General Ecology

General Ecology

Second Edition

David T. Krohne

Wabash College

BROOKS/COLE

THOMSON LEARNING

Australia • Canada • Mexico • Singapore • Spain • United Kingdom • United States

BROOKS/COLE

THOMSON LEARNING

Sponsoring Editor: *Nina Horne*
Project Development Editor: *Marie Carigma-Sambilay*
Marketing Team: *Rachel Alvelais, Mandie Houghan,*
  Carla Martin-Falcone
Assistant Editor: *Samuel Subity*
Editorial Assistant: *Rebecca Eisenman*
Production Editor: *Tom Novack*
Production Service, Manuscript Editing, Interior Design,
  Art Editing, Interior Illustrations, Photo Editing,
  Photo Research, Typesetting: *Thompson Steele, Inc.*

Permissions Editor: *Mary Kay Polsemen*
Cover Design: *Roy R. Neuhaus*
Cover Photo: *Frans Lanting/Minden Pictures*
Print Buyer: *Vena Dyer*
Cover Printing: *Phoenix Color Corporation*
Printing and Binding: *Quebecor/World—Dubuque*

For more information about this or any other Brooks/Cole products, contact:
BROOKS/COLE
511 Forest Lodge Road
Pacific Grove, CA 93950 USA
www.brookscole.com
1-800-423-0563 (Thomson Learning Academic Resource Center)

For permission to use material from this work, contact us by
www.thomsonrights.com
fax: 1-800-730-2215
phone: 1-800-730-2214

Printed in United States of America

10 9 8 7 6 5 4 3 2 1

Library of Congress Cataloging-in-Publication Data

Krohne, David T.
   General ecology / David T. Krohne.—2nd. ed.
          p. cm.
   Includes references (p. 459) and index
   ISBN 0-534-37528-6 (alk. paper)
   1. Ecology.   I. Title

   QH541 .K75 2001
   577—dc21   00-057952

To Sheryl and Pete

Contents in Brief

Contents

PART **I**

Introduction   2

PART II

Population Ecology 76

PART III

Community Ecology 216

PART IV
Ecosystems 354

Preface

My goal in the second edition of *General Ecology* is fundamentally the same as it was in the first edition: to produce a lucid, interesting explication of modern ecology. Accordingly, for the second edition I focused on two kinds of change. First, my revision of the text emphasized clarity of explanation. I believe this was one of the strengths of the first edition. But as every good teacher knows, the search for a new and clearer elucidation never ends. As in the first edition, the audience for this text is the student—the individual confronting this material for the first time. And so the writing, the new design, and the many new figures, were designed to facilitate their interaction with this text.

Second, it was important to update the text. The rate of production of new material in ecology continues to increase. New journals contribute to the explosion of information. Moreover, the Internet and its potential for rapid communication and informal publication has further increased the rate at which new knowledge accumulates. There has not been a fundamental paradigm shift in ecology in the three years since the publication of the first edition. However, our understanding in many areas has been refined by important new work and I've incorporated these changes throughout the book.

This leads me to one of the great dilemmas in the preparation of a textbook: when to replace the older classic examples with new ones. Some of the standard studies that have appeared in a generation of textbooks are flawed in one way or another. In some cases, new knowledge challenges the fundamental assumptions of the older studies. In others, modern experimental design, increased computing power, or the application of other new technologies almost renders the "classics" obsolete. Still, many of these studies are part of the culture and history of ecology. The serious student of ecology needs to be aware of them. The decision about when to replace an older study with a new example is a difficult judgment—one I've considered carefully in each case. Although the text cannot possibly include all new as well as the older examples, we can incorporate the new and the old in the classroom as a teaching device. I often present an older, perhaps flawed example to my class along with a new study that I think might replace it. This leads to a discussion of the nature of the two studies and an important lesson in analysis and critical thinking. And in the end, the students are aware of the historically important example *and* the newer literature.

Some aspects of the text have not changed in this edition. I have tried to retain the emphasis on analysis and experimental design. Therefore, in most cases I've included examples of the studies that demonstrate important ideas. Moreover, I have tried to retain the practice of showing the actual data in tables and graphs as much as possible. It is my hope that this leads students to question each study by looking carefully at the quality and variability of the actual data. It is, of course, the analysis and critical thinking that are crucial. In any science textbook, many principles will prove to be ephemeral; it is the *process* of science that is lasting.

In addition, the text still emphasizes an evolutionary approach to ecology. Today more than ever, it is important that students see the direct application of evolutionary principles in biology. It must be clear that evolution is not a side issue in biology, of interest only to paleontologists and taxonomists, and subject to attack from various outside forces. Rather, students should see it as a fundamental underpinning of biology in general, and ecology in particular. In this way it becomes a vital, operative part of our science. Finally, the application of ecological principles to environmental issues is as important now as ever. I have retained the boxes on environmental applications in each chapter. However, environmental applications appear more frequently in the main text as well.

Like the first edition, this text is designed to work for any student who has completed a general survey course in biology. Where does a course in ecology properly fit in biology curriculum? I think the student can benefit from this material early on in his or her career or later, perhaps as a capstone course. Early in the curriculum, ecology presents the student with the broad picture of whole organisms facing fundamental problems. So much of physiology, cell biology, molecular biology, and development underlie the solutions to these problems. In this position, ecology sets up the study of other parts of biology. The other viable option is late in the curriculum. At that point the student is proficient in many areas of biology, and ecology can serve as the basis for a synthetic capstone. In virtually any ecological interaction, one can remind the students of the relevant sub-organismal biology—C3/C4 photosynthesis, surface/volume ratios, neural networks, mutation—the list is endless.

There is some evidence of a somewhat disturbing trend in biology curricula. As molecular biology grows in importance, there appears to be a greater separation of cell and molecular biology from organismal biology and ecology. Specialization in one area or the other occurs earlier and earlier in the undergraduate curriculum. I would argue that precisely the opposite should occur. Molecular biology continues to transform all of biology, including ecology. Its application to ecological questions continues to expand, and the second edition reflects this. Even more important, however, is a philosophical issue. I believe that in the next decades, molecular biology and ecology will become philosophically more similar. The subject matter will, of course, be quite different, but there will be a philosophical convergence. The reason is that modern molecular biology is quickly moving past the stage of working with mapping the function of single genes. More often, the questions of interest are those of gene *interactions* and *systems* of genes. The traditional and beautifully reductionist approach that works so well with individual genes does not apply so well to complex systems. For example, developmental biologists are beginning to take a much more holistic analytical approach to cascades of genes in development. And, of course, this is precisely what ecologists have been doing for some time. I believe the end result is that students of ecology should be well versed in modern molecular biology—it will be part of their repertoire of approaches. And conversely, the student of molecular biology can be informed by the philosophical approach of ecologists.

Acknowledgments

This book would not have come to be without the input, direct and indirect, of a great many people. The true authors of this text are the many ecologists whose published data and insights have shaped the science of ecology. I am grateful to the fine group at Brooks/Cole who worked on this edition, especially Tom Novack, Nina Horne, and Marie Carigma-Sambilay. Marie was instrumental in helping me shape the second edition. I benefited greatly from the wisdom and suggestions of a number of outside reviewers. Many faculty who used the first edition volunteered ideas and suggestions. Andrea Fincke of Thompson Steele guided me though the production phase with great skill and good humor. Rochella Endicott of the Biology Department at Wabash College made invaluable technical contributions to the book. Greg Hoch of Kansas State University regularly stimulated new conversation about the science of ecology and shared his insights. Finally, I thank the biology students at Wabash College for teaching me about learning.

David T. Krohne

General Ecology

Introduction

The Science of Ecology

At a biological field station in the Great Plains, a fire races across a stand of native prairie grasses. The flames are particularly intense because a series of wet summers stimulated greater-than-normal growth of the grasses. The favorable conditions also permitted more species than normal to coexist in this stand, including a few rare species of plants not usually found here. At the margins of the stand, the flames kill seedlings of sumac shrubs, which in a wet year like this begin to invade the grass, eventually extending the forest into the grassland. Voles and deer mice flee ahead of the flames. After the fire, there will be no mice to consume the new growth of grass until the animals recolonize. Where the fire has already passed, birds congregate to consume ants from mounds no longer protected by dense grasses. The rains that finally extinguish the flames will wash phosphorus, which had been locked up in the grasses, from the ash back into the soil, where the minerals can again support new plant growth. Within a few weeks, the blackened earth is replaced by a carpet of lush new growth (Figure 1.1). Grasses and wildflowers emerge, fortified by the flush of phosphorus and the light that can now penetrate to the soil surface.

FIGURE 1.1 New growth of tallgrass prairie following fire. (*Photo by David Krohne*)

The Scope of Ecology

This simple scenario describes a web of *interactions*—a series of associations and relationships among the components of that stand of grass. An ecologist watching these events might be inspired to ask a number of questions about the relationships in this grassland:

1. What is the relationship between precipitation and plant growth or production?
2. How is an area's production related to the number of species it contains?
3. What factors cause changes in the populations of rare species?
4. How does the plant community develop and recover after a disturbance such as a fire?
5. What impact does grazing have on the plant community? On species diversity?
6. How does bird predation on ants affect the prey populations? The predator populations?
7. What are the patterns of movement of nutrients through the living and nonliving components of this system? Is this movement cyclic? What determines the rate at which nutrients move?

These questions illustrate some of the biological explorations and investigations that constitute the field of **ecology,** the study of the interactions between organisms and their environment. Strictly speaking, the word *ecology* refers to the study of these interactions, but in common usage it may refer to the interactions themselves, as in the phrase "an organism's ecology." Figure 1.2 is a simplified diagrammatic representation of some ecological interactions. Ecologists may focus on the environmental interactions of an entire species, of populations of the species, or of individual organisms. Note that the environment consists of both biotic (living) and abiotic (nonliving) components.

If we consider the interactions we might observe in other, very different systems—a tropical rain forest, a desert, the Galápagos Islands, or a meadow near timberline in the Rocky Mountains—we see that each system has its own complex web of interactions. By considering a variety of systems, we can begin to ask additional questions:

1. Are the interactions we observe in the prairie system fundamentally similar to or different from those we observe in these other systems?
2. Are there processes and interactions that are unique to each system?

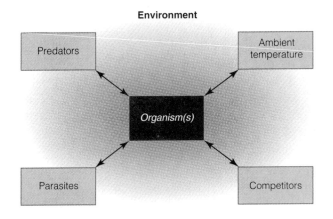

Environment

FIGURE 1.2 An example of the kinds of interactions that ecologists study. Note that the environment contains both biotic and abiotic components. Interactions between living components are connected by double-headed arrows because the interaction affects both participants.

3. What roles do other biological processes, such as genetics and evolution, play in the diversity of interactions we observe?

As you have no doubt concluded from this discussion, ecology is a tremendously diverse field that encompasses a wonderful variety of questions about biological phenomena. And in trying to answer these questions, it draws on other biological disciplines, including genetics, evolution, physiology, anatomy, and biochemistry.

Ecology and Levels of Biological Organization

Biologists tend to organize biological principles and phenomena into a hierarchy like that shown in Figure 1.3. Each level is more inclusive than the one below; cells are made of organelles, tissues are composed of cells, and so forth. Ecologists are primarily interested in the levels of this hierarchy from the organism through the ecosystem. Thus ecology is a science that examines the most complex levels of biological organization. This accounts for the tremendous diversity of interactions included in the field of ecology.

Subdisciplines in Ecology

Ecology can be subdivided according to the levels of the biological hierarchy. Thus the word **organisms** in Figure 1.2 might refer to any of the higher levels of the hierarchy—from the individual to the ecosystem.

The interactions between an individual organism and its environment are termed **autoecology.** Often the focus of autoecology is the physiological response of the individual to the abiotic environment.

Population ecology examines interactions that occur between a population and its environment. A **population** is a group of individuals that belong to the same species and inhabit a particular locale. We define a **species** as a group of actually or potentially interbreeding individuals. Thus the population ecologist is concerned with a portion of a species and the interactions in which it participates. A population ecologist studying our prairie stand, for example, might ask how the size of the vole population is regulated and controlled, or might wonder why some populations of prairie plants reproduce primarily by seed, whereas others reproduce vegetatively. These are ecological questions framed at the population level of organization.

Community ecology studies interactions among the populations of all species living in an area at a particular time, which together constitute the community. If a prairie stand does not burn frequently, forest eventually encroaches into the grassland. The progression of species replacements leading to forest is called succession. This web of interactions leading to forest is a community-level phenomenon. The number of plant species in the prairie is determined in part by the competitive interactions among the species of grass. Like succession, competition represents a community-level interaction.

Ecosystem ecology is the study of the most inclusive interactions, those among all the biotic and abiotic components of the system. An **ecosystem** thus includes both the community and its physical environment. The exchange of carbon dioxide among the atmosphere, the oceans, plants, and animals is an example of a global ecosystem-level interaction. Ecosystem interactions also occur on a smaller spatial scale. For example, our prairie system (including the grasses, voles, predators, atmosphere, and soil) also represents an ecosystem. The cycling of phosphorus—its release from grasses by the prairie fire and deposition into the soil, its incorporation into new grasses, its consumption by voles, and its return to the soil in feces—is also an ecosystem interaction.

Modern ecology has also developed a number of subdisciplines that focus on highly specialized facets of the science. For example, **physiological ecology** examines the ways that the bodily processes of organisms are adapted to the physical environment. **Genetic ecology** is the study of the ways in which an organism's ecology shapes its heredity and the ways in which genes influence ecological processes. Another approach to ecological interactions, **systems ecology,**

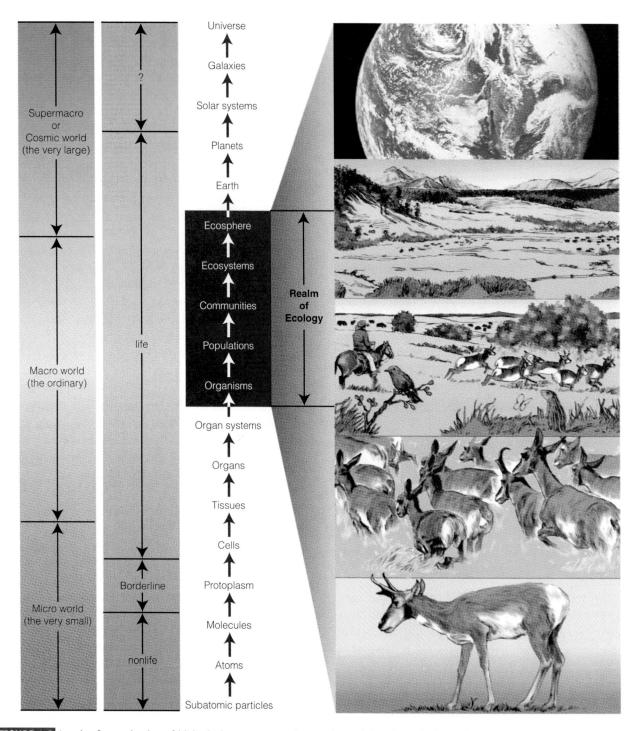

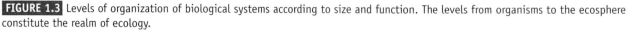

FIGURE 1.3 Levels of organization of biological systems according to size and function. The levels from organisms to the ecosphere constitute the realm of ecology.

emphasizes mathematical modeling of the interactions among the components of an ecological system, particularly the movement of energy and materials among the biotic and abiotic components of an ecosystem. A relatively new field, **landscape ecology,** focuses on the spatial patterns of ecological processes.

Sciences Allied to Ecology

Three fields of scientific study are closely related to ecology: natural history, environmental science, and the resource management sciences.

Natural history is the study of the habits, behaviors, and interactions of organisms in their natural environments. Although natural history and ecology are intimately related, they differ in that modern ecologists generally attempt to test an explicit hypothesis, whereas natural historians tend to focus on the descriptive study of natural phenomena. Thus, in natural history, the description itself is the goal. Often that description raises ecological questions or provides the background information needed for further ecological research.

Until the middle of the nineteenth century, when Charles Darwin published *On the Origin of Species*, much of the study of the natural world was natural history. In fact, much of the data Darwin accumulated during his voyage on the *Beagle* was descriptive natural history. His observations led Darwin to the theory of evolution by natural selection, which in turn led to a huge number of hypotheses about the ecological interactions that occur in nature. Thus natural history data preceded and paved the way for modern ecological work.

This is not to say that the science of natural history is dead or no longer useful; on the contrary, careful natural historical study is still crucial to modern ecology. As one example, consider our knowledge of the banner-tailed kangaroo rat (*Dipodomys spectabilis*). This small desert mammal has been the subject of many ecological studies of dispersal, genetics, population change, and social organization (Jones et al. 1988; Waser and Jones 1983). These studies have shown that in response to habitat saturation and a low probability of successful emigration, parents tend to be more tolerant of juveniles that remain in the natal area. These results enabled W. T. Jones and his colleagues to test hypotheses about such theoretical issues as the evolution of social systems and the role of dispersal in population regulation.

None of this research would have been possible, however, nor would the ecological principles behind it have been developed, without a background of careful natural history studies. These studies (Vorhies and Taylor 1922; Monson 1943) described the unusual fact that *D. spectabilis* inhabits conspicuous mounds containing extensive burrow systems that may take up to two years to construct and, once built, may last a decade or more. These vital burrows provide a cool, moist microclimate underground, serve as sites for seed caches, and offer protection from predators. Moreover, because the animals live together and only in these mounds, the ecologists were able to capture and mark all the individuals in the population. This natural history information was crucial to the subsequent experimental work, for it indicated that this species has unusual dependence on complex mounds, a fact that could be exploited by ecologists to test hypotheses about dispersal and social organization. Natural history studies play a similarly vital role in a wide variety of ecological analyses.

Environmental science is broadly defined as the study of the ecological effects of human activities on the environment. It is an interdisciplinary science that incorporates ecology, chemistry, geology, and even aspects of sociology and economics. Studies of acid rain, global warming, and oil spills are examples of environmental science research. The direct ecological effects of pollutants are the province of **environmental toxicology.**

In recent years, a field known as **conservation biology** has developed. This science uses the principles of ecology to maintain and manage biological diversity in both relatively natural systems and those more altered by human activity. The term *biological diversity,* or **biodiversity,** may refer to several of a system's features, including the number of species, the amount of genetic variation, and the complexity of interactions. Conservation biologists are concerned with identifying ways of maintaining or augmenting these sorts of variety in ecosystems. In this text we will examine numerous applications of fundamental ecological principles to conservation issues.

Although ecology and environmental science are intimately related, they are not identical. Ecology is a basic biological science in the same way that biochemistry and physiology are, whereas environmental science is an *applied* version of ecology. Thus the relationship between environmental science and ecology is similar to that between medicine and physiology. Just as medicine involves the application of physiological principles to human health, environmental science involves the application of ecological principles in studying the effects of human activities on the environment.

Even though our main focus here is on ecology as a basic science, the relationship between environmental science and ecology is so intimate and of such intrinsic interest that we will also consider some aspects of applied ecology. Moreover, every ecologist should have an inherent interest in environmental matters. Ecologists, like all scientists, have an obligation to use their knowledge to improve the human condition. And if ecologists (indeed, all of us) don't pay attention to the environmental health of the planet, they will have precious little left to study!

Several resource management sciences deal with human efforts to husband and manage natural populations in terms of economic, aesthetic, or other values. **Wildlife management** is the science of the control and manipulation of game and nongame wildlife populations to provide adequate numbers for hunters and other wildlife enthusiasts, as well as to ensure the long-term health of the populations and their habitats. Wildlife management makes extensive use of the work of basic ecologists who study the control and regulation of populations. In a similar way, **forestry, range management,** and **fisheries** biology apply ecological principles to the relevant resources to benefit both humans and the resources themselves. Like environmental science, they are applied versions of ecology.

The Conceptual Framework for Ecological Research

One goal of this text is to provide you with an understanding of the empirical basis of our ecological knowledge; thus we must become familiar with the concepts involved in conducting research. As you probably are aware, we must consider all scientific conclusions to be tentative. The factual information in any area of science changes so rapidly that virtually every scientific textbook is outdated by the time it is published. New information constantly forces us to revise our understanding of nature. Therefore, an understanding of *how to do science* is far more valuable than specific facts, at least some of which will be ephemeral. Thus this text will devote considerable attention to the kinds of experimental and analytical approaches used in ecology, including the following:

1. The types of analysis that are valuable in ecology
2. The kinds of ecological experiments that are most credible
3. How (and which) statistical analyses deepen our understanding of ecological data
4. The factors that confound ecological experiments

The Scientific Method

You have no doubt studied the **scientific method** and its application in a number of courses. The fundamental nature of the method is the same whether it is applied in ecology or in other sciences. One especially important feature of the scientific method is the nature of hypothesis testing. Good scientific experiments are designed to distinguish among alternative hypotheses. We can never show a hypothesis to be true; we can only demonstrate (or fail to demonstrate) that it is false. Thus science proceeds by the falsification of alternative explanations of an observed effect. One of these alternatives must be the **null hypothesis,** which states that the observed effect is simply the result of chance and is not due to some biological or other deterministic factor.

Let us illustrate the concept with an example. Suppose that in a study of predation on moths by bats, we notice that bats don't often catch and consume small moths. Our observation is consistent with the working hypothesis that bats prefer to prey upon large moths. We cannot accept this hypothesis, however, until we have rejected the null hypothesis, which in this case is that bats eat moths randomly with respect to size. To test the null hypothesis, we need new observations, so we measure the relative numbers of moths of various sizes that are flying while bats are hunting. We then measure the size distribution of moths actually captured by bats. When we look at the data, we find that bats do take more large moths but that this is because more large moths are available while bats are hunting (Figure 1.4). The bats are not choosing larger moths; they are taking all sizes of moths in the same proportion as they are available—that is, at random. Thus, because we cannot reject the null hypothesis, we cannot conclude that bats prefer large moths. This principle will be especially important in our consideration of community ecology in Part III.

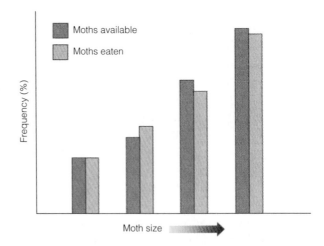

FIGURE 1.4 Hypothetical frequency distributions of the availability of moths of different sizes and predation on them by bats. The solid bars represent the proportion (percentage of the total) of moths of different size classes that are available to bats. The open bars represent the proportion of moths actually eaten by bats. There is no significant difference between the proportions of moths available and those actually eaten.

Reductionism versus Holism

We have defined ecology as the study of the interactions between an organism and its environment. For simplicity, we depicted only four such interactions in Figure 1.2. Of course, the number of interactions and the number of organisms and species involved can be huge. As the number of interactions under study increases, so does the complexity of the interactions. This complexity is both the fascination and the challenge of ecology.

The traditional scientific approach is based on **reductionism,** a process in which a system is reduced to its parts, each of which is described, studied, and understood in terms of its components. If we want to understand how a cell works, we take it apart and study its organelles and molecules. Experiments are then designed to isolate one component of the system, manipulate it, and observe the effect on the entire system.

According to our definition of ecology, we are interested in the web of interactions in which an organism participates. Ecology is thus a science based on **holism,** an approach that emphasizes the totality of the interactions.

The Limitations of Experimental Manipulation

Because ecology is fundamentally a holistic science, we face some potential difficulties when we attempt to conduct experiments. Suppose we want to study the role of wolf predation in controlling the moose population in a specific area (Figure 1.5). We can schematically depict the wolf/moose interaction as follows:

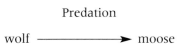

Using a reductionist approach, we would attempt to keep all other interactions constant and then experimentally manipulate predation so that we could observe its effect on moose abundance.

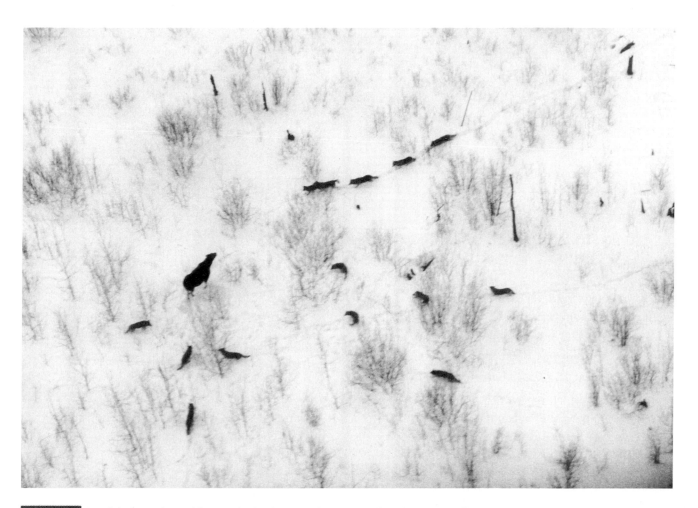

FIGURE 1.5 An adult moose hunted by a pack of wolves on Isle Royale. (*Photo by David Mech*)

At this level of the hierarchy, however, our approach encounters a fundamental problem. Our diagram of the system is oversimplified. The web of interactions is much more complex, perhaps more like the schematic in Figure 1.6. Now we can see that by doing an experiment to assess the role of wolf predation, we may also be affecting other interactions that affect the moose population. Thus, if we control the wolf population, the populations of other prey species such as deer or hares may increase. If these species compete with moose for food, the moose population may be affected. If we manipulate wolf predation and observe an effect on the moose, can we be sure the effect is a result of predation? No! It might have occurred via any of a number of indirect interactions, including competition. The complexity of interactions makes it difficult to ensure that any experiment we might do is not confounded by other interactions.

There is another problem with applying the reductionist approach at this level of the hierarchy. The complex set of ecological interactions that confound our experiments is underlain by other complex sets of interactions at lower levels. To understand fully the wolf–moose interaction, a reductionist approach demands that we also understand things like the physiological effects of parasites in moose, the biochemistry of digestion of woody plants by moose, the endocrinological effects of temperature stress on predator and prey, and so on. A simple-sounding ecological question (What is the effect of wolf predation on moose?) thus becomes complicated by the huge numbers of interactions at many levels of the biological hierarchy.

This is not to say that traditional kinds of experiments are doomed to fail in ecology. Indeed, some classic reductionist experiments have been crucial in developing our understanding of ecology. Given the complexity and potential pitfalls briefly outlined here, however, we must have detailed knowledge of the systems in which we work if we are to design a valid experiment and meaningfully interpret the results.

For example, R. T. Paine (1966) used manipulative experiments to demonstrate the role of predation in organizing the coexisting species of invertebrates and algae in a rocky intertidal community. He compared plots from which the predatory starfish, *Pisaster ochra-*

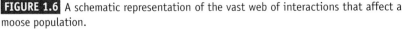

FIGURE 1.6 A schematic representation of the vast web of interactions that affect a moose population.

ceous, was removed with control plots. He found that in the absence of this predator, the number of species of algae declined dramatically, and the community was dominated by one species of mussel (*Mytilus californianus*).

Several factors account for the success of this experimental approach. First, because the basic ecology and natural history of the rocky intertidal habitat were relatively well known, unanticipated confounding interactions were unlikely. Second, the species involved were largely sessile invertebrates and plants, so the abundance of organisms could be readily and accurately determined. Finally, predation was clearly so pronounced relative to other interactions in this system that experimental manipulation produced obvious changes in the community.

Making Inferences in Ecological Research

At lower levels of the hierarchy, processes tend to be more universal. For example, the Krebs cycle, a central biochemical pathway in oxidative respiration, is affected by many interactions in the cell. Although it is by no means a simple mechanism, it functions in essentially the same way in yeasts as in whales. Put another way, its fundamental nature is universal among eukaryotes.

Now consider our ecological question concerning the role of predation in controlling prey populations, which is at a higher level of the hierarchy. It is not likely that the effect of wolf predation on moose is the same in northern Canada as it is in Maine. Differences in the severity of the weather, the presence of other species, the nature of the habitat, and so forth will almost surely lead to at least somewhat different results. And we have not even tried to make a truly general statement about predation; we have simply addressed the interaction between one pair of species. Would the results apply to lions and wildebeest? Probably not exactly.

But again, ecologists are not foiled by this difficulty. Rather, we rely on comparative studies. If common conclusions emerge from studies of wolf predation on moose in Canada and Maine, we may infer that we have discovered a more fundamental principle. When we find common themes in very different systems, we can be more confident that we have discovered a rather more general principle. Consequently, similar studies in different systems are of inestimable value in ecology.

Statistical Analyses and Mathematical Modeling

As we have seen, holism is an approach that emphasizes the workings of whole systems, rather than their individual parts. Phenomena that are observable only in the intact, complete system are known as **emergent properties.** Predation, for example, is a phenomenon that emerges from systems containing wolves, moose, woody plants, parasites, and a host of other organisms. Yet we cannot necessarily predict the nature of predation merely from our knowledge of the components of the system. Two important strategies for learning more about the nature of the interactions among those components are statistical analyses and mathematical modeling.

A number of statistical techniques can be applied to experiments in which interaction is a central focus. Suppose we are interested in the ecology of reproduction in a plant species. In particular, we are studying the environment's effect on the number of seeds each plant produces. We find that seed production increases both with increasing latitude and with the intensity of herbivory. By using a statistical method such as linear regression, we can quantify these relationships (Figure 1.7). The graphs reveal the essential nature of the interaction, which is that seed

production increases with latitude more rapidly when herbivory is high than when herbivory is low. The regression equations demonstrate the existence of an interaction between latitude and herbivory.

We can also use mathematical models to approach many ecological problems holistically. In developing a model, we attempt to formulate a set of equations that reflect the operation of the system. For example, we may wish to understand the pattern of movement of carbon in a salt marsh ecosystem. We can express mathematically the amount of carbon and its rate of transfer among the various biotic and abiotic components of the salt marsh. Or we may want to analyze the role of predation in controlling insect populations in a prairie ecosystem. The number of individuals of each insect species and their rates of change over time can be described by equations.

In any ecological model, we begin by making some fundamental assumptions about the operation of the system. Often these assumptions involve simplifications because we know that we will have difficulty obtaining enough information about the system to write a complete set of equations. Furthermore, ecological systems are so complex that any series of equations intended to describe the system completely is likely to be too complex to solve. Thus we make assumptions that will simplify the mathematics. For example, in the study of prairie insect populations, the numbers of

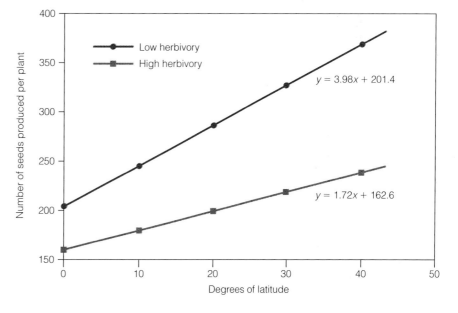

$$y = 3.98x + 201.4$$

$$y = 1.72x + 162.6$$

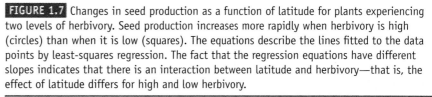

FIGURE 1.7 Changes in seed production as a function of latitude for plants experiencing two levels of herbivory. Seed production increases more rapidly when herbivory is high (circles) than when it is low (squares). The equations describe the lines fitted to the data points by least-squares regression. The fact that the regression equations have different slopes indicates that there is an interaction between latitude and herbivory—that is, the effect of latitude differs for high and low herbivory.

insects will be reduced by both predation and emigration, but to simplify the mathematics, we might assume that emigration is so small as to be insignificant. Of course, such assumptions also decrease the degree to which the model simulates the system.

The two basic approaches to modeling depend on the goal of the model. In the first approach, where the aim is to provide information for management decisions, we often incorporate a great deal of empirical data into a set of equations that predict the behavior of the system. For example, we might develop a mathematical model of carbon flow in a salt marsh by making many measurements of the carbon content of the components of the system and the rates of carbon transfer among them. We manipulate the equations so that the model accurately describes what our empirical data actually demonstrate in the system. This model can then be used to play "what if" games to make predictions. What happens if the carbon input into the system increases because of increased air pollution? We can plug greater carbon inputs into the model and see what the effects will be. Because the model is so heavily dominated by empirically based data, it will not provide great insight into the inner workings of the carbon cycling process; rather, it will be a mathematical description of a single aspect of nature.

The other approach to modeling is to simplify the mathematics so that we can describe the essential features of some ecological process. Precision is sacrificed so that we can use the model to elucidate the essential features of the system. Thus we would ask which of the components can be omitted, via simplifying assumptions, and the model still provide a reasonably accurate, qualitative description of carbon cycling. When we can answer this question, we have attained some general knowledge of the process of carbon cycling.

Proximate and Ultimate Explanations

We have seen that good natural history observations precede experimental ecological work. Often we observe something about nature that intrigues us—that birds avoid certain butterflies, that the tropics have many more species of animals and plants than the polar regions, that some desert plants are surrounded by rings of bare soil devoid of other plants. Natural history provides questions; ecology attempts to answer them.

Our current ecological explanations of nature are predicated on the concept of evolution. The great geneticist T. Dobzhansky said, "Nothing in biology makes sense except in the light of evolution," and this is particularly true in ecology. A cornerstone of Darwin's theory of natural selection is that ecological circumstances largely dictate the success or failure of certain individuals in a population.

Time and again throughout this text, we will see that ecological explanations are evolutionary explanations. Birds avoid certain butterflies because the butterflies have evolutionary adaptations that make them distasteful or toxic. Perhaps there are more species in the tropics because that environment provides more opportunities for speciation. Some desert plants have adapted to competition for water by producing chemicals that diffuse into the soil and prevent the growth of other plants.

Scientists distinguish between two kinds of explanations for ecological phenomena: A **proximate explanation** is the immediate cause of a phenomenon, whereas an **ultimate explanation** provides a more general reason, often one rooted in evolutionary principles.

For example, the clutch size of some species of birds correlates with the abundance of insects. When insects are plentiful, the birds lay more eggs; in years of insect scarcity, fewer eggs are laid. Biologists have found that if females are well fed and have stored much energy in the form of fat, they produce more eggs. This physiological effect—the amount of stored energy reserves—is the proximate explanation for the variations in clutch size; it explains the immediate cause of the observed effect. Perhaps this physiological process is an evolutionary response that ensures that females will have enough energy to produce eggs and gather food for the number of young in the nest. It would be disadvantageous for the birds to produce many young in years with poor food resources. This evolutionary explanation of clutch size is an ultimate explanation for the proximate mechanism we described; it explains in a broader sense the reason for or purpose of the clutch size phenomenon.

We need to be careful in using words like *ensures, purpose of,* and *reason for.* In evolutionary ecology, these words do not imply rational, conscious thought on the part of the organism, nor do they imply a grand design or overriding intention directing the phenomena we observe. Rather, they are bits of language customarily adopted by biologists to convey, in just a few words, all that the evolutionary process entails. They are a shorthand notation, if you will. In this example, we said that the "reason for" variations in clutch size with insect abundance was to "ensure" adequate energy for reproducing and rearing the young. This is a simplified way of saying, "Birds that produce small clutches in lean years and large clutches in good years, on average, leave more descendants in the next generation than those individuals that lack this physiological energy storage mechanism for adjusting clutch size. The adaptation increases the survival and reproductive fitness of those individuals who possess it." Certainly, we are not implying that the birds consciously make a decision about clutch size; the "decision" is physiological.

We began this chapter with a description of a fire in a grassland ecosystem. What may have initially appeared to be a simple natural event—flames racing through grass—is in fact a highly complex set of interactions. The ecological effects of the fire are manifold; they affect the plants, the animals, and the abiotic environment. Moreover, we can study these interactions at the population, community, or ecosystem level of organization.

But the effects are even more far-reaching. The fire and its ecological consequences also represent a selective force on the organisms, and it may play an important role in the evolution of grassland species. The interplay of ecology and evolution is a central feature of this text. In the next chapter, we consider the fundamental evolutionary processes and mechanisms that are central to the study of ecology.

Adaptation and Evolution

I n this chapter we begin our exploration of a theme that will continue throughout the text: the interaction between evolution and ecology. We start by considering the definitions of these terms and the nature of their relationship to each other.

The Interaction of Ecology and Evolution

In Chapter 1 we defined ecology as the study of the interaction between an organism and its environment. **Evolution,** in its broadest sense, is genetic change in a population of organisms over time. As we will see, a number of mechanisms can lead to such change. Thus ecology and evolution are intimately related because an organism's ecological situation directs its evolution, and the organism's response to its ecological situation may be evolutionary. Let us illuminate this statement with two examples.

A species of desert cactus facing a harsh climate with high temperatures and infrequent precipitation must compete with neighboring plants for water. The ecological situation favors plants that produce a web of fine roots near the soil surface instead of a single deep taproot. Plants with shallow roots are able to capture the water that occasionally falls in a thunderstorm before their deep-rooted neighbors can. These shallow-rooted plants can now grow more vigorously, may reach higher population densities, and may outcompete their neighbors for scarce water. Ecological relationships have been fundamentally changed. Many other species in this desert community are affected either directly or indirectly by these changes.

Thousands of years ago, canid progenitors of the modern wolf preyed mainly on small animals that could be captured and subdued by a single adult. Individuals lived and hunted alone, coming together only for mating. Over time, the canids gradually found themselves surrounded by abundant, large, herbivorous mammals that represented a bountiful food resource but were formidable opponents for a single predator. Gradually, the canids were able to exploit the new food resource by hunting in groups, in the process dramatically changing the ecological situation. To take advantage of the new food resources, the canids had to become more cooperative and more social; a whole new set of behaviors was required. The predator's population size probably increased, and the herbivores, facing a new preda-

tion threat, had to respond if they were to survive this new source of mortality. We can envision the effects of these changes rippling throughout the community.

These two examples no doubt represent a plethora of such responses that have occurred since life began on Earth. Each current ecological situation is a force that shapes an evolutionary solution to an ecological problem. Each evolutionary solution is called an **adaptation**—a genetically determined characteristic (whether behavioral, morphological, or physiological) that improves an organism's ability to survive and reproduce in a particular environment. (Note that we also use the term *adaptation* to refer to the evolutionary process whereby organisms become better suited to their environments.) Adaptations have had profound effects on the ecological relationships of systems. The title of a book by the renowned ecologist G. E. Hutchinson, *The Ecological Theater and the Evolutionary Play* (1965)—is an apt metaphor for the interaction between ecology and evolution. The ecological situation represents the stage on which evolution unfolds.

Throughout the remainder of this text, we will discuss the adaptations by which organisms have responded to their ecological context. Many of these adaptations are remarkable; indeed, one of the fascinations of biology is the wonderful adaptations that organisms evolve in response to their environments.

Evolution in Populations

To understand the nature of adaptation, it is important to begin with at least a brief discussion of genetic variation; we will consider this topic in greater detail in Chapter 6.

Genetic Variation

One of the most important observations we can make in nature is that for many of the characters we can measure or describe in natural populations—growth rates and litter sizes in mice, for example—some inherent variability exists among individual organisms (Figure 2.1). To put it another way, for any given character, different individuals exhibit different traits. The visible manifestation of a character in an organism is called the **phenotype.**

Variability such as that depicted in Figure 2.1 has two possible bases: Some traits are the result of differ-

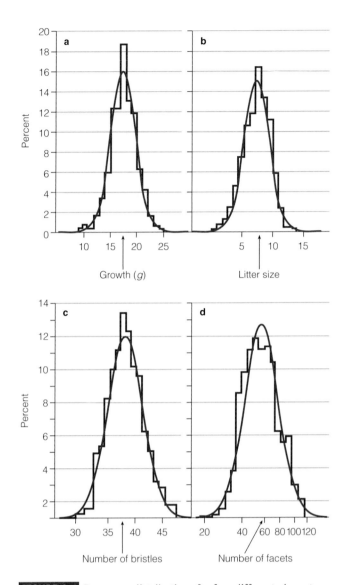

FIGURE 2.1 Frequency distributions for four different characters in different organisms. (**a**) Growth rate in mice from 3 to 6 weeks of age; (**b**) litter size in mice; (**c**) number of abdominal bristles in the fruit fly, *Drosophila melanogaster;* (**d**) number of eye facets in the eye mutant "bar" in *D. melanogaster.* The curves superimposed on the raw data in each graph show that each of these characters has a normal (bell-shaped) distribution. (*From Falconer 1981*)

Fundamentals of Population Genetics

The basic evolutionary unit is the *population,* a group of conspecifics (members of the same species) inhabiting a specified geographic area. The area is defined such that the individuals are in close enough proximity that any two have a sufficiently high probability of encountering each other and mating.

The sum total of all the genes in a population is the **gene pool.** In population genetics we are concerned with the nature of the gene pool and the ways it changes. In diploid organisms, each individual carries two copies of each gene. These copies are called alleles, and an individual's two copies of a particular gene may be the same or different. The total number of alleles in the gene pool is $2NG$, where N is the number of individuals in the population and G is the number of genes in each individual.

One way we characterize the gene pool is by measuring allele frequencies. Consider a simple character such as flower color in pea plants, which is controlled by a single gene with two alleles, R (the dominant) and r (the recessive). Homozygous dominant individuals (RR) and heterozygous individuals (Rr) have red flowers; homozygous recessive individuals (rr) have white flowers. The allele frequency for R is the frequency of this allele among all flower-color genes in the population. If the population contains 100 individuals, there are 200 alleles of the flower-color gene (each individual has two copies). If among those 100 individuals there are 40 copies of the R allele, then its allele frequency is 40/200, or 0.20. Because there are only two alleles, we know that the frequency of the r allele must be $1.00 - 0.20$, or 0.80. Population geneticists typically use the symbol p to denote the frequency of the dominant allele and the symbol q to denote the frequency of the recessive allele.

Earlier we defined evolution as genetic change in a population. Now we can define it more specifically as a change in the allele frequency in a population. Returning to the previous example, if the frequency of the R allele has increased from 0.20 in one generation to 0.30 in succeeding generations, then evolution has occurred.

Let us consider a simple case, a character for which there are two alleles (A and a), with A being dominant. Our population of 100 individuals has the following distribution of genotypes:

AA: 30 individuals

Aa: 20 individuals

aa: 50 individuals

ences in the **genotype,** or genetic constitution, of the individuals being measured, and other traits develop differently among individuals because of differences in their environments. In other words, some of the variability among individuals is a result of inherited differences, and some is environmentally induced. Here we are primarily concerned with genotypic differences among individuals in a population. Accordingly, we need to understand the genetic organization of populations, and to do so, we must consider some fundamental aspects of population genetics.

From these data we can calculate the value of p by adding all the A alleles and dividing that sum by the total number of alleles:

$$p = 2(30) + (20)/200 = 0.4 \qquad \textbf{(1)}$$

Because homozygous individuals carry two A alleles, they contribute 2×30, or 60, A alleles. Each heterozygote carries one A allele, contributing 20 in Equation 1. Thus the total number of A alleles (80) is divided by the number of alleles for that gene in the population (200).

We can calculate the value of q in the same way:

$$q = \frac{2(50) + 20}{200} = 0.6 \qquad \textbf{(2)}$$

Alternatively, because we know that $p + q = 1$, we can also calculate q by subtracting p from 1.0.

Now let us consider what will happen if the individuals in this population mate at random. We are concerned with the frequencies of alleles and genotypes in the entire gene pool. Because we know the frequencies of the two alleles, we can calculate the probabilities that different combinations of the two alleles will come together under random mating. This calculation is based on the product rule of probability, which states that the probability of two events occurring together is the product of their independent probabilities. Thus, if the probability that a given gamete (reproductive cell) contains the A allele is p, then the probability of a homozygous dominant (AA) genotype occurring is p^2; by analogous reasoning, the probability of a homozygous recessive genotype occurring is q^2:

AA: $p^2 = (0.4)^2 = 0.16$

aa: $q^2 = (0.6)^2 = 0.36$

The calculation for the heterozygote probability involves slightly different reasoning. The probability that an A allele and an a allele will join is pq. But this combination can occur in two ways: An A can join with an a, or an a can join with an A. Thus the probability of an Aa genotype occurring is $2pq$:

Aa: $2pq = 2(0.4)(0.6) = 0.48$

So now we can express the frequencies of the three possible genotypes as:

$$\begin{array}{ccccc} p^2 & + & 2pq & + q^2 & = 1.0 \\ 0.16 & + & 0.48 & + 0.36 & = 1.0 \qquad \textbf{(3)} \\ (AA) & & (Aa) & (aa) & \end{array}$$

The relationship among genotypes and allele frequencies in a population was described mathematically by two population geneticists, G. H. Hardy and W. Weinberg, working independently (Hardy 1908; Weinberg 1908). Equation 3, which is known as the Hardy–Weinberg equation, describes the genotype frequencies expected from the allele frequencies. A population whose genotype frequencies match the values predicted by this equation is said to be in Hardy–Weinberg equilibrium. Note that the population we began with was not in Hardy–Weinberg equilibrium because the genotype frequencies did not match those generated by the equation:

Genotype	AA	Aa	aa
Frequencies predicted by Hardy–Weinberg equation	0.16	0.48	0.36
Frequencies in population	0.30	0.20	0.50

Now let the population mate randomly. If we allow no forces other than the laws of probability to affect the frequency with which gametes combine to form new individuals, random mating will lead to a change in the genotype frequencies. Because A alleles will combine with a probability p^2, the frequency of homozygous dominants in the next generation will be 0.16. Similarly, the frequencies of the heterozygotes and homozygous recessives will match the predictions of the Hardy–Weinberg equation. Note that the values of p (0.4) and q (0.6) have not changed from the original population; continued random mating will result in no change in allele or genotype frequencies. A population in Hardy–Weinberg equilibrium does not evolve so long as no forces affect the random union of gametes.

Conditions Leading to Evolutionary Change

The Hardy–Weinberg equation describes an equilibrium situation for a population. Thus the frequencies of genotypes in Equation 3 will remain constant so long as certain conditions hold:

1. The population must be large. The reason for this condition is that the predictions of the Hardy–Weinberg equation are based on the laws of probability. If the population is small, chance events will cause deviations from the expected genotype frequencies.

2. There must be no net immigration of genotypes into, and no net emigration of genotypes from, the population. Such movements would cause a dis-

TABLE 2.1
Conditions and Violations of the Hardy–Weinberg Equation

Condition for Hardy–Weinberg Equilibrium	Corresponding Mechanism of Evolution If Violated
Infinitely large population	Genetic drift
No net immigration or emigration of alleles	Gene flow
No differential mortality or survival by genotype	Natural selection
No new mutations	Mutation pressure

crepancy between the expected genotype frequencies and those actually observed.

3. There must be no differential mortality or reproduction of the different genotypes. If the predictions of probability theory are to apply, there must be no external force raising or lowering the frequencies of the three genotypes.

4. There must be no new mutations. The appearance of a third allele would automatically alter the genotype frequencies, and thus they would no longer conform to the predictions of the laws of probability for two alleles.

These conditions lead us to the most important features of Hardy–Weinberg equilibrium:

1. A population that is in Hardy–Weinberg equilibrium will experience no change in either genotype frequency or allele frequency.

2. If one or more of the conditions is violated, genotype frequency and allele frequency will change.

Because we defined evolution as a change in the allele frequency in a population, a population that is in Hardy–Weinberg equilibrium is not evolving. Moreover, violations of the different conditions for maintaining Hardy–Weinberg equilibrium represent agents of evolutionary change (see Table 2.1). The terms *evolution* and *natural selection* are often used interchangeably. However, this is not correct. One of the most important conclusions we can draw from this table is that evolution proceeds by more than one mechanism.

Mechanisms of Evolution

The elucidation of natural selection in nature was, of course, Darwin's great insight, and this mechanism is still the process most commonly associated with evo-

lutionary change. We will find that in addition to their direct role in evolution, each of the other mechanisms has crucial ecological effects and consequences.

Genetic Drift

Genetic drift is defined as stochastic (random) shifts in allele frequencies. If the population is small (a violation of the Hardy–Weinberg conditions), the genotype frequencies expected from Mendel's laws and the laws of probability may not be achieved. Chance events take on greater importance. Examples of genetic drift are shown in Figure 2.2. The changes in

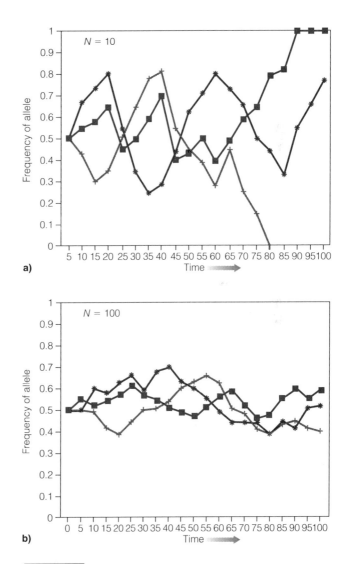

FIGURE 2.2 Examples of changes over time in allele frequency resulting from genetic drift. Each line plots allele frequency for a different population of the same species. (**a**) Populations containing 10 individuals; (**b**) populations containing 100 individuals. Note that the fluctuations in allele frequencies are more extreme in the smaller populations than in the larger populations.

allele frequency in these graphs are due entirely to chance. Note that drift occurs in both small and large populations but that the effect is exaggerated in smaller populations.

Drift may ultimately result in the complete loss of an allele or its increase to a frequency of 1.0. When the latter occurs, we say that the allele has become fixed. When several separate populations are undergoing the effects of drift, differences in allele frequencies gradually increase as chance loss and fixation occur in the separate populations. If the process continues, loss and fixation result in a bimodal allele frequency distribution as shown in Figure 2.3.

We have said that the effects of genetic drift are more pronounced in small populations than in large ones. The key variable in this phenomenon is the **effective population size, N_e,** which is defined as the number of individuals participating in random mating. Any factor that limits the number of interbreeding individuals in a population decreases N_e. For example, if an organism's social system is such that only a few males in the population actually mate, the effective population size will be significantly smaller than the total number of individuals. This will also happen if the sex ratio is skewed toward one sex. Other factors that decrease N_e include fluctuations in population size, variation in the number of offspring per parent, and overlap of generations; each decreases the actual number of randomly mating individuals relative to the total population size.

Ledig et al. (1999) documented an extreme bottleneck and its genetic effects in the rare Mexican pinyon pine (*Pinus maximartinezii*). This species is confined to a single population numbering fewer than 2500 trees in Mexico. Genetic analysis revealed that this species has a low percentage of polymorphic loci. In addition, these few polymorphic loci have at most two alleles—in contrast to most pines, which have many alleles per locus. Ledig et al. calculate that this genetic structure resulted from an extreme bottleneck in population size four or five generations (less than 1000 years) ago.

When we examine the various ways to measure and describe population size in Chapter 4, we will see that from an ecological standpoint, the population size may not be the same as N_e, the genetically relevant population size. The effective population size may have important consequences for conservation: If N_e declines too much, the loss of genetic variation resulting from drift may put species at risk of extinction (see the box on page 42).

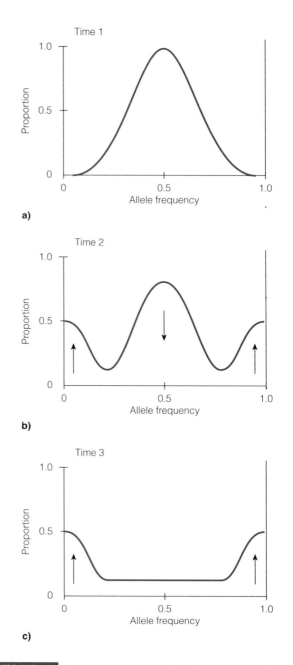

FIGURE 2.3 Changes over time in allele frequency distributions as a result of genetic drift. Each graph represents several populations simultaneously undergoing drift. (**a**) At time 1, the populations have a normal distribution of allele frequencies. (**b**) As time passes, drift causes loss or increase of alleles, (**c**) eventually resulting in a bimodal distribution in which populations with allele frequencies of either 1.0 or 0 predominate.

Gene Flow

Evolution can also proceed by the loss or gain of alleles via emigration or immigration, a process we call **gene flow**. If subpopulations differ in allele frequencies and the movement of alleles is nonrandom, the net effect will be changes in the allele frequencies of the subpopulations. By definition, evolution has occurred.

The movement of alleles can occur by various means. Dissemination of plant propagules such as spores and seeds may result in the movement of alleles. Gene flow may also result when animals establish themselves in a new population and successfully breed there.

In plants, the movement of pollen represents an important potential source of gene flow. For wind-pollinated plants, the direction in which the wind blows and the size and mass of the pollen largely determine the nature and extent of gene flow. For insect-pollinated plants, the foraging habits of the animals as they collect pollen and move from plant to plant determine the pattern of gene flow. It is apparent from the distribution of pollen dispersal distances for the yellow fawn lily (*Erythronium grandiflorum*) shown in Figure 2.4 that bees move pollen distances exceeding 20 meters. Progressive gene flow by pollen movement can move alleles far enough that they enter a new population and cause evolutionary change.

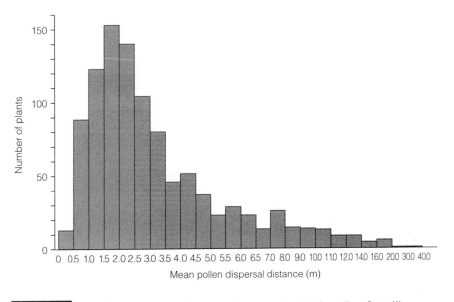

FIGURE 2.4 Mean distances over which bees disperse pollen in the yellow fawn lily, *Erythronium grandiflorum*. Note that the intervals on the horizontal axis are unequal. (*From Thomson and Thomson 1989*)

Natural Selection

Throughout this text, we will encounter examples of organisms whose genetic constitutions have been altered by the organisms' interaction with their environment. If allele frequencies are to remain in Hardy–Weinberg equilibrium, then there must be no differential mortality or reproduction associated with particular genotypes. This condition leads us to the concept of **natural selection:** the differential representation of genotypes in future generations, resulting from heritable differences among them in survival and reproduction.

Darwin's original formulation of the process of evolution by natural selection still stands as essentially correct today. He argued that

1. Inherited variation exists among individuals in the population; that is, for any character we measure or observe, not all individuals have the same phenotype. (These phenotypic differences reflect genotypic variation. Although Darwin did not understand the mechanisms of inheritance, he did recognize that the traits of interest must be heritable.)

2. Resources are limiting. Virtually any population is able to grow at a rate that greatly exceeds the rate of production of the resource that most critically limits it. Thus there will be competition for scarce resources.

3. As a result of points 1 and 2, those individuals whose phenotypes make them better able to garner scarce resources will, on average, leave more offspring than other members of the population. (Subsequently, our developing understanding of genetics led us to appreciate that because the advantageous phenotypes have a genetic basis, the genes for these traits will gradually increase in frequency in the population.)

We can clearly see how competition works when organisms are competing for food and a particular trait makes some of them more efficient at locating or consuming food. However, we define the concept of competition more broadly. If the phenotype of some individuals in the population makes them less likely to be captured and consumed by predators, then those genes (and phenotypes) will increase in the population as well. Thus the second point of Darwin's argument, the role of scarce resources, connects evolution and ecology in the most fundamental way.

One of the best examples of the selective pressure of competition has been described by H. L. Gibbs and P. R. Grant (1987) for Darwin's finches on the Galápagos Islands. One species in particular, *Geospiza fortis,* is apparently under intense selection for various aspects of body size, including bill dimensions. A granivorous species, it prefers small, soft seeds. In drought years these seeds are quickly depleted, leaving large, hard seeds as the only food resource available. Only birds with heavy, thick bills can crack such seeds. A drought increases the competition for seeds, and only those individuals with large bills are able to survive and reproduce. Thus, following droughts we observe a shift to larger bill sizes (Figure 2.5). After a year of abundant rainfall, there is a shift to smaller bills. A remarkable feature of this example is the rapidity of the evolutionary change in these populations.

The third tenet of Darwin's argument leads to the concept of **fitness,** the relative genetic contribution by an individual's descendants to future generations. Population geneticists distinguish between fitness measured in absolute and in relative terms. **Absolute fitness** is the expected number of offspring produced by a particular genotype. This measure incorporates both the survival ability of the genotype and its fecundity.

Relative fitness is based on the relative ability of a genotype to obtain representation in the next generation. If, for example, the relative fitnesses of three genotypes, *AA, Aa,* and *aa,* are 1.0, 0.8, and 0.6, respectively, then genotype *Aa* will be expected to contribute only 80 percent as many genes to the next generation as genotype *AA.*

There are three important properties of fitness:

1. Fitness is a property of a genotype, not of an individual or a population. Individuals with the same genotype share the same fitness.
2. Fitness is specific to a particular environment. As the environment changes, so do the fitness values of the genotypes. Again, this relation connects ecology and evolution.
3. Fitness is measured over one generation or more. We must know that the advantage persists and is stable over time.

Modeling Natural Selection

We can model natural selection using the same set of equations that led us to the Hardy–Weinberg equilibrium. To do so, we need to include a factor that incorporates the fitness of each genotype. We will consider the simplest case, a system with two alleles in which (1) there is complete

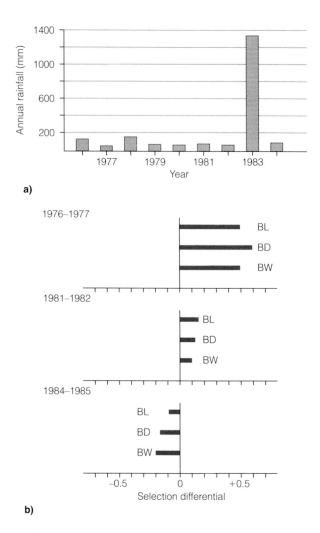

FIGURE 2.5 Selection on bill size in *Geospiza fortis* and its relationship to changes in food supply. (**a**) Annual rainfall on Daphne Island in the Galápagos. More small, soft seeds are available during years with high rainfall. (**b**) Selection differentials for bill traits during drought years (1976–1977 and 1981–1982) and following a season of very high rainfall (1984–1985). Selection differentials are an index of change in a trait before and after selection. A positive selection differential indicates that the trait has increased in size; a negative differential signifies that the trait has become smaller; a selection differential of zero indicates no change in the size of the character. BL = bill length; BD = bill depth; BW = bill width. (*Adapted from Gibbs and Grant 1987*)

dominance, and (2) the homozygous recessive genotype does not survive so well as the homozygous dominant and heterozygous genotypes.

The relative decrease in survival of the *aa* individuals is expressed by the selection coefficient, *s.* Thus, if the fitnesses of the homozygous dominant and the heterozygote (recall that dominance is complete) sum to 1, then the fitness of the homozygous recessive is $1 - s$. If *s* is very small, the homozygous recessive is almost as fit as the homozygous dominant. If $s = 1$, the allele is lethal: None of the homozygous recessives survive.

We can model the selection process algebraically in order to understand the change in allele frequency, Δp. The change in allele frequency is the difference between the original value of p and its value (p') after selection has occurred. The new allele frequency, p', is calculated in the same way we previously calculated p. Each homozygous dominant carries two A alleles, and each heterozygote carries one. We divide the sum of these values by the total number of alleles (or two times the population size), which is calculated as follows: We know that in the Hardy–Weinberg equation, the sum of the genotype frequencies equals 1. But some fraction of that total was eliminated by selection—this is calculated as the frequency of the homozygotes (q^2) times their selective disadvantage (s). Thus the new population is equal to $1 - sq^2$, and the total number of alleles is $2(1 - sq^2)$. Now we can calculate the value of p':

$$p' = \frac{2(p^2) + 2pq}{2(1 - sq^2)}$$

$$= \frac{p(p + q)}{(1 - sq^2)}$$

$$= \frac{p(1)}{1 - sq^2}$$

$$= \frac{p}{1 - sq^2}$$

Then Δp can be calculated:

$$\Delta p = p' - p$$

$$= \frac{p}{1 - sq^2} - p$$

We solve the equation to obtain

$$\Delta p = \frac{p - p(1 - sq^2)}{1 - sq^2} = \frac{p - p + spq^2}{1 - sq^2}$$

$$\Delta p = \frac{spq^2}{1 - sq^2}$$

From this derivation we see that the rate of change of allele frequency under this kind of selection depends not only on the intensity of selection (the magnitude of s) but also on the starting allele frequencies. We expect the rate of selection to depend on its intensity, but the effect of the initial allele frequencies is not intuitively obvious.

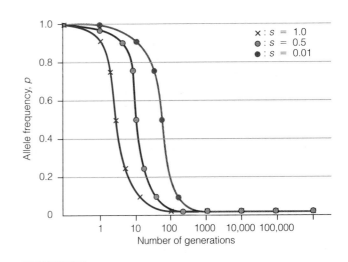

FIGURE 2.6 Changes in allele frequency (p) as a function of the number of generations, for three different selection coefficients (s). (*Data from Strickberger 1990*)

With this kind of analysis, we can model the changes in p and q under different intensities of selection. Results from such simulations are shown in Figure 2.6. Some interesting facets of the process emerge. Note that even if the recessive allele is lethal, after 9000 generations selection still has not eliminated it; it persists at very low frequency in the population. This occurs because in each generation, the recessive allele is protected from selection within the heterozygote. It is very difficult for selection alone to completely eliminate such an allele. However, some other agent of evolutionary change (such as genetic drift) may finally eliminate the allele once it reaches extremely low frequency.

Modes of Natural Selection Since Darwin's original formulation of the theory, the concept of natural selection has been expanded in important ways, so that we now define three fundamental modes of selection according to their effect on the population. In **directional selection** (Figure 2.7a), the phenotype at one extreme of the population distribution has a selective advantage; that is, it is favored by selection, and individuals with that phenotype leave more offspring. If the phenotype is heritable, the mean of the population will shift over time (to the right in this example). The changes in bill size in finches during drought on the Galápagos Islands are one example of directional selection.

When the intermediate phenotypes are advantageous (Figure 2.7b), **stabilizing selection** occurs. Individuals with intermediate phenotypes leave more offspring than those at either extreme, and over time the population will tend to have a narrower distribution centered around the original mean value. The birth weight of ungulates (hoofed animals such as deer and horses) is an example. Very small neonates may be at a disadvantage because they cannot keep up with their parents or regulate body heat as efficiently as larger young. Thus directional selection might increase the size of newborn young. At some point, however, large fetuses may have difficulty passing through the birth canal, which would set an upper limit to the size of a viable fetus. As a result, neonates of intermediate size will be the fittest.

Finally, there is **disruptive selection** (Figure 2.7c), in which the intermediate phenotype is at a disadvantage and is selected against. Clear examples of this kind of selection are relatively uncommon in the literature. One reason is that Mendelian genetics complicates the process. Genotypes that produce one or the other of the extreme phenotypes often produce intermediate phenotypes when they mate. If the two extreme types are still able to mate, we will be less likely to observe a population composed mostly of just the two extreme phenotypes, even if the intermediates are selected against.

T. B. Smith (1991) has discovered an example of disruptive selection operating on the African finches called black-bellied seed crackers, *Pyrenestes ostrinus*. In collections of this species, there are individuals with large and those with small bills but no individuals with intermediate-sized bills. Breeding experiments showed that crosses of large- and small-billed individuals produce no offspring with intermediate phenotypes. Smith also found that the seeds of two species of sedges are the main food source for these finches. One species produces hard seeds, the other soft. The finches prefer the soft seeds, but when soft seeds become scarce, the birds will eat the hard seeds. Smith hypothesized that disruptive selection favors either individuals with large bills that can crack

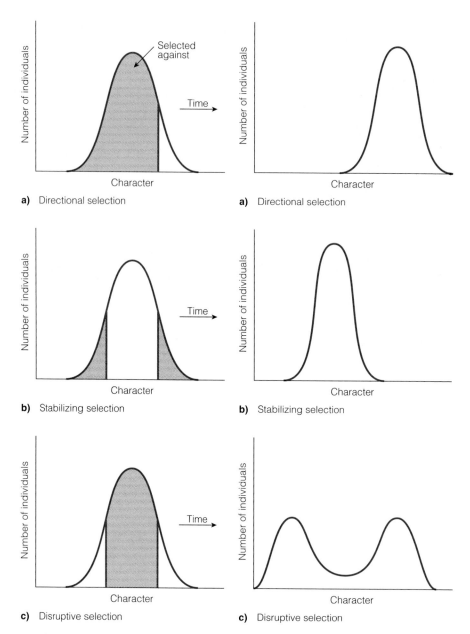

a) Directional selection

b) Stabilizing selection

c) Disruptive selection

a) Directional selection

b) Stabilizing selection

c) Disruptive selection

FIGURE 2.7 Changes in phenotypes under three modes of selection. The shaded regions represent phenotypes selected *against*.

the hard seeds or individuals with small bills that can more easily handle soft seeds.

The nature of an organism's interaction with its environment also determines the nature of selection. If certain genotypes have an advantage because they are able to obtain resources in competition with other members of the population, then more individuals with the best phenotype for obtaining the resource will survive. B. Wallace (1981) calls this **soft selection.** If, on the other hand, the interaction is directly connected to the physical environment, all individuals may die under certain circumstances (and under others, all may live). For example, if low temperatures kill

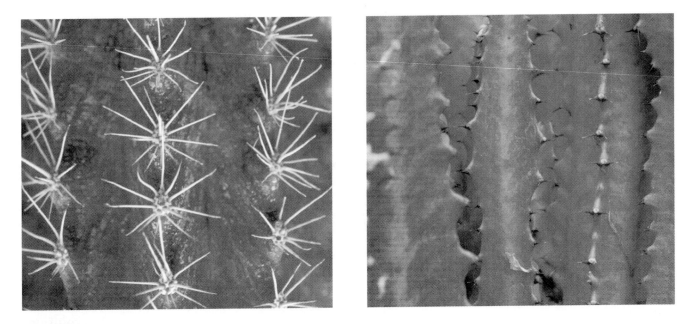

FIGURE 2.8 The spines of cacti and euphorbs represent convergent evolution. (*Photos by David Krohne*)

mosquitoes, an early fall frost may kill virtually every individual; the mortality rate is independent of population density. This is referred to as **hard selection.**

When the advantage of a phenotype changes with its frequency in the population, **frequency-dependent selection** is occurring. A number of factors can cause the fitness of an individual to change with the frequency of its phenotype. In some animal social systems, males with a rare phenotype are preferred by females (see Chapter 8); in such cases, selection will favor the rare phenotype.

The expression of sex may be influenced by frequency-dependent selection as well. In some mammals, genetic effects can alter the expression of sex. For example, in the wood lemming, *Myopus schisticolor,* a mutant allele on the Y chromosome, when combined with an X, produces some XY individuals that are female.

We might expect that if the allele that converts XY individuals into females increases in frequency, then at some point males will become rare. If the population is made up largely of females (both XX and XY females), a female that produces male offspring will then be at a selective advantage; her male offspring will enjoy a disproportionately large representation in the next generation, because many offspring will be sired by the few males in the population. Thus selection will favor the alleles that cause XY individuals to be male, and the sex ratio will shift back toward more nearly equal numbers of males and females. The advantage of the allele is purely a function of its rarity.

Natural Selection and Adaptation Adaptations are *heritable* features of organisms that increase their fitness. In our previous example, large bill size in *Geospiza* finches is an adaptation to food scarcity. The wings of birds and bats are adaptations that allow them to exploit resources more efficiently.

Adaptations can affect virtually any aspect of an organism: its morphology, physiology, biochemistry, or behavior. Of particular interest to biologists are the intricate matches between an organism's phenotype and its environment. For example, adaptations to cold climates in different animals involve a variety of aspects of the phenotype. The ears of arctic hares are short and thick relative to those of other hares; the reduced surface area decreases heat loss. Other adaptations, such as the hibernation of the arctic ground squirrel, are physiological. This species reduces its body temperature and metabolic rate to enter a torpid state during the coldest period of the year. Other small mammals respond behaviorally by confining their activities to the region under the snow, the subnivean space, where the insulative qualities of the snow result in a much warmer environment than the air above.

It is not unusual for selection to arrive at similar solutions in different organisms in response to the same ecological pressures. Similar adaptations in organisms with different phylogenetic histories are the result of **convergent evolution.** Examples of this process abound. The spines of cacti in the New World and of euphorbs in the Old World (Figure 2.8) represent convergent solutions to the problem of protecting succulent plants in arid environments. The placental and marsupial mammals diverged early in mammalian evolution. Although these groups have very different phylogenetic histories, many similar life-forms have evolved in similar environments (Figure 2.9).

Fitness and the Evolutionarily Stable Strategy

Natural selection acts to increase the average fitness of a population. As advantageous genotypes obtain proportionately higher representation, the average fitness of the population rises. As new genotypes and alleles enter the population through mutation and immigration, they affect the overall fitness of the population. A new genotype that is fitter than the current one will gradually replace it.

If the current genotype cannot be replaced by an invading one, it is said to represent the **evolutionarily stable strategy,** or **ESS** (Maynard Smith and Price 1973). If we plot the fitness of each possible genotype, we obtain a graph like that in Figure 2.10. If all the individuals in the population have genotype A, then a new genotype with greater fitness, genotype B, may be able to invade. The new genotype will eventually replace the old one. If the population contains only genotype C, however, neither A nor B can invade; each has lower fitness. Genotype C, then, is the ESS for this situation.

A number of ecological and evolutionary problems can be considered from the standpoint of ESS. The concept of the ESS has had an especially great impact on behavioral ecology, as we will see in Chapter 8. One reason for this is that in behavioral ecology, an organism must often make an evolutionary "decision"; that is, it must adopt a certain strategy. Should it reproduce sexually or asexually? Should it give an alarm call when a predator is sighted, thus warning other individuals but calling attention to itself? Should males remain with the females and provide more parental care, or should they abandon the females

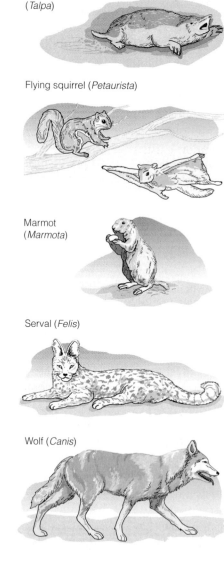

Marsupials

Pouched mouse (*Sminthopsis*)

Marsupial mole (*Notoryctes*)

Flying opossum (*Petaurus*)

Wombat (*Vombatus*)

Eastern native cat (*Dasyurus*)

Tasmanian wolf (*Thylacinus*)

Placental Mammals

Harvest mouse (*Mus*)

Common mole (*Talpa*)

Flying squirrel (*Petaurista*)

Marmot (*Marmota*)

Serval (*Felis*)

Wolf (*Canis*)

FIGURE 2.9

Similarities between pairs of Australian marsupial mammals and placental mammals represent examples of convergent evolution. (*Redrawn from Young 1981*)

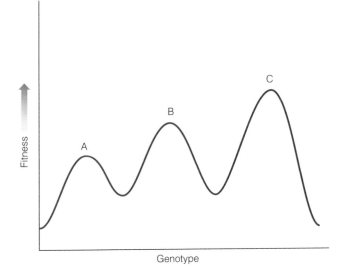

FIGURE 2.10 Fitness of three hypothetical genotypes, A, B, and C. Each genotype is composed of a large number of genetic loci. Because C has the greatest fitness, it represents the ESS. Genotypes A and B, whether they arise by mutation or immigration, cannot successfully invade a population containing genotype C.

to seek additional matings? Each of these dilemmas implies that some optimal strategy or solution exists. Phenotypic models of evolution using the concept of the ESS can be used to analyze such cases.

Optimality of Adaptation and Constraints on Selection

Biologists have been intrigued by the precise fit between phenotype and environment. Truly beautiful adaptations abound, and any student of biology has wondered at some point at the myriad "inventions" of evolution: the human eye, the speed and agility of the cheetah, the echolocation system of bats, the precise match between a flower and its pollinators. Consequently, questions concerning optimality will be important in any evolutionary/ecological study: Are the organisms optimally designed for the interactions in which they participate? Does a predator hunt in an optimally efficient manner? Are the wings of insects aerodynamically optimal? Is the breeding system of the hamadryas baboon optimal for its savanna habitat? Is the morphology of forest trees optimal for capturing light?

It is worth thinking about the meaning of the answers to these questions. We gain insight into evolutionary ecology when we consider cases of nonopti-

mal adaptations. There are at least three important explanations:

1. Perhaps our expectation is incorrect; what we envision as optimal may not in fact be the best solution.
2. Perhaps constraints on the evolution of the trait in question prevent an optimal solution.
3. Perhaps there simply has not been sufficient time for the adaptation to evolve.

The second and third answers to the question are the most important, for they suggest that selection is not omnipotent—that it has constraints. The constraints on evolution derive from the precise mechanism by which it occurs. There are four important sources of constraint.

First, selection can work only on the genetic variation that exists in a population. If all individuals in the population are genetically identical, it does not matter how intense selection is; the species cannot respond. This is implicit in the model of selection presented earlier. The rate of change in allele frequency depends not only on s, the selection coefficient, but also on the values of p and q.

Second, various aspects of the organism's biology may constrain adaptation. For example, developmental constraints may prevent adaptations. With a few exceptions, monocotyledonous trees such as palms are unbranched. Some dicots that live in the same habitats are branched, so there is clearly some adaptive advantage to branching. But monocots cannot branch because they are incapable of secondary thickening, a process that strengthens a twig enough to enable it to become a major branch. This developmental constraint limits the nature of morphological adaptation in monocots (Maynard Smith, 1989).

The fundamental organization of an organism is a product of its evolutionary history, and this history also constrains the adaptational options available. For example, vertebrates have evolved four limbs; consequently, we are not likely to see a bird with two legs, two wings, and a pair of arms with hands. Regardless of how strong the selection pressures might be, such a design is unlikely, given the historical organization of vertebrate morphology.

Given these constraints, we see that selection follows the path of least resistance. The adaptations it produces are, in the words of Stephen J. Gould, "contraptions" (Gould 1980). An adaptation begins as a modification of features already present in an organism. Thus we might envision some "optimal solution,"

FIGURE 2.11 The giant panda (*Ailuropoda melanoleuca*) shown with bamboo, its sole food source. (*Photo by Lynn Stone/Animals Animals*)

but because of the nature of the organism, a different adaptational pathway may be followed, or no adaptation may occur. The precise nature of the adaptation that results will be constrained by this principle.

Gould (1980) suggests that the evolution of the panda's "thumb" represents an example. Giant pandas (*Ailuropoda melanoleuca*) are specialists on bamboo; it is their sole food source (Figure 2.11). They strip the leaves from the bamboo with a "thumb" that

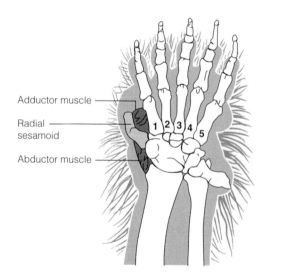

FIGURE 2.12 The anatomy of the panda's "thumb," which is not a true thumb but an extra digit derived from the radial sesamoid bone. (*From Gould 1980*)

is adapted for the purpose. If we examine the bones and muscles forming this sixth digit, however, we find that it is fundamentally different from the primate's opposable thumb. The panda's thumb is derived from the radial sesamoid bone of the wrist and its muscles (Figure 2.12). It may not be the optimal structure for stripping bamboo leaves, but, as Gould points out, it is functional and was "easier" to evolve than an opposable thumb because existing bones and muscles could be co-opted for this new function. It is a contraption, a contrivance that works. It could evolve because of the presence and nature of the radial sesamoid.

Third, physical laws constrain evolution. No matter how strong the selection, the laws of physics cannot be repealed. For example, the laws of aerodynamics determine the maximum weight an organism can have if it is to be able to fly with a given set of wings.

Finally, the nature of the selection process itself may constrain the adaptations it produces. Sewell Wright (1931) illustrated this point with his concept of the *adaptive landscape,* a topographic map of genotypes and their fitnesses.

In Figure 2.13 we have plotted the potential genotypes for a trait along the horizontal axes of a three-dimensional graph. The vertical axis represents the fitness of each genotype. A species at point a on this graph is reasonably fit; in the immediate vicinity are other, similar genotypes with *lower* fitness. A very different genotype (b) would result in even higher fitness, but to get from point a to point b, the population must travel down from its peak, across the landscape, and then up to the higher peak. Therein lies a problem. Selection, by its most fundamental nature, cannot move the population to lower fitness; selection will move the organism only to a higher fitness. Thus the population cannot make the initial move downhill, even though such a move would eventually allow it to move up to a higher peak.

How, then, does a population in this situation become fitter? Wright suggests that forces other than selection may be crucial. Recall from our consideration of the Hardy–Weinberg equilibrium that selection is but one of four mechanisms of evolutionary change. Random changes in allele frequency occur via the process of genetic drift; the population can move in *any* direction in genotype space by these stochastic changes. Wright suggests that drift is crucial to the adaptational process; it is the only force that can move the population downhill. If it does, selection can subsequently take the population uphill to a higher adaptive peak (Figure 2.13).

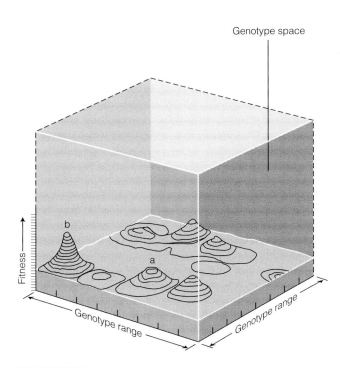

Genotype space

FIGURE 2.13 Wright's adaptive landscape. The two horizontal axes represent the range of different possible genotypes and define a region of "genotype space." The vertical axis depicts the fitness of each unique genotype. Points a and b represent two different genotypes and their fitnesses.

Two features of this system are important. First, genetic drift is more pronounced in small populations; hence adaptation may be more rapid when N_e is small. Second, changes produced by drift are unpredictable. Thus an adaptation that we expect might not occur simply because of these restrictions.

The four constraints outlined here limit the nature of organisms' adaptive responses to the environment. To expect every trait of every organism to be maximally efficient and adaptive is unrealistic (Gould and Lewontin 1979). We will examine numerous adaptations to the physical and biological environment in this text. One of the most difficult but most fascinating tasks for the evolutionary ecologist is to tease out the relationships among adaptation, optimization, and constraint.

We also noted that we may not observe some expected adaptations because there has not been enough time for them to appear in the course of evolution. It is crucial to understand that the evolutionary process is dynamic; at any moment, we see organisms at various stages of the evolutionary and adaptive process. The ecological interactions that we observe are colored by the different adaptive states of the species involved.

Sexual Selection and Kin Selection

Table 2.1 lists the primary mechanisms of evolution derived from the concept of Hardy–Weinberg equilibrium. We defined natural selection as the differential survival or reproduction of individuals according to their genotype. In this section, we examine two phenomena that were formerly thought to require fundamentally different adaptational processes. However, we will see that both are, in fact, variations on the fundamental process we have already discussed.

Sexual Selection

One phenomenon that fascinated Darwin was the appearance of bizarre and exaggerated adaptations in some species. He was aware that these adaptations often appear in only one sex—usually the male—and deduced that this was a clue to their origin.

Darwin understood that some of an organism's features are important in attracting mates. Most typically, it is the female that chooses a particular male from among a number of potential mates; this may be because males are more numerous than females (see Chapter 4 for a discussion of sex ratio). In addition, males can usually mate many times, whereas females can produce only a limited number of eggs. The result may be male–male competition for opportunities to mate with more than one female. Such additional matings by a relatively few males may preclude some males in the population from mating at all.

If females choose the males with which they will mate, then they constitute a selective force on males. Those males that attract mates are clearly fitter; any aspects of a male's phenotype that enhance his probability of being chosen by females will experience intense selection.

Darwin referred to this process as **sexual selection.** It operates in essentially the same way as natural selection: Alleles that increase fitness will increase in frequency. Darwin coined a separate term for the process, however, because, unlike natural selection, it can lead to traits that otherwise are maladaptive. Highly exaggerated morphological features that attract mates may be deleterious in other ways. The spectacular tail of the peacock is attractive to potential mates, but functionally it is hardly an optimal design for a tail. Indeed, because it is an energetically expensive structure that may hinder movement

a)

b)

c)

FIGURE 2.14 Examples of some exaggerated traits thought to have evolved by sexual selection. Each of the traits shown is found only in males of the species. (**a**) The male sage grouse has large, brightly colored inflatable sacs on the neck. (**b**) The peacock has an elaborate, brightly colored tail. (**c**) Males of the extinct Irish elk had huge antlers. (*From Gould 1973*)

and make males more vulnerable to predation, it may be a detriment. Many highly exaggerated traits presumably arose via sexual selection; some examples are depicted in Figure 2.14.

R. A. Fisher (1930) suggested that sexual selection operates in conflict with natural selection. He termed the former "runaway selection" because a sexually selected trait will become exaggerated, at some point becoming so embellished that it is a serious detriment to the overall fitness of the male. At this point, the trait stabilizes at an equilibrium phenotype that balances the two conflicting pressures.

Others have questioned this view. A. Zahavi (1975) suggests that the elaboration of sexually selected traits may in fact represent "truth in advertising." The argument is based on the assumption that females choose mates according to their phenotype as indicators of the general fitness of the male. In the case of the Irish elk (Figure 2.14c), males with large antlers may have originally been selected by females because genes for such an effective defensive weapon would be desirable in offspring. As the trait was exaggerated by sexual selection, such massive antlers gradually came to be less useful as a weapon or anti-predator adaptation. The trait may, however, have come to symbolize fitness in a new way: Only those males who were extremely fit in the general sense would be able to carry such an absurdly large set of antlers. By mating with such a male, females could ensure that their offspring would receive a set of highly adaptive genes.

Kin Selection

Behavioral data collected by some ecologists have extended the concept of adaptation to include components of the organism's behavior. For many kinds of behaviors, including predator avoidance, habitat selection, and courtship, the adaptive advantage to an organism is clear; a behavior that reduces the risk of predation is of obvious selective value. Genes that promote such behavior would surely increase in frequency in precisely the same way as genes that modify physiology to match the climate or genes that increase a male's chance of being chosen by a female.

These data reinforce the importance of natural selection because they explain how this process shapes yet another aspect of organisms. However, certain behaviors—in particular, behaviors classified as altruistic—seem more difficult to explain in terms of natural selection. **Altruistic behaviors** are behaviors that enhance the fitness of other individuals in the population at the expense of the individual performing them. A classic example of altruistic behavior is the alarm call. If an individual detects the presence of a predator and gives a warning call, the fitness of conspecifics increases because they have a better chance to escape, but the caller's fitness decreases if the call attracts the attention of the predator.

If such behaviors have a genetic basis, then elimination of altruists should cause the frequencies of the altruistic genes to decline. We would not expect alarm calling to be an ESS; it should always be invaded by genes for not calling. How, then, do these altruistic behaviors persist?

W. D. Hamilton (1964) suggested that we need to expand our definition of fitness and our notion of selection. He defined **inclusive fitness** as the relative ability of an organism to get its genes, or copies of them, into the next generation. The key feature is the notion that copies of the genes are of value. Hamilton suggested that it is irrelevant to the evolutionary process whether an individual passes on its own genes or enhances the transmission of copies of them in another individual.

Who carries copies of an individual's genes? Obviously, relatives are the most likely to carry copies. In fact, it is a simple matter to calculate the proportion of one's genes that are expected to be carried by relatives. Offspring receive half their genes from each parent; thus half an offspring's genes are maternal and half are paternal. The average proportion of genes shared by two individuals is called the **coefficient of relationship, *r*.** Values of *r* for different relatives are shown in Table 2.2.

To return to Hamilton's argument, an altruistic allele can increase in frequency if it increases the inclusive fitness of the individual; that is, it can increase in frequency if the individual transmits the gene or if copies of the gene are transmitted by relatives. According to Hamilton, we can quantify the probability that an altruistic allele will increase in frequency. It can increase if

$$k > 1/r$$

where *k* is the ratio of the increase in fitness of the recipient to the decrease in fitness of the altruist, and *r* is the average coefficient of relationship between the altruist and the beneficiaries of the altruistic act. Thus if I commit an altruistic act toward my brother (*r* = 0.5), the genes for this behavior will increase if I more than double his fitness:

$$k > 1/(0.5) = 2$$

The altruistic act may decrease the fitness of the altruist, but if the loss is offset by enhancing the fitness of enough relatives such that the same number of genes ultimately make it to the next generation, then the trait is adaptive. Hamilton once quipped, "I would lay down my life for two brothers or eight cousins!"

Does such **kin selection** occur in nature? One of the most convincing demonstrations that it does comes from a study of alarm calling in Belding's ground squirrels (*Spermophilus beldingi*) by P. W. Sherman (1977). This species, which lives in groups in high mountain meadows in the Sierra Nevada, gives a distinctive alarm call when a predator is sighted. Not all individuals call, however; females are much more likely to give alarm calls than males. This makes sense when we consider the genetic relationships common in the groups. Because males disperse from their natal group soon after they are weaned, the typical group composition is a matriline: a female, her female offspring, and other female relatives (her sisters, her mother, and so on). A female was found to be more likely to give an alarm call when female relatives were present than when they were not. These results are entirely consistent with kin selection. Males who disperse from the kin group are unlikely to call; they would be warning nonrelatives.

It is important to note that no conscious process or reasoning is involved here. Females do not need to be "aware" that relatives will hear their alarm calls; all that is necessary is that there be genes that cause females to call and males to refrain from calling. Over time, those genes will increase in frequency because the social organization is such that by calling, females enhance their inclusive fitness. Males' fitness is higher if they do not call.

The concept of kin selection has been an important addition to evolutionary theory. As we will see in Chapter 8, it can be used to explain a number of unusual behavioral phenomena.

TABLE 2.2

Values of *r*, the Coefficient of Relationship, for Different Relatives in a Diploid Species

Relationship	Average Fraction of Genes Shared, *r*
Mother/child	0.5
Father/child	0.5
Full siblings	0.5
Half siblings	0.25
Grandparent/grandchild	0.25
Aunt or uncle/nephew or niece	0.25
First cousins	0.125

The Evolution of Interactions among Species

Throughout this chapter, we have alluded to the fact that other organisms constitute an important part of an organism's environment and hence exert selective pressure. These selective effects and the interspecific interactions they spawn can become quite complex.

Mimicry

Mimicry is the physical resemblance of two or more species resulting from inherent advantages of similar appearance. Mimicry is different from convergent evolution, which we defined as the production of similar adaptive solutions to an ecological problem in different taxonomic groups. As a result, the two taxa may resemble each other (see Figure 2.9). The resemblance of cacti and euphorbs (see Figure 2.8) is a by-product of their independent adaptations to herbivory. In contrast, in mimicry, it is the similar appearance itself that is selected for.

We can usually detect which species has come to resemble another. The imitative species is called the **mimic;** the species it resembles is called the **model.** There are three fundamental types of mimicry: Batesian, Müllerian, and aggressive.

In **Batesian mimicry,** a benign species resembles a noxious or dangerous one. The colubrid snakes (genus *Pliocercus*) of Central America are Batesian mimics; they resemble the highly poisonous coral snakes (genus *Micrurus*) found in the same region (Figure 2.15). The degree to which the colubrids mimic the coral snakes is remarkable. Despite considerable geographic variation in the color patterns of coral snakes, in each region a race of *Pliocercus* mimics the local coral snake very precisely (Greene and McDiarmid 1981). Predators and other animals avoid snakes with the striking coloration patterns of coral snakes. Thus colubrids, which are not venomous, gain an advantage from their resemblance to the more dangerous coral snakes.

The operation of mimicry systems depends on both learned and innate responses of the other species that interact with the models and mimics. In sampling possible food items, young birds learn to avoid the striking, orange-and-black-patterned monarch butterfly (*Danaus plexipus*), which is distasteful (Brower 1969). Thus the butterfly's conspicuous **aposematic coloration** serves as a warning to predators. In the case of the coral snake/colubrid system, the avoidance of snakes with red, yellow, and black bands is probably

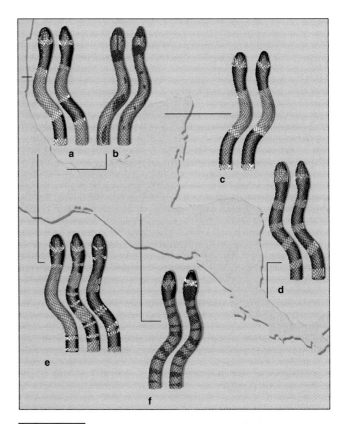

FIGURE 2.15 An example of Batesian mimicry: similarities in color patterns of colubrid snakes (*Pliocercus*) and coral snakes (*Micrurus*) in different Central American locations. The presumed models (coral snakes) are on the left; the mimics are on the right. In (**e**), two colubrids living in the same locality mimic one coral snake (center). (*From Greene and McDiarmid 1981*)

innate (Greene and McDiarmid 1981). Whether the avoidance is innate or learned, the mimics benefit from their resemblance to the noxious model.

One interesting feature of this kind of mimicry system is that the mimics must be less common than the models. If the mimic becomes more common than the model, its adaptive advantage will disappear. The same principle applies to innate avoidance mechanisms; selection will not favor the avoidance of a particular color pattern if it more frequently signifies a benign species.

Sometimes several noxious species come to resemble each other. This is called **Müllerian mimicry.** The yellow and black patterns of many bees and wasps exemplify this kind of mimicry. The advantage to each member of a group of Müllerian mimics is that the avoidance response that other species develop is reinforced when many species share the same aposematic coloration.

An apparent vertebrate example of Müllerian mimicry occurs among birds of the genus *Pitohui* in New Guinea. The skin and feathers of the hooded pitohui

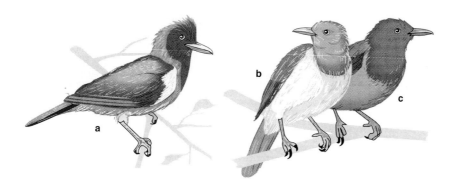

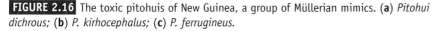

FIGURE 2.16 The toxic pitohuis of New Guinea, a group of Müllerian mimics. (**a**) *Pitohui dichrous*; (**b**) *P. kirhocephalus*; (**c**) *P. ferrugineus*.

(*P. dichrous*) have been found to contain homobatrachotoxin (Dumbacher et al. 1992), a highly toxic alkaloid compound that causes paralysis and convulsions. (It is closely related to the toxin found in the poison dart frog—a remarkable example of convergent evolution.) The striking orange and black plumage of *P. dichrous* (Figure 2.16) presumably constitutes aposematic coloration that acts to deter potential predators; four races of the variable pitohui (*P. kirhocephalus*) and the rusty pitohui (*P. ferrugineus*) resemble the hooded pitohui and also contain the toxin.

The third type of mimicry is **aggressive mimicry,** in which a noxious or dangerous species comes to resemble a benign one. Many predators adopt this strategy. For example, the zone-tailed hawk (*Butea albonotatus*) of the southwestern United States seeks its small-mammal prey while soaring. Whereas most hawks glide with their wings held horizontally, zone-tailed hawks glide with their wings in a V shape like the harmless turkey vulture, *Cathartes aura* (Willis 1963). Moreover, these hawks often soar among flocks of vultures, so their prey do not recognize them as dangerous.

Coevolution

In mimicry systems, one or more mimic species take adaptive advantage of the presence of a model species, but the model plays a passive role in the mimic's evolution. Other interactions between species are reciprocal—an evolutionary change in one species may constitute a new selective force on another. If the second species responds, then the environment has changed for the first species, and it may respond in turn.

P. R. Ehrlich and P. H. Raven (1964) coined the term **coevolution** to refer to one instance of this kind of interaction: the interactions between butterflies

and the plants on which they feed. Other authors use the term to refer to interactions in which each species has exerted a crucial selective force on the other; that is, they define coevolution as the interdependent evolution of two or more species that have an obvious ecological interaction (Lincoln, Boxshall, and Clark 1982). This second definition is more general and perhaps less precise than the original.

The current trend is to use the term in a much more restrictive sense that includes only those situations in which direct genetic change in one species is attributable to genetic change in the other (Janzen 1980; Futuyma and Slatkin 1983). Some authors emphasize that each species faces a host of selective pressures from other species, rather than having a simple one-to-one genetic correspondence with a coevolutionary partner (Futuyma 1986; Schluter and Grant 1984; Rummel and Roughgarden 1985).

Some of the most fascinating ecological and evolutionary phenomena occur among sets of interacting species. A number of kinds of potentially coevolutionary interactions among species are summarized in Table 2.3. In some interactions, such as parasitism, one species benefits to the other's detriment. In others, the species mutually rely on each other. In addition, the interaction may be facultative—that is, essential for neither species—or it may be obligate, in which case both species depend on it. For example, in some cases of the coevolutionary interaction between pollinators and flowers, the association is absolute: A particular pollinator is the only means of pollen transfer for that plant species. In other plants, insect pollination is required, but any of a number of species can accomplish the task. In fact, some of these interactions

TABLE 2.3

The Effects (Positive, Negative, or None) of Various Kinds of Interactions among Species

Association	Effects	
	Species A	Species B
Parasitism	+	−
Commensalism	+	0
Mutualism	+	+
Predation/herbivory	+	−
Competition	−	−

are so specialized that ecologists have sometimes discounted their general ecological importance. However, we are coming to understand the important role they play in shaping ecological systems and their role in biological control (Thompson 1999; Burdon and Thrall 1999).

Next we consider a few examples of evolutionary relationships among interacting species. Our discussion is by no means exhaustive; rather, it is intended simply to describe the nature of some of the interactions that occur. In some cases, such as predation and competition, a full discussion of the evolutionary relationships involved can be found in the chapters devoted to those topics.

Parasitism   **Parasitism** is an interaction in which individuals of one species—the parasite—derive their nutrition from the living tissues of a host species. The parasite benefits from the interaction, but its host's fitness is lowered. When we think of parasites, internal parasites come most frequently to mind. Two examples of internal parasites are the beef tapeworm of humans and the sheep liver fluke. Note that the life cycles of these parasites (Figure 2.17) have two features in common. First, each has a dispersal phase in which the parasite seeks a new host. The parasite's eggs must be relatively resistant to abiotic factors, for it is this stage that leaves the relatively benign environment of the host to travel to the next individual to be infected. In these cases, this process is facilitated by an intermediate host: cows for the tapeworm and snails for the fluke.

Second, both life cycles have a reproductive phase in which huge numbers of offspring are produced. This too is an adaptation to the risks involved in finding a new host. Because only a tiny fraction of the individuals dispersed from one host will ever reach another host, the parasite's reproductive strategy includes the production of huge numbers of offspring. Adult tapeworms are essentially egg-production machines. Because the adults inhabit the relatively benign and stable environment of a mammalian digestive tract, they do not need adaptations to protect them from abiotic factors. They do not need digestive systems, because they can simply absorb predigested nutrients from their hosts. They can devote the resulting energetic and morphological "savings" to reproduction.

The host species is also under intense selection imposed by the presence of the parasite. Because the parasite reduces the fitness of the host, anything the host can do to resist infection or destroy the parasite

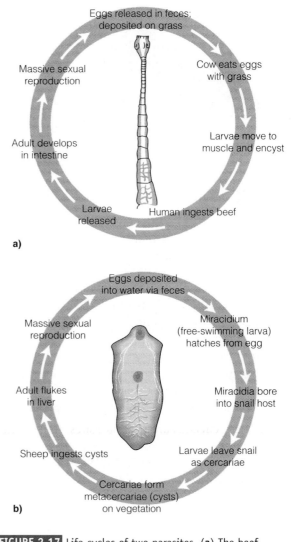

FIGURE 2.17 Life cycles of two parasites. (**a**) The beef tapeworm of humans (*Taenia saginata*). (**b**) The sheep liver fluke (*Fasciola hepatica*).

will be selected for. In vertebrates, the immune system is particularly important in this regard. The production of antibodies to the parasite constitutes an important selective pressure on the parasite, and evolution results. The parasite adopts strategies to avoid the host's immune system, including changing its external proteins frequently or producing external proteins similar to those found in the host and thus confusing the host's immune system.

Parasites are under selection to decrease their virulence. This may seem surprising, given that infecting hosts and usurping their resources are the essence of the parasites' subsistence. However, it is clearly disadvantageous for a parasite to kill its host quickly, because it could then perish before reproducing and disseminating. Thus the optimal strategy for the parasite is to use the host's resources but not kill it. Gradually, over evolutionary time, parasitism is perhaps replaced by **commensalism,** an interaction in which one species benefits but the other is neither helped nor harmed.

An example of this process can be seen in the interaction of myxomatosis virus and hares (Fenner and Ratcliffe 1965). When European hares (*Oryctolagus cuniculus*) were introduced into Australia, they multiplied so rapidly that they soon became a scourge on the landscape. One of the control methods instituted was the intentional introduction of myxomatosis virus, which is virulent to infected hares. After a few years, however, the mortality rate of infected hares declined, for two reasons. First, as expected, the hares developed resistance to the parasite; second, the virus became less virulent. That the virus had changed was clearly demonstrated when previously unexposed genetic lines of hares were experimentally infected; the once-lethal virus was no longer so virulent.

The range of interactions that constitute parasitism is vast. Internal parasites represent one end of the range; at the other end of the spectrum are external parasites such as the parasitic cuckoos in the genus *Cuculus.* These cuckoos are nest parasites; that is, they lay their eggs in the nests of their hosts. The hosts incubate the parasites' eggs and raise the hatchlings as their own. The negative impact on the hosts occurs because they invest energy in the offspring of another species at the expense of their own young. The host and parasite each constitute an important facet of the other's environment, and hence each exerts a strong selective influence on the other.

One of the most thoroughly studied nests parasites is the reed warbler cuckoo, a race of *Cuculus canorus,* which parasitizes the reed warbler (*Acrocephalus scirpaceus*). The cuckoo is highly adapted for parasitism. The female quietly watches the warbler's nest, and when both members of the pair are away from the nest, she glides in and deposits an egg. She also removes one warbler egg from the nest. The entire process requires less than 10 seconds. The cuckoo's egg closely resembles the warbler's; artificial eggs that do not match are quickly removed by the host (Davies and Brooke 1991). When the cuckoo chick hatches, it is fed the same food the warbler chicks receive. It grows extremely quickly and has its own adaptations to enhance its survival at the expense of the warbler chicks. Eventually, the cuckoo chick balances its warbler nestmates on its back and pushes them out of the nest, thus ensuring more food for itself.

Mutualism   When both species benefit from the interaction, we refer to it as **mutualism.** In some simple mutualistic relationships, neither species is highly adapted for the interaction, and often the relationship is facultative rather than obligate. Other mutualisms represent complex morphological, physiological, or behavioral adaptations to an association that is necessary for the survival of both species.

The relationship between hermit crabs and some sea anemones is an example of a simple, facultative mutualistic relationship (Vermeij 1983). Sea anemones attach themselves to hermit crab shells and then use their stinging spines to protect the crabs from some predators such as octopuses. The anemone presumably benefits from the interaction because the crab provides some mobility for this otherwise sessile creature.

One complex and highly evolved mutualistic interaction occurs between tanagers and mistletoe. Both species possess specialized morphological adaptations to the interaction. Mistletoe is a parasitic species that lives on trees. Its seeds are adapted to enhance its dispersal by tanagers (Figure 2.18a). The seed's outer pulp layer is highly nutritious, has a high sugar content, and can be easily digested by a feeding tanager. Inside the pulp is a layer of vicin, a substance that protects the seed from digestion and enhances its passage through the bird's gut. The vicin layer also sticks to the bird's cloaca, causing the bird to rub the seed off on a tree branch, where a new mistletoe plant can begin to grow.

The tanager's gut is also highly adapted for the interaction (Figure 2.18b). In most birds the gizzard is an important digestive structure that contains small stones and gravel that the bird ingests. Contraction of the organ's muscular walls churns the abrasive stones and grinds the bird's food. In some tanagers the gizzard lies to the side of the gut as a blind sack; its entrance can be opened or closed by a sphincter muscle. If the bird eats an insect, the sphincter opens and the insect is ground up in the gizzard; if a mistletoe seed is eaten, the sphincter closes off the gizzard so that the seed is not destroyed. The bird digests the nutritious pulp and eliminates the rest of the seed in the feces. In some tanager species that specialize on mistletoe, the gizzard has been almost entirely eliminated.

Another important class of mutualistic interactions involves plants and their pollinators. In the flowering plants, pollen produced in the male flower parts must be transferred to the female flower parts. Obviously, it is advantageous if the transfer takes the pollen to a

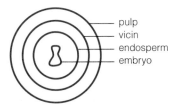

a) Mistletoe Seed

Typical insect-eating bird's gut

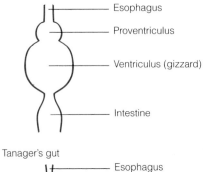

Tanager's gut

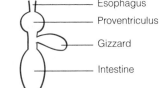

Mistletoe-specialist tanager's gut

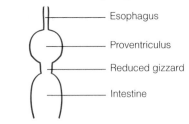

b) Avian Digestive Tracts

FIGURE 2.18 (**a**) Structure of a mistletoe seed. The vicin layer helps the seed pass intact through avian digestive tracts. (**b**) Anatomy of digestive tracts of various kinds of birds. In the mistletoe-specialist tanager, the gizzard has been almost entirely eliminated.

different plant, because cross-pollination produces genetic variation. Although many plants depend on the wind to transfer pollen, others use various strategies to attract animal pollinators. Flower color is a simple adaptation that indicates to the pollinator a potential reward for visiting the flower. The eyes of bumblebees are particularly sensitive in the blue range of the spectrum, and thus many bee-pollinated flowers are pigmented in that way. Many hummingbird-pollinated flowers are red tubular flowers.

Both nectar and excess pollen serve as food rewards for pollinators. These floral rewards can be quite pre-

cisely adjusted to the needs of the pollinator. In some nectar-producing plants, the sugar content of the nectar produces a viscosity appropriate to the diameter of the insect's proboscis, facilitating ingestion of the liquid. In other plants, special nutrients such as essential amino acids or oils are incorporated into the nectar. Some orchids even produce fragrances that the insect incorporates into its own sex pheromones.

In many cases, the degree of coevolution and mutualism between pollinator and flowering plant is truly remarkable. One of the most extraordinary examples of this is the relationship between the yucca (*Yucca* spp.) and its pollinator, the yucca moth (*Tegeticula yuccasella*) (Figure 2.19). These species have an obligate relationship—reach is dependent on the other. The yucca's large white flowers remain open at night, when they are pollinated by the nocturnal yucca moths. A female moth enters the flower, climbs up a stamen, and deliberately scrapes pollen from the anther, using a pair of specialized mouth parts called palps. She tucks the pollen under her chin, moves to other stamens, and accumulates a ball of pollen. She next flies to another flower on the same or a different plant, where she thrusts her ovipositor through the wall of the flower's ovary and lays a single egg in the ovary cavity near the yucca's ovules. She then walks to the top of the pistil and thrusts the pollen into the stigmatic opening, pushing the pollen inside with her tongue and effecting pollination. As the yucca seeds

FIGURE 2.19 The yucca and its moth pollinator, an example of an obligate mutualistic relationship. (*Photos by David Krohne*)

develop, the moth's egg hatches and a larva emerges. It eats its way through the middle row of seeds in the pod, gnaws a hole in the wall of the seed pod, and emerges. It then spins a thread of silk to lower itself to the ground, where it spins a cocoon around itself. The emergence of the adult moth coincides with the flowering of the yucca the next season.

Clearly, both species benefit. The yucca moth is a highly effective pollinator because it directly forces the pollen into the stigmatic opening. Later, the yucca seeds provide important nutrition for the newly hatched larva. The larva never eats all the seeds, however; it eats a few but leaves others intact and available to propagate the plant.

The coevolution between a flowering plant and its pollinator can drive both members of the interaction to extreme morphological specialization. The star orchids of Madagascar participate in such an interaction with the hawk moths that pollinate them. Like all orchids, star orchids have stamens that are fused to the style. The resulting structure, a column, is a bisexual organ. At the tip of the column, pollen grains are produced in cohesive groups called pollinia, which detach and are transferred en masse during pollination.

An intimate coevolution has taken place between one star orchid species, *Agraecum sesquipedale,* and its hawk moth pollinator, *Panagena lingens.* The orchid is characterized by an extremely long floral tube, and the hawk moth has an incredibly long proboscis. If pollination is to occur, the floral tube of the orchid must be longer than the moth's proboscis, so that the moth must press up against the column to reach the nectar reward at the base of the tube. In the effort, it contacts and picks up the pollinia on the tip of the column. If the proboscis were longer than the floral tube, the moth could more easily reach the reward without effecting pollination.

The coevolutionary aspect of this interaction is thought to proceed as follows: Selection on the orchid favors the lengthening of the floral tube, because pollination depends on the moth pushing against the column. Selection on the moth favors lengthening the proboscis so the nectar reward can be reached. The resulting system of mutually reinforcing selection pressures progressively increases the length of the tube and the proboscis and leads to the bizarre morphology of both species.

Conditions leading to mutualism The ecological distribution of mutualistic relationships suggests some of the conditions necessary for its evolution. Ecologists have long recognized that the examples of mutualisms

are more common at low latitude. For example, Orians (1974) indicates that many mutualistic systems end not far north or south of the equator. Ant–plant mutualisms do not extend north of 14°N. Nectarivorous and frugivorous bats do not occur north of 33°N. Finally, there are few orchid bees or plants with extrafloral nectaries (nectaries that provide nutrients to specialized mutualists such as ants) north of 24°N.

The general interpretation of this distribution is that the tropics, with its stable and benign climate and infrequent disturbance, nurtures the long-term evolutionary relationships between species necessary for mutualism. In more northern (or southern) latitudes, harsh climate, short growing season, and a relatively young ecology (due to recent glaciation) mitigate against the evolution of mutualistic relationships. Nevertheless, mutualism does occur in many ecosystems. Taxa differ considerably in the proportion of mutualistic relationships they maintain. For example, note in Figure 2.20 the variation among insect taxa

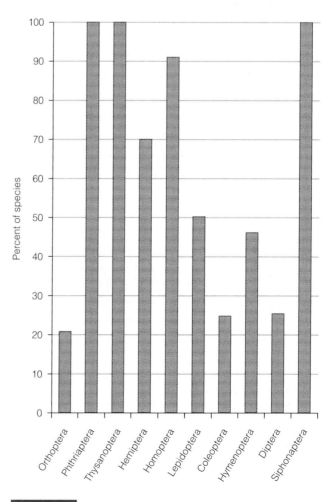

FIGURE 2.20 The proportions of various insect taxa in the British Isles that have mutualistic microorganisms. (*Data from Buchner, 1965*)

that maintain mutualisms with microorganisms in the British Isles.

Can we determine the evolutionary conditions under which mutualism will evolve? Keeler (1981) developed models to address this question. She began with the assumption of a population polymorphic for three variants. Some proportion of the population consists of successful mutualists; a second group contains unsuccessful mutualists, species that provide but do not receive benefits; and the rest are nonmutualists, species that do not provide or receive benefits. We can describe the selection coefficients of the three morphs in the population as follows:

Successful $s_{ms} = (H)(1 - A)(1 - D) + I_A + I_D$
mutualist:

Unsuccessful $s_{us} = (H)(1 - D) + I_A + I_D$
mutualist:

Nonmutualist: $s_{nm} = (H)(1 - D) + I_D$

where

> H = proportional loss of plant tissue if there is no defense
>
> D = the protection of plant tissue by defenses other than mutualists
>
> A = the amount of defense by mutualists
>
> I_A = the cost of benefits provided to the mutualists
>
> I_D = the cost of defenses

The model is based on a cost–benefit analysis in which mutualism can evolve only if the net benefits outweigh the costs. We can analyze this situation with selection models similar to those previously described in which we use fitness as a measure of benefit and cost. In this context, then, the conditions favoring mutualism are represented by the equation

$$p(1 - s_{ms}) - q(1 - s_{us}) > 1 - s_{nm} \qquad (4)$$

This equation represents the situation in which the total fitness of a population consisting of successful and unsuccessful mutualists exceeds that of a population of nonmutualists.

Competition Competition is the interaction between two species over a limiting resource that negatively affects their population growth rates (see Table 2.3). Because competition has been viewed for many years as one of the central interactions in ecology, we devote an entire chapter (Chapter 9) to it.

The process of natural selection is based on the relative ability of individuals to survive and reproduce. Because access to limiting resources constitutes an important determinant of fitness, competition has important evolutionary effects. Of interest to us here is the process known as character displacement.

Figure 2.21a depicts the pattern of resource use for two species whose use of the resources overlaps—that is, the species compete for the resources. Note that not every individual uses the same resources. If these intraspecific differences in resource use are heritable, natural selection can alter the pattern of resource use. We would expect individuals of species 1 on the far left of the curve to have more resources available to them because they do not compete for these resources with species 2. Similarly, the individuals of species 2 on the far right of the curve face no competition from species 1. The availability of additional resources should result in more offspring for the individuals that don't compete and, consequently, cause a shift, via natural selection, in the resource use by the two species (Figure 2.21b). Such an evolutionary shift in resource use to avoid competition is called **character displacement** and is examined more fully in Chapter 9.

Predator–Prey and Herbivore–Plant Evolution

Predator–prey and herbivore–plant interactions are among the most complex evolutionary and ecological systems. As with competition, we devote an entire chapter (Chapter 10) to these interactions. Because **predation** (the capture, killing, and consumption of an animal, usually by another animal) and **herbivory** (the consumption of plant tissue by an animal) are life-and-death matters for both parties in the interaction, they result in intense selective pressures. The predator faces intense selection to locate and capture its prey; its prey is under pressure to avoid detection or resist capture. The same is true for herbivores and the plants they consume.

Predator–prey and herbivore–plant interactions often result in a series of escalating adaptations. An adaptation by the predator provides it with an advantage, which then exerts pressure for a response from the prey. The predator responds to the prey's response with further adaptation, and so on. This pattern of escalations is referred to as a **coevolutionary arms race.** We will discuss examples of such interactions in considerable detail in Chapter 10. Here we will consider just two examples to illustrate the process.

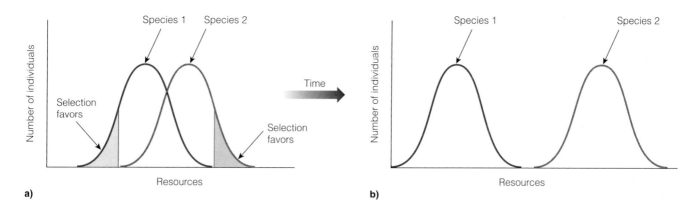

FIGURE 2.21 Diagram of the process of character displacement. In (**a**), the resource utilization curves for two species overlap. Selection favors those individuals in each species that do not use resources used by the other species. Over time, the resource utilization curves may separate (**b**).

In the New World tropics, the frog-eating bat, *Trachops cirrhosus,* is a predator that specializes on frogs (Tuttle 1982). These bats use their sophisticated hearing to home in on the mating calls of frogs in the dark. Some of the frogs on which these bats prey have responded to predatory selection pressure by producing toxins that make them unpalatable and poisonous to mammals. The adaptive response by the bats has been to discriminate between the calls of poisonous and nonpoisonous frogs. M. D. Tuttle demonstrated this by playing recordings of the calls of toxic and nontoxic frogs to bats in a darkened room. The bats attacked the source of the calls of nontoxic frogs but avoided those of the poisonous species.

Herbivores exert strong selection pressure on plants to deter grazing. A number of plant adaptations have resulted from this selection. Many defenses are morphological: Spines, hairs, and thick, rough foliage all deter grazing. Others are chemical: Plants produce a huge array of toxic and noxious compounds that deter herbivores. These compounds were once thought to be toxic by-products of metabolism that the plant isolated from its physiology by sequestering them in leaves or stems, where their presence secondarily served to deter herbivores. Because of this early interpretation, these compounds are still referred to as **secondary compounds.** The current view of them, however, is that they represent specific biochemical adaptations by plants to the selective pressure of herbivores (Ehrlich and Raven 1964). Mutations enhancing the de novo production of specific compounds are strongly selected.

As in predator–prey systems, herbivores and plants exert selective effects on one another. One example is the interaction of the milkweed (*Asclepias* spp.) and the larvae of the monarch butterfly (*Danaus plexipus*).

Milkweeds produce a number of noxious compounds to deter grazing, including cardiac glycosides that affect the rhythm of the heart in mammals. Milkweeds are the primary food source for monarch larvae, however, and the larvae have evolved the capacity to sequester the toxins in their bodies. In this way they are not only able to keep feeding on the milkweed without harm but are also able to co-opt the plant toxins for their own use: The sequestered chemicals are what makes the monarch butterfly noxious to its predators!

Another group of plants, the Umbelliferae or carrot family, have evolved furanocoumarins to thwart herbivores. These highly toxic compounds are capable of cross-linking and inactivating DNA. Some members of the moth family Oecophoridae and butterflies in the genus *Papilio* have evolved ways to detoxify these compounds. Some of these adaptations are biochemical, others behavioral. The caterpillars of one species feed only in rolled-up leaves. This protects them from the sun—and hence from the toxins—because the cross-linking of DNA by furanocoumarins is a photochemical reaction. Species that produce no furanocoumarins are fed upon by generalist herbivores, whereas those with a broad range of toxins (angular and linear furanocoumarins) are fed upon only by extreme specialists that have evolved the means to thwart the effects of the toxins.

L. VanValen (1973) has referred to the difficulty faced by species locked in a coevolutionary arms race as the "Red Queen Problem." This is, of course, a reference to the Red Queen in Lewis Carroll's *Through the Looking Glass,* who had to run faster and faster just to stay in the same place. This is an apt analogy: The environment constantly changes, and organisms must continue to evolve to survive. It also points out a significant feature of the evolutionary process: Fitness is

Loss of Genetic Variation

This chapter began with a discussion of the nature of variation in natural populations. We saw that for many traits, plotting the distribution of phenotypes produces a bell-shaped curve. This phenotypic distribution is a consequence of the underlying genetic variation in the population.

Heritable variation is essential for adaptation by natural selection. For some species threatened with extinction, loss of genetic variation is a significant problem. The adaptability, and thus the long-term survival, of some species may be threatened if genetic variation does not provide the raw material on which selection can operate. Moreover, the small populations that are characteristic of threatened species may lead to still more loss of genetic variation.

Small population size can lead to a loss of genetic variation in several ways. First, we have seen that the effects of genetic drift are greater in small populations and that one of the consequences of drift can be the loss or fixation of alleles (Figure 2.3). Thus small populations may lose genetic variation when allele frequencies change stochastically until one allele is lost.

Second, small populations are subject to inbreeding effects. Inbreeding results from the mating of relatives, who are individuals with a high proportion of identical genes. The formal definition of inbreeding hinges on the concept of genes that are identical by descent. Two copies of a gene are **identical by descent** if they can be traced to a copy of that gene in a common ancestor. When relatives mate, the probability of offspring containing genes that are identical by descent increases. We define a population as inbred if the probability that offspring receive copies of a gene that are identical by descent is greater than it would be in a purely randomly mating population. Obviously, a small population is more vulnerable to inbreeding than a large, randomly mating one.

Inbreeding does not change allele frequencies, but it does change the frequen-cies of the genotypes. In particular, it increases the frequency of homozygotes at the expense of heterozygotes at a rate of $0.5N$ per generation, where N is the population size. This phenomenon has an important effect: Deleterious homozygous recessive genotypes come together more frequently. The resulting decrease in fitness—an organism's ability to survive and reproduce—is called **inbreeding depression.** Species and populations respond very differently to inbreeding. Plants that self-pollinate regularly are clearly not negatively affected; other species show negative effects with only mild levels of inbreeding.

Newman and Pilson (1997) examined these phenomena experimentally by establishing populations of the annual plant *Clarkia pulchella* that differed in N_e. Populations with small effective sizes had lower fitness than large populations. Heschel and Paige (1994) demonstrated similar effects in field populations. They studied ten populations of the plant scarlet gilia (*Ipomopsis aggregata*). Populations of fewer than 100 individuals were found to have smaller seed size and lower germination success, and they were more susceptible to environmental stress. When these researchers introduced pollen containing new alleles from distant populations, seed size and germination success improved.

A third phenomenon that can result in loss of genetic variation is a population "bottleneck" in which the population undergoes a drastic decline and subsequent recovery. When the population experiences a severe reduction in size, not only does it become subject to the long-term effects of drift, but certain genotypes may be lost by chance in the population crash. When the population subsequently expands, some of its previous genetic variation may no longer be present.

Wildlife biologists have observed population bottlenecks; some of them result from human activities. For example, the northern elephant seal, *Mirounga angustirostris,* was hunted nearly to extinction along the west coast of North America in the nineteenth century. When the species finally became protected, fewer than 100 individuals remained. Today, they number more than 30,000. When M. L. Bonnel and R. K. Selander (1974) surveyed 21 protein loci using starch-gel electrophoresis, they found that none of the loci are polymorphic in this species—virtually all the genetic variation is gone.

Bottlenecks can occur for natural reasons as well. Recent research (O'Brien, Wildt, and Bush 1986) has shown that some populations of the cheetah (*Acinonyx jubata*) have extremely low genetic variation; 55 loci surveyed in South African populations of this species were monomorphic. The researchers found that the DNA sequences of individual cheetahs varied by less than 1 percent. By contrast, such sequences in humans vary as much as 32 percent. Skin grafts among seven cheetahs resulted in no tissue rejections! Clearly, these animals have lost a great deal of genetic variation. Even though the low genetic variation in African cheetah populations may largely be the result of past natural processes, current human effects on their habitat put this species at greater risk. Not only will cheetahs be less able to respond to environmental change should it occur, but some data suggest that they are also at risk from disease. In the Wildlife Safari Park in Oregon, 25 of the park's 42 cheetahs succumbed to feline infectious peritonitis. The more genetically variable lions in the park were also exposed, but none was affected. If we are to preserve populations at risk of extinction, we must take special steps to ensure that their genetic variation is conserved. Each species has a unique genetic milieu, so it is difficult to develop hard and fast rules about preserving genetic diversity. Nevertheless, it is clear that it is advantageous to maintain large populations for genetic reasons, and it is important that management programs assess genetic variation and, when necessary, enhance it.

relative to the environment. In other words, there is no such thing as fitness in an absolute sense; an organism (or genotype) is fitter or less fit than another *in a particular environment*.

In the introduction to this chapter, we stated that the ecological situation is the context within which evolution occurs. We have now developed that notion more fully: The concepts of fitness and adaptation are relevant only in a particular ecological context. Or, in Hutchinson's terms, the "evolutionary play" develops differently in different "theaters."

SUMMARY

1. Ecology and evolution are intimately related biological fields. An organism's ecological situation determines the kinds of evolutionary forces it faces. At the same time, a species's evolutionary history accounts for important aspects of its ecological relationships.

2. Not all individuals in a population are identical in phenotype. Many of these differences reflect differences in the genotypes of the individuals.

3. The sum total of alleles in a population is the gene pool. Evolution is defined as changes in the frequencies of alleles in a population.

4. The Hardy–Weinberg equilibrium defines the frequencies of genotypes in a population that is not evolving (is undergoing no change in allele frequencies). The processes that cause a population to deviate from Hardy–Weinberg equilibrium thus represent the mechanisms of evolutionary change: genetic drift, gene flow, natural selection, and mutation pressure.

5. Populations adapt in response to selection pressures from the environment. Although natural selection is a major force in adaptation, genetic drift can also play an important role in this process.

6. We do not always observe the adaptations we expect to see in organisms. This is because a number of factors can constrain or slow the process of adaptation.

7. When a female chooses the males with whom she mates, she constitutes a selective force on males. This selection may result in exaggerated traits in males, some of which may even be maladaptive. This process is sexual selection.

8. Kin selection theory explains altruistic behavior. Behaviors that reduce the fitness of an individual can evolve if they increase the fitness of relatives that carry copies of the altruist's genes.

9. The interactions among ecologically interacting species constitute important selective forces. These interactions may be detrimental to one or both species, or they may be mutually beneficial. In coevolution, adaptations in one species constitute a new selective force on the other species.

SELF-ASSESSMENT: CAN YOU ...?

1. Explain all the important mechanisms of evolutionary change within populations.

2. Explain the major reasons why we sometimes do *not* see the adaptations we expect organisms to have.

3. List and define all the forms selection can take.

4. Explain the significance of the Hardy–Weinberg equation.

5. Present and discuss several examples of the relationship between ecology and evolution.

6. Define and characterize the kinds of coevolutionary interactions that occur between species.

PROBLEMS AND STUDY QUESTIONS

1. Distinguish among the following terms:

hard selection
soft selection
frequency-dependent selection
directional selection
disruptive selection
stabilizing selection
sexual selection
kin selection

2. Consider the following scenario: A group of birds lives in dense colonies. Because most individuals disperse as young birds, few individuals in the colony are related. A new mutation arises that causes individuals to give alarm calls when predators are sighted.

   a. Is this an ESS? What would you expect to happen to the frequency of the gene causing this behavior?

   b. How would things change if another mutation arose that altered the alarm calls so that predators could not locate the caller?

3. Design a research protocol that would determine which member of a pair of species is the model and which the mimic.

4. Write a short, concise definition of fitness. Now, suppose you are conducting field research on the fitness advantages of certain adaptations. You need to measure the fitness of individuals. What could you measure in the field to measure fitness?

5. Aposematically colored species that are lethal present a special evolutionary problem. For example, if the first encounter with a coral snake is fatal, obviously no learning occurs. How might aposematic coloring work to warn other species in cases like this? How could you test your ideas?

6. Consider this scenario: In a species of diploid animals, each individual produces exactly two offspring. Predation is intense, and many young are eaten; when parents attempt to protect their young from predation,

they are eaten themselves. A mutation arises that causes some parents to abandon their own young and protect the young of their siblings. Can this be an ESS? Explain why or why not.

7. A population consists of 100 individuals with the following genotypes:

AA	Aa	aa
50	15	35

   a. What are the values of p and q?

   b. Is this population in Hardy–Weinberg equilibrium?

8. For the population in Question 7, imagine that the homozygous recessives are deleterious and that only 80 percent of these individuals survive to reproduce. If the other genotypes have a 100 percent survival rate, and if each surviving individual replaces itself through reproduction, what are the values of p and q after one generation?

9. In *Drosophila*, sepia eyes are caused by a recessive mutation. Suppose that you find that 16 percent of the flies in a population have sepia eyes.

   a. What is the frequency of the dominant allele?

   b. What must you assume to solve this problem?

10. The "thumb" of the giant panda was derived from the radial sesamoid. List and explain as many possible reasons as you can why this adaptation was based on that bone and not on the first digit, as in primates.

PROJECTS AND ADDITIONAL STUDY

1. Try to develop an algebraic model for selection like that on page 25 for a system with incomplete dominance.

2. Search the recent literature to review the use of the term *coevolution*. Identify the significant papers in the development of this concept. Analyze the ways in which different authors and researchers have used the term. (Thompson, J.N. 1999. *Oikos* 84:5–16 is a good place to start).

3. Rare or endangered species may suffer from the effects of low genetic variation. Choose one of the following and develop a literature review of recent studies of genetic variation and its effects.

desert pupfish

Royal catchfly (*Silene regia*)

northern elephant seal

North American wolves

bison

4. A recent article (Wells, J. 1999. Second thoughts about peppered moths. *The Scientist* 13:13) raises a number of questions about the validity of Kettlewell's studies of peppered moths. Read this article and discuss Wells's criticisms. Are they valid? What points favor Kettlewell's interpretations and conclusions? What do you conclude about peppered moths as an example of natural selection?

CHAPTER 3

Abiotic Factors and Limits

I n Chapter 1 we defined ecology as the interaction between an organism and its environment. One of the clearest examples of such an interaction is that between an organism and its physical, or abiotic, environment. Organisms inhabit a tremendous range of physical environments on Earth, including some that are inhospitable, at least from our human perspective. Furthermore, studying the relationship between organisms and their abiotic environment helps us understand the kinds of adaptational processes described in Chapter 2.

When organisms encounter harsh abiotic conditions, they may have one of two responses. Evolutionary adaptations in their morphology, physiology, or behavior may allow them to cope with the abiotic conditions and inhabit regions where such conditions prevail. Without such adaptations, their presence or absence in a region will be determined by the local abiotic environment. In this chapter we examine both responses, especially adaptations to the physical factors in the environment.

Physical Resources and Limiting Factors

We begin our discussion of abiotic factors by distinguishing between physical resources and limiting factors. **Physical resources** are those abiotic factors that an organism must assimilate if it is to live and prosper. A plant must assimilate light energy, water, and carbon dioxide if it is to photosynthesize; an animal must consume oxygen and water if it is to survive. Organisms require certain nutrients as well. All these components are properly called resources. The term **physical factor** denotes other kinds of abiotic parameters (such as salinity, pH, or temperature) whose physical or chemical effects may delimit a zone in which life is possible. When factors such as these determine the presence or absence of a species, we refer to them as **limiting factors.** In this chapter we focus on physical limiting factors. It is also possible for biotic factors to be limiting.

The distinction between a physical resource and a physical factor quickly blurs, however, when we think about an organism's response to its physical environment. As we will see in this chapter, one set of adaptations to the physical environment is the "choice" the organism makes about where it lives. If, say, a lizard cannot tolerate temperatures over 40°C, it will avoid places that exceed that temperature. The concept that all organisms, whether mobile or sessile, have habitat preferences—and the means to locate and colonize appropriate habitats—immediately implies that limiting physical *factors* may operate in essentially the same way as limiting *resources.* For the lizard, places where the temperature is below 40°C constitute a limiting resource. In the same way, places with appropriate salinity, pH, or any other physiologically important abiotic factor may be a limiting resource.

In this chapter we examine two important aspects of physical resources: their ecological effects and their importance as selective forces on organisms. Not only do physical factors define species' distribution and abundance, but they are also an important component of the environment—one that constitutes a selective force (see Chapter 2). Indeed, one of the most fundamental adaptations we can describe for an organism is its ability to cope with abiotic factors.

It is important to keep in mind that abiotic factors rarely operate on an organism independently of one another. If water is in the frozen state, it may be unavailable to plants. Thus temperature may play a role in water stress. If an animal uses the evaporation of water to cool its body, the availability of water becomes an important factor in the animal's ability to withstand high air temperatures.

Chapter 1 discussed the fundamental division of the environment into *abiotic* and *biotic* factors. Throughout much of the early history of ecology, the emphasis was on the abiotic factors important to organisms and their physiological responses to such factors. Only later did biotic interactions receive much attention. Indeed, as late as 1931, R. N. Chapman's *Animal Ecology* (1931) stated that "the interrelations of organisms in nature may be considered under the subject matter of ecology which is closely related to physiology, and in the minds of some biologists should be a subdivision of it."

The importance of the abiotic environment was codified more than 150 years ago. In 1840, the physiologist Justus von Liebig proposed his *law of the minimum,* which states that the success of an organism is determined by the one crucial ingredient in the environment that is in short supply. To put it another way, this principle holds that whereas the ability of an organism to live and reproduce depends on many abiotic factors, one such factor is usually critically limiting. This principle underlies much of the physiological ecology we will consider in this chapter. It also forms the basis for niche theory and competition (Chapters 10 and 11).

In 1913, Victor Shelford extended this concept by proposing his *law of tolerance.* According to this maxim, there are upper *and lower* bounds to the physical fac-

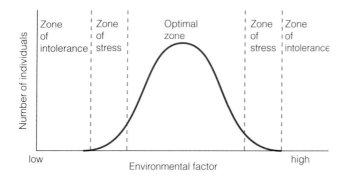

FIGURE 3.1 Shelford's law of tolerance. A plot of the number of individuals of a species as a function of some environmental factor (such as temperature) produces a bell-shaped curve that can be divided into various tolerance zones.

tors an organism can tolerate (Figure 3.1). For example, for any organism there is a maximum temperature that can be tolerated; higher temperatures are lethal. Similarly, there is a lower temperature bound, below which life is also impossible. The implication is that no organism can live everywhere. Although this notion is so obvious as to seem trivial, it nevertheless has profound consequences for ecology (see the box on page 72 for another application of the law of tolerance). A period of intense research to determine the abiotic limits for many organisms followed Shelford's publication.

Tolerance of abiotic conditions can take two nonexclusive forms. In adaptation by *natural selection,* organisms genetically adapt to the environment to which they are exposed. Phenotypic responses also occur as a result of exposure to the environment. Organisms that change their physiology or morphology in response to a particular environment are said to *acclimatize,* and the ability to do so may be an adaptation.

The Effects of Abiotic Factors on Species Distribution and Abundance

The effects of abiotic factors occur in ecological time—a change in the physical environment manifests itself in effects on the distribution and abundance of the current generation of plants and animals. The range of a species may expand or contract, or the abundance of the species may change locally.

As we have seen, the law of the minimum and the law of tolerance describe the role of abiotic factors in delimiting the places where an organism can live. These ideas are central to the community-level concept of the niche. One of the important formulations of the concept of the niche—an organism's place and role in the environment—is based on precisely these notions. An important component of the Hutchinsonian niche (Chapter 11) is a mathematical description of a species's tolerance limits for abiotic factors.

In the sections that follow, we consider a few examples of physical factors that determine where species can live: temperature, water abundance and salinity, and nutrient availability. Although we will consider these factors independently, it is important to keep in mind that many abiotic factors interact with one another. For example, Lloyd and Graumlich (1997) have recently shown that treeline, the highest altitude at which trees grow, in the Sierra Nevada has changed a number of times in the past 3500 years. Their climatic reconstructions show that neither temperature nor precipitation alone but, rather, the combination determines the treeline at any given time.

Temperature

One of the most important determinants of species distribution is temperature. Plants, by virtue of their sessile nature, provide some of the most dramatic examples of the effects of temperature on the geographic ranges that species occupy. Cold temperatures, a short growing season, desiccation resulting from low temperatures, and wind determine the latitudinal and altitudinal limits of trees (Enno 1997). Figure 3.2

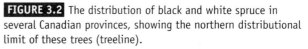

FIGURE 3.2 The distribution of black and white spruce in several Canadian provinces, showing the northern distributional limit of these trees (treeline).

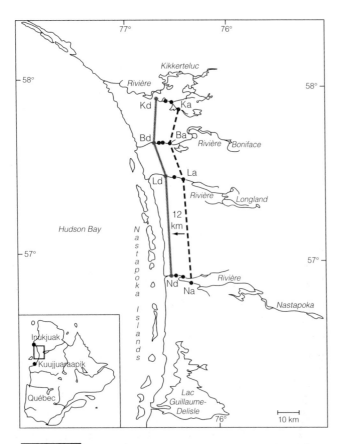

FIGURE 3.3 The shift in treeline along the eastern shore of Hudson's Bay. The dotted line represents the treeline at the end of the nineteenth century. The solid lines represents the current treeline, 12 kilometers closer to the coast (*From Lescop-Sinclair and Payette 1995*).

shows the approximate present position of the northern treeline in Canada, the boundary between the taiga and tundra. The black spruce (*Picea mariana*) and white spruce (*Picea glauca*) characteristic of the taiga and boreal forest reach their northern range limits at treeline.

That this boundary is determined by abiotic factors is supported by the fact that it is responsive to climatic change. On the basis of an analysis of pollen deposited in lakes, G. M. MacDonald and colleagues (1993) have shown that an episode of climatic warming that began in northern Canada approximately 5000 years ago created a transition from tundra to forest. Moreover, the change in vegetation was remarkably rapid, requiring only 150 years. Apparently, the northern treeline is particularly sensitive to temperature changes. Recent evidence documents this sensitivity. Lescop-Sinclair and Payette (1995) documented a rapid shift in treeline along the eastern shore of Hudson's Bay. In this region, conditions are most harsh near the bay; trees are relegated to protected valleys at some distance from the driving winds near shore. A warming trend beginning in the late nine-

teenth century has resulted in a 12 km shift closer to Hudson's Bay (Figure 3.3).

A classic example of the effect of temperature on the distribution of a species involves the saguaro cactus (*Cereus giganteus*). J. R. Hastings and R. M. Turner (1965) showed that the range limits of this Sonoran Desert species are precisely determined by a specific set of temperature requirements. This species dies if exposed to temperatures below freezing for more than 36 consecutive hours. If a thaw occurs within 36 hours, the plants survive. The range of this species corresponds closely to a map of sites for which there are no records of subfreezing temperatures for more than 36 hours (Figure 3.4).

The geographic distribution of some species of butterflies in the family Papilionidae is apparently determined by temperature extremes. To study the role of temperature in the distribution of papilionids, O. Kukal (1991) and co-workers exposed both *Papilio canadensis* from Alaska and *P. glaucus*, a related species from the southern United States, to northern winters. Pupae from populations of *P. glaucus* from Georgia and *P. canadensis* from Michigan and Alaska were placed in mesh bags on the ground (where they overwinter) in arboretums in Alaska and Michigan. Only 8 percent of *P. glaucus* individuals survived in Alaska, compared to

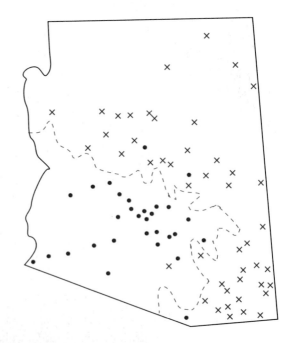

FIGURE 3.4 The northern and eastern boundaries of the range of saguaro cactus in Arizona. Dots represent sites where there are no records of periods longer than 36 hours without a thaw. Crosses indicate sites where such periods have been recorded. The dotted line represents the boundary of the Sonoran Desert. (*From Hastings and Turner 1965*)

more than 60 percent of both populations of *P. canadensis.*

Temperature effects on species distributions are not confined to plants and animals that do not maintain a constant body temperature; some mammal distributions are clearly related to temperature as well. P. Stapp and colleagues (1991) have shown that the northern limit of the distribution of the southern flying squirrel, *Glaucomys volans,* is related to temperature. Because of their small size, nocturnal habits, and gliding behavior, flying squirrels are particularly vulnerable to thermal stress and huddle together in a nest during colder months (Figure 3.5). This behavior allows the species to survive farther north than might otherwise be possible. As squirrels reach higher latitudes, however, the energy expenditure necessary to keep warm eventually becomes too great, and the northern range limit is reached (at about 45°N). At that latitude the energy expense of keeping warm exceeds 2.5 times the basal metabolic rate. Interestingly, many birds also reach their northern limit where their energy expenditures are 2.5 times the basal rate. Perhaps this is a general threshold for many species.

Local abundance of organisms is also affected by temperature. For example, some game birds such as quail are strongly affected by the temperature regime when the young hatch. Because hatchlings are highly vulnerable to thermal stress at this time, cold, rainy weather during hatching can depress local populations of quail.

The eastern reef-building coral, *Millepora* spp., provides another example. P. W. Glynn and W. H. de Weerdt (1991) described a local extinction caused by increased water temperature associated with an El Niño event. They studied populations of three species (*M. intricata, M. platyphylla,* and one undescribed species), all of which were abundant in Panama before 1983. When an El Niño event reached Panama in the winter of 1982–1983, the influx of warm water from the eastern Pacific raised the water temperature 2–3°C for five to six months. By May of 1983, only a few *M. intricata* were still alive; all the individuals of the other two species had perished.

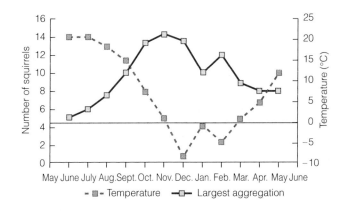

FIGURE 3.5 The relationship between huddling and ambient temperature inside a flying squirrel enclosure in New Hampshire. (*From Stapp, Peking, and Mautz 1991*)

Water Abundance and Salinity

Water also influences the distribution and abundance of many organisms. As with temperature, sessile organisms like plants provide some of the clearest examples. The western boundary of the distribution of the moisture-loving American beech, *Fagus grandifolia* (Figure 3.6), closely corresponds to sites where annual precipitation falls below 75 centimeters. Redwoods (*Sequoia sempervirens*) are restricted to the fog belt of

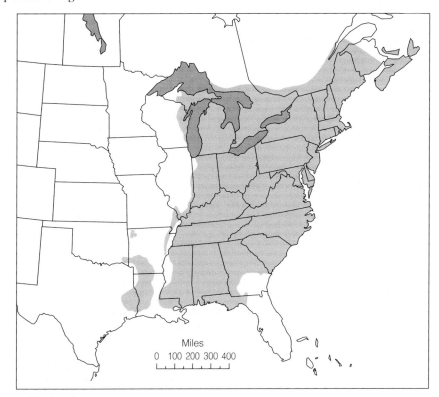

FIGURE 3.6 Geographic range of the American beech (*Fagus grandifolia*). The western limit is thought to be determined by moisture. (*From U.S. Department of Agriculture Handbook no. 271*)

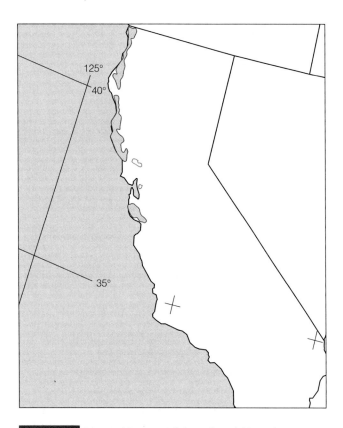

FIGURE 3.7 Geographic range of the redwood (*Sequoia sempervirens*). This species is restricted to a fog belt along the northwest coast of California and southwestern Oregon. (*From U.S. Department of Agriculture Handbook no. 271*)

FIGURE 3.8 Dense vegetation in a desert wash through which rainwater flows. (*Photo by David Krohne*)

northern California and southern Oregon (Figure 3.7), where precipitation is high and fog adds considerable summer moisture.

In habitats where water is scarce, moisture gradients are likely to have profound effects on plant distributions. In such locales, slight changes in water availability have exaggerated effects on the vegetation. In the desert, for example, a wash that channels the flow of the infrequent rains may support significantly different and more luxurious vegetation than sites just a few meters away (Figure 3.8). Similarly, in the tallgrass prairie, the various sites at which grass species occur separate along moisture gradients associated with slight changes in topography (Figure 3.9).

Soil salinity can affect plant distribution. In the deserts of the southwestern United States, many valleys formed from lakes that dried up at the end of the Pleistocene; salts were concentrated in the basins as the lakes receded. Often a gradient occurs from a relatively nonsaline substrate at higher elevations to highly saline soils in the valley bottoms. Many plant species respond to this gradient: Abundance is often greatly reduced in the saltier valley bottoms (Figure 3.10).

Salt marsh habitats provide an example of the combined effects of water and salinity on plant distributions. The zonation of plants in a salt marsh is often determined by the salinity gradient that occurs where freshwater meets seawater. S. C. Pennings and R. M. Callaway (1992) have shown that in one salt marsh in southern California, plant distribution is determined by two crucial physical factors: tidal flooding and soil salinity. Lower in the marsh, periodic tidal flooding constitutes an important abiotic factor because plants growing there must be able to tolerate inundation. Higher in the marsh, increasing salinity stresses plants that attempt to grow there. Thus the most benign conditions in the marsh are at intermediate elevations,

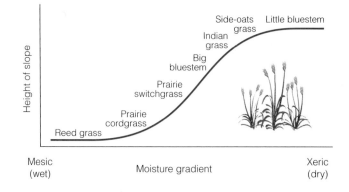

FIGURE 3.9 Distribution of prairie grasses along a moisture gradient associated with changes in topography. Species at the bottom of the slope experience mesic conditions compared to those at the top of the slope. Each of these species occupies a region along this gradient and may overlap with neighboring species.

FIGURE 3.10 Desert habitat in a saline valley in California. Note the changes in the vegetation between sites at the bottom of the valley (highest salinity) and those higher up the valley sides (lower salinity). (*Photo by David Krohne*)

and, not surprisingly, total plant biomass is greatest there.

At this site, pickleweed (*Salicornia virginica*) grows lower in the marsh than Parrish's glasswort (*Anthrocnemum subterminalis*) (Figure 3.11). When pickleweed and glasswort were experimentally transplanted throughout the tidal range shown in Figure 3.11, both species performed best in intermediate regions, where flooding and salinity were moderate. *Anthrocnemum* avoids both the hypersaline areas of the upper marsh and the flooded regions low in the

marsh. *Salicornia* can tolerate flooding and thus can survive in the lower region. Pennings and Callaway suggest that competition with *Anthrocnemum* probably excludes *Salicornia* from its preferred intermediate zone.

Nutrient Availability

Nutritional factors can have important ecological effects as well. Plants are frequently nutrient-limited, particularly with respect to nitrogen and phosphorus. Nutritional factors are clearly important for animals as well. It is known, for example, that many ungulates seek out mineral licks—areas with high soil concentrations of certain elements such as sodium or calcium. Deer (*Odocoileus* spp.), elk (*Cervus canadensis*), and mountain goats (*Oreamnos americana*) seek such areas and sometimes consume large quantities of soil. Small mammals such as mice and squirrels often chew shed antlers or the bones of dead ungulates. In both of these examples, the animals presumably are satisfying the need for certain nutrients, as opposed to calories, because they cannot obtain energy from these sources.

A. M. Schultz (1964; 1969) and F. A. Pitelka (1964) suggested that the population dynamics of lemmings (*Lemmus* spp.) in the Arctic is controlled by nutritional factors. The abundance of these small mammals is cyclical, with a four-year period. Schultz and Pitelka proposed the *nutrient recovery hypothesis* (Figure 3.12)

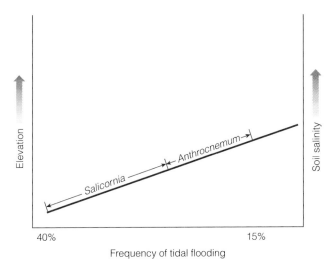

FIGURE 3.11 The effect of tidal flooding and salinity on the distributions of plant species in a southern California salt marsh. (*Data from Pennings and Callaway 1992*)

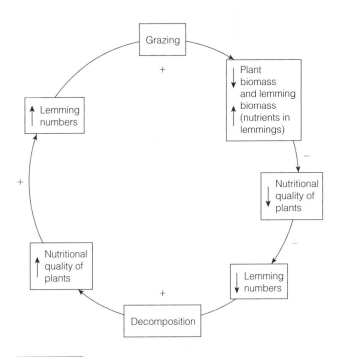

FIGURE 3.12 A schematic representation of the nutrient recovery hypothesis of lemming fluctuations in the Arctic. (*After Shultz 1969*)

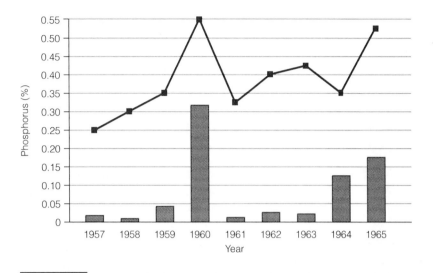

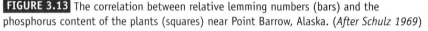

FIGURE 3.13 The correlation between relative lemming numbers (bars) and the phosphorus content of the plants (squares) near Point Barrow, Alaska. (*After Schulz 1969*)

to explain this interesting pattern of population fluctuation. According to the hypothesis, lemmings play a major role in the transfer of nutrients. Limited nutrient availability eventually slows their population growth; at some points of the cycle, large quantities of nutrients are "tied up" in lemming bodies. Only when decomposition returns this material to the soil can the vegetation, and eventually the lemming populations, recover.

The data supporting this hypothesis are correlational; that is, there is a relationship between lemming numbers and the phosphorus content in their forage (Figure 3.13). Without experimental studies, it is not possible to demonstrate cause and effect in such correlations or to prove that the nutrient drives the cycles. Lemming cycles are extremely complex phenomena, and it is unlikely that this simple model entirely explains their population dynamics. Nevertheless, the data relate an organism's noncaloric nutritional needs to its abundance.

Types of Adaptive Responses to the Abiotic Environment

Organisms' adaptive responses to the physical factors in the environment fall into two broad categories. Organisms may evolve physiological mechanisms to counteract the effects of the physical factors; these are referred to as **homeostatic mechanisms.** Alternatively, organisms may evolve ways of avoiding the stressful physical parameters. As we will see, avoidance strategies can involve behavioral or physiological mechanisms.

Homeostatic Mechanisms

The term *homeostasis,* coined by the physiologist Claude Bernard in 1878, refers to an organism's physiological mechanisms that work to maintain a constant internal environment in the face of external factors that vary over a broad range (Figure 3.14). For example, mammals maintain an internal temperature that varies only a few degrees Celsius while the external temperature may vary more than 80°C. Organisms can evolve homeostatic mechanisms for virtually any of the physical factors that can limit their growth and development.

It is important to note that by and large, homeostatic mechanisms require the organism to expend energy. Thus there is a direct or indirect energetic cost to homeostasis. For a mammal to maintain a body temperature of 37°C when the air temperature is below 0°C, it must expend energy. It can maintain its body temperature by shivering, but shivering requires energy. Similarly, at high ambient temperatures, keeping the body from overheating has an energetic cost. When a dog pants and uses the evaporation of water from its tongue to cool its body, the cost is the expenditure of water and energy.

An organism may not adopt a particular homeostatic mechanism for any of a number of reasons. Perhaps the energetic cost is too high, or some con-

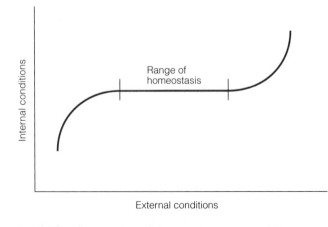

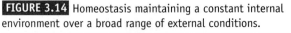

FIGURE 3.14 Homeostasis maintaining a constant internal environment over a broad range of external conditions.

straint (see Chapter 2) may prevent the adaptation we expect to observe. Ultimately, organisms must operate within the laws of physics in their responses to the abiotic environment; adaptations for, say, temperature regulation may be constrained by the laws governing heat transfer.

Avoidance Strategies

An organism can also adopt adaptive responses of the second broad category: adaptations that enable it to avoid the physical factor that stresses it. Consider the wide range of adaptations for avoiding stressful temperatures. Some animals have evolved the physiological ability to hibernate, thus avoiding the need to maintain a high body temperature during periods of low winter temperatures (and limited food availability as well). Other organisms have evolved resistant life cycle stages that are unaffected by temperature extremes. Still others avoid stressful conditions by migrating to more benign environments. Finally, some organisms' behavioral adaptations allow them to avoid certain stressful environments by carefully selecting a narrow range of sites that they inhabit or in which they concentrate their activities.

In the next section, we will consider the nature of the adaptive responses to various physical parameters of the environment. We will concentrate on four of the many abiotic factors that have an impact on organisms: temperature, water, light, and nutrients. For each abiotic factor, we will examine both the physiological adaptations that allow organisms to deal with the stresses imposed by that component of the environment and the physiological and behavioral adaptations that allow organisms to avoid those stresses. Temperature is examined in some detail, not because it is any more important than other factors, but as an example of the detailed understanding that has been developed by physiological ecologists.

The Effects of and Adaptations to Thermal Stress

Life on Earth is limited to a remarkably narrow range of temperatures, at least compared to the temperatures that might be encountered elsewhere in the universe. The temperature range of active life-forms extends from approximately −50°C to near the boiling point of water (100°C). Forms active at the upper extreme are very rare and, in fact, are limited to a few thermophilic bacteria inhabiting hot springs. Of course, resistant resting stages such as spores and seeds can tolerate even more extreme temperatures. Nevertheless, life-forms are confined to a relatively narrow range of temperatures.

The Effects of Temperature on Organisms

If we are to understand organisms' adaptations to temperature extremes, we must first consider the effects of temperature on living organisms. As with other chemical reactions, increases in temperature cause an increase in the rate of physiological reactions. The measure of this effect is termed Q_{10}, a parameter that represents the increase in the rate of a reaction for each 10°C increase in temperature (T). Thus, if increasing the temperature by 10°C results in a doubling of the reaction rate, then $Q_{10} = 2.0$.

We can illustrate the effect of temperature on physiological processes by examining the reaction rate as a function of temperature. For the hypothetical reaction data in Figure 3.15, we have plotted the rate of the reaction as a function of temperature when $Q_{10} = 2.0$. We see that the rate (R) increases exponentially.

The curve is described by the equation

$$R_2 = R_1 (Q_{10})^{\frac{T_2 - T_1}{10}}$$

In living things, the rates of reactions do not necessarily follow an exponential equation like this over

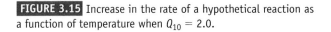

FIGURE 3.15 Increase in the rate of a hypothetical reaction as a function of temperature when $Q_{10} = 2.0$.

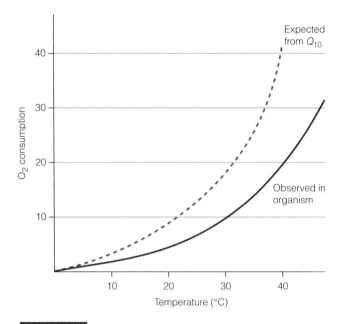

FIGURE 3.16 Increase in oxygen consumption of an animal as a function of ambient temperature (solid line) differs from that predicted from the physical effects (Q_{10}) of temperature on reactions (dashed line).

such a large temperature range. In Figure 3.16 we see that oxygen consumption increases more slowly than the equation predicts.

As an organism's temperature increases, death eventually results. The lethal maximum temperature varies widely across taxa and habitats; for some organisms it is as low as 6°C, whereas some bacteria can survive in hot springs above 100°C.

High temperatures can have a number of ultimately lethal effects on organisms. Proteins and DNA denature at temperatures above 40°C; the exact temperature depends on the chemical and physical structure of the molecule. In addition, an increase in temperature increases the rate of oxygen consumption. If the demand for oxygen outstrips an organism's ability to supply it to the tissues, death may occur. High temperatures may also disrupt normal cell membrane structure and function.

It may also be that temperature affects different metabolic processes differently. If complex metabolic pathways must interact but their reactions have vastly different Q_{10} values, thermal death can occur. In the Antarctic, fish of the genus *Trematomus* live in water very near freezing (Somero and DeVries 1967); death from high temperature occurs at 6°C! It is unlikely that any of the possible causes other than the disruptive effect of different Q_{10} values for different biochemical pathways could account for thermal death at only 6°C. Certainly, denaturation of proteins does not occur at so low a temperature.

Low temperatures may also be lethal. The temperature below which death occurs is as variable across taxa and habitats as upper lethal temperatures. Obviously, tissues that freeze are in jeopardy. Biochemical processes cease or slow greatly. In addition, ice crystals cause physical damage to membranes and organelles. For plants, frozen water in the soil is effectively unavailable for uptake by the roots, and death may occur via desiccation. Freezing is not the only cause of death associated with cold temperatures, however. In the guppy (*Lebistes reticulatus*), death occurs if fish that have been kept near room temperature (22°C) are cooled to 10°C (Pitkow 1960). Death apparently results from depression of the respiratory center by low temperature.

Principles of Heat Transfer

A number of physiological, biochemical, morphological, and behavioral mechanisms have evolved to prevent death from temperature extremes. Before we discuss them, we must first consider some aspects of the movement, or transfer, of heat.

It is important to understand the relationship between heat and temperature. Biologists measure temperature in degrees Celsius. The unit of heat is the calorie, the amount of heat required to raise the temperature of 1 gram of water 1°C. Different substances require different amounts of heat to raise their temperatures. The amount of heat required to increase the temperature of 1 gram of a substance by 1°C is its **specific heat capacity.**

Heat is transferred from one body to another by three processes: conduction, radiation, and evaporation of water. It is important to note that heat transfer between two objects occurs only if there is a temperature differential between them. Therefore, an organism loses heat to the environment only if its temperature is higher than that of its surroundings, and it gains heat only if its temperature is lower. Let us briefly consider each of the mechanisms of heat transfer.

Conduction is the transfer of heat between two bodies that are in physical contact with one another. It results from the transfer of kinetic energy in moving molecules. Conduction is dependent on several factors, including each body's thermal conductivity (a measure of its ability to transfer heat), the area of contact, and the temperatures of the two bodies.

When an object is in contact with a fluid (a liquid or a gas), heat can be transferred to the fluid by conduction. Heat can be transported in fluids by **convection,** the mass movement of fluids of different temperatures. Movement of warm fluid away from the surface of the object causes replacement by cooler fluid, which, in turn, is heated and also moves away.

Radiation is the transfer of heat between two objects that are not in physical contact. Any object with a temperature above absolute zero (0 K) emits electromagnetic radiation, which can pass through a vacuum (or through air, with only slight changes). Obviously, the temperature of the surface is very important in determining the rate of emission. The *emissivity* of an object is its propensity to emit radiation; *absorptivity* is its tendency to absorb the radiation that strikes it.

Emissivity and absorptivity have some interesting consequences for heat transfer in living organisms. Skin and fur have high emissivity in the infrared range, where essentially 100 percent of the radiation emitted from animals occurs. There is virtually no difference in emissivity between dark and light skins or furs; the differences that exist are related only to the visible portion of the spectrum. Thus, in considering heat emission, we must be careful not to base conclusions about the thermally adaptive value of animal coloration solely on an organism's appearance to our eyes.

Skin and fur colors do differ in the absorption of heat via solar radiation, however. Because approximately 50 percent of the energy in solar radiation is in the visible portion of the spectrum, the color differences we observe among animal furs and skins *are* physiologically important for the absorption of heat from sunlight.

The final avenue of heat transfer is evaporation, which requires large quantities of heat. The amount of heat required is the *heat of vaporization.* For water at 22°C, 584 calories are required to change 1 gram of water into vapor. For an organism, this means that water loss is a potentially important means of removing heat. Moreover, evaporation is the *only* means of lowering the temperature of a body if it is warmer than its environment. Consequently, animals have made extensive use of evaporation as a heat loss mechanism. The cost associated with this process is the detrimental effect of water loss.

Given these processes of heat transfer, we can generate an equation that describes the rate of heat gain or loss from an organism:

$$H_{tot} = \pm H_c \pm H_r - H_e \pm H_s$$

where

H_{tot} = total metabolic heat production (this must always be a positive quantity for a living organism)

H_c = conductive (and convective) heat exchange

H_r = radiation heat exchange

H_e = evaporative heat exchange

H_s = heat storage by the organism

This equation contains plus-or-minus signs for most components because, as previously noted, the direction of heat transfer depends on the temperature differential between the organism and the various components of the environment. The pathways of heat transfer included in this equation for an organism in its environment are depicted in Figure 3.17.

Animal Adaptations to Thermal Stress

Animals adapt to thermal stress via two general strategies: homeothermy and poikilothermy.

Homeothermy and Poikilothermy **Homeotherms** physiologically regulate their temperatures so as to maintain them within rather narrow bounds. In **poikilotherms,** body temperature fluctuates with that of the environment. Neither strategy is absolute; homeotherms cannot maintain a strictly constant temperature under all circumstances. Similarly, poikilotherms may adopt behavioral strategies to prevent their temperatures from falling too low or rising too high. Thus we should think of these terms as end-

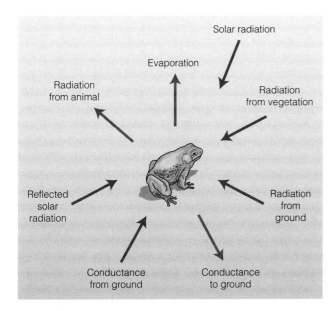

FIGURE 3.17 The pathways of an animal's heat gain from and heat loss to its environment.

points on a spectrum. They are more precise than the commonly used expressions *warm-blooded* and *cold-blooded*. Even though the phrase *cold-blooded* is sometimes applied to lizards, fish, and other poikilotherms, the tissues of these organisms are not necessarily cold; they simply fluctuate with changes in environmental temperatures.

We also distinguish between **endotherms**—animals that produce sufficient metabolic heat to maintain a high body temperature—and **ectotherms,** which obtain much of their heat from the environment. Although it is true that many homeotherms are endotherms and many poikilotherms are ectotherms, this correlation is not perfect. We should not use the terms *endothermy* and *homeothermy* (or *ectothermy* and *poikilothermy*) synonymously. For example, many insects derive much of their body heat from their surroundings (ectothermy and poikilothermy), but intense muscular contractions also contribute to body temperature (endothermy).

Homeotherms maintain a high and relatively constant body temperature that is essentially independent of that of the environment. One selective advantage of homeothermy is the ability to be active during cold periods. The ability to forage, seek mates, and avoid predation in the cold is of obvious selective advantage. Because poikilotherms become increasingly inactive as the temperature drops, resources no longer available to them are available to homeotherms.

In homeotherms, a characteristic relationship exists between the ambient (environmental) temperature and their metabolic rate. The **basal metabolic rate** is defined as the rate of consumption of oxygen by an organism at rest (usually measured in units of oxygen consumed per unit time per gram of body mass). The metabolic rate remains essentially constant over a range of ambient temperatures referred to as the **thermal neutral zone** (Figure 3.18). At temperatures below the lower critical temperature, energy must be expended to maintain body temperature. Above a temperature called the upper critical temperature, energy must be used to dissipate heat. In both cases the metabolic rate rises accordingly.

Adaptations to Hot Environments   In order to understand the mechanisms of heat regulation in homeotherms, we can return to the equation for heat balance:

$$H_{tot} = \pm H_c \pm H_r - H_e \pm H_s$$

The four terms on the right side of the equation represent the potential ways to dissipate heat when

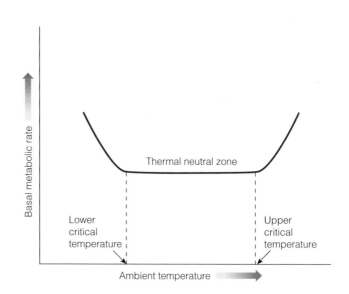

FIGURE 3.18 Plotting the basal metabolic rate (measured by oxygen consumption) as a function of ambient temperature makes it possible to identify the thermal neutral zone.

the ambient temperature is high. Clearly, evaporation and storage (H_e and H_s) are the most important options when the ambient temperature exceeds body temperature.

The two primary modes of evaporative water loss in animals are panting and sweating. Mammals (especially carnivores) cool by panting. One advantage of panting is that the animal facilitates evaporation by generating air flow over the moist nasal and oral surfaces. One disadvantage is that the excess respiration associated with panting can deplete the lungs of carbon dioxide, and its loss can leave the blood excessively alkaline. Panting obviously requires muscular movement and thus involves the expenditure of energy. The elasticity of the respiratory tract greatly reduces the energetic cost of panting, however. Birds also pant to regulate their temperature. In addition, they increase evaporation by rapid oscillations of portions of the throat, an action called *gular fluttering.*

Sweating is common in humans and in many ungulates, including horses, cattle, sheep, goats, and some antelope. Even in heavily furred mammals such as the camel, sweating is an effective heat loss mechanism because the dry desert air facilitates rapid evaporation.

A few homeotherms are able to inhabit very hot environments because of their ability to store heat. One of the classic examples is the antelope ground squirrel (*Ammospermophilus leucurus*) of the hot Mojave and Sonoran Deserts, where it is one of the few mammals active during the day. This small rodent is able to withstand considerable increases in its body temperature (Chappell and Bartholomew 1971). As it

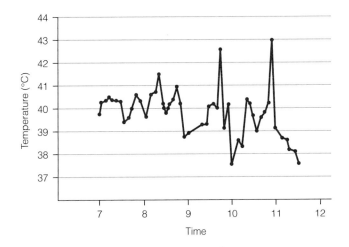

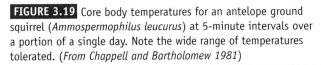

FIGURE 3.19 Core body temperatures for an antelope ground squirrel (*Ammospermophilus leucurus*) at 5-minute intervals over a portion of a single day. Note the wide range of temperatures tolerated. (*From Chappell and Bartholomew 1981*)

forages in the hot sun at temperatures up to 70°C, it accumulates heat, and its body temperature rises as high as 43.2°C. After a period of foraging, the ground squirrel retreats to its underground burrow, where the ambient temperature is many degrees lower than its body temperature. It sprawls on the cool, moist sand of its burrow and dumps its heat load via conductance. When its body temperature has dropped sufficiently, the ground squirrel emerges from its burrow to forage (and accumulate heat) again. Compared to desert species such as lizards and snakes, the antelope ground squirrel is a homeotherm, but its mechanism for dealing with thermal stress allows much wider fluctuations in body temperature than other mammals (Figure 3.19). The foraging patterns and duration of above-ground activity must be coordinated with the need to dissipate heat periodically (Hainsworth, 1995).

Adaptations to Cold Environments We noted previously that energy must be expended to maintain body temperature in the cold. Again, the heat balance equation illustrates an animal's options for maintaining a high body temperature in the face of low ambient temperatures. Increasing internal heat production and decreasing heat loss via conduction (H_c) and radiation (H_r) are the most important processes for animals in the cold.

One option that accounts for much of the increase in metabolic rate at temperatures below the thermal neutral zone is increased heat production by the animal. Muscular activity, exercise, and shivering contribute to metabolic heat production. Additionally, a process called nonshivering thermogenesis may oper-

ate. Animals that use this process store energy in the form of brown fat. This adipose tissue is high in cytochrome *c* and can consume oxygen at a high rate, making it useful in the generation of heat.

Another obvious strategy for maintaining high body temperatures in cold regions is to decrease conductance by increasing insulation. Fur is an excellent insulator whose value for that purpose increases linearly with thickness (Schmidt-Nielsen 1979). In addition, layers of subcutaneous fat constitute important insulation. The blubber of seals, walruses, and whales allows these marine mammals to range into arctic and antarctic waters. Although polar bear fur is not particularly effective as an insulator, a layer of subcutaneous fat enables this arctic mammal to swim in seawater below 0°C.

The first feathers probably arose as an insulating adaptation in birds' dinosaur ancestors; subsequently, they were co-opted as an aid to flight. Nevertheless, feathers still provide excellent insulation, particularly when the bird puffs them up to increase the amount of body-warmed air they trap next to the skin. Penguins are an excellent example of the insulation provided by feathers. Feather densities up to $70/in^2$ allow these birds to withstand extremely low temperatures on land and in the water.

Another adaptive option suggested by the heat balance equation is adjustment of radiative heat loss. Because radiative heat loss is proportional to the size of the radiative surface, one morphological adaptation available to an animal in a cold habitat is to reduce the relative size of its surface area.

This strategy manifests itself in two fundamental ways: body shape and body size. Animals in cold environments tend to have bodies and extremities that are thick and stocky. By contrast, animals in hot environments are more slender, with longer, thinner extremities that increase the relative surface area and facilitate radiative heat loss. Dall sheep (*Ovis dalli*), inhabitants of montane arctic habitats, are stocky with short limbs compared to the desert bighorn sheep (*O. canadensis nelsoni*), which has longer, more slender limbs.

Body size also affects radiative heat loss. Heat production is proportional to body mass (and hence volume), whereas radiative heat loss is proportional to surface area. Because the volume of an animal varies as the cube of its linear dimension, and its surface area varies as the square of the linear dimension, a small animal (small linear dimension) will have a higher surface-to-volume ratio than a large animal. As a result, small animals will lose heat more rapidly than large animals and thus are at greater risk of cold stress. Although many large animals inhabit cold habitats—

polar bears and whales in arctic and antarctic waters are two examples—very small animals are rare in cold climates.

The effects of these adaptations are shown in Figure 3.20. The slopes of the three lines indicate the *amount* of increase of metabolic rate required to maintain body temperature in response to a range of cold temperatures. Animals in Group 1, which include the arctic fox, very large mammals, and many marine mammals, do not need to increase their metabolic rate very much because they have small surface-to-volume ratios, small extremities, or thick layers of fat or blubber. The animals in Group 2 might include smaller mammals that lose heat rapidly and must increase their metabolic rate more to maintain a constant body temperature. This group includes species such as arctic ground squirrels, lemmings, and weasels. Group 3 represents tropical species such as sloths, coatis, and some small primates, which are poorly adapted to cold conditions. For these animals, a small decrease in temperature leads to a large increase in metabolic rate. Thus the energetic expense of keeping warm is high for this group.

For poikilothermic animals that inhabit cold climates, preventing damage from freezing may be an important process. Two options are available to such animals: supercooling and freezing tolerance. *Supercooling* occurs when the body temperature falls below the freezing point of pure water without freezing the body fluids. Some arctic and antarctic fishes and invertebrates live in water that is below 0°C. The presence in their tissues of solutes such as glycerol lowers the freezing point of their fluids, thereby preventing the damaging formation of ice crystals. *Freezing tolerance* refers to the prevention of damage when tissue fluids do freeze. The wood frog (*Rana sylvatica*) is one of five species of terrestrial frogs known to tolerate freezing of its body fluids during hibernation. Freezing tolerance is a result of the presence of glucose in sufficiently high concentrations to prevent physical cellular damage by tiny ice crystals (Storey 1990). Many insects use glycerol for the same purpose.

Strategies for Avoiding Thermal Stress   Some animals simply avoid thermal stress by behavioral or physiological mechanisms. Here we examine a few of the wide variety of such strategies that have evolved.

Reptiles are among the most adept behavioral thermoregulators of all poikilotherms. That each species has a preferred body temperature for activity can be demonstrated by providing captive reptiles with a gradient of temperatures and allowing them to choose a thermal comfort zone. In the field, they bask in the sun in the morning to raise their body temperature. They may orient themselves very precisely during this process to expose the maximum surface area to the solar radiation. They may also draw heat from warm rock surfaces.

Invertebrates may also choose thermal regimes appropriate for their life histories. A fascinating example involves populations of fruit flies, *Drosophila melanogaster*, inhabiting sites at high and low elevations. Because this species is able to develop over a rather narrow range of temperatures, populations living at high elevation may have difficulty completing development because of the low temperatures characteristic of such sites. To investigate, J. S. Jones and co-workers (1987) asked the question "Are *Drosophila* able to persist at high elevation by virtue of choosing oviposition sites with favorable thermal characteristics?"

It is very difficult to locate *Drosophila* egg-laying sites in the field, so the scientists adopted a clever approach to the problem. They used a temperature-sensitive eye mutation to identify the thermal environment used for oviposition. The eye color of these mutants depends on the temperature at which the pupae develop

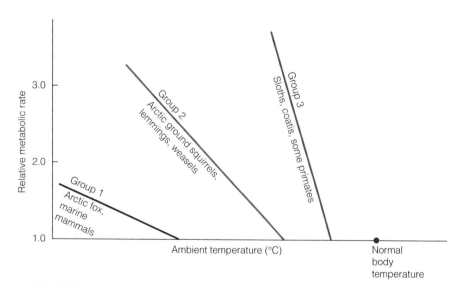

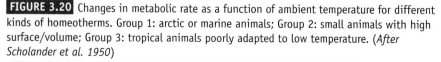

FIGURE 3.20   Changes in metabolic rate as a function of ambient temperature for different kinds of homeotherms. Group 1: arctic or marine animals; Group 2: small animals with high surface/volume; Group 3: tropical animals poorly adapted to low temperature. (*After Scholander et al. 1950*)

(pupae exposed to low temperatures have dark eyes). After the researchers introduced the mutant flies at both elevations and waited for the life cycle to proceed, they captured offspring and recorded their eye colors. They found that despite the large differences in air temperature between the low- and high-elevation sites, all the flies emerging in each area had virtually the same eye color. Therefore, they concluded, the flies at the high-elevation site must have chosen warmer microhabitats in which to lay their eggs.

Another behavioral thermoregulatory adaptation is huddling at low temperatures to reduce cold stress. Huddling helps maintain body temperature in the cold by decreasing the effective surface area of individuals. A huddling group exposes to the environment a smaller proportion of its total surface area than would any individual. A number of mammals adopt this strategy, including the southern flying squirrel (previously discussed) and many species of mice. Some birds, such as the emperor penguin (*Aptenodytes forsteri*), huddle as well. During the breeding season, which occurs during the antarctic winter, several thousand individuals may huddle together on the rookery. In this way, penguins can tolerate ambient temperatures as low as −40°C.

Another set of avoidance strategies entails using torpor to avoid unfavorable temperatures. Animals that adopt such a strategy are called **heterotherms;** animals that allow their body temperature to drop during the winter undergo **hibernation.** These animals enter a state of suspended animation, usually deep in a burrow. Their energetic demands are low because they are not attempting to maintain a high, relatively constant body temperature. Some species, such as arctic ground squirrels (*Spermophilus parryi*), have the added adaptation of incorporating natural antifreezes in their blood to allow the animals to tolerate subfreezing body temperatures during hibernation (Barnes 1989).

Hibernation is not an easy phenomenon to define precisely. Different species allow their body temperatures to drop by different amounts. Some species, including bears, do not enter deep hibernation. Their body temperatures drop only a few degrees, and some physiological functions, including gestation and birth of the young, continue. In addition, some animals enter a torpid state for months at a time, whereas others remain torpid only so long as temperatures stay below a certain level or food supplies are sufficiently limited. Hibernators are known among rodents, ungulates, carnivores, bats, and some Australian marsupials. Hummingbirds, swifts, and some mouse birds in the genus *Colius* also hibernate.

Another form of heterothermy is **estivation,** torpor during the summer months to avoid high temperatures or water stress. As we might expect, many desert animals (including some heteromyid rodents and some ground squirrels) estivate.

Finally, migration is also a means of avoiding temperature extremes. Like hibernation, migration is inextricably linked to avoiding energy shortages associated with winter. This strategy is particularly well developed among birds, although some mammals, including bats and certain marine mammals, migrate as well.

Plant Adaptations to Thermal Stress

By and large, plants take on the temperature of their environment. They face the obvious disadvantage that they are incapable of fleeing temperature extremes; options such as behavioral thermoregulation and migration are unavailable to them. Nevertheless, even though plants cannot change location, they are not completely helpless in the face of thermal stress because they are able to maintain temperature differentials between their tissues and the environment. The temperature of shoots and leaves may differ markedly from the ambient temperature. If they are in direct sunlight, their temperature may be several degrees higher than the air temperature. If they are transpiring rapidly, evaporative water loss may decrease plant temperatures relative to ambient temperature. The thickness of leaves can also be an important and adaptive temperature modulator: Thick leaves can be as much as 30°C warmer than the air, thin leaves 15°C cooler. Finally, the presence of small hairs on the plant surface modulates tissue temperatures by trapping a layer of air that buffers the plant from changes in air temperature.

High temperatures result in death in plants from the same kinds of processes as in animals. In addition, at high temperatures water loss may become severe (see the following section).

The effects of low temperature are more complex. Obviously, the freezing of plant tissue can result in death. Not only do ice crystals cause physical damage to organelles and membranes, but they may also cause the osmotic flow of water out of mitochondria, with lethal consequences. Aside from freezing effects, most species have minimum temperatures below which metabolic functions cease. Among the first tissues to be thus affected are the reproductive structures.

Cold-related mechanical injury may be important for trees. In the taiga and boreal forest, spruce trees

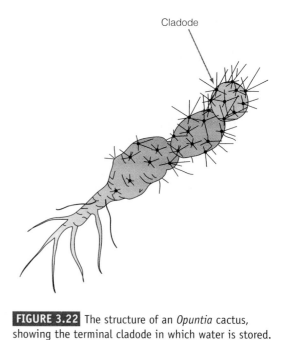

FIGURE 3.22 The structure of an *Opuntia* cactus, showing the terminal cladode in which water is stored.

FIGURE 3.21 The northern range limit of creosote bush (*Larrea tridentata*) in the Southwest (solid line) in relation to the −20°C minimum isotherm (shaded area). (*Adapted from Pockman and Sperry 1997*).

are subject to cracking at low temperatures. As the temperature drops at night, the outer layers of the trunk, including the bark, cool more rapidly than the inner portions, producing tangential stresses. If the temperature drops precipitously, these forces can be sufficient to split the tree along one radius of the trunk. Also, water transport can be disrupted at low temperatures when water freezes in the xylem. The resulting formation of air bubbles causes cavitation of the xylem—that is, disruption of the continuous column of water required for transport. Low temperature and cavitation determine the northern range limit of the desert shrub known as creosote bush (*Larrea tridentata*). Pockman and Sperry (1997) showed that minimum temperatures of −16°C to −20°C. result in the interruption of water transport in this species. The northern range limit of creosote bush corresponds to the isotherm at which the minimum temperature is −20°C (Figure 3.21).

Plants use a number of adaptive options under frigid conditions. Deciduous trees drop their leaves in the fall in order to avoid low temperatures and desiccation in winter. The cactus *Opuntia fragilis* is native to the Canadian prairies, where it experiences very cold winter temperatures. This species, unlike more southern cacti, is extremely cold-tolerant because it decreases the water content of its cladodes (Figure

3.22) to avoid tissue damage (Ishikawa and Gusta 1996). In order to do this, it must increase its drought tolerance because the water stored in the cladodes is used during dry periods. Some species switch from a perennial or biennial growth habit in the southern part of their range to an annual strategy in their northern habitats. They thus weather the low temperatures of winter as dormant seeds in the ground. In the vine *Menispermum canadense,* a thick woody stem is produced in the southern part of its range; in the north, however, the above-ground parts of the plant die back each winter (Daubenmire 1974).

A variety of adaptive options are available to plants for exerting control over low temperatures. For example, in the family Araceae, intense respiration in the inflorescence can raise the temperature of the spadix sufficiently to avoid frost damage. Among arctic plants, the flowers of *Dryas* are heliotropic; that is, throughout the day, the flower is consistently aimed at the sun. The parabolic shape of the inflorescence serves to concentrate the sun's rays, significantly raising the temperature of the reproductive organs (and of insect pollinators that frequent the plant).

Adaptations for Water Balance

Water is essential for all life; indeed, it is one of the most crucial limiting abiotic factors. Because life on Earth evolved in an aqueous environment, water plays a key role in many metabolic pathways. Consequently, all organisms must have sufficient amounts of water available to carry out their life functions. Our interest in this as ecologists stems from the fact that environments differ markedly in water availability and thus exert different potential water stresses on organisms.

The Physiological Ecology of Water

Water has unique chemical and physical properties that have profound effects on its role in physiological, and hence ecological, processes. First, water always contains gases, salts, and other compounds in solution. Some of these materials are essential for life, and water is the vehicle by which they enter the body. Others, such as certain ions or toxins, cause physiological problems for organisms. Second, water has both an extremely high heat capacity and a high heat of vaporization. As we saw previously, these features have important consequences for organisms with respect to temperature regulation. Third, water molecules are polar (Figure 3.23). One consequence of this property is cohesiveness; chains of water molecules can form. This property is important in the physiological ecology of water balance in plants because it provides a mechanism by which water can be drawn up the vascular tissue of plants. Indeed, the ability of a 100-meter redwood to pull water from its roots to its topmost leaves depends, in part, on the cohesive properties of water.

Osmotic effects are one of the most important aspects of the physiological ecology of water. **Osmosis** is the movement of water across a semipermeable membrane in relation to solute concentrations. Water flows from regions of lower solute concentration to regions of higher solute concentration. If the internal contents of an organism are more concentrated than the medium in which it lives, we say it is **hypertonic** relative to its medium. If the internal contents are less concentrated than the medium, we say the organism is **hypotonic** relative to its medium.

This process is ecologically relevant because of the variety of salt concentrations in the multitude of aquatic environments that organisms inhabit. In a highly saline environment, the organism is hypotonic. Water will tend to flow from the tissues of the organism. In a low-salinity environment such as a freshwater lake, water will tend to flow into the organism. In both cases, there must be physiological adaptations to modify these flows. If not, death will occur from the uncontrolled loss or gain of water. Organisms that can tolerate variation in the salt content of the water they inhabit are termed *euryhaline;* organisms that require a narrow range of salt concentrations are *stenohaline.*

Species respond to the osmotic problems presented by salt concentrations in two general ways. If they allow the osmotic conditions of their tissues to change in response to the environment, they are osmoconformers; if they attempt to maintain a specific solute concentration in their tissues despite fluctuations in the external environment, they are osmoregulators.

Animal Adaptations for Water Balance

Marine animals that live near land or at an interface with freshwater may encounter salt concentrations radically different from those of normal seawater. The distributions of many marine species that inhabit tide pools and estuaries, where they are likely to encounter fluctuations in environmental salt concentrations, depend on their physiological ability to cope with such changes. Some species, such as starfish and oysters, are osmoconformers; others, such as crabs and some annelids, are osmoregulators. Osmoregulators are generally better able to survive environmental changes in salt content.

The fact that osmoregulators tend to be hypertonic relative to their environment causes two problems. First, water tends to enter their tissues. The entry of water necessitates some mechanism for its removal, so very dilute urine is typically excreted. Second, this excretory process causes some important solutes to be lost. As a result, most osmoregulators have evolved mechanisms of active transport to move important ions against a concentration gradient and thus maintain appropriate tissue concentrations.

Marine vertebrates, which typically are osmoregulators, tend to use one of two strategies to deal with life in salt water. Some, such as the hagfish (*Myxine*),

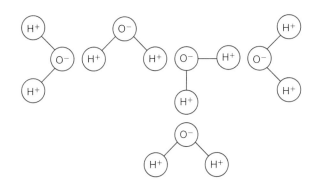

FIGURE 3.23 The polarity of water molecules, which accounts for water's cohesiveness.

are hypertonic or isotonic relative to seawater. These animals experience little osmotic flow into or out of their tissues (Schmidt-Nielsen 1979). Others, such as the teleost fishes that are hypotonic relative to seawater, drink seawater to counteract the inevitable loss of water from their tissues. One result is a buildup of salts, which must be eliminated. The gills are the primary organ responsible for this process. Active transport is required to move the salts against the concentration gradient from the tissue to the seawater.

The skin of most marine organisms is relatively impermeable, and thus some control of water flow in these species is achieved by limiting the surface area available for exchange. However, no skin is completely impermeable. In addition, the gills must be in contact with the water if gas exchange is to occur, and this contact necessarily leads to osmotic flow. The fact that the skin of many amphibians such as frogs and salamanders is highly permeable limits their ability to invade marine environments, an excellent example of an evolutionary constraint (see Chapter 2).

The crab-eating frog of Asia (*Rana cancrivora*) is an exception. This species lives in mangrove swamps, where it forages in seawater. As it evolved, this frog had two options for avoiding the osmotic loss of water: It could maintain high tissue solute concentrations (like the hagfish and other marine elasmobranchs), or it could excrete vast quantities of urine and drink seawater to maintain appropriate ion concentrations (like marine teleosts). The frog has adopted the former solution. It concentrates urea in its tissues up to 480 u mol urea per liter, thus preventing water loss (Gordon et al. 1961).

Freshwater animals face a different set of problems. Because the tissues of these organisms are hypertonic relative to their surroundings, the animals tend to gain water from the environment. Freshwater teleosts have evolved a set of adaptations to cope with these problems. First, they eliminate large volumes of a highly dilute urine, excreting up to one-third of their body weight in water each day. Second, to accommodate the loss of certain critical ions caused by this tremendous rate of urine excretion, the gills actively transport these ions from the water into the body. Similar adaptations are seen in freshwater amphibians. In frogs, for example, the skin is the major organ of osmoregulation.

The land was colonized by animals more than 425 million years ago. The organisms that were able to make this change benefited from the much higher concentration and availability of oxygen; higher metabolic rates and the increased activity they can support became possible. However, any organism making the transition to land must have been able to adapt to the potentially greater rates of water loss characteristic of a terrestrial existence.

Although some water loss occurs via excretion, evaporation is the primary avenue of water loss for terrestrial animals. This may occur through respiration or directly from the body surface if the organism sweats or has a water-permeable exterior. Amphibians and some invertebrates avoided this problem by remaining associated with moist environments. Other animals evolved integuments that are impervious to water to reduce the rate of evaporation; the keratinized skin of reptiles and the chitinous exoskeleton that covers many invertebrates are two examples. Some frogs have skin adaptations that allow them to leave moist habitats. Members of the South American genus *Phyllomedusa* have skin glands that excrete wax that coats the skin and prevents evaporative water loss. For all these animals, the respiratory surface, which must remain moist for oxygen exchange, remains the major surface from which water evaporates.

In addition to skin adaptations, a number of other strategies have evolved for reducing evaporative water loss. Some animals adopt behavioral strategies, such as nocturnality. Fossorial animals such as moles and gophers live underground in burrow systems in which the relative humidity is so high that evaporation is greatly reduced. This adaptation allows some pocket gophers in the genus *Thomomys* to inhabit desert regions.

Some mammals and birds further reduce evaporative water loss by recovering moisture from the air they exhale from the lungs. In the kangaroo rat (genus *Dipodomys*), inhaled air passing over the respiratory passages is warmed by conduction, which lowers the temperature of the passage walls. When warm moist air is exhaled from the lungs, it passes over the now cooler walls of the respiratory passages. Water vapor condenses there and is subsequently reabsorbed by the animal.

The other major source of water loss for terrestrial animals is urine and feces. Desert animals generally have long digestive tracts that absorb as much water from the feces as possible before it is excreted. Further, the nature of an animal's urine—specifically, what nitrogenous compound is excreted; see Table 3.1 and Figure 3.24—determines the amount of water lost. Each form of nitrogenous waste differs in toxicity and solubility, and the most toxic wastes are generally excreted in more dilute form. Thus species that excrete the more toxic ammonia or urea must invest more water in the elimination of nitrogen than do species that excrete uric acid. Adaptations that make it possible to excrete nitrogen without excessive water

TABLE 3.1		
The Nitrogenous Wastes Excreted by Various Terrestrial and Aquatic Organisms		
Organism	Habitat	Nitrogenous waste
Birds	Terrestrial	Uric acid
Snakes and lizards	Terrestrial	Uric acid
Gastropods	Terrestrial	Uric acid
Insects	Terrestrial	Uric acid
Turtles	Terrestrial	Uric acid and urea
Mammals	Terrestrial	Urea
Amphibians	Terrestrial/aquatic	Urea
Crocodiles	Terrestrial/aquatic	Ammonia and uric acid
Amphibians	Aquatic	Ammonia
Elasmobranch fishes	Aquatic	Ammonia
Teleost fishes	Aquatic	Ammonia and urea
Invertebrates	Aquatic	Ammonia

loss were crucial to the invasion of land. Birds and reptiles have an advantage in the desert because their excretion of uric acid requires a relatively small expenditure of water.

In terrestrial animals, the water needed to counteract evaporative and excretory losses is obtained by drinking, in the food, and via some metabolic pathways. The latter source is sometimes called metabolic water. The oxidation of organic energy sources releases a certain amount of water, as is apparent in the general equation for the oxidation of glucose:

$$C_6H_{12}O_6 + 6O_2 \rightarrow 6CO_2 + 6H_2O$$

Because the water produced metabolically is basically the result of the oxidation of hydrogen, the hydrogen content of the energy source determines the maximum amount of water that can be derived metabolically. For each gram of glucose oxidized, 0.6 gram of water can be produced. For starch the value is 0.56 gram, and for fat the average value is 1.07 gram (each of the many different kinds of fat has a slightly different hydrogen content). Metabolism of protein also produces water, but some of the hydrogen in protein combines with nitrogen and is thus unavailable to produce water. As a result, the digestion of protein typically produces less metabolic water than does that of fats and carbohydrates.

For some animals in extremely dry habitats, metabolic water is the primary source of water. Kangaroo rats, for example, can subsist on dry foods such as seeds without ever drinking free water, because as much as 90 percent of their water gain comes from metabolic water. However, a certain amount of water is lost (exhaled) during the respiration that is needed to obtain the oxygen required to metabolize the hydrogen in the food. The magnitude of this water cost depends on the relative humidity. Schmidt-Nielsen (1964) showed that if the relative humidity is above 10 percent (at 25°C), the kangaroo rat can survive on metabolic water and the small amount of water stored in seeds.

Plant Adaptations for Water Balance

Like their animal counterparts, terrestrial plants face the challenge of maintaining water balance in environments in which water may be a scarce resource. The ecological importance of this problem is confirmed by the fact that, as we have seen, the distributions of many plant species are correlated with habitat moisture gradients. Here we consider some of the adaptations plants have adopted to cope with the water balance problems imposed by a terrestrial existence.

Plants absorb water primarily via the roots (a little water is also absorbed by other organs). The important factors in determining the water intake of a plant are the amount and availability of water in the soil; the latter, in turn, depends on the nature and size of the soil particles and on the osmotic environment present.

When water enters a soil, it integrates into the soil matrix in the minute interstices (pores) between soil particles. In porous sandy soils, the interstices are relatively large, and the adhesive forces between the water and the soil are relatively small. Consequently, the loosely held water is drained away rapidly by

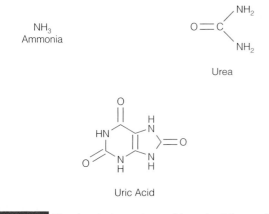

NH₃
Ammonia

Urea

Uric Acid

FIGURE 3.24 The chemical structures of important forms of nitrogenous waste.

gravity. In soils with finer particles, the pockets of water may be much smaller and more tightly held. The amount of water that stays in the interstices of a soil is called its **field capacity.**

The size and nature of the pores also determine the ease with which a plant's roots can pull water from the soil. In finer soils with small pores, water is held tightly by adhesive and capillary forces. As water is removed from the soil either by plants, drainage, or evaporation, at some point the only water left is in the smallest, most tightly binding pores and is unavailable to plants. This is called the **permanent wilting point.**

Osmotic relationships are also important. If a soil is hypertonic relative to the interior of the root, the plant will have difficulty taking in water. Because many desert soils are also saline, the water balance of desert plants is particularly precarious. In saline or other hypertonic soils, plants must expend energy to take water in against the osmotic gradient.

Obviously, root morphology and physiology constitute an important set of adaptations that determine a plant's water requirements. Plants in dry environments are called **xerophytes** and can adopt one of two strategies to ensure that they obtain sufficient water. Some produce taproots that penetrate deep into the soil, where they pick up water pulled to lower strata by gravity. Others, like the grasses, produce highly branched, shallow, fibrous roots. Having many highly branched roots near the surface accomplishes two things for a plant. First, the highly branched system helps ensure that the root will locate a region of soil that contains moisture above the permanent wilting point. This provides an advantage over a single taproot, which, once it locally depletes the soil moisture, has no other options for locating water. Second, the plant has the opportunity to capture water soon after it reaches the ground—before it can evaporate and before other plants can absorb it.

In the leaves and stems, loss of internal water occurs by evaporation. Most plants have a waxy covering on the surface of the leaves and stems, called the cuticle, that reduces the evaporative loss of water directly from the plant tissues. Xerophytes typically use additional strategies to reduce this form of water loss further. For example, many prairie plants have a rough texture due to thick cuticles that insulate the plant from water loss. Desert plants are particularly well adapted for avoiding evaporative water loss. Some, including cacti and paloverde (*Cercidium floridum*), have photosynthetic stems and no leaves, thus reducing evaporative surface area. Many desert plants have small surface hairs or spines that not only

deter grazing but also create a region of still air next to the cuticle, thereby decreasing evaporation.

Among the most important sites of water loss from a plant are the stomata on the leaves. When these structures are open to take in carbon dioxide and release oxygen during photosynthesis, the plant is highly vulnerable to evaporative water loss. When a plant faces hot, dry conditions, the guard cells that form the stomatal opening lose water; the resulting loss of turgor pressure in the cells causes changes in their shape, thereby closing the stomatal opening and reducing water loss. When this occurs, the stomata are no longer able to exchange gases, and the rate of photosynthetic activity may be inhibited.

Some morphological adaptations alleviate some of these difficulties. In certain species the stomata are more heavily concentrated on the underside of the leaves to reduce evaporative water loss; in others they are recessed deeply into the leaf surface. Leaf rolling in response to water loss, as seen in corn, probably also decreases evaporative water loss.

Clearly, then, plants must deal with the physiological consequences that arise because the mechanism of gas exchange needed for photosynthesis also increases water loss. In order to understand the physiological difficulties this causes, it is necessary to consider the biochemistry of photosynthesis.

The reactions of the typical, light-independent pathway of photosynthesis (the Calvin cycle) are depicted in Figure 3.25. The key enzyme in this pathway, RuBP carboxylase, catalyzes the reaction of CO_2 and ribulose bisphosphate (RuBP) to form phosphoglycerate, a 3-carbon molecule (hence this pathway is called the C_3 *pathway*). The activity of this enzyme is highly sensitive to CO_2 concentration; if CO_2 levels in the leaves fall, the enzyme will instead pick up oxygen. The result is a process called photorespiration, which not only fails to produce any energy for the plant but may also consume up to half the carbon fixed in the Calvin cycle. Therefore, it is important to a plant's energy balance that it keep the stomata open so that the CO_2 concentration in the leaves can be maintained at levels sufficiently high to prevent photorespiration.

The dilemma the plant faces during hot, dry weather is that if the stomata are opened to exchange CO_2 and O_2, water can evaporate readily. If the stomata are closed to prevent water loss, photorespiration becomes a problem.

One biochemical adaptation to resolve this dilemma is the C_4 pathway (Bjorkman and Berry 1973; Figure 3.26), which derives its name from the fact that CO_2 combines with phosphoenolpyruvate

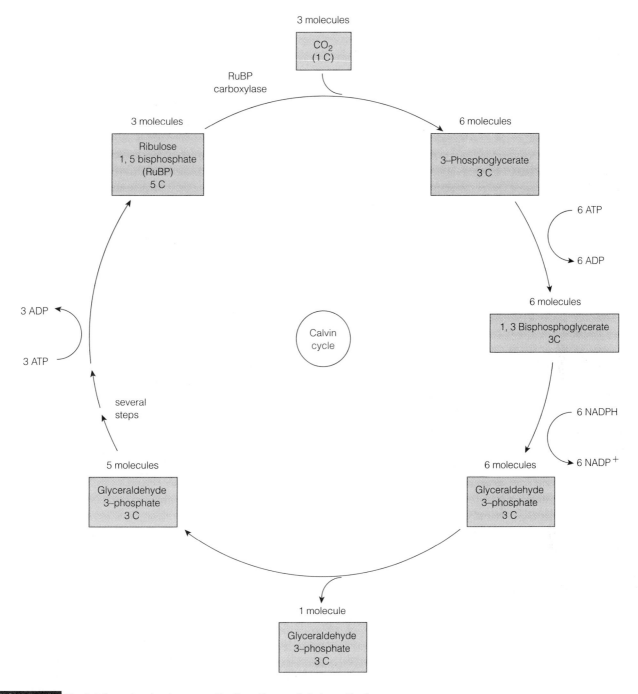

FIGURE 3.25 The Calvin cycle, also known as the C_3 pathway of photosynthesis.

(PEP), eventually producing malate, a four-carbon molecule. C_4 plants also have the C_3 pathway.

The enzyme that catalyzes the fixation of CO_2 in the C_4 pathway, PEP carboxylase, differs from the RuBP carboxylase of the C_3 pathway in that it has a much higher affinity for CO_2 and is not sensitive to the O_2 concentration. Thus it is able to fix carbon at lower CO_2 concentrations, and as a consequence, the C_4 pathway is not vulnerable to photorespiration.

C_4 and C_3 plants differ morphologically as well. C_4 plants have so-called **Krantz anatomy,** in which the chloroplasts are concentrated in bundle sheath cells around the leaf vein (Figure 3.27); C_3 plants have their chloroplasts scattered throughout the leaf. The malate produced in the mesophyll cell, site of the C_4 pathway, is transported into the bundle sheath cell, where CO_2 is released to enter the Calvin cycle. In effect, the plant separates the C_4 process from the C_3 in order to concentrate CO_2 at the site of the Calvin cycle and avoid photorespiration.

The C_4 pathway thus provides the plant with a potential solution to its dilemma. The stomata can

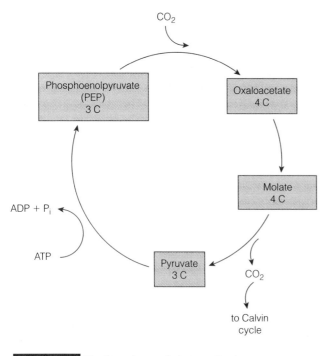

FIGURE 3.26 The C$_4$ pathway of photosynthesis.

remain closed for longer periods of time to conserve water because the plant can continue to fix carbon, even at low leaf CO$_2$ concentrations, thanks to the high affinity that PEP carboxylase has for CO$_2$.

The C$_4$ pathway is found in at least 17 plant families and apparently has evolved several times. Some species within a genus may be C$_3$ and others C$_4$. As we might expect, the C$_4$ pathway is common in plants that inhabit hot, dry climates. *Atriplex rosea* is a C$_3$ inhabitant of grasslands; its C$_4$ congener, *A. patula*, is a desert species. Some species can even switch from C$_3$ to C$_4$ under appropriate conditions.

Like most adaptations, the C$_4$ pathway comes at a price: C$_4$ plants require a great deal of light for photo-synthesis. Therefore, this adaptation is limited to plants inhabiting highly illuminated environments. Where seedlings must grow in intense shade, as in a deep forest, C$_3$ plants are at an advantage. Not surprisingly, many trees characteristic of moist deciduous forests are C$_3$ plants, including oaks (*Quercus* spp.) and the American beech (*Fagus grandifolia*).

Another adaptation to the dilemma posed by the stomata is crassulacean acid metabolism, or **CAM metabolism.** The name was derived from a modification of photosynthesis first described in the plant family Crassulaceae, which includes a number of desert succulents, among them cacti. The process has now been described in a number of other plant families as well. CAM plants assimilate CO$_2$ during the night, when evaporative water loss from the stomata is minimal. The carbon is stored in the four-carbon malate molecule. During the day, the stomata are closed, but photosynthesis proceeds by using the stored carbon. This process does not allow a plant to grow as fast as a C$_3$ plant provided with adequate water, but it does permit some plants to inhabit arid regions that they otherwise could not.

Adaptations to Light Stimuli

Light is an important abiotic factor for both animals and plants. A tremendous range of light regimes occurs in both terrestrial and aquatic habitats. Not only the amount of light, but also its quality, varies among habitats. For example, on a sunny summer day, plants and animals on the ground in deciduous forest experience a much lower light intensity than do organisms in a grassland. In addition to this quantitative difference, the wavelengths of light reaching the forest floor are not the same as full sunlight. Endler (1993) has shown that there are four major "light habitats" on the forest floor (Figure 3.28). The spectra of light available in these habitats differ significantly (Figure 3.29). In general, shade tends to be dominated by the yellow-green portion of the spectrum; woodland shade is blue-gray; in small gaps, red wavelengths predominate; and large gaps have white light similar to ambient sunlight. These different wavelengths have profound

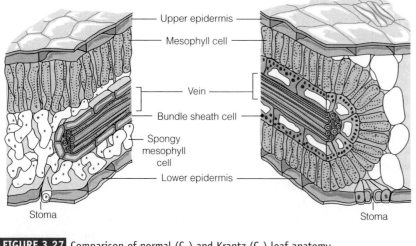

FIGURE 3.27 Comparison of normal (C$_3$) and Krantz (C$_4$) leaf anatomy.

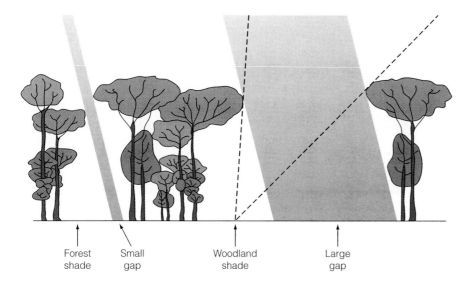

Forest shade — Small gap — Woodland shade — Large gap

FIGURE 3.28 The four light habitats in forest when sunlight is not blocked by clouds. In forest shade, nearly all light is reflected from or transmitted through the vegetation. In small gaps, there is direct sunlight from a small canopy gap. In woodland shade, a large portion of the incident light comes from the sky but not directly from the sun. In large gaps, most of the light is from direct sunlight. (*From Endler 1993*)

effects on plant growth and reproduction. For example, short-wavelength light affects pigment production and chloroplast development, germination, and the control of stomata. Long wavelengths influence seed germination, stem growth, and photosynthetic rates.

Animal Adaptations to Light Stimuli

Unlike plants, most of which require light to produce energy, animals do not have an absolute requirement for light. A number of animals live in completely dark environments, including fish that inhabit ocean depths beyond where any light penetrates. A number of cave-dwelling species never experience light. And,

of course, many animals are active nocturnally.

Animals do make use of light in indirect ways, however. Some species use light as a cue that a habitat is appropriate for other reasons. The light itself may not be required; it may instead signal an advantageous (or disadvantageous) environment and stimulate the animal to seek (or avoid) light. Such movements are referred to as **phototaxis**. If an animal moves toward light, it is said to be positively phototactic; if it moves away from light, it is negatively phototactic.

The larvae (maggots) of the common housefly (*Musca domestica*) use light as an environmental cue. Maggots have photoreceptors on both sides of the head. As a maggot moves, it turns its head from side to side. Light impinging on one of the photoreceptors stimulates the maggot to turn in the opposite direction. Eventually, this response leads it to a dark environment, where pupation takes place. Not only is the dark environment more likely to be cool and moist, but the vulnerable pupae are less likely to be seen by a predator.

Animals also use light as a stimulus for the timing of important life history events. When the natural light-and-dark cycle changes in regular fashion, animals can use the photoperiod to cue the initiation of events at certain times of year. Many vertebrates use photoperiod as a cue to initiate breeding. For example, a few teleost fish, such as the catfish (*Heterpneustes fossilis*), have photoperiodically induced reproduction. This phenomenon is rare in amphibians but common in birds and mammals, for which lengthening or shortening days may stimulate reproduction. Females of most species of mice, squirrels, and ferrets enter estrus, and males begin spermatogenesis, in response to increasing day length. Goats, sheep, and mink respond reproductively to decreasing day length. Invertebrates respond to photoperiodic cues also; in addition to stimulating reproduction, photoperiod may induce feeding, molting, or other life history events.

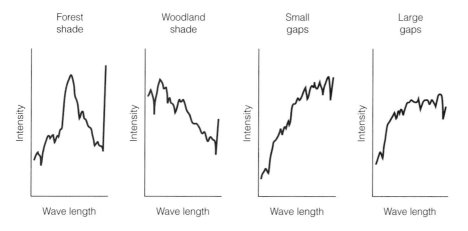

Forest shade — Woodland shade — Small gaps — Large gaps

Intensity / Wave length (for each panel)

FIGURE 3.29 The light spectra for the four main forest light habitats shown in Figure 3.28. (*From Endler 1993*)

Plant Adaptations to Light Stimuli

For plants, light is obviously a far more crucial abiotic factor, for it is their ultimate source of energy. Moreover, their sessile nature makes them vulnerable to changes in light intensity or quality. Not surprisingly, then, plants have evolved a number of adaptations to the variation in the light regimes they experience in nature. Among these adaptations are photosynthetic pigments that absorb those wavelengths to which the plant is commonly exposed in a particular habitat. Other adaptations include morphological changes to increase or decrease light absorption and movements of parts of individual plants (phototaxis) that optimize the light regime experienced by the individual.

Solar radiation in the range of 400–750 nanometers passes through Earth's atmosphere with the least reduction in energy. Accordingly, plants have geared their photosynthetic apparatus to light in this range: Their chlorophylls, flavins, and carotenoids have evolved to capture light in this portion of the spectrum. That the range of 400–750 nanometers also corresponds to the visible part of the spectrum may appear to be an interesting coincidence, until we recognize that evolving animal vision probably centered on those wavelengths for precisely the same reasons as did plant photosynthesis: the high-energy penetration of that part of the spectrum.

The exact portion of the spectrum that a plant receives depends on a number of factors. The wavelengths dominating incoming radiation vary over the course of the day. Early in the morning and late in the evening, the sun's rays pass through the atmosphere at a shallow angle, so they take a longer path through the atmosphere. More of the blue wavelengths are absorbed, and consequently, morning and evening sunlight has a higher red component. Filtration by airborne particulates (dust or volcanic ash), smog, or other plants' leaves can also change the wavelengths that reach a plant. The quality of light reaching aquatic plants changes markedly with water depth; shorter wavelengths penetrate far deeper than longer wavelengths. In aquatic habitats, differences in light quality are likely to be ecologically important. In these systems, algal zonation by depth is clearly mediated by changes in the wavelengths of available light. In terrestrial habitats, these kinds of effects are perhaps less important; there, the intensity and duration of light are more critical variables.

If a plant is to generate sufficient energy for maintenance and growth, it must absorb some minimal amount of light. Each leaf (or other photosynthetic surface) on the plant must produce enough energy to "pay for" the respiration it performs. In addition, sufficient excess energy must be available to support the activities of nonphotosynthetic tissues, such as roots and reproductive structures. When the plant receives insufficient amounts of light for these tasks, it experiences an energy deficit that, if sufficiently prolonged, results in death. The relationships among light, energy production, and respiration play important roles in the development of plant communities (succession), a topic discussed in Chapter 8.

The proportion of incident light required if the energy generated by photosynthesis is to match the energy expended in respiration is called the **light compensation point.** Light compensation points vary over two orders of magnitude in higher plants—from as little as 27 lux to as much as 4200 lux. For trees, the light compensation point varies from 3 percent to 30 percent of incident sunlight. Lower plants, particularly one-celled plants, have much lower compensation points because their photosynthetic apparatus must support so few cells.

Too much light can be as detrimental as too little. In certain habitats, excess light can be a limiting factor that determines species composition and abundance. Deserts are obvious examples. Cloudless skies expose plants to strong radiation that is intensified when the substrate is light in color and highly reflective. A prime example of such an environment is the White Sands desert of New Mexico, which is composed of gypsum sands. Only very light-tolerant species can survive there.

A number of adaptations ensure that plants can cope with the light regimes they encounter in their habitats. Some adaptations, such as the orientation of leaves toward the sun (Figure 3.30), have been known for over 100 years (Bay 1894). In plants subject to low light levels, more chloroplasts may be packed into the leaf tissue. Surface hairs and cuticles that might block light are generally absent from such plants, and the leaves tend to be thin and to have a relatively large surface area. Some algae that live at great water depths and some cave-dwelling mosses can photosynthesize at low light levels equivalent to a full moon (Daubenmire 1974).

High light intensity selects for a different set of adaptations. Leaf surface area may be significantly reduced, or leaves may be eliminated entirely in favor of photosynthetic stems. Fine hairs may shade the leaf surface, and thick leaves may protect inner chloroplasts from bleaching. Because sunlight eventually destroys chlorophyll, plants in highly illuminated habitats usu-

FIGURE 3.30 The leaves of this compass plant are oriented toward the sun to maximize the light striking them. (*Photo by David Krohne*)

ally regenerate chlorophyll rapidly compared to plants found in less illuminated environments.

These strategies reflect genetic adaptation of different species to the light regimes they most commonly experience. Desert species such as creosote bush (*Larrea tridentata*) have been modified by selection for

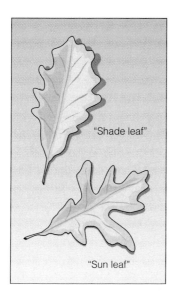

FIGURE 3.31 Difference in morphology of "sun leaves" and "shade leaves" in response to different light exposure can occur within the same individual or the same species, in this case white oak (*Quercus alba*).

that habitat, whereas species such as redwood sorrel (*Oxalis oregana*), which grows in the deep shade of redwood forests, are genetically adapted to low light levels. Acclimatization may also occur as a response to a particular light regime. For example, the distribution and number of chloroplasts can change appropriately in response to consistent exposure to high or low light levels. Leaf morphology may change as well. Sun-adapted plants have smaller, more finely dissected or lobed leaves (Figure 3.31); these morphological changes are determined by the light received by individual buds, so morphological variety may be manifest within an individual plant. Because leaves at different portions of a tree are exposed to different light intensities, so-called "shade leaves" develop lower on the tree, and "sun leaves" develop higher.

Adaptations to Nutrient Availability

Plants and animals must acquire, in addition to energy, a number of nutrients. Most of these are elements necessary for the structural components of cells or their biochemical processes. Among plants and animals, there is tremendous variation in the kinds and amounts of nutrients required.

Animal Nutritional Requirements

Precise information on required elements is available for only a handful of animals, most of them either laboratory strains or livestock. Because few complete studies of the nutrient requirements of wild animals exist, the physiological ecology of nutrient limitation in animals is an area ripe for research.

The elements known to be required for animals are listed in Table 3.2. Over 90 percent of the body weight of an animal is made up of only four elements: oxygen, carbon, hydrogen, and nitrogen. They are, of course, the building blocks of many organic molecules. Seven elements are needed in fairly high amounts: calcium, phosphorus, potassium, sulfur, sodium, chlorine, and magnesium. The rest of the required elements are needed only in trace amounts. Even though these elements are needed in very small amounts, their presence in the diet is crucial. For example, because iron is an important constituent of the hemoglobin molecule in higher animals, its presence in the diet of these animals is essential for the transport of oxygen to the cells. Sodium is limiting in some insects. Moths and butterflies employ a

TABLE 3.2
Essential Elements for Animals

Hydrogen	Magnesium	Molybdenum
Carbon	Phosphorus	Cesium
Nitrogen	Sulfur	Beryllium
Oxygen	Chlorine	Chromium*
Sodium	Iron	Selenium*
Potassium	Copper	Iodine‡
Calcium	Zinc	Vanadium‡

*Probably essential to higher animals.
‡Essential to tunicates and echinoderms.

behavioral mechanism known as "puddling" to obtain sufficient sodium. They gather at standing water and imbibe large amounts of water, thereby ingesting sodium. In some species, the males even transfer the valuable nutrient to the female during copulation so that it may be incorporated into the eggs (Smedley and Eisner 1995).

Plant Adaptations for Nutrient Uptake

Far more information is available on the nutritional needs and the ecological role of nutrition in plants. As we have seen, carbon, oxygen, and hydrogen are essential because CO_2 and H_2O are necessary for photosynthesis. In addition, all plants require relatively large amounts of elements referred to as **macronutrients** (Table 3.3). Others, called **micronutrients,** are needed only in small or trace amounts. To the extent that soils vary in their nutrient contents, nutritional elements play a role in plant distribution and abundance.

Some of the best information on the ecological role of plant nutrients comes from soils that are notably lacking in certain key elements. One of the most intensely studied soil types is known as *serpentine.* Serpentine soil, which is scattered throughout the world and is common in parts of the Sierra Nevada and the Coastal Range in California, is derived from metamorphic rock. It is characterized by a peculiar ion composition: It is low in calcium, nitrogen, and phosphorus but high in magnesium, nickel, aluminum, and iron (the latter three in toxic amounts). In addition, because serpentine soils have poor water-holding capacity, water tends to puddle (not penetrate) and then to evaporate rapidly. Thus some of the nutritional difficulties plants face on serpentine are exacerbated by low water availability.

A number of species have evolved the capacity to grow on serpentine soils. In fact, some, such as *Emenanthe rosea,* are endemic to serpentine (Tadros 1957). Many of these species are xerophytic; in addition, some have developed the capacity to take up magnesium and use it as a calcium substitute.

Although the role of nutrients in plant growth and development is readily apparent for serpentine soils, the problem is not limited to such peculiar soils. Limited nutrient availability is one of the most important physical factors affecting plant populations. In any plant community, either the natural scarcity of some nutrient or competition among plants for it results in nutrient limitation. Consequently, plants are under strong selection pressure to increase their ability to absorb whatever nutrients are present in soils.

Because plants are sessile, the growth pattern of their roots plays an essential role in the soil resources they are able to exploit. The more extensive the root system, the more opportunities for locating and taking up essential or limiting nutrients. The root systems of plants can be phenomenally extensive. The root mass of a forest tree often exceeds that of its crown. In tallgrass prairie, 80 percent of the community's biomass is underground in the form of roots. Thus the extensive ramification of root systems is a significant adaptation for ensuring adequate supplies of nutrients.

The soils that roots explore for nutrients are extraordinarily heterogeneous, and thus root growth must be plastic enough to locate and capitalize on pockets of high nutrient concentration. One developmental adaptation to this requirement is some plants' ability to modify the branching process according to the abundance of nutrients the root encounters. In regions with few nutrients, the root grows longer without branching; when it encounters a pocket of nutrients, its branching rate increases, providing a larger localized surface area that facilitates absorption.

TABLE 3.3
Elements Required by Plants

Macronutrients	Micronutrients
Hydrogen	Sodium
Carbon	Chloride
Nitrogen	Boron
Oxygen	Manganese
Sulfur	Copper
Phosphorus	Zinc
Potassium	Cobalt
Calcium	Molybdenum
Magnesium	Silicon

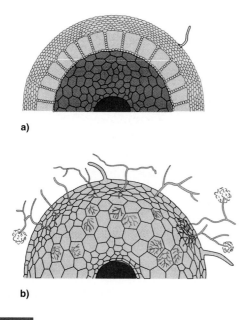

a)

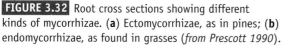

b)

FIGURE 3.32 Root cross sections showing different kinds of mycorrhizae. (**a**) Ectomycorrhizae, as in pines; (**b**) endomycorrhizae, as found in grasses (*from Prescott 1990*).

Many plants enter into symbiotic relationships with fungi to enhance their ability to take up nutrients. These mutualistic associations are called **mycorrhizae.** As many as 80 percent of all vascular plants form mycorrhizal associations, which can be either *ectomycorrhizal* or *endomycorrhizal,* depending on the nature of the penetration of the root by the fungus. In ecotmycorrhizae, the fungal mycelia extend great distances into the soil, thereby hugely increasing the absorptive surface area of the root. The plant, in turn, provides the fungus with some of its photosynthate. Thus the plant "trades" some carbohydrate to the fungus in exchange for an increased ability to take up nutrients.

In ectomycorrhizae, the fungus grows as a sheath around the exterior of the root (Figure 3.32a); some intercellular penetration by the fungus also occurs. This type of symbiosis is characteristic of beech, oak, birch, and pine trees. Endomycorrhizae penetrate into the root's cortical cells, where they coil and branch (Figure 3.32b). Most grasses and many agricultural crops form endomycorrhizal associations.

Other Abiotic Factors

The preceding discussion has focused on the abiotic factors that are most commonly limiting for plants and animals. However, virtually any physical factor in the environment can be limiting, in some cases because it

is in short supply, in others because it inhibits the growth or development of the organism. The final two abiotic factors we will consider here are oxygen and pH.

Oxygen

For most plants and animals, oxygen is necessary for energy metabolism and thus constitutes an important resource. For anaerobes such as cyanobacteria, however, oxygen is toxic, and they must live in habitats in which they are protected from oxygen.

For plants, oxygen is rarely a limiting factor. It is a by-product of photosynthesis, and most plants' roots can obtain the oxygen needed for respiration from the soil. Plants that inhabit wetlands are an exception; they may be oxygen-limited because their roots are in soil saturated with water.

Some of these plants have structures that rise above the surface of the water to absorb oxygen from the air. The pneumatophores of black mangroves (*Acicennia gerinans*) serve this function. Other plants use physiological mechanisms to bring oxygen to the roots. In rice, air is moved from the emergent upper parts of the plant to the roots. When consumption of oxygen in the roots and the dissolution of CO_2 into the surrounding water decrease the pressure in the network of internal air spaces in the submerged part of the plant, air moves by mass flow from the atmosphere through the stomata and down into the roots (Raskin and Kende 1985).

Animals must also assimilate oxygen from the environment. For land-dwelling animals, oxygen is abundant; indeed, the abundance of atmospheric oxygen is probably one of the selective factors leading to colonization of the land. For most terrestrial animals, the key to oxygen limitation is the ability to absorb oxygen and transport it to the tissues. Thus many of the relevant adaptations for land animals are associated with the nature of the absorptive surfaces (lungs or skin) and the circulatory mechanisms to deliver oxygen. For aquatic organisms, the oxygen content of the water determines the degree to which oxygen is limiting.

pH

For some organisms, pH is a crucial abiotic factor. Most species of plants and many aquatic animals have relatively wide tolerance limits for pH. Sometimes these limits are directly related to the effect of pH on metabolic physiology; for other organisms the effects of pH may be more indirect. For example, pH can

Tolerance of Pollution

The law of tolerance has direct application to the effects of pollutants on animals and plants. Much of the impact of pollution is attributable to direct mortality associated with anthropogenic changes in organisms' abiotic environments.

One of the most widespread environmental effects is the acidification of lakes and streams by acid precipitation (see the box "Subspecies, Races, and the Endangered Species Act" in Chapter 6). Many of the lakes and streams in the northeastern United States have experienced a significant drop in pH. The effect on the organisms that inhabit these lakes is complex, because each species has a unique tolerance curve for pH. For example, mussels are extremely sensitive to acid; they are unable to tolerate water with a pH below 6.0. In contrast, some invertebrates, such as water boatmen, tolerate pH values as low as 3.5. Fish also vary in their acid tolerance. Yellow perch and lake trout are relatively tolerant and can withstand pH values as low as 4.5, whereas smallmouth bass are eliminated when the pH drops below 5.5.

Consequently, acid rain has complex effects on communities. Even though acidified lakes are referred to as "dead," this may be far from the truth. After the loss of the most valued species such as gamefish, acidified lakes often come to be dominated by large numbers of a few less visible, acidophilic (acid-loving) species.

The tolerance curves of organisms, and thus their responses to disturbance, are rarely fixed. Natural selection can modify these curves to make organisms more tolerant of pollutants. One of the most thoroughly studied examples of this phenomenon is the response of sweet vernal grass (*Anthoxanthum odoratum*) to high concentrations of lead, zinc, and copper in mine tailings. J. Antonovics and A. D. Bradshaw (1970) have shown that populations of *Anthoxanthum* can evolve tolerance to these metals very rapidly. In one instance, zinc tolerance developed near a galvanized fence in only 25 years.

Two phenomena make predicting the effects of pollutants difficult. First, the effects of pollutants may be the product of interactions among several abiotic factors. For example, the effluent from some power plants results in thermal pollution of adjacent waterways. If the water temperature increases sufficiently, a variety of species may die because their maximum temperature tolerances were exceeded. Even if direct mortality from high temperatures does not result, the organisms' increased metabolic rates caused by elevated ambient temperatures may make finding sufficient food impossible. Furthermore, higher water temperatures lower the dissolved oxygen content of the water. As water temperature increases, the solubility of oxygen (and other) gases decreases (see the accompanying figure). Most fish require oxygen concentrations of at least 7 milligrams per liter in spawning areas; above 25 °C, many fish would be in peril. And this assumes that the water is saturated with oxygen, which is seldom the case.

The second complicating factor affecting the impact of pollutants is that their effects may be cumulative. In widespread areas of shallow-water tropical regions, particularly in the Caribbean, many corals are dying. The syndrome, characterized by whitening of the coral heads, is known as "coral bleaching." Mortality is associated with the expulsion of the zooxanthellae, microscopic algal symbionts that contribute to the nutrition of healthy corals.

A number of factors may be responsible for coral bleaching, including thermal pollution from global warming, increasing ultraviolet radiation from loss of ozone, decreasing seawater salinity, and nutrient pollution, especially increased phosphate concentrations. The bleaching problem occurs worldwide, and these factors are not equally prevalent in each area. Thus it has been suggested that several environmental insults may combine over time to produce the bleaching and mortality effects.

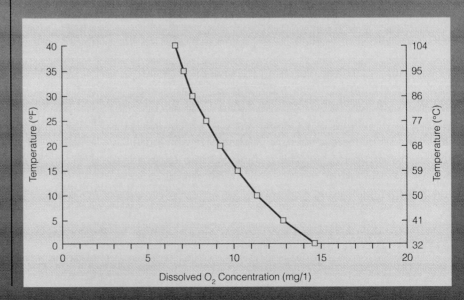

affect the ability of the crayfish (*Cambarus bartoni*) to locate food (Allison, Dunham, and Harvey 1992). This species prefers water with a pH of 7.5. Acid waters (pH below 4.5) affect its olfactory senses such that the crayfish has difficulty finding food. Although in this case the effects may be lethal, they are not directly attributable to pH. As acid precipitation becomes an increasingly significant environmental problem, the effects of pH on plants and animals will increase in importance.

In this chapter we have seen the role that abiotic components of the environment play in determining the distribution and abundance of animals and plants. Variation in the distribution of abiotic factors not only limits the distributions of organisms, but also constitutes an important selective force on them. Thus the examples considered in this chapter include instances of the application of Shelford's law of tolerance as well as discussions of many adaptive responses to abiotic conditions. The responses to abiotic conditions represent examples of the intimate interaction of ecology and evolution.

Much of the rest of this text is devoted to the variety of biotic interactions important in ecology. This emphasis reflects the current focus in ecology. It does not, however, imply that the physical factors discussed in this chapter are no longer thought to be significant. As you proceed through the text, it is important that you appreciate that the limits and adaptations described in this chapter underlie the biotic aspects of ecology.

SUMMARY

1. The abiotic environment limits the distribution and abundance of organisms. Some physical factors, such as light and water, constitute limiting factors for certain species. Other factors limit distribution because they delimit a zone in which life is physiologically possible. The distinction between abiotic factors and resources is a subtle one. For example, light for photosynthesis is a limiting resource for some plants. For others in very high-light environments, light may be a limiting factor.

2. There are two broad categories of responses to the abiotic environment. First, homeostatic mechanisms are adaptations that maintain a viable set of physical and chemical conditions in the organism despite a wide range of environmental conditions. Second, organisms can avoid stressful abiotic conditions physiologically or behaviorally. Whereas animals may move to more appropriate environments, plants can sometimes avoid harsh conditions with resistant seeds.

3. Thermal stress can occur because temperatures are too high or too low. In animals, physiological adaptations to temperature extremes depend on increasing or decreasing heat loss via conduction, convection, radiation, or evaporation. Extremes can also be avoided by hibernation or behavioral adaptations such as migration or habitat selection. Plants have a more limited set of adaptations to thermal stress. Many are associated with seasonal physiological changes that result in dormancy during stressful periods.

4. In animals, many aspects of water stress are associated with osmotic stress. Water and heat stress interact because some animals use evaporative cooling as a homeostatic adaptation to thermal stress. Plants face a special set of difficulties because the stomata must be open to exchange CO_2 and O_2 for photosynthesis but the open stomata also result in evaporative loss of water. C_4 photosynthesis is an important adaptation to this dilemma.

5. Animals do not have an absolute requirement for light. Consequently, their adaptations center on the use of light to locate other important resources or as stimuli for seasonal or daily activities. Because of the role of light in photosynthesis, the quantity and quality of light are a crucial limiting resource for plants. Their morphology and photosynthetic machinery are adapted to gather sufficient light of certain wavelengths and to protect the plant from excessive light energy.

6. Animals and plants require specific elemental nutrients. Both groups have a variety of adaptations to ensure the assimilation of these nutrients.

SELF-ASSESSMENT: CAN YOU ...?

1. Explain the physiological problems associated with the following:

low water availability

low temperature

high temperature

quantitative and qualitative differences in light

2. Explain the physiological adaptations of plants and animals to the physiological problem in Question 1.

3. Explain the ecological effects of the problems in Question 1.

4. Discuss the relationship between limiting factors and resources.

5. Discuss the ways in which plants and animals cope with harsh abiotic conditions by physiological adaptation versus avoidance.

PROBLEMS AND STUDY QUESTIONS

1. Compare and contrast the nature of avoidance mechanisms in plants and animals.

2. Discuss the osmotic stresses the following organisms would face:

a. A reptile that attempts to inhabit saline waters

b. A steelhead that forages in the ocean but returns to freshwater to spawn each year

c. A pelagic gull whose only source of water is the ocean

d. A desert plant inhabiting an alkali basin

3. Explain the adaptive advantage of C_4 plants in hot, dry climates.

4. If thermal stress is an important selective factor for animals in desert environments and water is limiting, what mechanisms of heat loss do you expect to be most common in desert animals? What would you predict about the general size and shape of desert animals?

5. Few amphibians inhabit marine environments. Using your understanding of the various constraints on adaptive evolution (Chapter 2), discuss why this might be.

6. C_4 plants can close their stomata and continue to photosynthesize. Why do you suppose C_3 plants persist? Shouldn't C_4 plants gradually outcompete them?

7. Figure 3.2 shows the northern limit of trees in Canada. This region was glaciated within the last 20,000 years. We know that conifers were pushed south ahead of the glaciers during the glacial advance. Thus it is possible that these trees are at the limit shown in this figure because that is as far north as they have moved since the retreat of the glaciers. How could one test this hypothesis?

8. You notice that a certain species of cactus reaches its southern range limit in central New Mexico. The geographic limit of the species could be due to temperature effects or water stress. How could you determine which is the limiting factor? Are these exclusive hypotheses?

9. For some abiotic *resources* such as light, organisms may compete directly for access. Abiotic *factors* generally are thought to set physiological limits. How is it possible that abiotic factors may lead to competition among individuals? What is the basis of the competition?

10. In the Arctic, where abiotic conditions are cold and dry, there are relatively few species, but at least some have large population sizes. In the tropics, where conditions are more benign, there are many species but often few individuals within a species. Speculate on why this might be.

PROJECTS AND ADDITIONAL STUDY

1. Perform a literature search to determine which species can tolerate the most extreme temperatures. Develop a summary of the physiological mechanisms these species use to tolerate extreme temperatures.

2. Search the literature to determine the photosynthetic pigments produced by green, brown, and red algae in marine ecosystems. Find the optimal wavelengths for these pigments and relate them to the penetration of light in seawater.

3. Endler (1993) shows that the light environment varies qualitatively in forest habitat. Examine the role of this variation on *animals* in forest environment.

Population Ecology

Demography and Population Growth

With this chapter we begin Part II: Population Ecology. As we noted in Chapter 1, this text focuses on the study of ecology at three hierarchical levels: the ecosystem, the community, and the population. Here we turn our attention to the interactions that occur among individuals of the same species—that is, within populations. Our focus is thus on the biotic portion of ecosystems. As we will see, however, we can never completely ignore the abiotic environment because it affects the biota at every level of the hierarchy.

At the population level of organization, ecologists ask questions such as the following:

1. What are the characteristics of populations? What population parameters can we measure? How do populations differ in such aspects as density, age distribution, and so on?
2. How do populations grow? What are the patterns of population increase (and decrease)? Are there consistent patterns in changes of abundance among species? What parameters can we use to describe quantitatively the changes in populations?
3. How are the numbers of individuals in populations controlled? What factors determine the limits of population size? Are there processes that stabilize populations?
4. Given that each individual has a finite amount of energy to devote to reproduction, how should this energy be apportioned among offspring? How many offspring should be produced? How often? How much energy should be invested in each?
5. What are the patterns of variation upon which natural selection acts among individuals of natural populations? How is variation generated and maintained?
6. How do members of a population interact behaviorally? What ecological factors determine the evolution of such interactions?

Chapters 4–8 will focus on these kinds of questions. We will first examine how we characterize natural populations and how their numbers are regulated. Then we will discuss variation in natural populations—the differences among individuals for a variety of characteristics, including reproduction. Finally, we will consider the behavioral and social interactions that occur in populations.

Population ecology has important applications in environmental issues. Humans have a vested interest in the populations of many species of animals and plants. Clearly, our activities have important impacts on many species; indeed, humans have been the basic cause of the extinction of hundreds of species around the world, and many thousands more are threatened. If we are to protect species from anthropogenic extinction, we must understand populations and the processes by which they are regulated. Humans also rely on many wild species for food and recreation. Our fishing fleets harvest millions of tons of fish from the sea, and sport fishers and hunters the world over tap populations of game fish and wildlife. Prudent management of commercially and recreationally important wild populations is rooted in the basic principles of population ecology. Finally, because many plants and animals are harmful, either directly to humans or to species we wish to cultivate, we try to control crop pests, weeds, parasites, and disease agents. Taking these actions, too, requires that we understand the basic principles of population ecology. We begin by considering one of the most fundamental aspects of a population, its demography.

Demography

Demography is the quantitative description of a population. Demographers are concerned with the "vital statistics" of a population: its size, its age and sex composition, its spatial distribution, and so forth. Before we can begin to examine the many ecological forces that act on populations, we must have a way to describe statistically or mathematically the important components of the population. And before we can make such descriptions, we must be sure we understand exactly what a population is.

Defining Populations

In the simplest sense, a **population** is a group of conspecifics inhabiting a specific place at a specific time. Thus we may refer to the population of *Paramecium* in a pond or to the population of spotted owls on the Olympic Peninsula. On the surface, identifying a population is a straightforward procedure. In practice, however, a number of difficulties arise.

Our definition of a population implies that we must define both its physical and its temporal boundaries. The latter is somewhat easier because the nature of the population of interest often defines the time scale over which we will study it. Thus, if we are interested in the patterns of population fluctuation of elephants in East Africa, an appropriate time scale would be years; on the other hand, if we are studying *Drosophila*, a consid-

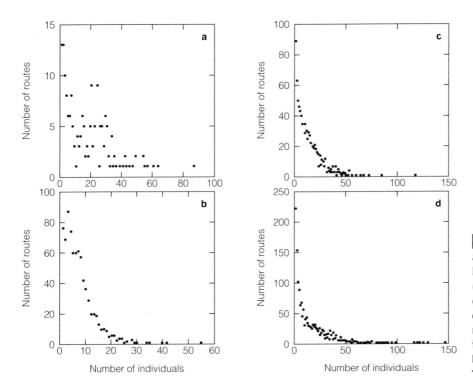

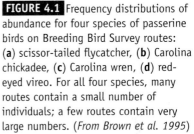

FIGURE 4.1 Frequency distributions of abundance for four species of passerine birds on Breeding Bird Survey routes: (**a**) scissor-tailed flycatcher, (**b**) Carolina chickadee, (**c**) Carolina wren, (**d**) red-eyed vireo. For all four species, many routes contain a small number of individuals; a few routes contain very large numbers. (*From Brown et al. 1995*)

erably shorter time scale would be more appropriate. Even though very long-term population studies are relatively rare, those few that exist reveal important population phenomena, such as cycles with long periods, that short-term studies may fail to reveal.

The physical boundary of a population is not always so clear-cut. Instead of making an arbitrary decision about how to define the spatial limits of a population, we try to find some ecological or genetic discontinuity associated with some physical or topographic feature of the landscape. We try to define a population in such a way that the population phenomena *within* the study area are more important than the processes that occur *between* the study area and other regions. However, two aspects of populations greatly complicate the identification of such discontinuities. First, the number of individuals is not uniformly distributed across the landscape. For example, an analysis by Brown and Stephens (1995) showed that there is great spatial variation in bird abundance (Figure 4.1). Indeed, for 77 of the 90 species analyzed, more than

50% of the total individuals were recorded in less than 25% of the samples.

Second, the movement of individuals greatly complicates the analysis of such discontinuities. Consider the situation depicted in Figure 4.2. A group of white-

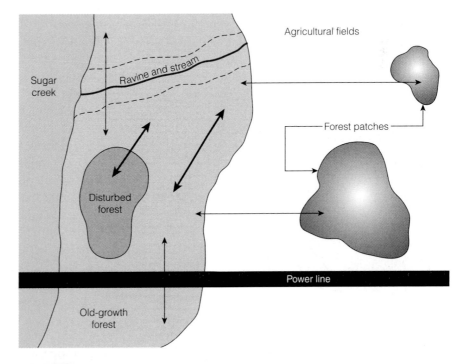

FIGURE 4.2 A complex group of "populations" of white-footed mice (*Peromyscus leucopus*) in a landscape composed of a large tract of old growth forest surrounding a disturbed forest and agricultural fields containing isolated forest patches. The ravine, power line, and agricultural fields represent barriers to movement of individuals. The thickness of the arrows thus indicates the frequency of movements of mice among regions.

footed mice (*Peromyscus leucopus*) inhabits a complex region of forest, agricultural fields, and small forest patches in the Midwest (Krohne et al. 1985; Krohne and Hoch 1999). Within some of these forests are disturbed areas with young, regenerating forest. Mice can move long distances within habitats. However, the ravines and agricultural fields are barriers to movement. Note in Figure 4.3 that the pattern of population fluctuation differs significantly among these regions:

1. The population in large, old-growth forest maintains moderate densities.
2. The population in the successional region goes extinct each winter and is at very low density at other times. Most animals in this region move there from the older, surrounding forest.
3. The density on the very small forest patch is extremely high.
4. The density on the larger forest patch is most like that in the large, old-growth forest.

How should we define a population of white-footed mice in this landscape? How do we know what part of this landscape is demographically independent? Do the inhabitants of each region (old growth, disturbed forest, each patch) constitute a separate population? Each has a physical or ecological boundary. Or should we consider all the mice in the entire region in Figure 4.2 a population? After all, the different regions are connected somewhat by the movement of individuals. This is not a trivial issue. If we want to study the

dynamics of this population, and especially the factors that cause the population to increase or decrease, we must define the population rationally and objectively. If we define our population too narrowly, we neglect the important influences of other nearby regions. If we define our population too broadly, we lump together units that may have separate regulatory mechanisms and effects.

This is a classic problem of scaling. We will encounter it again when we consider communities and ecosystems. Part of the problem stems from the fact that the idea of a "population" is, in fact, an arbitrary construct—a concept ecologists have invented to designate an ecologically discrete group of individuals. Unfortunately, few populations have unambiguous, natural boundaries. Even groups isolated on oceanic islands are connected to others.

We now understand that ecologically and genetically defined populations do not necessarily coincide. To return to the example of the white-footed mice, genetic studies show that in large forest regions, groups separated by ravines are demographically indistinguishable but genetically distinct (Krohne et al. 1985). In contrast, the groups in old-growth forest and disturbed forest are demographically very different (Figure 4.3) but genetically very similar. These results make sense given the patterns of movement. Where there are no barriers to movement, gene flow homogenizes the genetic population; barriers such as the ravines prevent gene flow and allow the demes on opposite sides to evolve independently.

If the habitat patch of interest is so large that practical difficulties make it hard to study, then it is necessary to sample the larger population. Thus, if we must study the mouse population in a forest of many thousands of acres, we establish numerous study plots that provide samples of what is almost certainly a very large population. We must endeavor to ensure that our samples are representative of the larger population as a whole.

Population geneticists recognize that there is spatial variation within and among populations, and like population ecologists, they have developed terms for the patterns of variation that focus on genetically homogenous groups physically separated from other such groups. A **deme** is a local genetically defined population characterized by random mating within the group (see Chapter 2 for the significance of random mating). Within a deme, mating occurs within the **genetic neighborhood.** Mating is random within the neighborhood and it contains $4\pi s^2 D$ individuals, where D is the density of the population, and

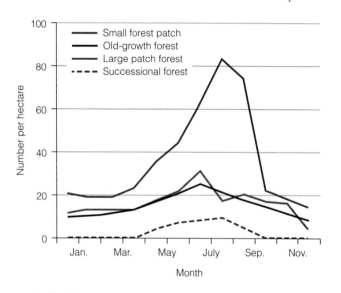

FIGURE 4.3 The patterns of population fluctuation for white-footed mice (*Peromyscus leucopus*) in the four different habitats depicted in Figure 4.2.

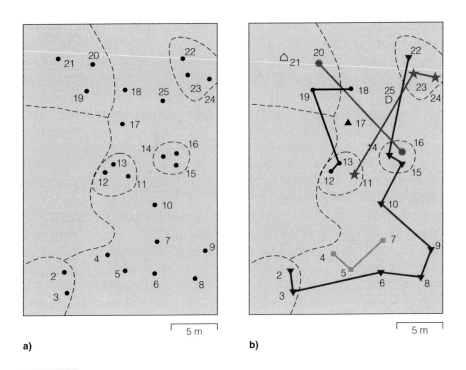

a)   b)

FIGURE 4.4 The distribution of *Rubus saxatilis* shrubs in a "mixed semi-open forest" plot in Sweden. The broken lines indicate limits of patches identified before the plants were genetically analyzed. In (**a**), locations of plants are plotted relative to the patches. In (**b**), in which different symbols, connected by solid lines, indicate different genets, the plant's genetic population distribution is seen to differ from its demographic distribution. (*Adapted from Eriksson and Bremmer 1993*)

s is the standard deviation of the distance individuals disperse between their birth and the birth of their offspring.

In practice, patterns of spatial genetic variation can be difficult to identify. Because the population is the level at which evolution is defined—that is, evolution is defined as changes in allele frequencies within populations—it is also appropriate to look at the nature of a population from a genetic or evolutionary point of view. Given the large number of new genetic techniques now available to population ecologists, we are able to investigate the genetic composition of populations and how it is related to other aspects of demography.

If we note the appearances of shrubs of *Rubus saxatilis* in one of its habitats, we might group the shrubs into patches that seem to contain like individuals (Figure 4.4a) and then decide to call each of these patches a population. But upon closer examination, we learn that this plant reproduces itself by cloning. The "individuals" we are observing are in fact ramets, independent members of clones called genets, and ramets from different genets can occur within a given patch (Figure 4.4b). Thus, when the distribution of genetic makeup of the shrubs is superimposed on its

demographic distribution, we find that the demographic and genetic populations are not necessarily the same.

A population's history can play a role in its genetic structure. Populations of the periodic cicada (*Magicauda* sp.) show similar demographic characteristics across this species's range (Figure 4.5), including the emergence of huge numbers of adults at intervals of either 13 or 17 years. However, the northern and southern parts of the range are genetically distinct, as determined by analysis of mitochondrial DNA (Martin and Simon 1990). The northern populations are characterized by only one genotype and by very little variation compared with the southern populations. Martin and Simon suggest that this population differentiation is due to the isolation of the northern group in a small refugium during the Pleistocene glaciation. This population "bottleneck" resulted in genetic drift, the loss of variation (see Chapter 2), and divergence from the southern group. There is no demographic distinction that corresponds to the genetic difference.

In the remainder of this chapter, we will discuss a number of parameters that quantify various aspects of populations. In those discussions we will assume that the populations have been chosen appropriately. We

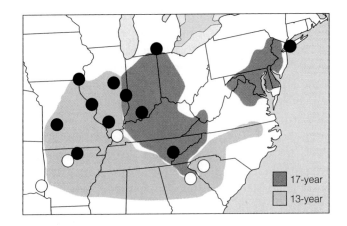

FIGURE 4.5 Ranges of 13- and 17-year cicadas (*Magicauda* sp.). Closed circles represent one genotype, open circles another genotype. (*Adapted from Martin and Simon 1990*)

will revisit this topic in Chapter 5 because for both practical and theoretical reasons, the identification of population boundaries is intimately connected to the issue of population regulation.

Population Density

The most fundamental demographic parameter is the number of individuals in the population. If we are to understand the ways and the degree to which a population changes, we need to know how many individuals constitute what we have defined as the population. We could express population size as the total number of individuals in the group, but it is often more meaningful to know the number of individuals *per unit area* (or volume, if that is appropriate). This quantity, called the **density** of the population, is calculated by dividing the total number of individuals by the total area they occupy. Thus expressions such as 200 mice per hectare, 1 grizzly per 100 square kilometers (km^2), and 10^6 diatoms per cubic centimeter (cm^3) are all expressions of density.

If we simply count the number of individuals per unit area across a study site without regard to the variation in the quality of the habitat, we have measured the **crude density.** However, not all parts of the habitat are equally appropriate living space. If we count the number of individuals per unit area of **appropriate habitat,** we have measured the **ecological density.** Obviously, if we are trying to measure a species's ecological density, we must have far more detailed knowledge of the species and its interactions with the habitat than if we are simply measuring crude density.

The densities exhibited among animal and plant species range over many orders of magnitude. Some large mammals are found at densities of only $10^{-2}/km^2$, or 1 animal per 100 km^2; some plant and insect densities range as high as $10^{12}/km^2$.

The techniques used to measure density vary widely with the kind of organism being studied. For some organisms, perhaps the most accurate technique is a census, a direct tally of the number of individuals in a study area. For plants and sessile animals, it is often possible to count individuals directly. Adult trees, for example, can be tallied to provide a very accurate total population count. A census may also be possible for large, conspicuous animals. Moose, for example, can be readily counted from the air, especially in winter, when they are visible against the white snow.

Other animals are less conspicuous. They may have cryptic coloration or secretive habits; nocturnality also makes censuses difficult. Other animals may move so frequently and so far that it is difficult to know whether an individual has already been counted. In aerial censuses of walrus (*Odobenus rosmarus*) in the Arctic, it takes so long to fly over the species's huge potential range, and the animals can move such great distances, that it is difficult to ensure that the census does not include multiple counts of some individuals.

Difficulties also arise with organisms for which it is hard to determine what an individual is. Among plants that reproduce vegetatively, should we count the ramets or the genets? Even though a genet might be the individual in a genetic sense, it may not be an ecologically meaningful unit. Not only is it difficult to decide how to define an individual in such cases, but practical difficulties are associated with finding and counting underground connections. For species such as lichens and mosses, in which an individual gradually expands over a large surface area, the number of "individuals" may not change, but certainly the extent of an individual may. For such species, *density* does not have the traditional meaning. Perhaps the most extreme case of this problem involves the fungus *Armillaria bulbosa,* which produces conspicuous above-ground mushrooms. Using DNA fingerprinting, M. L. Smith and co-workers (1992) showed that these mushrooms are in fact connected by a vast underground mat of mycelia, such that an individual actually extends over many hectares! The DNA profiles of the mushrooms inhabiting a vast area indicate that all of them are in fact parts of a single individual.

The appropriate means of censusing difficult species such as these depends on the kind of question being asked. If we are interested in genetic aspects of population changes, it may be important to identify genetically distinct individuals. If, on the other hand, we wish to determine the ecological success of a species, it may not be necessary to resolve the issue of the definition of an individual. For example, if we are interested in the success of a moss species under certain environmental conditions, we can measure abundance as the percentage of the available surface area covered by the species.

For animals for which direct counts are not feasible, other techniques allow researchers to estimate density. These techniques fall into two broad categories: estimates of absolute density and indices of relative abundance.

For *estimates of absolute density,* a sampling scheme is used to approximate the total number of organisms. One of the most widely used techniques is the Lincoln–Peterson method, which relies on capturing and marking some fraction of the total population and then using this fraction to estimate the actual population size. This method of estimation is based on the following logic. We know that we cannot capture all the individuals in the population. Thus, at t_1 we capture and mark some number of individuals. The proportion of the population that is marked is as follows:

Proportion of population marked =
$$\frac{\text{no. animals marked at } t_1}{\text{total no. animals in population}}$$

The marked animals are then released back into the population and allowed to mix thoroughly in it. If at t2 we capture another fraction of the population, this second sample will contain some marked and some unmarked animals. We can express the proportion of marked animals in the second sample as follows:

Proportion of sample marked =
$$\frac{\text{no. marked animals captured at } t_2}{\text{total no. animals captured at } t_2}$$

If the second sample is a random sample of the entire population, then we can set these two proportions equal to each other:

$$\frac{\text{no. animals marked at } t_1}{\text{total no. animals in population}} =$$

$$\frac{\text{no. marked animals captured at } t_2}{\text{total no. animals captured at } t_2}$$

Because we know three of the four quantities, the fourth—the total number of animals in the population—can be calculated.

The validity of this technique depends on a number of assumptions:

1. That the marking technique has no effect on the mortality of individuals (e.g., that marked individuals are not at higher risk of predation)
2. That the marking technique does not affect the probability of being recaptured (e.g., that marked individuals do not learn to avoid recapture)
3. That there is no net immigration or emigration of marked or unmarked individuals in the interval t_1 to t_2
4. That there is no mortality or reproduction in the interval t_1 to t_2

Violation of any of these assumptions would obviously affect the equivalence of the two sides of the final equation.

The other kind of technique that can be used on species that cannot be counted directly is an **index of relative abundance,** which frequently relies on indi-

rect evidence for making comparisons of the numbers of organisms. Such indices can answer such questions as whether there are more animals in habitat A or B and whether there are more this year than last year. For some ecological questions and for many purposes in game management, this is all we need to know. For example, if we are setting bag limits for a deer population for the coming season, we may need to know only that there are, say, twice as many deer in the area this year as last year.

Almost any evidence of the presence of animals can form the basis of an index of relative abundance. Counts of fecal pellet groups have been used with many big game animals. For some small mammals, the number of animals captured per day per trap can be an effective index of abundance. For some bird species, the number of nests constructed is related to population size; for others, call counts can be a very accurate index. For example, in some states the game department pays rural mail carriers to record the number of pheasant calls they hear on their daily routes. The number of calls heard per mile is highly correlated with pheasant abundance.

In some cases, indices of relative abundance can be converted into estimates of absolute density. It is possible, for example, to calculate the number of deer in a plot from fecal pellet counts if independent information on the rate of defecation is available. O. P. Pearson (1960) showed that as the density of the California vole (*Microtus californicus*) increases, it builds more runways through its grassland habitat (Figure 4.6). On the basis of this information, W. Z. Lidicker and P. K. Anderson (1962) used counts of runways

FIGURE 4.6 Aerial view of a runway of clipped vegetation made by the vole *Microtus californicus*. The abundance of such runways is highly correlated with vole abundance. (*Photo by David Krohne*)

along transects through the grassland as an index of the density of *Microtus*. The relationship between animal numbers and runway abundance is so strongly correlated that it is a simple matter to convert runway counts into density figures.

Dispersion

The second fundamental demographic characteristic of a population is its **dispersion,** a term that refers to the spatial distribution of individuals in the population. It should not be confused with **dispersal,** the term for the movement of individuals away from their home area. We will consider the ecology of dispersal in detail in Chapter 5.

Plant and animal populations can have three possible dispersion patterns (Figure 4.7). First, individuals may be randomly distributed across their habitat. In this case, the presence of an individual in a certain spot provides no information about the likelihood that another individual will be found in the vicinity. Secondly, the population may be aggregated, or clumped. Groups of individuals tend to occur together. Once an individual has been located, there is a high probability that another will be found nearby. Finally, a population can be hyperdispersed, or regular in dispersion. In this case the distances between adjacent individuals are roughly equal. Most species show aggregated rather than random or hyperdispersed distributions.

A species's dispersion pattern is a clue to important features of its ecology. The most common cause of an aggregated distribution is habitat heterogeneity; animals and plants congregate in patches of appropriate habitat and avoid unsuitable sites. Direct social interactions may also lead to either an aggregated distribution or hyperdispersion. Primates that travel together in troops are an example of the former. They may range widely across their habitat, but at any one time, the individuals in the population are likely to be highly aggregated. Hyperdispersion is a likely consequence of territoriality or avoidance of conspecifics.

A number of statistical procedures have been developed to identify the dispersion patterns of organisms (Pielou 1969). Most of these procedures test the observed dispersion against a purely random distribution. One frequently used procedure is based on the **Poisson distribution,** a mathematical description of infrequent, random events (for example, radioactive decay). If the mean rate of an event is known, we can calculate the Poisson probability (*P*) of any given number of occurrences by the formula

$$P_x = \frac{a^x e^{-a}}{x!}$$

where *x* is the number of occurrences, *a* is the mean number of occurrences, and *e* is the base of the natural logarithms. The assumptions for the Poisson distribution are small mean occurrence and that the mean number of occurrences equals the variance in the number of occurrences.

Figure 4.8 represents the hypothetical distribution of mosquito larvae in 25 small pools of water. The total population size is 37 larvae, and the mean number per pool is 1.5. The observed frequency distribution is given in the first two columns of Table 4.1. Using the foregoing formula, we can calculate the expected frequencies of pools containing each number

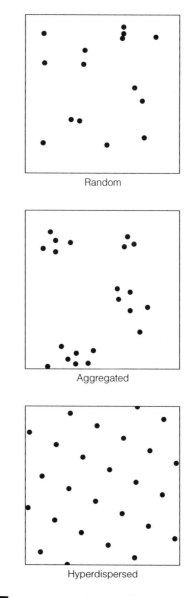

Random

Aggregated

Hyperdispersed

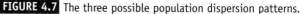

FIGURE 4.7 The three possible population dispersion patterns.

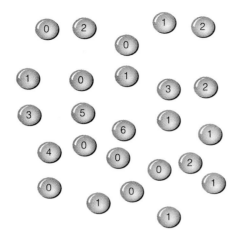

FIGURE 4.8 Hypothetical distribution of mosquito larvae among 25 pools of water. *(From Poole 1974)*

of larvae. For example, the expected frequency of pools with two larvae is

$$P_2 = \frac{1.5^2 e^{-1.5}}{2!}$$

$$= 0.251$$

Because there are 25 pools, we expect $0.251 \times 25 = 6.28$ to contain two larvae. Each of the expected values calculated in this way is listed in the third column of Table 4.1.

By comparing the two middle columns in Table 4.1, we can see that many more pools with five and six larvae were observed than were expected. But is the overall distribution of larvae in the pools random (in which case it would follow a Poisson distribution)? To determine this, we compare the observed and expected distributions using a statistical test called the chi square (χ^2) *test,* according to the formula in the last column of Table 4.1. In this instance, the chi square value is significant. We conclude that the distribution is not random.

Another property of the Poisson distribution that is useful in this context is that the mean and variance are equal; that is, if the dispersion follows a Poisson distribution, the mean number of larvae per pool is the same as the variance in the number of larvae per pool. Deviations from a mean/variance ratio of 1.0 indicate hyperdispersion or aggregation. A ratio larger than 1.0 implies that the variation among pools is rather small (relative to the mean). This would occur if most of the pools had the same number of larvae and would suggest hyperdispersion. If the mean/variance ratio is smaller than 1.0, it is because the variance is relatively large. This would occur if some pools had many larvae and some had very few—an aggregated distribution.

A test statistic—$(n-1)s^2/\bar{x}$, where n is the number of pools, s^2 is the variance, and $\bar{x}$ is the mean number of larvae per pool—is used to test whether the mean/variance ratio differs significantly from 1.0. This test statistic has a chi square distribution with $n-1$ degrees of freedom. For this example, $\bar{x}$ is 1.48, s^2 is 2.68, and thus the mean/variance ratio equals 0.55. Then we calculate the test statistic:

$$\frac{(n-1)s^2}{\bar{x}} = \frac{(24)(2.68)}{1.48}$$

$$= 43.5$$

This statistic has a χ^2 distribution with $n-1$ degrees of freedom ($25 - 1 = 24$ df.). For 24 df., the value 43.5 is significant at the 0.05 level. Thus, because the mean/variance ratio is less than 1.0, we conclude that the distribution is clumped.

Even though this procedure seems straightforward, in actual applications it may be more complicated. In our hypothetical example, the units we counted were individuals in clearly demarcated pools of water. Often we are applying this statistic to a distribution of individuals in a more continuous natural habitat, as represented in Figure 4.9a; in such cases we must overlay the distribution with a grid of quadrats and calculate the mean number of individuals per quadrat (each of which is analogous to a single pool). The choice of the size and number of quadrats is crucial and can affect the results.

In Figures 4.9b and 4.9c, two different quadrat sizes have been laid over the same distribution. In Figure 4.9b, the distribution appears aggregated; the quadrats are sufficiently large that some have large numbers of

TABLE 4.1

Poisson Analysis of Hypothetical Distribution of Mosquito Larvae in Pools

Number of Larvae in Pool	Observed number of Pools (O)	Expected number of Pools (E)	$\frac{(O-E)^2}{E}$
0	8	6.82	0.21
1	8	8.86	.08
2	4	6.28	0.82
3	2	2.49	0.10
4	1	0.82	0.04
5	1	0.21	2.97
6	1	0.05	18.05
	25	25.00	$\chi^2 = 22.27$

$\chi^2 = 22.27$; 6 df, $p < .001$.

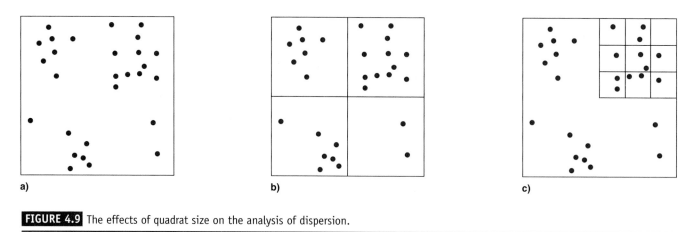

FIGURE 4.9 The effects of quadrat size on the analysis of dispersion.

individuals. When smaller quadrats are used (Figure 4.9c), the individuals appear to be randomly distributed. Determining which interpretation is correct can be difficult. That the Poisson distribution requires a small mean occurrence of individuals in a quadrat provides a clue: We should not choose the larger quadrat if the mean number of individuals turns out to be large. We can also base the decision on ecological considerations. It might, for example, be appropriate to choose a quadrat size that is approximately equal to the size of habitat patches.

An aspect of spatial distribution that is related to dispersion is the concept of grain (Pielou 1974). The term is derived from photography, where it refers to the size of clumps of silver in a photographic emulsion. In ecology, **grain** relates the size of an animal's habitat patches to the animal's vagility, or ability to move relatively long distances. A habitat is said to be **coarse-grained** for a given organism if the organism's vagility is low relative to the size of habitat patches. A habitat is considered **fine-grained** for a species if that organism has high vagility relative to the size of habitat patches. An organism living in a coarse-grained environment is likely to spend most of its time in one patch, whereas an organism in a fine-grained environment will range over many patches. Thus a hawk will experience the mosaic of agricultural fields and woodlots depicted in Figure 4.10 as a fine-grained habitat; it can readily hunt over both field and forest. How-

ever, a land snail will experience this same region as a coarse-grained habitat. Because of its low vagility relative to the distances between patches, a snail is likely to spend most, if not all, of its life in a single patch of forest.

Age Structure

The third important demographic feature of populations is the **age structure,** the distribution of individuals among age classes. Each individual in the population experiences a certain age-related risk of

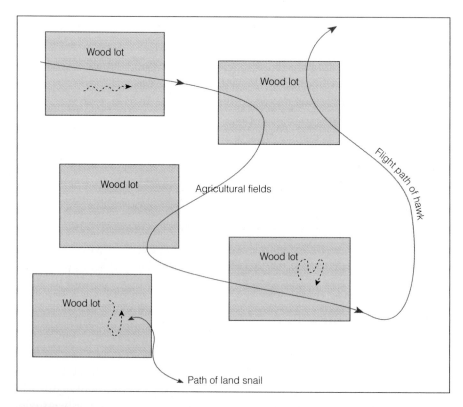

FIGURE 4.10 The concept of grain. This habitat mosaic of agricultural fields and woodlots is fine-grained for the far-ranging hawk but coarse-grained for the less mobile land snail.

dying. These probabilities determine the distribution of individuals in the various age classes. As we will see, a population's age structure contains important ecological information about that population.

If we are to study age distribution, we must have some means of determining the ages of individuals. In a number of organisms, age can be ascertained by examination of certain structures, depending on the species. A variety of plants and animals, for example,

produce annual growth rings in various structures. Counts of annual rings in the vascular tissue of temperate-zone trees, in the scales of many fishes, and in sheep horns can reveal the ages of individual organisms. The condition of dentition is used in determining age in some ungulates. In some species, annual rings of deposition in the teeth can be counted in cross sections. Deer can be placed in age classes on the basis of tooth wear (Figure 4.11). Some birds, including

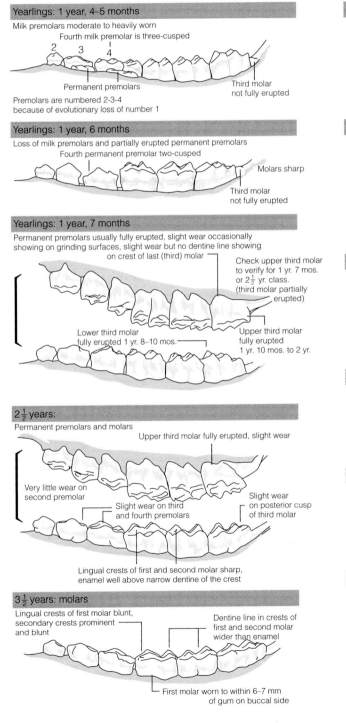

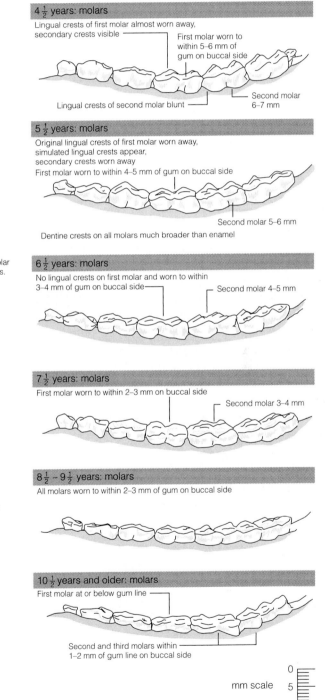

FIGURE 4.11 Patterns of tooth wear in deer enable biologists to place individuals into age classes. (*From Giles 1971*)

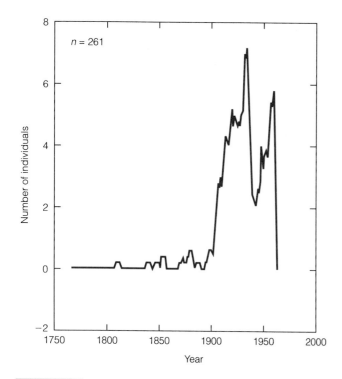

FIGURE 4.12 The observed recruitment of saguaro cacti seedlings in a population in Arizona. The figure represents the number of individuals that entered the population that year. Note the large fluctuations in the number of new individuals added. (*From Pierson and Turner 1998*).

TABLE 4.2

Life Table for Dall Sheep (*Ovis dalli*) Constructed from the Age at Death (as Determined by Annual Horn Rings) of 608 sheep in Denali National Park, Alaska

Age Interval (years)	Number Surviving at Beginning of Age Interval, n_x	Number Surviving as a Fraction of Newborn, l_x	Number Dying During Age Interval, d_x
0–1	608	1.000	121
1–2	487	0.801	7
2–3	480	0.789	8
3–4	472	0.776	7
4–5	465	0.764	18
5–6	447	0.734	28
6–7	419	0.688	29
7–8	390	0.640	42
8–9	348	0.571	80
9–10	268	0.441	114
10–11	154	0.252	95
11–12	59	0.096	55
12–13	4	0.06	2
13–14	2	0.003	2
14–15	0	0.000	0

Based on data in Murie 1944, as quoted by Deevey 1947.

TABLE 4.3

Life Table and Fecundity Table for the Thar, *Hemitragus jemlahicus* (Females Only)

1 Age in Years x	2 Frequency in Sample	3 Adjusted Frequency	4 Number of Female Live Births per Female at Age x (m_x)	5 $1000l_x$	6 $1000d_x$	7 $1000q_x$
0	—	205*	0.000	1000	533	533
1	94	95.83	0.005	467	6	13
2	97	94.43	0.135	461	28	61
3	107	88.69	0.440	433	46	106
4	68	79.41	0.420	387	56	145
5	70	67.81	0.465	331	62	187
6	47	55.20	0.425	269	60	223
7	37	42.85	0.460	209	54	258
8	35	31.71	0.485	155	46	297
9	24	22.37	0.500	109	36	330
10	16	15.04	0.500	73	26	356
11	11	9.64	0.470	47	18	382
12	6	5.90		29		
13	3		0.350			
14	4					
15	3					
16	0					
17	1					

*Calculated from adjusted frequencies of females other than kids (column 3) and m_x values (column 4). (*From Caughley 1966*)

grouse, cannot be placed into specific year classes, but first-year birds can be distinguished from all older birds.

Age structure data can provide insight into important population processes. In an 85-year study of saguaro cacti (*Carnegiea gigantea*) in Arizona, Pierson and Turner (1998) demonstrated the importance of episodic seedling establishment in maintaining the population (Figure 4.12). The occasional peaks in recruitment, shown by the numbers of individuals first appearing in each year, helped maintain the saguaro population in this region.

For any population of plants or animals we might study, individuals are distributed in age classes from newborn to the maximum age for that species. We compile a population's age distribution in a life table. There are two basic ways of generating data for a life table. In a static life table, we sample the population at a moment in time and record the numbers of individuals in each age class; from this sample we infer the pattern of mortality associated with each age class. In a cohort life table, we follow a single group, or cohort, of individuals from birth to death and record the number dying at each age.

Table 4.2, the life table for Dall sheep (*Ovis dalli*) in Denali National Park, is a classic static life table. E. S. Deevey (1947) developed it from age data gleaned from growth rings on the horns of dead sheep collected in the park by A. Murie (1944). The first column contains the age classes, from newborn to 14 years (no animal older than 14 years was discovered). The second column, the n_x column, contains the number of individuals found for each age class. For example, 608 individuals (the total number of skulls found) were alive in age class 0–1, 487 survived at least to age class 1–2, and so on. The third column, the l_x column, represents the age-specific survival rate. It is calculated as the proportion surviving that age interval. Thus for age class 0–1, $l_1 = 608/608 = 1.000$. For age class 2–3, $l_3 = 480/608 = 0.789$. The number dying during each age interval (d_x) is shown in the final column.

Table 4.3, another example of a static life table, shows the age structure of a population of the Himalayan thar (*Hemitragus jemlahicus*), an ungulate introduced into New Zealand (Caughley 1966). A total of 623 animals were shot, and the annual rings on the horns were counted. Note that this life table is based on thar females only because, as we will see, females provide the crucial information for some aspects of population growth. This table contains the same information as the life table for Dall sheep (Table 4.2), plus an additional column, m_x. Caughley estimated this parameter, the number of live births for

individuals of each age, by dividing the number of young by the number of females. The parameter measures the age-specific fecundity of the population.

Table 4.4, another static life table, shows the age-specific mortality rates for human males in the United States on the basis of census data. Note that this table contains an additional column, e_x, which gives the average number of years of life remaining for an individual of age x. In other words, e_x is the age-specific life expectancy.

The calculation of e_x is somewhat more complicated than those for other columns, so we will illustrate it with a simpler life table for a population of perennial plants growing in a field (Table 4.5). This table includes a new column, L_x, which represents the average number of individuals alive during each age interval: $L_x (n_x + n_{x11})/2$. Thus, from Table 4.5 we see that during the time interval from year 2 to year 3, 32 individuals entered the age class and 16 survived. On average, then, there were $(32 + 16)/2 = 24$ individuals alive during this period, so $L_2 = 24$.

The column T_x contains the total number of plant-years to be lived by all individuals of age x. (One individual plant living for 1 year is 1 plant-year). We want to calculate the total number of plant-years to be lived for each age interval. Each value of T_x in the table is calculated as follows:

$$T_x = \sum_{x}^{\infty} L_x$$

Thus $T_2 = L_2 + L_3 + L_4 + L_5 = 36$ plant-years. At age interval 2, all the individuals in the population will live 36 plant-years.

Now we can calculate the average number of years left to be lived by an individual by dividing T_x by the number of individuals alive at the beginning of that interval, n_x. Thus the mean expectation of life at each interval is given by

$$e_x = T_x/n_x$$

So, for e_2,

$$e_2 = 36/32 = 1.13$$

Now that we understand how e_x is calculated, let us return to the human life table in Table 4.4. Note that for a newborn male during 1929–1931, the expectation of life was 59.12 years. What determines the value of e_0? From the calculations we just outlined, it should be clear that two factors can increase e_0. First, if the maximum life span increases, so will e_0. If, for example, some individuals began living to 110 years

TABLE 4.4
Life Table for U.S. Males, 1929–1931

Age Interval of Life, x	Of 100,000 Males Born Alive		Mortality Rate	Complete Expectation of Life, x
	Number Alive at Beginning of Year of Age, l_x	Number Dying During Year of Age, d_x	Number Dying per 1000 at Beginning of Year, $1000q_x$	Average Number of Years of Life Remaining at Beginning of Year of Age, e_x
0–1	100,000	6,232	62.32	59.12
1–2	93,768	931	9.93	62.04
2–3	92,837	483	5.20	61.65
3–4	92,354	331	3.59	60.97
4–5	92,023	285	3.09	60.19
5–6	91,738	243	2.66	59.38
6–7	91,495	208	2.27	58.53
7–8	91,287	179	1.96	57.67
8–9	91,108	156	1.72	56.78
9–10	90,952	142	1.55	55.87
14–15	90,246	172	1.90	51.29
19–20	89,172	268	3.01	46.88
24–25	87,692	321	3.66	42.62
29–30	86,053	346	4.02	38.39
34–35	84,222	410	4.86	34.17
39–40	81,979	522	6.36	30.03
44–45	79,036	691	8.74	26.05
49–50	75,188	900	11.98	22.25
54–55	70,165	1,184	16.87	18.66
59–60	63,496	1,563	24.61	15.34
64–65	54,924	1,960	35.68	12.33
69–70	44,253	2,373	53.62	9.68
74–75	31,986	2,515	78.61	7.43
79–80	19,565	2,344	119.83	5.57
84–85	9,159	1,587	173.33	4.21
89–90	3,068	712	232.11	3.21
94–95	672	211	313.32	2.35
99–100	72	32	438.79	1.62
104–105	2	1	621.87	1.05
105–106	1	1	666.56	0.96

Data from Pearl 1940, cited in Allee et al. 1949.

TABLE 4.5
Life History Table for a Population of Perennial Plants Growing in a Field

x	n_x	l_x	L_x	T_x	e_x
0	125	1,000	91.5	172.5	1.38
1	58	464	45.0	81.0	1.40
2	32	256	24.0	6.0	1.13
3	16	128	10.0	12.0	0.75
4	4	32	2.0	2.0	0.50
5	0	0	0.0	0.0	0.00

TABLE 4.6
A Cohort Life Table for the Annual *Phlox drummondi* at Nixon, Texas

Age Interval, $x - x'$ (days)	Number Surviving to Day x, N_x	Survivorship l_x	Number Dying During Interval, d_x	Average Mortality Rate per Day, q_x
0–63	996	1.0000	328	.0052
63–124	668	.6707	373	.0092
124–184	295	.2962	105	.0059
184–215	190	.1908	14	.0024
215–231	176	.1767	2	.0007
231–247	174	.1747	1	.0004
247–264	173	.1737	1	.0003
264–271	172	.1727	2	.0017
271–278	170	.1707	3	.0025
278–285	167	.1677	2	.0017
285–292	165	.1657	6	.0052
292–299	159	.1596	1	.0009
299–306	158	.1586	4	.0036
306–313	154	.1546	3	.0028
313–320	151	.1516	4	.0038
320–327	147	.1476	11	.0107
327–334	136	.1365	31	.0325
334–341	105	.1054	31	.0422
341–348	74	.0743	52	.1004
348–355	22	.0221	22	.1428
355–362	0	.0000		

From Leverich and Levin 1979.

of age, e_0 would increase. However, e_0 will also increase if juvenile mortality decreases. Note in Table 4.4 that at age 1, e_x has actually *increased*. This is because an individual who survives the first year has avoided a period of high mortality and thus is more likely to live for the population's maximum life span (see the box on page 94 for more information on human populations).

An example of a cohort life table is shown in Table 4.6. These data are for the winter annual *Phlox drummondi*. It is easier to obtain cohort data for plants than for animals because their sessile nature enables us to mark individuals as seedlings and follow their fates over time.

Both static and cohort life tables have drawbacks. A number of assumptions are inherent to static life tables. One assumption is that the patterns of mortality remain relatively constant over the period of time included in each table. Any biases in collecting the individuals can skew the life table as well. With

respect to Table 4.3, for example, Caughley (1966) pointed out that certain age classes of thar may be more susceptible to collection by shooting, a condition that would bias the sample.

Cohort life tables are also not without disadvantages. If the year in which the cohort was born turns out to be unusual, or if mortality is abnormally high or low in a subsequent time period, the life table will not accurately reflect the patterns of mortality. Of course, there are also many instances in which it is impossible to follow a cohort throughout its life span. Even though it is possible to construct a static life table of instars of the fossil invertebrate *Beyrichia jonesi*, we obviously cannot generate a cohort table for these long-dead individuals!

Life table data provide important information on mortality for the various periods of an organism's life span, but mortality *trends* are easier to identify when life table data are plotted on a **survivorship curve,** typically a plot of the l_x column (log scale) versus age.

Human Demography

The demographic parameters described in this chapter also apply to humans and reveal important information about the current characteristics and future trends of human populations. Three demographic variables are particularly important: fertility rate, age structure, and life expectancy.

Fertility rates can be measured in a number of ways. The **general fertility rate** is the number of offspring produced per 1,000 females in the population who are of childbearing age. The **total fertility rate (TFR)** is the total number of children a female is expected to bear in her lifetime; it is exactly analogous to the value of R_0 described in this chapter and is calculated from the l_x and b_x data in human life tables.

If a woman produces exactly 2.0 offspring over her life span, she exactly replaces herself and her mate. However, because there is some mortality of offspring, the actual replacement rate is generally slightly higher than 2.0. In more developed countries, it is approximately 2.1; in less developed countries, it is closer to 2.5.

The TFR for the planet is 3.4, considerably higher than mere replacement. This value varies greatly among countries. For many developed countries, the TFR is less than 2.0; in some western European countries, it is 1.3. Before the political strife in Rwanda, the TFR there was 8.1.

Demographers describe the age structure of human populations in much the same way as for other organisms. They typically present the data in a graphical form that indicates the proportion of the population in each of several age classes; often the data are presented separately for males and females. Examples of age structure data for four groups of human populations are shown in Figure 4B.1.

Note in this figure that the age structures of the more developed countries differ significantly from that of Kenya, Nigeria, and Saudi Arabia. In particular, the latter countries have an age structure heavily skewed toward the younger age classes; nearly 40 percent of the female population is under age 44. This has great significance for future population growth in these countries. Because such a large fraction of the population is of reproductive age or soon will be, there is tremendous potential for population growth. In contrast, in the more developed countries, many individuals have passed beyond the childbearing years; consequently, these populations are likely to have smaller growth rates.

Even though this discussion does not include the TFRs for these populations, they obviously play a significant role as well. But given equal TFRs, the age structure is the pivotal factor in determining the rate of population growth.

Life expectancy plays only a relatively small role in determining the population growth rate, because once a female ceases reproducing, it is actuarially insignificant how much longer she lives. Rather, this parameter is often used as a measure of standard of living, as a reflection of the quality of health care, sanitation, and nutrition, and as an index of quality of life in general.

Human life expectancies are calculated in the same way as those indicated in Table 4.4. The nature of this calculation is significant for the interpretation of life expectancies. The life expectancy is e_0, the mean number of years a newborn infant is expected to live. Because this value is based on the summed age-specific life expectancies, juvenile mortality contributes importantly to it. Thus, when we read in the newspaper that the average life expectancy in the United States has increased, it is generally not because older individuals are living significantly longer (although that does make some contribution), but rather because infant mortality has declined.

Changes in age-specific mortality patterns thus have a significant effect on the age structure of the population—and hence on population growth. As we will see in Chapter 5, the timing of such changes plays a crucial role in the population processes in a developing country.

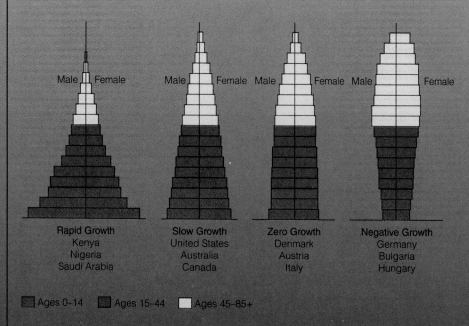

Rapid Growth	Slow Growth	Zero Growth	Negative Growth
Kenya	United States	Denmark	Germany
Nigeria	Australia	Austria	Bulgaria
Saudi Arabia	Canada	Italy	Hungary

■ Ages 0–14   ■ Ages 15–44   □ Ages 45–85+

FIGURE 4B.1 Population age structure diagrams for countries with rapid, slow, zero, and negative population growth rates. (*Data from Population Reference Bureau*)

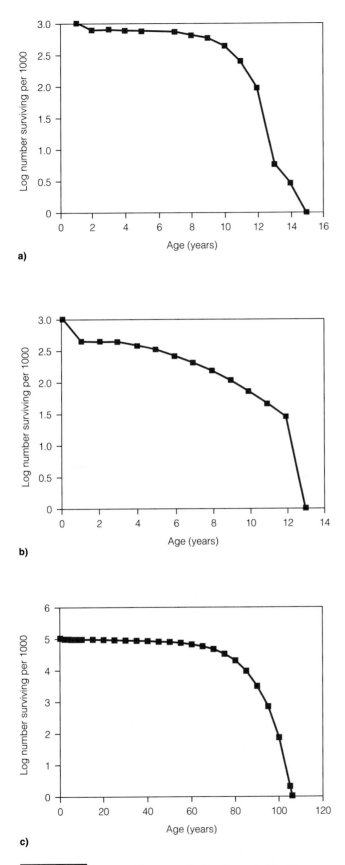

a)

b)

c)

FIGURE 4.13 Survivorship curves for the data tabulated in Tables 4.2–4.4. (**a**) Dall sheep (*data from Deevey 1947*); (**b**) female thar (*data from Caughley 1966*); and (**c**) U.S. males, 1929–1931 (*Data from Allee et al. 1949*).

Examples from the life tables presented in Tables 4.2–4.4 are shown in Figure 4.13. Note the general similarity in shape. After a dip in survivorship early in life, the curve flattens out for a long period. All three of these curves are for mammals; once a young mammal reaches a certain age, it has a high probability of living for the maximum life span. Finally, mortality increases significantly as the maximum life span is approached.

Even though this pattern is typical of mammals, variations occur. In Figure 4.14, survivorship curves for Columbian black-tailed deer (*Odocoileus hemionus*) and the roebuck (*Capreolus capreolus*) show the high mortality typical of young mammals; thereafter however,

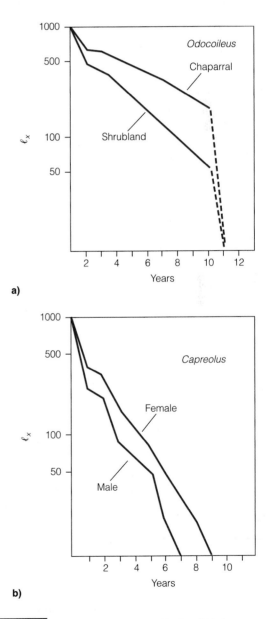

a)

b)

FIGURE 4.14 Survivorship curves for black-tailed deer (*Odocoileus hemionus*) and roebuck (*Capreolus capreolus*). (*Adapted from Hutchinson 1978*)

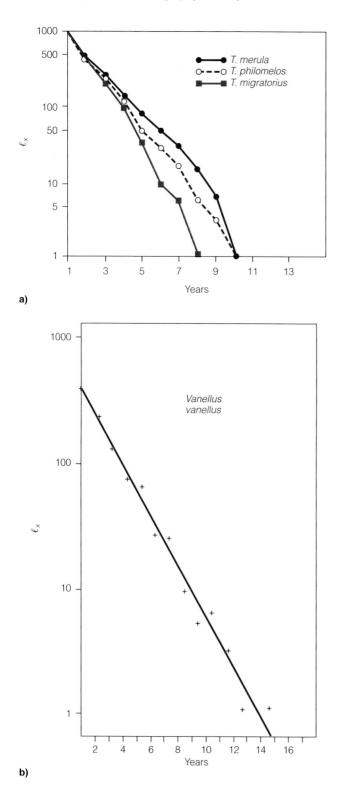

a)

b)

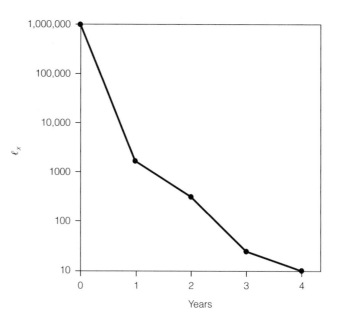

FIGURE 4.16 Survivorship curve for the prawn, *Leander squilla*. (*From Kurten 1953; cited in Hutchinson 1978*)

FIGURE 4.15 Examples of bird survivorship curves. (**a**) Survivorship curves for three species of *Turdus: T. merula,* the European blackbird; *T. philomelos,* the song thrush; and *T. migratorius,* the American robin. (*Data from Lack 1943a; 1943b; 1943c; cited in Hutchinson 1978*) (**b**) Survivorship curve for the lapwing, *Vanellus vanellus*. (*Data from Kraak, Rinkel, and Hoogenheide 1940, cited in Hutchinson 1978*) Due to inadequate estimates of mortality in the first year, all curves begin at age one.

they show greater mortality compared to the high survivorship that is typical in midlife.

Survivorship curves for various birds are shown in Figure 4.15. Compared to those of mammals, the survivorship curves of birds are essentially straight lines. Note that inadequate measures of early mortality require that the curves start at age 1. Even though mortality in the first year may have been higher, the crucial point is that after the first year, mortality among birds is typically constant across age classes.

A third type of survivorship curve, shown in Figure 4.16, is characterized by extremely high mortality among the youngest age classes. Curves of this shape are characteristic of many invertebrates, which typically produce a huge number of immature stages that are released into the environment, where they are very vulnerable to predation and abiotic stresses. A large proportion of them die before they are able to metamorphose into adults. After metamorphosis, survivorship is much higher.

Among animals, the basic shapes of survivorship curves are sufficiently stereotypical that ecologists have assigned numbers to the fundamental patterns (Figure 4.17). Many mammals exhibit a survivorship pattern referred to as a Type I curve, birds typically have a Type II survivorship curve, and many invertebrates show a Type III curve.

We can also gather useful information about differences between the mortality patterns of males and females by graphing their age-specific survivorships separately. Such data generate questions about the various sources of mortality at different ages for males

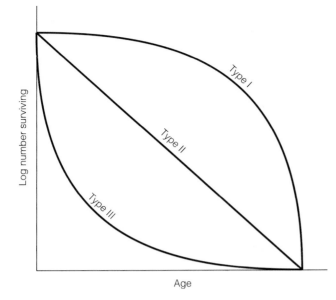

FIGURE 4.17 Three patterns of survivorship curves. Type I curves are typical for mammals, Type II curves for birds, and Type III curves for many invertebrates.

and females, questions that must be answered by field research projects.

Different habitats can be compared by examining survivorship curves. R. D. Taber and R. F. Dasmann (1958) compared survivorship of Columbian black-tailed deer in native chaparral and in shrubland that had been artificially enhanced to support more ani-

mals. Deer density in the shrubland was 34 square kilometers, compared to 10 square kilometers in chaparral. Note in Figure 4.14 that, paradoxically, deer survivorship was higher in the poorer habitat; in the shrubland, deer suffered significantly higher mortality during middle age. The reasons for this were not clear from the study.

The survivorship curves of plants, like those of animals, have a variety of shapes. Many forest trees have nearly rectangular curves, reflecting very high survivorship until a precipitous decline late in life. Herbs typically have survivorship patterns that combine some elements of Type I and Type II curves. Some grasses have the typical Type I curve, whereas others have the early mortality characteristic of mammals followed by a linear portion more like a Type II survivorship curve (Figure 4.18).

Sex Ratio

The fourth demographic parameter we will consider is the sex ratio. In the older literature, demographers expressed this parameter either as the ratio of males to females or as the number of males per 100 females. Most ecologists now express it as the proportion or percentage of the population that is male or female.

More refined measures of sex ratio can provide us with ecological insights, however. It may be useful or important to know the age structure of each of the

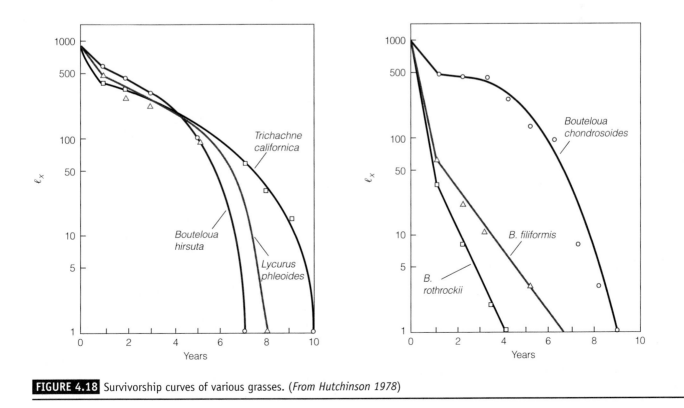

FIGURE 4.18 Survivorship curves of various grasses. (*From Hutchinson 1978*)

two sexes, from which we can derive a series of sex ratios based on age classifications:

Primary (1°) sex ratio:	sex ratio at fertilization
Secondary (2°) sex ratio:	sex ratio at birth or hatching
Tertiary (3°) sex ratio:	sex ratio at sexual maturity
Quaternary (4°) sex ratio:	sex ratio of adult population

It is common for these four sex ratios to differ significantly within a given population (Figure 4.19).

In addition to these sorts of demographic changes in sex ratio, important differences in the activities of individuals of each sex may modify the *effective sex ratio*. For example, in many species only a few males actually mate. Members of the grouse family epitomize this phenomenon. Mating takes place on a lek, a small piece of ground on which males display and females gather to choose males for mating. Only a few of the displaying males actually mate. Thus, in grouse populations, even though the numbers of adult males and females may be similar, from a genetic standpoint the sex ratio is effectively skewed by the disproportionate success of a few males.

Among plants, the concept of sex ratio is problematic because the sexes do not always occur separately on different individuals. The matter is further complicated by the fact that we can talk about the presence of sexual parts of flowers at various levels: in individ-

TABLE 4.7

Terms Used to Describe the Presence of Male and/or Female Flower Parts in Flowers, Individual Plants, and Plant Populations

Individual flower
Hermaphroditic—with both stamens and pistils

Staminate—with stamens only

Pistillate—with pistils only

Individual plant
Hermaphroditic—with hermaphroditic flowers only

Monoecious—with both staminate and pistillate flowers

Androecious—with staminate flowers only

Gynoecious—with pistillate flowers only

Andromonoecious—with both hermaphroditic and staminate flowers

Gynomonoecious—with both hermaphroditic and pistillate flowers

Trimonoecious—with hermaphroditic, staminate, and pistillate flowers

Groups of plants
Hermaphroditic—with hermaphroditic plants only

Monoecious—with monoecious plants only

Dioecious—with both androecious and gynoecious plants

Androdioecious—with both hermaphroditic and androecious plants

Gynodioecious—with both hermaphroditic and gynoecious plants

ual flowers, in individual plants, or in a population (Table 4.7). Thus flowers can have both male and female parts (hermaphrodites), or they may contain stamens only (staminate flowers) or pistils only (pistillate flowers). Individual plants may have many arrangements of sexual flower parts, and within a population, the sexes may be partially or completely separate.

For plant population biologists, then, the notion of sex ratio can become blurred. In strictly dioecious plants such as cottonwoods (*Populus deltoides*) and ginkos (*Ginkgo biloba*), the traditional concept of a sex ratio applies. We can easily count the number of individuals in the population that are male or female. In monoecious plants, however, the situation is not so straightforward. For example, in stinging nettles (*Urtica dioica*), individual plants bear varying numbers of male and female flowers. Consequently, it is customary to refer to maleness or femaleness in these plants in quantitative terms; that is, a monoecious *individual* will bear some proportion of male flowers, and the rest will be female flowers. In a dioecious population, some fraction of the plants will be male, and the remainder will be female.

Deviation from a 50:50 sex ratio in a population can result from either ecological or genetic factors—and, of course, the two interact. We have already seen evidence of one major cause of skewed sex ratios in our discussion of survivorship curves. Different age-

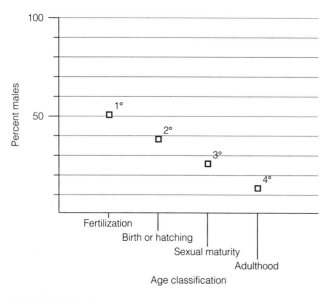

FIGURE 4.19 The change in primary, secondary, tertiary, and quaternary sex ratios, as expressed as percent males, in a given theoretical population.

specific mortality rates in males and females will skew tertiary or quaternary sex ratios in favor of the sex with higher survival. Because males generally experience higher mortality, sex ratios are often skewed toward females. For example, in a number of reptiles, the temperature at which the eggs incubate determines the sex of the hatchlings, and thus sex ratio can be altered by environmental conditions. In turtles, cool temperatures lead eggs to develop as males; warm temperatures lead to females. Janzen (1994) showed that in the western painted turtle (*Chrysemys pica bellili*), most nests produce hatchlings of all one sex. Nests with more vegetational cover on the western and southern edges were cooler and produced males (Figure 4.20).

In the boat-tailed grackle (*Quiscalus mexicanus*), nestlings have a 1:1 sex ratio, but by the time these juveniles are ready to breed the following year, the sex ratio is biased toward females, in a ratio of 1.42:1.00 (Selander 1965). Apparently, the exaggerated size of the tail in males, is an advantage in attracting mates but also constitutes a handicap that manifests itself in lower male survival.

The interaction between the demographic effects of sex ratio and a population's social system is complex and operates in both directions: Demographic effects influence the evolution of the social system, and the nature of the social system has demographic consequences for the population. We will explore this interaction further in Chapter 8.

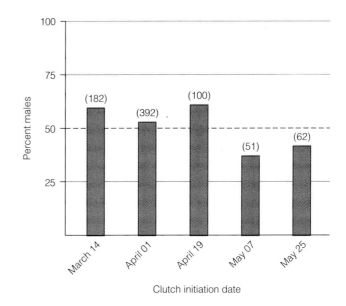

FIGURE 4.21 Sex ratio (proportion males) of American kestrel nestlings in relation to the date on which the clutch was initiated. Numbers in parentheses represent the number of nests sampled. (*Adapted from Smallwood and Smallwood 1998*)

Genetic mechanisms can also skew the primary sex ratio away from 50:50, and any deviations in this ratio will be manifest in subsequent sex ratios. As we saw in Chapter 2, the wood lemming (*Myopus schisticolor*) has a genetic mechanism that produces XY females; an allele at the gene for cell-surface antigens causes XY individuals to develop into functional females. The result is a population that is only 25 percent male (Kalela and Oksala 1966).

Because the primary sex ratio is under genetic control, it is possible for natural selection to modify it in ways that increase the fitness of an individual. In the American kestrel (*Falco sparverius paulus*) clutches laid early in the season are biased toward males; later clutches show a female bias (Figure 4.21; Smallwood and Smallwood 1998). These researchers suggest an "Early Bird Hypothesis" to explain these results. For males that breed as yearlings, all had fledged early the previous year. Males hatched early in the year seem to enjoy a competitive advantage in securing breeding sites after dispersal from the natal area. Consequently, a male bias early in the season is adaptive.

R. A. Fisher (1930) proposed a general theory of the evolution of sex ratio. In his model, the sex ratio adjusts such that the parents expend the same total amount of energy in rearing offspring of both sexes. This theory encompasses both the genetic aspects of frequency-dependent selection and the potential ecological differences in survivorship of the two sexes. If, for example, the two sexes experience the same mortality rates at birth, the sex ratio should be equal.

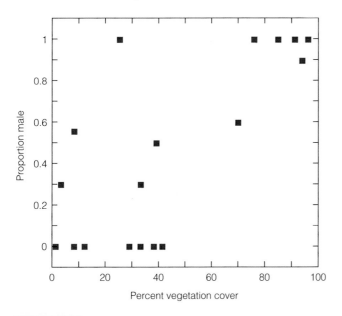

FIGURE 4.20 Sex ratio (proportion male) in nests of western painted turtles (*Chrysema picta*) as a function of the vegetational cover from southern exposure. Greater cover leads to cooler temperatures and a higher proportion of male young. (*From Janzen 1994*)

If, however, one sex experiences higher mortality or requires parents to invest more energy, then the sex ratio may be skewed.

For example, male common grackles (*Quiscalus quiscala*) are more active and grow more rapidly than females. Because of the greater parental investment of energy that males require, Fisher's model predicts that they should be less numerous than females. Indeed, at fledging the sex ratio of grackles is skewed toward females by a ratio of 1.65:1.00 (Howe 1977). In the blackbird-like oropendola (*Gymnostinops* spp.) of South America, the males are twice as large as the females at fledging. Females are 5–10 times more numerous than males in populations of this species (Smith 1968).

Fisher's theory predicts equal parental investment in male and female offspring *if all other things are equal.* Often, however, other selective factors affect the evolution of the sex ratio. For example, in some animal species, large females produce more offspring. In such cases, mothers that are physiologically capable of producing large offspring would be expected to produce more females than males, a mother's large daughters will benefit from their size advantage, and the increased number of their offspring will enhance the mother's fitness. J. J. Peterson and colleagues (1968) demonstrated precisely this effect in the nematode *Romamnomeris culicivorax*, which infects mosquito larvae. The worms are larger in larger hosts, and the sex ratios of cohorts of large worms are biased toward females.

R. L. Trivers and D. E. Willard (1973) hypothesized that for vertebrates, the sex ratio of offspring should vary with the physical condition of the mother. Their argument is based on the fact that in many vertebrates, male reproductive success is highly variable; a few dominant males achieve most of the matings, whereas nearly every female breeds. Therefore, a female capable of producing healthy, robust male offspring, which are likely to be dominant, will have high fitness; her male offspring will leave a large number of descendants. If she is in poor physical condition and is thus unlikely to be able to produce robust males, she is better off producing females. They will not have the same high fitness of a dominant male, but their fitness will be higher than that of a low-ranking male. Thus dominant females (those in excel-

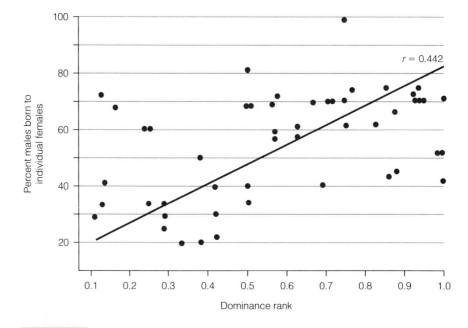

FIGURE 4.22 Sex ratio of red deer (*Cervus elaphus*) born to females in relation to maternal dominance rank. (*Adapted from Clutton-Brock, Albon, and Guiness 1986*)

lent condition) should produce litters biased toward males; females in poor condition should produce litters skewed toward females.

T. H. Clutton-Brock and co-workers (1982; 1986) demonstrated the existence of this effect in red deer (*Cervus elaphus*); dominant females produce a higher proportion of males than do females of low rank (Figure 4.22).

Similar kinds of skewed sex ratios are associated with growing conditions in plants. For example, J. K. Zimmerman (1991) has shown that for the orchid *Catasetum viridiflavum*, phenotypic gender is strongly bimodal: Populations tend to have very high or very low proportions of male flowers. In the orchid's tropical forest habitats, orchid sex ratio is related to light conditions, which are a function of forest age. In young forests where light is abundant, the sex ratio of flowers is skewed toward females; in older, darker forests, the sex ratio is 1:1. Thus investment in female tissue is greater under favorable light conditions. Experimental shading of plants in young forests confirmed that light is the determining factor.

Population Growth

We turn our attention now to the means by which the density of a population changes. Changes in population size are of fundamental importance in ecology, and the mechanisms by which change occurs and is controlled are of great interest. In addition, population growth, whether positive or negative, forms the basis of several important community-level phenomena (Part III). The development of a plant community after disturbance (succession) is in fact a process of differential population growth rates of different plant species under changing conditions. Competition manifests itself in increases and decreases of the populations of competing species. And, of course, predation has direct effects on the population growth rates of predator and prey alike.

Exponential Growth

If we introduce a species into a new and suitable habitat, the population will increase, probably rapidly. For simplicity, we begin our discussion with a species that has discrete generations, such as an annual plant or a univoltine insect. (Univoltine insects produce one brood per year.) Annual plants grow, flower, set seed, and die within a year; the seeds overwinter and germinate the following spring to produce a new and discrete generation. Univoltine insects reproduce once in a season and then overwinter in a resting stage. Imagine that each female plant or insect produces two offspring and then dies. With each generation, the population increases by a factor of 2. We call this factor the **net reproductive rate,** R_0. The population at any time is determined by the initial population size, the value of R_0, and the number of generations that have elapsed. For example, if $R_0 = 2.0$ and we begin with a population of 10 individuals, then the population size will be 20 after one generation, 40 after two, 80 after three, and so on. The general equation that describes this progression is

$$N_t = R_0^t N_0$$

where N_t is the population size after t generations, and N_0 is the initial population size. We see that a population growing in this way has the trajectory shown in Figure 4.18.

Ecologists define the net reproductive rate as the ratio of the number of females born to the number of females of the current generation. We simplify matters by considering only females; if each female exactly replaces herself, then $R_0 = 1.0$.

The information in a life table can be used to calculate the value of R_0. The number of females expected to be born to each female is the product of the age-specific mortality rate (l_x) and the age-specific birth rate (b_x). If all females always survived to reproductive age, we could simply sum the offspring each is expected to produce. If survival to reproductive age is not a certainty, however, the expected number of offspring will be proportionately less. Thus the value of R_0 is calculated as

$$R_o = \sum_0^\infty l_x m_x$$

Many organisms do not have discrete generations but rather breed continually, so that generations overlap. Exponential growth in such populations is described by

$$dN/dt = rN$$

where N is population size, t is time, r is a factor called the intrinsic rate of increase, and d indicates the first derivative. An important term in population biology, r refers to the instantaneous per capita rate of population growth. The value of r depends on the individual birth and death rates in the population:

$$r = b_0 - d_0$$

where b_0 and d_0 represent the average number of births and deaths per individual per unit time, respectively. The subscripts refer to the fact that these rates are measured when the population is very small. The actual value of r will vary with the quality of the environment. We refer to the value of r in the optimum environment as r_{max}. All other things being equal, we can use the value r as a measure of fitness, because it can be used to represent the rate of increase of a particular genotype. We will see this term again in the context of life history strategies (Chapter 12).

The differential equation $dN/dt = rN$ can be represented in integral form:

$$N_t = N_0 e^{rt}$$

where N_0 is the initial population size, e is the base of the natural logarithms (2.718), and t is time. Notice that in this model for overlapping generations, the values of t can be expressed in any time units, not just numbers of generations. This equation enables us to calculate the changing size of the population through time.

A species with overlapping generations can show exponential growth similar to that depicted in Figure 4.23 for a species with discrete generations. Caribou (*Rangifer tarandus*) were hunted to extinction on Southampton Island, Canada, by 1953. They were reintroduced in 1967 and began exponential growth that continues at present (Figure 4.24).

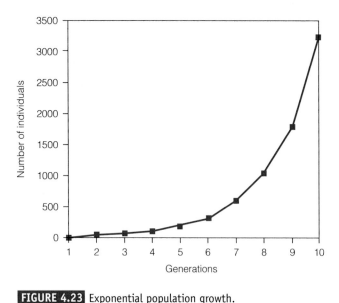

FIGURE 4.23 Exponential population growth.

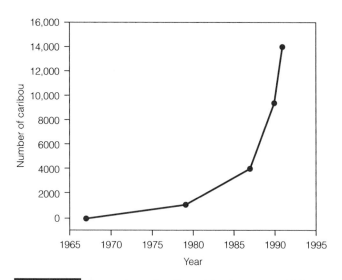

FIGURE 4.24 The numbers of caribou at least one year old from 1967 to 1991. (*From Heard and Ouellet, 1994*)

TABLE 4.8
Stable Age Structure over 10 Years for a Population with Three Age Classes and Constant Survivorship and Birth Rates

Age	l_x	b_x	Year									
			1	2	3	4	5	6	7	8	9	10
0	1	0	5	30	22	72	104	188	352	584	1080	1872
1	1	2	5	5	30	22	72	104	188	352	584	1080
2	0.5	4	5	3	3	15	11	36	52	94	176	292
3	0	0	0	0	0	0	0	0	0	0	á0	0
			N = 15	38	55	109	187	328	592	1030	1840	3244

An important and counterintuitive feature of population growth is that if a population has constant birth and death rates, it will gradually approach a **stable age distribution** (Lotka 1922). The population may continue to grow exponentially, but the *proportions* of the population *in each age class* will remain fixed. It is easiest to demonstrate this phenomenon with a specific example. For the hypothetical population shown in Table 4.8, the survivorship and birth rates remain constant over time, and the population size is projected through 10 years according to these parameters. Note that the population is growing but that the proportions of the population in each age class (Figure 4.25) become stable in the sixth year.

Sigmoid Growth

The equation $dN/dt = rN$ models the exponential growth of a population. However, we know that exponential growth cannot continue forever, because resources are not unlimited; at some point the popula-

tion size must level off. This situation, when graphed as in Figure 4.26, produces a generally S-shaped population curve called a **sigmoid growth curve.** Thus we need to modify our equation to account for the decrease in growth rate at higher population densities. In order to do this, we need merely to add a term, such that the equation becomes

$$dN/dt = rN\frac{(K - N)}{K}$$

The new variable in this equation, K, is the **carrying capacity,** the maximum population size the habitat can support.

This new equation describes the changes in the growth rate dN/dt as N changes. When N is very small, the term $(K - N)/K$ is very nearly 1.0, and thus the population's growth rate is essentially the exponential rate, rN. When $N = K$, the term $(K - N)/K = 0$ and dN/dt becomes 0. When N is greater than K, $(K - N)/K$ becomes negative, as does dN/dt, so the growth rate declines.

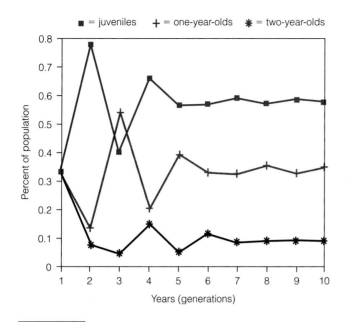

FIGURE 4.25 Approach to a stable age distribution for the population in Table 4.8. The percentage of the population in each age class stabilizes after six generations.

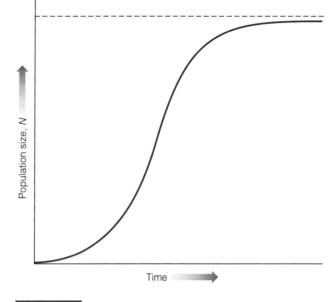

FIGURE 4.26 A sigmoid growth curve.

We thus have a mathematical description of the behavior of the population near the value of *K*. A population above *K* decreases, and a population below *K* increases. Whether biological processes in fact conform to the mathematical description we have developed here is discussed in Chapter 5.

Transition Matrices

If we know the probability of survival from one age category to the next, we can tabulate the changes in population age structure within a population. For instance, Horvitz and Schemske (1995) measured the probability of transition among several age (growth)

states in a tropical herb, *Calathea ovandensis* (Table 4.9). The probability of the herb's making the transition from seed at time *t* to seedling at time *t* + 1, for example, is 0.973.

We can analyze the pattern of growth of a population in yet another way by using the age-specific birth and survivorship rates to project the changes in the size and age composition of the population. This kind of analysis is called a **transition matrix**, or a **Leslie matrix** (Leslie 1945).

For a population with overlapping generations, the number of individuals in any age class is determined both by the survivorship of the individuals in the previous age group and by their reproductive output. To simplify the analysis, we assume that the age classes are discrete groups, each with an age-specific birth

TABLE 4.9

Transition Probabilities for the Herb *Calthea ovadensis* Among Various Growth Stages. Values on the diagonal represent the probability of remaining in a particular growth stage. Values above the diagonal represent the probability that an individual regresses in stage. (*From Horvitz and Schemske 1995*)

	Stage at Time *t*							
State at time *t* + *l*	Sd.	Sdlg.	Juv.	Pre.	Sm.	Med.	Lrg.	Xlrg.
Sd.	0.4983	0	0.5935	7.139	14.2715	24.6953	34.9027	40.5437
Sdlg.	0.0973	0.0110	0.0191	0	0	0	0	0
Juv.	0.0041	0.0442	0.3378	0.0698	0.0251	0.0065	0.0085	0
Pre.	0	0.0014	0.1355	0.4286	0.1736	0.0968	0.0427	0.0435
Sm.	0	0	0.0363	0.3841	0.6025	0.4258	0.2991	0.2174
Med.	0	0	0.0019	0.0254	0.113	0.2387	0.1709	0.2826
Lrg.	0	0	0	0.0095	0.0272	0.1548	0.3248	0.1957
Xlrg.	0	0	0	0.0032	0.0063	0.0452	0.1282	0.2391

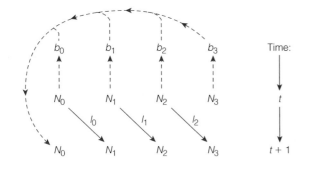

FIGURE 4.27 Schematic representation of the contributions of age-specific birth rates (b_x) and survival rates (l_x) to population size (N_x) in a subsequent time interval ($t + 1$). Note the similarities in locations of birth and survivorship rates in the figure to the locations of elements in the matrix in the text.

rate and survivorship rate. Obviously, real populations age continuously, and their birth and survivorship rates change accordingly. In such a system, the number of individuals in the next time interval can be derived as shown in Figure 4.27. Because newborn individuals come only from reproduction, the number of individuals in age class 0 (N_0) at time $t + 1$ is determined solely by the number of individuals born at time t (the contributions from b_0, b_1, b_2, and so on). The number of individuals in all other age classes is the result of survivorship from previous ages. Therefore, the number of individuals in age class N_1 at time $t + 1$ is the result of the survivorship of the individuals at time t. This survivorship value is l_0. Similarly, N_2 at time $t + 1$ is the result of the survival of N_1 at time t—that is, l_1.

We can represent the transition depicted in this figure by a series of **linear recurrence equations:**

$$N_{0t+1} = (N_{0t} \times b_0) + (N_{1t} \times b_1) + \qquad\qquad (N_{2t} \times b_2) + (N_{3t} \times b_3) \quad \textbf{(1)}$$
$$N_{1t+1} = (N_{0t} \times l_0) \quad\qquad\qquad \textbf{(2)}$$
$$N_{2t+1} = (N_{1t} \times l_1) \quad\qquad\qquad \textbf{(3)}$$
$$N_{3t+1} = (N_{2t} \times l_2) \quad\qquad\qquad \textbf{(4)}$$

Now let us consider a population of 660 individuals with the following demographic parameters at time t:

Age (x)	N_x	b_x	l_x
0	500	0	0.2
1	100	2	0.5
2	50	6	0.3
3	10	8	0.0

Using Equations 1–4, we calculate the values of N_x at time $t + 1$ as follows:

$$N_{0t+1} = (500 \times 0) + (100 \times 2) + \qquad\qquad (50 \times 6) + (10 \times 8) = 580$$
$$N_{1t+1} = (500 \times 0.2) = 100$$
$$N_{2t+1} = (100 \times 0.5) = 50$$
$$N_{3t+1} = (50 \times 0.2) = 10$$

The total population size is now 740 (580 + 100 + 50 + 10). We can continue these calculations to model how the population size and its age structure continue to change over time.

The notations for these calculations are simplified if we use matrix algebra to represent them. A brief introduction to this mathematics is presented in the chapter appendix.

We can construct a pair of matrices that enable us to perform the same calculations as in Equations 1–4. We devise a square matrix that consists of the age-specific birth and survivorship rates into which zeros are strategically placed so that when we multiply by the numbers of individuals in the age classes, the results are the numbers of individuals in those age classes in the next time period:

$$\begin{bmatrix} 0 & 2 & 6 & 8 \\ 0.2 & 0 & 0 & 0 \\ 0 & 0.5 & 0 & 0 \\ 0 & 0 & 0.2 & 0 \end{bmatrix} \times \begin{bmatrix} 500 \\ 100 \\ 50 \\ 10 \end{bmatrix} = \begin{bmatrix} 580 \\ 100 \\ 50 \\ 10 \end{bmatrix}$$

This hypothetical example follows the general form developed by Leslie, in which the matrix containing the age-specific reproduction and survival data (**A**) is multiplied by the matrix containing the numbers of individuals in each age class ($\mathbf{N}_t$) to generate a new matrix for time $t + 1$ ($\mathbf{N}_{t+1}$):

$$\mathbf{A} \times \mathbf{N}_t = \mathbf{N}_{t+1}$$

or

$$\begin{bmatrix} b_0 & b_1 & b_2 & \cdots & b_{n-1} & b_2 \\ l_0 & 0 & 0 & \cdots & 0 & 0 \\ 0 & l_1 & 0 & \cdots & 0 & 0 \\ 0 & 0 & l_2 & \cdots & 0 & 0 \\ \cdot & \cdot & \cdot & & \cdot & 0 \\ \cdot & \cdot & \cdot & & \cdot & 0 \\ 0 & 0 & 0 & & l_{n-1} & 0 \end{bmatrix} \times \begin{bmatrix} N_{0t} \\ N_{1t} \\ N_{2t} \\ \cdot \\ \cdot \\ \cdot \\ N_{nt} \end{bmatrix} = \begin{bmatrix} N_{0t+1} \\ N_{1t+1} \\ N_{2t+1} \\ \cdot \\ \cdot \\ \cdot \\ N_{nt+1} \end{bmatrix}$$

This kind of analysis has several interesting and useful properties. First, it is amenable to calculation by computer, which greatly facilitates demographic projections many generations into the future. If the survivorship and reproductive parameters are constant, the population will eventually attain a stable age

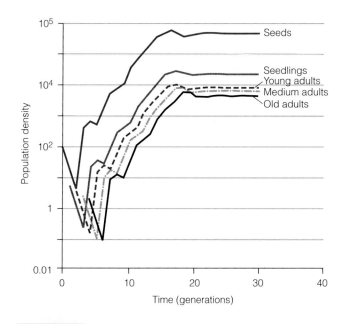

FIGURE 4.28 Results of transition matrix modeling for a population of the annual grass *Poa annua* over many generations. The population reached a stable age structure within 20 generations. (*From Law 1975*)

is depicted in Figure 4.28 for the annual grass *Poa annua*.

For organisms for which it is difficult or impossible to determine the age of individuals, it may be possible to make matrix projections by using size-class information in place of ages. In such cases, the matrix's size-class elements represent the *probabilities* of transition from one size class to the next; size-specific seed production replaces age-specific reproduction. From these data we can calculate the multiplicative growth rate (λ) of the population; λ values below 1.0 indicate declines in the population, and λ values above 1.0 indicate population growth.

From this kind of analysis, we can determine the rate at which a population may be harvested without causing its decline ($\lambda < 1.0$). It is also valuable for modeling the effects of various management strategies or environmental impacts on threatened or endangered populations. A number of "what if" questions may be addressed without experimental manipulations that might negatively impact a threatened population.

In this chapter we have discussed the ways in which we quantitatively describe populations and their growth. We turn our attention next to the mechanisms that control and regulate population density.

structure in which the proportion of the population in each age class no longer changes, despite the fact that the population may be growing. An example of using a transition matrix to model population age structure

SUMMARY

1. A population is a group of conspecifics inhabiting a specific place at a particular time. Demography, the quantitative description of that population, includes parameters such as density, age structure, sex ratio, and dispersion.

2. Density is measured by a number of means both direct and indirect. Indices of relative abundance are useful for comparisons of populations in different places or at different times.

3. Life tables contain information on the age-specific birth and death rates of individuals. From this information we can calculate other demographic parameters, such as age-specific life expectancy and the expected number of offspring produced by each female in the population (R_0).

4. Survivorship curves graphically depict the age-specific mortality of the population. Most organisms exhibit one of three main types of curves.

5. Populations introduced into new and suitable habitats will begin to grow exponentially. This exponential growth is described by the equation $N_t = N_0 e^{rt}$.

6. Because resources are finite, populations cannot grow exponentially forever. Eventually, they reach an asymptote at a value K, the carrying capacity. This pattern of population growth—from the exponential phase to K—is known as sigmoid growth. The equation $dN/dt = rN(K - N)/K$ describes this pattern.

7. Population growth can also be modeled with transition matrices that calculate the numbers of individuals in each age category through time from their age-specific mortality and natality rates.

8. Transition matrices are useful tools for population analysis because they can be used to predict the effects of various management or harvest patterns.

SELF-ASSESSMENT: CAN YOU ...?

1. Discuss the nature of a population, including its definition(s) and the relationship between the ecological population and the genetic population.

2. Calculate all columns of a standard life table from census and age structure information.

3. Depict the survivorship data in a life table as a survivorship curve.

4. Explain the origin and significance of the following equations:

$$N_t = N_0 e^{rt}$$

$$dN/dt = rN \frac{(K - N)}{K}$$

$$N_t = R_0{}^t N_0$$

5. Explain the ambiguity in determining the sex ratio of some species.

6. Explain the evolutionary ecology of sex ratio variation within and among species.

7. Compare the nature and mathematics of population growth for species with discrete versus overlapping generations.

PROBLEMS AND STUDY QUESTIONS

1. Discuss the problems inherent in defining a population for study. What are the effects of variation in time and space?

2. In a study of grasshopper density, you capture and mark 314 animals at t_1. One week later you capture 456 animals, 145 of which are marked. What is the population size?

3. In the method of analysis used in Problem 2, what is the importance or effect of the percentage of the total population that is captured and marked at t_1? Use hypothetical samples to illustrate the effect.

4. Using the following life table, answer the questions below.

Age	Number Alive at Start of Interval	Number Dying in Interval	Average Number of Female Young Born per Female
0	1000	890	0
1	110	70	0
2	40	19	5.1
3	21	14	3.2
4	7	7	0.0

a. Compute the values of l_x for this population.

b. What kind of survivorship curve does this population have?

5. Is the population in Problem 4 increasing or decreasing? Explain.

6. Why is the value of e_x not necessarily maximal at birth?

7. The average number of children per couple in the United States is less than 2.0, yet our population is still growing. How can this be?

8. Is this population random, clumped, or hyperdispersed?

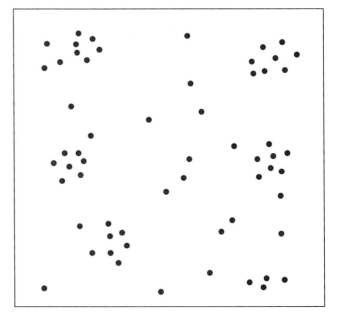

9. Use the Leslie matrix to calculate the population size and the age structure in the next time interval for the following population:

Age	N_x	m_x	l_x
0	800	0	1.0
1	400	3	0.5
2	200	2	0.1
3	20	0	0.5

10.   Explain the relationship between R_0 and r.

1. Search the data available via the Internet from the U.S. Bureau of the Census for age structure information on populations in the United States. Use these data to generate a life table and survivorship curve.

2. Visit a local cemetery and record the ages at death for as many individuals as possible. Use these data to develop a static life table for the population. What assumptions must be made? What factors might compromise the interpretation of the data?

3. Plot the data in Figure 4.23 using a logarithmic scale for *N*. What is the nature of the line that results? Analyze the meaning and significance of the slope of that line.

Appendix

Matrix Algebra: Multiplying a Matrix and a Column Vector

A matrix is a group of numbers (elements) arranged in rows and columns. Arrays are denoted by boldface letters.

One matrix, **M**, might contain the following values:

$$\mathbf{M} = \begin{bmatrix} 2 & 4 & 6 \\ 10 & 12 & 14 \\ 18 & 20 & 22 \end{bmatrix}$$

Another matrix, **N**, might be constructed thus:

$$\mathbf{A} = \begin{bmatrix} 0.2 \\ 0.4 \\ 0.6 \end{bmatrix}$$

A matrix containing a single column of numbers is referred to as a **column vector**. To multiply a column vector and a matrix, we begin by multiplying each element in the column vector by each of the elements of the first row of the matrix. We then add all these products and place the resulting sum in the first row of a new column vector. Next, the values in the column vector are multiplied by the values in the second row of the matrix, summed, and placed in the second row of the new column vector. This process is completed for all rows in the column vector. If we call the new column vector X, then the calculations for $\mathbf{M} \times \mathbf{A} = \mathbf{X}$ are as follows:

$$\mathbf{X} = \begin{bmatrix} (2 \times 0.2) + (4 \times 0.4) + (6 \times 0.6) \\ (10 \times 0.2) + (12 \times 0.4) + (14 \times 0.6) \\ (18 \times 0.2) + (20 \times 0.4) + (22 \times 0.6) \end{bmatrix} = \begin{bmatrix} 5.6 \\ 15.2 \\ 24.8 \end{bmatrix}$$

Note that in multiplying matrices, $\mathbf{M} \times \mathbf{A} \neq \mathbf{A} \times \mathbf{M}$. To be clear, we refer to pre- and post-multiplication. In $\mathbf{M} \times \mathbf{A}$, **M** is said to be *post-multiplied* by **A**.

Population Regulation

P *opulation regulation underlies most other ecological problems of interest . . . It is also integral to much that is of interest to evolutionary biologists. . . .* (Murdoch 1994)

In this chapter we consider the factors that regulate the densities of populations of animals and plants. We begin with two simple, empirical observations:

1. The population densities of animals and plants are not constant; they vary over time.
2. A population cannot grow forever; because of the finite nature of resources, unlimited growth is simply a physical impossibility.

Our goal in this chapter is to understand these two most fundamental observations. As we will see, the two observations may or may not be related. Two questions in particular are of interest: To what extent are the limitations of finite resources responsible for the patterns of fluctuation we observe? And to what extent are other factors such as predators, pathogens, and even chance important?

To address this fundamental issue in the science of population ecology, we return to the Lotka–Volterra equation for sigmoid growth presented in Chapter 4:

$$dN/dt = rN(K - N)/K$$

The equation provides a mathematical description of the change in population size over time shown in Figure 5.1. In this figure, a population grows exponentially until it approaches a maximum, K. At this point, the growth rate of the population decreases until it reaches zero at K.

This model represents an *equilibrium view* of a population. There is some equilibrium, K—carrying capacity—that is associated with resource levels. If perturbed away from this equilibrium value, the population returns to it. The equilibrium view of populations, whose roots are deep in the history of ecology, is associated with the concept of the "balance of nature," which implies that there is a normal state to which ecological systems return if left undisturbed.

Ecologists have recently questioned the validity of an equilibrium view of populations, however (Botkin 1990). A number of ecologists have suggested that populations fluctuate at random over time, without mechanisms or processes that move them to some equilibrium level. Associated with this nonequilibrium view is an emphasis on disturbance.

In this chapter, our approach to the issue of population regulation mirrors history. We begin with the development of models of population regulation

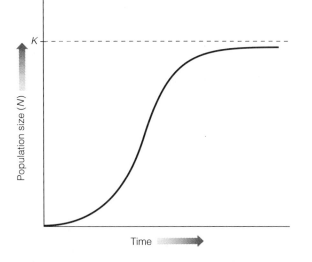

FIGURE 5.1 Sigmoid curve of population growth.

based on the equilibrium view of populations and then consider the implications of a nonequilibrium approach.

Some terms require definition at the outset. The process of **population regulation** can be characterized as bounded fluctuations in abundance (Murdoch 1994); implicit in this concept are feedback mechanisms that increase density when it is below a certain level and decrease it when it is too high. In other words, it is based on an equilibrium view of populations. **Population control** refers to the ecological mechanisms by which density is limited. Control factors may maintain densities below the carrying capacity; wolf predation, for example, may control elk density at levels below those that the habitat could actually support, or a farmer may control insect pests at levels below which crop damage is significant.

We may also ask what combination of factors determines the density of a given population at any given time, regardless of whether it is at the carrying capacity. In mathematical terms, we may seek an equation that allows us to predict the value of N (population number) from knowledge of a number of factors, such as food abundance, habitat quality, predator numbers, weather factors, and so forth. Even though no one factor by itself can be used to predict the population size, the combination predicts the value of N in any year and thus implies which factors determine population size. We can also note that the change in population in any time interval is described by the equation

$$\Delta N = (b + I) - (d + e)$$

where ΔN is the change in population size, b is the number of births, i is the number of immigrants, d is the number of deaths, and *e* is the number of emigrants. Although only four variables are needed to describe population change, we shall see that a number of biological factors contribute to their values and that they interact in complex ways.

Patterns of Population Fluctuation

The first step in explaining the ecological processes that regulate the densities of populations of animals and plants is identifying the patterns of fluctuation in natural populations. We saw in Chapter 4 that a huge range of population densities can exist for various species in different taxonomic groups.

There are four fundamental patterns of density fluctuations over time. Each is based on the extent to which population size changes and on the time scale involved.

Small-Magnitude Irregular Fluctuations

As the name of this pattern implies, species with this type of fluctuation do not experience large changes in density. We arbitrarily define "small-magnitude" fluctuations as year-to-year population changes of about one order of magnitude or less; high and low densities occur with apparent randomness. The population density curves for two populations of herons (*Ardea cinerea*) in Figure 5.2 exhibit this pattern. Heron numbers (expressed as the number of breeding pairs) fluctuate irregularly, but neither population changes by a factor of more than 3 over some 30 years of observations.

Large-Scale Irregular Fluctuations

This pattern is similar to the first, except that the range of variation is far greater. Peaks and lows again come without regularity, but the peaks may be several orders of magnitude higher than the lows (Figure 5.3). This pattern is typical for economically important insect pests that occasionally reach ex-

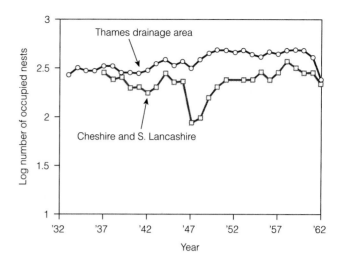

FIGURE 5.2 An example of small-magnitude irregular population fluctuations: the number of breeding pairs of the heron, *Ardea cinerea*, in two parts of England over a 30-year period. (*From Lack 1966*)

treme densities and cause major losses to crops or forest trees. Indeed, invertebrates, especially insects, are likely to have large-scale irregular fluctuations because their very high intrinsic rates of increase (*r*) endow them with the potential for explosive population growth.

Cycles

Among the most fascinating patterns of change in population density are those that occur with regularity—the population cycles. Cycles can show great ranges of density fluctuations; the population peaks

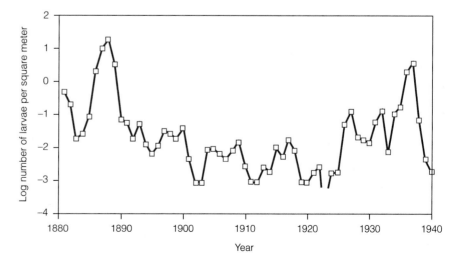

FIGURE 5.3 An example of a large-scale irregular population fluctuation. Winter counts of hibernating larvae of the moth *Dendrolimus* sp. per square meter of forest floor at Letzlingen, Germany, over 60 successive years. Note that densities are plotted on a logarithmic scale. (*Adapted from Varley 1949*)

for some species are more than an order of magnitude greater than the lows, whereas for others, population densities are less variable. The period of the cycle remains relatively constant for a given species, although it may be days for some species and years for others.

True cycles are relatively rare. None are known from the tropics or the Southern Hemisphere (Sinclair and Gosline, 1997). Two patterns of cycling have been studied especially intensely: the roughly 4-year cycles of microtine rodents (lemmings and voles) and 10-year cycles, such as those that occur in hare, grouse, and lynx populations.

The cyclical pattern of lemming abundance on one site in Manitoba is depicted in Figure 5.4. In some places in the Arctic, lemming cycles can be spectacular phenomena. Although the lemming population represented in Figure 5.4 reached a peak every three or four years at approximately 40 animals per hectare, peak densities in the Arctic can reach 1,000 lemmings per hectare one year and plummet to close to zero the next. The changes in amplitude of the 10-year cycle are rarely so dramatic as those for microtines (Figure 5.5). Data are simply not available in sufficient detail from enough sites across the continent for ecologists to assess whether statistically significant synchrony is actually a demographic feature of these species.

The regularity that defines a population cycle fascinates population ecologists because it implies that despite all the inherent year-to-year variations in the environment, some factor is controlling population density in a regular, clock-like fashion. The importance of this as yet unknown factor seems to override the stochastic variation inherent in the environment.

Another feature of cycles is also of special interest to ecologists: A cycling population experiences many population phenomena over the course of any one cycle. This regularity of change means that population biologists can study all the fundamental aspects and processes of density change—increase, decrease, carrying capacity, and so on—in a single population on a

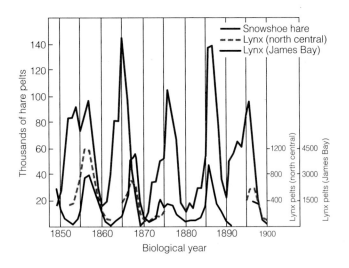

FIGURE 5.5 Population density patterns for snowshoe hares and lynx, as measured by Hudson's Bay Company pelt collections in the Hudson Bay region of Canada. (*From Keith 1963*)

predictable time scale. Consequently, population cycles, particularly those of lemmings and voles, are among the most intensively studied population phenomena.

Irruptions

Some species show patterns of density fluctuation that are called *irruptive.* This means that the population exists for much of the time at low density and with little fluctuation. Then occasionally, and perhaps unpredictably, the population explodes. The feral house mouse, *Mus musculus,* is typical (Figure 5.6). Following three years of very low densities, two irruptions of over 300 mice per acre occurred.

One of the remarkable things about irruptions is that they sometimes cover a wide geographic area. Most often they are the result of the convergence of a number of favorable conditions. For example, a year of favorable weather, abundant food resources, and the absence of predators might result in an irruption. However, because the likelihood that all those unusually favorable conditions will occur simultaneously over a wide area is small, it is important to ascertain whether a widespread irruption is a general population phenomenon or the result of the wide dispersal of a reproductively prodigious group of organisms.

Any general theory of population regulation must be able to explain these patterns of population density. We now proceed to

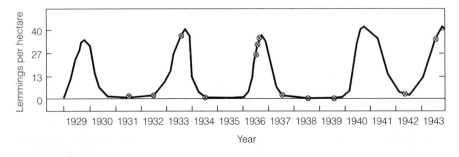

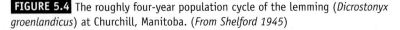

FIGURE 5.4 The roughly four-year population cycle of the lemming (*Dicrostonyx groenlandicus*) at Churchill, Manitoba. (*From Shelford 1945*)

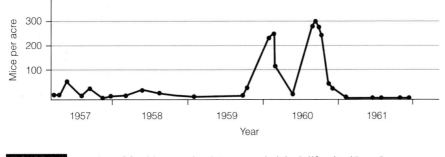

FIGURE 5.6 Irruption of feral house mice (*Mus musculus*) in California. (*From Pearson 1963*)

the theories of population regulation and control. In the context of those theories, explanations for each of these population patterns will emerge.

Equilibrium Theories of Population Regulation

Population ecologists began to focus intensely on the nature of population regulation in the 1930s. Subsequently, ecologists identified a number of specific mechanisms by which populations could be regulated. Between 1935 and about 1970, a rather intense debate raged over which processes are the most important for most species.

The fundamental dichotomy among ecologists with regard to population regulation centered on the relative importance of factors that operate in either a density-dependent or a density-independent fashion. **Density-dependent factors** are those whose effects on the birth rate or death rate change as a function of the population density—that is, they increase with increasing population density (Figure 5.7a). Intraspecific competition for a limiting resource, for example, can cause population growth to slow at high densities. As we will see, density-dependent effects may operate via a number of mechanisms. They may also include a time lag. That is, the effect on mortality or natality may be delayed by some period of time (Saitoh et al. 1999).

For **density-independent factors,** population birth rate and death rate do not change with the value of N (Figure 5.7b). Density-independent factors are usually associated with abiotic events—changes in the physical environment—that either promote or repress population growth, but their effects are independent of population density. A killing frost, for example, acts in a density-independent fashion; it kills the same proportion of the population no matter what the population size. The effect may be positive, too: Sufficient rainfall

may enable a high proportion of seeds to germinate and survive.

The Australian ecologist A. J. Nicholson (1933, 1954, 1957, 1958) introduced the notion that if population regulation is to occur, density-dependent factors must come into play; indeed, density dependence has come to define the concept of regulation (Murdoch 1994). Moreover, density-dependent regulation has come to be associated with the equilibrium view of populations. The connection is logical: The concept of density dependence leads directly to the notion of some equilibrium population density toward which the density-dependent factors move the population.

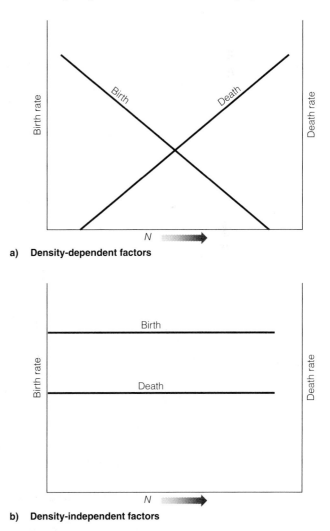

a) Density-dependent factors

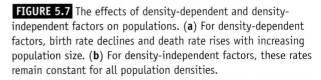

b) Density-independent factors

FIGURE 5.7 The effects of density-dependent and density-independent factors on populations. (**a**) For density-dependent factors, birth rate declines and death rate rises with increasing population size. (**b**) For density-independent factors, these rates remain constant for all population densities.

Two basic schools of thought incorporating density dependence have arisen. We will examine these first and defer a full discussion of density-independent factors until the section on nonequilibrium approaches to population dynamics. As you consider the examples and data pertaining to density-dependent regulation, evaluate the support they provide for the hypotheses. What factors confound the empirical studies? How strong is the evidence?

The Extrinsic Biotic School

This school of population regulation accepts the importance of density-dependent factors and emphasizes the significance of biotic factors external to the species being regulated. Among the important density-dependent extrinsic factors are the food supply, predation, and disease.

Food Supply and Population Regulation The importance of food supply as a population regulation factor has long been emphasized by ecologists such as David Lack and Frank Pitelka. Lack's 1954 book *The Natural Regulation of Animal Numbers* was particularly influential. It outlined the logical basis of the so-called food hypothesis as it is applied to birds. From surveys of a number of bird populations, Lack concluded that whereas some changes in reproductive rate occur in response to density, mortality is a more important density-dependent factor. Of the three major factors that result in increased mortality—starvation, predation, and disease (including parasitism)—Lack dismissed the latter two because there are so few observations of diseased or dying birds. On the other hand, birds are frequently observed to be more abundant in areas of high food density. Furthermore, during the winter, when food supplies are most limited, birds are very aggressive in their feeding behavior.

As evidence that starvation is an important source of mortality in birds, Lack cited data that compared two indirect measures of food supply with the mortality rate for wood pigeons (*Columbus palumbus*) in an agricultural area of England. Wood pigeons that weighed less than 450 grams were considered undernourished, and birds collected with empty crops were clearly having difficulty getting enough food. Note in Table 5.1 that when winter mortality rates were high (1960–1961 and 1962–1963), the proportions of birds with low weights and empty crops were also high.

There are basically two ways to test the hypothesis that food supply represents a density-dependent regulatory factor. The first is to look for correlations between density and food supply: A positive correla-

TABLE 5.1
Winter Mortality of Wood Pigeons and the Proportions of Undernourished Birds

Year	Percentage Change (Early December to Early March)	Percentage of Birds Shot at Less Than 450 g	Percentage of Birds Show with Empty Crops
1958–1959	–17	2	2
1959–1960	–45	25	—
1960–1961	–52	19	29
1961–1962	+25	5	5
1962–1963	–99	36	39
1963–1964	–20	14	3

Adapted from Lack 1966.

tion supports the hypothesis, whereas a negative correlation (or no correlation at all) results in rejection of the hypothesis. Of course, it must be remembered that correlation studies are weak tests of a hypothesis because the presence of a correlation cannot, by itself, prove causality.

One example of this type of analysis involves the lynx (*Felis lynx*) and its main food source, the snowshoe hare (*Lepus americanus*). Long-term records of the abundance of both species are available from fur-trapping records in Canada. Clearly, the patterns of density changes in the lynx and hare populations closely match over time (Figure 5.5).

The second approach to ascertaining whether food supply is a density-dependent factor is to supplement the food supply and see whether population density increases. The results of some food supplement studies have supported the food hypothesis. In one study, M. J. Taitt (1981) increased densities of the white-footed mouse, *Peromyscus leucopus,* by supplemental feeding; in another, R. G. Ford and Pitelka (1984) delayed the decline (relative to controls) of cyclic populations of the California vole, *Microtus californicus,* by supplying supplemental food (Figure 5.8).

Overall, however, food supplement studies have produced equivocal results. Despite the positive results just cited, other studies on the same species produced no effect (for *Peromyscus:* Wolff 1986; Hansen and Batzli 1979; for *Microtus:* Krebs and DeLong 1965). Among the difficulties with these studies is that food supplements may not be natural food sources and the supplements may be distributed in the habitat in an unnatural way. Thus it is difficult to draw meaningful conclusions from these studies.

Natural experiments in which the food supply dramatically increases without artificial supplements circumvent this problem. For example, the emergence of

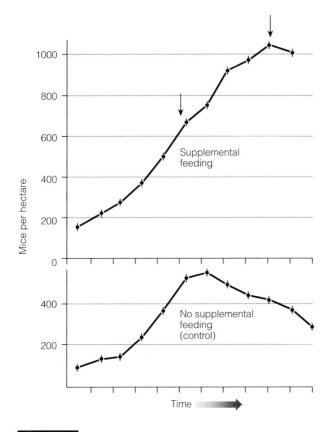

FIGURE 5.8 Effects of supplemental feeding in the California vole, *Microtus californicus*. Decline in the cyclical abundance of the voles given supplemental feed (above) was delayed relative to the control population (below). Arrows mark the onset and termination of supplemental feeding. (*Adapted from Ford and Pitelka 1984*)

periodic cicada adults (*Magicauda* spp.) provides spectacular, natural food supplements for certain species of insectivorous animals. After a period of development of either 13 or 17 years, during which nymphs feed on root xylem fluids in the forest floor, the adults emerge synchronously, flooding local areas of the woods with loud hordes of mating cicadas. Anderson (1977) reported that the reproductive success of the European tree sparrow (*Passer montanus*) increased significantly in a year during which 13-year cicada adults emerged. Similarly, the density of the short-tailed shrew (*Blarina brevicauda*), an insectivore, increased fourfold in a year in which cicadas emerged in Indiana (Krohne, Couillard, and Riddle 1991).

A comparison of irruptive and nonirruptive species of insect herbivores provides evidence of the importance of food supply in determining the nature of population dynamics. K. L. Dodge and P. W. Price (1991) compared the biology of two species of insects, the sawfly (*Euura exiguae*) and the leaf beetle (*Disonycha pluriligata*), both of which feed on the willow (*Salix exigua* Nuttall). The sawfly is nonirruptive; the leaf beetle occasionally exhibits huge population explosions. According to Dodge and Price, the difference between the population dynamics of these species is a consequence of their different feeding dynamics. The sawfly oviposits only on the long shoots of young willow plants. The larvae that emerge and then eat the shoots are very successful thanks to the high nutritional quality of this food resource. The young shoots are a rather limited food supply, however, and thus represent a constraint that prevents population explosions. In contrast, the leaf beetle oviposits in the ground, and the larvae emerge and feed on any willow, regardless of age or quality. Consequently, the carrying capacity for this species is far higher, and the willow plants can support significantly higher populations of the leaf beetle.

For plant populations, strong evidence exists to support the importance of nutritional resources in determining plant densities. Because of the sessile nature of plants, direct observation and experimental manipulation are more readily achieved. For example, J. W. Hughes (1992) examined the impact of the removal of competing vegetation on the spring herb *Erythronium americanum* in deciduous forests in New Hampshire. Removal of overstory trees, which increased the availability of nutrients, resulted in a 225 percent increase in density of the herb over a period of three years.

It is not uncommon for these kinds of effects to be particularly pronounced for juvenile plants that must compete for resources with well-established adults. In the case of longleaf pines (*Pinus palustris*) in the southeastern United States, adult trees, by virtue of the space they occupy, prevent the growth of juveniles; the only recruitment of juveniles occurs when adults die. Large adult trees have a loose, aggregated dispersion; juveniles are aggregated on sites that lack adult trees. Thus temporal variation in adult mortality, together with the coincident recruitment of new individuals, results in alternating phases of population growth and decline that vary in length and magnitude. The presence of adults constitutes an upper limit on density.

For plants growing in single-species stands, a phenomenon called **self-thinning** occurs. As a crowded population grows, the population density declines while the size of surviving individuals increases. Plant ecologists typically depict this phenomenon on a log–log plot of size versus density. Each stand is represented by a line that depicts a trajectory of the size-versus-density relationship for each stand. Each point on the trajectory represents a different point in time for the stand. Interestingly, for a number of species,

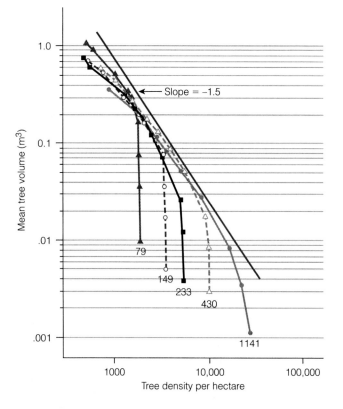

FIGURE 5.9 Self-thinning in a series of loblolly pine stands over a 50-year period. Each line represents a different stand. The initial densities of the stands are indicated by the numbers below the lines. Points on each line represent time points as each stand changes in density and tree size. All stands converge on the same density and tree size along a line of slope −1.5. (*Data from Peet and Christensen 1987*)

the pattern of change follows predictable paths. Regardless of the initial density of the stand, its trajectory eventually follows a line with a slope of −1.5.

An example is shown in Figure 5.9. A series of stands of the loblolly pine were studied by Peet and Christensen (1987). The initial stand densities are shown by the numbers below the initial point. Each line shows the time course of change in size as density declines for each stand. Note that all the lines approach a slope of −1.5 and eventually converge at a similar stand density and tree size.

Predation and Population Regulation The hypothesis that predation is another extrinsic biotic factor that can have density-dependent regulatory effects is a difficult one to test. Not only must we have information on the relative rates of predation at different densities, but we must also be able to relate that information to the carrying capacity of the prey species.

Many studies have indicated that predation depresses prey populations. For example, A. C. Fowler

and colleagues (1991) used experimental exclosures in a grassland habitat to show that birds depress grasshopper populations. Similar effects are known in plants as well. For example, M. A. Vinton and D. C. Hartnett (1992) showed that in tallgrass prairie, grazing of big bluestem grass (*Andropogon gerardii*) stimulated growth in the year in which grazing occurred. The long-term effect was negative, however: In the year following grazing, mean leaf length was significantly less in both grazed and artificially clipped populations, as was mean stem density.

Despite data such as these, demonstrating that predation regulates populations in a density-dependent fashion is far more difficult. F. H. Wagner and L. C. Stoddart (1972) present evidence that predation by coyotes (*Canis latrans*) plays a regulatory role in fluctuating black-tailed jackrabbit (*Lepus californicus*) populations in northern Utah. Data from a variety of sources indicate that coyote predation is an important source of mortality in jackrabbits. These authors found that the ratio of the numbers of coyotes and rabbits can be used as an index of the rate of coyote predation and that the jackrabbit mortality rate is negatively correlated with the coyote:rabbit ratio (Figure 5.10). Indeed, 69 percent of the variation in rabbit numbers is associated with that ratio.

Even though data such as these are consistent with density-dependent regulation, they do not unequivocally demonstrate that predation by coyotes regulates jackrabbit densities. The kinds of experiments that

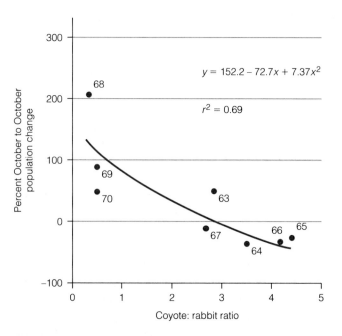

FIGURE 5.10 Correlation between annual rate of jackrabbit population change and coyote:rabbit ratio in northern Utah. (*From Wagner and Stoddart 1972*)

could unambiguously demonstrate density-dependent predation effects are difficult to conduct in field situations. In addition, the dynamics of the prey species and its food supply have important effects in predator–prey interactions. For our discussion we can consider these complex interactions in North American populations of moose. The regulation of moose populations is a matter of considerable interest because moose are important game animals. A major debate has centered on whether moose populations can be increased by removing or controlling their predators.

Of five studies in which wolves were experimentally removed to increase moose densities, only one showed an increase in the moose population (Boutin, 1992). W. C. Gasaway and co-workers (1992) suggest that it is not wolf predation by itself, but rather the combined effects of wolves, grizzlies, and black bears, that hold moose populations down.

The "top-down or bottom-up" controversy The emphasis by some ecologists on predation, and the emphasis by others on the prey's food supply, has evolved into a debate about "top-down" regulation (predation) versus "bottom-up" regulation (food supply). The population dynamics of moose on Isle Royale in Lake Superior represents an important example of the interacting effects of predation and food supply (Mech 1966).

Moose were not found on Isle Royale until the early 1900s, when a few individuals (or perhaps a single pregnant cow) swam the 32 kilometers from Ontario to the island. The boreal forest on the island had recently been decimated by fire, and the resultant rapid growth of early successional shrubs and small trees created an ideal habitat for the moose. Consequently, their population increased exponentially (Figure 5.11).

By 1930 the island was clearly overpopulated, and a crash of the moose population was imminent. Even though the moose population had begun to decline, wolves (*Canis lupus*) that arrived on the island in the mid-1940s found an abundant prey population, and their numbers increased, too. Subsequent to the arrival of wolves, the moose population stabilized at approximately 200–300 animals and coexisted with the wolf population in approximate balance for a

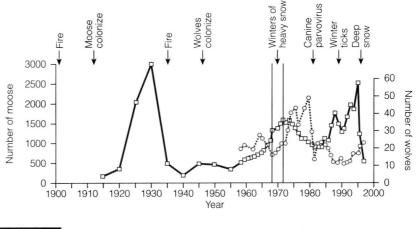

FIGURE 5.11 A century of changes in moose abundance on Isle Royale. (*Data from Mech 1966*)

time. Wolf predation provided a source of density-dependent regulation of moose numbers.

Recently, this balance has been disturbed. The quality of the moose habitat has deteriorated as succession proceeded to later, less usable stages; the wolf population has also declined, apparently as the rest of an imported canine disease. This example illustrates the difficulty of disentangling the interacting effects of predation and food supply.

Proposed explanations for the ten-year cycle of snowshoe hares also reflect the bottom-up and top-town controversy. L. B. Keith (1974, 1983) proposed a conceptual model of the snowshoe hare cycle in Alberta. Essentially, this model proposes that the cycle is generated biotically by the interaction of the hare and its food supply; overbrowsing in the winter initiates the decline in hare numbers. This is a crucial interaction because the highest hare densities occur just before winter, and the food supply (woody vegetation) does not regenerate during the winter months. Moreover, in the northern latitudes, where such cycles occur, the winter lasts seven to nine months. In contrast, K. Trostel and colleagues (1987) emphasize the importance of predation in driving hares to low population densities, from which recovery is slow. It is clear that the relative roles of predation and food supply are still not fully understood (Keith 1987).

Disease and Population Regulation The last regulatory factors to be considered in the extrinsic biotic school are disease and parasitism, which, like starvation and predation, represent potential density-dependent mortality factors. We might expect the rate of disease transmission to be related to population density, because more frequent encounters between individuals should increase the transfer of infectious disease agents. If so, disease could affect populations in a density-dependent manner. Dobson and Meagher (1996) have shown that the prevalence of brucellosis, a bacterial disease of ungulates, in Yellowstone National Park bison (*Bos bison*) is density-dependent (Figure 5.12).

We cannot demonstrate density dependence by simply showing a correlation between density and the rate of infection, however. Consider, for example, the case of red grouse (*Lagopus scoticus*) in Scotland infected with a nematode parasite (*Trichostrongylus tenuis*), which causes a debilitating condition called strongylosis (Jenkins, Watson, and Miller 1963, 1964). There is a strong positive correlation between grouse density and the percentage of birds infected (Figure 5.13). However, a number of problems of interpretation occur. First, this graph does not provide any information on the size of the parasite load in these individuals. Even if a high proportion of birds is infected, there will be little effect if the parasite load is low in most individuals. Second, it is not clear what the effects of the infections are, because it is possible for healthy individuals to contain high densities of nematodes. In addition, most outbreaks of strongylosis occur in the spring, when food is limited for grouse. Lovatt (1911) suggests that the infections are an effect rather than a cause; food shortages weaken the birds and predispose them to disease. Thus food supply, not disease itself, may be the real regulatory factor.

Further complicating the situation is the fact that not all species respond in the same way to pathogens. In Yellowstone, *Brucell aborta*, the agent of brucellosis, affects ungulates in very different ways. Bison show no pathological effects from infection, whereas infec-

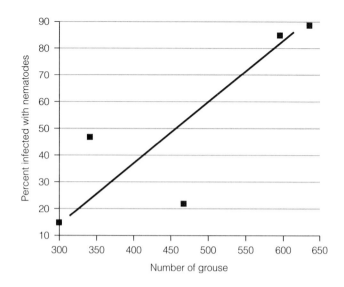

FIGURE 5.13 Correlation between red grouse (*Lagopus scoticus*) population density and the proportion of grouse infected with the nematode *Strongylus tenius* in Scotland. (*Data from Jenkins, Watson, and Miller 1963*)

tions of moose are virtually always fatal. Elk are intermediate—infected females often abort their first fetus (Dobson and Meagher 1996).

This is not to say that disease and parasitism do not influence populations. There are numerous instances of outbreaks of diseases like botulism and avian cholera decimating wintering flocks of waterfowl. These episodes of mortality typically occur when birds are concentrated at very high densities on wildlife refuges.

More recent data have supported the hypothesis that disease can be a fundamental regulatory or controlling factor. For example, tent caterpillars (*Malacosoma* spp.) produce cyclic outbreaks of defoliation throughout their ranges. The forest tent caterpillar (*M. disstria*) has population cycles in Ontario with a period of approximately 13 years (Figure 5.14).

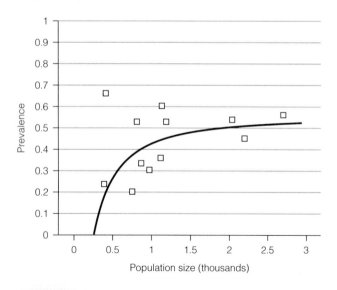

FIGURE 5.12 Prevalence of brucellosis in bison in Yellowstone National Park on the basis of seropositive tests for *Brucella aborta*. (*From Dobson and Meagher 1996*).

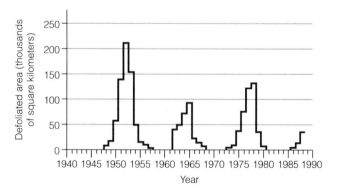

FIGURE 5.14 Population cycles of tent caterpillars, *Malacosoma* spp., in Ontario, as indicated by area of defoliated forest identified by aerial surveys. (*Adapted from Myers 1993*)

A number of unusual features characterize these cycles. It appears that food limitation does not explain the pattern. Even though the degree of defoliation (and thus the degree of food depletion) is highly variable across sites and between years, the populations remain cyclical, with very strong synchrony in different regions (Figure 5.15). When J. H. Myers (1993) conducted experiments in which she cropped some populations of the western tent caterpillar (*M. californicum*) to maintain them in the outbreak phase, the cropped population crashed just like the controls (Figure 5.16). In other experiments, she introduced populations to a new area using caterpillars from a population at the peak phase. These new populations crashed synchronously with the source populations, even though they were introduced into regions with abundant food supplies.

If food availability does not explain caterpillar population changes, what does? W. G. Wellington's (1960) description of caterpillars at different stages of the population cycle provided the clue. Expanding caterpillar populations are typically characterized by a preponderance of colonies of an active morph that disperses the caterpillars; at peak density, colonies of a sluggish, less hardy morph predominate.

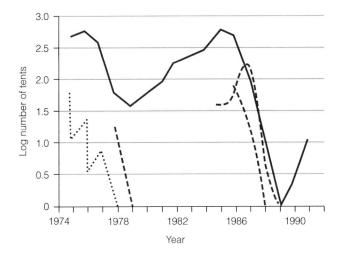

FIGURE 5.16 Experimental attempts to prolong a tent caterpillar outbreak by cropping (short dashed line) or to initiate outbreaks by introducing caterpillars into new areas (long dashed lines). Despite the continuing availability of food, the manipulated populations followed the trajectory of surrounding populations (controls, solid line). (*Adapted from Myers 1993*)

Myers suggests that these circumstances are consistent with the action of a baculovirus, which is ingested with the foliage on which the caterpillars feed. When the viral coat is digested, virions penetrate the insect's gut and replicate in cells. Death of the caterpillar follows and results in the release of millions of new virus particles onto the bark and leaves. The active morphs can disperse because they are not debilitated by the infection; they increase in density until the accumulation of virus particles in the caterpillar population eventually leads to a decline. The movement of virus particles, carried by dispersing caterpillars, explains the synchrony of outbreaks and the response of caterpillar populations to Myers's experimental manipulations.

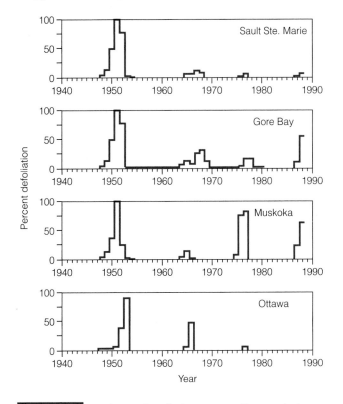

FIGURE 5.15 Synchrony of cyclical tent caterpillar populations in different regions of Ontario. The hypothesis that exhaustion of food supply (defoliation) triggers population collapse is not supported, because collapses often occur after only partial defoliation. (*Adapted from Myers 1993*)

The Intrinsic School

The second major school of population regulation is the intrinsic school, so named because it is based on mechanisms intrinsic to the population being regulated. In other words, the population is *self-regulated*. Over the years, a number of hypotheses have been developed within this framework. Here we will consider stress, territoriality, the genetic polymorphism hypothesis, and dispersal.

Stress and Territoriality   J. J. Christian and D. E. Davis (1971) proposed the social stress hypothesis, in which some mammalian populations show density-dependent effects on reproduction as a result of pathological effects of crowding. Increased population density leads to a large number of agonistic interactions among individuals, which stimulates hypertrophy of the adrenal glands. This pathological condition affects the production of reproductive hormones such that the reproductive rate decreases. Even though there are data to support this hypothesis, it may not apply over a wide range of taxa. Moreover, it is difficult to obtain pertinent field data; sampling hormones in wild animals is difficult, and it is hard to separate the physiological effects of crowding per se from effects associated with resource limitations.

Territorial behavior can regulate population density. When A. Watson and D. Jenkins (1968) artificially removed territorial individuals from a red grouse population to assess the effect of open territories on breeding by other birds, they found that only territory holders bred. Furthermore, overwinter survival of nonterritorial grouse was poor compared to that of territory holders. When territorial grouse were removed from the population, one of two things happened: Either their neighbors increased the size of their territories, or the vacant territory was quickly colonized by formerly nonterritorial grouse. The fact that the number of territory holders was relatively constant each spring (whereas the number of nonterritorial individuals varied) indicates that the ability to obtain and defend a territory effectively regulated grouse population size. Obviously, this intrinsic mechanism is at least indirectly related to an extrinsic biotic factor—the food supply—because a defense of territory is essentially a defense of food resources.

Genetic Polymorphism Hypothesis Another mechanism of intrinsic population regulation depends on genetic changes within individuals as population density changes. This **genetic polymorphism hypothesis,** first proposed by D. Pimentel (1968), suggests that some genetic feedback exists between, for example, a plant and the herbivores that consume it. Because increases in herbivore density constitute a selective force on the plant, genes for resistance to grazing increase in frequency. As a result of this resistance, the rate of grazing decreases. After several such cycles, the number of herbivores stabilizes, and the system attains equilibrium.

Similar feedback loops can occur between a parasite and its host. In attempts to demonstrate such genetic feedback in the laboratory, Pimentel maintained caged populations of the housefly (*Musca domestica*) and a parasitic wasp (*Nasonia vitripennis*). In the experimental cage, host (housefly) density was kept constant and parasite density was allowed to vary. In the control population, the hosts came from a population that had

not been exposed to the parasite. Over a period of 1004 days, the experimental host population became more resistant, and the experimental parasite population became less virulent. Concomitantly, the population density of the parasite in the experimental cage declined relative to parasite density in the control cage (Figure 5.17).

It has also been suggested that some population irruptions involve groups of genetically distinct individuals that are somehow more likely to increase at a high rate. In investigating irruptions of the crown-of-thorns starfish (*Acanthaster planci*) on the Great Barrier Reef in Australia, in which massive numbers of these predatory starfish decimate large areas of the reef, J. A. H. Benzie and J. A. Stoddart (1992) used electrophoresis of proteins to identify genetic loci from starfish from various outbreaks. The resulting genetic data could be used to distinguish between two competing hypotheses:

1. The outbreaks start at one site on the reef, and then starfish larvae genetically predisposed to producing outbreaks drift southward to start numerous secondary outbreaks.
2. Outbreaks have multiple origins, and the pattern of southward movement of outbreaks results from the movement of physical or biological factors.

The first hypothesis predicts that "outbreaking" and "nonoutbreaking" populations should have different levels of genetic differentiation; outbreaking populations should show a high degree of genetic similarity. The second hypothesis predicts that outbreaking and

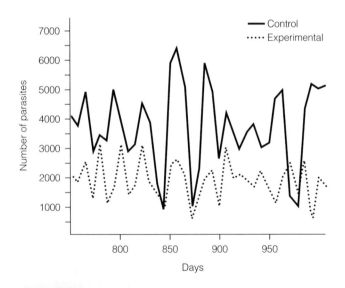

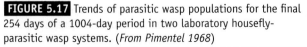

FIGURE 5.17 Trends of parasitic wasp populations for the final 254 days of a 1004-day period in two laboratory housefly-parasitic wasp systems. (*From Pimentel 1968*)

nonoutbreaking populations should be genetically similar.

Benzie and Stoddart analyzed eight different gene loci in starfish from 13 populations. From the allele frequency data they constructed a dendrogram (similar to a phylogenetic tree) based on Nei's D, an index of genetic distance among the populations. In such a diagram, populations with similar allele frequencies cluster together; genetically distinct groups are more distantly connected on the dendrogram. Note in Figure 5.18 that with one exception, the outbreaking populations are genetically very similar to one another and distinct from nonoutbreaking ones, a finding that supports the first hypothesis. Apparently, these starfish irruptions are a result of the southward movement of genetically similar populations.

C. J. Krebs and colleagues (1973) proposed a hypothesis to explain the four-year population cycle of microtines based on a modification of the genetic polymorphism hypothesis originally proposed by D. Chitty (1967). Their hypothesis is based on the empirical observation that the phenotype of voles changes over the course of the cycle; at the peak, voles are larger and more aggressive. Krebs and co-workers (1973) suggest that these phenotypic changes are driven by genetic change in the population over the course of the cycle and that during the high and low phases, the environment differs such that the selective pressures change. Selection causes genetic change—

and hence the phenotypic changes we observe. The cycle persists because of the regular alternation of the selective regimes.

In particular, when the population is low and forage has recovered from the previous cycle, microtines experience favorable conditions for population growth. In this circumstance, selection favors phenotypes that have high growth rates; competitive ability is of secondary importance. The population largely consists of small, nonaggressive, highly prolific individuals. As the population increases, resources become scarce and the selective pressures change. Now, large-bodied competitors are favored; the energy that previously could be devoted to reproduction must now be devoted to competing for food. The result is decreased population growth and the onset of population decline.

As more data have accumulated on the nature of microtine cycles, the Chitty–Krebs hypothesis has been further modified (Krebs et al. 1973); in particular, dispersal has been included in the model (Figure 5.19). At high population density, selection favors an aggressive, nondispersive morph; at low density, a nonaggressive, dispersive morph is favored. When density increases, interference among individuals causes emigration of the remaining dispersive morphs;

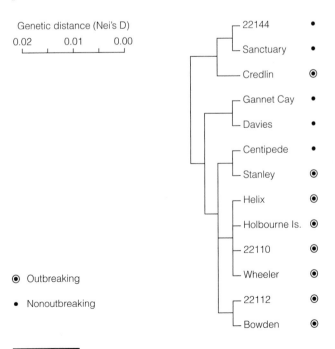

FIGURE 5.18 Dendrogram based on genetic distances among 13 populations of crown-of-thorns starfish (*Acanthaster planci*). Distances between branch points indicate genetic distance. (*From Benzie and Stoddart 1992*)

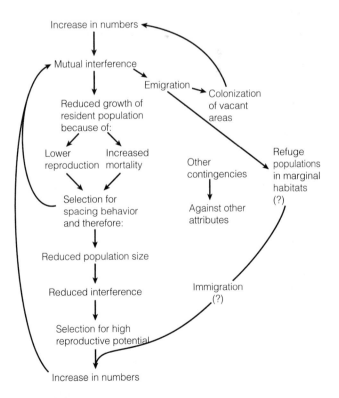

FIGURE 5.19 Schematic of the modified version of the Chitty–Krebs hypothesis to explain the cyclical population fluctuations of microtines. (*From Krebs et al. 1973*)

vacant areas become saturated as the population spreads into marginal habitat, where survival is poor. Selection increases the spacing behavior of individuals, and the population begins to decline. Immigration of dispersers from marginal habitats contributes to population increase toward the next peak.

Even though data that unequivocally demonstrate the operation of this model have been hard to come by, genotypes can indeed change as quickly as required by the model. R. H. Tamarin and Krebs (1969) and M. S. Gaines and Krebs (1971) used electrophoretic markers to demonstrate that significant shifts in allele frequency occur over the course of a population cycle of the meadow vole, *Microtus pennsylvanicus*. The interpretation of these shifts is unclear. They might be linked to other allelic changes that drive the phenotypic changes over the cycle, or perhaps they are simply associated with changes in population size. Other species of voles do not show any similar correspondence between allele frequencies and phase of the population cycle (Krebs 1979).

Dispersal   Up to this point, our discussion has focused almost exclusively on mortality and birth as factors that change population density, and much less on emigration and immigration. Even though external factors may stimulate it, the movement of individuals to and from a population is an intrinsic characteristic of the species, and thus it is appropriate to consider it here. We define **dispersal** as the movement of individuals from their natal area or their current home range. Can dispersal regulate population size?

For many years, population ecologists essentially ignored the role of dispersal as a potential population regulatory factor, at least in part because of the practical difficulties associated with studying it. If an animal disappears from the study site, how do we know whether it died or dispersed? When a new individual appears, was it born there, or did it immigrate from elsewhere? Furthermore, factoring in dispersal complicates the mathematics of population growth. Because of these difficulties, population ecologists generally assumed that emigration was balanced by immigration, resulting in no net population change attributable to dispersal.

In a study of the prairie vole (*Microtus ochrogaster*), Krebs (1969) evaluated the roles of emigration and immigration by comparing the changes in density over time of enclosed and unenclosed populations in the same field (Figure 5.20). Clearly, the density of the unenclosed population was affected by dispersal. W. Z. Lidicker (1986) reported similar findings for the California vole (*M. californicus*).

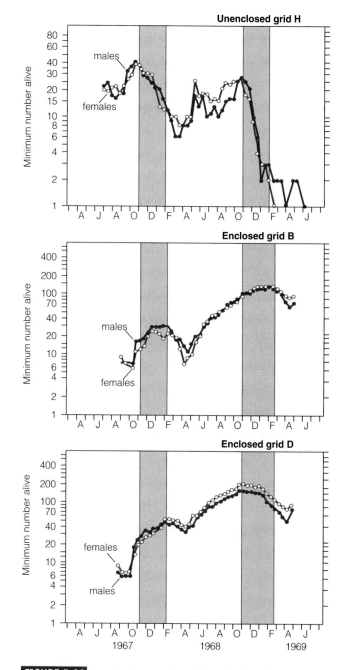

FIGURE 5.20 Density changes in prairie vole (*Microtus ochrogaster*) populations in unfenced field (grid H) and in fenced grids B and D. (*From Krebs 1970*)

Data such as these generated interest in the process of dispersal, and a number of studies on the frequency and effects of dispersal followed. Lidicker (1975) identified two basic forms of dispersal (Figure 5.21). **Saturation dispersal** occurs when the population has reached carrying capacity. The emigration of individuals at this time represents a sort of safety valve that releases the pressure the population exerts on deteriorating resources.

Presaturation dispersal occurs before the carrying capacity is reached. It may have a genetic basis;

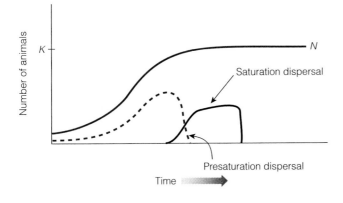

FIGURE 5.21 Relationship between changes in population size (*N*) and saturation and presaturation types of dispersal; *K* refers to carrying capacity. (*Adapted from Lidicker 1975*)

that is, individuals of certain age or sex classes may be programmed to emigrate. The evolution of this behavior suggests that some fitness advantage must accrue to presaturation dispersers. Lidicker suggests that because presaturation dispersers can be expected to be in better condition than saturation dispersers, they may gain in fitness from their ability to locate prime, unexploited habitat or to acquire other resources such as mates. In many species, presaturation dispersers are predominantly of one sex; such sex-biased dispersal might serve to reduce inbreeding in the population because it would make members of both sexes less likely to mate with close relatives (Greenwood 1983).

Both kinds of dispersal can affect population density, but the effects are achieved differently. Saturation dispersal can remove excess individuals at *K*; presaturation dispersal slows the rate at which the population approaches *K*. The key point is that if dispersal is to act as a density-dependent population regulation mechanism, the *rate* of dispersal must increase with increasing density. It may be that saturation dispersal is a manifestation of density-dependent resource limitation as defined by the extrinsic biotic school. Evidence for the hypothesis that presaturation dispersal operates in a density-dependent fashion, and is thus an example of an intrinsic mode of regulation, is not available for enough species to support meaningful conclusions.

Nonequilibrium Theories of Population Regulation

Early in the history of thinking about population regulation, a fundamental controversy arose. On one side were those, such as A. J. Nicholson (1954, 1957), who emphasized density-dependent factors that return populations to equilibrium values. The alternative view was initially advanced by two Australian entomologists, H. G. Andrewartha and L. C. Birch (1954), who emphasized the importance of density-independent factors such as weather. Their view de-emphasizes the concept of an equilibrium population, because the effects of weather are random and fundamentally independent of the population size.

There is currently great interest in the nonequilibrium view of populations (Botkin 1990; Krebs 1992). One reason for this change in emphasis is the accumulation of long-term data sets on population dynamics in very stable environments. Examples of populations that seem to fluctuate at random are now numerous enough to convince many ecologists that some populations do not in fact behave in equilibrium fashion. For example, in a 14-year study of insect populations in undisturbed rain forests in Panama, S. J. Wolda (1992) found that whereas some species show relative stability, others fluctuate wildly. Mean abundance of 22 percent of the species studied had mean annual fluctuations of more than 10 percent; some 4 percent of species fluctuated more than 20 percent per year. Thus, even in stable and undisturbed populations in the tropics, apparently random, large-scale fluctuations are evident.

Examples from a number of disparate taxa and systems lend support to the general importance of nonequilibrium dynamics. For instance, P. Doherty and T. Fowler (1994) report that coral reef fish dynamics are explicable as a nonequilibrium system. For the damselfish (*Pomacentrus mouccensis*) on the Great Barrier Reef in Australia, local abundance is largely determined by the recruitment rate of larvae; recruitment, the addition of new individuals, accounts for more than 90 percent of the variation in density among populations. Density-dependent factors do not compensate for the rate of recruitment among populations; rather, the differences interact with density-independent mortality to account for the local abundance of these fish.

Nonequilibrium theories of regulation are applied to populations that do not appear to return to an equilibrium value related to carrying capacity. Such populations may be controlled by abiotic factors or by a complex set of abiotic and biotic factors that give at least the appearance of random fluctuations.

Abiotic Extrinsic Regulation

The abiotic extrinsic school is the direct descendant of Andrewartha and Birch's ideas about the importance of density-independent factors. Their ideas arose out of their long-term studies of the insect pest *Thrips imaginis*, the apple blossom thrip, in Australia. The populations of this important pest of a number of agriculturally and horticulturally important plants show large-scale irregular fluctuations. After finding that the density of thrips was related to several weather variables, including temperature and rainfall, Andrewartha and Birch used multiple regression to formulate an equation for predicting thrip density in any given year. The equation they generated incorporates several abiotic factors to predict the log number of thrips in the spring of each year:

$$\log y = -2.390 + 0.125a + 0.2019b + 0.1866c + 0.0850d$$

where a is winter temperature experienced by a group of annual plants on which the thrips depend, calculated as (maximal daily temperature 248°F)/2; c is the same variable, but for the spring (September and October in Australia); b is amount of spring rainfall; and d is size of overwintering population.

Figure 5.22 shows that the equation does a good job of predicting the density of thrips in any given year; indeed, the equation accounted for 78 percent of the variance in thrip population size over the time period shown. Andrewartha and Birch contend that the abiotic factors they included in the equation are sufficient to explain the density fluctuations of this species. According to their model, the irregular fluctuations observed were the product of stochastic changes in the physical environment: benign conditions led to high densities, whereas severe weather led to low densities. Although some analyses of these data show evidence of density dependence, Andrewartha and Birch attempted to refute these claims.

The abiotic extrinsic school emphasizes that even though density-dependent effects may occur, density fluctuations can be explained without them. According to this school, the fundamental process of regulation is based on two premises:

1. Most species have a high innate capacity for increase, and thus, under favorable conditions, the population can grow rapidly.
2. The vagaries of weather and other abiotic factors are so great that no population experiences favorable growth periods for very long. Severe conditions always reduce densities before density-dependent factors come into play.

A number of studies, many of them on insects or other invertebrates, have revealed correlations between abiotic factors and population density. It seems logical to assume that small poikilotherms might be particularly vulnerable to abiotic stress; because they cannot internally regulate their body temperatures and because their small size provides little buffer against the environment (see Chapter 3), they are sometimes at the mercy of the weather.

Multiple regression equations can also accurately predict density for some populations of vertebrates. W. J. Francis (1970) measured a series of abiotic factors (and some biotic ones) and developed a model for the California quail (*Lophortyx californicus*). The productivity of quail, expressed as the ratio of juveniles to adults in fall samples, is predicted by the following equation:

$$y = 0.929H + 0.021A - 0.120C - 0.975$$

where H is soil moisture, A is the proportion of females that were adults in the previous fall, and C is total seasonal rainfall. The ratios predicted by this equation very closely correspond with the observed ratios. Almost 99 percent of the variation observed in the natural population during this period is explained by this equation. Natural historical observations indicate that cold, wet weather negatively affects chick survival.

The order of the variables in this equation is their order of importance. Soil moisture accounts for 83.1 percent of the variation in quail productivity. Adding the proportion of females that were adults into the equation explains another 12.4 percent of the variation. Seasonal rainfall accounts for another 3.3 per-

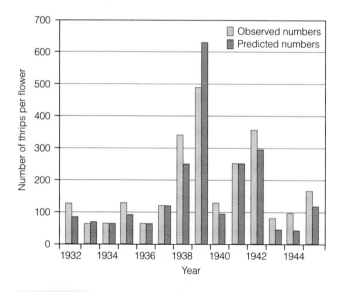

FIGURE 5.22 Comparison of observed annual population densities of thrips (*Thrips imaginis*) and densities predicted by an abiotic extrinsic model. (*Data from Andrewartha and Birch 1954*)

cent. The clear conclusion is that soil moisture, an abiotic factor, is a crucial determinant of quail production.

It seems reasonable to expect that abiotic factors would have the greatest effects in extreme environments such as deserts and the tundra. This is the case for the creosote bush (*Larrea divaricata*) in the Mojave Desert of California (Woddell, Mooney, and Hill 1969). The density of this species is positively correlated with rainfall (Figure 5.23). Apparently this shrub is relatively unaffected by indirect biotic factors, because the density of creosote bush does not decline as the density of competing species increases (Figure 5.24).

P. R. Grant and B. R. Grant (1992) provide another example of the importance of abiotic factors in rather extreme environments with their long-term studies of Darwin's finches on the Galápagos Islands, where rainfall fluctuates wildly. On Isla Daphne Major between 1976 and 1991, the yearly numbers of breeding finches of two species, *Geospiza fortis* and *G. scandens*, closely corresponded to annual amounts of rain, which is needed to produce the seeds on which these birds rely for food (Figure 5.25). Other population parameters also appear to vary with abiotic factors. Depending on the conditions, annual adult mortality varied from zero to 50 percent, and fledgling production ranged from zero to 10 chicks per breeding pair.

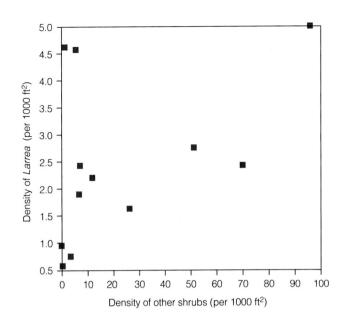

FIGURE 5.24 Density of the desert shrub *Larrea* as a function of density of competing shrubs. (*Data from Woddell, Mooney, and Hill 1969*)

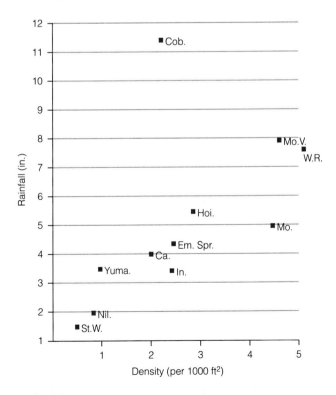

FIGURE 5.23 Scatter diagram demonstrating increasing densities of the desert shrub *Larrea* with increasing rainfall at 11 sites in the Mojave Desert. (*From Woddell, Mooney, and Hill 1969*)

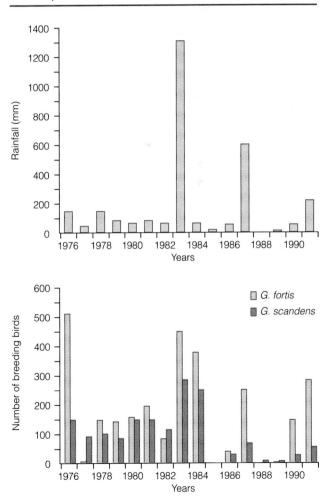

FIGURE 5.25 Annual rainfall (**a**) and numbers of breeding finches, *Geospiza* spp. (**b**) on Isla Daphne Major, 1976–1991. (*Adapted from Grant and Grant 1992*)

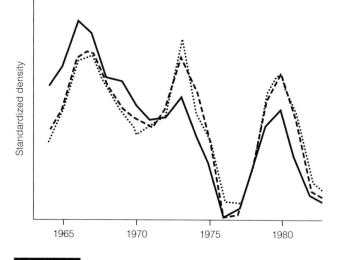

FIGURE 5.26 Synchronous changes in populations of three species of grouse in Finland. Solid lines = Capercaillie; dashed line = black grouse; dotted line = hazel grouse. (*Adapted from Linström et al. 1996*).

Population synchrony on a large spatial scale has been used to argue for the importance of abiotic factors such as climate. Three species of birds, capercaille (*Tetrao urogallus*), black grouse (*T. tetrix*), and the hazel grouse (*Bonasa bonasia*) have highly synchronized populations (Figure 5.26) across 11 provinces of Finland. Linström et al. (1996) conclude that spatial synchrony in climate is an important factor responsible for the synchrony.

Snowshoe hare populations, which cycle with a ten-year period, show remarkable synchrony across northern North America (Sinclair and Gosline 1997). Figure 5.27 shows the percentage of sites in Canada

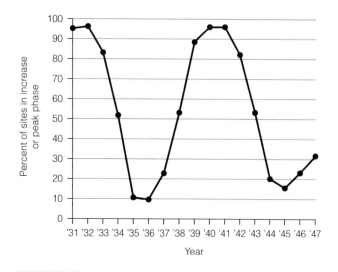

FIGURE 5.27 Synchrony in Canadian populations of snowshoe hares. For each year, the percentage of sites in the increase or peak phase of the hare population cycle is plotted. (*From Sinclair and Gosline 1997*)

south of the treeline in which the population is increasing or at its peak. Note that in the two peak years, 1931 and 1941, there is nearly 100 percent synchrony across this region. Sinclair and Gosline (1997) also show that the hare cycle is strongly associated with high-amplitude sunspot cycles. Although sunspots are unlikely to affect hare population directly, they may entrain climatic cycles that do.

When the boundaries of a population are not so well defined as on islands, the problem of identifying the regulatory factors is much more difficult. When this is the case, it is necessary to consider more than just the local population.

The Concept of the Metapopulation

In Chapter 4 we defined a population as a group of conspecifics inhabiting a specific place. This definition assumes that the boundaries of a population are relatively discrete. This assumption is more likely to be valid when there are sharp discontinuities in the habitat and for a species that is strongly associated with a particular habitat type. Even though we have seen that dispersal is an important demographic process, the definition of *population* that we have been using assumes—is in fact based on—little or no interaction with other groups via dispersal. Recently, considerable interest has developed in a broader, more inclusive concept of population that incorporates both dispersal and spatial variation in habitat type and quality.

R. Levins (1971) introduced the concept of the **metapopulation** as a collection of subpopulations interconnected by dispersal (Figure 5.28). This kind of population structure may apply to groups of populations that inhabit a landscape composed of habitat of varying quality that occurs in discrete patches. The fates of populations depend upon habitat quality. Some patches in optimal habitat produce large populations that may exceed *K*. Some of these individuals disperse to other patches, looking for space and resources. The probability of dispersal from one patch to another depends on the distance between patches and the nature of habitat corridors linking them. Other patches, in suboptimal habitat, may hold smaller populations or occasionally go extinct. These patches are recolonized by immigrants from high-quality habitat. The suboptimal patches rarely (if ever) send dispersers to quality habitat where densities are higher. The concept of metapopulations is applicable to nonequilibrium population dynamics because the nature of equilibrium within any habitat patch is irrelevant to the dynamics of the metapopulation. In

Levins's metapopulation, subpopulations are not in equilibrium. Indeed, many go extinct on a regular basis. The metapopulation persists, not because it achieves a state of equilibrium, but because the interactions among patches prevent extinction of the entire metapopulation (Murdoch 1994). Recall our discussion in Chapter 4 of the definition and nature of the "population." Note the similarity in structure between the metapopulation depicted in Figure 5.28 and the complex set of populations of white-footed mice in Figure 4.2.

The dynamics of a metapopulation depend primarily on both the quality of the habitat supporting the subpopulations and the quantitative aspects of dispersal among subpopulations. Habitat quality and size determine a subpopulation's extinction probability and its carrying capacity; dispersal rate is determined by the species's vagility and the distance separating the subpopulations. The forest herb, *Primula vulgaris*, has a metapopulation structure (Valverde and Silverton, 1997). This species cannot survive in mature forest because of the low light intensities. It colonizes gaps in the canopy of the forest when disturbance or death of a tree creates an opening. These openings form randomly in the forest, then gradually close in (see Chapter 13). The result is a metapopulation, a series of populations increasing or decreasing in gaps of various ages (Figure 5.29). As gaps close, pop-

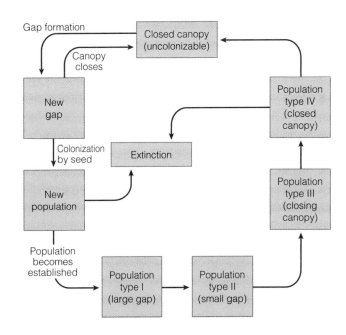

FIGURE 5.29 A metapopulation model for the forest herb, *Primula vulgaris*. Four population types occur in forest canopy gaps of different age. Persistence is lowest in gaps nearly closed. Populations can become established only in new gaps. (*Adapted from Valverde and Silvertown 1997*)

ulations go extinct; new populations are formed when seeds disperse to new gaps.

Metapopulation models are especially applicable for species that inhabit fragmented habitats. To put it another way, wherever human activity has reduced large expenses of continuous habitat to a series of smaller habitat patches separated to various degrees by lower-quality habitat, we may expect to see metapopulation dynamics. One example is the European nuthatch (*Sitta europea*), which inhabits mature deciduous forest, a habitat type that currently is highly fragmented in western Europe (Verboom et al. 1991).

J. Verboom and colleagues found that the distribution of this species is dynamic in space and time. The extinction and colonization frequencies of habitat patches are determined by the carrying capacity and dispersal rates, respectively (Figure 5.30). For the European nuthatch, the connection of the subpopulations via dispersal results in a metapopulation

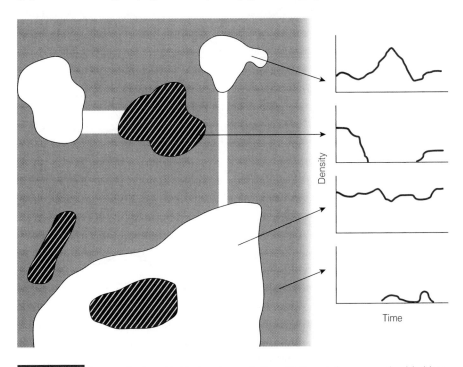

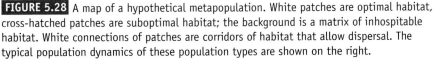

FIGURE 5.28 A map of a hypothetical metapopulation. White patches are optimal habitat, cross-hatched patches are suboptimal habitat; the background is a matrix of inhospitable habitat. White connections of patches are corridors of habitat that allow dispersal. The typical population dynamics of these population types are shown on the right.

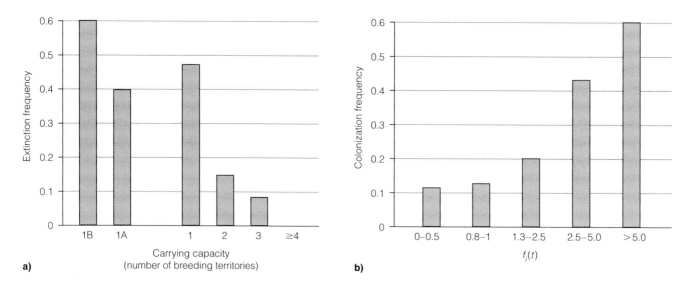

a)

b)

FIGURE 5.30 Characteristics of a European nuthatch (*Sitta europea*) metapopulation. (**a**) Extinction frequency as a function of woodlot carrying capacity (expressed as number of breeding territories). For woodlots containing only a single territory, optimal habitat (A) is distinguished from suboptimal habitat (B). (**b**) Colonization frequency as a function of dispersal influx index $f_i(t)$, which reflects the number of breeding pairs adjacent to empty habitat patches; low values correspond to a high degree of isolation. (*From Verboom et al. 1991*)

that is more persistent than any individual subpopulation.

Lidicker (1985) used the metapopulation concept to propose a multifactorial model explaining the four-year cycles of microtines. According to this model, habitat can be classified into "survival habitat," patches in which survival is possible in any season, and "colonizing habitat," patches that can support animals only during favorable periods. Although thinking in terms of this dichotomy of habitat types is useful in understanding the concepts of the model, in nature some gradation from survival to colonizing habitat is likely to occur. When organisms are prevented from dispersing by physical, biotic, or social barriers, they become so-called *frustrated dispersers*. Frustrated dispersal causes density to rise rapidly and is often followed by a severe crash or local extinction. Populations of several species of *Microtus* exhibit this effect (Lidicker 1985).

The key features of Lidicker's model are shown in Figure 5.31 for an idealized microtine cycle. At low density after a decline (year 2), the population is confined to survival habitat. Some population growth and dispersal to other patches occur during year 3. Because mild conditions permit high survival during the nonbreeding season that year, rapid population growth occurs during the next breeding season (year 4); "dispersal sinks" in colonizing habitat are rapidly filled by dispersers. Additional dispersal becomes frustrated and extreme densities are achieved, followed by a crash. Note that what appears to be a cycle for this

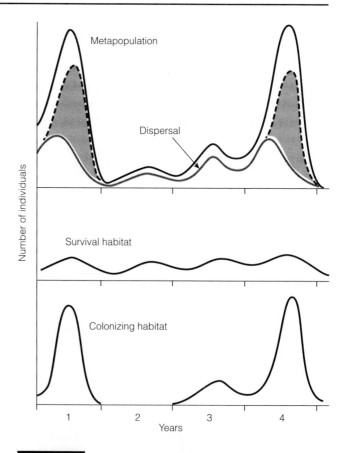

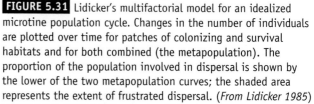

FIGURE 5.31 Lidicker's multifactorial model for an idealized microtine population cycle. Changes in the number of individuals are plotted over time for patches of colonizing and survival habitats and for both combined (the metapopulation). The proportion of the population involved in dispersal is shown by the lower of the two metapopulation curves; the shaded area represents the extent of frustrated dispersal. (*From Lidicker 1985*)

metapopulation is actually a reflection of distinct events that occur in the subpopulations. Specifically, densities remain relatively stable in the survival habitat, but radical fluctuations in numbers characterize the colonizing habitat.

Chaos Theory

Another potential explanation for the dynamics of populations that do not seem to be based on equilibrium processes was advanced by Robert May (1974). His seminal theoretical paper on the dynamics of populations explored the effects of modeling population fluctuations by using nonlinear difference equations. The fundamental equation used in these models is

$$N_{t+1} = N_t + N_t r(1 - N_t/K)$$

May examined the nature of population fluctuations generated by a range of different values of r (rates of increase) in this equation and found that a variety of dynamics resulted, some of which are shown in Figure 5.32. For some values of r, the population quickly achieves an equilibrium from which it no longer deviates; for others, stable cycles appear. The most interesting are those generated by values of r greater than 2.692, for which populations fluctuate unpredictably and without ever repeating a pattern. These patterns are called **chaotic.** Even though chaotic patterns of fluctuation appear to be random, they are not. They are generated by a simple *deterministic* equation; there are no stochastic factors in the model.

The patterns of fluctuation of May's chaotic populations are incredibly sensitive to the initial conditions. A minute change in the starting parameters results in completely different dynamics. This sensitivity to initial conditions is a defining characteristic of chaos.

When May plotted N_{t+1} versus N_t for various values of r, some interesting features of nonlinear dynamics emerged. In one case, the population approached equilibrium (Figure 5.33a). A second population oscillated back and forth (Figure 5.33b), a pattern called a *cyclical attractor.* In a third pattern, which on other plots had appeared to be random, the population was found to move back and forth across a parabola called a *strange attractor* (Figure 5.33c). Thus, apparent randomness was found to have a pattern: chaos.

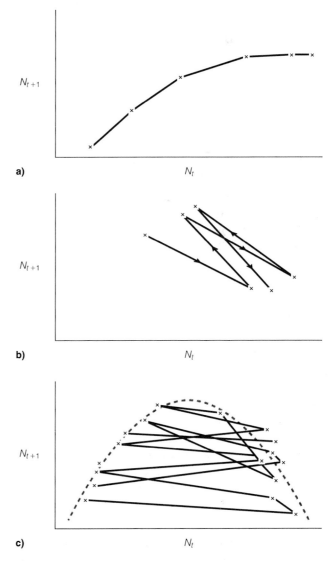

FIGURE 5.33 Plots of N_{t+1} versus N_t for populations with different values of r. (**a**) Approach to equilibrium; (**b**) a cycle; (**c**) chaos.

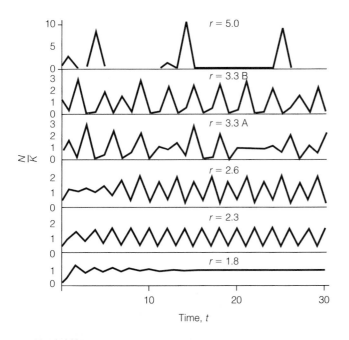

FIGURE 5.32 Population dynamics for six values of r; population density is expressed as a proportion of carrying capacity. (*From May 1974*)

How does this mathematics help us understand population regulation? The patterns of fluctuation that appear random, particularly the large-scale irregular fluctuations (see Figure 5.3), at least superficially resemble chaotic patterns; perhaps some natural populations exhibit complex nonlinear dynamics. If so, attempts to explain their behavior with simple hypotheses, such as those that invoke food supply or predation, will fail.

Ecologists have been slow to apply May's mathematics to natural populations, at least in part because the mathematics of chaos is perceived as formidable. Furthermore, it has not been entirely clear just what field biologists in search of chaos should look for.

In one attempt, M. P. Hassell and colleagues (1976) surveyed 4 laboratory and 24 natural populations of insects for chaotic behavior. Their model included a component to account for competitive interactions. The majority of the species they studied showed a return to equilibrium density; a few showed cyclic dynamics, and only one of the species exhibited chaotic behavior. The cyclic and chaotic dynamics were found primarily in laboratory populations. These investigators suggested that multispecies systems are more likely to have chaotic behavior. Unfortunately, a great deal of detailed empirical information is needed to test these ideas for systems containing several species.

Even though this research dampened the enthusiasm of ecologists for seeking complex nonlinear dynamics in natural populations, theoretical work by F. Takens (1981) and by W. M. Shaffer and M. Kot (1986a, 1986b) suggests some practical approaches to the problem of insufficient data. The application of chaos theory remains an area of interest among many ecologists. Whether it will help explain the behavior of some populations, or whether the conditions for its operation are so unusual that it will prove to be rare, remains to be seen.

A Synthetic View of Regulation

Fifty years of research into the mechanisms of population regulation have demonstrated the complexity of population dynamics. Even a straightforward concept like the carrying capacity, K, turns out to be rather complex. As shown in Figure 5.1, it is a constant, the maximum population that a particular habitat can support. But no reason inherent in the concept determines that K must be constant. Indeed, McLeod (1997) has shown that for environments characterized by a high degree of unpredictable variation, the concept of the carrying capacity does not apply to plant–herbivore dynamics.

Some researchers, such as Wolda (1995), have suggested that "the regulation versus nonregulation controversy should be quietly buried and quickly forgotten." But Murray (1999) argues that "the debate regarding the determination of population size is not going to disappear . . . ." Apparently, there will even be debate about the debate!

Given the complexity of ecological systems, it is not surprising that the debate persists. However, Turchin (1999) has suggested that, in fact, a consensus is emerging among ecologists. Among the points of agreement he outlines are the following:

1. The per capita rate of change, r, is affected by both exogenous and endogenous factors. Exogenous factors are those that change the population but are not affected by the population, such as abiotic factors. Endogenous factors are dynamic feedbacks that affect population numbers.
2. Exogenous factors are not "noise" that simply complicates the identification of endogenous factors. They are important factors that affect population densities.
3. Some negative feedback between r and population density (density dependence) is necessary (but not sufficient) for regulation.

We have seen evidence that supports the two basic schools of thought—the equilibrium and nonequilibrium views. Moreover, within each of these categories, we have seen evidence for a number of different mechanisms that determine population density. How do we reconcile these data and these theories, as well as Turchin's dual assertions about exogenous and endogenous factors? Initially the debate arose because we were searching for *the* mechanism of population regulation. Perhaps, given the tremendous variety of habitats, communities of producers, predators, parasites and pathogens, patterns of abiotic factors, and so on, the search for a single mechanism was naïve.

Consider again the dynamics of moose on Isle Royale (Figure 5.11). Clearly, the dynamics of moose on this island can be explained only as the combined effects of a number of agents (Table 5.2). There have been periods of relative stability and equilibrium; at other times the population was growing without control or crashing. Abiotic and biotic factors play a role. Top-down and bottom-up factors play a role. Chance events and deterministic factors play a role. It is likely that moose on Isle Royale represent an extremely

TABLE 5.2
Summary of the Factors Affecting the Population of Moose on Isle Royale

Abiotic Factors

Fire Initially led to increase in woody browse and increased *K* on Isle Royale. Second fire in 1936 created additional forage (a bottom-up influence on moose).

Isolation Isle Royale is not readily recolonized if moose (or wolves) go extinct. Dispersal to and from the island has low probability.

Snowfall Several winters of high snowfall (1969–1972) led to reduced population growth. Deep snow in 1996 led to starvation.

Biotic factors

Wolf predation Limited the growth rate of moose when numbers where sufficient (a top-down influence on moose).

Succession Reduced *K* as the vegetation of the island matured (a bottom-up influence on moose).

Winter ticks In 1989 a severe outbreak of this parasite led to exceptional moose mortality that year.

Canine parvovirus This disease decimated the wolf population in 1980–1982 and reduced the role of predation in moose population dynamics.

Genetics The very small size of the wolf population has led to significant inbreeding. This may also contribute to the decline of the wolf population and thus to the release of moose from control by wolf predation.

well-studied example of the complex sets of factors that determine population size. Even so, it is a simplified example, because neither moose nor their wolf predators are significantly connected to other populations by dispersal. You can imagine the web of interactions in an extensive metapopulation.

The equilibrium and nonequilibrium mechanisms also interact in important ways. For example, Saether (1997) has shown that in ungulates, stochastic changes in populations associated with changes in the environment are more pronounced when predators are absent. A stable equilibrium between food supply and ungulates is less likely when predators are absent, which leads to a greater role for stochastic environmental effects.

The metapopulation concept also helps synthesize the different schools of thought on population regulation. If the metapopulation is a more biologically meaningful concept than the single, isolated population, we can more easily integrate the various mechanisms of regulation. Subpopulations in different habitat types (separated to varying degrees by inhospitable habitat) and complex patterns of connection by dispersal lead to a system in which a number of regulatory mechanisms and agents may all be operating. The dynamics of the entire metapopulation is the result of these many interacting systems.

It would be easy to view the population regulation debate as naïve now that we see the accumulated evidence for the multitude of agents that affect population dynamics. However, the search for a single mechanism of over-riding importance was a valid one—it is an important goal of science to find general answers to fundamental questions. The kinds of data we have reviewed here led Lawton (1992) to ask rhetorically, "Are there ten million kinds of population dynamics?" As Lawton points out, we can devise a huge number of species-specific and location-specific models of regulation; indeed, each species has a unique set of agents and forces shaping its population dynamics. Still, there are only a few kinds of fundamental mechanisms that we must understand. The challenge is to elucidate their relative importance in specific cases and to continue the search for general laws of regulation.

Population Invasions

Thus far we have considered explanations for the dynamics of established and persistent populations. Early in the history of a population, however, its dynamics may more closely resemble exponential growth than the patterns of fluctuation we described at the beginning of this chapter (see the box on page 000).

One of the most significant effects that humans have on the ecology of populations results from their penchant for moving species to new regions. Sometimes this is done accidentally; at other times it is done intentionally for some expected benefit. The result is that the native biota is invaded by alien species—usually with significant consequences.

When a new species invades a region that contains appropriate habitat, we have the opportunity to observe exponential growth. In 1940 the house finch (*Carpodacus mexicanus*) was intentionally introduced to eastern North America (Veit and Lewis 1996). The species spread rapidly (Figure 5.34a). The population near the release site showed typical sigmoid growth: an exponential phase followed by fluctuations. In this case, the fluctuations were of the small-scale irregular type (Figure 5.34b).

In the course of an invasion, the spatial distribution of a species typically expands greatly. As the population explodes in the vicinity of the introduction, intraspecific competition drives individuals to extend the range of the species, often quite rapidly. For example, after the European starling (*Sturnus vulgaris*) was introduced to Central Park in New York City in 1890, it spread rapidly thereafter (Figure 5.35). By 1954 it had reached the West Coast of the United States and northern Mexico. Its current North American population is over 100 million. Five North American muskrats (*Ondatra zibithecus*) introduced into Czechoslovakia in 1905 spread so rapidly that their population size quickly exceeded one million individuals.

It is not uncommon for invading species to become pests in their new range. One of the most notorious is Klamath weed (*Hypericum perforatum*), which was introduced into California from Eurasia early in the twentieth century. It prefers overgrazed habitat and thus quickly spread throughout the West, and by the end of World War II it covered more than 800,000 hectares. This species is a scourge to ranchers and wildlife managers: It outcompetes native grasses and is mildly toxic to livestock and wildlife.

Australia has a long history of disastrous invasions. One of the most devastating was the introduction of the cactus *Opuntia*. Within a few decades, and until it was finally controlled via introduction of the *Cactoblastis* moth, this species covered many millions of hectares, choking out range grasses and forming impenetrable thickets.

The zebra mussel (*Dreissena polymorpha*) is a recent invader of North America that is still in the process of increasing its range and population densities. Accidentally introduced into the Great Lakes from the bilge of European and Asian ships, its population is growing exponentially in North America. This species can quickly cover all available surfaces and is

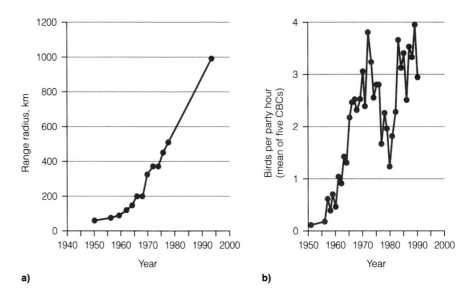

a)   b)

FIGURE 5.34 (**a**) Rapid range expansion of the house finch following its introduction into North America. (**b**) Sigmoid growth of an introduced population of the house finch. Note the exponential phase followed by small-scale, irregular fluctuations. (*From Veit and Lewis 1996*) Bird numbers are measured by an index of relative abundance: the number of birds noted per party of observers per hour in the annual Christmas bird count.

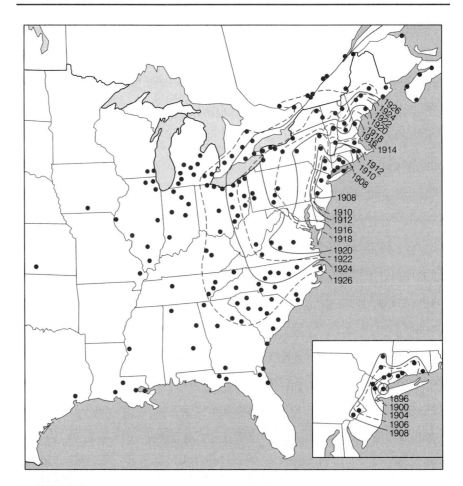

FIGURE 5.35 An example of the rapidity of spread of an introduced species, in this case the European starling (*Sturnus vulgaris*) from its introduction in New York City in 1890 until 1926. Dots outside the 1926 lines are chiefly winter records of pioneer spread. (*From Cooke 1928*)

already causing problems by clogging water intakes in the Great Lakes.

All such invasions have two common features: exponential growth in the vicinity of the introduction and then rapid range expansion as high densities are achieved in the original locale. Clearly, these population explosions occur because a species taken out of its natural system escapes the regulatory mechanisms with which it evolved in its native range. The fact that most such species do not reach these kinds of extreme densities in their natural range suggests that invasions essentially represent an escape from density-dependent regulatory systems.

The technique of **key factor analysis** (Morris 1957; Varley and Gradwell 1960) uses the information in a life table to identify those life cycle stages that are most vulnerable to mortality and thus are crucial to achieving high densities of invaders. We can use this technique to help identify the mortality factors that, if relieved, may lead to extreme densities of economically important invaders such as insect pests.

In the example presented in Table 5.2, the maximum potential offspring of an insect is calculated as the product of the number of reproductive females and the maximum mean fecundity of each. The mortality occurring over a generation, K, (this symbol is *not* the carrying capacity discussed previously) is then calculated from the sum of the mortality rates associated with each stage of the life history:

$$K = k_1 + k_2 + k_3 + k_4 + \cdots$$

We calculate the stage-specific mortalities known as instantaneous mortality coefficients (k_1, etc.) from the log of the number of individuals present at the start and finish of each life history stage:

$$k = \log (N_x) - \log (N_y)$$

where N_x is the number of individuals in a particular life history stage (eggs, larvae, etc.) at the start of the stage, and N_y is the number of individuals at the end of the stage. The values of k are shown in Table 5.3. With this information, we can determine the key factor in the population by visually correlating K with all the k values (Figure 5.36). The factor strongly associated with K is likely to be the key factor. The values of k_1, winter mortality, are highly correlated with the value of K. We conclude that winter mortality contributes more than any other factor to the overall mortality rate. We can then ask other questions about how that factor regulates the population, such as whether it operates in a density-dependent fashion.

TABLE 5.3

Summary of Natality and Mortality Factors Operating over One Insect Generation

	Number per 10 m²	Log Number per 10 m²	k
Potential natality (k_0)	1500	3.176	0.076
Eggs laid (k_1)	1260	3.100	0.171
Eggs hatching (k_2)	850	2.929	0.498
Third stage, larvae (k_3)	270	2.230	0.201
Larvae surviving parasitism (k_4)	170	2.230	0.276
Pupae surviving winter (k_5)	90	1.954	0.301
Adults emerging (k_6)	45	1.653	0.352
Adults reproducing	20	1.301	

From Southwood 1966.

This kind of analysis is valuable because it allows us to determine which life stages are most important in determining the density in a given year. It also enables us to understand something of the nature of the crucial mortality factors, in particular whether they are density-dependent and thus potentially regulating. This kind of information is especially useful if we are attempting to control outbreaks of invading insect pests, because knowledge of the conditions that might lead to an irruption can help us decide which stages to target with control measures.

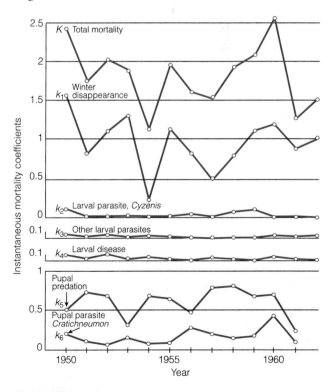

FIGURE 5.36 Plots of log N per 10 m² for the visual correlation of various k values with K. (*From Southwood 1966*)

Human Demographic History

Primitive humans were hunter-gatherers who lived in small, nomadic bands. The development of agriculture, beginning about 10,000 years ago (Figure 5B.1), led to the first major expansion of the human population. Still, the human population experienced a long lag phase and did not reach 200 million until approximately the advent of the Christian Era. A gradual increase continued, with some slight downturns associated with the plagues of the Middle Ages.

Only in the last 200–300 years has the human population entered the log phase of growth, where the population is growing exponentially. Most demographers associate this increase with the Industrial Revolution. During this time, agricultural production increased greatly, improved sanitation reduced the prevalence of infectious diseases, and modern medicine contributed significantly to a decrease in the mortality rate. During this period of exponential growth, the doubling time of the human population steadily declined to about 40 years.

Table 5B.1 shows that the doubling times vary considerably among countries, because different populations fall into different phases of a distinctive pattern of human demographic history (Figure 5B.2). Note that early in the history of a society, both birth rates and death rates are high; such a preindustrial population does not grow significantly because of the high mortality rate. As a country or society develops, however, the death rate declines while the birth rate remains high, producing rapid growth. This is what occurred in the Western world as a result of the Industrial Revolution.

As countries develop further, the birth rate begins to decline, for a number of reasons. In an industrial society, large families are not so important as they are in an agricultural society, where children are needed to help raise crops. In addition, more education and greater employment opportunities for women tend to decrease the birth rate. The decline in the birth rate to match the death rate is called the **demographic transition.** When this occurs, the total fertility rate declines, and the growth rate of the population declines and may even become negative.

Less developed countries experience tremendous population pressure because they have not yet undergone the demographic transition; until that point, the total growth rate of the population can be tremendous. A further challenge is that the industrialization that leads to the decline of the birth rate in the demographic transition is also associated with significant environmental effects from pollution and resource depletion.

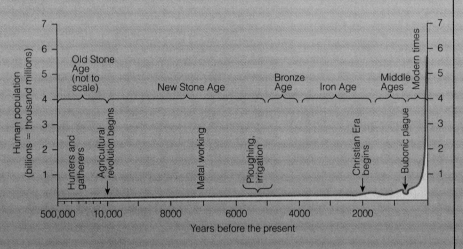

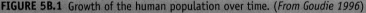

FIGURE 5B.1 Growth of the human population over time. (*From Goudie 1996*)

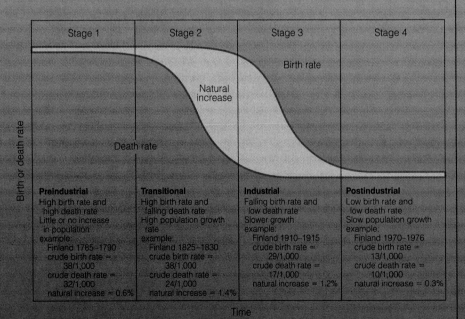

FIGURE 5B.2
The demographic transition. As a country develops, its death rate declines first, leading to a period of rapid population growth (shaded region). The demographic transition is complete when the birth rate declines to meet the death rate.

Human Demographic History, continued

Basic Demographic Data, Selected Countries

	Population Estimate, Mid-1991 (millions)	Crude Birth Rate	Crude Death Rate	Annual Natural Increase (%)	Population Doubling Time at Current Rate (years)	Population Projection, 2010 (millions)	Population Projection, 2020 (millions)
World	5384	27	9	1.7	40	7189	8645
More developed countries	1219	14	9	0.5	137	1345	1412
Less developed countries	4165	30	9	2.1	33	5844	7234
Less developed countries, excluding China	3014	34	10	2.3	30	4424	5643
Afghanistan	16.6	48	22	2.6	27	32.7	44.5
Australia	17.5	15	7	0.8	88	18.7	21.9
Bangladesh	116.6	37	13	2.4	28	176.6	226.4
Bolivia	7.5	38	12	2.6	27	11.4	14.3
China	1151.3	21	7	1.4	48	1420.3	1590.8
Denmark	5.1	12	12	0.0	1732	5.1	4.8
El Salvador	5.4	35	8	2.8	25	7.6	9.4
Haiti	6.3	45	16	2.9	24	9.4	12.3
Hungary	10.4	12	14	−0.2	—	10.5	10.1
India	859.2	31	10	2.0	34	1157.8	1365.5
Italy	57.7	10	9	0.1	1155	55.9	52.3
Japan	123.8	10	7	0.3	210	135.8	134.6
Kenya	25.2	46	7	3.8	18	45.5	63.2
Mexico	85.7	29	6	2.3	30	119.5	143.3
Peru	22.0	31	8	2.3	30	31.0	37.4
Sierra Leone	4.3	48	22	2.7	26	7.0	10.0
Sweden	8.5	13	11	0.2	367	8.3	8.0
Thailand	58.8	20	7	1.3	53	70.7	78.1
Turkey	58.5	30	8	2.2	32	83.4	102.7
United States	252.8	17	9	0.8	88	299.0	333.7
Former Soviet Union	292.0	18	10	0.8	91	333.0	363.0
Yemen	10.1	51	16	3.5	20	19.0	29.9
Zaire	37.8	46	14	3.1	22	67.5	101.1

Note: Figures for growth rate are sometimes rounded and thus do not exactly equal (birth rate − death rate).

From Population Reference Bureau, 1991. World Population Data Sheet.

Extinction and Risk Analysis

We saw in Chapter 4 that if R_0, the number of female offspring a female is expected to produce in her lifetime, is less than 1.0, then on average the females are not replacing themselves. R_0 values persistently less than 1.0 put the population at risk of extinction.

Even though humans have artificially raised the extinction rate for populations and entire species, it is important to remember that extinction is also an important natural process. Indeed, it is a major component of the metapopulation model of populations. As we will see in subsequent chapters, a number of community processes are driven by natural extinction of populations. The succession of a plant community after disturbance is essentially a series of local extinction events. As we will see in Chapter 12, the major mechanism determining species diversity on islands depends on extinction.

Obviously, a population is vulnerable to extinction if its mortality rate exceeds its birth rate for very long. Thus extinction can result from either a major decrease in the reproductive rate or an increase in the mortality rate, or both. The ultimate causes of such changes can be many. Foremost among them is loss of habitat, which may operate on both mortality and reproduction. Without suitable living space, individuals may not breed, and loss of habitat may make individuals more vulnerable to predation.

Our models of population growth assert that typically, the growth rate of a population is high at low density. Recall from Chapter 4 that populations at low density show exponential population growth, which slows as K (carrying capacity) is approached. However, some species behave differently: At very low density, the growth rate actually becomes negative, a phenomenon called the **Allee effect.** A population showing the Allee effect is depicted in Figure 5.37. In this figure, density increases to the right on the horizontal axis. As a population increases and decreases in size, it moves right or left along that axis, and its growth rate varies accordingly. Note that at low N, the growth rate is negative. For a population like this, extinction is likely when the population falls below some threshold low density, because at that point it will have a negative growth rate.

Why should this occur? Generally, social factors are implicated. For example, it may be that a certain minimum number of individuals must be present in the population if breeding is to occur. If such a "critical mass" is required to stimulate reproductive behavior,

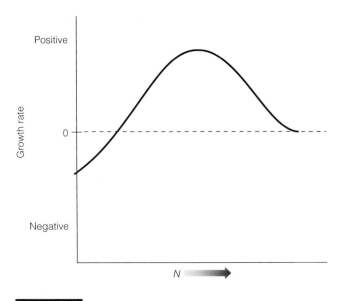

FIGURE 5.37 The Allee effect, in which population growth rate is negative at low population densities (N).

then the population might fail to reproduce at low densities.

In the case of terns, for example, small populations reproduce less successfully (proportionally) than large ones (Allee et al. 1949). Similarly, very small populations of the herring gull (*Larus argentatus*) do not breed when the population reaches small numbers (Darling 1938). The same phenomenon has been reported in the yellow-headed blackbird (*Xanthocephalus xanthocephalus*) by Fautin (1941). A number of mechanisms might cause the Allee effect. Allee and colleagues (1949) attributed it to the inability of small mobs of birds to drive off predatory species. However, few experimental studies of the actual process involved have been done.

With increasing frequency, population ecologists are called upon to make recommendations concerning management issues, particularly with respect to threatened or endangered species. Obviously, the topics we have discussed in this chapter bear directly on the probability that a population of interest will persist. Population regulation mechanisms are studied not only to determine the risk of extinction, but also to identify the parameters to which the probability of extinction is most sensitive. Our discussion of the complex nature of population regulation and the controversies surrounding the relevance of equilibrium and nonequilibrium models should alert you that all these models are difficult to develop.

The most important anthropogenic factor placing populations of plants and animals at risk of extinction is habitat loss. Three important consequences follow from the loss of habitat: (1) Decreased population size leads to a higher probability that demographic accidents will cause extinction. (2) Habitat alteration may result in fragmentation of large tracts of habitat into a series of smaller habitat islands or fragments. (3) Small populations tend to have lower genetic variation and higher levels of homozygosity than large populations. For the rest of this chapter, we will discuss these consequences of habitat loss.

Demographic Accidents

As total population size decreases, the risk of accidental extinction increases (Lande 1993); small populations are more vulnerable to catastrophic declines and extinction than are large ones. This is simply a result of the fact that chance events have a greater impact in small populations. A particularly severe winter, an influx of predators, or an epidemic can easily wipe out a small population. In addition, "demographic accidents" are more likely and more profound in their consequences. For example, if the sex ratio of a small population becomes skewed by chance, the population may go extinct because of a lack of mates for the remaining individuals. As previously noted, these kinds of events are assumed to occur with regularity in species with a metapopulation structure.

Stochastic effects such as these probably played a role in the extinction of the heath hen (*Tympanuchus cupido cupido*) on Martha's Vineyard (Figure 5.38). Overhunting caused a major decline in the population until 1907, when management regulations were imposed. Although the population increased there-

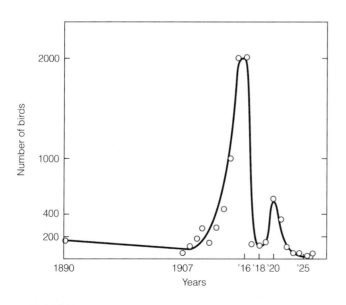

FIGURE 5.38 Population changes of heath hens (*Tympanuchus cupido cupido*) on Martha's Vineyard, Massachusetts. (*From Allee et al. 1949*)

after, it still attained only modest size. In 1916 a fire, a major storm, a cold winter, and an invasion of goshawks (*Accipiter gentilis*) combined to reduce the population to only 50 pairs. In subsequent years, the sex ratio was skewed toward males, further contributing to the decline (Gross 1928). By 1932 the population was extinct. Even though the cause of the initial decline was overhunting, stochastic effects that occurred when the population was small pushed it to extinction.

Habitat Fragmentation

Anthropogenic habitat changes often result in a mosaic landscape of habitat patches (fragments) surrounded by regions of more or less inhospitable habitat. Fragmentation can be expected to affect populations in one of two ways. First, a decline in total area of suitable habitat can result in smaller populations, for which demographic accidents may assume a greater role. Second, the formation of habitat islands results in effects on population size that are related to the size of the fragments and their degree of isolation. Andren (1994) reviewed well-studied cases of fragmentation and found that when the landscape contained more than 30 percent suitable habitat, the main effects on population sizes were due directly to loss of habitat; that is, the total population declined in proportion to the amount of lost habitat. If less than 30 percent of the habitat remained, however, population losses were greater than those attributable to habitat loss alone. The magnitude of the effects was more pronounced when very small, isolated fragments remained. We will discuss size and isolation effects of habitat islands in Chapter 12.

One effect of fragmentation is the formation of a regional metapopulation whose subpopulations may be at high risk of extinction. The vulnerability of the entire metapopulation to extinction depends on the probabilities that dispersal from other subpopulations will "rescue" extinct regions.

A metapopulation structure can lead to the persistence of species living in fragmented landscapes if four conditions can be shown to hold (Hanski et al. 1995). First, the habitat patches must be of sufficient size and quality that they can support breeding populations. Second, no single population can be large enough to ensure long-term survival. (If this were so, a metapopulation structure would not exist.) Third, the patches must be sufficiently connected via dispersal, either through corridors of habitat or by virtue of the vagility of the species, such that extinct patches can be recolonized. Finally, the patterns of local population dynamics must be asynchronous, ensuring that the simultaneous extinction of all patches is improbable. Models for estimating the probability of long-term survival of species with metapopulation dynamics must account for all four parameters.

Metapopulation models have been used to identify the risk of extinction for a number of endangered or threatened species, including the spotted owl (*Strix occidentalis;* Figure 5.39). In addition to the well-known controversy surrounding the survival of this species in the Pacific Northwest, where logging of old-growth forest has placed the species on the threatened species list, other populations are at risk as well. W. S. Lahaye and co-workers (1994) examined the risks faced by the California subspecies (*S. occidentalis occidentalis*), which, like its northern counterpart, is associated with late-seral-stage forests. In southern California, its habitat has been highly fragmented (Figure 5.40). In developing a metapopulation model for this species based on demographic data obtained from marked birds, these researchers found that some subpopulations are declining dramatically. If this demographic trend is

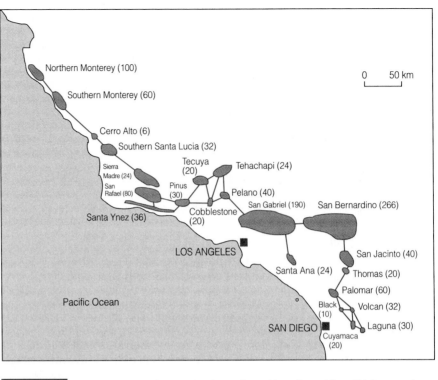

FIGURE 5.40 Metapopulation of spotted owls (*Strix occidentalis occidentalis*) in central and southern California. Lines indicate possible dispersal routes; numbers in parentheses are estimated carrying capacities. (*From Lahaye et al. 1994*)

FIGURE 5.39 Populations of the spotted owl (*Strix occidentalis*) are threatened in many parts of the species's range. (*Photo by Joe McDonald/Animals Animals*)

occurring in the other subpopulations as well, the metapopulation has a very high risk of extinction in the next 30–40 years. The exact nature of the potential decline (and thus the risk of extinction) depends on several demographic features of the model, which can predict the probabilities of extinction under purely deterministic conditions and if the environment fluctuates.

Another species whose population dynamics can be modeled in this way is the peregrine falcon (*Falco peregrinus anatum;* Figure 5.41). J. T. Wooten and D. A. Bell (1992) developed a model of this species in California, using a matrix approach (see Chapter 4). They modeled the dynamics of the population as a single unstructured population and by incorporating spatial (metapopulation) structure. Note that their metapopulation model, schematically depicted in Figure 5.42, incorporates the kinds of demographic features we have identified as important determinants of population dynamics: age-specific survival, age-specific fecundity, and dispersal between subpopulations.

Their metapopulation model indicates that the population cannot sustain itself without continued introductions of captive-reared birds. The predicted dynamics of the population depends heavily on the parameters incorporated in the model—specifically, on whether or not the population is structured and on the number of captive-bred birds released to sup-

FIGURE 5.41 Populations of the peregrine falcon (*Falco peregrinus anatum*) can be modeled as metapopulations. (*Photo by Darek Karp/Animals Animals*)

plement the population. One of the advantages of such a model is that we can identify workable management options by modifying the model. For example, Wooten and Bell found that enhancing adult survivorship is more effective than enhancing fledgling success in achieving a viable population. The fact that the model for a single unstructured population results in greater falcon abundance suggests that managers should concentrate on the healthier, northern California populations.

Genetic Risks to Small Populations

We know that small populations face some genetic risks as well (see Chapter 2). As population size declines, the importance of genetic drift and inbreeding increases; both processes increase the homozygosity of the population. Small populations in which drift has been operating for some time tend to have lower genetic variation than larger populations.

The theory of population genetics predicts just how homozygosity and genetic variation change as populations shrink. We know that increased homozygosity may have detrimental fitness effects as deleterious recessive alleles come together. In addition, we expect that populations with lower genetic variation may have lower fitness if the environment (and hence the selection pressures) changes. Of course, the actual effects we observe will vary according to the genetic and ecological attributes of the species.

M. S. Heschel and K. N. Paige (1995) demonstrated the effects of these kinds of genetic changes for populations of the scarlet gilia, *Ipomopsis aggregata*, a biennial or perennial herb found in montane regions of the West. Heschel and Paige located 10 populations with a wide range of densities—from 12 individuals to more than 4,500—and studied the fitness characters of the individuals from this wide range of population sizes. They found that seed size and germination success were significantly lower in the small populations. In addition, the small populations were more susceptible to stress from simulated herbivory. When new alleles were artificially introduced into the small populations, these effects were reversed.

As we have seen, the control and regulation of populations involve complex phenomena that depend on a number of features of the species and its environment. This chapter has focused on variation in the environment and its effect on regulation of numbers within a species. In the next chapter we consider other ways in which the spatial variation in the environment affects species.

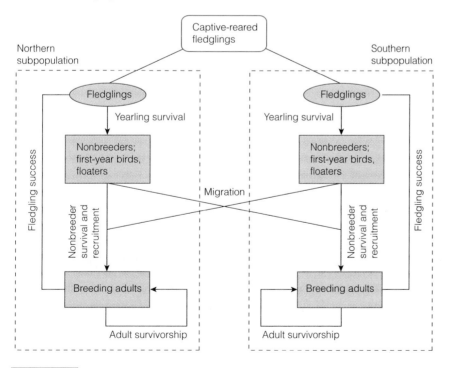

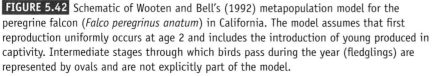

FIGURE 5.42 Schematic of Wooten and Bell's (1992) metapopulation model for the peregrine falcon (*Falco peregrinus anatum*) in California. The model assumes that first reproduction uniformly occurs at age 2 and includes the introduction of young produced in captivity. Intermediate stages through which birds pass during the year (fledglings) are represented by ovals and are not explicitly part of the model.

SUMMARY

1. Population growth is ultimately limited by resources. Consequently, populations cannot grow indefinitely. Eventually, the mortality rate will increase and the growth rate will decrease. Some populations achieve an equilibrium density near the carrying capacity.

2. Most populations show one of four patterns of fluctuation: small-magnitude irregular fluctuations, large-magnitude irregular fluctuations, cycles, or irruptions. It is the goal of population regulation theory to explain these patterns of fluctuation.

3. Equilibrium theories of regulation assume that populations will eventually attain an equilibrium value. Equilibrium theories of regulation are based on the concept of density-dependent changes in birth and death rates that cause the population to approach its equilibrium value.

4. Equilibrium regulation may occur through the effects of extrinsic biological factors such as food supply, predation, and disease, which can operate in a density-dependent fashion. It may also operate via intrinsic mechanisms, including density-dependent physiological and genetic factors and the dispersal behavior of individuals.

5. Nonequilibrium theories of population regulation assume that populations are rarely at equilibrium because a number of density-independent stochastic factors affect the population. Populations that are strongly affected by extrinsic abiotic factors such as weather may fluctuate stochastically.

6. Populations entering new habitat may expand rapidly. Invasions of new or introduced species are often characterized by exponential growth in the new habitat, by range expansion, and by extremely high densities in the absence of the species's natural predators, competitors, or parasites and pathogens.

7. When populations reach very low density for natural or anthropogenic reasons, they may be at risk of extinction. As the population declines to small size, stochastic changes in its composition may result in further declines. In addition, small populations face genetic problems as a result of the increased effects of inbreeding and genetic drift.

SELF-ASSESSMENT: CAN YOU ...?

1. Use material from Chapters 4 and 5 to explain the complexities associated with the concept of the population.

2. List and explain all the important mechanisms of population regulation.

3. Discuss our current understanding of population regulation, given these many potential mechanisms.

4. Relate the nature of population regulation to the concepts of proximate and ultimate factors.

5. For each of the four major patterns of population fluctuation, present a set of regulatory mechanisms that could explain it.

6. Discuss the mathematical representation of the growth of an invading population.

PROBLEMS AND STUDY QUESTIONS

1. Plot the following data to determine the nature of the population fluctuation. Does it matter whether you use a log or an arithmetic scale?

Year	Number	Year	Number
1900	3200	1904	4234
1901	2765	1905	5923
1902	5789	1906	4190
1903	3459	1907	9491

Year	Number	Year	Number
1908	13,999	1914	7913
1909	15,993	1915	5014
1910	132,000	1916	3908
1911	89,000	1917	2301
1912	78,992	1918	3290
1913	19,023	1919	9834

2. The concept of the metapopulation was presented as part of the discussion of nonequilibrium theories. Is it possible for this model to apply to equilibrium regulation? Explain how this might work.

3. Which of the theories of population regulation are mutually exclusive?

4. Explain the difference between an irruption and an invasion.

5. Outline the adaptive advantages that dispersers might realize. Are these advantages different for presaturation and saturation dispersal?

6. How can we tell whether a population achieves population equilibrium? What data would be necessary for this determination?

7. For equilibrium theories of regulation, is it necessary for K to be constant? Discuss the importance of the effect of changes in K for understanding population regulation.

8. The following data represent the effect of a particular factor on mortality. Is this a density-dependent or a density-independent factor? What kind of biological situation could result in data like these?

Density (per hectare)	Mortality Rate
1,092	.102
3,910	.117
4,513	.097
6,812	.243
7,134	.289
9,456	.456
13,890	.409
15,990	.512
16,549	.475

9. Discuss the following statement: Population regulation is a characteristic of the species; it is not a characteristic of the environment.

10. Choose one of the theories of population regulation. Design an experiment to test this hypothesis on a natural population. What predictions would the hypothesis make? What data would be necessary to test the predictions? What result would lead you to reject the hypothesis?

PROJECTS AND ADDITIONAL STUDY

1. Use the Internet to search for data on the population dynamics, regulation, and management of the following species:

snow goose

mallard

northern pintail

black duck

Compare and contrast their population trends and the essential regulatory features of each.

2. Prepare an analysis of the following statement: "Large ungulate populations are regulated and limited by predators."

3. Perform a literature search and analysis of the role of genetics and inbreeding in the status and management of plants and animals. Pay particular attention to the differences between plants and animals.

Intraspecific Variation

e begin with two simple observations of the natural world. First, the physical environment is neither homogeneous nor constant; the abiotic factors important to organisms vary both spatially (geographically) and temporally. Second, there is variation among individuals and populations of living things. These two observations seem almost trivially obvious, yet they are profoundly important to ecology. Together they represent an important aspect of evolutionary ecology.

In Chapter 2 we outlined the primary processes by which populations evolve—the fundamental mechanisms of change in allele frequency. Implicit in that discussion was the assumption that within the population there is genetic variation on which the mechanisms can act. If all individuals in the population have exactly the same genotype, there is no opportunity for selection to change the population. Similarly, drift cannot cause changes in allele frequency if only one allele is present. Thus the amount of genetic variation in a population is a crucial genetic parameter: It determines the possible adaptive responses to the ecological situation faced by the organism. These responses in turn affect the ecological relationships in which the organism participates.

Until recently, direct measurement of the amount of genetic variation in natural populations was difficult. In the past 20 years, however, the revolution in molecular biology that has transformed much of biology has also had an impact on the study of ecology and evolution. By applying molecular techniques, we can now study the interactions between ecological processes and their genetic and evolutionary bases. In addition, these techniques allow ecologists to address practical questions about the preservation of biodiversity. As we will see, for endangered species and for very small populations, the relationship between genetics and ecology is particularly relevant to making management decisions and designing recovery programs.

As we will see in this chapter, the genetics and ecology of an organism are so intimately related that disentangling cause and effect is sometimes difficult. Are the genetic patterns a result of the ecology of the species, or is the ecology of the organism a consequence of its genetic structure? We will see examples of both relationships. We begin with a discussion of the genetic mechanisms affecting variation and the techniques that allow ecologists to measure it. We then proceed to a discussion of the ecological factors that interact with population genetics.

Genetic Mechanisms That Affect Variation

Our discussion of evolution in Chapter 2 began with examples of continuous variation in natural populations. Figure 2.1 showed the frequency distributions of four phenotypic characters: litter size (mice), growth rate (mice), bristle number (*Drosophila*), and eye facets in the bar mutant (*Drosophila*). The three fundamental forms of selection—directional, stabilizing, and disruptive—all begin with phenotypic variation. What is the source of this kind of variation? According to the central dogma of molecular biology (Figure 6.1), the phenotype of an organism is derived from its genotype. As this diagram makes clear, phenotypic variation is derived from differences in the genetic material. Ultimately, genetic difference is derived from mutation of the DNA.

Mutations affect the phenotype in two ways: by an altered gene product or by a change in the regulation of other genes. Changes in the nucleotide sequence of the DNA can have either effect, depending on whether a structural gene, which is responsible for a protein product, or a regulatory gene, which directs the rate at which structural genes are transcribed and translated, is altered. Changes in regulatory genes alter the phenotype by changing the patterns or rates of gene expression and thus the developmental pattern of the organism.

Mutations may also affect the structure of the chromosomes. Occasionally, random breaks and reorganization of the chromosomes occur. These mutations can have profound effects on the phenotype, because the physical location of a gene plays a role in its regulation.

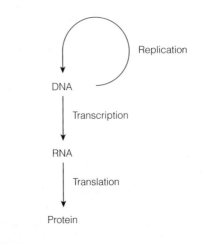

FIGURE 6.1 The central dogma of molecular biology.

Other genetic processes also contribute to the generation of new phenotypes in the population. Recombination results in new combinations of genes. The alternative alleles must be derived from mutation, but recombination shuffles them into new associations. Crossing over between homologous chromosome pairs also generates new gene combinations.

In Chapter 2 we argued that certain phenotypes are advantageous in a particular environment and leave more offspring. Gradually, those phenotypes (and the genotypes that produce them) increase in frequency. We might expect that over time, all genetic variation in the population would eventually disappear as selection weeded out inappropriate genotypes. This is not the case, however, because other mechanisms maintain genetic variation. We discussed one of these in Chapter 2, where we saw that a deleterious recessive allele persists in the population, despite selection against it, because the allele is protected in the heterozygote class. Thus diploidy and dominance maintain variation.

A second mechanism that maintains variation is the phenomenon of *heterozygote advantage,* or *heterosis.* If the heterozygote has higher fitness than either homozygote, the heterozygote class protects deleterious recessives from elimination by selection. The classic example of this phenomenon is the sickle-cell trait. At one of the loci that controls the production of hemoglobin, the dominant allele (*S*) produces "normal" hemoglobin. The recessive (*s*) is highly deleterious, however. Individuals that are *ss* have structurally altered hemoglobin that causes characteristic sickle-shaped red blood cells. These individuals suffer from a number of afflictions, including poor oxygen-carrying capacity of the blood, and they normally have short life expectancies. Although one would expect selection to rid the population of the *s* allele quickly, the heterozygotes (*Ss*) are more resistant to malaria than either homozygote. Consequently, in regions where malaria is endemic, heterosis ensures the maintenance of both alleles.

Genetic drift has the effect of decreasing the total amount of genetic variation in a population. Its primary effect is random change in allele frequencies. If drift proceeds long enough, loss or fixation of alleles is the inevitable result. If we begin with a normal distribution of allele frequencies, eventually the distribution will become perfectly bimodal (see Figure 2.3).

As we know, drift has more pronounced effects in small populations than in large ones (see Figure 2.2). Population geneticists use the concept of *effective population size* (N_e) to describe the effect of numbers on drift. The effective population size is the number of individuals, a subset of the total population, that mate randomly. If only a subset of the total population actually breeds, the *effective number* of individuals in the population is less than the total population size. Skews in the sex ratio and variance in the reproductive success of individuals, among other factors, lower the effective population size. Effective population sizes for populations typically range from a few tens of individuals to a few thousand.

Inbreeding also affects variation. The Hardy–Weinberg law (Chapter 2) assumes that mating is random. In small populations, this may be the case, but because of the small number of individuals, relatives may mate with high frequency. When relatives mate, the probability increases that an individual inherits two alleles that are **identical by descent**—that is, derived from the same copy of the allele in a past generation. The probability of an offspring receiving two alleles identical by descent is measured by the inbreeding coefficient, *F.*

The major genetic effect of inbreeding is an increase in homozygosity. The genotype frequencies of a population in Hardy–Weinberg equilibrium are given by the equation

$$p^2 + 2pq + q^2 = 1.0$$

If inbreeding is occurring, the genotype frequencies are described by the equation

$$(p^2 + Fpq) + [2pq(1 - F)] + (q^2 - Fpq) = 1.0$$

Obviously, the frequency of heterozygotes decreases while the frequencies of both homozygotes increase. The frequencies of the alleles do *not* change, however, even though the genotype frequencies change. For example, the homozygous dominant class increases by an amount *Fpq,* and the heterozygote class (containing some dominant alleles) declines by the amount $1 - F$. If we calculate the value of the dominant allele by algebraically reducing that portion of the equation that describes the homozygous dominants and heterozygotes, we see that *p* is unchanged:

$$p^2 + Fpq + (1/2)(2pq)(1 - F)$$
$$= p^2 + Fpq + pq - Fpq = p^2 + pq$$
$$= p(p + q) = p$$

Because the allele frequencies do not change, there is no direct effect on total genetic variation. *Genotypic* diversity decreases, however. Also, because

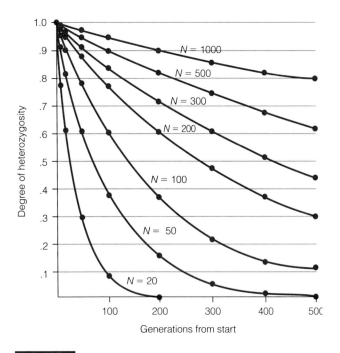

FIGURE 6.2 The proportion of heterozygoisty remaining in populations of different sizes, given the generations of random mating. (*From Strickberger 1990*)

homozygosity increases and heterozygosity decreases, alleles are exposed more directly to selection, and this can lead to a decrease in genotypic variation. In smaller populations, the probability of mating among related individuals increases. The rate of decay of heterozygosity for populations of various sizes is shown in Figure 6.2.

Measuring Genetic Variation Within and Among Populations

Having briefly examined the ways in which genetic mechanisms affect variation, we now consider the ways in which ecologists quantify genetic variation in populations. For the ecologist interested in genetic variation in natural populations, the first concern is the measurement of that variation. Several means of assessing variation are available to the field biologist. This section describes some of the most important.

Morphological Variation

The nature of geographic variation in morphology has been of interest to biologists for a very long time. Indeed, Darwin's theory of natural selection stemmed in part from his observation of visible differences among phenotypes in natural populations. Darwin, and those who followed him, concentrated first on morphological variation—observable differences in the phenotype.

Virtually any aspect of the morphological phenotype can show geographic variation. The scale over which that variation is apparent can vary from a small habitat patch to an entire continent.

If, in the opinion of the taxonomist working on the group, the variation is significant, the variants may be recognized as distinct **subspecies** or **races.** Theoretically, individuals from different subspecies can still interbreed, but each group has responded to local conditions with the evolution of a recognizable (and presumably adaptive) genotype.

The degree to which species are subdivided into races varies greatly. For example, turkey vultures, *Cathartes aura,* are widespread in North America but show little morphological differentiation from place to place. In contrast, the deer mouse, *Peromyscus maniculatus,* is also found throughout most of the continent but exhibits extensive subspecific variation. *Peromyscus maniculatus bairdii* is a grassland subspecies characterized by a short tail and short hind feet. It is very different in appearance from *Peromyscus maniculatus gracilis,* a forest-dwelling race with larger feet and a longer tail (Figure 6.3).

One of the advantages of morphological variation is that its measurement and quantification are relatively straightforward. Moreover, we can readily relate the morphological differences among groups to

FIGURE 6.3 The deer mouse (*Peromyscus maniculatus*) exhibits extensive intraspecific variation. A number of morphologically distinct subspecies or races have been described. (*Photo by Rod Planck/Photo Researchers, Inc.*)

environmental differences in an attempt to ascertain the ecological significance of each morphological trait.

Ecological studies of morphological variation generally make the fundamental assumption that the variants are genetically distinct from one another. We need to be careful about this assumption, however, because it is quite possible for geographic variation within a species to be a phenotypic response induced by the environment. In other words, a trait may be a developmental effect imposed on the organism by the environment. Distinguishing between genetic variation in morphology and phenotypic responses to the environment is not always a simple matter. Here we will consider two specific examples of geographic variation in morphology, one representing a phenotypic response to the environment and the other representing a genetically based variant.

Two Examples of Geographic Variation

One characteristic of the pocket gopher, *Thomomys bottae,* is its extensive subspecific differentiation. This mammal is known as a fossorial herbivore—an animal that lives primarily underground and eats plant material. It inhabits desert and grassland habitats in the West and Southwest. More than 200 races have been described for the *Thomomys bottae* complex, 47 of them in California alone (Patton and Brylski 1987). These local races are generally restricted to small regions isolated by the complex topography—mountains and valleys—of the West. Races are differentiated on the basis of coat color, morphology, and size.

Populations of this species have a long history of association with agriculture, particularly alfalfa fields, which provide abundant forage. Populations living in adjacent plots of natural vegetation and alfalfa fields differ markedly in size (Table 6.1).

J. L. Patton and P. V. Brylski (1987) analyzed the genetic and phenotypic basis of these size differences. Litters from field-caught animals from both types of populations were raised in the laboratory. Thus Patton and Brylski could assess the effect of optimal nutrition on the size relationship between the animals derived from alfalfa fields and natural vegetation. The results (Table 6.2) indicate that the growth rates of animals from both populations increased in the laboratory relative to the rates in the field. This significant increase in growth rate in the lab suggests that growth rate is plastic.

One of the classic studies of the basis of morphological differentiation was performed by J. Clausen, D. D. Keck, and W. M. Hiesey (1948). Yarrow (*Achillea lanulosa*) grows over a wide range of environments in California, including along a transect across the Sierra Nevada (Figure 6.4). Plants collected along such a

TABLE 6.1

Weight Differences of Pocket Gophers (*Thomomys Bottae*) in Different Habitats

Locality		Percentage of Weight Increase	
Alfalfa	Natural	Males	Females
Rose Valley Ranch	Freeman Canyon	90.1	53.6
Harvard	Mojave River	86.6	84.1
2-Hawk Ranch	Kingston Range	85.7	86.1
Coso Junction Ranch	Freeman Canyon	49.1	27.8
Butterworth Ranch	Coso Mts.	46.1	25.9
Benton	White/Inyo Mts.	43.4	44.2

For each locality, values are precent differences between alfalfa monocultures and adjacent natural vegetation communities.

Data from Patton and Brylski 1987.

TABLE 6.2

Changes in Growth Rates (g/day) of Field-Caught and Laboratory-Reared Pocket Gophers (*Thomomys bottae*) in the Laboratory

Sample	N	Field		Laboratory	N
Freeman Canyon (natural vegetation)	15	0.9474	***	1.1290	20
		***		**	
Rose Valley Ranch (alfalfa monoculture)	19	1.2994	***	1.6130	7

Asterisks denote significant differences between the two values separated by the asterisks.(**p <.01; ***p <.001) The Freeman Canyon population in the lab was as large as the Rose Valley Ranch population in the field.

From Patton and Brylski 1987.

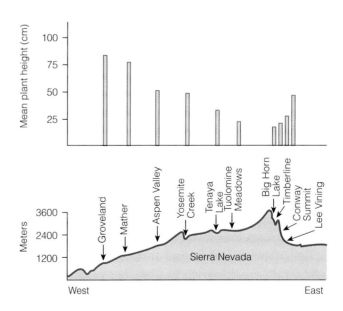

FIGURE 6.4 Variation in mean height of *Achillea* across a transect of the Sierra Nevada. (*From Clausen, Keck, and Heisey 1948*)

transect differ significantly in height. High-altitude plants are short in stature, like many species that grow in alpine tundra, whereas plants in the foothills grow significantly taller.

To analyze the basis of the variation in *Achillea*, Clausen and colleagues relied on common garden experiments. In such an experiment, individuals (from different places) that are suspected to differ genetically are grown under similar conditions in a garden or laboratory. If the phenotypes still differ, one can infer that the differences are genetically based. If they do not, it suggests that the differences originally observed were induced by the environment. This is a powerful tool to assess qualitatively the nature of the variation between populations.

Clausen and colleagues grew plants collected from the Sierra Nevada transect in three gardens: one located at Stanford on the coast, another at Mather at about 1300 meters, and a third at timberline at approximately 3400 meters. The results (Table 6.3) provide evidence of both genetic differentiation of populations and phenotypic plasticity. Note that plants from the high-altitude population (timberline) remain short in stature regardless of where they were grown. Even in the mild conditions of Stanford, they are dwarfed. On the other hand, the lower-elevation populations (such as Groveland) exhibit some plasticity in growth; they are dwarfed at high elevation. Clearly, they cannot cope with the harsh conditions of high altitude, and their phenotypes are adversely affected.

There are both advantages and disadvantages to genotypic and phenotypic adaptation. The advantages of genetic differentiation are twofold: It is biologically simple, and it enables the organism to be shaped directly by the selective effects of the environment.

Phenotypic changes are more complex. They may represent the results of stress or developmental changes that adapt the organism to the particular environment. Developmental plasticity allows the organism to respond appropriately to a number of different environments in that it can modify its phenotype as needed. This would be advantageous in a fine-grained species, which is likely to experience more than one kind of environment in its lifetime. It would also be advantageous in a species whose offspring disperse widely and thus may encounter an environment very different from that experienced by the parents. The disadvantage of this kind of response is its physiological and developmental complexity. The organism needs some mechanism to sense the nature of the environment, as well as a developmental switch to elicit the appropriate response.

TABLE 6.3

Heights (cm) of Yarrow (*Achillea lanulosa*) Collected from Different Localities and Grown in Common Gardens

Garden	Population Origin		
	Groveland	Mather	Timberline
Stanford	83.6	58.2	15.5
Mather	79.6	82.4	34.3
Timberline	21.2	31.6	23.7

Interaction of Genetic and Phenotypic Responses The experiments of Clausen and colleagues showed that the morphological variation in *Achillea* represents a complex interaction of both genetic and phenotypic responses to the broad range of climates experienced by populations of this species. In species that experience a range of environmental conditions, we would expect the interaction between these two kinds of responses to be complex.

Consider the example of the intertidal snail, *Littoria obtusata*. This species is exposed to significant wave action in some habitats but is protected from waves in others. The probability that a snail will be pushed off the rock by waves is determined by shell size and shape and by the strength of the foot. Snails from wave-exposed sites had larger feet and could resist greater shear forces than animals from protected sites. Lab and field studies by Trussell (1997) showed that there is a complex phenotypic response to waves. Animals from protected sites that were transplanted to exposed sites, or were exposed to wave action in a laboratory flume, grew larger feet. In contrast, animals from exposed sites subject to both high and low wave action showed no change in foot size. Thus the two populations exhibited different degrees of plasticity: Animals from protected sites adjusted their phenotypes more readily than animals from exposed sites. Trussell suggests that if natural selection favors phenotypic plasticity in heterogeneous environments, an asymmetry in the response may be advantageous too. An animal that does not increase its foot size when exposed to strong wave action may be washed away. However, the cost to a snail that does not reduce its foot size when exposed to low-energy waves is only the excess energy devoted to the size of its foot.

Understanding the relative roles of genetic variation and environmental effects is crucial because many of the topics important to evolutionary ecologists are approached in terms of adaptations by species to particular biotic or abiotic environments. We typically take this kind of approach to life history strategies, predation, and competitive effects. Many of the studies in these areas implicitly assume that the differences in the characters of interest are genetically based.

For instance, although seed size is clearly an important component of the reproductive strategy, we cannot necessarily assume that seed size is under simple genetic control such that it can be modified by natural selection (Wolfe 1995). In the biennial plant *Hydrophyllum appendiculatum*, in which seed size varies by at least a factor of 2, L. M. Wolfe found that there is no genetic variation for seed size. Thus there is no variation on which natural selection can act to modify seed size. Instead, seed size is affected by the light regime. Also, plants on which between 1 and 5 inflorescences were pollinated had higher seed weights than those on which 10 to 20 were pollinated. The only genetic effect was that seeds produced by self-pollination were smaller.

Chromosomal Variation

One kind of genetic variation that we can measure is chromosomal variation among individuals. Chromosomes are large enough so that with special staining techniques, we not only can see them but also identify chromosomal mutations. This enables us to describe an individual's **karyotype,** or the number and morphology of its chromosomes.

One of the first people to examine genetic variation revealed by chromosomal variation was T. Dobzhansky (1970), who was able to identify four major gene arrangements in the third chromosome of the fruit fly *Drosophila pseudoobscura.* Dobzhansky believed these rearrangements represent co-adapted gene complexes—groups of adaptive genes protected from recombination by inversion mutations of the chromosome.

There is considerable geographic variation in the distribution of the chromosome types of *D. pseudoobscura* (Figure 6.5). On the coast of California, the standard type is most common. Farther east the Arrowhead arrangement increases in frequency, only to be replaced by the Pikes Peak arrangement in Texas. In some instances, particularly where the environment changes rapidly, one chromosome type is abruptly replaced by another. In the Sierra Nevada, the frequency of the standard chromosome decreases from 45 percent to only 10 percent as the elevation rises from 280 meters to 3300 meters. Concomitantly, the Arrowhead arrangement increases from 25 to 50 percent.

Many other species are characterized by geographic chromosomal variation. Most do not show clines in

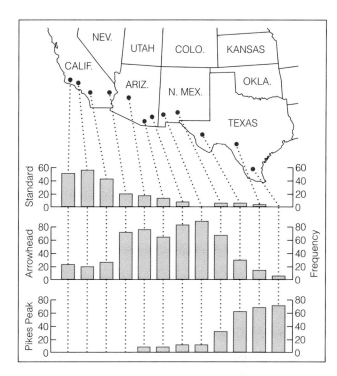

FIGURE 6.5 Geographic variation in *Drosophila* chromosomes. The names of the different arrangements are shown on the left. The bars represent frequencies of the various arrangements in the localities shown on the map. (*From Dobzhansky 1970*)

chromosome types of the kind described by Dobzhansky. More often, broad geographic regions are characterized by a particular karyotype. For example, the white-footed mouse, *Peromyscus leucopus,* can be divided into two chromosomal races (Figure 6.6; Baker et al. 1983), distinguishable on the basis of

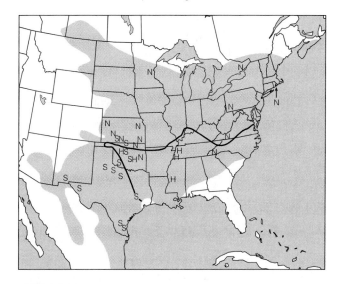

FIGURE 6.6 Chromosomal races of *Peromyscus leucopus* in eastern North America. N = the northeastern race; S = the southwestern race; H = localities where intermediate individuals are found. The heavy black line denotes the boundaries between three morphologically defined subspecies. (*From Baker et al. 1983*)

G-banded karyotypes (so called because Geimsa stain reveals patterns of bands in the chromosomes). The southwestern race is distinguished from the northeastern race by the presence of acrocentric (single-armed) chromosomes in pairs number 5 and number 20; these chromosomes are bi-armed in the southwestern race.

Allozymes

Another source of genetic information is the study of allozymes, genetic variants of proteins coded for by different alleles at a locus. Allozymes can be detected by the process of electrophoresis as follows. A sample of tissue is homogenized and placed on a gel made of starch, acrylamide, or other material. When an electric current is passed through the gel, the proteins in the sample migrate in the electric field according to their size and charge; small, highly charged proteins move faster than large molecules with small charge. (Of course, the size and charge of the protein are determined by its amino acid sequence, which, in turn, is determined by the sequence of nucleotides in the gene.)

After the proteins have migrated, the gel is sliced in the plane parallel to the direction of movement. Each slice is then stained for a particular protein. The presence of the protein is indicated by a stained band on the gel (Figure 6.7). The positions of the stained proteins can then be used to compare individuals in a population. If their proteins have moved different distances, the individuals are scored as genetically different.

The genetics for many protein systems have been worked out so that we know that each band actually represents an allele at a particular locus. By slicing the gel into many thin slices and staining each slice for a different protein, we can assess a number of different genetic loci. This allows us to measure the extent of variation within a population. We can quantify parameters such as the number of loci that are polymorphic (have more than one allele). We can measure the average heterozygosity of individuals in the population, which is also a measure of the amount of genetic variation. And we can compare populations by comparing the allele frequencies of the various proteins.

Protein electrophoresis has been used extensively to measure genetic variation. It offers two advantages over morphological measurements. One advantage is that a large number of loci can be sampled; we are limited only by the number of gel slices we can obtain and the number of loci that can be stained. Furthermore, we are measuring variation at the level of the cell's proteins—one step closer to the DNA. Thus the protein mobilities we measure are less likely to be the result of environmental effects (although such effects are not unknown in allozymes).

Extensive allozyme studies in a variety of species have revealed that most organisms maintain a great deal of genetic variation. Table 6.4 and Figures 6.8 and

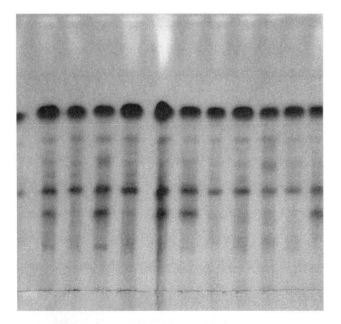

FIGURE 6.7 An electrophoretic gel. Each lane contains proteins from a different individual. Bands represent the allozymes present in that individual. (*Photo by David Krohne*)

TABLE 6.4

Measures of Genetic Variation (Proportion of Loci That Are Polymorphic and Average Heterozygosity) for Different Taxonomic Groups

	Polymorphism (mean)	Heterozygosity (mean)
Nine Groups		
1. Plants	0.259	0.0706
2. Invertebrata (excluding insecta)	0.399	0.1001
3. Insecta (excluding *Drosophila*)	0.329	0.0743
4. *Drosophila*	0.431	0.1402
5. Osteichthyes	0.152	0.0513
6. Amphibia	0.269	0.0788
7. Reptilia	0.219	0.0471
8. Aves	0.150	0.0473
9. Mammalia	0.147	0.0359
Three groups		
I. Plants (1)	0.259	0.0706
II. Invertebrates (2–4)	0.397	0.1123
III. Vertebrates (5–9)	0.173	0.0494

Data from Nevo 1978

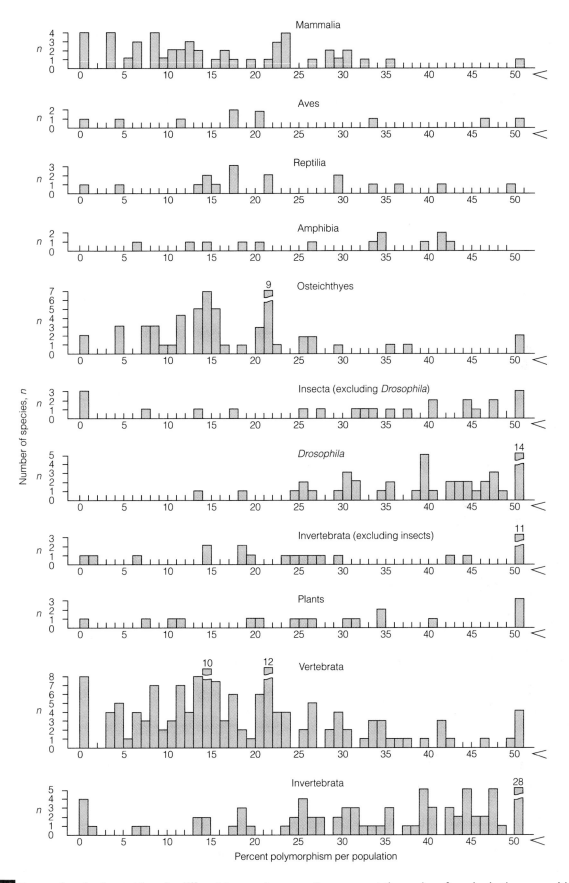

Number of species, *n*

Percent polymorphism per population

FIGURE 6.8 Frequencies of polymorphisms for different taxonomic groups. Bars represent the number of species in the group with that level of polymorphism for electrophoretic alleles. (*From Nevo 1978*)

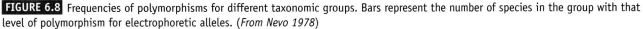

6.9 summarize these data. Note that the amount of variation differs among groups. When such data first began to appear, population biologists were surprised by the amount of variation maintained in most groups. Today, hundreds of such surveys of thousands of populations have confirmed the data presented here. In fact, allozyme studies may underestimate the amount of variation. If two proteins have different amino acid compositions (resulting from different DNA sequences) but by chance these morphs have the same electrophoretic mobility, we will not recognize them as different. The extent to which this factor results in underestimated values is not accurately known for most groups.

Allozyme data can also be used to measure differences between populations. M. Nei (1972) developed an index of **genetic similarity** based on allozyme data. The index, I_N is calculated as follows:

$$I_N = \frac{\sum_{i=1}^{m}(p_{ix}\,p_{iy})}{\sqrt{\left(\sum_{i=1}^{m}p_{ix}^{2}\right)\left(\sum_{i=1}^{m}p_{iy}^{2}\right)}}$$

where p_{ix} is the frequency of allele i in population X, p_{iy} is the frequency of the same allele in population Y, and m is the number of loci. In this equation, the numerator is the arithmetic mean of the products of the allele frequencies. The denominator is the geometric mean of the homozygote frequencies. The genetic similarity value can also be used to calculate the **genetic distance** (D_N) between populations according to Nei's formula:

$$D_N = -\ln I_N$$

We can use these kinds of measures of genetic distance to compare populations on virtually any spatial scale from a few meters to the planet. In Figure 6.10, the genetic distances among worldwide human populations have been represented on a phenogram. Populations that are genetically very similar are grouped together; very different populations are widely separated on the graph. The position of splits relative to the horizontal axis is a measure of genetic distance.

Another commonly used measure of genetic differentiation is a set of statistics developed by S. Wright (1965). This procedure is analogous to an analysis of variance—the total variation is partitioned into its component parts. The general concept, which is illustrated in Figure 6.11, is based on partitioning the

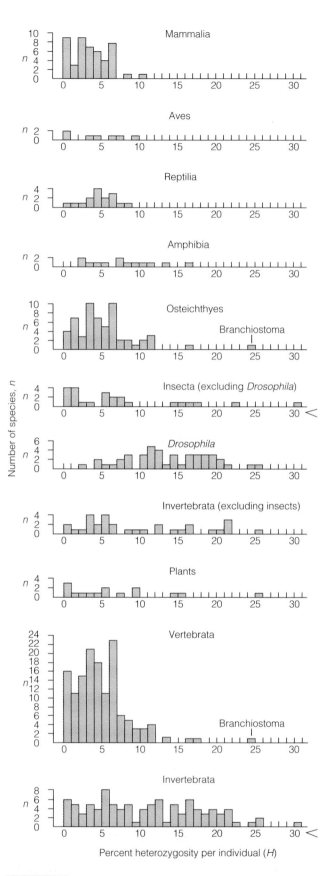

FIGURE 6.9 Frequencies of heterozygosity values for different taxonomic groups. Bars represent the number of species in the group with that proportion of heterozygous alleles. (*From Nevo 1978*)

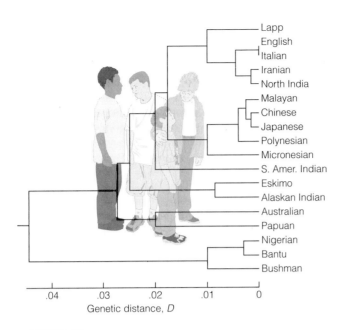

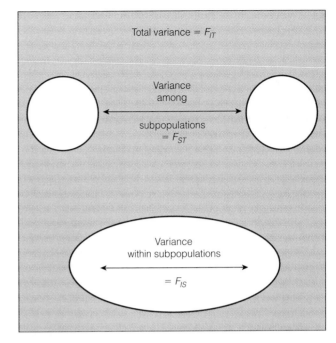

FIGURE 6.10 Phylogenetic relationships among 18 human populations, calculated using Nei's distance equations with data from electrophoretic alleles. (*From Nei and Roychoudhury 1982*)

FIGURE 6.11 The concept of Wright's *F*-statistics. The total nonrandom mating variance of the population is represented by F_{IT}. That variation is composed of portions due to variance within subpopulations (F_{IS}) and variance among subpopulations (F_{ST}).

components of nonrandom mating (inbreeding) in the population. Within a metapopulation, a subspecies, a species, and so on, a certain total amount of nonrandom mating (F_{IT}) occurs. A portion of this nonrandom mating is due to nonrandom mating within subpopulations and is designated F_{IS}, whereas some is due to variation in allele frequency among subpopulations (F_{ST}). These three measures are related:

$$F_{IT} = F_{ST} + (1 - F_{ST})F_{IS}$$

When one is studying patterns of genetic variation, it can be useful to identify the relative amount of the total variation that is due to variation among populations compared to that due to variation within populations.

W. F. Eanes and R. K. Koehn (1978) applied this kind of analysis to populations of the monarch butterfly (*Danaus plexippus*) in eastern North America. Nonmigratory butterflies were collected from Florida to Ontario and as far west as Missouri. Of 20 allozyme systems, 6 were polymorphic. The apportionment of genetic variation at those six loci is shown in Table 6.5. About half (0.008/0.017) of the total variation measured (for the pooled samples) is derived from variation within populations. The rest is associated with variation among populations. It is possible to test statistically whether an *F* value is significantly different from 0. Note in Table 6.5 that the values of F_{ST} for four loci were statistically significant, indicating significant genetic differences among subpopulations.

TABLE 6.5

Variation Among Nonmigratory and Migratory Populations of the Monarch Butterfly for Six Different Alleles, Using Wright's *F*-Statistics

Allele	Nonmigratory F_{ST}	Migratory F_{ST}
Phi-33	0.008	0.004
Pgm-100	0.06	0.005
Mpi-100	0.012*	20.006
MNdh-100	0.010*	0.012
Got-100	0.010*	0.007
Adh-100	0.009*	0.068
Mean	0.009	0.004

* Values are statistically significantly different from zero at the 0.25 level.

From Eanes and Koehn 1978.

DNA Fingerprinting

DNA fingerprinting (Jeffreys, Wilson, and Thein 1985; Jeffreys 1987) has recently emerged as an additional technique of genetic analysis that holds great promise for ecological genetics. This technique is based on the presence in DNA of so-called *minisatellites*, hypervariable regions that are composed of short segments (less than 20,000 base pairs) containing multiple copies of a short sequence (less than 65 base pairs). It is thought that minisatellites are noncoding regions. The variability of individual minisatellite loci is the result of mutations that cause the gain or loss of the repeat units.

A. J. Jeffreys and colleagues found that human DNA has a family of minisatellites with a common core sequence of a dozen nucleotides. This core sequence can be used as a probe to locate the hypervariable minisatellites, as outlined in Figure 6.12. The result is a series of bands associated with each individual—its fin-gerprint. Because the minisatellite loci are among the most polymorphic sequences ever described, the pattern of bands is essentially unique to each individual.

This technique can be used in ecological studies of genetic variation to compare the levels of variation among closely related populations. For example, D. A. Gilbert and colleagues (1990) used the technique to compare populations of the fox (*Urocyon littoralis*) on the Channel Islands off the coast of California. By comparing the numbers of bands shared by individuals within and among islands, the researchers could assess both the level of variation within populations and the degree of interisland genetic differentiation.

On some islands there was no variation. On others, the average percentage difference in band sharing was as high as 25.3 percent. Among islands, the average difference ranged from 43.8 percent to 84.4 percent. The similarity relationships among populations are depicted in the phenogram in Figure 6.13. Because subpopulations were sampled on some islands, this figure illustrates the degree of genetic similarity within and among populations.

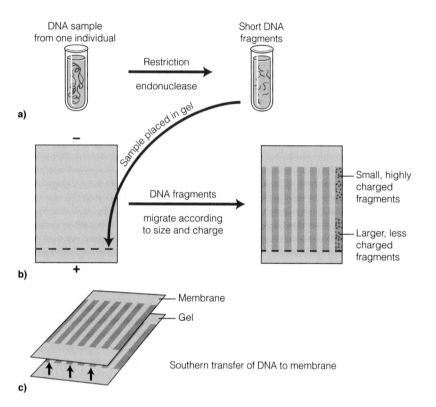

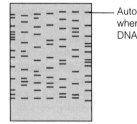

FIGURE 6.12 The process of DNA fingerprinting. (**a**) DNA is isolated from cells and cleaved at specific sites with an endonuclease. (**b**) The sample containing DNA fragments from each individual is placed in an electrophoretic gel where the fragments are separated by size and charge, producing a streak of fragments of different sizes in each lane of the gel. (**c**) The DNA fragments are then denatured into single-stranded segments and transferred to a nylon or nitrocellulose membrane by a Southern transfer. (**d**) The membrane is then washed with a solution containing single-stranded, radioactively labeled probes for the minisatellite DNA. The probe hybridizes with homologous fragments on the filter. (**e**) A piece of X-ray film is placed over the membrane and exposed. Each labeled hybrid fragment exposes the film and, upon development, shows up as a band. The pattern of bands comprises the individual's unique DNA fingerprint.

Mitochondrial DNA

Population biologists have also made use of the unique properties of mitochondrial DNA (mtDNA) to measure genetic variation in natural populations. Mitochondria contain a small set of genes that are separate from the nuclear genes and thus do not undergo segregation at meiosis. Instead, they are inherited entirely from the maternal side via the mitochondria in the egg. Because the mtDNA evolves especially rapidly, we can use the differences among individuals and populations to measure relatively recent genetic differentiation of populations.

Analysis of mtDNA for population genetics purposes generally uses either sequence information

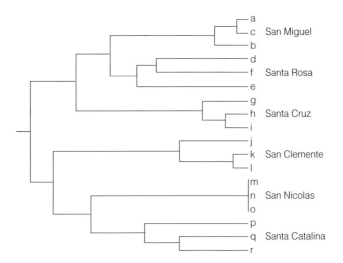

FIGURE 6.13 Phenogram of genetic relationships among Channel Island foxes based on DNA fingerprinting. The scale is proportional to the differences among subpopulations. Lower-case letters represent distinct populations. (*From Gilbert et al. 1990*)

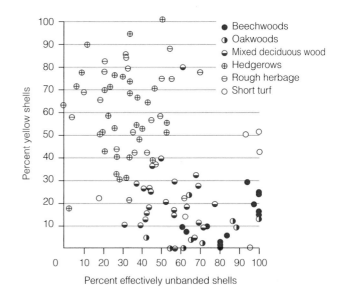

FIGURE 6.14 Shell color and banding pattern in relation to habitat in the snail *Cepaea nemoralis*. (*From Cain and Sheppard 1954*)

or restriction mapping. The advent of automated sequencing machines has made it practical to compare mtDNA sequences. Restriction mapping involves cutting the DNA with restriction enzymes and analyzing the size distribution of the resulting DNA fragments, which can be separated electrophoretically and the resulting banding patterns compared among individuals.

The Interaction of Genetic and Ecological Processes

The relationship between population ecological processes and genetic variation is complex. As we will see, a number of population processes influence patterns of genetic variation or differences among populations. The direction of causality is not simple or straightforward, however. The patterns of genetic variation also have an impact on demography and population processes. In this section we will look at examples of three fundamental interactions that affect variation: selection, dispersal, and population size.

Selection

As we saw in Chapter 2, the process of selection represents a very direct interaction between genetics and ecology: It is the ecological environment, whether abiotic or biotic, that constitutes the selective force on a population.

A classic example of this kind of interaction is known for the land snail, *Cepaea nemoralis,* in which intraspecific variation is a product of selection for color matching to avoid predation. Colors of the shells show great variety, ranging from brown to yellow to pink, and the degree of banding also varies greatly. A. J. Cain and P. M. Sheppard (1952, 1954) found a strong association between particular shell patterns and certain habitat types (Figure 6.14). This association is related to the ability of the banding pattern to render the snail difficult for predators to see.

Birds, especially thrushes, are the snail's most important predators. When a thrush captures a snail, it carries its prey to a stone and uses the stone as an anvil against which to peck the shell open. Cain and Sheppard compared the frequency distributions of the banding patterns of snails killed by predators to the frequency distribution of the patterns in the entire population. They found that "nonmatching" shells were preyed upon significantly more often than those that were difficult to detect against the background in a particular habitat.

If there is significant spatial variation in the selective pressures imposed by the environment, one result may be the development of genetically different populations. Geographic variants that result from adaptation to local conditions are called **ecotypes.** For example, tickle grass (*Agrostis scabra*) grows at several sites near Sudbury, Ontario, that are contaminated by heavy metals. Plants from contaminated sites show higher tolerance of heavy metals (Archambault and Winterhalder 1994). Seed germination and root growth in the presence of heavy metals was higher in plants from contaminated sites than in plants from uncontaminated sites (Figure 6.15).

The wild oat, *Avena barbata*, exemplifies the relationship between local and regional ecotypic differentiation. Ecotypes associated with specific abiotic conditions across a broad geographic scale are also detectable when similar abiotic conditions change on a much smaller scale.

In the foothills of the Sierra Nevada, Clegg and Allard (1972) identified two electrophoretic genotypes associated with extreme mesic or xeric sites found across this region. Each genotype is based on the allozymes present at five loci. The populations from which these types were identified are monomorphic for the genotypes.

Hamrick and Allard (1972) looked for the same ecotypic differentiation on a much smaller scale by comparing the allozymes of seven subpopulations located at different places on a single hillside. Conditions ranged from mesic grassland and savanna at the bottom to xeric evergreen forest at the top. For each subpopulation on the transect, they measured the genetic distance from the mesic genotype and from the xeric genotype identified in the earlier study. The results are presented in Table 6.6. The populations in xeric sites (higher sites, indicated by the higher numbers) on the hillside have very high similarity to the xeric genotype, and the populations from the mesic sites on the hillside (lower sites, indicated by the lower numbers) have high similarity to the mesic genotype. These data

TABLE 6.6
Avena Genotypes along a Moisture Gradient

	Site						
	Mesic ←					→	Xeric
	1	2	3	4	5	6	7
Mesic genotype	0.864	0.209	0.186	0.499	0.00	0.356	0.018
Xeric genotype	0.451	0.979	0.978	0.864	1.00	0.927	0.9998

Values are similar to the xeric or mesic genotypes identified by Clegg and Allard (1972).

Data from Hamrick and Allard 1972.

indicate that the mesic and xeric ecotypes can occur on a much smaller geographic scale.

Even more complex ecological interactions such as competitive ability can be modified by local selection. L. A. Mehrhoff and R. Turkington (1990) demonstrated that the competitive ability of two species of grass changed according to the origin and history of the populations. In some cases, reversals of competitive dominance occurred. For example, production of the grass *Lolium perenne* from newer pastures with a shorter history of competition was equal to or greater than that of *Holcus lanatus*. In older pastures where competition had been a selective force for some time, *Holcus* had higher production.

Dispersal

In Chapter 2 we noted that one important mechanism of allele frequency change in populations is gene flow. In Chapter 5 we discussed the nature of dispersal movements—emigration and immigration—from a population. If dispersal and breeding occur, gene flow has taken place. This process can have significant effects on the degree of genetic differentiation of populations. If a group of subpopulations experiences little gene flow between them, each will tend to be genetically distinct, and selection and drift will cause them to diverge. High rates of gene flow homogenize subpopulations, reducing the differences between them.

Let us consider two examples of the role of dispersal and gene flow from very different spatial scales. Earlier in this chapter we described the values of F_{ST} for nonmigratory populations of the monarch butterfly. W. F. Eanes and R. K. Koehn (1978) also examined the genetic differentiation of migratory populations of this species.

Some populations of monarchs undertake long-distance migrations. Monarchs in the eastern United States migrate south to a single site in Mexico where

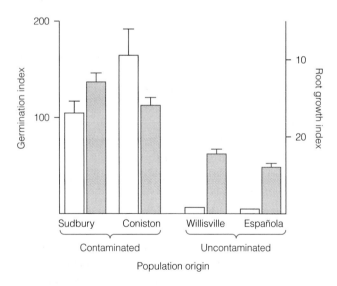

FIGURE 6.15 Tolerance indices for seed germination (open bars) and root growth (solid bars) in heavy metal–contaminated soil for plants from contaminated and uncontaminated sites near Sudbury, Ontario. Lines above the graphs indicate standard errors. (*Adapted from Archambault and Winterhalder 1994*)

they overwinter. In the spring they return northward. There is no evidence of a tendency to return to the specific summer range from which they originally came, however. As they return, they mate and lay eggs in the summer range. Thus the populations in eastern North America mix on the winter range before migrating north and breeding. The F values of the migratory populations are shown in Table 6.5. Note that the values of F_{ST} are much smaller in the migratory populations than in the nonmigratory population.

Compare these data to the genetic structure of a population of the pocket gopher (*Thomomys bottae*) examined by J. L. Patton and J. H. Feder (1981) on a study site of only a few hundred hectares. Wright's F-statistics were applied to allozyme data from this population. Fields of grass separated by chaparral constitute the subpopulations for the analysis. Note in Table 6.7 that in both years of the study, the mean values of F_{ST} are significant and are considerably larger than those for migratory butterflies in Table 6.5, despite the far smaller geographic scale involved.

The relevant difference between these two organisms and the genetic structures they manifest is dispersal. Because pocket gophers are fossorial, most individuals move a maximum of only a few hundred meters over the course of their entire lives. Thus the population size is relatively small and gene flow is low. As a result, population differentiation occurs on a finer scale than in the migratory butterflies studied by Eanes and Koehn.

The barriers to dispersal need not be large to affect the pattern of population structure. For example, a

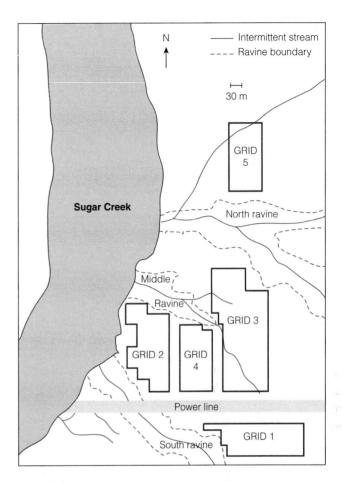

FIGURE 6.16 Map of forested habitat for a study of a *Peromyscus leucopus* population, showing the location of sampling grids in relation to barriers to dispersal—ravines and a power line right-of-way kept clear of trees. (*From Krohne and Baccus 1985*)

TABLE 6.7

F_{ST} Values for *Thomomys bottae* Subpopulations in 1976 and 1977

Gene Locus	1976	1977
Idh-1	0.065*	0.029
Idh-2	0.097*	0.121*
αGpd	0.072*	0.066*
Pept-1	0.061*	0.015
Pept-2	0.099*	0.035*
6Pgd	0.098*	0.050*
Est-4	0.039	0.053*
PreAlb	0.023	0.015
Trf	0.082*	0.067*
Ada	0.026	0.044*
Ga3Pdh	0.063*	0.045*
Mean	0.066*	0.049*

*Values are significantly different from zero.

From Patton and Feder 1981.

study by Krohne and Baccus (1985) showed that ravines and power line rights-of-way dissecting forest habitat act as barriers to dispersal in the white-footed mouse (*Peromyscus leucopus*), a forest-dwelling mammal. These researchers followed the movements of animals among subpopulations on study plots in forest habitat, some of which were separated by barriers to movement (Figure 6.16). Note in Table 6.8 that the frequency of dispersal movements among subpopulations separated by a power line or by a ravine is significantly less than among subpopulations without such barriers.

Even though the entire mouse population occupied an area of only 75 hectares, significant genetic differentiation among subpopulations was observed. Table 6.9 gives F_{ST} values calculated across subpopulations. In both years, there was significant variation among subpopulations. For the natural units separated by barriers (middle column for each year), significant F_{ST} values were again recorded. For subpopulations 2, 3, and 4 only, which were not isolated from dispersal, F_{ST}

TABLE 6.8

Frequencies of Movements of *Peromyscus leucopus* Among Grids Shown in Figure 6.16

Movement from	Movement to					
	G1	G2	G3	G4	G5	Total
G1		0.10	0.08	0.03	0.03	0.23
G2	0.05		0.05	0.26	0.05	0.41
G3	0.00	0.03		0.13	0.05	0.21
G4	0.00	0.05	0.05		0.00	0.10
G5	0.03	0.00	0.00	0.03		0.05
Total	0.08	0.18	0.18	0.44	0.13	1.00

Some columns and rows do not sum to totals because of rounding.

Rows are movements to a grid; columns are movements from a grid.

From Krohne and Baccus 1985.

TABLE 6.9

F_{ST} Values for *Peromyscus leucopus* in 1980 and 1981 in the Study Area Shown in Figure 6.16

Gene Loci	1980			1981		
	12345	1-234-5	234	12345	1-234-5	234
Trf	0.003	0.005	0.000	0.018	0.051*	0.022
CA-1	0.025*	0.025*	0.024	0.024*	0.013*	0.030*
ADA	0.626*	0.083*	0.011	0.021*	0.015*	0.026
PP1	0.009	0.003	0.017	0.024*	0.023*	0.036*
$\bar{X}$	0.166*	0.029*	0.013	0.023*	0.026*	0.029

Values are based on different subdivisions (subpopulations). 12345 represents each grid as a potential subpopulation; 1-234-5 represents three subpopulations separated by barriers to dispersal (grid 1; grids 2, 3, and 4; and grid 5); 234 represents the three subpopulations (grids 2,3, and 4) that are not separated by barriers to dispersal. Asterisks denote values significantly different from 0.

From Krohne and Baccus 1985.

was not significantly different from zero (third column in each year). Again we see that genetic differentiation can occur on a relatively fine spatial scale and that patterns of dispersal can play a significant role in determining the pattern of genetic variation.

A perennial plant, the blazing star (*Liatris cylindracea*), exhibits genetic subdivision on an extremely fine spatial scale (Schall 1975). As in the two previous examples, this pattern is attributable to the peculiarities of dispersal in the species. B. A. Schall examined the patterns of genetic structure in a rather small population of this species from a single hillside in Illinois. Even though the entire population occupied only 600 square meters, it was quite dense—as many as 175 plants per square meter. The study population was divided into a series of 66 quadrats, each 3 meters

square. Plants were sampled from these quadrats, and 27 allozymes were analyzed for each plant.

Schall found extreme genetic subdivision in this population. Adjacent quadrats differed by as much as 20 percent in allele frequencies. An analysis of F_{ST} values showed that the variance among subpopulations (the quadrats) was significant (Table 6.10). The values of F_{ST}, measured on a scale of only 600 square meters, were greater than those of the monarch butterfly measured across all of eastern North America (Table 6.5)!

What causes this unusual pattern? If local selection pressures were causing these differences, one would expect an association between allele frequencies and soil or microclimatic features. Schall could eliminate this explanation because attempts to correlate allele frequencies with edaphic factors failed.

Another possibility is that the local differentiation is a result of restricted gene flow. Two factors result in extremely small dispersal distances: pollinator behavior and seed dispersal distances. Pollination is accomplished primarily by bees, which tend to move between nearest-neighbor plants (Levin and Kerster 1969). Consequently, gene flow is greatly restricted. Although *Liatris* seeds are wind-dispersed, they are relatively heavy seeds and often become ensnared by nearby vegetation. The mean dispersal distance of a closely related species, *Liatris aspera,* is only 2.49 meters (Levin and Kerster 1969). Thus the restricted

TABLE 6.10

F-Statistics for *Liatris cylindracea*

Locus	F_{ST}
GOT	.0184
MDH	.0903
ADH	.0452
AP_1	.2240
AP_2	.0438
Es_1	.0464
Es_2	.2190
P_{ep}	.0256
G-GPDH	.0677
6-PDH	.0756
P_{er}^-	.0138
P_{er}^+	.0395
PGI	.0767
AUP	.0361
Es_3	.0091
$\bar{X}$	.0687

From Schall 1975.

movement of genes in *Liatris cylindracea* results in extreme local subdivision of this population. The effect of restricted gene flow is apparent in Figure 6.17, which shows the relationship between Nei's index of genetic distance and the geographic distance between samples (Schall 1975). Clearly, as geographic distance increases, genetic distance increases as well.

C. F. Williams and R. P. Guries (1994) have demonstrated the genetic consequences of differences in seed dispersal mechanism. They studied three species of forest herbs, *Cryptotaenia canadensis, Osmorhiza claytonii,* and *Sanicula odorata,* the seeds of which differ significantly in their ability to adhere to the fur of animals (Figure 6.18) and thus in their dispersal distances. *Sanicula* seeds attach easily and tightly to mammalian fur, whereas those of *Cryptotaenia* fall relatively near the parent plant because they are not transported on fur. Electrophoretic genetic analysis and *F*-statistics indicate that the greatest degree of variation among subpopulations (within drainages) occurs in *Cryptotaenia,* the species with the shortest seed dispersal distances. *Osmorhiza* is intermediate, and *Sanicula* had the least variation among subpopulations. Analysis of genetic differences among populations of these species yields similar results: The species with the highest dispersal ability showed the highest genetic similarity, even when measured over increasingly larger spatial scales, such as between drainages or between regions (Figure 6.19).

The pattern of gene flow can be different from that predicted by the potential for long-distance seed dispersal. The mangrove (*Avicennia marina*) is a species found on the shore of a number of subtropical regions in the western Pacific. This plant produces seeds that germinate on the tree (see Figure 17.21) and fall into

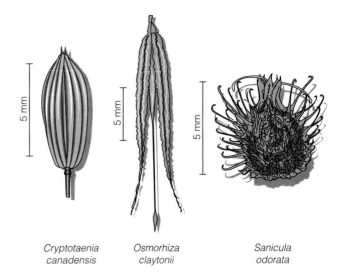

*Cryptotaenia
canadensis* *Osmorhiza
claytonii* *Sanicula
odorata*

FIGURE 6.18 Structure of seeds of three species. *Cryptotaenia* is least likely to adhere to fur; *Sanicula* is most likely to adhere. (*From Williams and Guries 1994*)

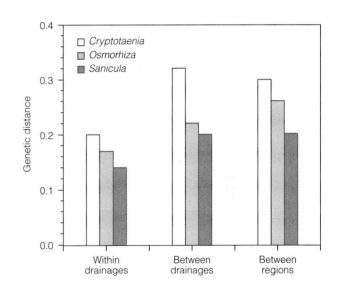

FIGURE 6.19 Genetic distance among species at three spatial scales: (1) within drainages; (2) between populations in different drainages of the same region; and (3) between populations in different regions. (*From Williams and Guries 1994*)

the water where tides and currents disperse the seeds, potentially for great distances. A new plant develops when the seed comes to rest in shallow water in appropriate coastal habitat. From this information, one might predict that the potential for long-distance seed dispersal in this species would lead to high gene flow and little local genetic differentiation. However, Duke et al. (1998) used protein electrophoresis to show that genetic structuring in the Indo-West Pacific is characterized by sharp discontinuities in allele frequencies. Gene flow among distant populations was

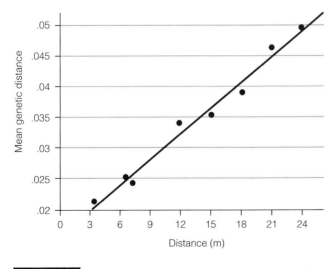

FIGURE 6.17 Genetic distance in *Liatris* as a function of spatial distance between samples. (*Data from Schall 1975*)

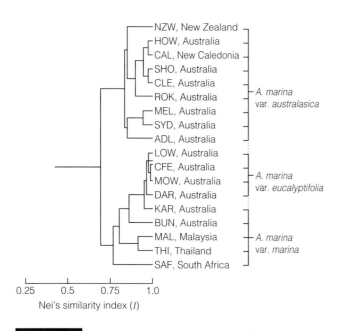

FIGURE 6.20 Dendrogram indicating genetic similarity among populations of *Avicennia marina* in the western Pacific. (*From Duke et al. 1998*)

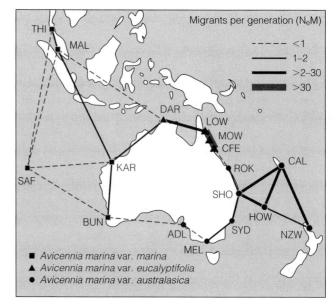

FIGURE 6.21 Patterns of gene flow in *Avicennia marina* among a series of sampling sites. The thickness of the lines represents the number of migrants per generation. (*From Duke et al., 1998*)

relatively low; most occurred among continuous populations. Note the correspondence between the pattern of gene flow in Figure 6.20 and the pattern of genetic similarity in Figure 6.21.

The spatial distribution of populations also plays a role in the movement of alleles within and between populations. In particular, populations in small, isolated habitats experience different patterns of gene flow than those in large, continuous habitats. Electrophoretic analysis of allozymes conducted by Young and Merriam (1994) showed that the genetic structure of populations of sugar maples (*Acer saccharum*) in patches of forest differs from the pattern in populations in large tracts of forest. The researchers analyzed genetic variation on two spatial scales. At the finer scale, in which variation was assessed between 100-square-meter subplots, there was less mixing of genotypes in the patches than in the control populations. They attribute this finding to less overlap of seed shadows in patches (Figure 6.22). Because subpopulations in the patches tended to be dominated by one or a few trees, the subpopulations were more differentiated there. When the genotype distributions were compared on a larger spatial scale (10,000 square meters), the results were reversed: Patch populations showed more genotypic variation than did large tracts of forest. The authors also hypothesize that patches of forest are subject to increased immigrant pollen and hence gene flow. Because the edge or perimeter is larger relative to the total area for a patch than for a large forest tract, there is more opportunity for the influx of foreign pollen.

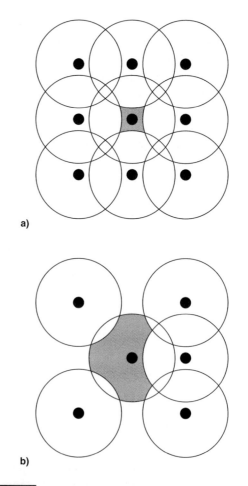

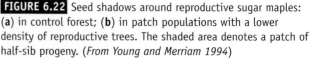

FIGURE 6.22 Seed shadows around reproductive sugar maples: (**a**) in control forest; (**b**) in patch populations with a lower density of reproductive trees. The shaded area denotes a patch of half-sib progeny. (*From Young and Merriam 1994*)

Population Size

B. S. Bowen (1982) demonstrated the effect that microtine population cycles have on the genetic structure of the California vole, *Microtus californicus.* As we saw in Chapter 5, microtine cycles are characterized by tremendous changes in population density over a four-year period. In the high phase of the cycle, dispersal results in wide movement of voles and in the colonization of suboptimal habitats. Then, after the population crashes, the few surviving animals are restricted to a few widely scattered patches of optimal habitat.

Bowen examined the changes in genetic structure over the course of a four-year cycle and found that F_{ST} changed as shown in Figure 6.23. During the peak of the population cycle, when dispersal and extensive gene flow have homogenized the population, F_{ST} is low; there is little differentiation between subpopulations. During low-population phases, however, the isolation of survivors and the low numbers result in a high degree of differentiation of subpopulations—and thus high values of F_{ST}.

The pattern of population fluctuation can have profound effects on the nature of genetic variation in populations. We saw in Chapter 2 that evolution occurs rapidly in small populations in which selection can change allele frequencies more rapidly than in large populations. In addition, small populations may be subject to loss of genetic variation by drift or inbreeding. Isolated populations may not have sufficient gene flow to maintain genetic variation.

The probability of extinction in small populations is higher as a result of the negative effects of these genetic processes. The loss of variation has three main detrimental effects (Franklin 1980; Lande and Barrowclough 1987; Lynch and Lande 1993; Lande 1994):

1. Reduction in the capacity of the population to adapt to environmental changes. If genetic variation is the raw material on which natural selection operates, the adaptive ability of small populations may be compromised.
2. Inbreeding depression. The homozygosity in small populations may generally depress the fitness of individuals by combining and expressing deleterious recessive alleles.
3. Fixation of new deleterious mutations. Recent models (Lande 1994) indicate that even for moderately large populations, the fixation of new, even mildly deleterious mutations may result in a substantial risk of extinction.

The principles of intraspecific variation that we have outlined have considerable practical application

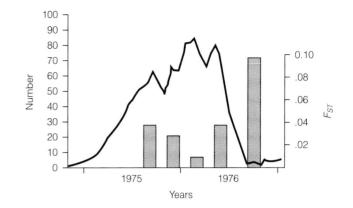

FIGURE 6.23 Values of F_{ST} for voles over time. The line denotes population changes; the bars represent mean F_{ST} at different points in time. *(From Bowen 1982)*

to the field of conservation biology. Specifically, our efforts to conserve rare or endangered species must be informed by our understanding of intraspecific variation (see the box on page 163).

The population sizes of many species have decreased because of human activities. In some cases, direct anthropogenic mortality from overharvesting is the root cause of the population decline. In others, population sizes are decreasing as a result of fragmentation of habitats by human activity. Faced with these declining populations, ecologists are recognizing that the interaction between genetics and ecology is crucial to biological conservation (Lande 1988). Our discussion of this interaction will focus to a great extent on populations and species that have been affected by human activity in ways that make these processes significant.

The marsh gentian (*Gentiana pneumonanthe*) is an example of a plant whose populations have been reduced by the effects of human activity, with clear genetic consequences. In the Netherlands, this species was once common in heathland habitats, but pollution of the soil and habitat loss have resulted in significant population declines, in some cases by as much as 40 percent (Mennema 1985). When Raijman and colleagues (1994) examined allozymes from populations ranging from 1 to more than 50,000 individuals, they discovered that most of the populations did not contain any rare alleles, a result that suggests an overall loss of variation. The number of polymorphic loci (loci at which there is more than one allele) is a measure of the overall level of genetic variation in a population. Both this parameter and the average number of alleles per locus were positively correlated with *G. pneumonanthe* population size (Figure 6.24), confirming the loss of variation in small populations.

The northern elephant seal (*Mirounga angustirostris*) presents one of the most extreme examples of the

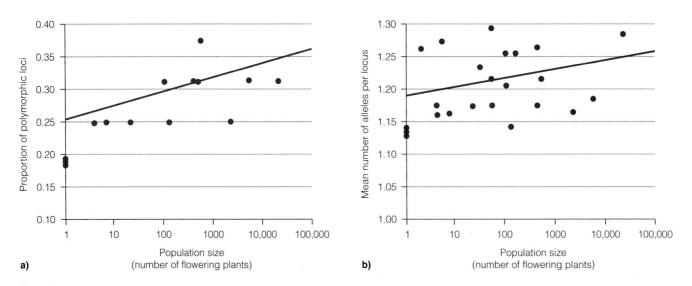

FIGURE 6.24 Genetic variation and population size in *Gentiana*. (**a**) Proportion of polymorphic loci as a function of population size. (**b**) Mean number of alleles per locus as a function of population size. (*From Raijman et al. 1994*)

effects of very small population size on genetic variation. This species was hunted nearly to extinction in the nineteenth century—the population was reduced to only about 20 individuals. In such a small population one might expect the loss of variation through genetic drift and inbreeding. Bonnell and Selander (1974) confirmed this by showing that although the species had recovered to more than 30,000 individuals by the middle of this century, there was no genetic variation at any of 24 electrophoretic loci they analyzed. This kind of population history is called a **bottleneck;** the population size narrowed, then expanded again.

Two species with very different recent population histories in North America, the gray wolf (*Canis lupus*) and the coyote (*C. latrans*), illustrate the effects of demography on genetic variation. Restriction analysis of mtDNA indicates that the continent-wide coyote population is composed of a larger number of genotypes (32) than is the wolf population (18) (Lehman and Wayne 1991; Wayne et al. 1992). Coyote populations are not highly genetically differentiated; their F_{ST} values were not significantly different from zero. F_{ST} values for the wolf, however, were large (0.76) and highly significant, which indicates that the North American population is highly subdivided into significantly different subpopulations.

A likely explanation for these differences is the recent demographic history of the two species. Coyotes have greatly expanded both their range and their population size (and thus their gene flow) as a result of habitat alterations in North America (Voight and Berg 1987). In contrast, the wolf has experienced shrinking and more highly fragmented populations (Mech 1970; Carbyn 1987) with less gene flow between them.

Studies of other species are consistent with this interpretation. The wild turkey (*Meleagris gallopavo*) populations were drastically reduced by the clearing of forest land in the eastern United States in the nineteenth and early twentieth centuries. By the 1940s, the eastern populations had been extirpated from more than 80 percent of their original range (Mosby 1949). The remaining populations inhabit a highly fragmented forest landscape. An electrophoretic survey of turkeys from Arkansas, Tennessee, and Kentucky revealed extremely high F_{ST} values (Leberg 1991). In fact, the mean F_{ST} value of 0.102 is one of the highest reported for any avian population: Nearly 89 percent of the genetic variation in this species is due to differences among populations. Populations of turkeys that did not suffer significant declines and fragmentation do not show this extreme genetic subdivision. Clearly, the absence of gene flow among fragmented populations and their small size significantly affect their genetic structure.

One difficulty with all these studies is that we generally do not have population genetic data from the period prior to the reduction in population size. This makes it difficult to assign the current genetic pattern unequivocally to a historical bottleneck. Bouzat et al. (1998) recently found an ingenious way around this problem. In a study of genetic variation in the greater prairie chicken (*Tympanuchus cupido*), these researchers were able to extract and analyze DNA from the feather roots of museum specimens of this species collected before the population reduction. In Illinois, the population of this species declined from estimates in the millions of individuals in the mid-nineteenth century to fewer than 50 by 1993. Western populations in

Subspecies, Races, and the Endangered Species Act

In 1973 the U.S. Congress passed a piece of landmark conservation legislation: the Endangered Species Act. This legislation mandated that the Fish and Wildlife Service determine which species in the United States are at risk of extinction. When a species is determined to be endangered, the government is required not only to protect the remaining individuals but also to develop and implement a recovery plan designed to move the species from the brink of extinction. In addition, the Endangered Species Act recognizes the importance of protecting distinct subspecies, and even populations, in order to maintain the naturally occurring genetic diversity so important to species survival.

This last requirement has led to some practical problems in the implementation of the act. Part of the policy that has developed prohibits creating hybrids of species or even of subspecies in order to facilitate recovery. The logic of this position is sound: The goal of the act is to preserve naturally occurring diversity; destroying that diversity through hybridization would defeat the fundamental purpose of the legislation. Because of the complex nature of natural variation within and among species, however, the application of this policy is problematic. S. J. O'Brien and E. Mayr (1991) point out some examples of the problems that can arise.

A unique subspecies of the mountain lion, *Felis concolor coryi,* once occupied much of the southeastern United States. Today this subspecies, also known as the Florida panther, is critically endangered. Less than 40 individuals remain in the swamps of southern Florida. Recent genetic analysis of the Florida panther has revealed that the subspecies is contaminated with genes from a South American subspecies, a few individuals of which were released into the Everglades in the late 1950s or 1960s. As a result of this genetic contamination, the Florida panther is no longer a pure subspecies. Under a strict interpretation of the hybrid policy, the Florida panther might be ineligible for protection.

The gray wolf, *Canis lupus,* represents another example. Recent genetic analyses of this species have shown that it is contaminated by genes from coyotes (*Canis latrans*). These closely related species have very different ecological niches, but matings between them can produce viable, fertile hybrids. Such hybrids were rare in presettlement times, when habitat separation made contact between the species relatively unlikely. As the coyote's habitat has been modified by humans, however, its range has expanded, and the result has been an increased frequency of hybridizations.

Since the discovery of the impurity of the wolf genome, ranchers in Wyoming, Montana, and Idaho, worried about possible losses of livestock to wolf predation, have petitioned the Fish and Wildlife Service to remove the gray wolf from the Endangered Species list. This step would not only prevent the development of wolf recovery plans but also permit the destruction of the few wolves remaining in those areas.

The goal of the Endangered Species Act is noble: to protect the remaining biological diversity of the United States. The act is based on the ecological and evolutionary importance of genetic variation. However, the great diversity in the patterns of intraspecific variation complicates our efforts to preserve it.

The issue of races and subspecies is only part of the problem. Hughes et al. (1997) recently attempted to estimate the number of extant *populations* in the world. Using literature data and population range maps, they calculate that there are approximately 220 populations per species. Depending on the estimate of the number of species on earth (see Chapter 12), this leads to an estimate of between 1.1 and 6.6 billion populations. If population extinction increases linearly with habitat loss, the global rate of population extinction is 1800 populations per hour in tropical forest alone! As you've seen in this chapter, important local genetic differentiation and differences exist among populations. Although this study makes a number of crucial and untested assumptions, it provides a sobering estimate of the magnitude of the problem.

TABLE 6.11

Alleles Found at Six Microsatellite Loci for Four Extant Populations and Museum Specimens from Illinois. Each letter is an allele. Bold letters are those found in the museum specimens that have been lost from Illinois. Italic letters represent those unique to the Illinois population prior to the bottleneck. (*From Bouzat et al. 1998*)

| Population | Loci | | | | | |
	ADL42	ADL23	ADL44	ADL146	ADL162	ADL230
Illinois	ABC	ABCD	A D F H	ABC	B E	C EFGHI
Kansas	ABCD	BCDEF	A CDEFGH	ABCDE	ABCDE	ABCDEFGHI
Minnesota	ABCD	ABCD	ABCDEFGH	ABC E	BCDE	BCDEFGHI
Nebraska	ABCD	ABCDE	ABCDEFGH	BC EF	BCDE	BCDEFGHIJK
Museum specimens	AB	BCD**E**	A D F	ABC **E**	**BCDE***FG*	**BCD**EG I *L*

Kansas, Minnesota, and Nebraska remained relatively large—up to 100,000 birds—during this period (Bouzat et al. 1998).

These researchers used a relatively new technique of genetic analysis, microsatellite DNA, to measure genetic variation. This technique is similar to DNA fingerprinting using minisatellite DNA. However, microsatellite loci, which are composed of a variable number of tandem repeat loci, are small and highly polymorphic. Moreover, they are known to be inherited in Mendelian fashion, making population genetic analysis more straightforward. The results are presented in Table 6.11. All alleles present in extant Illinois populations are also found in larger western populations, which suggests that they were present in Illinois prior to the bottleneck. Most significantly, the museum specimens contained nine alleles not currently present in the Illinois populations (Table 6.11). Comparison with the western populations shows that the alleles that were lost tended to be the rarer alleles, as one might predict.

Natural patterns of population change are known to affect the patterns of genetic variation as well. The cheetah, *Acinonyx jubatus,* is also extremely depauperate in genetic variation. When S. J. O'Brien and colleagues (1983) surveyed 47 allozyme systems in samples from populations from South Africa, they found that the 55 cheetahs in their samples were monomorphic for all 47 allozymes. Such extreme genetic uniformity has since been found to include the major histocompatibility complex (MHC), the genes that control cell-surface antigens responsible for cell-mediated rejection of tissue grafts. Even though the MHC is one of the most variable of all genetic systems, the cheetah is monomorphic for this system as well (O'Brien et al. 1985). Unlike the northern elephant seal, however, the monomorphism is apparently the result of a natural, rather than an anthropogenic, bottleneck in the recent past. This situation makes the cheetah extremely vulnerable to disease or to a change in the environment that would require selection to operate.

SUMMARY

1. The mechanisms of evolution are fundamentally mechanisms of genetic change within a population. The ecology of an organism is the context within which its evolution occurs. Consequently, there is a vital interaction between genetics and ecology.

2. Natural selection, genetic drift, and gene flow are key mechanisms of evolutionary change. Each acts on genetic variation within the population. Thus genetic variation is an important parameter that determines the nature of evolutionary change in a population.

3. Genetic variation arises by mutation. Variation is maintained by dominance and diploidy, recombination, and heterozygote advantage. Variation is decreased by genetic drift and inbreeding.

4. Population geneticists have used a number of measures of genetic variation. One of the most important has been morphological variation. Subspecies or races are identified primarily on the basis of morphological variation. This measure has drawbacks, however, including the small number of loci that can be examined and the potential for environmental effects to confound the interpretation of morphological differences.

5. Some variation within species has a genetic basis. Some is affected by the environment. For an organism adapting to variation in its environment, there are advantages to responding genetically and advantages to responding phenotypically.

6. A number of molecular techniques have recently been adapted for use in measuring genetic variation. They include allozymes, DNA fingerprinting, and analysis of mitochondrial DNA. These techniques avoid some of the difficulties associated with measuring morphological variation.

7. The ecology of an organism plays a key role in determining the pattern of genetic variation within and among populations. The environment, both biotic and abiotic, constitutes an important selective force that operates on the variation within a population. The patterns of dispersal and gene flow determine the scale over which populations vary genetically. The population size, through its impact on the process of drift, also affects the amount of genetic variation within and among populations.

SELF-ASSESSMENT: CAN YOU ...?

1. Discuss the nature of genetic and phenotypic variation in species and their interaction. What are the selective advantages and disadvantages of each?

2. Discuss the advantages and disadvantages of the various methods of measuring genetic variation and differences among populations or species.

3. Discuss the relationship between each of the following and genetic variation or genetic differentiation.

Selection
Population size
Dispersal

4. Relate the ecological effects discussed in this chapter to the Hardy–Weinberg law.

5. Compare the nature and results of gene flow and selection in plants and animals. How are they different? How are they the same?

PROBLEMS AND STUDY QUESTIONS

1. The technique of DNA fingerprinting does not measure allele variation because the minisatellite variants do not correspond to Mendelian alleles. Consider how this might affect the application, usefulness, and interpretation of DNA fingerprinting data.

2. For the population of *Peromyscus leucopus* described by Krohne and Baccus (1985; Figure 6.16), there are significant genetic differences between populations separated by barriers to dispersal. What evolutionary mechanisms might cause these populations to differ genetically? How could you test your hypothesis? What kinds of data would be needed?

3. Using the information on genetics in this chapter and the information on population regulation in Chapter 5, discuss the factors that cause small populations of endangered species to be at risk of extinction.

4. In a study of birds along the coast of South America, you find significant morphological differences among populations that lead you to describe several subspecies. A subsequent genetic analysis of these subspecies using allozymes finds no significant genetic differences among the groups. How can you explain these results?

5. The red mangrove is a tree that inhabits tropical coastal waters. It produces seeds that fall into the water and can survive long periods of immersion in salt water until they grow into adult trees. The black mangrove, on the other hand, lives farther from the shoreline. It reproduces vegetatively and by seeds that fall to the soil near the parent plant. What patterns of genetic variation would you expect to see in each species on a local and regional scale?

6. One of the assumptions of most studies using allozymes or mtDNA markers is that these markers are selectively neutral; that is, the selective coefficients of all the alleles are equal. Why is this assumption important? What would be the effect if it were not true? What if it were true for some alleles but not for others?

7. Explain verbally or algebraically how inbreeding can affect genotype frequencies without affecting allele frequencies.

8. In the example of the migratory populations of the monarch butterfly, the individuals that migrate back north and breed do not necessarily return to the area from which they came. What difference would it make if they did? Does it matter whether they mate on the wintering grounds or where they lay their eggs?

9. In Chapter 5 we pointed out that human activities are causing many wildlife habitats to become fragmented into separate habitat patches. Discuss the potential effect of this process on the genetic variation of a species. What mechanisms account for your predictions?

10. You sample two species of insects along a 6000-meter transect. Every 1000 meters you take a sample and analyze the allozymes in the two species. For each adjacent pair of populations for each species, you calculate Nei's index of genetic similarity:

Distance:	1000 m	2000 m	3000 m	4000 m	5000 m
Species 1:	0.93	0.89	0.88	0.94	
Species 2:	0.91	0.93	0.45	0.89	

Also note that a comparison of the populations of species 2 at 3000 meters and 5000 meters results in a Nei similarity of 0.41.

Using your knowledge of the factors that affect genetic variation, suggest a scenario to explain these results. It might include the nature of the species as well as the nature of the habitat or other aspects of the environment. Is there only one possible explanation?

PROJECTS AND ADDITIONAL STUDY

1. Choose one of the following organisms. Search the literature for data on genetic differentiation. Analyze those data with respect to

 a. The kinds of genetic techniques used to measure variation

 b. The concordance or similarity in results using different techniques

   Gray wolves

   Humans

   Pocket gophers (*Thomomys bottae*)

   Drosophila

2. Search the literature for information on genetic variation in the threatened prairie plant, *Silene regia*. Analyze and discuss the utility of electrophoretic markers in this species and plants in general.

3. Write a careful analysis of the nature of the concept of a "population" based on the information in this chapter and in Chapter 4.

4. Analyze the paper by Hughes et al. (1997) discussed at the end of this chapter with respect to the assumptions they made.

Life History Strategies

onsider the following "life histories":

1. A salmon, having spent five years feeding voraciously in the North Pacific, enters the Yukon River and swims upstream some 2000 miles to a small tributary. Over the course of the journey, during which it does not eat, the fish gradually digests its muscles and organs. By the time it reaches the small stream where its parents mated, it is near death. Finally, it spawns and dies.

2. A female red kangaroo in the desert of Australia cares simultaneously for three young at different stages of development. The oldest has left its mother's pouch and lives independently, although it remains near its mother's side. The second, a newborn, is attached to a teat in the pouch. It is helpless and incompletely developed. The third is a fertilized egg in the uterus, where it will remain, unattached to the placenta, for 204 days.

3. A mayfly egg hatches in a small stream in the Rockies. After feeding under the surface of the water for a few weeks, the nymph swims to the surface and hatches into the first adult stage. This winged form flies off and conceals itself in the vegetation along the stream. In a few hours, it sheds its skin and becomes a sexually mature adult. After males and females fly over the water and mate, the females lay eggs on the surface, and both sexes die.

4. A bamboo plant in Patagonia reproduces vegetatively for 100 years. Along with other individuals, it forms a dense stand of plants. Then, in one season, all the individuals in the population flower simultaneously, reproduce sexually, and die. Another 100 years later, the process is repeated.

5. A dandelion seed lands in a well-manicured lawn and germinates that same day. Within a week, the plant has a small rosette of leaves and has produced a flower only a few inches tall. The flower asexually produces a huge number of seeds that are scattered by the wind. A few days later, the plant flowers again.

This short list encompasses a wide range of life histories, but it is only a sample of the huge array of life history patterns in animals and plants. What is the source of this tremendous variety? What selective factors operate on these different organisms? Can we predict which life histories will be common in particular environments? Are there useful ways of organizing the variety of patterns we observe? This chapter will explore the answers to these and related questions.

The term **life history** has come to mean a number of things in ecology. In the broadest sense, it refers to any aspect of the developmental pattern and mode of reproduction of an organism. Here we will explore five fundamental aspects of the life history of a species:

1. *Size.* The mass and dimensions typical of adult individuals of a species.
2. *Metamorphosis.* The presence of a major developmental change in shape or form from the juvenile to the adult.
3. *Diapause.* The presence of a resting stage in the life history.
4. *Senescence.* The process and timing of aging, degeneration, and death.
5. *Reproductive patterns.* The magnitude and timing of reproductive events (clutch size, age at reproductive maturity, size of young, number of reproductive events in a lifetime, amount of parental investment and care, and the like).

Some authors use the term more narrowly, limiting its use to the reproductive pattern of a species. The last part of this chapter examines the reproductive components of life histories, an area of particularly active research in ecology.

In the five life history examples presented above, it is obvious that the environments faced by the five organisms differ markedly. We assume that these environmental differences have selected for different "tactics" for growth, development, metamorphosis, and reproduction. Indeed, life histories represent one of the clearest examples of the interaction of evolution and ecology.

We need to be careful in using such terms as *strategy* and *tactics,* however. In this context they do *not* imply a conscious decision-making process on the part of an organism. Rather, they are a shorthand way of describing the evolutionary basis for the patterns we see. For example, when we say that bamboo has the reproductive strategy described above, we mean that over many thousands of years, those bamboo plants with a long period of vegetative growth followed by simultaneous flowering, sexual reproduction, and death left more offspring than did individuals with other reproductive characteristics. In that particular environment, that pattern of development and reproduction was adaptive and gradually came to predominate in the population. This underlying process is what we mean by the term *strategy.*

Let us now consider each of the fundamental components of life histories. We begin with body size, one of the most easily observed aspects of the life history and one in which there is a great range of variation among species.

The Effects of Body Size

The body sizes of organisms span a tremendous range. In linear dimension this scale extends from bacteria 1 micrometer in length to redwoods 100 meters tall (not including the roots), a span of eight orders of magnitude. Even within a taxon, there is a great range of body sizes (Figure 7.1). For many taxa, the distribution of body sizes is right-skewed, as in the examples in Figure 7.1 (Gardezi and Silva 1999). In addition, mean body size is often negatively correlated with the number of species in the taxon, which suggests that more niches (see Chapter 11) are available for small-bodied taxa.

Presumably, the various sizes we observe in nature are adaptive in important ways, for clearly the size of an organism affects its ecology in many ways. For example, the great height of redwoods gives them access to light unavailable to smaller plants, but at the same time it makes them vulnerable to the destructive effects of wind and lightning in ways that smaller plants may not be. Bacteria are so small that the random bombardment of water molecules may affect their movement (known as Brownian motion), but

their small size confers advantages as well, allowing vital nutrients to enter the organism rapidly by simple diffusion, for example.

For virtually any organism we can consider, body size has an important influence on life. These effects can be ecological, physiological, or both. A water strider can exploit a unique ecological niche, for example, because its mass is small enough to enable the animal to "walk" on the surface of calm water without breaking the surface tension. Physiological effects of size in part determine organisms' food habits and the suitability of habitats. An 80,000-kilogram blue whale requires a huge amount of food to sustain life, but because its surface area (from which heat is lost) is so small relative to its mass (which generates heat), it can inhabit the cold waters of the deep ocean. In contrast, a 10-gram shrew requires a far smaller absolute quantity of food, but its surface area is so large relative to its mass that it loses heat at a tremendous rate and must therefore eat almost continuously.

We can make some generalizations about the effects of body size on other aspects of the biology of species. An organism's total food requirements increase with increasing size, while per-gram food requirements decrease. Larger organisms have somewhat lower risks of predation. Vulnerability to physical factors also varies with size. Certainly, a tall tree is more susceptible to lightning damage, but increased girth may protect it from other effects, such as wind or ice damage. Larger organisms generally have longer life spans and thus longer generation times, which affect the potential rate of evolution via natural selection. Organisms that go through their life cycles very quickly (insects, for example) can often adapt more rapidly to environmental changes.

It has been suggested that within a particular taxonomic group, size tends to increase over evolutionary time; this notion is referred to as Cope's law (Cope 1896). For example, Gotelli and colleagues (1991) showed that in minnows, those species closest to the phylogenetic root of the family are the smallest.

Consistent evidence for Cope's law is difficult to find in the fossil record, however. Although some groups of mammals and birds show increases in size over time, in no group is this trend absolute (Futuyma 1986). For example, the common ancestor of equines, *Hyracotherium*, the dawn horse, was a small animal. Even though the fossil record indicates that the line leading to modern horses increased steadily in size, other lines of horse-like mammals did not show this trend. Thus it would be incorrect to conclude from these data that a general phylogenetic increase in size occurred among equines.

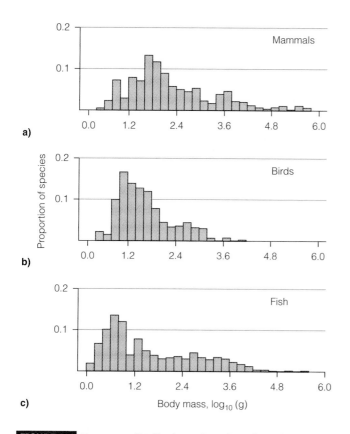

FIGURE 7.1 Frequency distributions of number of species with respect to log body mass for North America mammals (**a**), birds (**b**), and freshwater fish (**c**). (*From Brown, Marquet, and Taper 1993*)

J. H. Brown, P. A. Marquet, and M. L. Taper (1993) have suggested an interesting new hypothesis to account for the evolution of body size in mammals. The distribution of mammalian body sizes has a mode at about 100 grams. According to these researchers, 100 grams represents an optimal body size in this group. In their model, fitness is defined as reproductive power, or the conversion of energy into offspring. Reproductive power is limited by two processes: (1) the acquisition of energy, which increases with mass raised to the 0.75 power, and (2) the rate of conversion of energy to offspring, which changes as a function of mass to the -0.25 power. The model predicts that optimal body size is a balance between these two limitations. Even though the smallest individuals can convert energy to reproductive output at a high rate, their high metabolic rate requires them to spend a lot of their time foraging. In contrast, the largest individuals can obtain resources at a great rate but are constrained by the rate at which the resources are transformed into offspring.

Using the two energy relations described above, Brown, Marquet, and Taper calculate the reproductive power for mammals of different sizes. Reproductive power is maximal for animals of about 100 grams (Figure 7.2), which interestingly is also the mode of the distribution of mammalian body sizes.

In the evolution of size, different parts of the organism may be under different selective pressures, resulting in different patterns of growth for different structures. The situation in which different morpho-

logical characters change at different rates is referred to as **allometry** (literally, "different measures"). The general equation for allometric relationships is $Y = aX^b$ where x and y are measures of some aspect of the organism and b does not equal 1. In Figure 7.3 the relationship between body weight and brain weight for mammals is plotted on a log–log scale. The relationship is described by the equation

$$Y = 0.16X^{0.67}$$

where Y is brain mass and X is body mass. Even though larger mammals have larger brains in an absolute sense, their brains are smaller relative to their body size. For a 10,000-kilogram animal, the brain/body mass ratio is 0.0056. For a 100-gram animal, the ratio is 0.018. In other words, brain weight and body weight change at different rates—that is, allometrically. As shown in Figure 7.3, a log–log plot of these variables yields a straight line (log y = log b + a log x) where log y is the intercept and a is the slope. When $a < 1$, y changes more slowly than x. When $a > 1$, the reverse is true. Sometimes the allometric relationship between two variables is simply a product of geometry. For example, as we have mentioned several times, the surface-to-volume ratio of a small organism is greater than that of a large organism. This is simply a result of the geometric fact that surface area varies as the square of the linear dimension, whereas the volume varies as its cube. In this case, the allometric equation would be $y = bX^{2/3}$, where y is the surface area and x is the volume.

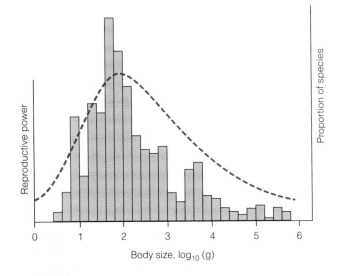

FIGURE 7.2 The curve of reproductive power (dashed line) as a function of log body mass in the model of Brown and colleagues. Note that the distribution predicted by the model closely matches the observed distribution (bars). (*From Brown, Marquet, and Taper 1993*)

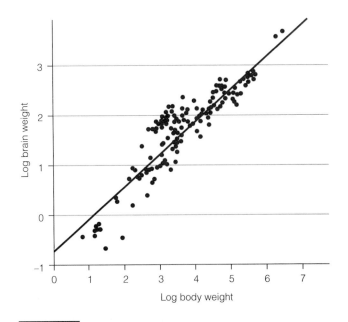

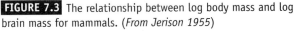

FIGURE 7.3 The relationship between log body mass and log brain mass for mammals. (*From Jerison 1955*)

Interestingly, mammals show a consistent allometric relationship between metabolic rate and body mass in which the slope is 0.75, a phenomenon known as Kleiber's law.

Physiologists have tried to find general patterns and rules for the allometric relationships they observe, and ecologists have tried to use these rules to explain ecological phenomena. For example, in many ecological and physiological allometries, the value of b is 0.25 or 0.75, a phenomenon known as **quarter power scaling.** Enquist et al. (1998) showed that in trees, the basal stem diameter scales to mass to the 0.375 power, a fascinating result because in mammals, the diameter of the trachea and aorta scale to the same power. This lead West et al. (1997) to devise a general allometric model for transport. They reasoned that transport systems should have three properties: (1) The system must reach all parts of the organism. (2) The terminal branches should be approximately the same size in different species, perhaps because of an optimal size for delivery. (3) These systems should be optimally efficient at fluid transport. It turned out that not all branching patterns satisfy these requirements. The optimal design was fractal in nature—that is, the length and radius of vessels change at the same ratio at each branch. The model they developed could then be used to predict the energy use patterns of organisms from their transport ability. West et al. found that their model predicted the 0.75 scaling of metabolic rate and mass. Their model has ecological application as well. A number of ecological factors are predictable directly from their energy/transport model, including the observation that plant and animal population density scales as the -0.75 power of body mass.

These results suggest that there are perhaps some general patterns of scaling according to body size. Because reproduction and body size are related, we will return to allometric relationships when we consider the reproductive aspects of life histories. We have seen in this section that the size attained by an adult is an adaptive feature of the life history. Ultimately, it is a product of selection; proximally, it is a result of the organism's development. In the next section we consider organisms that undergo metamorphic changes that lead to reorganization of the body plan.

Metamorphosis

Organisms that metamorphose undergo radical changes in morphology, physiology, and ecology over the course of their life cycle. A frog egg hatches a tadpole that metamorphoses into an adult frog within a few days or weeks. A fruit fly egg hatches a larva that feeds for a few hours or days and then pupates. In a few days, an adult, winged fly emerges from the pupal case. The changes encompassed by the term **metamorphosis** may be so great that it is reasonable to describe two separate and very different niches (see Chapter 9) for the species. In fact, the larvae of two species may be more like each other than they are like adults of either species.

Organisms that utilize different resources at different stages face an unusual evolutionary problem: Exploiting different niches may be difficult with a single body plan. The solution is a juvenile form specialized for one niche, followed by metamorphosis to an entirely new body plan, adapted to a different niche, in the adult.

Clearly, species that metamorphose must undertake complex genetic and physiological processes in the transformation. These changes require complex regulatory mechanisms that involve turning on and off many genes at appropriate times. In addition, the reorganization of the body plan in a metamorphic species entails considerable energy costs. What sorts of ecological advantages could outweigh these complications?

One prevailing hypothesis is that metamorphic species specialize so as to exploit habitats with high but transient productivity—and hence high potential for growth (Wilbur 1980). Part of this strategy is that specializations for feeding, dispersal, and reproduction are separated across stages. A frog tadpole occupies an aquatic environment (such as a pond) with extremely high potential for growth; the existence of the pond or its high production may be transient, however. Whereas an aquatic larva is not capable of dispersal to new ponds if its habitat becomes unsuitable, the adult frog is. In this case, rapid growth in the larva is separated from dispersal and reproduction in the adult. Although the adult feeds, its growth rate is far less than that of the tadpole. The energy that adults obtain from feeding is dedicated to dispersal and reproduction.

Many insects benefit from the same strategy. Although a butterfly larva feeds voraciously, often on a very specific set of host plant species, the adult does not grow. If it feeds, it does so only to maintain the energy reserves required for dispersal and reproduction. The monarch butterfly exemplifies this strategy (Urquhart 1987). Larvae feed specifically on milkweeds. Pupation also occurs on this host plant. The emerging adults migrate long distances—from all over eastern North America to nine small sites in the Sierra

Madre of Mexico. There, females become sexually mature and migrate north, mating along the way and feeding only to maintain energy reserves. In this example, the feeding-specialist stage is again separated from the dispersal and reproductive stages.

In many marine invertebrates, the pattern is reversed: The larvae are specialized for dispersal, whereas the adults grow and reproduce. A barnacle produces planktonic larvae that move passively in the water column, eventually settling onto the substrate. If the larvae settle in a suitable place, they metamorphose into small adults that are sessile, attached to the substrate. They feed by filter feeding and grow in proportion to the energy they can obtain. Adults reproduce by the release into the water column gametes, which fuse and develop into new planktonic larvae. Clearly, in this case, sessile filter-feeding adults cannot disperse. This aspect of the life cycle is accomplished by the larva.

In all the previous examples, the reproductive function is delegated to the adult. Under certain ecological conditions, however, it is apparently advantageous for reproduction to occur in the larval stage. Thus even the reproductive function typically fulfilled by the adult can be modified by selection under certain circumstances. Species that show this modification of a metamorphic life cycle are said to demonstrate **neoteny,** a life cycle in which the larvae of some populations or races become sexually mature and no longer metamorphose into adults. Neoteny is an example of a secondary adaptation because it could evolve only secondarily—after metamorphosis evolved. To put it another way, neoteny results from the modification of another adaptation, in this case a metamorphic cycle. Some populations of the salamander *Ambystoma maculatum* show this trait. In fact, the larvae of this species were originally classified as a separate species, *Sirenodon mexicanum,* before it was recognized that they are neotenic forms.

The selective factors leading to neoteny are not well understood. We know, however, that neotenic forms are more frequently found in extreme environments, often high altitudes or latitudes. High-altitude populations of *Ambystoma gracile* and *A. tigrinum* have higher frequencies of neoteny than low-elevation populations (Sprules 1974). If the larval environment is rich compared to the harsh adult environment, selection may favor neoteny (Alford and Harris 1988). Licht (1992) has ruled out simple food effects: Supplemental food did not increase the frequency of metamorphosis. Thus it appears that neoteny is a genetically determined feature of some amphibian life histories.

The process of metamorphosis allows the organism to specialize its form and function for life history tasks according to the selective demands of its environment. If, however, it sometimes faces extremely harsh environmental conditions, the developmental process may also include resistant resting stages.

Diapause and Resting Stages

If conditions occasionally or regularly become harsh, it may be advantageous for the organism to have a resistant stage built into the life cycle. In such a life history strategy, the organism forgoes any growth, reproduction, or other activities for a while so that they may occur at a later, more hospitable time. This genetically determined resting stage, characterized by cessation of development and protein synthesis and by suppression of the metabolic rate, is called **diapause.** Many other kinds of resting stages, with different levels of suppression of physiological activities, are known.

The variety of resistant stages among living organisms is huge. Bacteria form highly resistant spores in response to desiccation, heat, and certain chemical environments. The spores of fungi are likewise highly resistant. Of course, many plant seeds fulfill this function as well. Many insects pupate in the fall. The pupae are resistant to the adverse conditions of winter, and the life cycle is simply delayed until favorable conditions return in the spring.

Some of these resistant stages can be extremely long-lived. In one case, seeds of the arctic lupine (*Lupinus arcticus*), a member of the pea family, recovered from ancient lemming burrows in the Arctic, germinated in three days even though they were carbon-dated at more than 10,000 years old (Porsild, Harrington, and Mollisu 1967)!

Organisms face two important abiotic problems. The *severity* of the abiotic regime, including temperature extremes, insufficient or excess water, wind, and so forth, obviously affect organisms' ability to grow and survive. The *unpredictability* of the abiotic conditions is also an important factor. Indeed, unpredictable conditions probably pose greater difficulties for organisms than harsh, predictable conditions. Adaptations to the regular change of seasons in the temperate and polar regions may be relatively simple. For example, many seeds require a period of *stratification*, exposure to low temperatures for some minimum period, before they will germinate. This is a simple adaptation to ensure that germination occurs *following* the winter conditions rather than immediately prior to their onset. Mammalian hibernation is clearly a response to the same problem.

Unfavorable conditions that occur unpredictably pose considerable problems for organisms. Two general strategies are associated with this kind of environment. First, many adaptations are based on a resting stage that awaits favorable conditions. In addition, vulnerable aspects of the life history may be compressed into a short period of favorable conditions. How can organisms cope with the unpredictable onset of good or poor conditions?

We will consider two examples from the vertebrates. The first is the red kangaroo (*Megaleia rufa*). This marsupial inhabits the deserts of central Australia, where the onset of rains and the resulting flush of vegetation are extremely unpredictable. Obviously, it is advantageous for a kangaroo female to produce young at a time when plant productivity is sufficient to support her offspring. For such a relatively large mammal, however, gestation is so long that if a female waited to mate and carry the young until after the rains came, the favorable period might be past. The kangaroo's life history adaptation to this problem involves the use of embryonic diapause during gestation (Figure 7.4).

After a 31-day gestation period, the female gives birth to a tiny, helpless young typical of marsupials. The newborn crawls into the pouch and attaches to a teat, where it continues to grow and develop. After 235 days it leaves the pouch but remains with the mother and obtains milk from her. Two days after parturition, the female goes into estrus and mates again. The fertilized egg enters a 204-day period of embryonic diapause during which it remains in the uterus but does not attach. It then implants, and 31 days later, parturition of the second young occurs. Note that the first young leaves the pouch at just this time. Again the female enters estrus, fertilization occurs, and another diapause follows. The eventual result is that at any one time, the female has three young at various stages of development: one in diapause, one in the pouch, and one on the hoof. If conditions are unfavorable and one of the young dies, another is immediately available to take its place. If conditions suddenly improve, a juvenile is nearly always ready to take advantage of the ephemeral favorable situation.

A similar strategy—accelerated development combined with a resting stage—has also allowed amphibians to inhabit deserts. The spadefoot toads such as Couch's spadefoot toad, *Scaphiopus couchi*, inhabit some of the most severe deserts in North America.

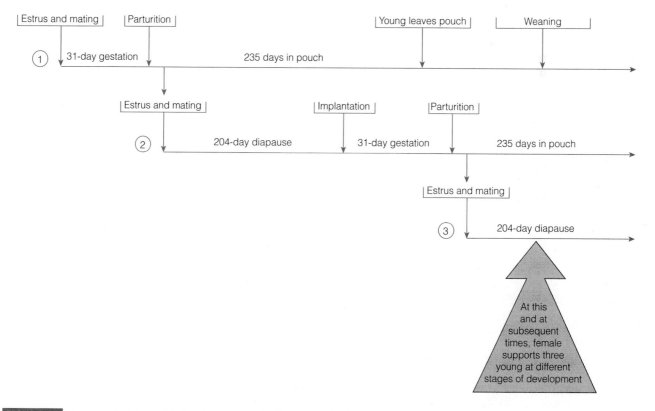

FIGURE 7.4 Timelines (not to scale) showing embryonic diapause and other reproductive events for a single red kangaroo (*Megaleia rufa*) female. After a female's initial pregnancy, estrus occurs two days after parturition. Note that this strategy eventually results in the simultaneous presence of three young at different stages of development.

Adults of this species burrow deeply into the substrate where it is cooler and perhaps more moist. Here they enter a resting state in which they are covered with a protective layer of dead skin. When it rains, the adults emerge and congregate to mate at temporary ponds. Development is greatly accelerated: The eggs hatch within 48 hours, and the tadpoles metamorphose at 16–18 days. Consequently, they can complete the life cycle during the brief "window" of favorable conditions and then return to the resistant resting stage to await the next rainfall.

Many plants of North American deserts are also adapted to unpredictable conditions. In the Mojave and Sonoran Deserts, rainfall is infrequent, and even though it may occur in a predictable season, it often falls over a narrow and unpredictable geographic area in any one year. Many of the plants in these regions have adopted an annual habit: They complete their life cycle in a single year (or season) and die. The seeds are highly resistant to desiccation and can remain dormant in the desert for many years. When rainfall does occur, germination is rapid, and the life cycle is completed quickly while soil moisture is high enough to support growth and flowering (Figure 7.5). Most documented examples of resting stages are adaptations to photoperiod, temperature, or moisture conditions. Recently, Hunter and McNeil (1997) showed that host-plant quality affects larval diapause in a moth, *Choristoneura raceana*. When this species feeds on host plants of low nutritional quality, such as red maple (*Acer rubrum*) or black ash (*Fraxinus nigra*), larvae are more likely to enter diapause than if they feed on chokecherries (*Prunus virginaiana*), a higher-quality species.

Resting stages thus constitute a series of adaptations that allow the species to avoid the most difficult abiotic conditions. Regardless of how well adapted the developmental pattern may be, however, the life span of all organisms is finite. We consider next the underlying basis of mortality.

Senescence

According to the old saying, two things cannot be avoided: death and taxes. We will leave the matter of the inevitability of taxes to social scientists. Here we are concerned with other questions: Why must organisms die? Is death truly inevitable? Could the timing of death evolve? Why do some organisms live so much longer than others? Each of these questions is currently attracting much interest in life history ecology (Avise 1993). Recent findings suggest the possibility that life span has some genetic basis and may, in fact, evolve. For example, C. E. Finch (1990) selectively bred strains of *Drosophila* with life spans 30 percent longer than wild-type flies. Also, newly discovered mutants of the nematode, *Caenorhabditis elegans*, live twice as long as individuals with the wild-type gene (Kenyon et al. 1993).

Biologists define **senescence** as the degenerative changes that result in an increase in expected mortality with age. Eventually, the probability of survival reaches zero (see Figure 4.16). The timing of senescent changes determines the life span. The range of life spans of various kinds of organisms reveals just how fertile an area for inquiry senescence is. At one extreme are desert annual plants that germinate, grow, flower, and die all within a few days. At the other end of the spectrum are plants with life spans measured in thousands of years, such as the bristlecone pine, *Pinus aristata*. The oldest known individual bristlecone pine is more than 5000 years old. Even among species far from these maxima, some organisms have remarkably long life spans. Sturgeon are known to live as

FIGURE 7.5 Desert annuals blooming after winter rains. Long-lived seeds of many desert annuals remain viable in the ground until rain occurs. (*Photo by David Krohne*)

long as 150 years. The lowly earthworm can live for 10 years. Captive tarantulas have lived for several decades.

Species that clone themselves are especially interesting, although it is difficult to identify the life span of a single clonal individual. For example, creosote bushes in the Mojave Desert live at least as long as bristlecone pines. As a single individual grows, however, it forms a clonal ring in which the older parts of the plant in the center of the ring die. A single such genet may be many thousands of years old, even though the living parts of the plant are not that old. Recently, huge living fungi of the species *Armillaria bulbosa*, which are perhaps 10,000 or more years old, have been discovered growing in northern forests (Smith, Bruhn, and Anderson 1992). Among animals, soft corals have life spans that rival these plants.

The problem of interpretation is that in clonal species the clone is long-lived, whereas parts of the individual may die back and be replaced. In other words, the ramet may die but the genet lives on (Orive 1995). But which is important for understanding the evolution of senescence? As M. E. Orive points out, most of the theories emphasize the genetic unit, and thus the genet, but it is the ramet that has specific life history characteristics.

Does life span evolve? It was long thought that life span is determined by how "intensely" or how "fast" the organism lives. Small animals are known to have shorter life spans than large ones. Because the per-gram metabolic rate, heart rate, and so on are much higher in small mammals, their shorter life spans were believed to be a consequence of this rapid physiology—the organism simply wears out sooner. This theory had to be discarded, however, when researchers demonstrated that birds generally live longer than mammals of comparable size, even though birds have higher metabolic rates, heat production, and heart rate (Ricklefs, 1998). Other contrary examples are also known. For example, small breeds of dogs generally outlive larger breeds.

Currently, there are two main hypotheses to explain the process of senescence: the mutation accumulation hypothesis and the evolutionary senescence hypothesis. According to the mutation accumulation hypothesis, each cell in an organism is subject to the deleterious effects of natural environmental insults. Ultraviolet radiation, chemicals, and oxygen free radicals produced by the cell degrade the cellular machinery and cause somatic mutations (Gensler and Bernstein 1981). The accumulation of damage ultimately results in a decrease in survivorship with age. According to this hypothesis, senescence *per se* does not evolve; rather, it is the inevitable result of exposure to the environment.

The second hypothesis posits that the pattern of senescence evolves in organisms. Specifically, it holds that the decline in survivorship with age is a result of mutations whose negative effects are felt later in life (Williams 1957). If these mutations act after either all or most reproduction has occurred, then natural selection cannot remove them from the population. They will gradually accumulate, and their effect will be to increase mortality in the older age classes. A version of this hypothesis suggests that the patterns we observe may be caused by genes with pleiotropic effects—in this case, beneficial effects before or during reproduction but negative effects afterward. Such genes would increase in frequency and cause mortality at specific postreproductive ages.

Note a fundamental difference between these two hypotheses: The first is a proximate explanation (see Chapter 1) based on immediate environmental effects on the organism. The second is an ultimate explanation derived from consideration of the evolutionary and selective forces acting on the population.

Central to this argument is the concept of **reproductive value.** The concept of reproductive value provides a selective basis for different life history strategies. We have assumed that species can apportion the energy available to reproduction in many ways, each of which will result in a different-shaped V_x curve. (See Figure 7.6.) Selection favors organisms with the greatest area under the V_x curve—that is, those with the greatest total reproductive value over all ages will leave the greatest proportional number of offspring. We thus have a way to compare the trade-

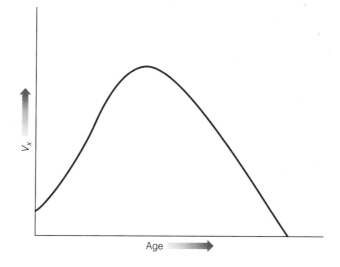

FIGURE 7.6 Reproductive value (V_x) as a function of age for a hypothetical species.

offs (costs and benefits) associated with different life histories.

The reproductive value of an individual of age x, V_x, is defined as

$$V = \sum_{t=x}^{\infty} (l_t/l_x)(m_x)(N_t/N_x) \tag{1}$$

where l_x is the age-specific survivorship, m_x is the age-specific birth rate, N is population size, and x is the current age. The term l_t/l_x represents the probability of surviving from age x to age t. This equation measures the relative reproductive (genetic) contribution of an individual at age x. This "value" is dependent on the probability of survival to future age classes and on the age-specific expected reproduction. Clearly, a low probability of survival should result in a low reproductive value—the organism is not likely to live to reproduce. Similarly, a low reproductive rate means that the individual will not contribute many genes. The effect of population size is less obvious. If the population is declining, N_t/N_x will be greater than 1, which will cause the reproductive value to increase. Conversely, in an increasing population, N_t/N_x will be less than 1, thus reducing V_x. A female that contributes young to an expanding population will make a smaller proportional contribution than one that contributes to a declining population. Note that V_x at age 0 equals R_0 (see Chapter 4).

The age-specific survival and natality rates for the sparrowhawk (*Accipiter nisus*) are shown in Figure 7.7. Note that survival peaks at approximately four years, whereas the number of young produced peaks at age six. The reproductive value, derived from the combined effects of survival and reproduction, is also shown. Natural selection will have less impact on deleterious mutations whose effects are manifest after the peak in reproductive value. Consequently, senes-

cence may result from the accumulation of such mutations.

There is evidence to support both hypotheses of senescence. Cells contain certain enzymes, such as superoxide dismutase, that scavenge oxygen radicals and thus protect cells from damage. When W. C. Orr and R. S. Sohal (1994) developed transgenic *Drosophila* that contained three copies of one of the genes for these enzymes, the flies with extra copies of the gene had 33 percent longer life spans than normal flies.

L. Partridge and K. Fowler (1992) manipulated populations of *Drosophila* to test the mutation accumulation hypothesis. They used artificial selection to generate two genetic lines: an "old line" generated by allowing only the oldest individuals to mate, and a "young line" generated by mating only young individuals. Individuals from the "old line" had longer life spans (Figure 7.8). Partridge and Fowler interpret this finding as evidence that supports the mutation accumulation hypothesis. By selecting older individuals that were capable of breeding, they eliminated the deleterious genes that previously were expressed in older individuals. Thus the life span before the selection experiment was limited by the expression of deleterious alleles.

One line of support for the evolutionary hypothesis has come from comparisons of sexual and asexual species, particularly species that reproduce by binary fission (Bell 1984). In such fissiparous species, two "daughter" individuals are produced. Both are of equal age; there is neither a "parent" nor an "offspring." If the evolutionary hypothesis is correct, postreproductive mutations should not accumulate in such species, at least as compared with sexual species. When G. Bell compared fissiparous oligochaete worms with other sexual invertebrates of similar size, he found that the fissiparous species did not show

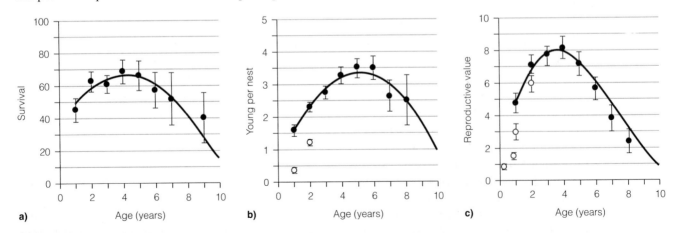

FIGURE 7.7 Age-specific changes in (**a**) mean annual survival (**b**) mean annual production of young, and (**c**) reproductive value of female sparrowhawks. (*From Newton and Rother 1997*)

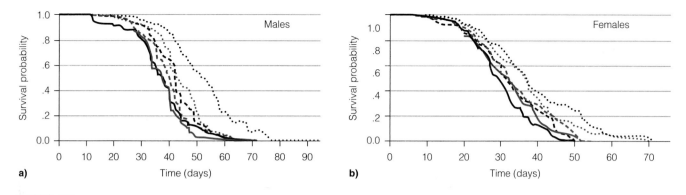

FIGURE 7.8 Survivorship curves for (**a**) male and (**b**) female *Drosophila* from "old" (dashed lines) and "young" (solid lines) selected strains. (*From Partridge and Fowler 1992*)

senescence; that is, there were no significant age-related changes in the probability of survival among "daughter" cells.

Tatar et al. (1997) documented altitudinal variation in senescence in *Melanoplus* grasshoppers in the Sierra Nevada. The evolutionary hypothesis predicts that earlier senescence should evolve at high elevations where late-age reproductive opportunities are limited by the shorter season and harsh conditions. Indeed, low-elevation populations had lower senescence, as depicted by the survivorship curves in Figure 7.9. These data were generated from populations in laboratory cultures under the same conditions, which indicates that the differences in senescence between high- and low-altitude populations have a genetic basis.

These experiments do not conclusively prove the evolutionary hypothesis, because another variable complicates all studies of the evolution of senescence: the act of reproduction. Because reproduction costs energy and may also expose individuals to increased predation or other mortality risks, it is difficult to dis-

entangle the role of reproduction from the potential influence of mutations with age-specific effects. An interesting experiment to demonstrate the effect of reproduction on mortality was devised by W. A. Vanvoorhies (1992). He compared the age-specific mortality rates of two mutant nematodes—one that does not produce mature sperm and one that does not mate—with wild-type animals. The wild-type strain had higher age-specific mortality and a shorter life span than both mutants (Figure 7.10).

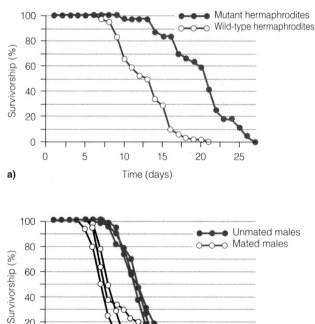

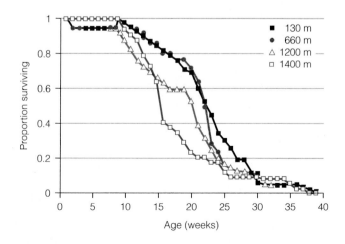

FIGURE 7.9 Survivorship curves for male grasshoppers (*Melanoplus*) from different elevations in the Sierra Nevada. (*From Tatar et al. 1997*)

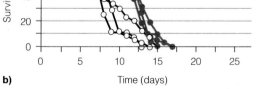

FIGURE 7.10 (**a**) Survivorship curves for nematodes. Open circles are wild-type males; closed circles are mutant males that do not produce sperm. (**b**) Survivorship curves for mated wild-type males (open circles) and mutant males that do not mate (closed circles). (*From Vanvoorhies 1992*)

The genetic basis for senescence and its possible adaptive significance is an area of current research interest among evolutionary ecologists. Mortality, especially if correlated with certain genotypes, is, of course, one of the prime forces behind natural selection. Another is the reproductive contribution of the individual. In the next section we consider the evolution of reproductive patterns.

Reproductive Strategies

The two most fundamental forms of propagation are sexual and asexual reproduction. Asexual reproduction occurs by a variety of means in different groups. Bacteria reproduce by simple binary fission. Other organisms, such as yeasts and many invertebrates, produce new individuals by budding or from fragments of the parent individual. Among the plants, stolons and rhizomes leading from a parent plant may generate a new individual. In other species, such as dandelions, asexual seeds with no fusion of gametes are produced in flowers. In some animals parthenogenetic reproduction occurs. For example, some lizard species in the genus *Cnemidophorus* are composed of entirely female lines. They reproduce via the development of unfertilized eggs.

Clearly, both sexual and asexual strategies are successful, for they are found among virtually all taxa of both plants and animals. From an ecological perspective, asexual reproduction can have advantages. First, asexual reproduction produces exact genetic copies of the organism. In a uniform or unchanging environment, this may be advantageous because there is no need for shuffling of genes by sexual recombination. Second, the organism does not need to find a mate. In certain ecological situations—for example, colonization of new habitats by one or a few individuals—this may be an advantage.

Our discussion of reproduction as a life history strategy focuses on the timing and magnitude of the reproductive events. The fundamental principles associated with this aspect of the life history strategy apply to both sexual and asexual species.

Energetics of Reproduction

We begin with a fundamental property of organisms—that they have available to them a finite amount of energy to devote to reproduction. In Chapter 14 we will consider the factors that determine how much energy is available for this purpose. For now, we can simply assert that this limitation affects every organism. In our consideration of the evolution of reproductive life history strategies, we make two basic assumptions:

1. Because the amount of energy for reproduction is finite, the species must make an evolutionary "decision" about how to apportion that energy in the reproductive process. Reproduction has many aspects, and each consumes energy in a different way. All other things being equal, large clutches or litters require more energy than small ones. Intense or prolonged parental care requires more energy than simply producing fertilized eggs and then abandoning them. To reach sexual maturity at a young age, an organism must expend energy in a different way than if reproduction is delayed until later in life. Thus each organism faces a series of trade-offs in which adopting one sort of reproductive activity precludes adopting another. The advantages of producing a large clutch of eggs must outweigh its energetic costs, because the energy invested in a large clutch is unavailable for other aspects of reproduction.

2. A relationship exists between the demography of a species, particularly its mortality schedule, and its reproductive pattern. As we will see in greater detail later in this chapter, reproduction and mortality interact in a complex way. Each reproductive effort may be expected to increase the mortality rate. The energy invested in reproduction may come from reserves that are important in maintaining life: energy for hibernation, predator avoidance, and so forth. Moreover, nesting and protection of young may have direct mortality costs. In addition to these direct mortality costs, the pattern of age-specific mortality (see Chapter 4) may affect the optimal pattern of reproduction. The reproductive strategy may evolve so as to ensure that reproduction occurs before death.

Trade-Offs in Reproductive Strategies

In this section we address some of the fundamental trade-offs that shape the evolution of reproductive life histories: the number of offspring per reproductive effort, the number of reproductive events, and the age at reproductive maturity. Keep in mind that each of these trade-offs interacts with the others, so changes in one probably affect the others as well.

Number of Offspring per Reproductive Event

A great deal of attention has been paid to clutch size or litter size as a component of the reproductive strategy. This feature is relatively easy to measure in the field—at least compared with some other parameters, such as subtle changes in age at sexual maturity or the number of reproductive events in a lifetime. Moreover, the natural variation both within and among species is great enough so that one can readily test hypotheses by comparing clutch sizes in different environments.

There are two basic strategies for the number of offspring produced per reproductive event: Females can apportion reproductive energy either to a few large offspring or to a large number of smaller individuals.

One might assume that larger litters or clutches are always advantageous—after all, the greater numbers of young produced should increase fitness. However, there are data that indicate that there is an upper bound to the optimal number in a clutch. For example, Jacobsen and Erikstad (1995) studied the effects of artificially increased clutches in the kittiwake (*Rissa tridactyla*). When they increased the naturally occurring broods of two chicks to three chicks, no parents were able to raise the enlarged broods to fledging. Clearly, certain factors that limit how many offspring parents can raise, such as higher feeding costs for the parents and increased conspicuousness to predators, select for clutches smaller than the maximum number of eggs a female might be able to lay.

Other data support the notion that large litters or clutches are too costly in important ways. In the collared lemming, *Dicrostonyx groenlandicus,* the birth weights of pups in large litters are similar to those in small litters. However, the weight at weaning, probably an important factor in survivorship, is significantly lower in large litters (Figure 7.11; Hasler and Banks 1975). In the wood rat, *Neotoma lepida,* mean birth weight declines with increasing litter size (Table 7.1). Even though body weights are no longer significantly different at the time of weaning, a pup from a litter of five takes 2.5 times as long to reach this weight as a pup from a litter of one (Cameron 1973). The cost is sufficiently great in large litters that a smaller proportion of young survives to weaning. Total pup production *declines* for large litters.

Geographic comparisons of litter and clutch sizes reveal variations that may be attributable to similar trade-offs. For many mammals and birds, litter or clutch size increases with increasing latitude (Figure 7.12). Notable exceptions to this "rule" are hibernating mammals and hole-nesting birds, which have litter or clutch sizes smaller than expected on the basis of the general latitudinal trend.

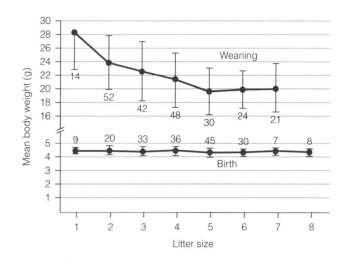

FIGURE 7.11 Mean weaning and birth weights for litters of different sizes in the lemming, *Dicrostonyx*. Numbers refer to number of pups. (*From Hasler and Banks 1975*)

TABLE 7.1

Characteristics of Litters of Different Size in the Wood Rat, *Neotoma*

Litter Size	Number of Litters	Mean Birth Weight (g)	Mean Weaning Weight (g)	Mean Weaning Age (Days)	Percentage of Young to Weaning	Production/ Litter
1	6	10.5	34.4	16.0	100	1
2	3	10.7	34.5	21.8	100	2
3	5	8.8	32.0	26.0	100	3
4	3	7.8	34.0	24.2	75	3
5	2	6.3	35.5	41.8	42	2

From Cameron 1973.

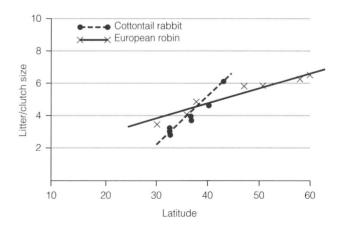

FIGURE 7.12 Litter sizes of the cottontail rabbit (*Sylvilagus floridanus*) and clutch sizes of the European robin (*Erithacus rubecula*) as functions of latitude. (*Data for robins from Lack 1947, 1948a; data for rabbits from Barkalow 1972, Majors 1955, Lowe 1958, Brimley 1923, Llewellyn and Handley 1945, and Dalke 1942*)

Several hypotheses have been suggested to explain this trend. D. Lack (1948b, 1956) proposed that the longer days at higher latitudes allow parents more time to gather food to feed more young, enabling them to support larger clutches. However, this hypothesis does not account for the exceptions, the hibernators and hole-nesters. More important, it is contrary to expectations for mammals, many of which forage nocturnally. M. H. Smith and J. T. McGinnis (1968) and A. W. Spencer and H. W. Steinhoff (1968) hypothesized that a shorter breeding season at high latitudes selects for larger clutches because fewer can be produced. This hypothesis also fails to explain the exceptions. R. D. Lord (1960) and R. E. Ricklefs (1968) offered a third explanation: The higher mortality associated with the more severe northern climates selects for larger clutches or litters. This hypothesis has the advantage of accounting for small clutch sizes in the exceptions. Selection does not favor larger clutches in these species because hibernators avoid much of the climatic severity by being safe underground for many months of the year, and hole-nesting birds are presumably safer from both predation and the elements than other species. It is important to note that all of these explanations are correlative; they rely on correlations among latitude, litter size, and demographic features to provide an explanation of the phenomenon.

Present versus Future Reproduction

We have seen that the reproductive value of an individual is the product of its reproductive output and its probability of survival. If the act of reproduction itself decreases the probability of survival, there is necessarily a trade-off between present reproduction and the possibility of future reproductive events. Thus an optimal reproductive strategy balances the benefits of current reproduction against the costs of future reproduction. We can separate the current and future reproductive values of an individual by modifying the reproductive value in Equation 1:

$$V_x = m_x + \sum_{t=x}^{x} l_t / l_x (m_t) \qquad (2)$$

Here m_x represents the current reproductive value, whereas the rest of the equation characterizes the residual reproductive value. The latter measures the relative contribution of a female of age x over the remainder of her life. We can use the equation to quantify the relative fitness of different strategies.

When we again examine the survivorship curves discussed in Chapter 4 (Figure 4.17), we see that an organism with a Type I survivorship curve has a high probability of survival during its early years. In contrast, species with a Type III survivorship curve experience very high mortality in the young age classes. If sexual maturity occurs during the steep part of the curve for such a species, selection will favor those individuals that reproduce maximally at the earliest opportunity. If sexual maturity occurs later, they run the risk of dying before reproducing. In contrast, organisms with a Type I curve do not have to reproduce maximally as early as possible; reproduction can be spread out over the adult life span because mortality of adults is low enough that survival to the next reproductive event is likely.

As we saw in Chapter 4, Type I survivorship curves are typical of mammals. Dall sheep ewes, for example, are likely to have several reproductive efforts over the course of their life span. We refer to this strategy as **iteroparity.** Species with a Type III curve are more likely to have a single, large reproductive event in their lifetime, a strategy referred to as **semelparity.** Massive semelparous reproduction is sometimes referred to as **big-bang reproduction.**

Note that the terms *semelparity* and *annual* are not strictly synonymous. *Semelparity* refers to the type of reproductive event, whereas *annual* refers to the life span. Similarly, the term *perennial* refers to organisms (most often plants) that live for more than one year, whereas *iteroparity* refers to the number of reproductive events in the life span.

A desert annual plant is an example of an organism with a semelparous life history strategy. The probability of its survival after a brief period of water availability is sufficiently low that selection favors a single, massive reproductive event before death.

Note that the trade-offs discussed above represent correlations; that is, organisms with Type I survivorship curves tend to be iteroparous, and species with Type III curves tend to be semelparous. As with many correlations, disentangling cause and effect is difficult. We have described a survivorship curve that is associated with a particular reproductive strategy. But as we have also seen, the act of reproduction has a survivorship cost, and a Type III curve may actually be a product of a single, massive reproductive effort, not a cause of it.

The classic example of a semelparous reproductive strategy is the sockeye salmon (*Oncorhynchus nerka*). These fish have an anadromous life cycle: They are hatched in freshwater, and after a few weeks there, the juveniles migrate to salt water. There they exploit the tremendous productivity of the temperate and arctic marine environment. After several years in the open ocean, the salmon begin a grueling migration back to the very stream in which their parents reproduced. These migrations can cover more than a thousand miles and require the fish to scale numerous rapids and waterfalls. As the fish re-enter freshwater, a number of physiological and morphological changes occur. The jaws become more hooked and thus less useful for feeding, and their color changes from bright silver to red (Figure 7.13). The energy demands of the upstream migration are so great that the salmon begin to digest their muscles and organs en route and are physiologically depleted by the time they reach the spawning beds. A massive spawning is followed by death. The life history of this species exemplifies the interaction of reproduction and mortality. The journey to the breeding site is so demanding that the resulting high mortality cost dictates semelparity.

This particular life history strategy appears to be based on the way these fish use the extremely productive waters of the ocean. M. R. Gross, R. M. Coleman, and R. M. McDowell (1988) showed that anadromous fish such as the salmon tend to occur in temperate lat-

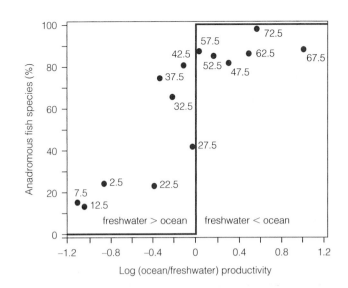

FIGURE 7.14 The frequency of anadromous fish as a function of the log ratio of ocean to freshwater productivity. Log (ocean to freshwater) productivity is zero when the productivity of neighboring oceans and freshwater is equivalent. The heavy bar is the theoretically predicted frequency of anadromy. Freshwater productivity is greater than ocean productivity to the left of the vertical bar, and ocean productivity is greater than freshwater on the right. Numbers by each point indicate latitude. (*From Gross, Coleman, and McDowell 1988*)

itudes, where the oceans are more productive than the freshwater systems. The frequency of anadromous fish increases abruptly when the oceans surpass freshwater in productivity (Figure 7.14).

What is the quantitative nature of the trade-off between semelparity and iteroparity? In 1954, L. C. Cole demonstrated mathematically that the difference between these life history strategies is surprisingly small (Cole 1954). He showed that "for an annual species, the absolute gain in intrinsic population growth which could be achieved by changing to the perennial habit would be exactly equivalent to adding one individual to the clutch size." In other words, $b_a = b_p + 1$ (where b_a and b_p represent the clutch sizes of annual and perennial species, respectively). Subsequent research has shown that the relative values of adult and juvenile survivorship modify the difference between semelparity and iteroparity. Nevertheless, even if the difference is not exactly one individual per reproductive event, the fundamental result—that the differences in clutch size between the two strategies are small—seems robust.

Age at Sexual Maturity

It seems reasonable that selection would push the age at which sexual maturity occurs to the lowest possible age. As R. C. Lewontin (1965) pointed out, the situation is analogous to depositing money in a bank: If money grows by

FIGURE 7.13 Migrating sockeye salmon (**a**). Death follows a massive spawning event (**b**). (*Photos by David Krohne*)

compound interest, then the earlier one deposits, the greater the ultimate return. Because populations (and the proportional contribution of individual females) grow in the same way, directional selection may favor a young age at sexual maturity. Cole (1954) showed that the advantage of early sexual maturity is even greater for semelparous species.

This aspect of the reproductive strategy also involves a trade-off, however, because delaying reproduction may have advantages as well. For example, if juvenile survival is greater than adult survival and there is a huge cost to reproduction, delaying reproduction (reaching sexual maturity later) may be favored. In addition, if reproductive success depends on age, size, status, or experience, delayed reproduction may be advantageous. In such cases, the organism trades a higher probability of not surviving to reproduce for a greater potential reproductive success.

The Theory of *r* and *K* Selection

In the preceding section, we examined a series of trade-offs and considered each as an independent result of selective effects. However, these aspects of the reproductive strategy are clearly interconnected in complex ways. R. H. MacArthur and E. O. Wilson (1967) developed the theory of *r* and *K* selection to provide a comprehensive explanation of the forms that reproductive strategies can take. The theory was originally developed as part of their theory of island biogeography (see Chapter 12).

MacArthur and Wilson suggested that organisms can be classified into two fundamental groups on the basis of their position on the sigmoid growth curve (Figure 7.15) and the resulting life histories. Some species, called ***r*-selected species,** are commonly found at low population densities, where growth is exponential. These species are regularly decimated by catastrophes in the physical environment, such as storms, fire, or drought. Other species are commonly found at much higher population densities, near the carrying capacity (*K*); these so-called ***K*-selected species** face intense intraspecific competition for scarce resources.

MacArthur and Wilson proposed that for these two fundamentally different kinds of organisms, distinct life history strategies will evolve. For species near *K*, selection puts a premium on the production of large, competitive offspring that can gain access to limited resources. For species commonly in exponential growth, on the other hand, selection favors maximum energetic investment in reproduction at the expense of competitive ability (Table 7.2). Reproduction must

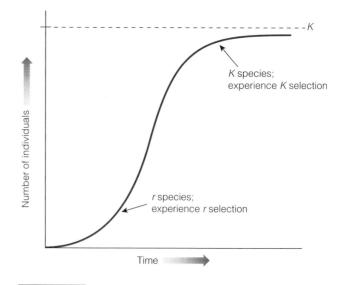

FIGURE 7.15 The positions of *r* and *K* species on the sigmoid population growth curve.

occur quickly and at high volume, because the environment's ability to support the population will soon deteriorate. The reproductive strategy for this species includes some or all of the features that serve to increase the value of *r*: increased clutch or litter size, small size of offspring at birth, little or no parental care, and early age at sexual maturity. The emphasis on rapid production before mortality intervenes selects for semelparity.

Do life histories allow the predictions of the *r* and *K* model? A gross comparison of organisms suggests that the theory applies. Elephants have demographic attributes of species closer to the carrying capacity

TABLE 7.2
Characteristics of *r*-selected and *K*-selected Species

Feature	*r*-selected Species	*K*-selected Species
Population size	Highly variable; often in log phase	Constant, high near *K*
Mortality	Unpredictable; sometimes high	Constant
Survivorship curve	Generally Type I	Generally Type II or III
Competition	Weak	Intense (especially intraspecific)
Reproduction and development	Low age at maturity Rapid growth Semelparity Large litter or clutch size Little or no parental care	High age at maturity Slower growth Iteroparity Small litter or clutch size Much parental care
Emphasis	Production Quality of offspring	Efficiency Quality of offspring

than do, say, *Drosophila,* which are more sensitive to the vagaries of the abiotic environment. Moreover, the reproductive patterns of the two groups differ as predicted by the theory. Elephants are iteroparous and have delayed sexual maturity: A single, large young is born, and intensive and extensive parental care is provided. In contrast, *Drosophila* have a short life span. Sexual maturity is achieved in a matter of days, and a huge number of small eggs are laid and abandoned.

In addition to these rather gross comparisons, the theory is supported by comparisons of closely related species with more subtle demographic differences. For example, two vole species, *Microtus pennsylvanicus* and *M. breweri,* can be distinguished according to the *r* and *K* theory. These species of herbivorous grassland rodents are very closely related. *M. breweri* is endemic to Muskeget Island off the coast of Massachusetts. Genetic data indicate that it evolved recently from the mainland species, *M. pennsylvanicus*

TABLE 7.3

Comparison of Three Characteristics of the *K*-selected *M. breweri* and the *r*-selected *M. pennsylvanicus*

Characteristic	*M. breweri*	*M. pennsylvanicus*
Mean body weight (g)	54.0	44.7
Mean litter size	3.5	4.5
Mean life span (weeks)	13.5	9.5

From Tamarin 1978

(Tamarin 1978). The island form does not experience the high-amplitude cycles that are characteristic of *M. pennsylvanicus* and many other microtines (Figure 7.16). Instead, it remains at relatively high density. *M. breweri* has characteristics associated with *K* selection, whereas *M. pennsylvanicus* exhibits a more *r*-selected strategy (Table 7.3).

A series of elegant studies by M. Gadgil and O. T. Solbrig (1972) demonstrated that *r* and *K* selection can occur intraspecifically as well. Gadgil and Solbrig studied the common dandelion, *Taraxacum officianale,* on the grounds of the Concord Field Station in Massachusetts. The species was found in three distinct environments: a well-traveled pathway, a lawn that was regularly mowed, and a successional field that was mowed once a year. The three habitats represented different degrees of density-independent disturbance. Electrophoretic studies on plants collected from the three environments revealed four distinct genotypes that were associated with particular environments. Genotype A was most common in the pathway population, whereas genotype D was most common in the successional field (Table 7.4).

Comparisons of the reproductive strategies of the genotypes revealed that the genotypes common in the highly disturbed pathway were *r*-selected, whereas those in the field were *K*-selected. Genotype D was characterized by adaptations that favored competition over reproduction: fewer flowering heads per plant and

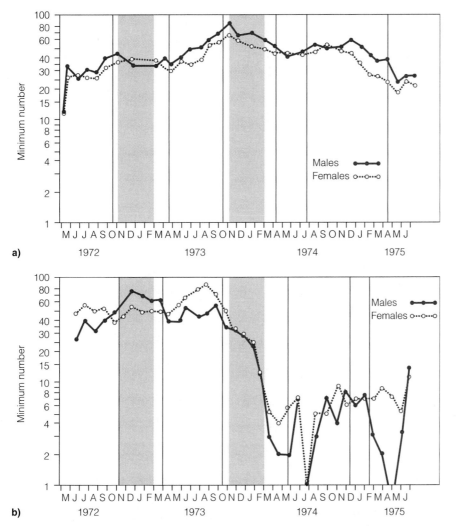

a) 1972 1973 1974 1975

b) 1972 1973 1974 1975

FIGURE 7.16 Log population size over time for **(a)** *M. breweri* and **(b)** *M. pennsylvanicus.* Note that *M. breweri* remains at high density; its population does not experience high-amplitude cycles. *(From Tamarin 1978)*

Life Histories and Conservation

Conservation biologists warn that many species are at risk of extinction as a result of human activities. But are all species equally vulnerable? As we manage ecosystems and predict the impact of our activities, can we assess the likelihood that a species will be threatened with extinction?

In this chapter we have described the essential elements of what ecologists call the life history of a species. Included in this concept are features such as the size, developmental pattern, and reproductive biology of the species. It turns out that these parameters provide important insights because they are key determinants of the vulnerability of a species to extinction.

Let us consider size first. The size of an organism is correlated with a number of features related to its probability of extinction, including generation time (Figure 7B.1). This is true for both plants and animals. The age at sexual maturity is also positively correlated with body size. All other things being equal, large species respond more slowly to environmental change than small ones do. Although larger individuals may be better buffered against abiotic factors, they also require greater total resources to maintain themselves. This leads to the observation (Chapter 8) that home range size increases as a function of body size. One consequence of this is that large organisms generally cannot achieve the population densities that small ones can. As a result, their populations are more vulnerable to accidents or chance decreases in population size that may lead to extinction.

The reproductive strategy employed by an organism also affects its susceptibility to extinction. We saw in Chapter 4 that R_0 measures the average number of female offspring expected over the life span of a female. Recall that the generation time (G) for an organism is the amount of time required for the population to increase by R_0. We derive the relationship of G to R_0 as

follows: If the initial population size is N_0 and after one generation it has grown to N_G, the ratio of the two is R_0:

$$N_G/N_0 = R_0$$

From the relationship in Chapter 4,

$$N_G/N_0 = e^{rG} = R_0$$

Solving for r yields

$$rG = \ln R_0$$
$$r = \ln R_0 /G$$

These equations demonstrate that the intrinsic rate of increase (r) is negatively associated with the generation time. Because there is a positive relationship between generation time and size, we expect large organisms to have a low value of r. Indeed, this is the case. Consequently, large organisms face the added difficulty that should their population size decrease, their low reproductive rate will prevent them from recovering rapidly.

The r and K dichotomy is relevant here. We expect K-selected species to be vulnerable because of a number of interacting factors. Being large, and thus requiring greater

resources, they find themselves near K. Consequently, they compete intensely for limited resources. The optimal reproductive strategy for K-selected species emphasizes quality of offspring rather than quantity: delayed sexual maturity and reproduction, extended parental care, small clutch size, and long interbirth intervals. This array of characteristics makes these species more vulnerable to extinction.

There are abundant examples of K-selected species at considerable risk of extinction. Grizzlies, condors, and elephants all share this combination of features, including large size and very low reproductive rate. Grizzlies do not reach sexual maturity until 4 or 5 years of age; condors produce only a single chick every other year; the gestation period in the elephant is 22 months. Clearly, these species, all of which are threatened or endangered, do not have the reproductive capacity to recover readily from diminished population sizes.

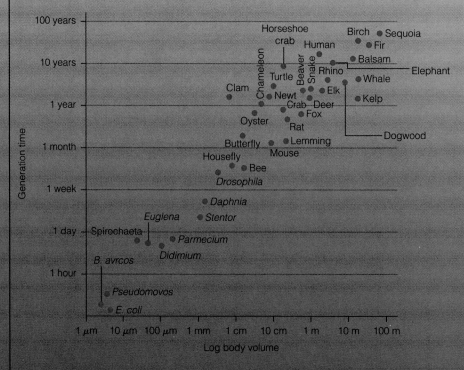

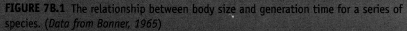

FIGURE 7B.1 The relationship between body size and generation time for a series of species. (*Data from Bonner, 1965*)

TABLE 7.4
Frequencies of the Four Dandelion Genotypes in Populations from Three Habitats and the Number of Flowering Heads Produced by Each

Population	Genotype				Number of Flowering Heads per Plant			
	A	B	C	D	A	B	C	D
1 (pathway)	73	13	14	0	3.6	2.3	1.5	—
2 (lawn)	53	32	14	1	2.6	2.1	1.9	—
3 (field)	17	8	11	64	3.8	2.3	0.5	1.2

From Gadgil and Solbrig 1972.

larger photosynthetic surface area (Figure 7.17). These results are significant in two respects. Not only do they support the general predictions of MacArthur and Wilson's theory, but they also demonstrate that *r* and *K* selection can occur intraspecifically, even for populations that are in close physical proximity.

The many comparisons of this kind that have been made have generally supported the notion of *r* and *K* selection. The dichotomy inherent in the theory is useful as a general classification scheme, although it is generally more useful to think in terms of an *r*–*K* continuum rather than a dichotomy. Species can generally be placed somewhere along this continuum rather than classified as absolute *r* or *K* strategists with all the predicted attributes. The classification of *r* and *K* species also has important conservation implications (see the box on page 184).

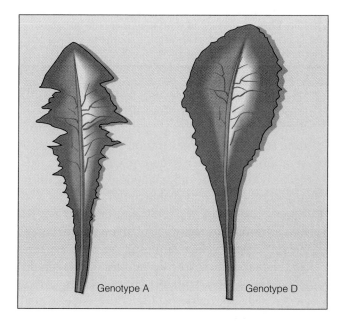

FIGURE 7.17 Leaf shapes of dandelion genotypes A and D grown under identical conditions. Note that the genotype D leaf has a larger photosynthetic surface area, which enhances its competitive ability. (*From Gadgil and Solbrig 1972*)

Genotype A   Genotype D

Not all species fall neatly onto this continuum, however. G. I. Murphy (1968) and W. M. Schaeffer (1974) described what has come to be called a "**bet-hedging**" strategy (Stearns 1976) that combines elements of *r* and *K* selection. According to this theory, if juvenile mortality is variable and occasionally high, neither a classic *r* nor a classic *K* strategy is optimal. An *r* strategy would fail because of the relatively high probability that a massive reproductive effort would coincide with a period of poor juvenile survivorship, and the entire effort would be wasted with little chance of another. Instead, an iteroparous strategy farther toward the *K* end of the spectrum would be optimal: Although such individuals might not be as successful as *r* strategists that produce a large clutch at just the right time (a period of high juvenile survival), they would be less likely to experience a total reproductive failure. Thus the organism "hedges its bet" to ensure that *some* reproduction occurs successfully.

M. S. Boyce and C. M. Perrins (1987) described what may be an example of such a strategy in passerine birds. In Great Britain, great tits (*Parus major*) face large annual variation in juvenile mortality. The average clutch size is 8.5, even though the most productive clutch size—the one that produces the most offspring under optimal conditions—is 12. In response to selective pressures, tits have apparently reduced the mean clutch size as a bet-hedging strategy.

Thus far, we have approached life history strategies from an evolutionary perspective, as adaptations to the ecological situation. The study of these adaptations is complicated by the many interacting facets of the life history. We conclude this chapter with a discussion of some of the difficulties that arise in the study of life histories as adaptations.

Constraints and Ambiguities in the Study of Life History Strategies

The theories of the evolution of life history strategies are optimization theories; that is, they attempt to explain the patterns of life history by determining the optimal life history solution to a particular ecological problem. If an organism matches the hypothesized optimal solution, we conclude that the hypothesis was correct. For example, the theory of *r* and *K* selection hypothesizes that delayed reproduction, small clutches of large offspring, and so on are optimal for species near *K*. We test the hypothesis by analyzing the reproductive strategies of species that face intense competition near *K*.

What do we conclude if the life history does not match the predictions? We discussed the general issue of constraints on the process of evolutionary adaptation in Chapter 2. Theories of optimization represent a good example of the role of such constraints. If the organism's characteristics do not match the predictions of a theory, we may conclude that we should reject the theory as an incorrect explanation of the relevant factors and processes. However, because we are dealing with the evolutionary response of organisms, it is also possible that the poor match results from one or more of the following constraints:

1. Evolution is an ongoing process. Each species or taxonomic group is of a different evolutionary age; no species has completed the evolutionary process. Perhaps our hypothesis is correct, but the species simply has not yet arrived at the predicted solution. It may be that insufficient time, the lack of appropriate random mutations, or other constraints have prevented the evolution of the predicted traits. Life history traits do not evolve in a vacuum; they must be compatible with the other aspects of the physiology, morphology, and biochemistry of a species. For example, in the sand lizard (*Lacerta agilis*), the maximum clutch volume is constrained by the size of the abdominal space available to hold the clutch (Olsson and Shine, 1997).

2. Fitness is a relative property. Although our theory might suggest the optimal life history, selection cannot choose the solution with absolute fitness; it can only choose the fittest of the options available. Our theory may predict the perfect solution, but for the reasons discussed in the previous constraint, selection can only choose the fittest option among those available. (See Chapter 2 for a discussion of fitness and Wright's adaptive landscape.)

3. It may be difficult for us to ascertain the appropriate time scale over which to consider an organism's life history (Stearns 1977). This constraint may make comparisons between theoretical predictions and empirical observations difficult. An example of this situation is depicted in Figure 7.18, where the life histories of two intertidal barnacles are represented by the product of age-specific birth and mortality rates ($l_x b_x$). The curves for the species look very different, depending on whether they are plotted on an absolute or a relative time scale. One could make reasonable arguments for choosing either. We might choose the absolute

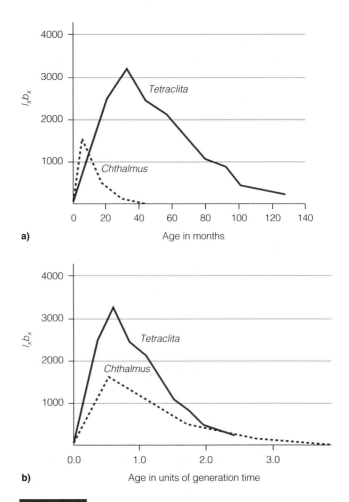

a)

b)

FIGURE 7.18 Demonstration of how the choice of time scale can lead to ambiguity in life history depicted by the product of age-specific birth and mortality rates ($l_x b_x$). **(a)** The life histories of two intertidal barnacles graphed in absolute time. **(b)** The same life histories graphed in generation time. The first graph emphasizes the differences between the two species; the second, the similarities. (*From Stearns 1977*)

scale because abiotic factors change on a yearly, seasonal basis. On the other hand, because evolution works over several generations, a relative time scale (units of generation time) may also be appropriate. In order to choose correctly, we need detailed knowledge of the biology of the organism, and this is often unavailable to us.

4. Allometric relationships may confound our analysis of life history strategies (Kaplan and Salthe 1979). Figure 7.19 plots the relationship between egg mass and maternal mass for two species of salamander. Even though the proportion of maternal mass devoted to egg mass (0.113) is the same in both species, one species, *Ambystoma opacum*, falls exactly on the allometric line for all salamanders (dashed line), whereas the other, *A. tigrinum*, lies above the line. *A. opacum* has a life history like other salamanders; *A. tigrinum* is investing differently than other

salamanders. Which analysis is correct? Our decision might affect whether or not we reject a particular hypothesis about life history evolution.

Even though other constraints and ambiguities complicate our analysis of the evolution of life history strategies, we should not be discouraged from analyzing life histories. Rather, we must take the constraints carefully into account in evaluating empirical studies of this fascinating area of evolutionary ecology.

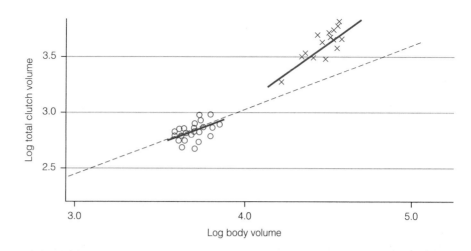

FIGURE 7.19 Allometric relationship between log body volume and log clutch volume in two species of salamanders. The dashed line represents the allometric relationship for all species of salamanders. The *x*'s represent *Ambystoma tigrinum;* the open circles represent *A. opacum.* (*From Kaplan and Salthe 1979*)

SUMMARY

1. The life history strategy of a species consists of four fundamental aspects: size, developmental pattern, senescence, and reproduction. The developmental pattern may include metamorphosis and resting stages. A species's reproductive life history strategy consists of the magnitude and timing of reproductive events, including clutch or litter size, age at reproductive maturity, size of the young, number of reproductive events in the lifetime, and amount of parental care.

2. The life history strategy is an evolved response to the ecological conditions the species faces. Because most life history adaptations have both advantages and disadvantages, the life history strategy represents a series of trade-offs optimized by natural selection.

3. Body sizes of various species span a large range. The potential advantages of small size and of large size vary with the nature of the species. Large organisms may be less vulnerable to predation and to the effects of the abiotic environment (the latter may not be true for large trees), but their absolute energy (food) requirements are large. Small organisms require less total energy but may be affected strongly by abiotic factors.

4. The sizes of different parts of a species may change at different rates, a phenomenon known as allometry.

5. Metamorphosis is an adaptation that allows the individual to exploit different niches and environments in the course of its life span. One important advantage of metamorphosis is specialization of developmental stages for specific tasks such as feeding, dispersal, and reproduction. The disadvantage of metamorphosis is its genetic and physiological complexity.

6. Resting stages allow the organism to survive periods of harsh abiotic conditions or poor resources. Resting stages take a wide variety of forms, including long-lived seeds, resistant spores, and embryonic diapause.

7. There are currently two main hypotheses that have been proposed to explain the process of senescence. One maintains that the life span of an organism is determined by the deleterious effects of the environment, including ultraviolet radiation and oxygen free radicals. The other postulates that aging is an evolved response resulting from deleterious mutations expressed late in life.

8. A finite amount of energy is available to an organism for reproduction: Energy devoted to one aspect of reproduction is unavailable for others. Consequently, each species makes an "evolutionary decision" about how to apportion that energy in its reproductive pattern. The trade-offs in reproductive components include number and size of offspring, number of reproductive events in the lifetime, and age at reproductive maturity.

9. The evolution of the reproductive strategy is also affected by the interaction between mortality and reproduction. Each reproductive event has an effect on the age-specific mortality rate.

10. The theory of *r* and *K* selection has been a useful heuristic model for understanding the reproductive trade-offs that species face. This model proposes that organisms fall into two broad categories: *r*-selected species face density-independent and often catastrophic mortality. Their reproductive strategy emphasizes the production of large numbers of offspring with little energetic investment in each. Components of *r* strategies typically include large clutches, early age at maturity, little parental care, and fewer reproductive events. *K*-selected species are typically at high density near the carrying capacity, where they face intense intraspecific competition. Their reproductive strategy emphasizes producing a few competitive offspring.

SELF-ASSESSMENT: CAN YOU...?

1. Explain the concept of reproductive value and its significance in the evolution of life histories.

2. Explain the nature of allometric relationships and their effect on measurements of life history traits.

3. Discuss the trade-offs associated with the following life history traits:

   metamorphosis
   small/large body size
   diapause
   large/small clutches
   present and future reproduction

   semelparity/iteroparity
   age at sexual maturity

4. Discuss the interaction of demography, especially survivorship curves, and life history strategies.

5. Discuss ambiguities in experimental design that result from the many interactions among life history traits.

6. Discuss the evolution of life history strategies in light of your knowledge of the mechanisms of microevolution (Chapter 2).

PROBLEMS AND STUDY QUESTIONS

1. Explain the relationship between the concept of reproductive value and the evolution of senescence.

2. You are presented with an organism that has the following characteristics: Type III survivorship curve, desert habitat, inhabitant of desert washes (isolated streambeds that are sometimes scoured clean by flash floods), and low vagility (unable to move long distances). Predict the characteristics of the life history strategy of this species.

3. In Chapter 5 we discussed the evolution of dispersal in animals. Discuss whether the pattern of dispersal should be considered a component of the life history strategy.

4. A species of minnow inhabits waters where either sunfish or bass dominate as predators. The sunfish prey intensely on adult, sexually mature minnows; juveniles are largely immune to predation. Where bass are the main predators, predation on immature minnows is sometimes intense, but it is also much more variable in intensity than that by sunfish. Predation on adults is relaxed.

   a. What would you expect the life history of minnows in waters with sunfish to be? (Consider age at maturity, number of clutches, clutch size, and so on.) Explain.

   b. How would you expect the life history strategy to change if minnows were transplanted to waters in which bass predominate? Explain.

5. The accompanying figure shows the allometric relationship between clutch size and body size in salamanders. The solid line represents the relationship seen for *all species of salamanders*. The x's represent a population of one species in its common low-elevation habitat. The circles represent a newly discovered population at high elevation. Can you tell whether this is a genetic adaptation to high elevation? Why or why not? What are the issues?

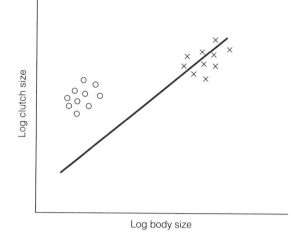

6. Explain why metamorphosis and resting stages are not part of every species's life history.

7. A strong association exists between a species's survivorship curve and its reproductive strategy. Determining the direction of causality is difficult, however: Does the survivorship curve select for a particular reproductive pattern, or does the reproductive pattern determine the survivorship curve?

    a. Explain the basis of this association.

    b. Devise an experiment that could enable you to determine the direction of causality.

8. Discuss the evolution of life history strategies in light of your understanding of the constraints on evolution and adaptation (Chapter 2). How do we interpret a life history strategy that is different from what we expect to be optimal?

9. Draw the relationship between V_x and age that you would expect for an r-selected species and for a K-selected species.

10. Discuss the reasons why organisms may evolve life spans that extend beyond the last age at which reproduction occurs.

PROJECTS AND ADDITIONAL STUDY

1. Design a hypothetical Leslie Matrix (Chapter 4) to examine the effects of changes in litter size, age at sexual maturity, or number of reproductive events on the fitness of individuals.

2. Perform a literature search and analyze the literature on the hypotheses for the latitudinal increase in clutch and litter size. Include a review of the hypotheses to explain hole-nesters as exceptions.

3. Use literature reports to measure the relationship between body mass and seed mass in selected taxonomic groups of plants.

4. Use U.S. Census data from the Internet or elsewhere to plot the reproductive value curve for humans in the United States.

Behavioral Ecology

Thus far we have discussed the essential features of population ecology—population dynamics and regulation, the patterns of intraspecific variation, and the evolution of life history strategies. Each of these topics represents an example of the evolutionary response of a population to a particular ecological situation. In this chapter, which concludes Part II on population ecology, we turn our attention to another form that the response may take: behavior.

The study of behavior in and of itself is so broad that we cannot hope to cover its entire scope. Our purpose in this chapter is not to summarize the entire field but, rather, to focus on the aspects of behavior most directly involved with the ecological interactions in which a population participates. The behavioral elements we will consider fall under the rubric of **behavioral ecology,** the study of behavior in an ecological context. Some behaviors that we will address, such as communication and aggression, represent interactions among conspecifics in the population. Others, such as territoriality and mating systems, are consequences of the effects of abiotic and biotic factors external to the population.

We begin our discussion with a consideration of the genetic and environmental components of behavior. Next we turn to two ecologically important mechanisms of interaction: communication and aggression. Finally, we examine three aspects of social systems that have ecological importance: territoriality, mating systems, and cooperation.

Genetic and Environmental Components of Behavior

We begin with a discussion of the processes by which behavior and social systems arise. In Chapter 2 we argued that behavior, like other components of the phenotype, is subject to the influence of natural selection. We discussed mechanisms by which adaptive behaviors, including sexual selection and kin selection, can arise.

Like all considerations of evolution, that discussion assumed that the traits of interest have a genetic component. If selection is to operate on a behavioral character, there must be a gene or genes that influence the phenotype for that trait—it must be heritable. Behavioral ecology has long involved a dichotomy: We have tended to characterize behavioral traits as either environmentally determined or genetically programmed. This is the classic *nature*-versus-*nurture* argument. Although the current view is that this is a

false dichotomy, it is instructive to review the innate and environmental bases of behavior.

Innate Behavior

Certain kinds of behavior are shared by virtually all members of a species. These behaviors appear in response to a certain stimulus and from the first are performed in their typical stereotypical fashion. Ethologists such as Niko Tinbergen (1951) characterized such behaviors as **innate** or **instinctual.** The specific behavior is referred to as a **fixed action pattern (FAP).** FAPs can be remarkably consistent in form. For example, the courtship behavior of the goldeneye duck (*Bucephala clangula*) includes a movement called the "simple head throw." Within an individual, the mean duration of this FAP never deviates by more than a few hundredths of a second (Dane and vander Kloot 1964).

The FAP is initiated by a sign stimulus, a behavioral or visual trigger generated by another individual. In the common flicker (*Colaptes auratus*), the "mustache patch" of adult males is a sign stimulus that elicits aggression from other males.

The response to a sign stimulus sometimes increases if the stimulus is greater than normal in some way—larger, more frequent, or more intense. A magnified sign stimulus that provokes an exaggerated response is called a **superstimulus.** Superstimuli are most often revealed by the use of artificially enhanced stimuli provided by the experimenter. For example, D. B. E. Magnus (1958) showed that for the male fritillary butterfly (*Argynnis paphia*), the flashing of the female's orange wings is a sign stimulus for mating behavior. When Magnus showed males patterns of mechanically produced orange flashes, he could elicit more intense responses by increasing the flashing rate beyond that observed in nature.

Natural selection has exaggerated many traits to achieve the same effect; many ornate or elaborate structures and behaviors no doubt arose because of their effect as superstimuli. For example, because many avian nest parasites (see Chapter 2) are larger than their hosts, their chicks benefit from the superstimulus effect. Their larger size makes the behavior that stimulates parents to feed their offspring—neck stretching upward, mouth agape, wings fluttering, head bobbing—appear to be exaggerated. Consequently, the parasite chicks receive a disproportionate share of parental care.

Another line of evidence for a genetic component to behavior comes from recent studies of geographic variation in behavioral patterns. Much of the early work in this area assumed that behavior, especially reproductive behavior, is relatively invariant. However, a number of studies have now revealed significant geographic variation in behavior among a wide variety of taxa (Foster, 1999). For example, calling, courtship elements, schooling behavior, migration, and many defensive behaviors show geographic variation across a wide range of taxa. These data imply the existence of underlying genetic variation (and control) for some traits.

Learning

At the other end of the spectrum is learned behavior. One of the most important forms of learning results from the association between certain stimuli in the environment (including interactions among conspecifics) and the response of the individual. This is classical conditioning. We encountered an example of this process in our discussion of mimicry (Chapter 2). Young birds sample visual patterns from the environment. If they attempt to eat an orange and black butterfly and find it unpalatable, they will subsequently associate the color pattern with the bad taste and will avoid such items.

The ability to learn is of obvious adaptive value. We see it in a great variety of organisms, including those we don't usually associate with great mental power. Bees, for example, can learn the locations of their nests and food sources with great accuracy. Clark's nutcrackers (*Nucifraga caryocatactes*) have the ability to learn and remember the locations of up to 9000 caches of pine seeds, each of which may have as many as 10 seeds in it. The information learned while storing the seeds is retained for many months. Experiments in an aviary by R. O. Balda (1980) showed that the birds actually remember the locations of caches rather than using other stimuli (landmarks or olfactory cues) to help in locating them. The birds were allowed to cache food on the floor of the aviary. Then the birds were moved to another cage, the seeds were removed, and the floor was cleaned of all signs of seed storage. When the birds were returned to the aviary after a week, they probed in the locations where they had buried seeds.

The Interaction of Genetic and Environmental Factors

It is becoming increasingly clear that the separation of the innate and environmental bases of behavior is a false dichotomy. The development of virtually any behavioral trait we can imagine has both environmental and genetic components. Even behaviors that we call instinctive (and thus assume to have a large genetic component) involve a complex interaction between the innate and learned elements. We will demonstrate this interaction with two examples.

In Chapter 2 we discussed how alarm calling in the Belding's ground squirrel has evolved by kin selection. Females, who because of their philopatry tend to be surrounded by relatives, are more likely to give warning calls than males who disperse from the family unit. The behavioral interactions among individuals in this species exemplify the interaction between genes and the environment. That interaction complements the information we already have about the role of kin selection in this species.

W. G. Holmes and P. W. Sherman (1982) studied the ontogeny of kin recognition in this species. At birth, some pups were switched to different females in combinations that created four groups of offspring: sibs reared apart, sibs reared together, nonsibs reared apart, and nonsibs reared together. At weaning, the frequencies of aggressive interactions were measured among individuals of the four groups. Holmes and Sherman found that whether they were sibs or not, individuals reared together demonstrated less aggressive behavior toward each other than individuals reared apart. The pups apparently learn to treat the individuals with whom they grow up as kin.

An interesting difference between males and females, however, reveals some genetic effects as well. Female sibs reared apart showed lower rates of aggression toward each other than did male sibs reared apart. In fact, males were aggressive toward any individual with whom they did not share a nest as a juvenile. Some innate mechanism allows females to recognize their female kin and behave less aggressively toward them. This makes sense if you recall that female kin tend to live together and warn each other of predators by using alarm calls. Because males disperse and thus are unlikely to interact with any relatives, male or female, selection for the ability to recognize or behave differently toward kin is probably not very strong.

Extensive studies of the development of song in the white-crowned sparrow (*Zonotrycha leucophrys*) have demonstrated the complexity of the interaction between genes and the environment. P. Marler and M. Tamura (1964) showed that males of this species have song dialects characteristic of the region where they were raised (Figure 8.1).

A series of elegant experiments by M. Konishi (1965) elucidated the interplay between genetics and learning in the development of the male song. The results of

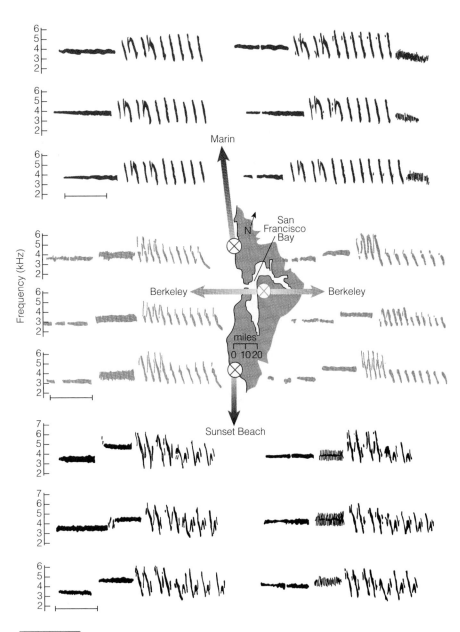

FIGURE 8.1 Song dialects in white-crowned sparrows from Marin, Berkeley, and Sunset Beach, California. Males in each location have their own distinctive song, as revealed in these sonograms of six singers from each location. (*From Alcock 1989*)

In the normal development of song, the male begins singing what is called a *subsong* at about 150 days. This has elements of the full song but lacks distinctive elements, which are acquired by refinement and practice between 150 and 200 days of age. If the male does not hear a white-crowned sparrow song during the critical interval from 10 to 50 days, he develops an abnormal subsong that is never refined by practice into a dialect-specific full song. If the bird is deafened at five months, the full song loses the species-specific characteristics of normal song.

Marler (1970) interpreted the results of these experiments as follows. There is a genetically based template of the basic elements of the song, which means that white-crowned sparrows can learn only a species-specific song. To develop a full song (with a local dialect), however, the male must hear white-crowned sparrow song as a model during the critical period. Then, when he begins to sing, he must be able to hear himself so that he can match his song to the refined template in his memory.

The developmental process is, in fact, far more complex than these experiments suggest. The nature of the social interactions in which the young birds participate also influences the outcome of the song-learning phase. If, for example, a young white-crowned sparrow is housed near another species (say, a finch) in an aviary in which other white-crowned sparrows are singing, it will learn the song of the finch (Baptista and Petrinovich 1984). These experiments are indicative of the complex interaction between the innate and the learned.

Thus far we have seen that the development of behavior is clearly not simply innate or environmental but, rather, a complex interaction between the two. Now we turn our attention to a series of important mechanisms of behavioral interaction.

these experiments are summarized in Figure 8.2. Eggs gathered in the field were hatched in the laboratory, and the individuals were divided among three groups:

1. Chicks maintained in soundproof chambers in which they were not permitted to hear songs of any kind.
2. Chicks allowed to hear songs of their species until they were five months old, at which point they were surgically deafened.
3. A control group of chicks that were permitted to hear white-crowned sparrow song and were not surgically manipulated.

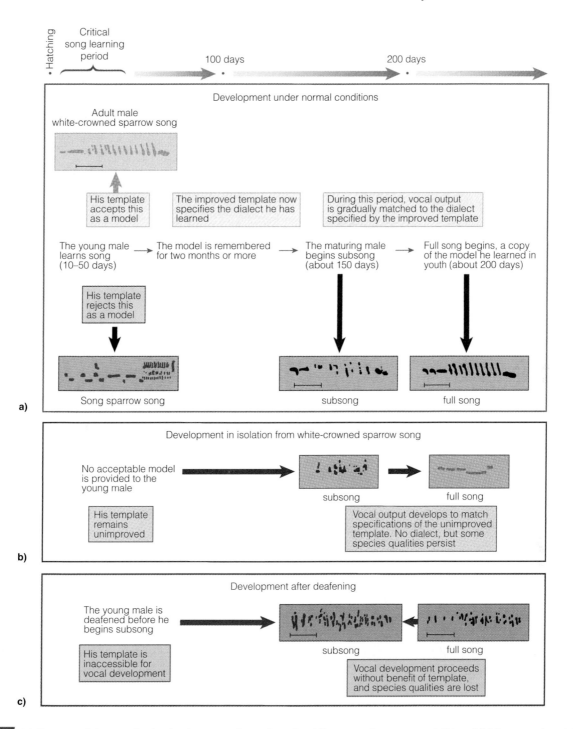

FIGURE 8.2 (a) The essential events in the development of song in male white-crowned sparrows and (b) and (c) two experimental manipulations of song learning. (*From Wilson 1975*)

Mechanisms of Behavioral Interaction

Our purpose here is to consider the kinds of interactions that have particular ecological relevance. Thus our treatment will not be as inclusive as one would find in a behavior text. We will focus on communication, aggression, and territoriality.

Communication

Behavioral ecologists define **communication** as any action on the part of one organism that alters the probability of a behavior in another individual. The fitness of both individuals increases as a result. Thus the signal itself is not communication until it is received and acted on by the other organism. The sending, the receiving, and the elicited behavior are all components of the communication system.

TABLE 8.1
Functions of Communication Among Conspecifics

Type of Communication	Description
Monitoring	Assessment of the presence, demeanor, or behavioral state of neighbors
Contact	Signals that keep individuals aware of each other's presence—often important in dense habitats
Individual and class recognition	Recognition of castes or particular age and sex classes
Status	Identification of rank
Begging	Solicitation of parental care by offspring
Invitation to groom	Solicitation of behavior
Alarm	Warning of the presence of a predator
Distress	Calls for aid
Threat	Show of imminent aggression
Appeasement	Attempts to dissuade aggression
Sexual receptivity	Identification of a female's (or male's) physiology and behavioral readiness to mate
Courtship	Communication of species identification

TABLE 8.2
Properties of Major Channels of Communication

Signal Characteristic	Channel			
Ability to Reach Receiver	Chemical	Acoustic	Visual	Tactile
Range	Long	Long	Medium	Very Short
Transmission rate	Slow	Fast	Fast	Fast
Flow around barrier?	Yes	Yes	No	No
Night use	Yes	Yes	No*	Yes
Information Content Fadeout time	Slow	Fast	Fast	Fast
Locatability of sender	Difficult	Fairly easy	Easy	Easy
Cost to Sender Energy expense	Low	High	Low to moderate	Low

•Except bioluminescent signals
From Alcock 1989.

The functions of animal communication are myriad. A few of the important types are summarized in Table 8.1. Many of the functions outlined in this table have clear ecological roles. For example, in animals that travel in groups in dense vegetation, communicating one's position or even simply one's presence is adaptive. For marine mammals such as whales that travel over vast stretches of the ocean, keeping contact via long-distance signals is adaptive. Providing information to conspecifics, especially the kin group, about the presence and identity of predators is a clear example of the significance of behavior in an ecological context.

In general, the nature of communication signals is correlated with the ecology of the species. A vast array of sensory modes is used for communication. Virtually every mode that allows a signal to be emitted and received has been adapted for communication in animals (Table 8.2). Habitat plays a major role in the kinds of channels used by a species. For a whale trying to communicate with conspecifics over several hundred miles of open ocean, chemical signals would be too slow and would become too diffuse; acoustic signals are more adaptive. For a moth communicating sexual readiness, auditory signals might provide a predatory bat with information on its location; chemical communication is safer. The nature of the signal mechanism is fine-tuned by evolution.

The sounds emitted for acoustic communication are highly adapted for the purpose and for the habitat. For example, to warn of a predator, some birds have evolved alarm calls that are difficult to localize. C. H. Brown and P. M. Waser (1984) have shown that the vocalizations of blue monkeys (*Cercopithecus mitis*), which live in the rain forest canopy, seem to be adapted to transmission over great distances. In that habitat, sounds in the range of 125–200 hertz show less attenuation than those of other frequencies, and there is also less background noise in that frequency range. The vocalizations of the blue monkey fall in the optimal frequency range suggested by these data.

Chemical signals can likewise be fine-tuned by selection. Many insects communicate chemically via pheromones. Selection has modified the signaling system to take advantage of the chemical properties of different kinds of pheromones. For example, some insects use pheromones as sex attractants and as warning signals. To be effective as a sex attractant, a pheromone must be highly specific; that is, a female should attract only males from her own species. Sex pheromones tend to be high-molecular-weight compounds because these molecules have the potential variability, and hence the potential specificity, to be useful as attractants. Alarm pheromones, on the other hand, need to travel far and fast to warn conspecifics as widely as possible; specificity is much less important. Consequently, alarm pheromones tend to be volatile, low-molecular-weight compounds.

Communication signals can be divided into two basic types: discrete signals and graded signals (Wilson 1975). Discrete signals provide simple, digital—yes or no, on or off—kinds of information. Graded signals fall along a continuum of intensity or complexity, and their variability allows the signaler to provide richer, analog information.

FIGURE 8.3 Fireflies (*Photinus* sp.) produce flashes of light that deliver discrete signals. (*Photo by Patti Murray/Animals Animals*)

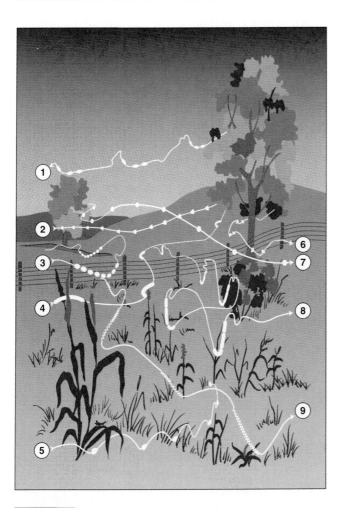

FIGURE 8.4 Discrete signals in the sexual communication of fireflies. The flashing and flight paths of the males belonging to nine species of *Photinus* are shown as they would look in a time-lapse photograph. Each species has a distinct, relatively invariable (and hence discrete) flashing pattern. When a female on the ground observes the pattern of her own species, she flashes in response, attracting the male to her. (*From Lloyd 1966*)

Fireflies in the genus *Photinus* (Figure 8.3) produce discrete signals. Each species has a characteristic flashing pattern that is relatively invariant (Figure 8.4). The flashing patterns of both males and females are highly stereotyped. Males fly across the habitat and flash. The female communicates her receptivity and location by responding with her own flashing, but the flashing provides no information about *how receptive* she is.

In contrast, the waggle dance of the honeybee (*Apis mellifera*) is a series of graded communication signals (Frisch 1967). When a honeybee locates a new food source, she returns to the hive and communicates the direction, distance, and quantity of the food to her nest mates. In performing the waggle dance, depicted in Figure 8.5, the worker bee repeatedly moves in a figure-eight pattern on a flat surface. In the center of the figure eight, she performs a so-called *straight run,* a straight-line movement during which she vibrates (waggles) her abdomen. It is the straight run that contains the graded information.

The direction of the straight run signals the direction to the food source. If the worker is dancing outside the hive on a horizontal surface, the straight run

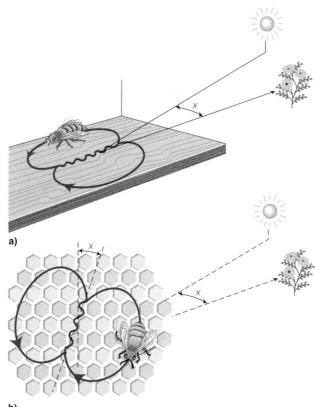

FIGURE 8.5 The waggle dance of the honeybee, which is performed both (**a**) outside the hive and (**b**) inside the hive on a vertical surface. (*From Frisch 1967*)

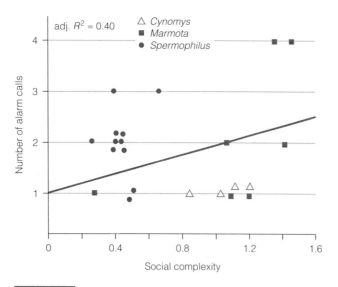

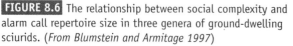

FIGURE 8.6 The relationship between social complexity and alarm call repertoire size in three genera of ground-dwelling sciurids. (*From Blumstein and Armitage 1997*)

is oriented toward the food source. If the dance occurs in the hive on a vertical surface, the straight run occurs at an angle from vertical that corresponds to the angle between the food source and the sun. The distance to the source is communicated by the length of the straight sun, and the quantity of food is indicated by the intensity of the waggle during the straight run. Thus a very complicated and variable set of information is transferred.

Graded information can be transmitted in many ways. Levels of aggression, for example, can be communicated by signals that either change in form or increase in intensity, as when a monkey stops simply staring, moves to a standing position, and finally opens its mouth to show its canines to convey increased hostility.

One might assume that a species's communication repertoire is related to the complexity of the social system. In more complex social systems characterized by cooperation and division of labor in organized groups of kin (Philips and Austad 1990; Hauser 1996), one might expect more complex patterns of communication. Blumstein and Armitage (1997) tested this assumption in ground-dwelling sciurids by comparing alarm call complexity with social complexity. This taxonomic group, which consists of ground squirrels (*Spermophilus* sp.), marmots (*Marmota* sp.), and prairie dogs (*Cynomys* sp.) exhibits a wide range of social systems from nearly solitary species to highly organized cooperating groups. There is a significant positive relationship between the complexity of the social system and the size of the alarm call repertoire (Figure 8.6). Note in this figure that the strongest relationship holds

for the genus *Marmota*. For the other two genera, the size of the communicative repertoire was not as strongly related to social complexity.

Aggression

Aggressive behavior is another means of interaction among conspecifics. Biologists define **aggression** as an act by which one individual achieves dominance over another by physical violence or the threat of it. Here we ignore aggression toward other species. Such behavior is typically considered an aspect of predation, defense against predators, or competition.

Aggression within species is observed in a number of contexts. Territories may be guarded aggressively to ensure sole access to their resources. Dominance of one individual over another may be enforced by aggression. In some instances, males threaten or attack females to force them to mate or to prevent them from leaving for other males. Parents commonly use aggressive behavior to control their offspring and to force weaning or fledging at the appropriate time. In all these instances, the aggressor's actions have the potential to increase his or her fitness.

Generally, aggression is reserved for responding to the most direct threats to fitness. For example, wolves (*Canis lupus*) have a strong dominance hierarchy. Only the alpha, or dominant, male mates. Clearly, an individual's status in the pack has a direct bearing on genetic fitness. In some pinnipeds (Figure 8.7), such as fur seals (*Arctocephalus*) and sea lions (*Otaria, Zalophus*), males aggressively guard harems of females against intrusion of other males and desertion. In both of these instances, important genetic values are at stake.

FIGURE 8.7 Many species of pinnipeds form harems such as the one shown here. A single dominant bull guards a group of females from other males. (*Photo by Eastcott/Momatiuk/Animals Animals*)

In other cases, the effects on fitness are less direct but no less important. As resources become scarce, for example, intraspecific aggression may increase significantly. The incidence of wounding in the voles *Microtus pennsylvanicus, M. ochrogaster,* and *M. californicus* increases during the winter months when resources are scarcest (Rose 1979; Rose and Gaines 1976; Lidicker 1973, respectively).

For some time, the conventional wisdom held that most aggressive interactions actually are highly ritualized—that fighting is more ceremonial than lethal. White-tailed deer (*Odocoileus virginianus*) bucks fighting over an estrus doe do little more than spar. To understand that the bucks' antlers and hooves can be used as much more formidable weapons than they are in these "battles," one need only witness their use when a deer is threatened by a coyote or wolf. The argument was made that this ritualized pattern evolved because the fitness of the species depends on protecting important members from lethal combat that would deprive them of the opportunity to pass on their genes.

To make evolutionary sense, however, this hypothesis perhaps requires a "group selection argument," because the benefits of this behavior accrue to the group. Groups whose members possess this trait have an advantage over other groups that fight to the death. The problem with a group selection argument arises when the advantage for the group is different from what is advantageous for the individual. Because males who fought a little harder would have higher fitness, especially in groups in which most individuals were programmed for ritual fighting, genes for more aggressive combat would increase in frequency and destroy any fitness advantage that the group enjoyed.

It is more likely that the adaptive basis for ritualization of such behavior involves individual fitness. It may be to both antagonists' advantage to recognize quickly, and with minimum trauma, which is the superior animal, so that both the winner and the loser can live to fight (and breed) again. The younger or smaller individual often loses such an encounter, but given time, that individual may well become large enough to dominate others. Thus his fitness will be greater if he lives long enough to realize this advantage.

Under these circumstances, individuals may adopt strategies that take advantage of the reluctance of inferior individuals to challenge. For example, one result of the great tit (*Parus major*) having as many as eight different territorial songs, each of which is sung from a different perch on its territory, is that a nonterritorial male may "conclude" that he faces too many territory holders and that his best strategy is to seek a less fully occupied area (Krebs, Ashcroft, and Webber 1978). A similar phenomenon has been described in red-winged blackbirds (*Agelaius phoeniceus*). When recording equipment was used to broadcast a variety of red-winged blackbird songs from a territory, fewer intruders were detected than in experiments in which a single song was broadcast.

These examples of ritualized aggression aside, there is ample evidence of serious aggression among conspecifics under natural conditions. In spotted hyenas (*Crocuta crocuta*), which are among the most violently aggressive mammals known (Kruuk 1972), intense competition and aggression occur over kills that the pack has made. In response to selective pressure, females are larger and more aggressive than males, presumably to ensure access to kills for themselves and their offspring (Frank 1986). They also are anatomically masculinized: The clitoris is enlarged, and the vaginal labia are folded to form a scrotum. Presumably, this masculinization of females is a by-product of the selection for increased size and aggressiveness, which was achieved via increased androgen levels in females (Kruuk 1972).

The high androgen levels and masculinization of females have apparently led to extremely high levels of aggression among neonates as well (Frank 1991). Hyena neonates have elevated androgen levels, precocial motor development, and fully erupted front teeth. Most litters are twins. In same-sex litters, neonates fight to the death of one of the twins, allowing the surviving pup to grow larger and thus have greater access to food. Because female hyenas do not disperse, death of a female sib results in less competition for status in the clan. For males, who do disperse, the larger size resulting from the death of a competing brother may increase the chances of integration into a new clan after dispersal.

In other species, infanticide (Hrdy 1979) represents another example of lethal aggression. The genetic benefits of infanticidal behavior are well known. For example, G. W. Schaller (1972) observed that male lions (*Panthera leo*) kill cubs when they take over a pride. This is almost certainly done to gain genetic advantage. After taking over a pride, the males kill the cubs and mate with the females, thus ensuring that they share genes with the cubs they care for in the future.

J. O. Wolff and D. M. Cicerello (1989) have suggested that infanticide in the white-footed mouse, *Peromyscus leucopus,* is rooted in both genetic and environmental reasons. It appears that in the males, infanticide is associated with confidence of paternity just as it is in lions. Males that have not sired pups or that have dispersed are infanticidal; males that have sired pups or have home ranges overlapping those of lactating females do not kill pups. In females, by contrast, infanticide appears to be related to crowding. Resident females kill any pups on their territories, related or unrelated, if the pups are outside the female's nest, particularly if the pups are far away from the nest. This suggests to Wolff and Cicerello that the appearance of pups near the edge of the territory indicates high-density conditions and elicits infanticidal behavior by females.

We have seen that aggression is an important means by which an individual can gain access to resources. One important way in which an animal can ensure access to critical resources is by having exclusive use of a particular region—a territory.

Territoriality

Individuals use the available habitat nonrandomly: Each individual uses a subset of all available habitat. The area an individual normally traverses in its daily activities is called the **home range** (Figure 8.8). If the animal has exclusive use of the area by means of aggression or advertisement, we refer to it as a **territory.**

Generally, territorial behavior evolves to ensure access to crucial resources. The kinds of resources thus protected are quite variable among species and include food supply, shelter, access to females, nest

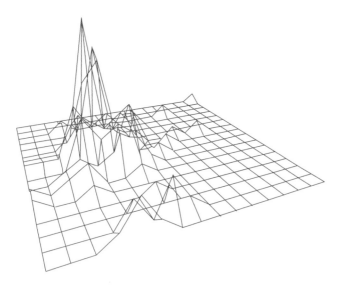

FIGURE 8.8 A three-dimensional depiction of the home range of an elk. The horizontal axes indicate the spatial extent of the home range. The vertical axis represents the frequency of occurrence in each part of the range. (*Adapted from Anderson 1982*)

site, spawning sites, sexual display sites, hibernation sites, and resting sites. Wilson (1975) has classified territories according to their purposes and characteristics as follows:

Type A: Large defended area; courtship, mating, nesting, and food gathering.

Type B: Large defended area; all breeding but not foraging.

Type C: Small defended area around nest.

Type D: Mating territories; for courtship activities only.

Type E: Roosting or shelter positions.

The type of territoriality seen for a given species clearly represents an interaction between the evolution of behavior and the ecology of the species. The concept of a territory is based on the importance of the resource being defended. In a colony of breeding seabirds, for example, maintaining safe space near the nest is crucial for reproductive success. Thus these birds have type C territories. For insectivorous forest birds, an adequate food supply may be the critical limiting resource, and thus they develop type A territories.

If holding a territory is to be advantageous, then the costs of defense must be outweighed by the advantages of exclusive use, and this suggests that territoriality can be explained by a hypothesis of economic defensibility (Brown 1964). Defense of a territory entails several costs. The energy expended for defense is not available for other activities such as reproduction. Defense also exposes the territory holder to higher mortality because the risk of predation is higher; furthermore, the energy costs of defense may increase mortality rates. Thus it seems reasonable to assume that where we observe territoriality, the benefits exceed the costs.

Several lines of empirical observation suggest that this hypothesis is correct. First, for type A territories, a strong negative correlation exists between the size of the territory and the abundance of the food supply. For example, C. C. Smith (1968) studied the sizes of territories of squirrels in the genus *Tamiasciurus*. At his study site, the cones of the lodgepole pine (*Pinus contorta*) are the primary food source for this animal. When Smith assessed the potential food supply of a territory and measured the rate of food consumption by the squirrels, he found that the mean ratio of food available to that consumed was 1.3, indicating that, on average, territories contained a slight excess of food. In poor habitat, squirrels maintained larger territories.

Other data consistent with the economic defensibility hypothesis come from a very precise energy budget for territorial African golden-winged sunbirds (*Nectarina reichenowi*) developed by F. B. Gill and L. L. Wolf (1975a, 1975b). This species sometimes defends patches of a mint species that produces nectar, the sunbirds' source of nutrition. Gill and Wolf measured the range of nectar production over all mint patches and determined how long sunbirds must forage to collect enough nectar each day. From known values for the energy costs of perching, foraging for nectar, and rapid flying, and from detailed time budgets for birds defending territories, the researchers could calculate the energy costs of territoriality.

The economic balance sheet is complex (Table 8.3). As the nectar production of a patch increases, the time required to obtain enough calories to sustain the bird for one day declines. At the bottom of the table is a cost–benefit analysis for defending territories with different nectar production levels. If the production rate of a defended territory is twice that of an undefended site, the bird gains 2400 calories per day by maintaining the territory. This advantage declines to zero if a territory is no more productive than an undefended site. At this level of productivity, territoriality is abandoned.

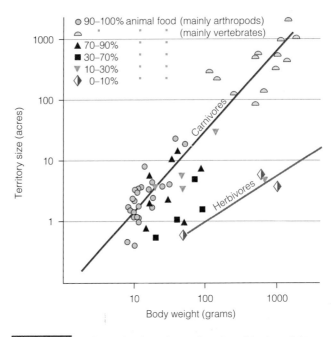

FIGURE 8.9 Territory size (acres) as a function of body weight (grams) for birds of various feeding categories. Each point represents a different species. The different symbols indicate the amount of animal matter in the diet. (*From Schoener 1968*)

The behavior of the birds in the field indicates that they respond to the economics of the situation. When production is high everywhere, sunbirds do not defend territories. Any heterogeneity in nectar production among patches, however, leads to territoriality. If bird density increases such that the costs of defense increase, territoriality is again abandoned.

Another line of evidence in support of the economic defensibility hypothesis is the observation that territory size increases as a function of body mass (McNab 1963; Figure 8.9). This relationship is based on the notion that the energy requirements of an animal increase with its mass. Thus larger animals have higher energy requirements and need larger feeding territories to sustain them.

Note in Figure 8.9 that the relationship is different for herbivores and carnivores. The slope is greater for carnivores. This makes intuitive sense, because carnivores are higher in the food chain and the numbers of their prey are smaller (compared to the "prey" of herbivores). Thus a larger territory is required to maintain a carnivore.

Social Systems

Like behavioral mechanisms, social systems form a broad field of study. Again we will concentrate on those aspects of social systems that are derived from

TABLE 8.3
Energy Balance Sheet for Sunbirds in Territories of Different Quality

Activities	Expenses/Hour (calories)
Resting on perch	400
Foraging for nectar	1000
Chasing intruders from territory	3000

Nectar Production (microliters per blossom per day)	*Hours Required to Collect Sufficient Nectar for 1 Day*
1	8
2	4
3	2.7
4	2

Nectar Production

In Undefended Site	*In Territory*	*Hours of Resting Gained*	*Calories Saved**
1	2	8 − 4 = 4	2400
2	3	4 − 2.7 = 1.3	780
4	4	2 − 2 = 0	0

*For each hour spent resting instead of foraging, a bird expends 400 instead of 1000 calories, which equals 600 calories saved per hour. Total calories saved = 600 × number of hours spent resting instead of foraging.

From Gill and Wolf 1975b.

the ecological and evolutionary interactions within species.

Animal social interactions encompass a wide range of systems. In some species, individuals are solitary most of the time and come together briefly only for mating. At the other end of the spectrum are species with complex social interactions, including group living and cooperation. We define a **social system** as the characteristic size and composition (if any) of a group and the nature of the behavioral interactions among its members.

The social system exhibited by a species can be described in terms of three fundamental components:

1. *Group size and structure.* The number of individuals that live in proximity to one another is a basic element of the social system. Obviously, group size determines the potential for interactions, both positive and negative, among individuals. The age and sex structure of the group constrains other aspects of the social system. Because group size is clearly related to the nature of the resources—abundant or scarce, randomly distributed or clumped, and so on—this component has strong ecological correlates.

2. *The mating system.* The mating system comprises the relationships between males and females during reproductive activity. The duration of the association between males and females is one component of the mating system; another is the relative number of each sex that participate in breeding, both in terms of the proportion of the population and in terms of the number of partners each individual has.

3. *Cooperation and helping.* In species with large group sizes, the opportunity exists for some level of cooperative interaction among members of the group. We generally categorize as most advanced those social systems with the highest levels of cooperation. As we will see, the ecological situation plays a major role in the development of cooperation.

Our approach makes the explicit assumptions that a social system is a product of evolution and that the evolution of a social system is guided by the particulars of an organism's ecology. A species's trophic position, its habitat, and its demography all constitute selective pressures that impinge on its social biology. These ecological factors interact with the physiology and genetics of the species to determine the course of the evolution of the social system. Next we examine each of the fundamental components of social systems in more detail.

Group Size and Structure

Group size can be a difficult parameter to characterize. Any species that reproduces sexually must have a group of at least one male and one female at some time in the life cycle, even if the individuals live completely solitary lives except for one brief mating period. Thus there is a temporal component to the description of group size.

It is also important to understand the spatial and habitat dynamics of a species in order to characterize its group size. Does a group of ground squirrels living together in an alpine meadow constitute a socially cohesive colony, or is this a local population that is simply restricted to a grassland habitat? Assessment of "group size" requires some information on the nature of the interactions among the group—on its structure. We might define group size differently depending on whether the meadow population is (1) a group of independent individuals who happen to have chosen the same habitat or (2) a group of related females and their offspring.

At one extreme of the group size spectrum are animals that are truly solitary. They live their lives essentially independently of other conspecifics. Although males and females come together briefly to mate and the young interact with the mother for a period before independence, they otherwise carry out their life tasks on their own.

A large proportion of mammalian carnivores are solitary. Species such as the raccoon (*Procyon lotor*), the ring-tailed cat (*Bassariscus astutus*), the mountain lion (*Felis concolor*), and some members of the weasel family such as the American badger (*Taxidea taxus*) and the mink (*Mustela vison*) are truly solitary.

Many of these species are top carnivores. It may be that the scarcity of prey for an animal so high in the food chain demands solitary living. Home range sizes might not be large enough to support even a pair of animals. Those carnivores commonly found in groups, such as wolves (*Canis lupus*) and hyenas (*Crocuta crocuta*), typically must cooperate to take prey larger than any individual could kill.

For animals that live among their conspecifics, the nature of the group can vary considerably. We define several kinds of groups:

Aggregation. An aggregation is a group of conspecifics that includes more than one family unit (parents and offspring) and is localized in a particular area. There is no internal organization or cooperative behavior. An aggregation of seabirds feeding together on a school of herring is an example.

Colony. A colony is a group of conspecifics that form a society of highly integrated individuals. There may be division of labor and/or organization of individuals into specialized castes. A honeybee hive is a colony.

Group. This term has two meanings. We have already used it as a generic term to mean a band, assemblage, or party of individuals without regard to the degree of social organization. It also is sometimes used to refer to a specific kind of assemblage that is intermediate between an aggregation and a colony. In this usage it is the most flexible term associated with sets of individuals found in the same place at the same time. Even though interaction may occur among nonrelated members of the group, such a group lacks the level of organization of a colony. Moreover, the temporal nature of the association is highly variable: The group may remain in contact for some time or break apart often. Various species of primates form groups for foraging or resting, and considerable behavioral interaction occurs among group members.

The evolutionary "decision" about whether to join a group (in the generic sense) is based on the relative advantages of solitary living and group living. We can assume that each lifestyle will entail costs. In a group the individual may receive protection and warning from predators. It may be able to find or capture prey that would otherwise remain unavailable to it, and it may receive protection from abiotic factors via huddling or other forms of shelter. On the other hand, because individuals in groups also face intraspecific competition for resources, they must be able to participate in the complex behavioral interactions that must occur in the group.

An example of the balance that must be struck in the evolution of group living—the optimization of group size—is shown in Figure 8.10. Here we see that as the number of individuals in the group increases, the energy derived from the habitat increases, but so does the total amount of energy required to support the group. The optimal group size is found at the intersection of those two functions. Beyond that point, the energy requirement of the group outstrips the habitat's energy yield.

A more general model is shown in Figure 8.11. If we plot two social contributors to fitness, A and B, as a function of group size, we see that the inclusive fitness of each changes with group size. For example, A might be defense against predators. At small group sizes, there are not enough eyes and ears to provide warning of predators, so there is no advantage to a small

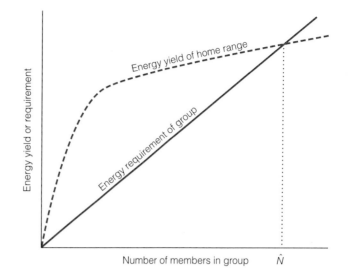

FIGURE 8.10 A model of optimization of group size, expressed as a function of the habitat's energy yield and the group's energy requirements. If unaltered by other selection pressures, the optimal group size ($\hat{N}$) should change via evolution to a point at which the habitat's energy yield is fully utilized.

group. At intermediate group sizes, enough individuals are present to detect predators, and the individuals benefit from the group defense. At large group sizes, fitness declines again, perhaps because large groups are more conspicuous to predators and thus are attacked more often. B might represent group foraging. At small group sizes, there is a net loss in fitness. Perhaps there are insufficient prey items to support the group. But at larger group sizes, cooperative group

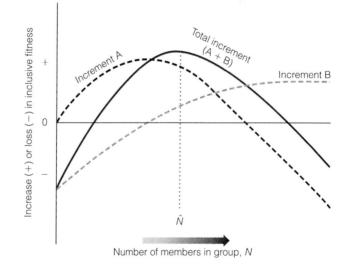

FIGURE 8.11 A more general model of optimal group size, given as a function of the maximum sumed components of group fitness. Two social contributors to fitness, indicated by A and B, could represent, for example, the increments due to superior group foraging and the increment due to superior group defenses against predators.

foraging becomes possible, and fitness increases. The optimal group size is where the sum of these two functions is maximally positive.

The evolution of group size is an example of an optimization process in which the species balances the costs and benefits of group size and structure specific to its ecological situation. Both solitary and group-living species must come together to mate, however. In the next section we consider the evolutionary ecology of mating systems.

Mating System

We describe the nature of the relationships among males and females during breeding as the **mating system.** The mating system traditionally has three components: (1) the number of mates an individual takes; (2) whether the male and female form a pair bond—that is, remain together for at least one mating; and (3) the duration of the pair bond. These components vary tremendously among animals.

The evolution of the three components of the mating system is driven by two related processes: sexual selection and the demography of the species. In Chapter 2, we described the fundamental process of sexual selection. Underlying this phenomenon is the fact that energy investment in reproduction is greater in females. Males can reproduce by contributing only sperm, whereas females invest more heavily in egg production, gestation or incubation, and (in some cases) care of the young. Consequently, once a female mates, her commitment to her offspring precludes additional matings. The result is that for males, females are a limiting resource. The evolutionary result is that sexual selection is driven by female choice of mates and by male–male competition for females.

Recent genetic evidence provides additional information on the basis for sexual selection. It was long assumed that the Y chromosome contains only the sex-determining SRY gene. However, new evidence shows that the Y chromosome contains at least 20 genes or gene families (Roldan and Gomendio (1999). Moreover, this material does not recombine with genes on the X chromosome.

The nature of the Y genes leads some to argue that they are important in sexual selection. In mammals, the male is the heterogametic sex. Some of the genes on the Y chromosome code for testicular development and function. They may play a role in sperm competition if females mate with more than one male. Others affect body size and tooth development, traits that might play a role in male–male contests over females.

Because there is no recombination with the X chromosome, the genes on the Y are inherited paternally in mammals. Traits beneficial to males but detrimental to females can develop in such a system. However, in birds the female is the heterogametic sex. Thus sexual selection for traits on the Y chromosome spread only through the female line and can favor traits beneficial only to females. Roldan and Gomenio (1999) suggest that this explains why female choice of phenotypically costly traits is more common in birds than in mammals.

In addition to sexual selection, there is a feedback interaction between sexual selection and demography. Recall from Chapter 4 the lower age-specific survivorship for males. The competition among males feeds back negatively to affect their survivorship curves. The mating system that evolves is the product of the peculiarities of this feedback interaction and the ecological context of the species.

Polygyny   In polygynous mating systems, males mate with more than one female. This implies (correctly) that not all males in the population necessarily mate—there is variance in male reproductive success. S. T. Emlen and L. W. Oring (1977) suggest that ecological factors play a significant role in the origin of polygyny. If resources are clumped, the potential exists for one male to control a disproportionate share of the resources, leading to intense competition among the males. Because females choose males who control resources, there is a high variance in reproductive success among males, and the potential for polygyny arises.

The other key factor in Emlen and Oring's model is the temporal distribution of potential mates. As the degree of synchrony in sexual receptivity of females decreases, the opportunity for multiple matings increases; if all females are receptive at the same time, it will be difficult for a male to service more than one. There is a limit to this, however. At high levels of asynchrony, the cost of maintaining exclusive use of a resource—in this case, attracting a mate—exceeds the potential reproductive benefits. The tendency toward polygyny decreases again. These effects are depicted in Figure 8.12.

Polygyny can assume several forms, depending on the specific ecological factors that lead to its evolution. **Female defense polygyny** occurs when females associate with each other in groups. Most often the selective pressure for female associations is predation. In gorillas (*Gorilla gorilla*), females travel in groups to avoid leopard predation (Harcourt et al. 1981); males defend such groups from other males. In many pin-

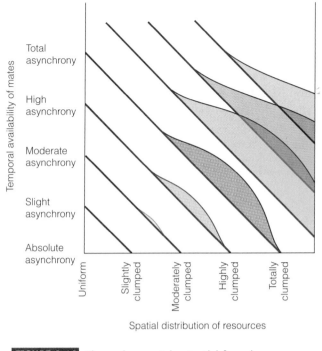

FIGURE 8.12 The environmental potential for polygyny (indicated by the vertical height of the shaded areas) as determined by the temporal availability of mates and the distribution of resources. (*From Emlen and Oring 1977*)

FIGURE 8.13 Sage grouse (*Centrocercus urophasianus*) are an example of a polygynous lekking species. (*Photo by Leonard Lee Rue III/Animals Animals*)

nipeds, such as the elephant seal (*Mirounga angustirostris*), females come ashore en masse to give birth, which is rapidly followed by new matings. These aggregations of females have given rise to the formation of harems of females that are aggressively defended by males (LeBoeuf 1972, 1974; see Figure 8.7).

In **resource defense polygyny,** some environmental resource is clumped, and males defend the resource. Females are attracted to the territory for its resource, and males gain access to multiple females in this way. The key to this system is the existence of some resource that is highly localized and thus highly defensible by a male. For example, the orange-rumped honeyguide (*Indicator xanthonotus*), a bird of Africa and the Indian subcontinent, feeds entirely on beeswax. Males defend bee colonies from other males. Females are attracted to the colonies as a food source and copulate with the male defending a particular colony (Cronin and Sherman 1977). Extreme polygyny results. In one instance, a single territorial male copulated 46 times with at least 18 different females. Males without territories rarely copulated.

Marmots (*Marmota flaviventris*) also exhibit resource defense polygyny. For this mammal, overwintering burrow sites constitute an important ecological limiting factor that can be defended. Males defend such

sites against other males. Females and their offspring congregate at such sites to form a colony associated with a male (Armitage 1962; Downhower and Armitage 1971).

One of the more unusual polygynous systems is the **lek.** In lek systems, the males defend only a tiny portion of a communal display area. Females approach the display area and choose among the males that are displaying there. For example, on sage grouse (*Centrocercus urophasianus;* Figure 8.13) leks in prairie and sage habitats (Wiley 1973), as many as 400 males may display over an area of more than a hectare. On each lek is a mating center where copulation occurs. Only those males that have display territories adjacent to the mating center copulate. Indeed, fewer than 10 percent of the males achieve 75 percent of the matings. Possession of these sites depends on age and experience. Young males move from the periphery toward the center as they mature.

The ecological force that drives the evolution of such a system is apparently a wide distribution of resources, which scatters females over a large area. For example, in lekking sage grouse, the home range of the female may be more than 1000 hectares (Gibson and Bradbury 1986). For a short time, males attract these widely dispersed females to a lek for mating. Otherwise, the energetic costs of seeking and attracting a mate would be prohibitive for both sexes.

Monogamy In monogamous systems, one male mates with one female. The pair bond between them can be brief or lasting; this parameter varies greatly among species. The frequency of monogamy is far higher in birds than in mammals; approximately 90 percent of all bird species are monogamous (Lack

1968). In contrast, fewer than 5 percent of mammals are monogamous (Orians 1969).

The ecological context plays a role in the evolution of monogamy as well. The conditions that favor monogamy are essentially those that prevent polygyny. The fitness advantage to males of multiple matings is so great that essentially the only features that lead to monogamy are those that render polygyny impossible. Thus monogamy is favored by certain uniform to moderately clumped distributions of resources over various degrees of synchrony of female mating receptivity (see Figure 8.12).

Differences between birds and mammals illustrate these conditions. In birds, the timing of resource availability, such as seasonal availability of insects, selects for reproductive synchrony among females. The result is little opportunity for polygyny. Under these circumstances, a male gains more by forming a monogamous pair bond with a female and providing parental care for his offspring. His contribution to their success increases his fitness more than seeking additional copulations could. In mammals, the reproductive physiology of females is the prime reason why monogamy is rare. Because females carry the young for a long period in utero and then nourish them via lactation after birth, males can contribute only indirectly to the care of the young. A male's fitness can best be increased by seeking additional copulations: polygyny.

Analyzing the role of ecological factors in the evolution of monogamy has been difficult, because to determine whether a species is monogamous or polygynous, we require good information about the actual distribution of matings and the numbers of mates per individual of each sex. Historically, obtaining such data has been difficult, particularly for secretive or nocturnal species as many mammals are. Recent advances in genetic techniques, however, now allow ecologists to determine unambiguously the mating system of a species.

For the California deer mouse, *Peromyscus californicus,* for example, researchers had only circumstantial evidence that this rodent is monogamous. In the field, associations of males and females appear to last a long time (McCabe and Blanchard 1950), and in the laboratory, males show some tendencies to care for the young (Gubernick and Alberts 1987). Any ambiguity in the matter was removed when D. O. Ribble (1991) used DNA fingerprinting techniques (Chapter 6) to determine conclusively that males mate with a single female. This procedure has important applications to such questions not only because each individual's "fingerprint" is unique but also because offspring can be unequivocally assigned to parents on the basis of shared elements of the parental banding patterns. Ribble found that in *P. californicus,* males mated with only a single female.

In the case of *P. californicus,* it seems that males defend resources rather than a female. Ribble (1991) found that males will remain on a home range for up to eight months without a mate. Apparently, like marmots, they are gaining access to females by garnering critical resources. The reasons why monogamy resulted rather than resource defense polygyny are not clear.

Polyandry   As we have seen, the ecology and physiology of many species lead to polygyny. The explanations for the evolution of polygyny suggest that the converse system should be less common. Indeed, polyandry, in which females mate with more than one male, is the rarest of the mating systems. In some cases of polyandry, the sex roles are reversed, and males provide most of the parental care.

The greater investment that females typically make in producing and rearing the young makes it unusual for a female to be able to gain in fitness by mating with more than one male. Once a female is inseminated, she generally has sufficient sperm to fertilize her eggs, and additional matings are of little potential fitness value. The female might gain by having multiple males in attendance to provide parental care, but selection generally would not favor males participating in such a system.

Consequently, few species are polyandrous. Some examples of females mating multiple times (with different males) over the course of a breeding season are known among insects, including fruit flies, butterflies, and flies. Among birds, the American jacana (*Jacana spinosa*), a wading bird, exhibits an extreme form of polyandry. Females defend large "superterritories" that contain multiple nests attended by different males.

The ecological situations leading to polyandry may be unique in each species. For one well-studied species, the spotted sandpiper (*Actitis macularia*), extensive studies by L. W. Oring and M. L. Knudson (Oring and Knudson 1973; Oring 1985) have elucidated the key features of this system. Females compete for mating territories, into which two or three males settle. Each male defends a single nest in a portion of the female's territory, providing parental care for the young, while the female defends her territory and attempts to attract still other mates.

According to Oring, a unique accident of the evolutionary history of this bird contributes to polyandry: No matter how much resource abundance varies, each

Behavioral Ecology and Captive Breeding

One of the most important techniques available to conservation biologists is captive breeding of rare or endangered species. When a species becomes so rare that extinction is imminent, the decision is often made to bring some or all of the remaining individuals into captivity so that they can be completely protected; then an attempt is made to increase their numbers through a captive breeding program. This step is taken as a last resort, when the remaining wild individuals are in obvious jeopardy. This is the approach that was taken with the California condor. A number of other species, such as the black-footed ferret and the golden lion tamarin of the Brazilian rain forests, have been saved in this way. When captive breeding is attempted, it is vital to the success of the program that biologists fully understand the behavioral ecology of the species they are trying to help. Without this kind of knowledge, captive breeding may simply increase the probability of extinction. A number of different behavioral and social factors must be understood if captive propagation is to be successful.

Kleiman (1980) illustrates the potential role of social systems in a comparison of the social systems of two mammals, the wolf and the lion. Both are group-living animals—wolves live in packs, lions in prides. The groups of both species are similar in size and composition. At the same time, significant differences exist that could alter the outcome of attempts to rear each in captivity. Wolves are basically monogamous, and only the alpha male and alpha female mate. Other adults participate in rearing the offspring but do not breed. Female wolves are very aggressive to other females. In contrast, lion prides contain several breeding females. The males are usually related to each other. The breeding males are often displaced by others. They do not participate in parental care.

Consequently, the optimal strategy for rearing these species in captivity differs. Whereas a wolf pack must be established by a single pair, a pride of lions can be initiated with a group of related females and a core of related males. Considerable aggression may be associated with the development of a wolf pack; much less aggression occurs in a developing lion pride.

In a number of species, particularly primates, complex relationships and interactions are associated with mating. The considerable male–male fighting in the wild for access to females can be very disruptive and even lethal in captivity and may thus subvert the goals of captive breeding. In some species, such as langurs, breeding units may be invaded and taken over by all-male groups. The result is great social upheaval and sometimes even mortality.

Although it may be possible to eliminate such aggressive interactions in captivity, this course of action may have negative consequences as well. In the rhesus monkey, for example, aggressive behavior and display help males achieve the androgen levels needed for breeding. A captive breeding program that either fails to recognize this system or in some way undermines it may be doomed to failure.

In addition to the mating system, the structure and size of groups may be vitally important to success in captive breeding. For example, talapoin monkeys are polygamous, but males and females tend to segregate; indeed, reproduction is inhibited if the sexes are forced into close contact. If the sexes are not separated in captivity, aggression by females may cause mortality of males. Group size is crucial in some species, such as the short-tailed leaf-nosed bat. In this species, reproduction is higher in groups of 10 to 20 females than in groups of 3 or 4.

female lays only four eggs per clutch. Consequently, a female cannot gain fitness by increasing her clutch size. Instead, she can gain fitness only by defending a high-quality territory and attracting additional mates.

For this bird species, a physiological constraint (fixed clutch size) dictates polyandry, but the ecological situation allows it to be successful. Its food resource (mayflies) is so abundant that a single male can adequately provision a clutch. Furthermore, the demography of the species is such that the sex ratio is skewed toward males, thus providing a reservoir of additional potential mates for the female.

To this point, we have seen the great variety of mating systems and group structures among animals (see the box on page 207 for a discussion of how these systems affect captive breeding). Thus far, our discussion of the evolution of these systems has emphasized the direct fitness interests of individuals. We now focus on the ways in which these same interests lead to cooperative behavior in some species.

Cooperation and Helping

The final component of social systems that we will consider is the degree of cooperation and helping among individuals in the group. According to Wilson (1975), cooperation and division of labor are the hallmarks of highly social species. Some species have truly remarkable levels of cooperation—to the point where they exhibit high levels of altruism. *Helping* is a term sometimes applied to vertebrates who help raise young other than their own.

We will see that some of these arrangements are a result of unusual genetic systems, whereas others are derived from special ecological circumstances. Our treatment of highly integrated and cooperative societies cannot be complete, for every social system is the product of a unique set of circumstances. We will, however, present examples of genetically and ecologically derived systems.

Genetically Based Cooperation   One of the most fascinating examples of cooperation is seen in the hymenopterous insects—the bees and wasps. A colony in a typical hive of the honeybee (*Apis mellifera*) consists of a single fertile female (the queen), relatively few males (drones), and many thousands of sterile females (workers). Eggs, which are produced only by the queen, are fertilized by the males and cared for entirely by the sterile workers. Because all the workers are the female offspring of the queen, all

care for the young is provided by sisters to sisters. They forgo their own reproduction and instead forage, defend the hive, and raise new broods of young. How could this curious system evolve?

The answer is provided by the theory of kin selection (Hamilton 1964; Wilson 1971). The chromosomal sex-determining mechanism in hymenopterous insects differs from that in other animals in that the insects are **haplodiploid,** which means that females are diploid ($2N$) while males are haploid ($1N$). An unfertilized egg develops into a male; a fertilized egg develops into a female. The significance of haplodiploidy is that it skews the typical coefficients of relationship among members of a kin group. In a typical diploid organism, parents and offspring are related by $r = \frac{1}{2}$. Similarly, full sibs are related by $r = \frac{1}{2}$. In haplodiploids, however, sisters are more closely related to each other than parents and offspring are.

The logic is as follows. In a diploid organism, each parent contributes half the genes to the offspring. Each has two sets of genes that it might give, and it is equally likely that either set will be transmitted to the offspring. Thus, the coefficient of relationship between sibs is

$$r = \underbrace{\left(\frac{1}{2} \times \frac{1}{2}\right)}_{\text{male contribution}} + \underbrace{\left(\frac{1}{2} \times \frac{1}{2}\right)}_{\text{female contribution}}$$

$$r = \frac{1}{4} + \frac{1}{4} = \frac{1}{2}$$

The male contribution is derived from the fact that half the offspring's genes are paternal, and it is a 50:50 chance which of his two sets he donates to the offspring. The calculation for the female is the same. By similar logic, the value of r for parents and offspring is $\frac{1}{2}$.

For a haplodiploid, however, the situation is different because the male has only one set of genes to contribute. Thus the value of r for sibs is

$$r = \underbrace{\left(\frac{1}{2} \times 1\right)}_{\text{male contribution}} + \underbrace{\left(\frac{1}{2} \times \frac{1}{2}\right)}_{\text{female contribution}}$$

$$r = \frac{1}{2} + \frac{1}{4} = \frac{3}{4}$$

The parent–offspring relationship is not affected for females. A daughter receives half of her genes from each parent, so $r = \frac{1}{2}$. Therefore, in a haplodiploid, sibs are more closely related ($r = \frac{3}{4}$) than are parents and offspring ($r = \frac{1}{2}$). According to the theory of kin selection, a sister is more valuable than a daughter. This strange calculus explains what amounts to the ultimate in altruism—forgoing one's own reproduction to care for one's sisters.

Even though most biologists accept this explanation, it is not complete because it fails to account for some facts. First, not all hymenopterous insects are social; some live solitary lives with no hint of the extreme altruism shown by some bees and wasps. Second, in some of the highly social bees and wasps, the queen mates with more than one male. This changes the coefficients of relationship such that the difference in degree of relatedness between sisters and mother–daughter pairs decreases. Third, some diploid animals also have social, altruistic societies. For example, termites exhibit the same level of altruism and division of labor as bees, yet they do not have the haplodiploid sex-determining mechanism. One example of such a society is even known among mammals (see below). Thus, even though we can accept kin selection as a partial explanation of the extreme altruism in the hymenoptera, some additional factors remain to be elucidated.

Habitat Saturation and Food Supply

A number of examples of cooperative breeding have evolved among the birds. One of the most thoroughly studied is the acorn woodpecker, *Melanerpes formicivorous* (Koenig and Mumme 1987; Stacey and Koenig 1984; Mumme, Koenig, and Pitelka 1988). The social unit of this species is a territorial group of up to a dozen members (both males and females) that gather and store acorns in thousands of holes drilled in a tree called a granary. This cache provides the major source of food over the winter. Granary trees are developed, maintained, and defended by all members of the group.

This social system has three unusual features:

1. *Division of labor and helpers.* The group is divided into breeders and helpers. Helpers provide care for the young and assist in defending the territory, but they do not breed.
2. *Mate sharing.* There is only one active nest in the group at any one time. Among the breeders, any female may breed with any male, and vice versa. Electrophoretic studies of paternity show that these matings actually do result in *genetic* contributions from the shared mates (Mumme et al. 1985).
3. *Infanticide.* Females kill the first eggs laid by other females. Whenever birds enter a group to fill a vacancy, they kill all the young.

How do we explain the evolution of this social system? First, the habitat is saturated; all good territories are occupied. Consequently, opportunities for young to disperse are extremely limited. Groups actively resist immigration; in fact, only if all the breeders of one sex die is a new individual of that sex allowed to immigrate. Thus, because the young have little opportunity to leave and begin reproducing, they remain in the natal area, where their best fitness option is to contribute to the reproductive success of the group. The structure of the group, especially its size and proportion of helpers, is determined by the amount of mast (nuts fallen to the ground) that has been stored (Figure 8.14).

The importance of habitat saturation is demonstrated by comparative studies of this species in habitats of different productivity. At the Hastings Reservation in California, there is sufficient mast to last the winter. At Water Canyon in New Mexico, the production of nuts is more variable. When the stored food is exhausted, the group abandons its territory. Vacant territories are eventually recolonized. As expected under the habitat saturation hypothesis, fewer birds remain in their natal group at Water Canyon than at Hastings. In southeastern Arizona, acorn production is sporadic and insufficient to support a group through the winter. Granaries are not constructed at this site. At such low productivity, birds must forage widely for many kinds of food. Colonies with helpers do not develop at such sites.

Second, mate sharing results because it provides advantages to the individual breeder. Woodpeckers that share mates raise more young per individual than do monogamous pairs. For example, in groups of three (one female and two males), the per-male production of young is 1.16; in groups of two (one female and one male), it is only 0.92. Moreover, annual mortality is 30 percent for nonsharing birds but only 21 percent and 14 percent for sharing females and males, respectively.

Finally, the infanticide is apparently adaptive because it provides a genetic advantage over other members of the group. Female laying is not precisely synchronous. When the first egg is laid, it is almost always removed and killed by another female. The process is repeated with each egg laid until the second female begins laying eggs. This behavior helps the infanticidal female ensure that the group care is invested in eggs that are predominantly hers. She stops killing eggs when she begins laying because she can no longer be sure whose egg she is killing. The infanticide committed by immigrants is also associated with ensuring that care is preferentially directed to one's own offspring.

One of the most fascinating social systems known among mammals is that of the naked mole-rat,

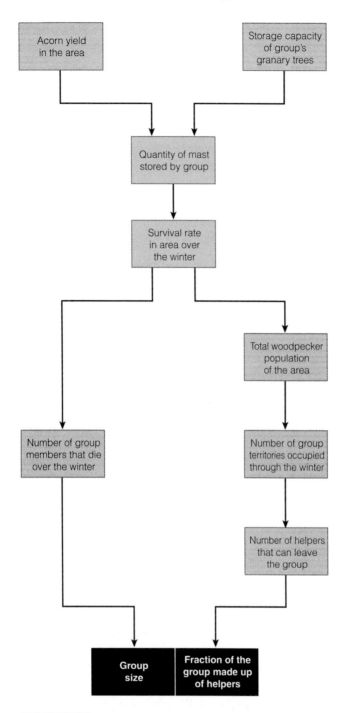

FIGURE 8.14 The structure of the acorn woodpecker group is strongly influenced by the quantity of mast the group can store in the fall. The quantity of store mast is itself determined by the yield from the trees in the group's territory and the storage capacity of the granaries held by the group. The quantity of mast affects how many birds survive the winter. The survival rate influences the total woodpecker population of the area and also the number of desirable territories that remain occupied all year.

If the bird population is large and many territories are occupied, few young can leave their parental groups to breed on their own As a result, the groups are large, and a substantial fraction of each group consists of nonbreeding helpers. The quantity of stored mast also influences the number of young raised successfully by the group. (*From Stacey and Koenig 1984*)

Heterocephalus glaber; Figure 8.15), which is social in ways reminiscent of the honeybees, including extensive division of labor and cooperation (Sherman, Jarvis, and Braude 1992; Sherman, Jarvis, and Alexander 1991). The mole rat is an example of a species whose social system is shaped by the environment and by its effects on the genetic relationships within the group. This rodent species inhabits harsh desert-like habitats in eastern Africa. Naked mole-rats are fossorial and burrow extensively through very hard substrate for the underground plant parts on which they feed.

Mole-rats have been termed **eusocial**—truly social—mammals. Their social system consists of the following unusual elements:

1. *Extensive cooperation.* The group digs cooperatively. Individuals form long chains and push dirt from one individual to another until it is finally kicked out of the burrow entrance.
2. *Division of labor.* Breeders in the colony produce young and provide them with nutrition. The nonbreeders defend the tunnel system and clean and carry pups. Among nonbreeders, there is division of labor according to size. Small individuals maintain the tunnels. Large ones are important in defending against snakes and other intruders, including mole-rats from other colonies.
3. *The presence of a queen.* The queen, who is the largest individual in the colony, is the only female that breeds. The queen is very large and can bear as many as 27 pups in a single litter! She is a despot

FIGURE 8.15 The naked mole rat (*Heterocephalus glaber*) is thought to be one of the few eusocial mammals. (*Photo by Raymond A. Mendez/Animals Animals*)

who patrols the tunnels and shoves the nonbreeding members of the colony to stimulate them to increase their work efforts. The nonbreeding females engage in battles to determine which will replace a dead or sick queen.

This unusual social system affects the genetic relationships within the colony. The restriction on the number of individuals mating results in very high inbreeding within the colony. DNA fingerprinting studies (Faulkes, Abbott, and Mellor 1990; Reeve et al. 1990) have shown that the coefficients of relationship among colony mates are as high as 0.81. Thus they are more closely related than full sibs ($r = 0.50$). Consequently, there is ample opportunity for kin selection to affect the evolution of the behavior of colony mates.

The details of the genetic relationships within a colony support the importance of kin selection. H. K. Reeve (1992) has shown that some of the workers in a colony are lazier than others and require more aggressive shoving by the queen. These individuals tend to be less closely related to the queen than are workers whom she shoves less frequently and aggressively.

P. W. Sherman, J. U. M. Jarvis, and S. H. Braude (1992) and B. G. Lovegrove (1991) hypothesize that, very much as in the case of the acorn woodpecker, habitat saturation and a harsh climate are the key ecological features that have led to the evolution of this social system. Because the food supply is patchily distributed, a colony can monopolize a patch. Because all patches are occupied, dispersal to new patches is not possible. In addition, intense predation makes aboveground travel and dispersal very risky. Thus the young must stay in the natal colony. Digging and foraging are so difficult that cooperation and group digging are advantageous.

The result is selection for the formation of cooperative groups (Figure 8.16). The reproductive suppression of others by the queen leads to a very high genetic

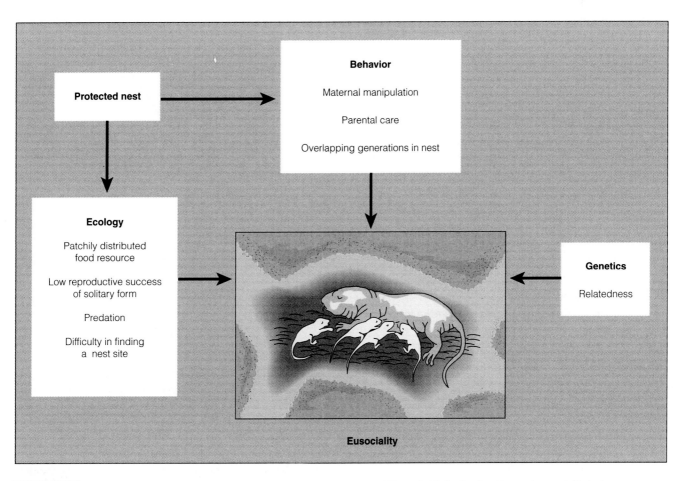

FIGURE 8.16 Genetic, ecological, and behavioral factors that combine to promote the evolution of eusociality in naked mole-rats. Inbreeding due to limited dispersal and underground colonies promotes cooperation and helping. In the harsh desert ecosystem, the cost of dispersal is so high that most individuals are philopatric. Behavioral patterns also contribute to group living. The protected nest tends to reinforce the effect of the harsh habitat and the behavior of naked mole-rats. (*From Honeycutt 1992*)

relatedness among colony members. Consequently, their inclusive fitness is enhanced more by helping to care for the group offspring than by breeding themselves. The specialization of the queen for reproduction results in more offspring for the colony than an individual could expect to produce on its own.

Thus, like that of the acorn woodpecker, the social system of the naked mole-rat is a product of the interaction of the ecological situation and the genetic patterns it generates. It has recently been shown that another member of the same family, the Damaraland mole-rat (*Cryptomys damarensis*), an inhabitant of harsh deserts in South Africa, is also eusocial (Jarvis and Bennett 1993).

The fundamental evolutionary basis of cooperative group living is the relative advantages of living in a family unit. According to S. T. Emlen (1994), such advantages could derive from the benefits of philopatry—that is, remaining in the natal area—or from some ecological constraint. Emlen suggests that initially some limitation on breeding opportunities causes individuals to live in family units. The benefits of philopatry arise secondarily. Over time, individuals modify their behavior to take advantage of the group situation.

Not all examples of cooperative group living fit this notion of habitat saturation, however. Two lines of evidence argue against the role of habitat saturation for the Seychelles warbler, *Acrocephalus seychellensis,* a cooperative breeder (Komdeur 1992). Vacant territories tend to be filled by pre-reproductive birds from equal or lower-quality territories. Furthermore, individuals that delay reproduction in a high-quality territory have higher fitness when they eventually breed there than do individuals that disperse and breed immediately in lower-quality territories. These data favor the idea that the benefits of philopatry drive the change to cooperative breeding.

Still, even in this system there is evidence for the role of habitat saturation. When J. Komdeur moved birds to unoccupied islands, no cooperative breeding occurred until all the high-quality territories were occupied.

The previous example suggests that the food base may be an important factor in cooperative breeding. Recent data and models confirm this (Hatchwell and Wrege 1996; Hatchwell 1999). These researchers developed graphical models to examine the effect of food supply on cooperative breeding. There are fitness effects of the amount of parental care (Figure 8.17). Note that as care increases, so do the reproductive costs and benefits for the parents. A reduction in

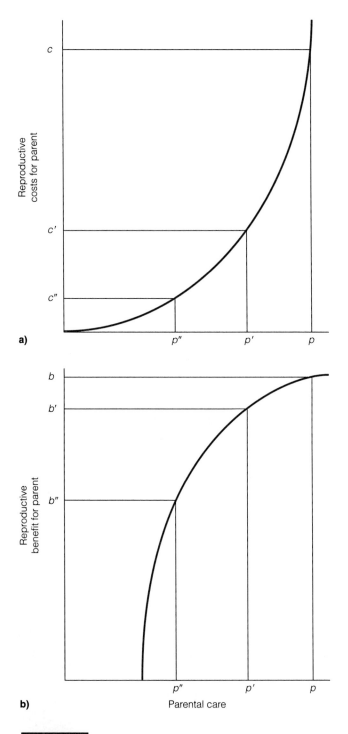

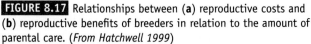

FIGURE 8.17 Relationships between (**a**) reproductive costs and (**b**) reproductive benefits of breeders in relation to the amount of parental care. (*From Hatchwell 1999*)

parental care (*p* to *p'*) leads to decreased cost (*c* to *c'*). Note that further reduction (*p'* to *p"*) leads to a smaller reduction in cost (*c'* to *c"*). Now consider the benefits. Reduction in care from *p* to *p'* has a small effect on the benefit (*b* to *b'*), whereas reduction to *p"* (*p'* to *p"*) has a large effect (*b'* to *b"*).

How should this affect cooperative breeders? The effect of helpers in a cooperative system can be additive or compensatory. In *additive care,* the effect of helpers is added to that given by the parents, yielding a larger total amount of care. In *compensatory care,* any help given by nonparents reduces the amount of care the parents must give. The models tell us how the system should respond to food availability (Figure 8.18). When starvation is frequent, helper care should be additive (hatched area). The reason for this is that benefit reduction is minimal only at the upper end of the benefit curve (Figure 8.17). In other words, when parental care is high, slight reductions in parental care have minimal effects on their benefits. On the other hand, when parental care is lower, slight reductions in care lead to larger reductions in benefit (Figure 8.17). In this case, helper care should be compensatory. In ecological terms, this means that when brood starvation is common, parental care should remain high and helper effects will be additive. When brood starvation is rare, parental care is lower and helper effects will be compensatory. Hatchwell (1999) compared 27 species of cooperative breeders across a range of ecological conditions and showed that the predictions of the model are supported. Thus the effect of helpers in a cooperative breeding system is sensitive to the ecological context.

This chapter completes our consideration of population ecology. We turn our attention now to the next higher level of organization, the community.

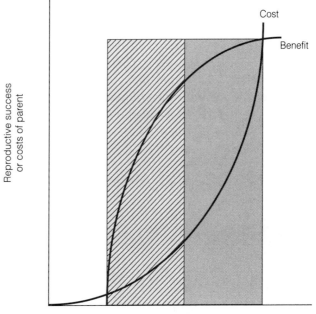

FIGURE 8.18 The relationship between reproductive costs and benefits (see Figure 8.17) as a function of parental care. The shaded area indicates the region in which parental care could be compensatory. The hatched region indicates the area of additive parental care. (*From Hatchwell 1999*)

SUMMARY

1. Behavior represents an adaptation to the environment. It is shaped by interactions with conspecifics and by the distribution of resources.

2. Behavior is shaped by the interaction of genetic and environmental factors.

3. Communication occurs via a number of sensory modalities in different taxa. The signals may convey discrete information or more complex, graded data.

4. Aggression is generally adaptive in protecting important resources, including food, territories, mates, and the individual's genetic contribution to the next generation.

5. Territories evolve where the energy invested in their defense is balanced by the returns resulting from exclusive use of the resource.

6. The nature of a social system is determined by the group size and structure, the mating system, and the degree of cooperation among members of the group.

7. The size of a group is determined by the balance between the advantages of association with conspecifics and the effect of such association on access to resources. For some species, such as top carnivores, energy limitations lead to a solitary habit.

8. The mating system comprises the nature of the pair bond and the number of mates an individual has. Polygynous systems are associated with variance in male reproductive success and sexual selection. Monogamous systems evolve when males cannot mate successfully with more than one female. This occurs if female reproduction is highly synchronized or if male parental care is required for successful reproduction.

9. Cooperation is associated with altruistic behavior because the cooperating individual generally forgoes reproduction or other opportunities in order to help the group. Kin selection is an important factor in the evolution of cooperation in some groups, such as bees and wasps. In others, habitat saturation limits the opportunities for members of the group, and they gain in fitness only by helping.

SELF-ASSESSMENT: CAN YOU ...?

1. Explain why the concept of "nature versus nurture" is a false dichotomy.

2. Outline the ecological factors that impinge on the evolution of communication signals.

3. Discuss the evolutionary ecology of

   territoriality

   aggression

   group size

4. Explain the genetic factors associated with haplodiploidy that lead to altruism.

5. Explain the essence of the argument made about additive versus compensatory parental care in Figures 8.17 and 8.18.

PROBLEMS AND STUDY QUESTIONS

1. Communication among animals has a straightforward biological definition. Language is more difficult to define. Discuss possible biological definitions of language. How could we determine whether an animal is using language?

2. Explain the interaction between the evolution of group size and territoriality.

3. Discuss the role of kin selection in the evolution of behavioral mechanisms and social systems.

4. Develop a flow chart showing the ways in which ecological resource distributions determine the nature of the social system.

5. How is the evolution of altruism among the workers in a honeybee colony affected by doubling the typical number of breeding males in the colony?

6. Explain how altruism and cooperation can evolve in species that are not haplodiploid.

7. A conflict arises between parents and offspring over the age at which weaning should occur. Explain the effect of each of the following on this conflict: (a) the species is semelparous; (b) the species is iteroparous, but in each mating the female mates with a different male.

8. Many species of seabirds forage on schools of fish in the open ocean. Why do these birds not defend feeding territories?

9. Explain why grizzlies and mountain lions are solitary, whereas wolves live in groups.

10. Any mode of communication potentially alerts predators to the individual's presence. What kinds of adaptations can avoid this problem?

11. Discuss the relationship between superstimuli and sexual selection.

PROJECTS AND ADDITIONAL STUDY

1. Relate the information in this chapter concerning the genes on the Y chromosome to the important theories of sexual selection outlined in Chapter 2.

2. Divide the class into groups. Each group should choose one of the following taxonomic groups and (a) survey the literature to determine the social system, (b) develop an explanation for the ecological factors leading to the social system, and (c) assess the amount of variation in the social system within the taxon.

    pinnepeds

    whales

    grouse

    pelagic seabirds

    water striders

    primates

    marmots

3. Choose a locally available, conspicuous species. Conduct observations to determine basic aspects of the social system, including group size, territoriality, mating system (if possible), nature of communication, and any cooperative feeding. Waterfowl, cranes, park squirrels, ground squirrels, and gulls all make good subjects.

Community Ecology

Competition

This chapter is the beginning of Part III of the text, which introduces the fundamental interactions of community ecology. Here we broaden our focus from populations of a single species to the next most inclusive level of organization—groups of interacting species. We start, in Chapters 9 and 10, with two fundamental interactions among species, competition and predation, and then, in Chapters 11 and 12, we consider how these forces combine to organize communities. Finally, we examine the patterns and mechanisms of community development following a disturbance, a process known as succession (Chapter 13).

A cornerstone of Darwin's theory of evolution is the notion that important resources are sometimes in critically short supply. The implications of this fact were impressed upon Darwin by an essay of Thomas Malthus (1826), who argued that the human population increases exponentially while its food supply increases linearly. Malthus concluded that human populations will ultimately be limited by food because beyond the point where graphs of these two quantities intersect (Figure 9.1), the food supply will always be insufficient for the population size. Darwin generalized this relationship to plants and animals on the basis of his own observations of natural systems. He realized that such limitations lead to the "struggle for existence" that drives the process of natural selection. Any variants in the population with characteristics that enhance their ability to garner scarce resources will therefore be more frequent in the next generation (Chapter 2).

In Chapter 2 we saw the wide range of adaptations that result from natural selection and observed that virtually any aspect of the organism can be affected. Many adaptations directly or indirectly are related to the ability to obtain resources. As they occur, the competitive ability of the population gradually increases. The tremendous success of Darwin's theory, and its basis in competition for scarce resources, led ecologists to emphasize the fundamental importance of competition in community ecology. As we will see in this chapter, competition still holds an important place in the way ecologists think about the nature of communities. In fact, for many years competition was considered to be *the* underlying force structuring communities of plants and animals. Recently, that view has been modified—a topic we will cover in Chapter 11.

In Chapter 3 we saw that each species has limits of tolerance for various abiotic factors such as temperature and light. A graph of the number of individuals that can tolerate different abiotic conditions is often a bell-shaped curve. Similarly, not all individuals in a population use biotic resources in precisely the same way. For example, Figure 9.2 shows that most individuals use prey about 30 millimeters in size. A few individuals, however, can use very large prey, and a few others rely on very small items. This figure refers to prey size, an important aspect of the food supply, but a number of other resources important to the organism can be depicted in the same way. Soil moisture might be an important resource for a desert plant, for example, whereas the size of areas of unoccupied substrate might be crucial for a sessile barnacle in a marine intertidal system.

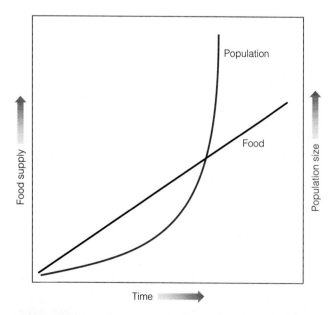

FIGURE 9.1 Changes in population size and food supply over time. According to Malthus, food supply increases arithmetically, whereas population size increases exponentially.

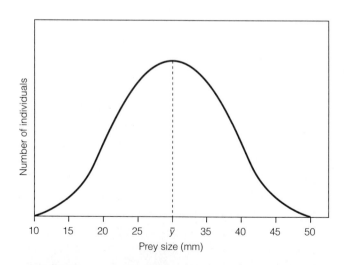

FIGURE 9.2 Normal distribution of resource use by individuals in a population.

When we describe the patterns of resource use for a species, we have defined the niche of the species. Because this concept is fundamental to understanding the process of competition, we explore it first.

The Concept of the Niche

The term *niche* first appeared in the ecological literature in 1917 when Joseph Grinnell used it to describe an organism's physical location in the environment, a concept we now express with the term **habitat.** The first to use the term *niche* in its modern context was Charles Elton (1927), who described the ecological *role* of a species as its **ecological niche.**

Modern ecologists refer to Elton's concept of the role an organism plays in the community as the **functional niche.** For example, a bat such as the little brown bat (*Myotus lucifugus*) plays a particular role in an ecological community as a nocturnal, aerial insectivore. This description by no means includes all the ecological interactions that constitute this bat's role, but it certainly includes some of the potentially important ones: the bat's trophic position and where and when it forages.

We must be careful to avoid using the terms *ecological niche* and *habitat* interchangeably. The niche is the role the organism plays, whereas the habitat is a description of where the organism is found. The Mexican free-tailed bat (*Tadarida braziliensis*) hunts in the desert, whereas the little brown bat (*Myotus lucifugus*) preys on insects in deciduous forest. These two species have different habitats but similar niches.

Even though the functional niche is clearly useful as a description of some important facets of the ecology of a species, the concept has some limitations as well. First, it is too broad: The precise kinds of interactions it involves are not clearly stated. Second, it is not amenable to quantitative measurement or mathematical analysis.

The concept of the **Hutchinsonian niche** solves these problems. G. E. Hutchinson (1957) suggested that for any species, we can identify a series of resources for which the species has a range of tolerance. Figure 9.3 shows a hypothetical example of the use of two resources by a fish. On the *x*-axis is plotted the frequency distribution of the sizes of prey taken by the fish. It forms a bell-shaped curve. On the *y*-axis is plotted the frequency distribution of oxygen concentrations at sites where the species is found in a stream. The area of intersection of these two resource utilization curves is a quantitative description of two of the

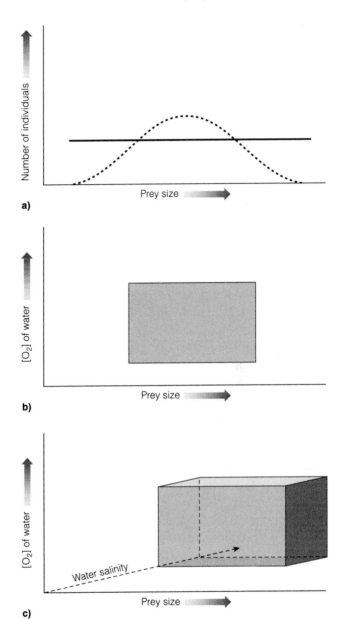

FIGURE 9.3 The dimensionality of the Hutchinsonian niche. (**a**) A fish's niche described in one dimension: prey size. The range of acceptable prey sizes is indicated by the line. (**b**) When a second dimension, oxygen concentration, is added, the niche is represented by an area. (**c**) Including a third dimension, salinity, produces a three-dimensional niche represented by the volume indicated.

resources used by that species. If we include a third axis, we have defined a volume that describes the pattern of use of three resources. We can, in fact, define *n* biotic and abiotic resource axes, each of which will be utilized with a certain frequency distribution. The set of *n* resource axes defines the Hutchinsonian niche, a hypervolume in multidimensional space. Even though we can depict and visualize only three resource axes, an *n*-dimensional hypervolume is amenable to mathematical and statistical analysis.

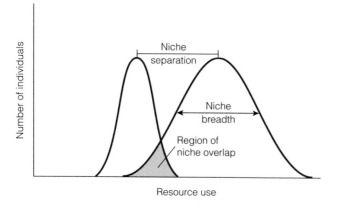

FIGURE 9.4 Important parameters of a niche. Each curve represents the niche in one dimension for one species.

Mathematical description of resource use quantitatively describes the functional niche of the organism.

In practice, it is difficult or impossible to measure all the resources utilized by a species. However, because much of the theory surrounding competition and the niche is based on the *limiting* resources, we can limit our attention to a smaller number of resource axes. The important mathematical descriptions of a niche based on a single limiting resource are depicted in Figure 9.4. We know that the resource utilization curve tends to be bell-shaped, indicating that not all individuals use resources in the same way. A measure of this variation is the **niche breadth,** a value that quantifies the variance in resource use. The **niche separation** is the distance between the mean resource use curves for two species. The hatched region indicates the region of resource overlap where competition is expected.

A species with a very narrow niche breadth is a specialist (the curve on the left in Figure 9.4), whereas one with a broad niche is a generalist (the curve on the right in Figure 9.4). For example, the giant panda (*Ailuripoda melanoleuca*), which has an extremely narrow food niche—it eats only bamboo—is an extreme specialist. In contrast, a relative, the raccoon (*Procyon lotor*), has an omnivorous diet and is a generalist.

The concept of the niche provides the basis for understanding competition among species for resources. Niche overlap among species may result in competition among them.

Types of Competition Defined

If a resource is a limiting resource—that is, one that ultimately limits the rate of population growth in a species—then the use of that resource by another species will have ecologically important consequences. The other species may not use the resource in exactly the same way, but any use of a resource renders a portion of it unavailable. Such resource depletion may limit the growth of one or both populations. We refer to interactions such as these as **resource competition.** Specifically, we define **competition** as any interaction between two or more species over a limiting resource that causes a decrease in population growth of one of the species.

Virtually any resource in short supply can lead to competition. For most plants, for example, light, water, and nutrients are the basic resource requirements. Consequently, the kinds of niche parameters typically applied to animals, such as prey size, are not as applicable to plants. Most plant ecologists recognize that the fundamental resource for which plants compete is probably space in which seeds can germinate and seedlings can survive and grow. Thus plant ecologists refer to **preemptive competition** for space.

The predominant definition of competition in this text describes **interspecific competition**—that is, competition that occurs between two species. The term **intraspecific competition** refers to competition between members of a single species. In this case, the activities of some individuals of a species make resources unavailable for other members of the same population. As we saw in Chapter 5, according to the extrinsic biotic school, resource limitation is the basis of population regulation.

Competition can be classified according to the mechanism of operation: whether organisms use up resources or instead prevent other organisms from using them. These two ways in which organisms compete are called exploitation competition and interference competition. In **exploitation competition,** the actions of one species (or individual, in the case of intraspecific competition) significantly reduce the amount or availability of the resource for another. The two species may never come in direct contact with each other. In desert ecosystems, for example, ants and rodents compete for seeds, an important source of food (Brown and Davidson 1977), even though there is no direct interaction between them. Any seeds eaten by ants are unavailable to rodents, and vice versa. Another example of exploitation competition occurs in tropical rain forests, which support extensive growth of epiphytic plants—that is, species that grow on other plants. Epiphytes are not parasitic; they do not rob the supporting plants directly as a true parasite would. Rather, they use the host plant simply for support. However, many epiphytes have elaborate adaptations for trapping nutrients that fall through the canopy in rain or that run down the stem of the supporting plant (Daubenmire 1974). By absorbing these nutrients, the epiphytes render them unavailable to plants rooted in the forest floor, even though the epiphytes do not actively interfere with the absorption activities of other species. Thus, in exploitation competition, use of the resource by one species renders it less available for the other.

In **interference competition,** one species actively interferes with the ability of the other to use the resource. An example of this sort of direct interference is the phenomenon of **allelopathy,** in which organisms produce and release chemicals that have a deleterious effect on other, nearby organisms. In plants, in which resource preemption is an important competitive force, allelopathy can interfere with the ability of other species to acquire space. Black walnut trees are known for the production of juglone, an allelopathic chemical that interferes with the ability of other plants to establish themselves nearby (Massey 1925). In Caribbean coral reef systems, the liver sponge, *Plakortis halichondroides,* inhibits the growth of the sheet coral, *Agaricia lamarcki* (Porter and Targett 1988). For sessile organisms in this habitat, space is a limiting resource. Porter and Targett showed that *Plakortis* creates a zone of dead coral around its base, apparently by releasing waterborne chemicals that decrease the number of zooxanthellae (and their chlorophyll content) in the coral. The effect is seen even if there is no direct contact between the species.

In animals, territoriality can also exemplify exploitation competition. Near the crest of the Cascade Mountains in Washington, two similar species of tree squirrels, the red squirrel (*Tamiasciurus hudsonicus*) and the chickaree (*T. douglasii*), are sympatric—they occupy the same range at the same time. Both feed on the cones of lodgepole pines and other conifer species. In this region of sympatry, the two species maintain interspecific territories, an arrangement that ensures that both species have access to crucial food supplies (Smith 1968). Defense of the territories actively prevents use of the resource by others.

Often competition is not a matter of interactions between only two species. Sometimes several species in a community are involved in competition to varying degrees. A species that is affected by the combined competitive effects of several species is said to face **diffuse competition.** The seedling of a canopy tree in a deciduous forest faces diffuse competition for light (and for other resources as well) from the many species in the canopy that are absorbing light before it reaches the forest floor. The summed effects of competition can have profound effects on the composition of a community, as we will see in Chapter 11.

Methods for Obtaining Evidence of Competition

How do we demonstrate that competition is occurring between two species? According to the definition of interspecific competition, the relevant criterion is a negative impact on the growth of one species as a result of the presence of the other when both use the same limiting resource. Thus, if competition is occurring, the removal of one species from the system should result in an increase in the population of the other.

Experimentation

A classic example of this type of experiment was performed by J. H. Connell (1961) in his study of two species of barnacles, *Chthalamus stellatus* and *Balanus balanoides,* that inhabit the rocky intertidal zone off Scotland. In the intertidal zone (Figure 9.5), tides determine the exposure of the habitat to air. Different parts of the intertidal zone are exposed for different lengths of time, depending on their position, the

FIGURE 9.5 An example of an intertidal zone in Southeast Alaska. At low tide, as shown in this figure, large areas are exposed to the air. (*Photo by Anne W. Rosenfeld/Animals Animals*)

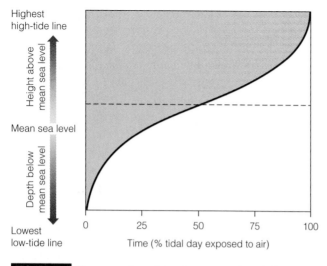

FIGURE 9.6 The proportion of the time exposed to air for various portions of the intertidal zone. (*From Segar 1998*)

height of the tide, and the steepness of the slope; (Figure 9.6). Both species of barnacles produce pelagic larvae that drift about in the water column and eventually settle onto the substrate to metamorphose and begin a sessile adult life. Settlement is random, and larvae of both species settle throughout the intertidal zone. Still, Connell noted that adult *Chthalamus* are always found higher in the intertidal zone than are *Balanus*. Although this distribution could be a result of competition, it could also result from different tolerances of abiotic factors, because species that inhabit the higher portions of the intertidal zone must withstand desiccation at low tide. If *Chthalamus* is simply more resistant to desiccation than *Balanus*, then competition may not be a factor in their different distributions.

Two important features of this system allowed Connell to undertake a series of elegant experiments on the importance of competition. First, the substrate in this system is composed of many rocks and boulders, so Connell could move rocks bearing barnacles from one part of the intertidal zone to another. Second, because both species are sessile as adults, Connell knew he could find specific individuals again to document their fates. Connell performed the following experiments:

1. He moved rocks bearing *Chthalamus* from the upper to the lower intertidal to see whether this species could survive in the lower zone. He also did the reciprocal experiment, moving *Balanus* higher in the intertidal to determine whether it could survive the desiccation it would face there.

2. He physically removed *Balanus* from rocks in the lower intertidal and *Chthalamus* from rocks in the upper intertidal. These experiments were designed

to determine whether *Chthalamus* could grow in the lower zone in the absence of its putative competitor, *Balanus*, and vice versa.

He also photographed rocks over the course of the experiments to ascertain the fates of individuals of both species.

Connell found that whereas *Chthalamus* is capable of living lower in the intertidal zone than it normally does, *Balanus* does not fare well in the upper intertidal zone, where it experiences more frequent periods of desiccation. This result led Connell to conclude that the absence of *Balanus* from the upper intertidal zone is due to its inability to tolerate the abiotic factors it encounters there. In the experiments in which *Balanus* was removed from the rocks in the lower zone, *Chthalamus* was able to persist in the lower zone, whereas in the controls in which *Balanus* was not removed, *Chthalamus* gradually decreased in numbers (Figure 9.7). The photographic records from the controls showed clear evidence that as *Balanus* grows, its expanding base literally pries *Chthalamus* off the rocks. Thus Connell concluded that *Chthalamus* is relegated to the upper intertidal by the competitive effects of *Balanus*. *Balanus* cannot outcompete *Chthalamus* in the upper zone because it does not tolerate desiccation as well. Space is the limiting factor for which these sessile organisms compete.

Light and nutrients are often limiting for plants. Competitive effects under light limitation are well documented. Although underground competition for nutrients is difficult to study, ecologists are coming to recognize the importance of below-ground competitive interactions (Casper and Jackson 1997). When

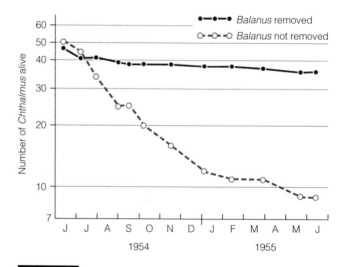

FIGURE 9.7 Survivorship of *Chthalamus* in the lower intertidal zone when *Balanus* was removed, compared to control. (*From Connell 1961*)

light is not limiting, nutrient resources may limit population growth. Experiments that prevent other species from gaining access to local nutrients can be used to demonstrate competitive effects in plants. Coomes and Grubb (1998) showed that tropical forest trees had higher growth rates when trenches were used to block neighboring trees' access to local nutrients.

It seems logical that competition should be most intense between similar organisms. After all, they are most likely to need similar resources. Still, J. H. Brown and D. W. Davidson (1977) demonstrated that competition can occur between animals as taxonomically unrelated as rodents and ants. Annual plants, important primary producers in many deserts, take advantage of infrequent rains by flowering and producing huge numbers of seeds. Because these seeds can lie dormant on the desert floor for long periods of time, desert soils contain a large reservoir of seeds that can be exploited by granivorous animals. After showing that the size distributions of seeds utilized by ants and rodents overlap extensively, Brown and Davidson used removal experiments to determine whether these taxa compete for seeds. They set up four experimental conditions: (1) Two plots were fenced with wire mesh to exclude rodents, and all rodents within the exclosures were removed; (2) on other plots, granivorous ants were removed by treating colonies with insecticide; (3) plots were established in which both groups were removed; and (4) two plots were unmanipulated controls. Both rodents and ants increased in density when the other group was eliminated. The 71 percent increase in ants over the control plots (Table 9.1) is strong evidence for competition between ants and rodents.

These examples illustrate the nature of ideal experiments for demonstrating that two species compete for an important resource. Such experiments allow us to test the very essence of our definition of competition:

the negative impact of one species on the growth rate of another. If we can hold all other variables equal, the removal or addition of a species is a powerful experimental test of the extent of competition between two species.

Observation and Inference

The experimental approach just described can only be used with a system in which manipulation is practical. Moreover, removals or additions may affect other interactions besides competition. For example, if the removal of a species has effects on predators or parasites in the system, these effects may confound our conclusions about competition. Thus, for some ecosystems and some pairs of organisms, manipulation may not be possible. In such cases we *may* be able to *infer* that competition is occurring from our observations of patterns of species distribution. In one example, J. M. Diamond (1975) noted that some closely related species of birds in the Bismarck Archipelago near New Guinea have a "checkerboard" distribution pattern. Two closely related species are not found to coexist on the same island; some islands have one species, some have the other, but no island has both species (Figure 9.8). In this case, no experiments were done. The detailed observation of distribution patterns was used to infer logically that competition is occurring.

This process can sometimes lead us astray, however. The species of lagomorphs that originally occurred on the island of Newfoundland once differed from those found on the mainland. Whereas the mainland has

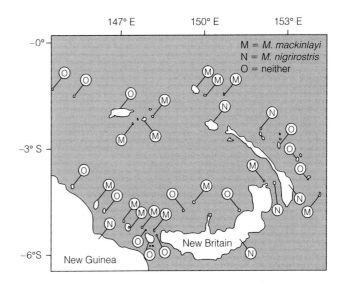

FIGURE 9.8 Checkerboard distribution of cuckoo dove species (*Macropygia*) in the Bismarck Archipelago. Most islands have only one species. No island has both, and some have neither. (*From Diamond 1975*)

TABLE 9.1

Comparisons of Ant Colonies and Rodent Numbers in Plots from Which Potential Competitors Were Removed

Taxon	Treatment			
	Rodents Removed	Ants Removed	Control	Increase Relative to Control (%)
Ant colonies	543	—	318	70.8
Rodents				
Number	—	144.00	122.00	18.0
Biomass (kg)	—	5.12	4.13	24.0

Adapted from Brown and Davidson 1977.

FIGURE 9.9 (a) The arctic hare (*Lepus arcticus*), an inhabitant of tundra. (*Photo by Brian Milne/Animals Animals*) (b) The snowshoe hare (*Lepus americana*) is found in taiga. (*Photo by Joe McDonald/Animals Animals*)

snowshoe hares in taiga habitats and arctic hares in the tundra (Figure 9.9), both habitats in Newfoundland were occupied by arctic hares, the only species on the island. When the snowshoe hare was introduced to the island, arctic hares were soon found only in tundra habitats. The conclusion was that each species is relegated to its preferred habitat by competition from the other.

This interpretation turned out to be incorrect, however. Instead, a predator, the lynx, plays a critical role. When the snowshoe hare was introduced to Newfoundland, lynx numbers increased because of the greater abundance of prey. The arctic hare, which for unknown reasons is far more vulnerable to predation in taiga than is the snowshoe hare, rapidly decreased in taiga. Its disappearance was the result of increased predation in that habitat, not competition from snowshoe hares (Bergerud 1967).

As this example illustrates, if we are to demonstrate competition between two species, our understanding of the natural history of the system must be fairly complete. The presence of other interactions can affect the population of either species, confounding our conclusions regarding competition. Studies extending the work of Brown and Davidson (1977) on competition between rodents and ants have shown that the interaction among these groups is mediated by an indirect effect: the correlation of some ant species populations with grass cover (Valone et al. 1994). Even when direct competitive effects are demonstrated, other interactions can modify the ultimate result. In a study of competitive effects on regeneration of northern red oaks (*Quercus rubra*), Buckley et al. (1998) showed that although removal of plant competitors has a positive effect on oak regeneration, removal effects were compromised by the effects of browsing by deer.

Competition has been such an important paradigm in ecology that observations *consistent with competition* have too often been confused with *proof of competition*. Science should proceed by rejecting competing hypotheses rather than by attempting to confirm a preferred idea. We will see in Chapter 11 that many ecologists believe competition is a fundamental force that determines the assembly of communities. As we will also see, the difficulties associated with demonstrating competition between two species are compounded when one attempts to make inferences about the role of competition in a complex community.

The Relationship Between Intraspecific and Interspecific Competition

We should expect that individuals of the same species will be more similar in resource use than will members of another species. Thus conspecifics may compete more intensely with one another than with members of other species. One indication of this is that the intensity of territorial defense is generally greater within species than between them. For animal populations, intraspecific competition is the basis of the extrinsic biotic school of population regulation (Chapter 5). According to this school, competition within a species for food ultimately limits the growth of the population. We have seen several examples of the regulation of animal populations that can be attributed to intraspecific competition for limited food resources, a density-dependent situation.

In plants, a phenomenon known as self-thinning attests to the importance of intraspecific competition. The term is derived from the horticultural practice of thinning crops or other plants in order to achieve higher production of a few individuals by removing some of their intraspecific competitors. For example, plantations of commercially valuable timber are often thinned to promote growth of the remaining individuals. Natural and laboratory populations are also known to "self-thin." This means that the higher the initial density of seedlings, the higher the mortality rate, causing the population density to decline and the size of individuals to increase.

The rye grass *Lolium perenne*, exemplifies this phenomenon. W. M. Lonsdale and A. R. Watkinson (1983) sampled a series of populations of different initial densities at various times after planting. Each line in Figure 9.10 represents the combination of density and weight per plant over the course of time. In the dense populations, both mortality and the mean size of each plant increased. In other words, as the population "thinned" in terms of number of individuals, the size of each individual increased. The straight line associated with the decline in density over time has a slope of $-3/2$. Interestingly, this slope is found in many plant self-thinning lines (Figure 9.11). These data indicate that for many populations, intraspecific competition is an important factor that can result in mortality at very high densities.

We have seen how difficult it is to demonstrate the existence of interspecific competition. How then can we assess the relative intensities of intra- and interspecific competition? This is indeed a difficult task. The best information we have comes from plants because of their sessile nature and because their densities can readily be manipulated in the greenhouse.

One such interaction that has been extensively studied by D. R. Marshall and S. K. Jain (1969) involves two very similar species of wild oats, *Avena barbata* and *A. fatua*, which are found sympatrically in the grasslands of California. The relative effects of intra- and interspecific competition were studied in

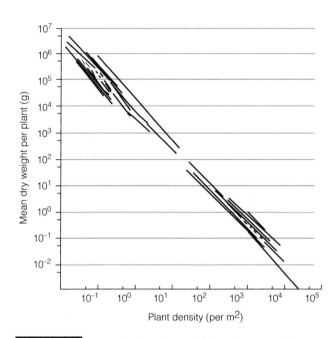

FIGURE 9.11 Regression lines from self-thinning curves for 31 stands of different species of plants. (*From White 1980*)

the laboratory in mixed plantings of the two species. The researchers used a technique called a replacement series, in which a series of pots is sown with seeds of the two species in varying proportions. For example, in one experiment 128 seeds were planted in each pot, but the relative numbers of seeds of the two species varied in the series of pots. At one extreme, all 128 seeds were *A. fatua;* at the other extreme, all were *A. barbata.* Other pots were sown with 16 *fatua* and 112 *barbata,* 32 *fatua* and 96 *barbata,* and so forth. The replacement series thus varied from pure intraspecific competition in *A. fatua* (128 *A. fatua* seeds and 0 of *A. barbata*) through equal numbers of the two, and then to pure intraspecific competition for *A. barbata.* At the intermediate value of 64 seeds of each species, the intensity of intraspecific competition was least, and the number of interspecific competitors was greatest.

Marshall and Jain measured both survival of seedlings and eventual seed production for both species in the replacement series. From observations on pure stands of each species, they knew that approximately 75 percent of all seedlings survive under pure intraspecific competition. This formed the basis of the null hypothesis for their experiment. If intra- and interspecific competition were equivalent for these species, we would expect 75 percent of the seedlings to survive regardless of whether they were competing against conspecifics or the other species; that is, we should see 75 percent survival of each species throughout the replacement series. For survival, this is indeed what they found (Figure 9.12a). Thus there is

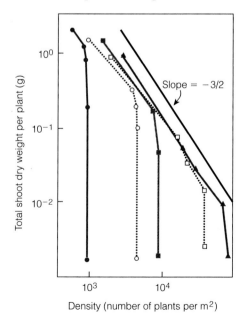

FIGURE 9.10 Self-thinning in the rye grass *Lolium*. Each line represents the trajectory of one population. (*From Lonsdale and Watkinson 1983*)

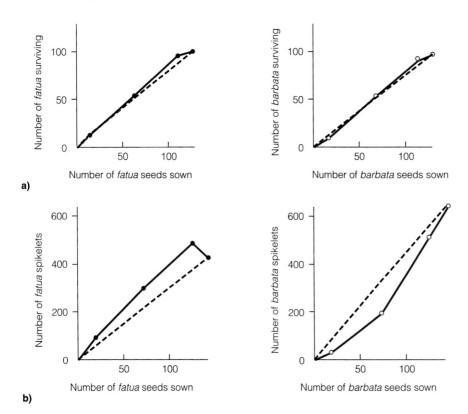

FIGURE 9.12 Results of replacement series competition experiments in *Avena barbata* and *A. fatua*. (**a**) Survival of seedlings. (**b**) Seed production. The dotted lines represent the expected result based on pure intraspecfic competition. The solid lines and data points represent the actual performance in competition. (*From Marshall and Jain 1969*)

The Effects of Competition

We can distinguish between long-term and short-term effects of competition. The long-term effect, known as **character displacement,** operates on an evolutionary time scale—that is, over many generations. The short-term effect, called **competitive exclusion,** occurs in ecological time—that is, within a single or a few generations. It is rather difficult to predict which will occur. We discuss the long-term effects of competition first.

Character Displacement

Character displacement is the gradual separation of two species in morphology or physiology as a consequence of competition. To the extent that these aspects of an organism determine its resource use, the niches of the two species

no significant difference between the effects of intra- and interspecific competition on survival.

The results differed, however, when Marshall and Jain considered reproduction. Now the dependent variable in the replacement series graph was the number of seeds produced, measured by counting the number of flowering spikelets. In this case, *A. fatua* produced more seeds than expected in intense interspecific competition (Figure 9.12b). *A. barbata*, on the other hand, produced fewer seeds than expected in interspecific competition. Consequently, the presence of *A. fatua* would ultimately be expected to decrease the population size of *A. barbata* because of lower seed production in the latter. The researchers concluded that for *A. fatua*, intraspecific competition is more intense than interspecific competition and that for *A. barbata*, the reverse is true. Thus we would expect *A. fatua* gradually to outcompete *A. barbata* if the two were placed in direct competition.

What would be the end result of such an interaction? Our definition of competition is based on a change in the growth rate of one species's population. If competition proceeds for a time, the negative effect on one species will lead to important changes in the community.

separate. Imagine two species with the frequency distributions for a character or trait shown in Figure 9.13. If this character is related to resource use, there will be considerable overlap in resource use. Now consider this in evolutionary terms. If we assume that the character (and thus the resource use) has a genetic basis (that individuals at different points on the bell curve differ genetically), we can predict what natural selection will do. Those individuals of species 1 that fall on the left tail will have an advantage because they do not compete with individuals of species 2. Directional selection (see Chapter 2) will move the curve for species 1 to the left. Similarly, the individuals of species 2 on the right tail will have a selective advantage, and directional selection will shift *that* curve to the right. Individuals of both species face the most intense competition at intermediate phenotypes and will have the lowest fitness. The result, after many generations, is the character divergence of the two populations—and hence divergence with respect to their use of this resource. Clear demonstrations of character displacement in the field are relatively rare, because so many factors other than competition can modify characters associated with resource use.

The changes in body size of sympatric weasel species (Figure 9.14) are an example of character displacement. The long-tailed weasel (*Mustela frenata*) is smaller at those latitudes where it is sympatric with the least weasel (*M. nivalis*). In the same way, the ermine (*M. erminea*) is smaller in the part of its range that it shares with the long-tailed weasel. Note that the changes in size for a given species correspond rather sharply to the appearance of another sympatric species. Assuming that prey size is dependent on body size, these data are consistent with the predictions of character displacement (McNab 1971). Note that in the absence of sympatric competitors, body size increases. This is probably because a large weasel can capture and subdue both large and small prey. On the other hand, a small weasel is restricted to smaller prey items. Thus, when a species is released from competition, its body size increases, enabling it to take a wider range of prey items. When faced with competition, one of the species specializes on smaller items.

Although this interpretation may be correct, it depends on certain important assumptions: that the size of the animal is a direct reflection of the sizes of its prey, and that prey size is the primary factor influencing the evolution of body size. Clearly, many other factors may also be important (see Chapter 3).

One of the most convincing examples of character displacement comes from detailed studies of coexisting species of desert ants (Davidson 1978). D. W. Davidson demonstrated that in the ant *Veromessor pergandei*, the assumption that character size reflects resource use is correct. For a given species of ant, mandible size is highly correlated with the size of

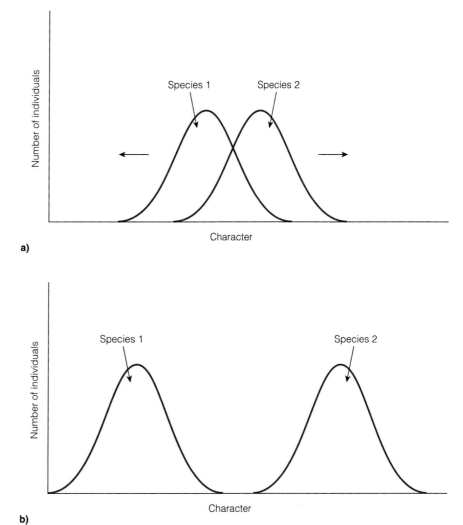

a)

b)

FIGURE 9.13 The process of character displacement. (**a**) Individuals of one species that use resources in regions that do not overlap with the other species have a selective advantage. (**b**) Over time, selection will separate the niches of the two species.

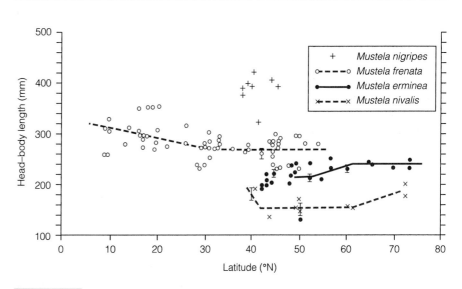

FIGURE 9.14 Head–body lengths of male weasels as a function of latitude. (*From McNab 1971*)

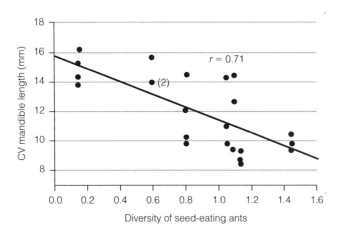

FIGURE 9.15 The coefficient of variation (CV) in mandible length in *Veromessor pergandei* as a function of the number of granivorous ants in the community. (*From Davidson 1978*)

the seeds used as food. Within ant colonies, the variation in mandible size is negatively correlated with the number of potentially competing sympatric species (Figure 9.15). Thus, as the number of competitor species increases, *Veromessor* becomes more specialized on smaller seeds.

In different deserts, *Veromessor* faces various potential competitors with different mandible sizes (Figure 9.16). Note in this figure that when faced with only one very rare larger competitor in Mojave, California, *Veromessor* colonies have a uniform distribution of mandible sizes. Species with large mandible sizes are more common at the other sites, and the proportion of large morphs in *Veromessor* declines. In the Arizona sites, the presence of smaller-mandible competitors also reduces the proportion of small-mandible *Veromessor*. Davidson was able to rule out the effects of seed size availability as a causative factor for the distribution of mandible sizes in *Veromessor*.

N. Eldridge (1974) has described a convincing case for character displacement in two species of trilobites, *Phacops rana* and *P. iowensis*. The fossil record for both of these marine arthropods is quite complete. Over much of the Middle Devonian period, the two

species had allopatric distributions; from no strata are fossils of both species known. During this long period of allopatry, neither species varied greatly in size or in other aspects of morphology; the size range of *P. iowensis* consistently fell within that of *P. rana*. Toward the end of this period, as the size of the North American sea shrank rather abruptly, the two species became sympatric. In fact, this is beautifully documented by a specimen with both species in the same rock. Sympatry resulted in a marked increase in the size of *P. iowensis* such that specimens from sympatric populations are larger than any other known examples of *P. iowensis*.

Eldridge's studies demonstrate the amount of divergence after long periods of time; other studies illustrate the process of divergence in progress. M. M. Martin and J. Harding (1981) studied populations of two species of annual plants called filaree. Two species, *Erodium cicutarium* and *E. obtusiplicatum*, are distributed in the grasslands of California such that

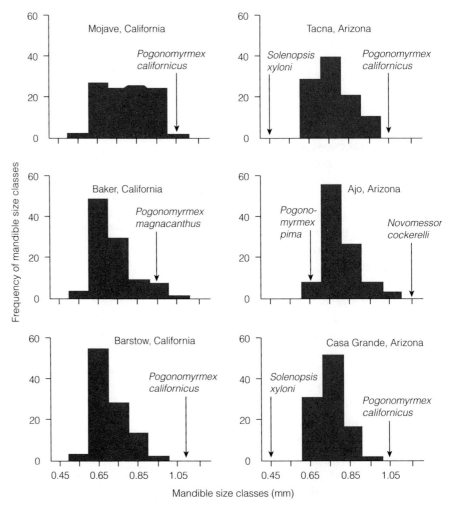

FIGURE 9.16 Frequency distributions of mandible sizes at different sites. The mean mandible lengths of the most similar competitors of *Veromessor pergandei* are indicated by the arrows. (*From Davidson 1978*)

there are both sympatric and allopatric populations of these species. In studying the intensity of competition between the two species as a function of the history of sympatry or allopatry, Martin and Harding used replacement series like those Marshall and Jain used with the two species of *Avena* as previously discussed. They compared the intensity of competition between *E. cicutarium* and *E. obtusiplicatum* in three sets of situations involving four sites (Figure 9.17):

1. *Sympatric populations*. Replacement series were established with individuals of both species from populations on site 1 where the two are sympatric.
2. *Allopatric populations*. Replacement series were established with individuals of both species from allopatric populations (*E. cicutarium* from site 4 and *E. obtusiplicatum* from site 3).
3. *Transposed populations*. In this comparison, individuals of E. cicutarium from a population sympatric with *E. obtusiplicatum* on site 1 were paired with *E. obtusiplicatum* individuals from site 2. In other words, the populations being compared were from sympatric populations, but the particular populations that were paired had not encountered each other before.

The results of these experiments, shown in Figure 9.18, revealed subtle but important differences among the three situations. The most intense competition occurred between allopatric populations—that is, populations that had not been in contact in the recent past; the least intense competition occurred between currently sympatric populations. The transposed pop-

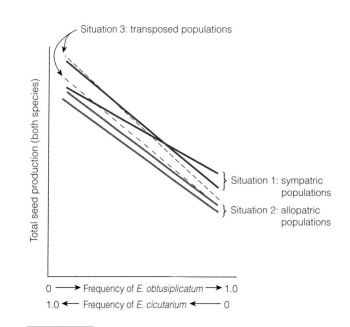

FIGURE 9.18 Total seed production at different levels of intra- and interspecific competition for populations of two species of *Erodium*. (*From Martin and Harding 1981*)

ulations were intermediate. The sympatric populations showed the least intense competition because during their time together, character displacement had worked to decrease the overlap in resource use. The allopatric populations competed most intensely because in the recent past they were not under pressure to diverge in their use of resources. Character displacement had not reduced competition. The transposed populations were intermediate because they had experienced some character displacement but not with the exact population with which they were tested. This last case is very interesting because it shows that competitive effects and character displacement occur locally and to specific populations.

Character displacement also reduces intraspecific competition. In some cases, the sexes differ slightly in the resources they use. In many of the raptors (hawks and owls), for example, there is significant sexual dimorphism in size. Given that predators of different sizes specialize on prey of slightly different sizes, intraspecific competition may thus be reduced. This may be particularly important for carnivores feeding high on the food chain where food resources are scarce. Different age classes of a given species may also use different resources. In animals that metamorphose, such as amphibians and insects, different life history stages may utilize completely different resources and thus avoid competition. Tadpoles feed very differently from adult frogs, and almost certainly these different life stages do not compete with each other for food resources. There were probably many

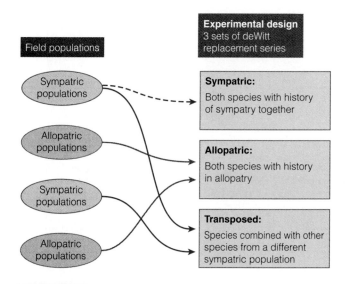

FIGURE 9.17 The experimental design used by Martin and Harding (1981) to examine the evolution of competitive interactions.

reasons for the evolution of metamorphosis, but one consequence may have been the reduction of intraspecific competition.

Competitive Exclusion

The short-term effect of competition is called **competitive exclusion,** which is defined as the local extinction of a species as a result of interspecific competition. It was first described by the Russian protozoologist G. F. Gause, whose experiments with *Paramecium* provided the first clear example of extinction caused by the presence of another species. Gause plotted the pattern of population growth of *P. caudatum* and *P. aurelia* in separate cultures (that is, in allopatry). Both species showed typical sigmoid growth curves when grown individually (Figure 9.19), approaching an asymptote that corresponds to our concept of carrying capacity. When the two species were placed in the same culture, however, *P. aurelia* always approached an asymptote (albeit a lower one), whereas *P. caudatum* eventually died out. Gause concluded that *P. caudatum* lost out as a result of interspecific competition. These results led Gause to postulate that no two species can long coexist on the same limiting resource, a concept that has come to be known as the **competitive exclusion principle,** or **Gause's law.** We will examine this idea in detail in Chapter 10.

Competitive exclusion may have been important in the extinction of species over the course of evolution. One example, the extinction of the marsupial family *Borhyaenidae* in South America, is generally attributed to competitive exclusion. The fossil record indicates that South America once had a splendid diversity of marsupial mammals. The family *Borhyaenidae* were carnivores that showed extreme convergence with placental mammals in the order Carnivora from North America. For example, the marsupial cat *Thylacosmilus* strongly resembled the North American saber-toothed cat *Smilodon* (Figure 9.20). *Borhyaena* was a predator with a remarkably dog-like skull. Other species attained the size of bears and a similar morphology. Beginning about 5 million years ago, newly formed connections between North and South America permitted North American forms to enter South America. The result was intense competition and the extinction of many South American marsupials, apparently as a result of competitive exclusion (Simpson 1965). The box on p. 233 discusses the effects of some other invading species.

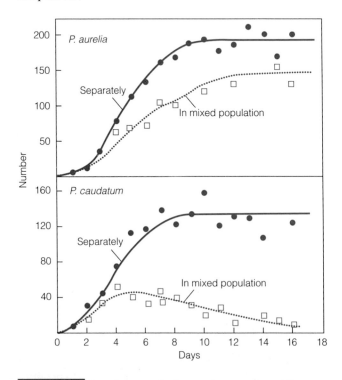

FIGURE 9.19 Growth curves for *Paramecium aurelia* and *P. caudatum* in separate and mixed cultures. (*From Gause 1934*)

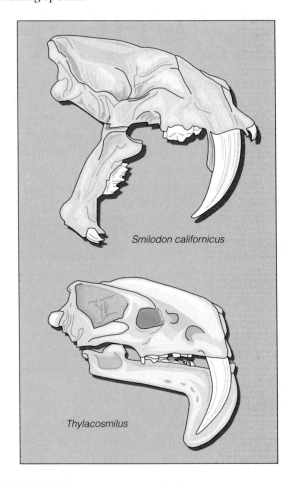

FIGURE 9.20 The skulls of the saber-toothed cat *Smilodon californicus*, a placental mammal, and the marsupial cat *Thylacosmilus*. (*From Vaughan 1972*)

Introduced Species: Potentially Devastating Competitors

Which of the following are exotic invaders: bluegrass in Kentucky, zebra mussels in the Great Lakes, mountain goats in Olympic National Park, or tumbleweed in Wyoming? Answer: All of them! Bluegrass is a Eurasian grass (*Poa pratensis*). Zebra mussels were released into North America from the ballast of ships from Asia and Europe. Mountain goats were released into the Olympic Mountains from native populations in the Rockies. Tumbleweed is another Eurasian weed that was imported in the nineteenth century along with grains from Europe.

The movement of species from their native habitats to new areas is one of the most common and insidious human effects on the natural world. Some introduced species cause much more harm than others. Bluegrass is relatively well controlled and has only minor adverse effects on natural habitats. Zebra mussels, on the other hand, are wreaking havoc with pipes and other submerged structures. Millions of dollars are spent annually to remove them from the intakes to water supplies. Mountain goats in the Olympics are drastically reducing the populations of many alpine plants that are not adapted to such intense grazing pressure.

Some invaders arrive by accident, as was the case with the zebra mussel. Others represent deliberate introductions, often for economic reasons. In the 1870s, Spencer F. Baird of the U.S. Fish Commission imported carp from Asia as a food fish. For a while, the fish was a culinary fad, and members of Congress demanded that the fish be introduced into their home waters. The result was an invasion by a nonnative species that destroys benthic habitat and outcompetes native fish.

The most infamous introductions are those that result in a population explosion of the invading species. Islands such as Australia seem particularly vulnerable. Two major invasions have had serious impacts there. The European rabbit (*Oryctolagus cuniculus*) was imported from Great Britain and soon reached such numbers that it destroyed immense tracts of rangeland. Estimates made in the 1950s put the population at over 1 billion rabbits. Another import to Australia, *Opuntia*, which was introduced as a garden ornamental, also underwent a population explosion in its new home, covering thousands of square kilometers with impenetrable stands of cactus. Kudzu was imported from China into the southern United States in the 1930s to control erosion. In the benign climate of the South, the plant can grow up to a foot a day. Today it covers more than 2 million acres there, costing as much as $50 million a year in lost farm and timber production.

Why do these alien species cause such problems? Not all introductions are so devastating—many species that are introduced inadvertently or intentionally never even establish populations. Two factors probably account for the more intractable exotics: (1) the species's ability to colonize new habitats and (2) its ability to outcompete native biota. Because so many North American crops came to this continent from Europe and Asia, weeds have flowed primarily *to* North America as stowaways in seeds of those crops. However, the movement of aliens has not been unidirectional. A number of North American species have invaded Europe, where they are now pests. For example, a plant native to the North American

tallgrass prairies, tall goldenrod (*Solidago altissima*), is now a noxious weed in Europe (Weber and Schmidt 1993; Zwölfer 1976).

Among the "weedy" species that are good colonizers, purple loosestrife is an example. This plant produces tough, long-lived, wind-dispersed seeds in prodigious numbers—up to 2.5 million per plant. Under favorable conditions it can grow as much as a centimeter a day, and it can tolerate intense intraspecific competition; seedling densities sometimes reach 20,000 per square meter. These kinds of *r*-selected adaptations can lead to population explosions in a new habitat. Indeed, purple loosestrife is rapidly invading North American wetlands.

It was once thought that successful invaders appropriated "empty niches" and that such opportunism accounted for their spectacular success. Recently, B. Herbold and P. B. Moyle (1986) argued that exotics outcompete natives instead. As evidence, they point out that in California, 48 of 137 species of freshwater fish are nonnative. Of these 48 introduced species, only 6 are found in undisturbed or pristine waters, and 4 of those are species of trout or salmon introduced into fishless lakes in the high Sierra. Of the 26 species of exotics for which data exist, 24 are known to have a negative impact on native fish. Thus the successful, and often most devastating, introductions are organisms that have strong competitive abilities. Rather than entering empty niches, they displace native species.

Because it is difficult or impossible to predict which species will display such explosive abilities in new habitats, it is very important that we use extreme caution when intentionally introducing species and that we take all precautions to prevent accidental movements of species into areas where they are not yet established.

Fugitive Species

Species that inhabit transient environments are often very susceptible to competitive exclusion. Called **fugitive species** (Hutchinson 1951), these organisms typically colonize habitats immediately after a disturbance. As succession proceeds, the nature of the habitat changes until it is no longer hospitable to them. They are called fugitives because they or their offspring must always be in search of the next disturbance site for colonization. In salt marshes, disturbance creates patches of bare soil. Exposure to the direct sun leads to higher salinity due to increased evaporation. The plants that colonize these disturbances are fugitives; the community will eventually fill in the patches. Greenhouse experiments show that these fugitive species are much more tolerant of hypersaline conditions than other salt marsh plants (Bertness et al. 1992).

Fugitive species must have a reproductive and dispersal strategy that allows their propagules or offspring to reach the next disturbance site, because their current habitat is always disappearing. Thus they tend to be *r* strategists (Chapter 7) that put more energy into producing a large number of mobile offspring than into competitive ability. Consequently, fugitive species generally are known for their poor competitive ability. Figure 9.21 shows the sharp boundary between a population of fireweed (*Epilobium angustifolium*) in a burned forest and an undisturbed meadow community. Fireweed is a fugitive species: Its many wind-dispersed seeds rapidly colonize recent burns

FIGURE 9.21 Blossoms of fireweed (*Epilobium angustifolium*), a fugitive species that depends on fire disturbance. Note that the population of fireweed abruptly ends at the edge of the burned forest. It cannot invade the undisturbed meadow of tall grass. (*Photo by David Krohne*)

TABLE 9.2

Number of *Mirabilis hirsuta* Seedlings Surviving at Different Densities (number/0.2 m²) on Badger Mounds and in Undisturbed Prairies

	Location of Seedlings	
Density of Seedlings	Badger Disturbances	Undisturbed Prairie
1	0.8±0.2	0.2±0.2
2	1.4±0.4	0.2±0.2
10	0.4±0.2	0
20	0	0

From Platt 1976.

where competition from other plants is low. Note in the figure, however, that fireweed cannot invade the undisturbed meadow. In many cases it is not clear which is the cause and which is the effect: Do poor competitors tend to be colonists of disturbed habitat, or do colonists tend to be poor competitors because there is little selection for that ability?

One well-documented example of a fugitive species is a prairie forb called the hairy four-o'clock (*Mirabilis hirsuta*), which colonizes small disturbed areas caused by badger excavations of ground squirrel burrows. Platt (1976) showed that seedlings of this species survive to maturity only if they are located on such disturbed sites; they cannot compete successfully with other prairie plants in undisturbed prairie (Table 9.2).

According to Gause's law, species with very similar resource utilization curves generally do not coexist for long. If competitive exclusion always results in the extinction of one of the species, character displacement can never evolve. Because the two species must be able to coexist, even if precariously, for selection to modify the niches, the conditions that permit coexistence are of interest.

Conditions for Coexistence

One aspect of Gause's principle is central to the matter of coexistence: the degree to which species' resource utilization curves are "similar." Although similarity obviously is related to the degree of resource overlap, we have not yet quantitatively described what is meant by this term. The Lotka–Volterra models describe a method for doing so.

The Lotka–Volterra Graphical Models

A. J. Lotka (1925) and V. Volterra (1926) independently developed a set of graphical models that help answer these questions. Their graphical models were based on the equations we have used to describe population growth of a single species. Recall from Chapter 4 that we can model a species's population growth with the following equation:

$$\frac{dN}{dt} = rN\frac{(K - N)}{K} \qquad (1)$$

where r is the intrinsic rate of population growth and K is the carrying capacity. We can modify this equation to account for the effects of competition by another species. We know that by definition, competition decreases the growth rate of the two species. In addition, the intensity of the effect depends on the density of the competing species and the amount of overlap in resource use. Moreover, a large population of competitors has a greater effect than does a small population. The more similar the resource use by the competitor, the more intense the competition should be.

The following modification of Equation 1 accounts for these effects of interspecific competition on species 1:

$$\frac{dN_1}{dt} = r_1N_1\frac{(K_1 - N_1 - \alpha_{12}N_2)}{K_1} \qquad (2)$$

where α_{12} is the intensity of competition from species 2 on species 1, N_1 is the population size of species 1, and N_2 is the population size of species 2. The value of α ranges from 0 to 1. One way to think about this component of the equation is to consider it a factor that converts the number of individuals of species 2 into an ecologically equivalent number of individuals of species 1. The more similar the two species, the more nearly an individual of species 2 depletes the resource to the same extent as an individual of species 1. If species 1 and 2 were exactly identical, α would equal 1.0. Both N_1 and $\alpha_{12}N_2$ are subtracted from the carrying capacity, K_1, to simulate the effect of resource depletion on dN/dt. Note also in Equation 2 that we have added subscripts to identify the species to which each variable applies. Thus N_1 is the population size of species 1, and K_1 refers to the carrying capacity for species 1. We can write a similar equation for species 2 and its competitive interaction with species 1:

$$\frac{dN_2}{dt} = r_2N_2\frac{(K_2 - N_2 - \alpha_{21}N_1)}{K_2} \qquad (3)$$

Note the differences in subscripts between Equation 3 and Equation 2, especially that in Equation 2, α_{12} denotes the competitive effect of *species 2 on species 1,*

whereas in Equation 3, α_{21} refers to the competitive effect of *species 1 on species 2.*

Now let us consider the conditions that might allow species 1 and species 2 to coexist. In the simplest sort of coexistence to model, both species have nonzero populations that are no longer changing. Thus we are interested in combinations of densities of the two species such that $dN_1/dt = 0$ *and* $dN_2/dt = 0$, where N_1 and N_2 are nonzero. To satisfy these conditions, we must look for ways to make the right sides of Equations 2 and 3 equal to zero. In fact, all we need to do is make the numerators of the right sides of Equations 2 and 3 equal to 0. In other words, if $(K_1 - N_1 - \alpha_{12}N_2)$ and $(K_2 - N_2 - \alpha_{21}N_1) = 0$, dN_1/dt and dN_2/dt both equal zero. Thus, for species 1 we can set

$$(K_1 - N_1 - \alpha_{12}N_2) = 0 \qquad (4)$$

and for species 2 we can set

$$(K_2 - N_2 - \alpha_{21}N_1) = 0 \qquad (5)$$

These equations describe straight lines. At each point on each line, $dN/dt = 0$ for that species. Each of these lines is referred to as an **isocline.**

We can plot these two lines on a graph of N_1 versus N_2 (Figure 9.22). Consider the isocline for species 1. Note that the population of species 1 cannot exceed K_1 on the x-axis. We can show this by substituting 0 for N_2 in Equation 4. When we do this, we see that $N_1 = K_1$. In other words, in the absence of competitors, species 1 reaches its carrying capacity as we would expect. The y-intercept is derived by substituting 0 for N_1 in Equation 4. When we do this, we see that the y-intercept is K_1/α_{12}. Similarly, from Equation 5 we show that species 2 cannot exceed K_2 on the y-axis, and we determine that the x-intercept for species 2 is K_2/α_{21}. Combinations of species 1 and species 2 that fall on this line result in no growth of species 2.

Now let us consider what happens if we establish populations of various sizes of these two species. The isoclines tell us where each population is not growing. For either species, a population density beyond the isocline (away from the origin) results in a decrease in the population. For example, note from Equations 2 and 4 that species 1 will decline if the numerator on the right side of the equation is negative. Thus species 1 will decline if:

$$N_1 > K_1 - \alpha_{12}N_2 \qquad (6)$$

A similar analysis shows the conditions under which species 2 declines. Figure 9.22 shows these situations

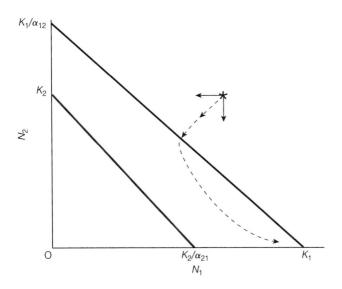

FIGURE 9.22 Isoclines in a Lotka–Volterra graph of interspecific competition. The asterisk represents the population densities of two species living in sympatry or in a single culture; it represents an experiment begun at a particular set of densities. The solid arrows represent the trajectories of each population; the broken arrow represents the combined trajectory (summed vectors) of the two populations. In this case, species 1 always outcompetes species 2.

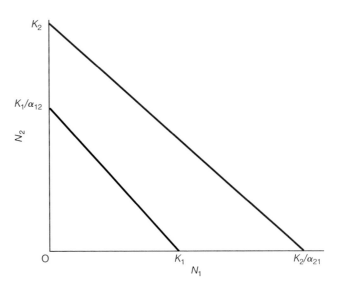

FIGURE 9.23 A Lotka–Volterra graph in which the locations of the isoclines dictate that species 2 always outcompetes species 1.

graphically. In the regions beyond the isocline of each species (away from the origin), numbers decline. In the region inside the isocline (toward the origin), numbers increase.

With this information, we can analyze the consequences of these relationships for the two species. Three outcomes are possible: competitive exclusion, unstable equilibrium, or stable equilibrium.

Competitive Exclusion In Figure 9.22 we have arranged the two isoclines such that the isocline for species 1 lies beyond that for species 2. Imagine that we are establishing laboratory cultures of two species of *Paramecium*. We establish a culture with numbers of the two species corresponding to the point indicated by the asterisk on the graph. This single point depicts the densities of the two species. Because the point lies beyond the isoclines of both species, we expect both species to decline. In Figure 9.22, the trajectory of species 1 is toward lower density (toward $x = 0$); the trajectory of species 2 is also toward lower density (toward $y = 0$). The trajectory of the combined densities is a vector that is the sum of the two vectors representing the individual species. In this case, it points toward the origin. If we followed this culture over time, we would see the densities decline in this way.

Eventually, the densities will reach the species 1 isocline. Again, we can reason what will happen next. Species 1 is now on its isocline, and in our analysis it is given that $dN/dt = 0$ for this population. Species 2 still lies beyond its isocline, however, so it continues to decline. The result is that the densities of the two species move into the region between the isoclines. Here species 2 is still declining, but because species 1 is now below its isocline, it begins to increase. The trajectory of the two populations will move steadily toward $N_2 = 0$ and $N_1 = K_1$ (see Figure 9.22). Thus, in this particular case, species 1 drives species 2 to extinction. We would expect species 1 to reach its carrying capacity, K_1, in the absence of the other species, and this is precisely what happens. Convince yourself that regardless of the initial densities of the two species, species 1 will always outcompete species 2 when the isoclines are arranged as in Figure 9.22. Figure 9.23 shows the reverse situation; species 2 always outcompetes species 1.

Unstable Equilibrium Now consider the arrangement of isoclines shown in Figure 9.24. Here the outcome is not so easy to predict. Again, let us begin a hypothetical culture of two species at the combination of densities indicated by the asterisk. The same logic indicates the trajectory of the populations. Again, both species decline toward the origin. If they happen to hit the intersection of the two isoclines, the numbers of both species stop changing; at this point both $dN_1/dt = 0$ and $dN_2/dt = 0$, and the two species coexist. What if the culture is begun at other population densities? If the populations are begun at point a, the trajectory is toward K_1 because we are in a region where species 1 increases but species 2 declines to

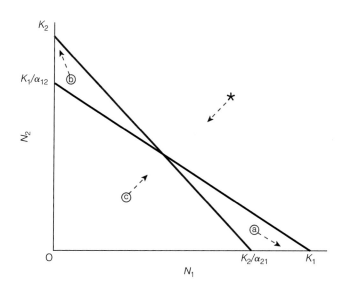

FIGURE 9.24 A Lotka–Volterra graph in which unstable equilibrium occurs. The asterisk and points a through c represent four different combinations of population densities. Note the different population trajectories in each of the four regions of the graph.

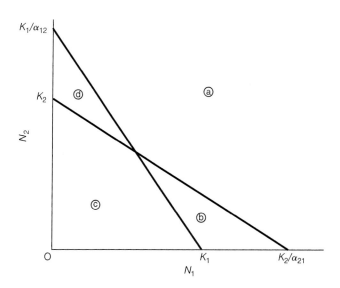

FIGURE 9.25 A Lotka–Volterra graph depicting stable equilibrium between two competing species. The populations converge on the intersection of the isoclines regardless of initial population densities.

extinction. If the populations are begun at point b, species 2 always wins. If the initial population densities are at point c, both species increase until one of the isoclines is reached. Depending on which isocline the vector hits, the populations enter the region described for point b or point a, with results as described for those conditions. Thus, in this configuration of isoclines, the outcome of interaction between the species varies according to the initial population sizes. The point of intersection of the isoclines represents an equilibrium point. If the populations reach that point, they no longer change because both are on their isocline. This is an **unstable equilibrium** point, however, because any perturbation away from that exact point results in the extinction of one of the species.

Stable Equilibrium   It is also possible for the isoclines to intersect as in Figure 9.25. In this situation, no matter what the initial population densities, the trajectories of the populations converge on the intersection of the two isoclines. Determine the trajectories for cultures begun at points a through d and convince yourself that this is so. We refer to the intersection of the isoclines in this example as a **stable equilibrium** because the populations will always reach that point and, if perturbed away from it, will always return. Note also that both species can increase from low population densities, even in the presence of large numbers of the other species.

Recall that we began this discussion by asking how species might coexist without competitive exclusion

occurring. We can see that Figures 9.22 and 9.23 represent classic competitive exclusion. A biological interpretation of the situation in Figure 9.22 suggests that because the carrying capacity of species 1 is so high, this species is not limited by this resource to the extent to which species 2 seems to be. Perhaps species 1 can use other resources not used by species 2. Thus, if it loses some of this resource to species 2, the effect is less devastating. Or perhaps species 1 is more efficient at obtaining the resource, so the resource's effective availability is higher.

Stable coexistence occurs when the isoclines are arranged as in Figure 9.25. How do we interpret these conditions biologically? In order for the isoclines to cross as in Figure 9.25, $K_1 < K_2/\alpha_{21}$ and $K_2 < K_1/\alpha_{12}$. Rearranging these inequalities, we see that $\alpha_{21} < K_2/K_1$ and $\alpha_{12} < K_1/K_2$. For this to be the case, the competition coefficients—the α's—must be small. In particular, for both species the effect of the competitor (measured by α) must be small relative to the ratio of its carrying capacity to that of the other species. These conditions must hold for both species simultaneously, and this is possible only if the carrying capacities of the two species are similar such that their ratio, regardless of which is the numerator, is close to 1. Thus any great disparity in the carrying capacities of the two species will result in competitive exclusion, or a nonstable equilibrium. In summary, stable coexistence in this graphical analysis occurs under conditions that make intuitive sense: small values for α and large carrying capacities for both species.

Empirical Examples of the Models' Predictions

How well do these graphical models describe real populations? Certainly, we can use the demography of species in the absence of competition to predict the results of interspecific competition. This procedure was demonstrated by an elegant set of experiments on the competitive interactions of four species of phytoplankon (Huisman et al. 1999). For aquatic algae, light is a critically limiting factor. In the laboratory one can determine the *critical light intensity* for a species, the minimum amount of light required to sustain population growth. Huisman and his co-workers measured the critical light intensities for the four species.

They predicted that in pairwise interactions, the species with the lower critical light intensity would be the stronger competitor; the species with the higher critical light intensity would be more likely to lose in direct competition. According to their measurements, the predicted competitive abilities of the four species were as follows: *Chlorella* > *Aphanizomenon* > *Microcystis* > *Scenedesmus*. These predictions were then tested in cultures in which two species were in direct competition. The results are shown in Figure 9.26. Note that in each experiment, the species with the lower critical light intensity reached higher density than its competitor and that in most cases, the poorer competitor was driven virtually to extinction.

However, when we can directly measure the parameters of the Lotka–Volterra models, we see that the models are not perfect representations of the results of competition between natural or laboratory populations.

For example, even though $K_1 = K_2 = 1.0$ and $\alpha_{12} = \alpha_{21}$ for populations of *Drosophila melanogaster* and *D. simulans* in the laboratory, these populations do not exhibit competitive exclusion, as the Lotka–Volterra models predict for these values (Miller 1964). Another example of the imperfection of these models was obtained for laboratory populations of *Drosophila*

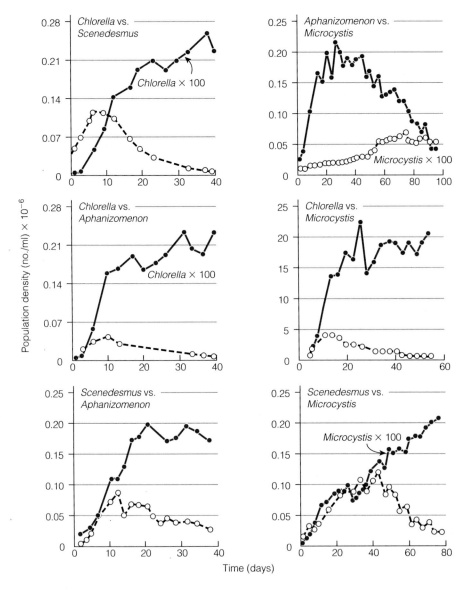

FIGURE 9.26 Results of competition between pairs of phytoplankton species. The solid lines and filled circles represent the species with the lower critical light intensity of the pair. (*Adapted from Huisman et al. 1999*)

pseudoobscura and *D. serrata*, and for *D. pseudoobscura* and *D. willistoni* (Ayala 1969, 1972). These species coexist despite the fact that they violate the conditions for coexistence set by the models. When the coexisting densities of *D. pseudoobscura* and *D. serrata* are plotted, as in Figure 9.27, they lie below the density predicted by the graphical model. F. J. Ayala suggests that perhaps the isoclines need to be "bent" downward. Such a deformation of the isocline might occur if α changes as a function of density. There is evidence that competition may be density-dependent in this fashion. For example, S. J. Smith-Gill and D. E. Gill (1978) measured values of the competition coefficient for two species of frogs, *Rana pipiens* and *R. sylvatica,* in laboratory populations of various densities. They

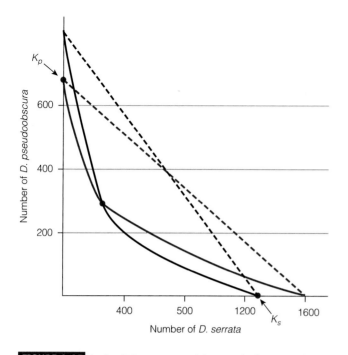

FIGURE 9.27 Lotka–Volterra competition graphs for *Drosophila serrata* and *D pseudoobscura*. The solid lines are observed isoclines; the dashed lines are isoclines predicted by the model. K_p and K_s represent the carrying capacities (K_2 and K_1) of *D. pseudoobscura* and *D. serrata*, respectively. The observed population densities "bent" the isoclines into curves. (*Data from Ayala 1969*)

found that both species' α changed in complex ways as a function of the density of the two species.

A series of classic laboratory experiments by T. Park (1954, 1962) on competition between two species of flour beetles revealed two important features of competition not borne out in the Lotka–Volterra models: (1) The abiotic conditions under which the interaction takes place affect the results of competition, and (2) The outcome of competition is not absolutely predictable. Park used two species of flour beetles, *Tribolium castaneum* and *T. confusum*, to study competi-

tion in laboratory populations cultured on wheat flour. A series of cultures was maintained at different combinations of temperature and relative humidity. The results of these experiments are shown in Table 9.3. Which species won in direct interspecific competition depended on the conditions. *T. confusum* was superior under cool, dry conditions whereas *T. castaneum* won more frequently under warm, moist conditions. The data in Table 9.3 are the frequencies of wins by *T. castaneum* and *T. confusum*. Each species shows a tendency to win under certain conditions.

The experiments by Ayala and Park suggest that a number of additional factors can affect the outcome of competitive interactions. In fact, one can conceive of a large number of complex interactions that can determine the outcome of competition, including interactions with a third species. N. A. Moran and T. G. Whitham (1990) have shown that two species of aphids, *Pemphigus betae* and *Hayhurstia atriplicis*, feed on plants in the genus *Chenopodium* (pigweed). *Pemphigus* feeds on the roots, whereas *Hayhurstia* feeds above ground and forms galls. Both use the same resource, the plant's phloem, but the two never physically encounter one another in nature. *Hayhurstia* has a particularly severe effect on host plants, decreasing their mass by up to 54 percent and the seed set by 60 percent. Moran and Whitham found that the competitive interaction between the two aphids depends in part on the degree of the host plant's resistance to *Hayhurstia*. On plants susceptible to *Hayhurstia*, the incidence of *Pemphigus* is reduced by 91 percent by the extensive presence of its competitor. On plants resistant to *Hayhurstia*, however, populations of *Pemphigus* are large. In this case, the evolved resistance of a third species, the host plant, ultimately determines the outcome of competition.

Habitat type may also affect the outcome of competitive interactions. T. P. Livdahl and M. S. Willey (1991) have documented an example of this in two

TABLE 9.3
Percent Wins by *Tribolium castaneum* and *T. confusum* in Competition Under Different Laboratory Conditions

Temperature (°C)	Relative Humidity (%)	Climate	Single Species Numbers	Mixed Species (%Wins)	
				T. confusum	*T. castaneum*
34	70	Hot-moist	*confusum = castaneum*	0	100
34	30	Hot-dry	*confusum > castaneum*	90	10
29	70	Temperate-moist	*confusum < castaneum*	14	86
29	30	Temperate-dry	*confusum > castaneum*	87	13
24	70	Cold-moist	*confusum > castaneum*	71	29
24	30	Cold-dry	*confusum > castaneum*	100	0

Values (percent wins) are the proportions of cultures in which one or the other species won in competition.

From Park 1954.

mosquito species. The native North American treehole mosquito, *Aedes triseriatus*, faces competition from the recently introduced Asian species, *A. albopictus*. Both species breed in small volumes of standing water, such as occur in tree holes or other cavities and even in abandoned tires. Livdahl and Willey showed that the predicted outcome of competition based on Lotka–Volterra graphical analysis depends on the exact habitat type. In Figure 9.28 we see that stable coexistence is expected in tree holes, whereas extinction of *A. triseriatus* is predicted in abandoned tires containing small pools of water.

Complex life cycles may also affect the potential for two species to coexist. S. C. Walls (1990) has shown that whereas larvae of the salamander *Ambystoma talpoideum* have become competitively superior to larvae of *A. maculatus* by inducing high mortality in the latter, the reverse is true for the adults. In this case, the differential competitive abilities of the larvae and adults probably result in the continued coexistence of the two species. Under these circumstances, neither can outcompete the other.

Finally, it may be important to consider the time scale over which competitive interactions are viewed. Two pairs of Caribbean corals, *Agaricia agaricites* and *Porites astreoides*, and *Agaricia agaricites* and *Montastraea annularis*, compete for space in the same way as the barnacles *Balanus* and *Chthalamus* that Connell studied. In this case, one species attacks by extending

digestive filaments to contact the other. Long-term photographic studies of these pairs of competitors showed that over time, repeated reversals of competitive advantage occur (Charnesky 1989). E. A. Charnesky points out that the pattern of repeated reversals may in fact reflect a long-term coexistence not evident in short-term studies that suggest that one species is winning.

These examples indicate that the Lotka–Volterra models, like most mathematical models, are an oversimplification of what actually occurs in nature. Nevertheless, they are useful for visualizing the relationships between the competition coefficients and carrying capacities in determining the outcome of competitive interactions between two species. They demonstrate that the critical variables are the carrying capacity, K, and the competition coefficient, α. Given the importance of the competition coefficient, it is useful to be able to measure it in nature.

In this chapter we have reviewed a number of examples of evidence for the existence of competition in communities. Because competition can lead to extinction of one species or to evolutionary change in the niche of a species, it is thought to be a central organizing force in community ecology. The consumption of one organism by another also has important ecological and evolutionary effects. Predation, as this interaction is called, is the subject of the next chapter.

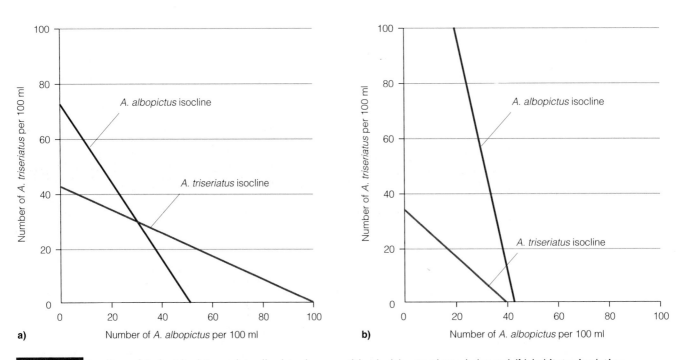

FIGURE 9.28 Isoclines of *Aedes triseriatus* and *A. albopictus* in competition in (**a**) natural tree holes and (**b**) habitats simulating abandoned tires. (*From Livdahl and Willey 1991*)

SUMMARY

1. The niche is a property of the species. The functional niche, a descriptive term, is the specie's *ecological role* in the community. The Hutchinsonian niche is defined by a hypervolume in *n* resource dimensions.

2. In practice, community ecologists often focus on a single, limiting resource because those resources are the sources of competition.

3. Competition is the negative effect on the growth rates of species whose niches overlap. This definition forms the basis for the empirical demonstration of competition: There must be a negative demographic effect on at least one species of a pair of potential competitors.

4. Species can compete for virtually any resource that limits population growth. Resources may be biotic factors such as prey items of certain sizes or abiotic resources such as regions of appropriate oxygen concentration, pH, and so forth. For many sessile organisms, space is a critically limiting resource.

5. Intraspecific competition may be particularly intense because conspecifics have the same niche and use the same set of resources. Intraspecific competition often limits populations.

6. Interspecific competition has two fundamental effects: The short-term effect (ecological time scale) is competitive exclusion in which one species declines to extinction as a result of competition with another. The long-term effect (evolutionary time scale) is character displacement, an evolutionary shift in the niche to decrease overlap with another species and reduce competition.

7. The Lotka–Volterra models provide a graphical analysis of the conditions leading to competitive exclusion or coexistence. Coexistence is more likely when the niche overlaps of the species are small relative to the ratios of the carrying capacities.

SELF-ASSESSMENT: CAN YOU ...?

1. Discuss the methodological issues associated with demonstrating competition in the field or in the laboratory.

2. List and distinguish among the various forms that competition can take.

3. Explain the relationship between intra- and interspecific competition.

4. Discuss character displacement in the context of microevolution as discussed in Chapter 2.

5. Derive the algebraic representation of competition in the Lotka–Volterra models. Using these models, explain the basis of competitive exclusion, unstable coexistence, and stable coexistence.

6. Explain the biological significance of the Lotka–Volterra conditions for coexistence and our empirical attempts to demonstrate them.

PROBLEMS AND STUDY QUESTIONS

1. Discuss the potential effects of intraspecific competition on niche parameters such as niche breadth.

2. Explain why α_{12} does not necessarily equal α_{21}. What kinds of niche parameters lead to this situation? What are the implications for competition between these two species?

3. Using Lotka–Volterra graphs, explain the relationship between niche overlap and carrying capacity that leads to competitive exclusion.

4. Discuss the reasons why a pair of species might undergo character displacement rather than competitive exclusion. Incorporate your understanding of the constraints on natural selection and evolution.

5. You observe three species of oaks living in the same stand of a deciduous forest. Their dispersion within the stand appears to be random. Does this violate Gause's law? Explain.

6. Design a set of experiments to test the hypothesis that the three species in Problem 5 are competing.

7. Using library literature and references, compile descriptions of the functional niches for the following:

    Everglades kite (*Rostrhamus socialbilis*)

    Common bladderwort (*Utricularia vulgaris*)

    Naked mole-rat (*Heterocephalus glaber*)

    Horseshoe crab (*Limulus limulus*)

    Yucca moth (*Tegeticula yucaasella*)

8. Species with broad niches are known as generalists. Some species have large geographic ranges. Discuss the relationship between geographic range and niche generalists. How is the range related to the niche? Does a large geographic range indicate a broad niche? How is the idea of a species's habitat related to these questions?

9. What features of Connell's system made his classic study of competition in *Balanus* and *Chthalamus* possible?

10. The bird distributions in Figure 9.8 are interpreted as evidence of competition. Discuss alternative explanations for these data.

PROJECTS AND ADDITIONAL STUDY

1. Search the literature for a putative example of competition between two species. Analyze this work in terms of (a) experimental design and (b) possible confounding factors.

2. Try to develop an analytical and experimental method for demonstrating that competition is an important mechanism in a *community of several species.* Compare your ideas to the systems discussed in Chapter 11. (*Hint:* Consider how competition should affect the Hutchinsonian niches through character displacement.)

3. Use laboratory cultures of protozoa (or algae) or greenhouse populations of fast-growing plants to

    a. Measure the population growth rate in pure culture

    b. Measure the population growth rate in the presence of one or more other species

    c. Test the effects of resource limitation (or excess) on competition

4. In this and previous chapters, the *size* of an organism has been shown to be important in a number of ecological contexts. Write a careful analysis of the ecological importance of size on the basis of Chapters 1–9.

Predation

e all have an intuitive notion of what predation is. The word conjures up images of a lion
stalking and finally rushing a wildebeest on the
Serengeti in Africa or a pack of wolves testing a moose
on a frozen lake in Alaska, pressing closer and closer
to detect any weakness or unwillingness to fight back.
Our mental image of predation is nature "red in tooth
and claw." Moreover, the lay person often sees the
predator as a villain, an image perpetrated by fairy
tales and cartoons in which ominous music signals the
approach of the cruel, bloodthirsty predator. Even
some nature shows designed to inform the public
about natural history anthropomorphize predators as
evil. Not only are such images biologically incorrect

(predators are neither evil nor good), but they have
led to some unfortunate public attitudes and policies.

The relationships between predators and their prey
are among the most fascinating in all of ecology and
are characterized by wonderful adaptations on the
part of both. Predation also has rich and far-reaching
ecological effects. After all, it is the consumption of
one species by another that drives the flow of energy
in ecosystems. We have already seen the role that predation can play in the regulation of animal populations.

Predation is difficult to define precisely because of
the wide range of ways in which one organism can
consume another (Figure 10.1). Some ecologists

FIGURE 10.1 Predation takes many forms. (**a**) A double crested cormorant (*Phalacrocorax auritys*) actively seeks and chases its fish prey. (**b**) A little blue heron (*Florida caerulea*), also a fish predator, waits motionless to ambush its prey. (**c**) A Minke whale (*Balaenoptera acutorostrata*) feeds on vast numbers of small crustaceans in the Antarctic Ocean. (**d**) A pitcher plant in a bog (*Serracenia* sp.) traps insects in its highly modified leaves. (**e**) Insect larvae graze on the stems and leaves of fireweeds. (**f**) A moose (*Alces alcres*) feeds in the same patch of fireweed. (*All photos by David Krohne*)

define predation broadly to include virtually all acts of consumption, including herbivory, detritivory, parasitism, and carnivory. Others restrict the term to a subset of these acts. The difficulty derives from our need to define arbitrary categories in a spectrum of activities that overlap considerably. In this text, we define **predation** simply as the consumption of one organism by another that attacks its prey while the prey is still alive. Thus our definition excludes detritivores. We define **heterotrophs** as organisms that obtain energy from other organisms rather than from inorganic sources. Thus all predators are heterotrophs, but not all heterotrophs are predators (some are detritivores). But note that the definition says nothing about the kind of organism that eats or is eaten, whether the entire organism is eaten, or even whether the predation kills the prey. Consequently, our concept of predation includes the processes of herbivory and parasitism, as well as carnivory.

Our discussion will focus on two major forms of predation, carnivory and herbivory, and on the adaptations of the predators and prey. Less attention is devoted to parasitism, not because it is unimportant—indeed, it is ubiquitous in nature and of great importance—but because relatively less is known about it. We then turn our attention to some theoretical aspects of predation: optimal foraging theory and the ecological effects of predation.

Carnivory is the predatory interaction in which an organism captures, kills, and consumes an animal. The vast majority of carnivores are animals, but a few plants, such as sundews, the Venus's flytrap, and bladderworts, are carnivorous as well. **Herbivory** is a general term that refers to the consumption of plant material by an animal. We recognize several different kinds of herbivores. **Grazers** (also called **folivores**) generally consume leafy material, whereas **browsers** eat mostly woody material and bark. Grazers and browsers may or may not consume the whole plant. **Granivores** consume plant seeds. Because they usually eat the entire seed and embryo, they consume and kill a potential entire plant. **Frugivores** consume fruit.

Parasites are organisms that form an association with individuals of another species (called hosts) in order to obtain nutrients or energy from them. We consider parasites separately from grazers because they are likely to form these associations with only one or a few hosts over their life spans, whereas grazers generally consume parts of many individuals. In actuality, though, grazers and parasites overlap: Some grazers, such as an insect on an oak tree, may spend their entire lives on a single individual.

Parasitic forms are found in virtually every phylum of animals and most divisions of plants. In addition, there are myriad parasitic bacteria, fungi, and viruses. This area is highly important for both commercial and health reasons. Parasites affect everything from our food crops to our own health. Parasitic infections and diseases are major causes of human mortality around the world. Although we know a great deal about some of these parasitic species, we know precious little about the roles of parasites in ecological systems (Minchella and Scott 1991). A few examples of the role of parasites in community ecology have recently emerged (see Chapter 11), but the ecology of parasitism is still a rich area for future research.

Another group of organisms that fall under the rubric of predation are the **parasitoids.** These organisms are very similar to parasites in that they consume nutrients and energy from a host, but we classify them separately because their parasitic activities are limited to their larval stages. The adults generally are free-living and do not form an association with a host. The adult females lay their eggs on or near an animal host. Upon hatching, the larvae consume the host, either partially or completely. This form of predation is confined to the insects in the orders Hymenoptera and Diptera.

Predation, in its many forms, is an important interaction at all levels of the ecological hierarchy. It drives the movement of energy and nutrients in ecosystems. It plays a role in organizing communities. It affects the patterns of population dynamics. Consequently, we discuss predation and its effects throughout this text. In this chapter, our focus is on the evolutionary ecology of predator and prey species as an important aspect of community ecology.

The predator, be it carnivore, parasite, or grazer, constitutes an important part of the environment of the prey. And the prey, with its anti-predator adaptations, exerts strong selective effects on the predator. Some authors (Dawkins and Krebs 1979) reason that in most cases, the intensity of selection on the predator and the prey is not symmetrical; that is, the cost to the predator of losing a meal is not so great as the cost to the prey of being consumed. R. Dawkins and J. R. Krebs refer to this as the "life-dinner principle." The predator risks its dinner; the prey risks its life. However, in other systems, the prey represent a clear danger to the predator. If so, there is the potential for a "coevolutionary arms race" (Brodie and Brodie 1999). In such a system, there is reciprocal selection on the predator and prey: Each adaptation by one side constitutes a new selective force that leads to a counteradaptation by the other. It is important to note that in coevolutionary systems, the units of interest are lineages or populations (not individuals) and the time scale is multi-generational (evolutionary time).

Nevertheless, the fact remains that each group, predator and prey, constitutes an important part of the other's environment and hence exerts selective pressure on the other. In the next section, we begin to explore these evolutionary interactions through a discussion of the adaptations associated with carnivorous predators and their prey.

Carnivorous Predation

Predator Adaptations

A predator must find, recognize, capture, and consume its prey. In this section we will consider the adaptations by which predators accomplish these tasks. First, we will look at the sensory adaptations that help predators detect and recognize their prey; then we will consider the various strategies for hunting and foraging.

Prey Detection and Recognition   Predators use a stunning array of sensory modes to locate potential prey items. In addition to the modes that quickly come to mind, such as sight and smell, some predators have adopted more exotic mechanisms, such as the use of electric fields and echolocation. But regardless of the sense used, the predator must distinguish a prey item's input from the general background input of the environment. This is accomplished by the development of a **search image.** Although this term probably suggests a visual image, it can be applied to any sensory mode. A search image develops via a simple learning process. As the animal samples sensory patterns from the environment, it learns which patterns provide a potential food reward. The search image is the pattern of input that comes to be associated with prey. With experience over time, conditioning improves and refines the search image. This concept implies that the predator undergoes a perceptual change in which it "learns to see" prey items that were previously undetected.

It is easiest to think of this process in a visually hunting predator. Indeed, most of the work on this topic has been done with birds that hunt by *sight.* For a bluejay hunting insects, certain visual patterns such as wings, eyes, or legs come to be associated with food. For a hawk hunting small mammals, a certain pattern of color and shape comes to be associated with a prey item. An analogous process occurs with other sensory

modes. An owl learns that a certain sound pattern indicates a foraging mouse on the forest floor. A hyena quickly learns to distinguish the odors of potential prey items from the myriad other odors it detects.

Predators that rely on sight are familiar to us. We can relate more easily to their mode of locating prey because humans are such a visually oriented species. Hawks and eagles, whose eyes are highly adapted for detecting prey at great distances, are the most obvious examples of visually hunting predators. It has been estimated that the resolving power of a bald eagle's eyes is approximately that of a pair of 7× binoculars. Because many such predators are particularly sensitive to motion, a cryptic animal may avoid detection until it gives away its presence by moving. Large birds and mammals are not the only animals that make use of this sensory mode. Jumping spiders use their eyes to detect and capture their insect prey. Dragonflies are aerial insect predators that use sight to hunt.

Smell (chemoreception) is also commonly utilized by predators. This mode of prey detection is more difficult for humans to appreciate because our olfactory powers are so poorly developed compared to those of many other animals. Many mammalian carnivores, including wolves, coyotes, badgers, and bears, make extensive use of their sense of smell in locating prey. Among freshwater and marine fish such as sharks, chemoreception is a principal sensory mode.

The ability of some animals to detect the odor of prey is truly remarkable. Many of the canids (members of the dog family) can detect odors in concentrations of just a few parts per billion, which enables them to locate prey by following even faint concentration gradients. Reptiles have also taken these abilities to high levels of sensitivity with a special set of adaptations. After a snake samples molecules in the air with its forked tongue, it places the tips of the tongue into two pits in the roof of the mouth (the vomeronasal organ), thereby depositing the molecules directly onto olfactory nerve endings.

Sound is another common means of detecting prey. Among the vertebrates, owls have taken this adaptation to high levels of sensitivity. Because they are nocturnal predators, sight is of little value to them, and like birds in general, they have poor olfactory capabilities. The auditory centers of the brain, however, are highly developed in owls. In addition, an owl's face includes remarkable morphological adaptations that enhance its hearing. The circular patterns that surround the eyes (Figure 10.2) actually have functional significance: The feathers work as parabolic dishes to gather sound. Tiny feather tracts from the outside edge of each parabola lead directly to the ear, so that

Barred owl

Barn owl

Great gray owl

Snowy owl

FIGURE 10.2 The circles around the eyes of many species of owls are parabolic reflectors that help gather sound and channel it to the ears.

when an own turns its head from side to side, it is actually gathering sound information on the direction and distance of an object.

The bats, whales, and porpoises have evolved extremely sophisticated auditory sensation. They have combined highly sensitive and discriminating hearing ability with the ability to produce high-frequency sounds to form an echolocation system for finding prey. In bats, high-frequency sound is emitted from the mouth and nose. The bizarre shapes of the faces of many species of bats are apparently important in directing sound emissions. The sound strikes objects in the environment, and the echoes from those objects are perceived by a highly sensitive auditory system. A similar system evolved in cetaceans. In this case, sound is emitted through a "lens" of fat in the forehead, and the echoes from objects in the environment are likewise interpreted.

The sensitivity and precision of this system, particularly in bats, are astounding. Bats can detect objects as small as 0.1 millimeter, determine the direction of an

object to within less than 1 degree of arc, and distinguish textures. The direction of an object is determined by differences in the timing and intensity of echoes detected by the two ears. If the right ear detects an echo louder and sooner than the left ear does, the bat senses that the object is to the right. The distance to an object is sensed from the time the echo takes to return; the degree of pulse/echo overlap is greater for very near objects than for distant objects (Figure 10.3). Bats are able to modify the frequency of the sound they emit to match the size of the object they are tracking; smaller objects can be resolved better with sound of shorter wavelengths. To obtain detailed information more rapidly, bats can also increase the rate of sound emission in the last phases of approach before capture.

A few animals use information from *touch,* or at least nonauditory vibrations, to gather information about prey. Many web-building spiders detect vibrations through their legs. When an insect hits a spider's web, the spider sitting in the center of the web feels the vibration. To determine the location of the insect, the spider sequentially plucks the radial strands of the web. The insect caught in the strands dampens the vibrations along that radius, and the spider moves out to secure its prey with silk.

A group of snakes, the pit vipers, have developed a unique sensory mode: *heat detection.* Pit vipers derive their name from two pits located on the head, which are heat sensors that enable them to detect the presence of homeotherms. These snakes, most of which are nocturnal predators of desert areas, take advantage

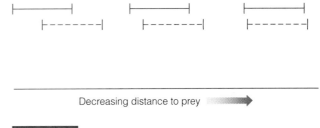

Decreasing distance to prey ➡️

FIGURE 10.3 Pulses (solid lines) emitted by a bat and echoes (dashed lines) received by the bat as it approaches an object. As the distance to the object decreases, the echo overlaps with more and more of the original sound impulse.

of the fact that the desert typically cools down rapidly at night. Under these conditions, the search image of a warm-bodied animal stands out against the cooler ambient background.

Finally, some fish have developed the ability to use *electric fields* to detect prey. Some of these species live at great depths where there is little or no light. The presence of a prey item is detected by sensing characteristic distortions in a weak electric field emitted by the predatory fish.

Tactics for Capturing Prey
Detecting a prey item is, of course, only part of the task. The predator must then capture, subdue, and consume the prey item. Some prey species are formidable opponents for predators either because they are large and well armed or because they have developed special noxious adaptations. A full-grown moose is quite capable of defending itself by virtue of its mass, antlers, and sharp hooves. The darkling beetle (*Eleodes*) appears far less imposing. Nevertheless, it has evolved a potent chemical defense. When attacked by a predator, the beetle lifts its abdomen into the air and excretes a noxious blast of chemicals.

Predators have responded with a number of adaptations of their own to facilitate the process of capturing and killing their prey. The grasshopper mouse, for example, counters the chemical defense of *Eleodes* by grabbing the insect and jamming its abdomen into the sand so the blast can do the mouse no harm. It then proceeds to eat the beetle from the head down. Wolves hunt in cooperative groups that negate the superior fighting ability of the moose.

Predators can be broadly classified into several groups on the basis of their hunting style. One such classification is based on the movement pattern of the predator—whether it waits in hiding to ambush its prey or actively seeks them. Another grouping is based on whether the predator hunts singly or in groups.

Ambush predators (sometimes called sit-and-wait predators) remain quiet and hidden in hopes of ambushing an unsuspecting prey item. The Venus's flytrap and the great blue heron epitomize this strategy. Some ambush predators increase the chances that a prey item will come near enough for a strike by providing some kind of lure or attractant, such as the projection on the head of the angler fish. In a few cases, the sex hormone of the prey acts as an attractive lure. The bolas spider secretes a chemical that mimics the sex pheromones of its moth prey and thus attracts males to their deaths. It has recently been shown that some web-building spiders also are attractors (Craig

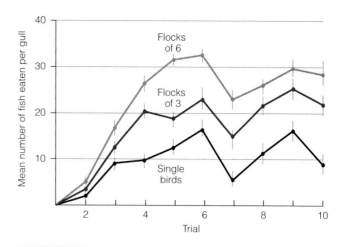

FIGURE 10.4 Foraging success of captive black-headed gulls in flocks of different sizes. Gulls in flocks of six captured more prey per 3-minute trial than did single birds or flocks of three. (*From Gotmark, Winkler, and Anderson 1986*)

and Bernard 1990). The webs of garden spiders in the genus *Argiope* are decorated with special strands of ultraviolet-reflecting silk. This silk reflects the same wavelengths that some flowers use to attract insect pollinators. Thus the insects are attracted to a "flower" only to find themselves caught in a spider web. The reflectance patterns of spider silk are highly adapted. The tropical orb-spinning spider, *Nephila clavipes*, spins silks with different reflective properties in different light environments (Craig et al. 1996). In dim light, they produce white-reflecting silks that are difficult for insects to detect. In bright light, they produce pigments that make their silks appear golden yellow. Experiments with bees have shown not only that bees are attracted to gold webs but also that they have difficulty learning to avoid them.

Other predators encounter their prey via their own movements. Parasitoids, many parasites, virtually all herbivores, and many carnivores fall into this group. Both active and passive means of movement can be involved. Female parasitoids search actively for host larvae on which to lay eggs; mountain lions roam their territories in search of the sight or smell of prey. Parasites tend to use more passive means of locating their hosts. The malarial parasite, *Plasmodium*, is transferred from host to host via the bites of mosquitoes. Ticks acquire new hosts by waiting on vegetation and then moving toward them in response to carbon dioxide gradients resulting from their respiration.

Cooperative groups of predators are distinguished from those that simply forage in a group by their need for help from conspecifics in actually capturing the prey. Bison, for example, forage together but not cooperatively—they do not require groups to subdue the grass! Cooperative hunting in terrestrial mammals is found in lions, several species of canids (wolves, coyotes, hunting dogs), and hyenas. Less is known of aquatic group foraging, but many reports indicate that killer whales hunt cooperatively. For all these hunters, the cooperative group can take much larger prey than could any lone individual. There is a limit to the advantages to individuals that forage in a group, however. In lions the relative food intake per individual is highest for hunting pairs but declines in larger groups (Packer 1986); in other species, such as gulls, there are energetic advantages to larger groups (Figure 10.4).

Prey Adaptations

Prey species develop adaptations to counter those of their predators. These adaptations consist of avoiding one or more of the three categories of the predator's efforts: detection, capture, or handling.

Avoiding Detection   A prey species that avoids predation by avoiding detection is said to be **cryptic.** Even though we are most familiar with crypticity for visually hunting predators, it applies equally well to all sensory modalities used by a predator. For prey, the general principle of crypticity is precisely the opposite from that of the predator seeking to locate prey in the environment: To avoid detection, the prey should be indistinguishable from a *random sample of the background* (Figure 10.5). If the prey cannot be distinguished from a random sample of the background, the predator cannot develop a search image for it. In gen-eral, a prey animal is less visible in any of the following circumstances:

Its color matches the background.

Its shape is asymmetrical.

It is countershaded (the lower part of the body is light and the upper part is dark, so that in bright sunlight the contrast between the animal's shaded and unshaded areas is reduced).

Its outline blends into the background.

Its eyes are hidden.

It remains motionless.

Mimicry of objects in the environment, such as sticks or leaves, is a result of this same set of adaptations.

As we have said, the same principle of crypticity applies to other sensory modalities. To foil echolocation by hunting bats, some moths in the family Noctuidae have evolved "auditory crypticity" by emitting sounds in the same frequency range as the bat's sonar. These sounds confuse the echolocation data the bat receives by making the moth "sound like" a random part of the background.

Avoiding Capture   Once a prey item has been detected, its only remaining options are to flee or resist capture. Both strategies are used extensively. We are all familiar with the fleetness of certain animals such as antelope, whose speed advantage is so great that predators have little chance of running them down unless they are very old, very young, or somehow impaired. Some species can escape predation if they are warned in time. Alarm calls in ground squirrels function in this way. (The evolution of alarm calls is discussed in Chapter 2.) In some species these calls are quite sophisticated. C. N. Slobodchikoff, C. Fisher, and J. Shapiro (1986) demonstrated that the alarm calls of Gunnison's prairie dog can communicate a great deal of information. Not only do the prairie dogs have different calls for different kinds of predators, but recent experiments have shown that they can distinguish between individual humans and can even communicate which particular human is approaching (Slobodchikoff et al. 1991).

FIGURE 10.5 (**a**) A white-tailed ptarmigan (*Lagopus leucurus*) (*Photo by David Fritts/Animals Animals*) and (**b**) the American bittern (*Botaurus lentiginosus*) are examples of cryptic coloration. (*Photo by Ray Richardson/Animals Animals*)

Predation can be deterred if the predator is confused by the activity of the prey. A number of species use some behavior to startle and freeze the predator at the critical moment and thus avoid capture. Anyone who has had a nearby pheasant suddenly take to the air in a whir of wings understands the startle effect. Some moths suddenly reveal eye spots on their wings when disturbed. The moment of hesitation induced in a predator may be sufficient to allow escape. Some species provide false targets for the predator. The white spots on or near the tails of many birds may well be adaptations that increase the likelihood that a fox will come away with only a mouthful of tail feathers. Some marine invertebrates apparently use flashes of bioluminescence to startle predators.

For other species, avoiding capture involves safety in numbers. Some prey are safer in large groups, especially when there is some form of group defense. Adult musk oxen are known for thwarting wolf predation by pressing closely together in a circle with their horns facing outward. D. C. Heard (1992) has shown that the size of musk ox groups increases with the density of wolves (Figure 10.6), particularly in winter when other prey are less available to wolves. Larger groups mean more eyes for vigilance and stronger defense groups.

The concept of the selfish herd is associated with a safety-in-numbers strategy. The idea of the selfish herd, a concept developed by W. D. Hamilton (1971), is that if animals form herds, predators will be more likely to attack only peripheral individuals. In fact, Hamilton proposed that predation on the marginal individuals is an important driving force that maintains aggregative behavior. This perhaps is the reason why a number of animal species form schools, herds, or flocks.

Data bearing directly on the anti-predator advantages of schooling in fish have been difficult to obtain. J. K. Parrish (1989) studied the risk of predation for individuals in different positions in a school of Atlantic silversides (*Menidia menidia*) being preyed on by black sea bass (*Centropristis striata*). Individuals that straggled behind the school were at greater risk—it was clearly safer to be part of the school. However, the relative risk of various positions in the school depended on the type of attack (Figure 10.7). When the attacking fish launched from below, the central fish were safer; when the attack came from above (passer), it tended to divide the school in half. The central fish were then at risk because they suddenly found themselves at the back of the school.

Another form of safety in numbers, referred to as predator swamping, is based on the fact that when prey densities are very great, the predators cannot

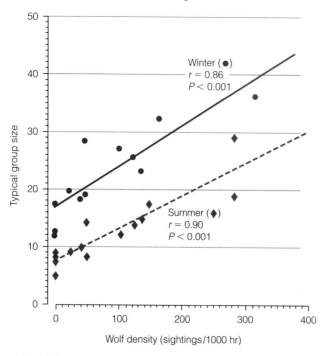

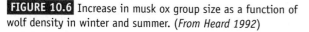

FIGURE 10.6 Increase in musk ox group size as a function of wolf density in winter and summer. (*From Heard 1992*)

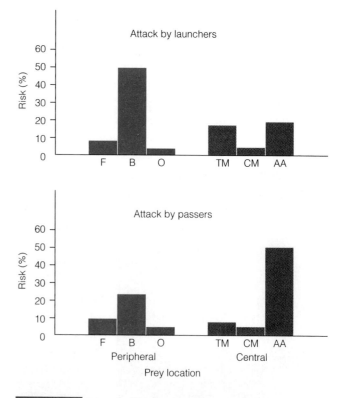

FIGURE 10.7 Relative amount of the average total risk of predation incurred by Atlantic silversides (*Menidia menidia*) in six positions within a school under attack by launchers and passers. F = front; B = back; O = outside; TM = true middle; CM = confused middle; AA = after attack. (*Adapted from Parrish 1969*)

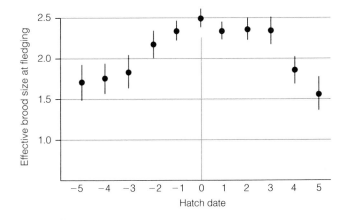

FIGURE 10.8 The relationship between average (±1 standard error) effective brood size at fledging and hatch date in lesser snow geese in Manitoba. Values are derived from a pooled (1973–1979) sample of successful females. (*From Findlay and Cooke 1982*)

possibly consume all individuals. Even though some will certainly be eaten, many more will escape predation. Periodical cicadas may rely on this strategy. They lie dormant for many years (13 or 17 years in North America). Then they emerge all at once locally and probably swamp their predators.

Lesser snow geese have adopted a similar swamping strategy for egg laying, although the time scale is much shorter. Most individuals in the population lay eggs synchronously. Fox and gull predation on eggs is intense on any clutches laid prior to the peak. The probability of raising a large brood is greatest for those individual snow geese that lay eggs synchronously (Figure 10.8).

Disrupting Predator Handling   Many prey species are readily detected by predators and are unable to escape from their predators either by fleeing or by seeking the shelter of a group. For them, some means of resistance or deterrence is the only option. This strategy has taken many forms among prey, including the development of significant defensive weapons such as antlers, horns, shells, and chemical deterrents.

A variety of morphological features are effective anti-predator adaptations. The shells of many invertebrates are simply passive barriers to predator attack; obvious examples are the hinged shells of clams and mussels. Other features are employed more actively in defense. Sea urchins produce long spines bearing toxin- or irritant-filled sacs to discourage potential predators.

Among the most interesting prey defenses are those induced directly by the predators. Harvell (1986) documented an example of inducible defense in the marine bryozoan *Membranipora membranacea*. When this species is preyed on by nudibranchs, it responds to unwitting chemical cues from the nudibranch by rapidly producing new chitinous spines that reduce predation rates by as much as 60 percent. As we will see, inducible defenses are known in plants as well.

In the newt, *Taricha torosa*, a potent neurotoxin called tetrodotoxin acts as a chemical deterrent to predation. This compound is among the most potent toxins known in nature; a tiny piece of newt tail contains enough tetrodotoxin to kill garter snakes (*Thamnophis* spp.). Most species of garter snakes avoid the aposematically colored *Taricha*. However, E. D. Brodie III and E. D. Brodie Jr. (1999) showed that among individuals and broods in *Thamnophis sirtalis*, there is heritable variation in the ability to tolerate tetrodotoxin. Populations of this species that are sympatric with newts are more resistant to the toxin than those that have not experienced it. Thus the advantage of the toxin is countered by the evolving resistance in the garter snake.

An anti-predator toxin in another species, the bulber grasshopper (*Romalea guttata*), is the focal point in another example of the state of flux in many predator-prey relationships (Yosef and Whitman 1992). This large grasshopper (it is up to 7 centimeters long) is aposematically colored and produces toxins that make it unpalatable to 21 species of birds and lizards. Still, one avian predator, the loggerhead shrike, has developed an effective counteradaptation. Shrikes characteristically impale their captured prey to kill them. The loggerhead shrike dispatches a bulber grasshopper by impaling it on thorns or barbed wire and leaving it there for two to three days. During this time the toxins degrade, and the shrike can then consume the grasshopper with no ill effects. The shrikes are careful to impale the grasshopper through the thorax, where the toxins are most concentrated. As this example illustrates, counteradaptations in the coevolutionary race need not involve complex morphological or biochemical changes; in this case, a simple behavioral change countered the toxin's effectiveness.

For some prey species, the strategy for avoiding capture involves simply *appearing* to be larger or more formidable than they actually are. The spoonwing lacewing (*Palmipena aeloeptera*), which is preyed on by robber flies, apparently deters predation because its huge underwings make it appear much larger. Experiments by M. D. Picker, B. Leon, and J. G. H. Londt (1991) on tethered lacewings with their wings either intact or amputated showed that robber flies were far more likely to attack the lacewings without the large appendages.

Evolutionary adaptations and counteradaptations of predator and prey are not limited to carnivores and their prey. We will see this in the next section, where we turn our attention to those predators that consume plant material.

Herbivorous Predation

Like all predator–prey systems, plant–herbivore systems constitute evolutionary interactions in which the plants and their grazers exert selective influences on each other. In this section we will follow the same pattern as in the previous section, focusing first on the adaptations of herbivores and then on the counteradaptations of the plants. As we saw, carnivorous predators sometimes have difficulty detecting their prey, but this is not typically a problem for herbivores. Instead, herbivores are often faced with the task of countering the defense mechanisms of the plants, which may include morphological adaptations such as spines and thick cuticles or chemical defenses known as secondary compounds. Our discussion will focus first on the adaptations that herbivores employ to counter these defenses.

Herbivore Adaptations

One might not think that locating prey would be so difficult for an herbivore. However, there is a great range in nutritional quality among and within species that grazers must recognize. Plants change seasonally in energy and mineral content. Interspecific differences are significant. For example, in the Serengeti there are significant seasonal changes in the quality of the C4 grasses associated with the pattern of precipitation. As a result of seasonal rains, there is a "green wave" of plant production to which wildebeest (*Connochaetes* sp.), zebra (*Equus* sp.), and eland (*Taurotragus oryx*) respond (Frank et al. 1998). In Yellowstone National Park, elk (*Cervus canadensis*) and bison (*Bos bison*) follow a similar "green wave" as the vegetation emerges after snowmelt. In the early spring, these species migrate to higher altitudes to track the phenologically younger and more nutritious vegetation (Frank et al. 1998; Frank and McNaughton 1992).

The second problem herbivores face is the suite of plant deterrents to grazing: structural and chemical defenses. Among mammalian herbivores, the koala (*Phascolarctos cinereus*) exemplifies a suite of adaptations to undermine the anti-grazing strategy of the

plant. The koala feeds entirely on the leaves of eucalyptus trees (*Eucalyptus* spp.), which are highly protected: They are low in nutrients and high concentrations of lignin and tannin reduce the digestibility of the nutrients. The koala's adaptive response includes anatomical, physiological, and behavioral traits (Lee and Martin 1990). The gastrointestinal tract of the koala is very long for its body size. In addition, the koala has a lengthy cecum where food is held for as long as eight days to ensure that as much nutrition as possible is derived from the leaves. Koalas also have very slow metabolic rates, which enables them to subsist on such poor-quality food. At rest they expend only 74 percent of the calories used by an average marsupial of similar size and only 50 percent of that used by a placental mammal. In addition, koalas eat only one leaf at a time and chew slowly and meticulously.

Other herbivores have developed direct counteradaptations to the chemical defenses of plants. For example, the monarch butterfly sequesters the cardiac glycosides it ingests in milkweeds, thus making the adult butterfly toxic to its predators. Other counteradaptations to plant chemical defenses include detoxification of the secondary compounds and their use as feeding cues and stimulants. In this way, insects may use the plant defense mechanisms to direct them specifically to plants to which they are adapted and to help them avoid plants whose defenses they cannot circumvent.

A number of insects use behavioral means to avoid plants' chemical defenses. The squash beetle (*Epilanchua tredecimnata*) feeds in a very distinctive way—by cutting a circular trench around part of the leaf until it is held by only a few veins and the lower epidermis. This prevents the plant from sending its chemical defenses to the part of the plant under attack. Similarly, alder sawflies (*Eriocampa ovata*) sever the veins of the leaf before feeding (Rhoades 1985).

D. F. Rhoades (1985) has proposed that overall, herbivores fall into two fundamental groups based on their strategies. On the one hand are so-called *stealthy herbivores,* species that attempt to obtain their plant nutrients without mobilizing the active plant defense systems, such as local production of toxins or hormone mimics. Herbivores that adopt the other strategy are called *opportunistic herbivores.* These species take advantage of plants that for various reasons are otherwise stressed and thus unable to mount a capable defense. Plants that are under abiotic, competitive, or nutritional stress are especially vulnerable to preda-

TABLE 10.1
Typical Attributes of Stealthy and Opportunistic Herbivores

Stealthy Herbivores	Opportunistic Herbivores
Relatively stable population size	Variable population size
Conservative use of plant prey	Profligate use of prey
High efficiency of conversion of plant material	Low efficiency of conversion of plant material
High digestibility of food	Food typically poorly digestible
Low feeding rate	High feeding rate
Low metabolic rate	High metabolic rate
Territorial	Colonial
Low r	High r

Adapted from Rhoades 1985.

tion, and opportunistic herbivores limit themselves to such plants. Herbivores themselves may contribute to the stress by launching massive attacks that weaken the plant and thus subvert its defenses. The general characteristics of stealthy and opportunistic herbivores are listed in Table 10.1.

Plant Adaptations to Herbivores

Like carnivorous predators, herbivores constitute an important selective force on their prey. As is the case for animal predators and their prey, plants and the animals that feed on them interact in an evolutionary relationship.

The strategies available to plants to deter predation are more limited than those available to most prey animals because of the sessile nature of plants. Some adaptations, however, are similar to those found in prey animals. For example, Janzen (1976) proposed that the synchronous flowering of bamboo is a predator-swamping adaptation. Many species of bamboo are known to grow vegetatively for many years (some species for as long as 100 years). Then, suddenly, all the individuals in the population flower simultaneously and set seed. So many seeds are produced that granivores are attracted and consume great numbers of seeds. No predator can specialize on bamboo seeds, however, because the interval between flowering is so long. When flowering *does* occur, the predators that respond are swamped, and many individuals escape predation.

Still, plants typically have fewer options for anti-herbivore adaptation. Their basic strategies can be divided into structural and chemical adaptations.

Structural and Chemical Adaptations Structural elements of plants can decrease the availability of nutrients to grazers or can deter them by making the plant less palatable. One of the most obvious structural adaptations for deterring grazing is the production of spines or thorns. These structures are most effective against vertebrate herbivores; insects and other small grazers may be able to work around the spines.

Structural elements of the plant tissues may also deter grazing. Many grasses incorporate silica in their stem and leaf tissue (McNaughton et al. 1985), which not only makes the plant tissue less palatable but also wears down the teeth of the grazers. Note that this adaptation is fundamentally different from thorns and spines, which over ecological time deter grazing on individuals. Silica acts as a selective agent on the grazer: Over time, herbivores will evolve an aversion to plants with large amounts of silica. This particular plant evolutionary adaptation is very precise; populations that have been subjected to heavy grazing pressure have higher silica concentrations (McNaughton et al. 1985).

The other major form of plant defense is chemical deterrence. It has long been known that many plants contain compounds that deter grazers. They are called **secondary compounds** because they were thought to represent by-products of plant biochemical pathways that are toxic to herbivores. You are probably familiar with some of these compounds, because many of them are used for medicinal or other purposes. Morphine, caffeine, and nicotine are secondary plant compounds that belong to a large group of compounds called alkaloids. The unique tastes of many spices are a result of secondary compounds that deter grazing by insects.

There are two fundamental views of the origin of secondary compounds in plants. C. H. Muller (1970) suggested that many of these compounds are secondary metabolites, compounds produced as by-products of plant metabolism and co-opted for defense because of their toxic nature. If a plant can sequester these toxic compounds in its tissues without harm to itself, the chemical can be used for the plant's defense. Some plants make some biochemical modifications that render the compounds toxic to animals but not to themselves.

The alternative view was suggested by P. R. Ehrlich and P. H. Raven (1964), who hypothesized that these compounds are not derived from metabolic by-products but rather are evolved *de novo* in response to grazing as a selection pressure. According to this view, the production of these compounds has a higher energetic and metabolic cost to the plant than Muller predicted. Most ecologists have come to accept the Ehrlich and Raven hypothesis (Stamp 1992), because biochemical evidence indicates that these secondary compounds are produced and degraded regularly in the cell and are not by-products of other pathways.

Plants have adopted a wide array of chemical compounds for defense and a wide range of strategies for thwarting their predators. We can categorize these strategies into three main groups: repellents, toxins, and hormone/pheromone mimics.

Repellents are compounds that discourage predators from feeding or laying eggs on the plant. The thistle *Parthenium hysterophorus,* for example, produces a compound that repels the larvae of its main insect predators. Tannins are common secondary compounds used by many plant species to render leaves unpalatable and less digestible to herbivores.

Toxins go a step further. Rather than simply repelling the herbivore, they prevent grazing by causing mortality of the predator. An example is the cyanogenic white clover (*Trifolium repens*), a plant whose tissues contain cyanogenic glycosides, which are sugar compounds bound to cyanide. When the tissue is damaged by grazing, two special enzymes located in the tissue are released and separate the cyanide residues from the sugars, thereby releasing this highly toxic compound into the digestive system of the grazer.

Bird's-foot trefoil (*Lotus corniculatus*) in Europe exemplifies how precisely such chemical defense systems are matched to the degree of herbivory. Where winters are more severe, and thus mortality of insect predators is higher, the proportion of cyanogenic forms of this plant is lower; where the winters are milder, most plants are cyanogenic because they face higher rates of attack by insects (Figure 10.9). These data are consistent with the hypothesis that these secondary compounds are produced as a specific metabolic end product to deter grazing, not as a by-product of other pathways. Only where the plants face intense predation do they incur the metabolic costs of producing these compounds.

Some species have the ability to produce compounds that mimic important hormones of their attackers. In one example, the potato (*Solanum berthaulthii*) synthesizes a component of an alarm pheromone that aphids release when attacked by a predator. Release of this compound by the potato elicits a flight response in the aphids. Another plant, the flossflower (*Ageratum houstonianum*), produces a chemical mimic of the insect molting hormone ecdysone. Insect larvae grazing on this plant are killed because the compound causes them to molt prema-

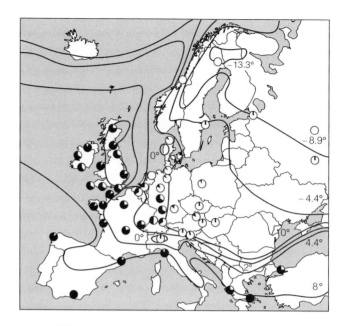

FIGURE 10.9 Ability to produce free cyanide varies among individual bird's-foot trefoil plants. A large percentage of plants in warmer regions manufacture both cyanogenic glycosides and the enzymes necessary to liberate cyanide from them. (The percentage of plants in a region that are cyanogenic is indicated by the darkened part of the circle in that region.) In areas with colder winters (contour lines are labeled with the January mean temperature in degrees Celsius), far fewer plants are cyanogenic. Plants in cold regions, where winter weather is an effective predator control, have less need to defend themselves chemically. (*From Rosenthal and Kotanen 1994*)

turely. Other plant species produce hormones that cause abnormal development that results in mortality. In Africa, bugleweed (*Ajuga remota*) produces a phytoecdysteroid that causes developmental abnormalities, including two heads and abnormal mouthparts, in locusts that prey on it.

Secondary compounds to deter herbivory are found not only in angiosperms and gymnosperms but in many lower plants as well. Chemical defense is known to occur in some very ancient organisms, including the cyanobacteria. One cyanobacterium, *Anabaena affinis*, produces an endotoxin that helps protect it from grazing by zooplankton (Kirk and Gilbert 1992). Coral reef algae face such intense grazing pressure that between 50 and 100 percent of total plant production may be consumed by herbivores (Hatcher and Larkin 1983; Carpenter 1986). Experiments by M. E. Hay, W. Fenical, and K. Gustafson (1987) demonstrated that a number of effective secondary compounds produced by coral reef algae deter herbivory in the coral reef sea grass *Thalassia* (Figure 10.10).

If plant secondary compounds evolved de novo as grazing deterrents, one might expect adaptive mecha-

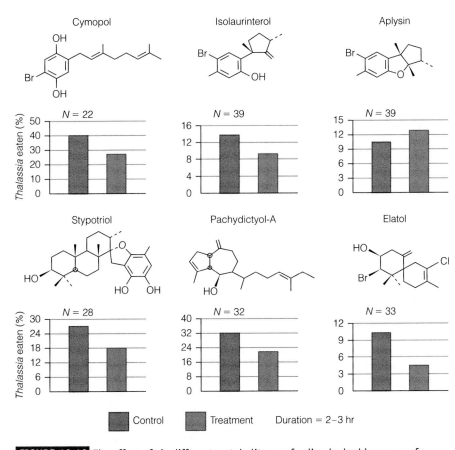

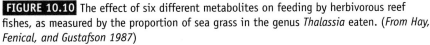

FIGURE 10.10 The effect of six different metabolites on feeding by herbivorous reef fishes, as measured by the proportion of sea grass in the genus *Thalassia* eaten. (*From Hay, Fenical, and Gustafson 1987*)

stand is attacked by tent caterpillars, it *and other trees* decrease the nutritional quality of their leaves. In growth chamber experiments, he found that once the first individual is attacked, other individuals in close proximity, but without any physical connection, also decrease leaf nutritional quality. Perhaps some sort of pheromone is released to signal the local activity of grazers, but neither the mechanism nor the extent of this phenomenon in other plants is known.

E. Haukioja and S. Neuvonen (1985) examined the basis of inducible resistance to insect defoliation in the white birch (*Betula pubescens*). To explain why larvae of attacking insects grow poorly on leaves from trees that had been defoliated the previous year, they tested two hypotheses: (1) that the insect attack lowers the nutrient content of the plant such that subsequent growth is of poor nutritional quality for the herbivore, and (2) that the effect is an intensification of the defensive chemistry of the foliage. Their results supported the second hypothesis. When damaged plants were fertilized, they still produced foliage of poor nutritional quality—a result contrary to what is predicted by the first hypothesis. In addition, insect damage was a more effective inducer of the changes than was mechanical damage.

nisms to fine-tune their production to the intensity of grazing. Selection should favor plants that produce energetically expensive deterrents only when grazed. A number of examples of inducible secondary compounds are known. For instance, Cronin and Hay (1996) showed that amphipod grazing causes the brown alga, *Disyota manstrualis,* to produce high concentrations of the secondary compounds known as diterpenoid dictyols. Nearby ungrazed plants did not produce the compounds. Similar patterns are evident from taxonomic comparisons of species facing different grazing intensities. For seven genera of algae, tropical species that face diverse and intense grazing pressure are significantly less palatable than their temperate relatives that did not evolve under such selection (Bolser and Hay 1996). These researchers showed that the differences were due not to differences in nutritional quality but, rather, to increased chemical defense in the species subject to more intense grazing.

Among the most intriguing findings regarding plant defenses against herbivory is the suggestion that plants may, in fact, communicate information about herbivore attack. D. F. Rhoades (1985) reports that when a single Sitka willow (*Salix sitchensis*) tree in a

Allocation of Resources to Defense in Plants

Because defense against herbivores is assumed to have metabolic costs, it would seem likely that evolutionary decisions must be made about how and when to apportion energy to defense. We have seen that some plant defenses are inducible and that not all plants are equally subject to grazing pressure. Selection pressure should thus dictate which species devote more resources to defense and how those resources are allocated. Root (1996) has shown, for example, that the response of goldenrods (*Solidago altissima*) depends on the intensity of the herbivore pressure. This species favors maintenance of established individuals over seed production. When faced with herbivore pressure, goldenrods are likely to reduce flower and seed production to permit investment of resources on existing stems.

The basic hypotheses that have been advanced to explain how plants should respond to this variation in predation are typically based either on the spatial and temporal vulnerability of the plant to grazing or on the differences in the costs of deterring grazing for a particular plant species. P. Feeney (1976) proposed that plants that are highly "apparent" in space should be expected to invest more heavily in defense. They should have "quantitative" chemical defenses that make them nutritionally inferior. P. D. Coley, J. P. Bryant, and F. S. Chapin (1985) hypothesized that defense should be at a premium in plants inhabiting nutrient-poor sites because they are less able to recover if attacked. P. J. Edwards (1989) suggested that plant defenses can be categorized as either neutral resistance or defense. The former includes structural elements such as thick epidermal cell walls that provide protection against any outside element. Defenses are features, such as chemicals, whose primary functions are deterrence of animal attack. Neutral resistance is more typical of primitive plants; true chemical defense is characteristic of more advanced plants.

Recently, P. J. Grubb (1992) has suggested four additional, though related, kinds of factors involved in the allocation of resources to morphological plant defense:

1. *Availability of plants to herbivores.* Many species found in well-lit openings in forests have spines, as do many species that inhabit permanently open sites. Grubb suggests that in such sites, the most productive plants in the habitat are close to the ground and thus more readily available to grazers. Because herbivores concentrate on these species, selection has favored the evolution of defense mechanisms in them.
2. *Plant architecture.* The architecture of the plant is also important. A high proportion of palms and tree ferns have spines that protect one or a few apical meristems from herbivory. Selection favors increased protection in such plants.
3. *Seasonal scarcity.* Some 450 species of evergreen holly exist worldwide. Of these, 13 have spines to protect the plant from grazing, and 12 of the 13 are found in deciduous forests, where the unavailability of leafy plant material puts the evergreen hollies at increased risk from herbivores during winter.
4. *Involvement of specialized herbivores.* Although vegetation of the heathlands of South Africa superficially resembles that of a similar habitat in Australia, the Australian heath flora has a high proportion of spiny species. Grubb suggests that this difference is a result of strong selective pressure exerted in Australia by a diverse marsupial fauna highly adapted as herbivores.

The Effects of Herbivory

Our discussion thus far has been based on the assumption that the negative effects of herbivory constitute a selective force on plants. The prevalence and complexity of plant defense mechanisms suggest that this view is largely correct, and direct measures of the impact of herbivory reveal that this is clearly the case for many instances of herbivory. If a specialist seed predator consumes and digests large numbers of seeds from an individual plant, it is not only killing a large number of potential new individuals, but also decreasing the fitness of the parent plant.

Consumption of plant parts is also known to have negative effects. When an herbivore eats even part of a leaf, it decreases the photosynthetic area of the plant. Replacement of this leaf tissue requires mobilization of reserves from elsewhere in the plant, including food stores and minerals whose use may well decrease the plant's total production and upset its energy balance. The direct extraction of food reserves by insects such as aphids has also been shown to have negative effects on plant production (Dixon 1971). In the arctic, grazing by geese leads to lower belowground nonstructural carbon stores in the grasses *Dupontia fisheri* and *Eriophorum scheuchzeri* (Beaulieu et al. 1996).

Some other evidence, however, suggests that the activity of herbivores has beneficial effects. Grazing increases seed production in some species (Hendrix 1979; Janzen 1979), increases biomass production in others (McNaughton 1976), and increases the nutrient content of the plant in still others (Owen and Wiegert 1976). These effects may be mediated by such things as the removal of old leaves, alteration in hormone distributions, or increased production to compensate for the losses. Nevertheless, even though such effects are known, most of the evidence suggests that for most plants, the effects of grazing are more negative than positive (Belsky 1986).

Even if we can document some positive effects of grazing on the physiology of the plant, such effects do not necessarily reflect fitness benefits to the plant. K. N. Paige and T. G. Whitham (1987) were, however, able to document fitness benefits of grazing on scarlet gilia (*Gilia aggregata;* Figure 10.11). Grazing pressure by

FIGURE 10.11 Scarlet gilia (*Gilia aggregata*) may be an example of a plant that benefits from grazing. (*Photo by David Krohne*)

mule deer and elk was intense on their Arizona study sites: The grazers fed on nearly 75 percent of all plants and removed as much as 95 percent of the above-ground biomass. When a single stalk was eaten, however, it was replaced with four new stalks. These plants produced 3.05 times as many fruits as unbrowsed plants. Experimental cropping showed that when all aspects of a plant's fitness were accounted for, including total fruit production and seed germination, browsed plants realized a 2.4:1 fitness advantage over unbrowsed plants. Recent studies have confirmed this effect with a more sophisticated experimental design

that matched grazed and ungrazed plants for size (Paige 1999). Naturally grazed plants also produced significantly more fruits than ungrazed plants of similar size (Figure 10.12).

Data such as these have generated considerable controversy among plant ecologists, provoking two opposing views: that herbivores exert a negative impact on plants (Belsky 1986) and that herbivores benefit their prey (McNaughton 1976). The life span of the plant may prove to be an important variable in this debate. D. F. Doak (1992) points out that short-lived species such as aquatics and terrestrial annuals tend to have decreased survival and fecundity when grazed. On the other hand, long-lived perennial plants are probably more likely to show positive effects.

J. Maschinski and T. G. Whitham (1989) emphasize that the controversy may be resolved as we come to understand the complexities of grazing patterns and intensities. These researchers found that the nature of herbivores' impact on the biennial herb *Ipomopsis arizonica* depends on plant associations, nutrient availability, and the timing of grazing. Similarly, Doak (1992) followed individuals of the perennial dwarf fireweed (*Epilobium latifolium*) grazed on by the larvae of the butterfly *Mompha albapalpella*. When the plants were subjected to low-intensity attack, he noted minor decreases in growth rate even if attacks were frequent. Under high-intensity grazing, growth was suppressed even if attacks were infrequent. Thus the effects of herbivory depend on a number of interacting ecological variables, such that on balance, we probably cannot characterize herbivory in general as either positive or negative.

Thus far, we have seen a number of predator adaptations for locating and consuming prey. In addition to these traits, predators must develop optimal tactics for determining how specialized the diet should be, how long to continue foraging in a particular spot, and so forth. Many of these tactics represent behavioral adaptations to optimize the foraging pattern of the predator.

Optimal Foraging Theory

The life-dinner principle states that the evolutionary relationship between predator and prey may not be symmetrical in intensity. Nevertheless, predators are clearly under intense selection to find and consume their prey. The elaborate adaptations cataloged earlier in this chapter testify to the rigors imposed by selective pressure on predators seeking a meal. We touched

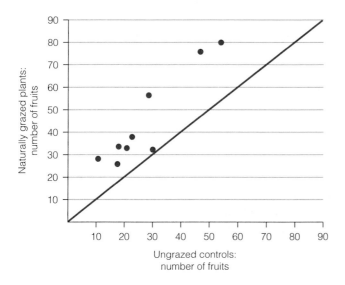

FIGURE 10.12 Fruit production in matched pairs of plants of scarlet gilia (*Ipomopsis aggregata*). Each point represents a single pair. The diagonal line is the expectation if grazed and control plants have identical fruit production. (*From Paige 1999*)

on just a small subset of the morphological and physiological adaptations of predators.

In the past 30 years, ecologists have attempted to understand precisely how selection also shapes the *behavior* of predators. The theory that has developed in this field is known as optimal foraging theory, and many of the important principles derived from it may be applied to carnivores and herbivores alike. Optimal foraging theory emphasizes the similarities among many kinds of consumers; its principles predict the optimal foraging behavior of predators as different as insects specializing on nectar and carnivorous predators.

Optimal foraging theory relates the actual behavior of a predator as it searches for and consumes its prey to some "optimal" foraging pattern predicted by the ecological models. Most of these models assume at the outset that other environmental variables and interactions are held constant. Optimal foraging is defined as the maximum possible energy return under a given set of foraging conditions. Species for which obtaining the maximum amount of energy is crucial are referred to as **energy maximizers.** For other species, obtaining a given amount of energy in the minimum amount of time is vital, perhaps because the animal has other pressing tasks (such as defending a territory or seeking mates) that put a premium on obtaining energy quickly. Such species are called **time maximizers.** Recall the antelope ground squirrel discussed in Chapter 3. This desert rodent accumulates a heat load while foraging and must periodically return to its burrow to "dump" heat to the cool, moist soil. The foraging strategy of this species must optimize the time available for this crucial physiological activity (Hainsworth 1995).

The fundamental assumptions of this theory are (1) that foraging behavior has a genetic basis that can be molded by selection and (2) that fitness (with respect to a foraging mode) is related to the net energy intake (the total energy consumed minus any energy costs associated with the foraging mode).

Our discussion of optimal foraging theory, which has become a large subdiscipline in behavioral and community ecology, considers two examples of optimal foraging models: the optimal diet and the optimal use of food patches.

Optimal Diet

The optimal diet model we will consider was designed to explain the breadth of the diet consumed by a predator—that is, the diversity of items included in the diet. Predators can be specialists with a narrow range of prey items or generalists with catholic tastes. R. H. MacArthur and E. R. Pianka (1966) developed a simple model to explain when a predator should specialize and when it should broaden its diet.

MacArthur and Pianka reasoned that any forager faces a trade-off between a specialized and a broad diet. In the former case, the items consumed will be of high value, but the predator may spend a great deal of time or energy searching for those items. By contrast, a generalist will spend less energy in search of items, but not all such items will be of equal (or high) value.

In any given diet, each item consumed contributes to the average energy input. A better diet is one that increases this average energy input. The behavioral decision a predator faces is whether to expand its diet to include other prey items. Adding an item decreases the predator's overall search time but increases the variation in the quality of prey items. What is the optimal solution?

For a given diet, the average energy input per item can be expressed as

$$\bar{E}/\bar{h}$$

where E is the energy (caloric) content of each item in the diet and h is the handling time (or energy) required to capture, subdue, and consume each item. Now, if a new and less profitable diet item, i, is added to the diet, the additional energy input is given by

$$E_i/h_i$$

Is the new diet more favorable than the old diet? To answer this question, we need to know their relative energy values. If the predator chose not to eat the new item, it would have to continue searching for another acceptable item. This additional search time would be the average search time for the old diet, which we designate as $\bar{s}$. Thus the energy input for this item is given by

$$\bar{E}/(\bar{s}+h)$$

Now we have energy inputs for both the new diet and the old diet. Because the optimal diet maximizes the energetic input, the diet should expand to include the new item if

$$E_i/h_i \geq \bar{E}/(\bar{s}+h) \tag{1}$$

This decision is made for each new potential item in the diet. The diet will expand only so long as adding new items increases the average energy input.

How does the biology of the predator–prey interaction fit this model? The model gives clues to which features of the relationship are important. Obviously, the relative values of the search and handling times are crucial. In the simplest case, the handling times for a few items are very small. In this situation, E_i/h_i will be large, which will select for a broad diet.

This contrasts with the situation in which the search time is very small. Under these circumstances, $(\overline{E}/(\overline{s} + \overline{h})$ becomes approximately equal to $(\overline{E}/\overline{h})$. Only new diet items with smaller handling times are added to the diet. The result will be a relatively more specialized diet. In other words, predators with long search times relative to their handling times should be generalists. Predators with long handling times relative to their search time should be specialists.

These predictions make sound intuitive sense. In the first instance, it makes sense that if handling a prey item does not take too long, a predator might as well eat the prey when it runs across the item. The energy expenditure is so small that any energy derived from the item will be of net value. In the second case, search time is small, but handling time determines whether an item is consumed. This too makes sense. If little energy is invested in finding items, but they cost a great deal to consume, then a predator should choose only those that are easy to handle.

Imagine two hawks, one a small kestrel, the other a large red-tailed hawk. Both are hunting a field that contains an abundance of mammal prey of various sizes, including mice, ground squirrels, rabbits, and woodchucks. For these two hawks, the handling times for the various prey items vary considerably. For the kestrel, a rabbit or even a ground squirrel would be a formidable enemy, so the handling times for these animals would be extremely high. For the red-tailed hawk, the handling times would be rather small for all items except perhaps the woodchuck. Search times are similar for both hawks. We would expect the kestrel to specialize and the red-tailed hawk to generalize. This is precisely what happens: Of the mammals in the field, the kestrel will eat only the mice, whereas the red-tailed hawk will include all the species in its diet.

So far in our discussion of this model, it has explained mathematically various aspects of foraging that we might logically predict. The model also makes some interesting predictions that are not so intuitively obvious. For example, consider the effect of prey density on the breadth of the predator's diet. If prey densities increase, search times will decrease. Note that search time is included only on the right side of Relation (1). Thus, whether or not a new item is added to the diet is independent of its abundance but highly dependent on the overall prey density (because this determines $\overline{s}$). The equation tells us that if prey are generally abundant, s will be small, and this will cause the right side of the inequality to be large. New items should not be added to the diet; that is, the predator should be specialized *even if the new potential item is very abundant.*

This prediction has been borne out in experiments by E. J. Werner and D. J. Hall (1974) on bluegills feeding on *Daphnia.* When the overall prey density was increased, the bluegills became more specialized and fed only on the largest *Daphnia,* even though they were encountering many small *Daphnia.*

Optimal Use of Food Patches

A second optimal foraging theory makes predictions about how long a predator should stay and forage in a given patch of food (Charnov 1976; Stephens and Krebs 1986). Consider a predator that forages on a food source that is patchily distributed (not an unrealistic situation). As the predator forages, it depletes the available food, and its rate of energy gain begins to decline; if the predator remains and feeds long enough, it will completely deplete the prey. Whenever the predator leaves a patch, however, it must seek out a new one and incur the costs of travel, during which no energy is gained. When should the predator leave a patch and seek a new one?

E. L. Charnov hypothesized that a predator should base its "decision" about leaving a patch on the marginal value of the patch—that is, the rate of extraction of energy at the time the predator leaves the patch. If we plot the amount of energy gain as a function of time, we obtain a curve like that in Figure 10.13. The time before arrival in the patch is the travel time between patches, t_t; the time after arrival is the staying time, t_s. For any amount of time in the patch, the rate of energy extraction is the slope of the line where it intersects the curve. The maximum slope occurs at the value of t_s where the line forms a tangent to the curve. Figure 10.13 shows the difference in rates of energy gain between two different staying times. At t_{s2} the predator has stayed longer in the patch than at t_{s1}, and the rate of gain is less at t_{s2} than at t_{s1}. Because the tangent at t_{s1} has the maximum slope, t_{s1} represents the optimal staying time. This value is derived by taking the derivative of the function describing the energy gain per unit time and solving for the maximum.

This theory also enables us to make predictions about how the predator should forage under different conditions. If the patches are far apart, the travel time t_t increases from T_{t1} to T_{t2}. This changes the optimal

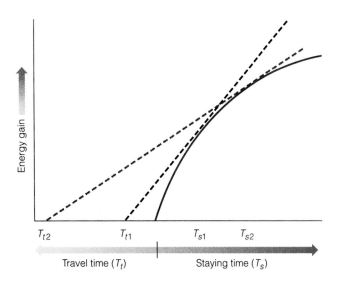

FIGURE 10.13 Optimal patch theory. The curve depicts the cumulative amount of energy gained as a function of the amount of time a predator stays in a patch. Tangents represent rates of energy extraction. The influence of travel time between patches on optimal staying time in a patch is apparent: The greater the distance between patches, the longer the predator should stay in each patch.

staying time from T_{s1} to T_{s2}. The theory predicts that when travel time is greater, the predator should spend longer in a patch before moving on. This makes good intuitive sense: If patches are far apart, the costs of travel increase, and the predator should forage longer in a patch before moving on and incurring these greater costs.

Species must, of course, have a means of assessing the crucial variables in this model. For many species, the rate at which a predator encounters prey can provide the necessary information on the marginal value of a patch. But as in other aspects of predation, selection can fine-tune the ability of predators to assess prey populations. For example, kestrels in northern Europe track local rodent populations. Crashes of the vole populations cause emigration of kestrels as regions become locally depleted of prey. Viitala et al. (1995) demonstrated that kestrels use an interesting mechanism to assess prey density. Voles scent-mark their runways (see Figure 4.6) with urine and feces that reflect in the ultraviolet range. Apparently, kestrels can detect these marks and assess recent vole activity and prey density.

Relatively few direct field tests of the marginal value theory exist because it is difficult to obtain the kind of energy data needed to test the model. However, J. M. Pleasants (1989) recently examined the application of the theory to bees foraging for nectar on larkspurs (*Delphinium nelsonii*). In this plant, each flower in an inflorescence contains a nectar

reward. A bee must decide when to leave a patch (that is, an inflorescence) and travel to another. Once Pleasants had assembled data on the production of nectar in flowers, he used computer simulations to assess the prediction of staying times from the marginal value theory. He found that the theory indeed predicted the energy gain by bees as a function of staying time. When he compared these data with field measurements of bee foraging, he found that the foraging pattern of actual bees was indistinguishable from the theoretical predictions.

We have seen in the preceding discussions that predators and prey exert selective pressures on each other. Thus, in a sense we have already considered one aspect of the effects of predation, the evolutionary consequences of the process. In the next section, we turn our attention to the ecological effects of predation.

Community-Level Effects of Predation

Predation is fundamentally a community-level effect: It represents an interaction between two or more species in a community. However, its effects have ramifications at all levels of the ecological hierarchy—population, community, and ecosystem. Consequently, throughout this text we discuss aspects of consumption and predation. In Chapter 5 we described the important role some predators play in regulating the density of prey populations. One of the most important effects of predation is its role in determining the structure of communities, a topic explored in Chapter 11, where we consider the structure of communities in general and discuss how predation interacts with competition. In Chapter 14 we see how energy flows through ecosystems as a result of the activities of consumers.

Functional and Numerical Responses to Predation

When we discussed the role of predation in population regulation in Chapter 5, we treated predators simply as an external force that increased a population's mortality. We did not concern ourselves with either the mechanisms by which mortality changes or the details of the predator response to changes in prey densities. It is appropriate here to turn our attention to these phenomena.

As prey densities increase, predators can respond in either of two ways: Each predator can consume more prey, or the predators can increase in number (and those larger numbers of predators will eat more prey). We call the former the **functional response,** the latter the **numerical response.**

A classic study by C. S. Holling (1959) demonstrated the nature of these two responses for a community in which small-mammal predators consume (among other things) pine sawfly pupae. After the pine sawfly lays its eggs on pine trees, the larvae emerge, consume the needles, and then migrate down the trunk and burrow into the soil to pupate. The pupae remain dormant in the soil for some months. Three species of small mammals prey on the sawfly pupae in this system: the short-tailed shrew (*Blarina brevicauda*), the white-footed mouse (*Peromyscus leucopus*), and the masked shrew (*Sorex cinereus*). Because each of these small mammals opens and consumes the pupae in a distinctive manner, the intensity of predation by each of the three species can be estimated by counting opened pupae.

In addition to predation intensity, Holling estimated prey density by sampling the soil for pupae. He also trapped a sample of small mammals to estimate their density in comparison to the density of pupae.

The numerical and functional responses of the predators to changes in prey density are shown in Figures 10.14 and 10.15, respectively. Note that in this example there is an inverse relationship between the numerical and functional response for any one species. *Blarina* has a small numerical response but a large functional response, *Sorex* shows just the opposite pattern, and *Peromyscus* is intermediate.

Holling also combined functional and numerical responses of predators, expressed as percent of predation as a function of prey density (Figure 10.16). For

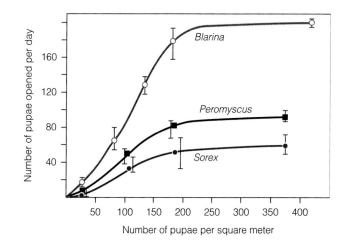

FIGURE 10.15 The functional response: changes in the rate of predation as a function of prey density. (*From Holling 1959*)

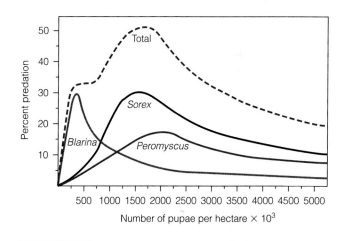

FIGURE 10.16 Functional and numerical responses combined to show the relationship between percent predation and prey density. (*From Holling 1959*)

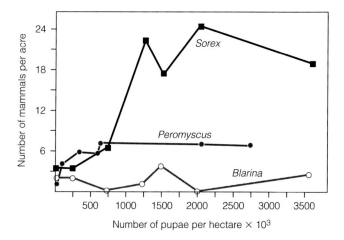

FIGURE 10.14 The numerical response: changes in the numbers of three predators (small mammals) as a function of prey density. (*From Holling 1959*)

each predator species, percent of predation is derived by multiplying the number of pupae eaten by the number of predators present and then dividing this product by the density of pupae present. In other words, Figure 10.16 shows the proportions of pupae eaten by each species (and the sum of all three) at each prey density.

The pattern of change in this figure is interesting. Note that the proportion of prey that each species consumes peaks at a slightly different pupal density. This result has important implications for the regulation of this insect. If the sawfly increases its density rapidly to a point beyond the range where its pupae will be most heavily preyed on by *Blarina*, it enters the range of pupal densities at which *Sorex* predation is highest. The same is true if pupal density further increases to the range in which *Peromyscus* consumes its largest fraction of prey. Thus, this three-species community is more likely to limit sawfly abundance than one in which the three species reach maximal consumption at the same pupal density. At very high densities, however, the sawfly can escape the controlling effect of small-mammal predation.

The nature of the functional responses for any species is a function of the nature of the predator-prey coevolutionary interactions and the reproductive biology of the predator. Holling predicted that three forms of functional responses would occur (Figure 10.17). A Type I functional response is a linear response between the number of prey consumed and the prey density. In a Type II response, the consumption of prey reaches an asymptote. This is also true of the Type III response, but in this instance the curve is sigmoid. The asymptote predicted in the Type II and Type III responses is a result of either satiation of the predator or the increase in the amount of time spent handling prey as they are being consumed at a high rate.

The role of predation in controlling prey will depend on the nature of the functional response. This is apparent if we consider the proportion of prey eaten over the range of prey densities for each kind of functional response (Figure 10.18). Note that the Type I response is density-independent; as prey densities increase, there is no corresponding increase in the proportion of prey eaten. In a Type II response, the proportion of prey eaten decreases with increasing prey density. In a Type III response, however, the proportion consumed increases with prey density: true density dependence. When we recall from Chapter 5

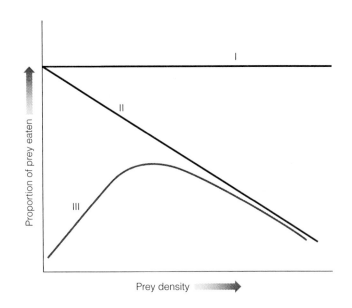

FIGURE 10.18 The three types of functional responses, expressed as the *proportion* of prey eaten as a function of prey density.

that regulation of population density requires density-dependent mechanisms, we see that the Type III response is most likely to regulate prey populations.

The Type III response can result from any number of factors that decrease hunting efficiency at low prey densities. These might include a failure to develop a search image without sufficient positive reinforcement or the presence of refuges in which prey can hide at low population densities. Understanding the functional response is crucial for biological control where predators are used to control pests (see the box on pages 264–265).

The prevalence of predator–prey interactions in nature indicates that the net result is rarely the extinction of prey by virtue of the actions of predators. As a result of the intense coevolutionary interactions between predators and prey, some form of coexistence seems to be the norm. What factors lead to this situation?

Conditions for Coexistence of Predator and Prey

A. J. Lotka (1925) and V. Volterra (1926) developed simple graphical models to examine the conditions under which predator and prey coexist. These models are similar in concept to the models we used to describe competition in Chapter 9.

We begin with expressions for the rate of growth of the predator and prey populations. For the predator,

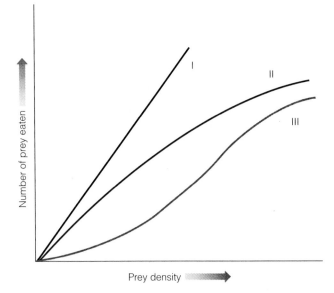

FIGURE 10.17 The three types of functional responses, expressed as the *number* of prey consumed at various prey densities.

$$dP/dt = apHP - dP \qquad (2)$$

where dP/dt is the instantaneous growth rate of the predator population (P); a is the production efficiency of the predator (see Chapter 14), the proportion of energy assimilated by the predator that is converted into new predator tissue; p is the ingestion efficiency of the predator, the proportion of available prey actually consumed; H is the density of the prey (H stands for herbivore); and d is the death rate of the predator. According to this equation, the rate of increase of the predator is equal to its growth rate minus its death rate. The per capita growth rate ($apHP$) is determined by the efficiency of predation, the efficiency of conversion of prey material, and the number of prey items in the environment.

A similar equation describes the prey growth rate:

$$dH/dt = rH - pHP \qquad (3)$$

where dH/dt is the instantaneous change in the prey population; r is the intrinsic rate of growth of the prey (see Chapter 4); and p, H, and P are the same variables as in Equation 2. In this case, the birth rate of the prey (rH) is decreased by the death rate (pHP). The latter term is analogous to the death rate in Equation 2 (assuming that all death is from predation). The death rate from predation is a function of predator efficiency and the frequency of encounters between predators and prey (which will be proportional to HP).

Recall that in our graphical model of coexistence between competitors in Chapter 9, our goal was to identify the conditions under which neither population changes. In a fashion analogous to that situation, predator and prey populations no longer change when $dP/dt = 0$ and $dH/dt = 0$. Thus, to identify the conditions under which the predator population no longer changes, we set the two equations equal to zero and solve for H:

$$dP/dt = 0 = apHP - dP$$
$$apHP = dP$$
$$apH = d$$
$$H = d/ap \qquad (4)$$

When the density of prey items (H) is equal to $d/$ap, the predator population remains stable $(dP/dt = 0)$. Similarly, for the prey population,

$$dH/dt = 0 = rH - pHP$$
$$rH = pHP$$
$$r = pP$$
$$P = r/p \qquad (5)$$

which expresses the density of predators at which the prey population no longer grows.

When we plot prey and predator population sizes on a graph (Figure 10.19), the prey isocline is described by some number of predators ($P = r/p$). If there are more predators than this, they drive the prey to lower numbers; that is, in the region above the prey isocline in Figure 10.19, the prey population is declining ($dH/dt < 0$). The predator isocline is described in terms of the number of prey ($H = d/ap$); if there are fewer prey than this, the predator population declines ($dP/dt < 0$).

Let us consider the dynamics of the predators and prey generated by this relationship. In each quadrant of Figure 10.19, the trajectories of the predators and prey are represented by the arrows. When prey are abundant—that is, beyond the predator isocline—the predator population increases (quadrants I and IV); when prey are scarce (less than the predator isocline), predators decline (quadrants II and III). Similarly, when predator numbers are greater than the prey isocline (quadrants I and II), prey numbers decrease; when predator numbers are below the isocline (quadrants III and IV), prey numbers increase. The trajectory of the two populations will be the summed vector describing the predator population and the prey population. The result in this example is that the two populations will move around the graph as shown by the circle. Convince yourself that this is so.

We can use Figure 10.19 to plot the changes in predator and prey populations over time. At each

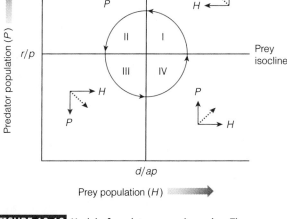

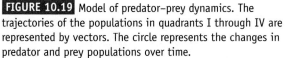

FIGURE 10.19 Model of predator–prey dynamics. The trajectories of the populations in quadrants I through IV are represented by vectors. The circle represents the changes in predator and prey populations over time.

Biological Control of Pests: Putting Predators to Work

Pests and weeds cause tremendous damage to both natural and agricultural habitats. Some of these pests are exotic species, accidentally or intentionally introduced to regions beyond their natural range (see Chapter 9). At one time, resource managers hoped that the development of herbicides and pesticides would eradicate or control unwanted species. Two fundamental problems mar this approach, however. First, the targets of the chemicals develop resistance; their short generation times and large population sizes enable them to adapt quickly, rendering chemical control ineffective. Second, as we saw in Chapters 5 and 6, these toxins may themselves pose serious environmental problems.

Thus there is great interest in **biological control**—the use of one or more natural means of control. In the classic sense, biological control involves the use of natural predators or parasites to control pests, but it may also involve measures such as

manipulating the reproductive biology of the pest. Insect pheromones or sex attractants can be used to lure pests to traps, or sterile males can be released into the environment to render large numbers of matings infertile. When biological control is based on some form of predation, we can use the concepts of predation to understand the ecological bases of using predators, parasites, parasitoids, or pathogens to control pests.

We know that predators can decrease the population size of their prey. For biological control purposes, we attempt to reduce the prey population to a new equilibrium level below the level at which it has economic impact (Figure 10B.1). Several conditions must hold for this approach to work. The predator must be highly specialized on the pest. To be able to respond numerically to increases in pest numbers, the predator must have a high r value and a generation time similar to that of the pest. It also must be able to persist when the prey

(pest) is scarce. Unfortunately, relatively few predators satisfy these conditions. Because parasites and parasitoids are more likely to have these characteristics, much biological control is based on them.

This theory also assumes that predators come to equilibrium with their prey species. Recently, W. W. Murdoch, J. Chesson, and P. L. Chesson (1985) have suggested that biological control by predators may more commonly represent a non-equilibrium situation in that pest populations, like many species, are more likely to exist in a metapopulation structure. Predators may cause local extinction of prey, followed by recolonization from other subpopulations; prey numbers may fluctuate greatly in different subpopulations, eliciting local numerical or functional responses of the predators. The net result is control, but not of the equilibrium type depicted in Figure 10B.1.

Resource managers are enthusiastic about the use of **integrated pest management**

FIGURE 10B.1 A classic type of biological control in which the average abundance of an insect pest is reduced after the introduction of a predator. The economic threshold is determined by humans' activities, and its position is not changed by biological control programs. (*After VandenBosch, Messenger, and Gutierrez 1982*)

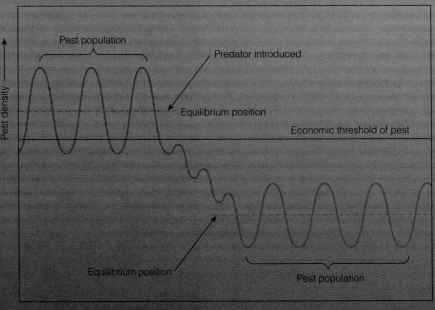

Biological Control of Pests:
Putting Predators to Work, continued

(IPM), a program of carefully selected control techniques tailored to address each particular insect pest problem. Insect numbers are monitored carefully, and population models are used to predict irruptions of the pest. Insecticides are used only sparingly and locally in response to this information. In addition, natural predators or parasites are released widely in areas of local infestation. Finally, sterilization procedures or sex pheromones may be used as well.

Measures to control the gypsy moth are an example of the use of IPM. After this species was accidentally released in Massachusetts in 1869, the population exploded across the northeastern United States, defoliating many hectares of forest. Currently, it is gradually spreading south and west. For this problem, IPM involves the use of pesticides, parasites, and sex pheromones. Insecticides are applied wherever local infestations erupt. Two parasitic flies produce moth mortality. One species lays its eggs on the gypsy moth caterpillar, and the emerging larvae bore into the caterpillar and kill it. The other species lays its eggs on foliage, where they are ingested by caterpillars. In addition, sex pheromones are used in several ways. Glyplure, a synthetic version of a sex attractant that female gypsy moths produce, attracts males into traps full of sticky material. Alternatively, glyplure can be scattered in the environment; males are confused by the prevalence of attractants and fail to mate.

Plants can also be controlled by biological means. Among the most important organisms for plant control are parasitic fungi, such as the rusts, which obtain access to plant nutrients by secreting chemicals, phytotoxins, that cause lesions on the plant surface. Extensive infestation eventually kills the plant. The fungus *Colletrichum gloeosporoides* is used to control northern

jointvetch, a weed in rice fields. This fungus, which is native to the American South, produces phytotoxins that kill the weed but do not harm the rice.

Another plant pest that can be controlled by biological means is spotted knapweed. Introduced into North America from Europe, probably along with alfalfa, it has become an important weed, particularly in the northwestern United States and southwest-

ern Canada, where it has taken over millions of acres of rangeland. A promising means of control has been discovered by A. Stierle and G. Strobel, who found that the pathogenic fungus *Alternaria alternata*, produces a phytotoxin, maculosin, that apparently is specific for knapweed. The special promise of this means of control is its specificity, a relative rarity among fungal phytotoxins.

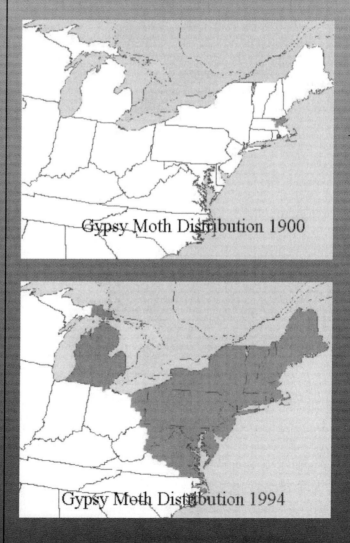

FIGURE 10B.2
The rapid increase in the distribution of the gypsy moth after its introduction to North America. (*Data from U.S. Forest Service laboratory, West Virginia State University*)

place in the figure that the circle intersects an isocline, we can characterize each population as small, medium, or large. Arbitrarily beginning at the left of the circle, we see that at the start (t_1), P is medium and H is small. At t_2, P is small and H is medium. Continuing around the circle and plotting both population levels over time produces the graph in Figure 10.20. At t_5, the populations are at the same levels as at t_1. We can see that the predator and prey populations cycle out of phase with each other.

In using this model to identify the conditions under which the predator and prey populations cycle in this manner and do not go extinct, we have oversimplified the situation in nature. For example, we have not included the carrying capacity of either species; doing so will make the model more realistic (Rosenzweig 1969, 1973). For the predator, including a carrying capacity will mean that at very high values of P, its growth rate will decline, and the predator population will level off. We can depict this graphically by bending the predator isocline to the right at high P (Figure 10.21). This modification decreases the size of the region in which predator populations grow. In a similar fashion, we can depict a prey carrying capacity. At high H, we bend the prey isocline down. This means that at high H it will be more difficult for the prey population to grow. We further increase the reality of the model by including the Allee effect (see Chapter 5). This effect is represented by bending the prey isocline down at low prey densities (Figure 10.21).

What effects do these changes in isoclines have on the relationship between predator and prey? The answer depends on the position of the predator isocline relative to the "hump" in the prey isocline. If the predator isocline intersects the prey isocline to the left of the hump (Figure 10.22), the trajectory of the two populations is an expanding outward spiral. This is an

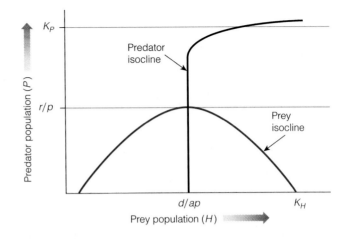

FIGURE 10.21 The predator–prey model with modified isoclines to account for predator and prey carrying capacities and for the Allee effect for small prey populations.

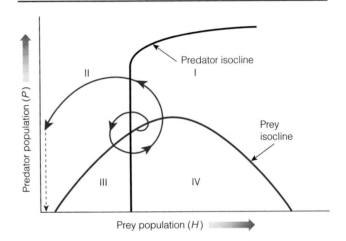

FIGURE 10.22 The trajectory of predator and prey populations when the predator isocline is to the left of the prey isocline peak. The eventual effect is the extinction of the prey population.

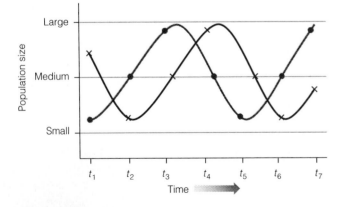

FIGURE 10.20 Changes in population sizes of predator (x) and prey (•) over time. Population sizes at t_1 are those at the left of the circle in Figure 10.19.

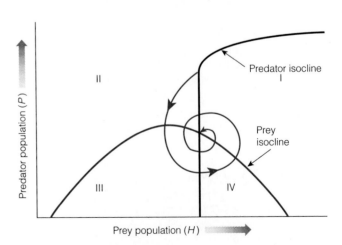

FIGURE 10.23 The trajectory of predator and prey populations when the predator isocline is to the right of the prey isocline hump. The result is coexistence of the two populations.

unstable situation: Extinction will result, first for the prey and perhaps for the predator as well. If the predator isocline is located to the right of the prey hump, however, the result is an inward spiral and coexistence (Figure 10.23).

What biological conditions are represented by these shifts in the predatory isocline? When the isocline shifts to the right, in effect the quadrat in which the predator population does not grow (dP/dt) has expanded; very high prey densities are required for P to increase. This shift could result from either crypticity on the part of the prey or inefficient foraging by the predator. When the predator isocline shifts to the left, the predator population is capable of growing even at very low prey densities—this predator is an efficient hunter. As the predator is less adept at capturing prey (as the isocline shifts to the right), the likelihood of coexistence between predator and prey increases.

Suppose that the prey have some refuge in which they can always hide and escape predation—say they are safe from predation if they can hide in a crevice. But there can be only so many crevices; if the prey population is large, not all individuals will be able to hide. If, on the other hand, the prey population is small, every individual will be able to hide, and predators will not be able to capture any (and the predator population will not be able to grow). This situation is depicted in Figure 10.24, in which the prey isocline is turned upward at low prey densities. The result is that no amount of predators can cause dH/dt to be negative. Even if the predator isocline is to the left of the prey hump, the outward spiral will hit the prey iso-

cline, which will prevent extinction of the prey. Coexistence is again possible.

These models have been manipulated extensively to make them more realistic. Nevertheless, a number of assumptions still limit them. For example, these models assume a Type I functional response by the predator, which is probably unrealistic in most situations. These models thus represent the kinds of heuristic models discussed in Chapter 1. We do not expect them to make precise predictions about the numbers of predators and prey; instead, we use them to describe and illustrate the essential features of stability in predator–prey systems. We should also note that without other density-dependent mechanisms, extinction of predator or prey would be likely at low population densities simply as a result of stochastic demographic effects.

In a classic empirical study that supports the qualitative predictions of these models, C. B. Huffaker (1958) examined the interaction between the six-spotted mite (*Eotetranychus sexmaculatus*) and its predator, another mite (*Typhlodromus occidentalis*). Laboratory populations of the two species were established in chambers containing oranges as a food source for the six-spotted mite. Huffaker varied the number of available oranges and the initial population sizes of the two mites.

In the absence of the predator, the six-spotted mite reached stable population levels. When the predators were introduced, the prey populations were quickly eaten to extinction, and the extinction of the predator soon followed (Figure 10.25). Huffaker could delay

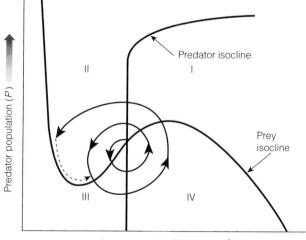

FIGURE 10.24 The trajectory of predator and prey populations when the prey items have a refuge where they can avoid predation. Even when the predator isocline is to the left of the prey hump, coexistence of predator and prey populations results.

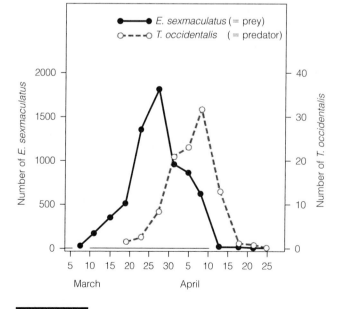

FIGURE 10.25 The effects of the introduction of predatory mites into a relatively simple laboratory system containing a population of prey mites. (*From Huffaker 1958*)

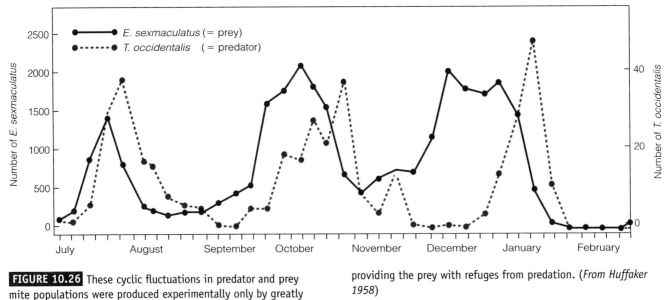

FIGURE 10.26 These cyclic fluctuations in predator and prey mite populations were produced experimentally only by greatly increasing the structural complexity of the system—in effect, providing the prey with refuges from predation. (*From Huffaker 1958*)

extinction of the prey by increasing the distances between oranges so that the six-spotted mites dispersed to new areas of the cage. Only by greatly increasing the complexity of the system was Huffaker able to obtain cyclic fluctuations of both species (Figure 10.26). He greatly increased the dispersion of oranges and slowed the dispersal of the predator by erecting barriers between the oranges and covering them with Vaseline. (The predators could not cross these barriers, but the prey could surmount them with the aid of silk, in much the way a spider might.) Wooden pegs scattered throughout the cage also provided the prey with platforms from which to launch themselves on their silks. In effect, Huffaker created refuges for the prey. Clearly, he had to go to great lengths to obtain the kinds of cycles predicted by the Lotka–Volterra models.

Few examples of long-term predator–prey cycles have been described in nature. Still, because we know that predators do not continually drive their prey to extinction, we can conclude that coexistence in some form must be the rule. Predator–prey systems are probably best characterized as another example of nonequilibrium ecological interactions. Local fluctuations, extinctions, recolonizations, and temporary cycles probably characterize most predator–prey systems—in other words, they are dynamic interactions.

Clearly, predation is an important ecological factor that affects populations and communities. In the next chapter, we investigate the ways in which predation and competition interact to organize the structure of communities.

SUMMARY

1. The term *predation* encompasses a wide range of feeding interactions, including carnivory, herbivory, detritivory, and parasitism. Despite the great variety among these interactions, they have in common the consumption of one individual by another. Consequently, predation, broadly defined, includes common evolutionary and ecological features.

2. Predation has both ecological and evolutionary consequences for both the predator and the prey. Predation leads to important and complex sets of adaptations for both.

3. Predators face strong selection pressures to enhance two aspects of their ability: the sensitivity and specificity of their ability to sense prey and the adaptations that help them capture and subdue their prey.

4. Prey face strong selection pressures to counter these predator adaptations. They evolve means of subverting the sensory modes of their predators and of escaping capture once detected.

5. Because plants are typically conspicuous to herbivores, their adaptations emphasize structural or chemical deterrents to grazing.

6. The behavior of predators is thought to evolve according to the tenets of optimal foraging theory. This body of theoretical models is based on the concept that predator "decisions," such as how diverse the diet should be and how long to forage in a patch, are adapted to maximize either the total energy input or the energy input per unit time.

7. Predators respond to increases in prey density by increasing their feeding rate (functional response) or by increasing their numbers (numerical response). The relationship between these two responses and the ability of the prey to increase in numbers significantly affects the ability of the predator to control the prey population.

8. The Lotka–Volterra graphical models of predation describe the conditions in which predators and prey coexist. These models suggest that inefficient predators, prey adept at avoiding predation, and refuges in which the prey are not vulnerable to predation lead to stable coexistence.

SELF-ASSESSMENT: CAN YOU...?

1. Describe and compare the many forms that predation can take.

2. Outline and discuss the major forms of predator and prey adaptations. Describe the fundamental "problem" each faces and the general forms of the adaptive solutions.

3. Explain the basic concepts of an optimal foraging strategy. Discuss the specific examples of diet breadth and the marginal value theorem.

4. Explain the fundamental ecological effects of predation.

5. Explain the application of Lotka–Volterra models to predation. Compare them to the models for competition in Chapter 9.

PROBLEMS AND STUDY QUESTIONS

1. Explain how the patterns of functional and numerical responses affect the potential for predators to control prey populations.

2. If a predator shifts to a larger or more dangerous prey, what would you expect to happen to the diversity of its diet? Why?

3. Explain the fundamental principle behind prey adaptations to avoid detection by predators.

4. Why should the handling time increase as a function of the number of prey items included in the diet?

5. Explain the fundamental difference between the isoclines used in the Lotka–Volterra competition (Chapter 9) and predation models.

6. As an anti-predator adaptation, a prey species develops a potent toxin that kills its predators. However, the predator must consume a whole individual to be affected by the toxin. If the prey individual is killed, how can this be an effective anti-predator adaptation?

7. Relate the factors that lead predators to forage in groups to the general factors that affect the evolution of sociality (Chapter 8).

8. Would you expect one effect of predation to be a mortality pattern that leads to *r* selection (Chapter 7)? Why or why not?

9. Discuss the reasons for and against defining predation broadly to include carnivory, all types of herbivory, detritivory, and parasitism.

PROJECTS AND ADDITIONAL STUDY

1. Controversy in wildlife management often arises around the control of predators. Survey the literature for the control of predators of either large mammals (deer, moose, elk, and caribou) or waterfowl (ducks and geese). Write a position paper on the role of predator control in the management of these species. Use your understanding of the evolutionary ecology of predation to support your position.

2. Field studies can be devised that use domestic eggs (chicken, domestic quail, etc.) to measure the potential intensity of predation on ground-nesting birds. Artificial nests with such eggs can be monitored for the frequency and intensity of predation. Devise and carry out such experiments to measure the intensity of predation in various kinds of habitats in your region (such as large versus small patches of forest, grassland versus woodland, etc.)

3. Use laboratory cultures of protozoan prey (such as *Paramecium*) and their protozoan predators (such as *Didinium*) to study the effects of prey density on the rate of predation.

4. Survey the literature on plant secondary compounds to analyze the association between certain types of secondary compounds and plant families. Analyze how secondary compounds are distributed taxonomically.

Community Structure

I n this chapter we examine the forces that structure communities—the ecological factors that organize the number of species in a community and form their relationships. In Chapter 13 we will discuss the process of community development after a disturbance, but here we are concerned primarily with the relationships in mature communities. As we will see, however, there is currently considerable interest in the role of disturbance as the predominant process governing community structure. Thus, succession and community development are intimately related to the topics of this chapter.

In Chapter 9 we defined the community as a set of interacting species. To understand community structure fully, it will be helpful to amplify that definition: A **community** is the entire ensemble of species in some area delimited by the practical extent of interspecific interactions. The second part of that definition is problematic in the same way as our definition of the population. At some level, all species on the planet interact, but certainly it is impractical to define only a single global community. If we delimit the community too narrowly, important interactions extending across our boundaries will confound our understanding of the

processes within. Some authors discuss taxon-specific groups or functional groups as communities, referring, for example, to "the community of lizards" or the "plankton community." Strictly speaking, these are not communities according to our definition. The former is a taxonomic association; the latter is a **guild,** a group of coexisting, ecologically similar organisms. Nevertheless, the interactions within such groups are important components of community structure.

As an introduction to the notion of community structure, consider two communities. In the African rift lakes formed by the separation of two tectonic plates in East Africa is a community inhabited by a large number of fish species in the family Cichlidae. These fish, most of which are in the genus *Haplochromis,* have undergone massive adaptive radiations in the rift lakes. More than 250 species of these closely related fish coexist in Lake Victoria alone. They have diverged into a number of different niches (Figure 11.1) in at least 10 trophic groups, including molluscivores, epiphytic algae-scrapers, insectivores, macrophytivores, detritivores, phytoplanktivores, prawn-eaters, piscivores, zooplanktivores, and even one species that eats crabs. Even though great diversi-

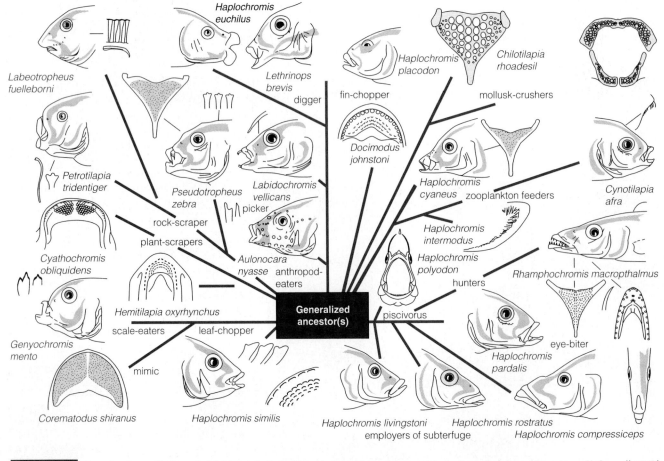

FIGURE 11.1 The adaptive radiation of cichlid fish in Lake Malawi, East Africa, into a vast number of niches. (*From Lowe-McConnell 1987*)

fication of food niches has occurred among these fish, the huge number of sympatric species suggests that many of them overlap greatly in food resources. Indeed, many species within these trophic guilds have extremely similar niches. Even though there is evidence of slight microhabitat variation, it is clear that many species of African cichlids coexist in these lakes despite extensive niche overlap (Witte 1984; Lowe-McConnell 1987).

Contrast the organization of this aquatic community with that of three desert rodent communities in the American Southwest. Studies by M. A. Bowers and J. H. Brown (1982) of the community composition of three desert sites revealed a clear pattern of coexistence among the granivorous rodents in the three deserts (Figure 11.2). Although the coexisting species differ, the size relationships are very regular.

Each of these communities is intriguing in its own right. In the African lakes, we might wonder how so many species manage to coexist in sympatry; in the desert system, we might ask what forces organize the community so precisely. In attempting to answer these questions, we will be searching for the general rules that explain the organization of these two different communities.

Ecologists focus on two major categories of explanations for community structure: equilibrium and nonequilibrium hypotheses. According to the equilibrium view, certain forces and processes—specifically, competition and predation—organize communities. When these processes have reached equilibrium, we observe community structures that exhibit characteristics derived from the effects of predation, competitive exclusion, and character displacement. The nonequilibrium hypothesis holds that communities rarely reach an equilibrium state and that disturbance and its effects underlie much of what we observe in the structure of communities. We will examine both of these basic approaches to community structure in turn, along with the processes each deems important.

The equilibrium and nonequilibrium views of communities have their historical roots in a debate among ecologists early in the twentieth century. According to F. E. Clements (1916, 1936), communities are highly organized groups of species that persist together over long periods of time. After many generations of competitive interactions, the coexisting species and the community they constitute reach equilibrium. In contrast, H. A. Gleason (1926, 1939) argued that communities are the result of individual species' responses to the environment in ecological time. In Gleason's view, the community is not a highly organized entity; instead, it is a nonequilibrium, random assemblage of species individually responding to environmental changes.

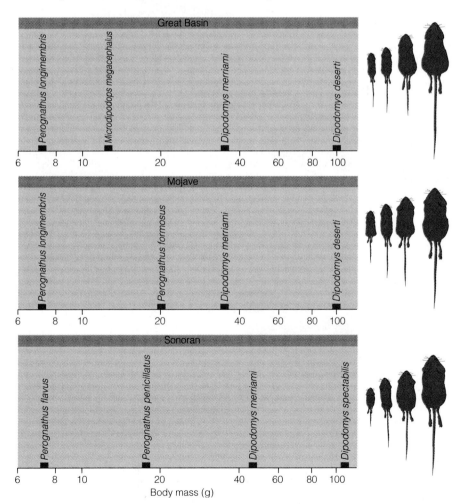

Equilibrium Views of Community Structure

The effect of both competition and predation is to reduce the densities of some species in a community. In Chapters 9 and 10, we described

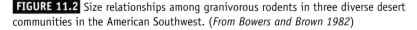

FIGURE 11.2 Size relationships among granivorous rodents in three diverse desert communities in the American Southwest. (*From Bowers and Brown 1982*)

the conditions that lead to equilibrium population densities. Now we consider the possibility that these equilibrium densities are important determinants of community organization.

Interspecific Competition as a Structuring Force

In Chapter 9 we discussed the mechanisms of competition, its effects on populations, and its fundamental importance in ecology. We left unanswered the question of how coexisting species in a community are ecologically related to one another. Our analysis was limited to the simplest case of two competing species. The questions of interest to us here are the following: How are niches arranged in highly diverse communities? How does the interspecific competition we considered in Chapter 9 affect entire communities? We begin with discussions of two phenomena that suggest an important role for competition in shaping community structure: changes in niche dimensions and patterns of species distributions.

Changes in Niche Dimensions

Let us return to the concept of the Hutchinsonian niche from Chapter 9. When we describe the Hutchinsonian niche for a species in the absence of any potential competitors, we have described its **fundamental niche.** Competitors modify this fundamental niche because their presence makes certain resources unavailable. The subset of the fundamental niche used when competitors are present is called the **realized niche.**

An example of the relationship between fundamental and realized niches in the presence of competition was provided by R. S. Miller's (1964) classic studies of four species of pocket gophers. The functional niche of pocket gophers is as a fossorial herbivore; for such burrowing animals, the nature of the soil is of critical importance. The soil resource axis ranges from deep, fine soils (high quality) to coarse, shallow ones (low quality). When the four species are allopatric, their fundamental niches overlap extensively.

But in sympatry, these fundamental niches are greatly reduced. Each species is relegated to a specific subset of the range of soil conditions. These data suggest that competition plays a role in organizing the relationships among species in this community.

Many examples of very fine subdivision of resource use are known, strengthening the conviction of some that competition plays a significant role in community structure. In northern coniferous forests, for example, one can observe five similar species of warblers foraging in the same tree (Figure 11.3). They have very similar functional niches: All are insectivores that glean insects from conifer trees. With detailed observational studies, MacArthur (1958) showed that each of the several species is actually foraging in a slightly different microhabitat within the tree (Figure 11.4). Thus the five species are not in fact occupying the same niche.

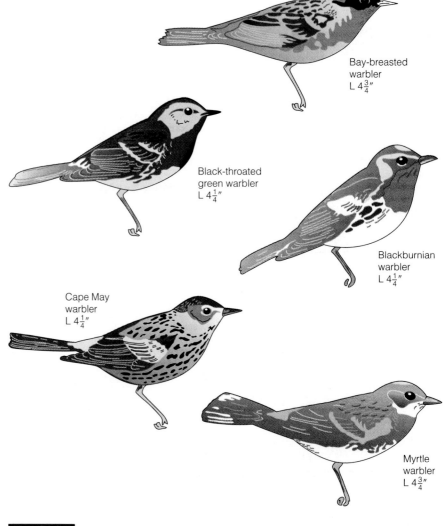

FIGURE 11.3 The five species of warblers that MacArthur studied. Note the similarity in size and bill shape.

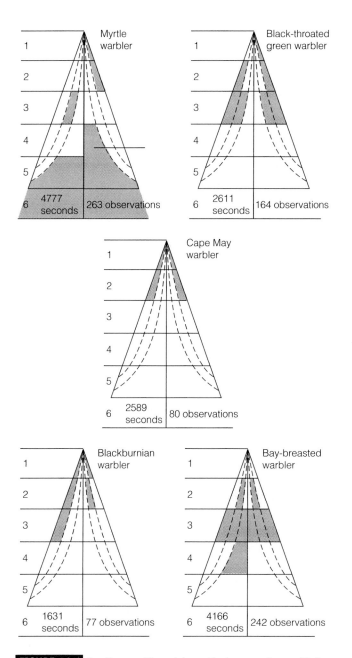

FIGURE 11.4 Feeding positions (six vertical zones of a conifer) of five sympatric warblers based on percentage of observations. Dotted lines represent the zones from exterior to interior at which species presence was recorded. Solid lines represent vertical zones. The shaded zones represent those regions where at least 50 percent of the activity is concentrated. The five species in fact occupy separate niches. (*Adapted from MacArthur 1958*)

In Chapter 9 we saw that the niche of the granivorous ant, *Veromessor*, is affected by the number of competing species. Several other studies have found similar effects of competition on the niche. We infer from them that because the number of competing species affects the niche dimensions, competition indeed plays a role in organizing the community. Martin (1996) has provided evidence of the direct fit-

ness costs associated with niche overlap in a group of sympatric ground and shrub-nesting birds in Arizona. This researcher measured the overlap in nest microhabitat choices in seven species and the predation rate experienced by each species. He showed that predation risk, and hence the fitness cost, was greater for species with high overlap in nest microhabitat. Competition for predation-safe sites is an important selective force in this community.

These data suggest that niches in high-diversity communities must be modified to avoid such fitness costs. There are three possible patterns of niche relationships in high-diversity communities (Figure 11.5):

1. Increased niche overlap
2. Decreased niche breadth
3. Increased range of resources utilized

Let us consider examples of each.

Few studies show a continuous increase in niche overlap as species diversity increases. In a study of small mammals in heath habitats in Australia, B. Fox (1981) examined the microhabitat niches of the

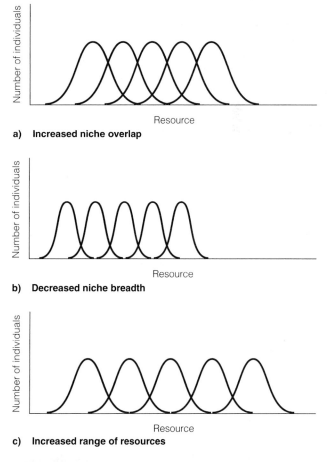

a) **Increased niche overlap**

b) **Decreased niche breadth**

c) **Increased range of resources**

FIGURE 11.5 Possible niche relationships in high-diversity communities.

species coexisting in a habitat mosaic by sampling mammals along transects laid across microhabitat types. For any species, the use of the microhabitat resource was a normal curve (Figure 11.6). Fox found that as the number of species increased, the overlap between pairs of species increased and then declined again (Figure 11.7).

It is possible to pack niches together in high-diversity communities by decreasing the niche breadth of each species (Figure 11.4b). If we represent the niche of an organism as a frequency distribution along a limiting-resource gradient, the width of that normal curve measured by the variance or the range represents the niche breadth of the species. High-diversity communities may be structured by changes in niche breadth resulting from the effects of competition.

K. A. Rusterholz (1981) described changes in niche breadth associated with different levels of competition for a guild of foliage-gleaning insectivorous birds. He studied a group of six sympatric warblers and vireos in montane forests of the American Southwest. The microhabitat niche of each species was described by detailed observations of the foraging height, perch size, tree species, and so on. An index of the intensity of the competition faced by each species was calculated as follows:

$$L_i = \sum_{j=1}^{n} a_{ij} x_j$$

where L_i is intensity of competition felt by species i, n is the number of species in the guild, a_{ij} is the competitive effect of species j on species i, and x_j is the abun-

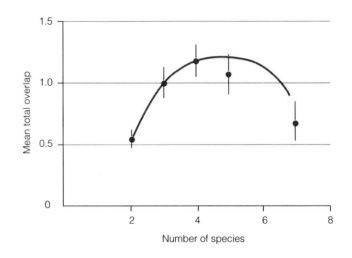

FIGURE 11.7 Mean microhabitat niche overlap as a function of the number of coexisting species from three heath study sites. Error bars represent mean value ±1 standard error. The curve is the best fit to the data points. (*From Fox 1981*)

dance of species j. This index is the mathematical product of the extent of overlap of foraging microhabitats and the population size of the competing species. The intensity of competition between two species increases as both the amount of overlap and the number of competitors increase. The six species were ranked by the intensity of the competition they faced.

Rusterholz then measured the niche breadth of each species by measuring the ratio of culmen length to bill depth (Figure 11.8) for each species—a measure of the shape of the bill and a feature of the bird intimately related to the insect prey it can capture and eat. By measuring the variation in bill shape in each species, he was actually measuring the niche breadth for prey type and size. Rusterholz found a significant negative correlation between the intensity of competi-

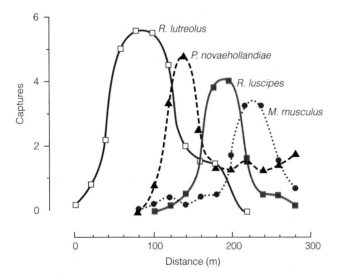

FIGURE 11.6 Partitioning of heath microhabitats by the four most abundant small-mammal species captured along a 300-meter transect. (*From Fox 1981*)

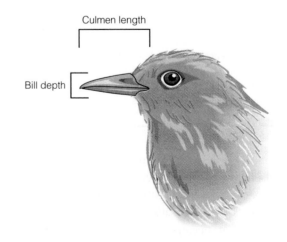

FIGURE 11.8 Bill dimensions used to evaluate bill shape as an indicator of niche breadth.

TABLE 11.1

Rankings of Intensity of Competition and Morphological (Niche) Variability (as Measured by the Ratio of Culmen Length to Bill Depth) in Six Insectivorous Birds

Species	Intensity of Competition	Coefficient of Variation for Culmen Length/Bill Depth
Olive warbler	74.8	7.88
Grace's warbler	30.8	5.21
Yellow-rumped warbler	19.2	5.60
Solitary vireo	12.1	7.93
Red-faced warbler	5.6	10.73
Mountain chickadee	4.4	11.86
Spearman rank correlation: = − 0.83		

Adapted from Rusterholz 1981.

tion and niche breadth (Table 11.1). Thus as competition intensified, the birds responded with narrower niches. This is the same result that D. W. Davidson (1978) reported for the niches of ants facing more competitors (Figure 9.15). (For a discussion of the relationship between highly specialized niches and extinction, see the box on page 278.)

E. R. Pianka (1973, 1975) compared the niche structures of high- and low-diversity lizard communities to ascertain the range of resource utilization. North American lizard communities typically contain between 4 and 11 species, whereas the Kalahari

Desert of Africa is more diverse, containing 12–18 species. The Australian deserts are the most diverse, with 18–40 species of lizards, depending on the site. Pianka was interested in how lizard niches are "packed" together in the more diverse communities. Reasoning that the three most critical niche components are time of activity, microhabitat, and diet, he made detailed measurements of all three for lizards in the three deserts, using direct observation and analysis of stomach contents. He found neither a decrease in niche breadth nor an increase in overlap, as in the previous examples (Figure 11.9), but rather a *decrease* in overlap. When he compared these communities with respect to prey sizes, he found that the Australian guilds utilized much larger prey than either of the other two guilds. Thus additional lizard species were able to "fit" into the Australian guilds by extending the range of resources used relative to their North American counterparts—specifically, by taking very large prey items. Large Australian lizards were eating vertebrates, including other lizards, a part of the resource axis not utilized in North America and the Kalahari.

One of the difficulties with this kind of analysis is that the tendency to emphasize a single niche dimension distorts the situation in nature. Figure 11.10

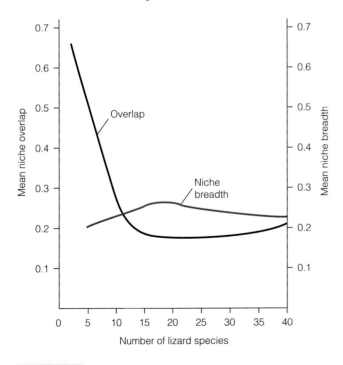

FIGURE 11.9 Niche overlap and niche breadth in North American, African, and Australian lizards as a function of lizard species number. (*From Pianka 1975*)

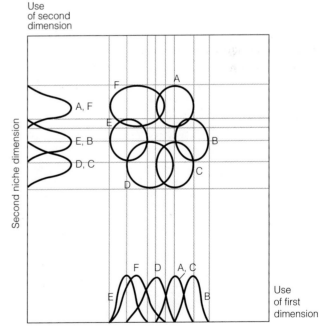

FIGURE 11.10 Hypothetical niches of six species in two dimensions. In each niche dimension, there is extensive overlap among species. However, when the niches are projected into two-dimensional niche space, the niche overlap is much less. (*From Pianka 1988*)

Niche Specialization and Extinction

Species extinction and the loss of biodiversity are serious concerns. In Chapter 7 we saw that the life history strategy plays a significant role in determining a species's vulnerability to extinction. With the concept of the niche, we can expand our understanding of the factors that make some species more vulnerable than others. One important generalization we can make is that species that occupy highly specialized niches are susceptible to extinction, particularly if resources are scarce. The reasons for this should be obvious. First, the species will probably exist at rather low densities because of the scarcity of some of the relevant resource. This is related to the issue of size and resource requirements considered in Chapter 7. The resource in short supply does not have to be a food resource. S. R. Humphrey (1985) points out, for instance, that the Mexican free-tailed bat (*Tadarida brasiliensis*) ranges widely for relatively abundant insect prey and that the more limiting resource is caves offering appropriate temperature and humidity conditions. In addition, a highly specialized niche makes a species vulnerable to competitive effects, particularly from exotic species introduced into the habitat.

The bird faunas of island systems like the Hawaiian Islands exemplify this problem. The Hawaiian Islands are among the most isolated islands on Earth—they sit some 3200 kilometers from the nearest islands and 4000 kilometers from the mainland.

Originally, the Hawaiian Islands supported 84 species of endemic birds derived from only 20 colonizations over the great expanse of ocean. Among the various groups of birds, the honeycreepers have undergone a major adaptive radiation, filling a number of niches on the islands (Figure 11B.1), in much the same way as Darwin's finches have done on the Galápagos. Of the 47 species of honeycreepers that evolved on the islands, 18 went extinct prehistorically. Eight more disappeared after the islands were visited by Captain Cook. There are currently 20

species, and they are at considerable risk. Of these 20 remaining species, 10 are dependent on a single population; 6 have populations of fewer than 500 individuals, and 3 number less than 50 individuals.

These species represent a classic example of a fauna at risk as a result of highly specialized niches, relatively scarce resources, and limited geographic ranges. Because of the small size and isolation of the Hawaiian Islands, population sizes are naturally small. Local extinction cannot be followed by recolonization because of the isolation of the archipelago. Disturbance, in the form of habitat loss or predation by exotics, has a significant impact. Moreover, the introduction of exotic birds results in competition for already scarce resources. These problems are not confined to the honeycreepers. Only 5 percent of the seabirds of Hawaii now nest on the main islands. Predation by introduced cats, rats, mongooses, and barn owls has eliminated most breeding colonies. On tiny Laysan Island, introduced rabbits so devastated the vegetation that three endemic birds went extinct: the Laysan rail, the Laysan miller

bird, and the Laysan honeycreeper.

These effects are certainly not restricted to island habitats. Any species that has a narrow geographic range, a narrow habitat tolerance, and a specialized niche will be at risk when human impacts threaten its requirements. A prime example is the Everglades kite (*Rostrhámus sociábilis*). This endangered inhabitant of sawgrass wetlands is highly specialized; it feeds almost exclusively on apple snails. The abundance of its prey is highly dependent on the quality of the Everglades habitat—particularly water levels, which have been greatly reduced by the development of south Florida.

One of the most celebrated endangered species, the giant panda (*Ailuropoda melanoleuca*), exemplifies the role of specialization in extinction. This native of the bamboo forests of China eats only bamboo. Because of human population pressure in China and its attendant development, bamboo forest is disappearing at an alarming rate. Concomitantly, the panda population is shrinking. It is estimated that only 1000 remain in the wild.

FIGURE 11B.1 Some representative examples of the diversity of honeycreeper species found on the Hawaiian Islands.

shows the resource utilization curves of six species in each of two niche dimensions. On the basis of the degree of overlap for either niche dimension alone, we might expect coexistence of these species to be difficult. However, when we project the niche breadths to the center of the graph to create two-dimensional Hutchinsonian niches, we find relatively little niche overlap for the six species. How should we interpret this situation? Do we concentrate on the relatively little overlap in two dimensions or on the relatively extensive overlap in one? We typically focus on the single resource dimension that is most critically limiting, because this enables us to measure the extent of the niche overlap that seems most likely to lead to competitive exclusion. But we should remember that this is often an oversimplification of the potential effects of competition, particularly when several resources are important.

Patterns of Species Distributions

Another traditional way to examine the role of competition in community organization is to observe the patterns of species distributions. J. M. Diamond (1975) conducted extensive studies of the bird distributions in the islands of the Bismarck Archipelago near New Guinea. After Diamond accumulated extensive data on the geographic ranges of many species on these islands, he organized this information in various ways to test the hypothesis that competition plays a crucial role in species distributions.

One of the most straightforward analyses involved the mapping of the distributions of similar species on islands. As we saw in Figure 9.8, the distribution of two species of small cuckoo doves, *Macropygia mackinlayi* and *M. nigrirostris,* has a "checkerboard" pattern; that is, each island has one or the other (or neither), but not both species of these ecologically similar birds. Diamond takes this finding as evidence that the bird communities on these islands are determined at least in part by competition via competitive exclusion. For other guilds, he was able to document the extinction of invading species shortly after their arrival on islands that already supported similar species. For example, Diamond observed that when islands on which the cuckoo dove *M. mackinlayi* was able to coexist with *M. amboinensis* were invaded by a third species, *M. nigrirostris,* the invader declined to extinction.

Another kind of biogeographic evidence is consistent with the role of competition. For each species of bird studied, Diamond constructed an *incidence function,* a graph of the proportion of islands on which the species is found plotted as a function of the total number of species on the island (Figure 11.11). Many of these functions have an S-shape: At low diversity a

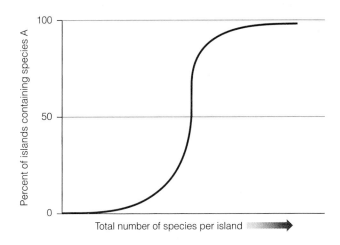

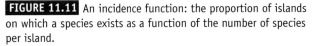

FIGURE 11.11 An incidence function: the proportion of islands on which a species exists as a function of the number of species per island.

given species is likely to be absent, but as species accumulate, that species is more likely to be found in the community. Each species has a slightly different incidence function. Most are S-shaped, although the position of the curve may shift to the right or left along the *x*-axis. Species with incidence functions shifted to the left tend to be good colonists; they are among the first species to be found on an island of small size or low diversity. Species with incidence functions shifted to the right tend to be found only on species-rich islands (often larger islands).

A few species are what Diamond referred to as "supertramps"—species with typical S-shaped incidence functions only on islands with low species diversity (Figure 11.12a). On islands with high species diversity (and thus greater potential competition), the incidence of supertramps peaks and declines to zero instead of remaining at 100 percent (Figure 11.12b). Diamond infers the importance of competition in community structure from the different incidence functions of supertramps for different levels of species diversity. On crowded islands, supertramps obviously cannot tolerate (compete with) too many other species, but on islands where potential competition is less, supertramps can persist.

Critiques of the Role of Competition

Even though the competition paradigm has been an important one in ecology for many years, some important critiques of this paradigm have forced ecologists to temper their emphasis on competition and design more careful experiments to document its role.

In contrast to the fine subdivision of niches among warblers demonstrated by MacArthur, some species apparently coexist without subdividing the use of

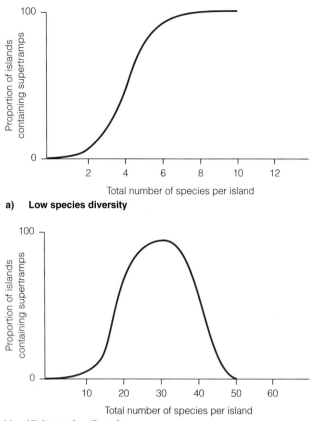

a)   **Low species diversity**

b)   **High species diversity**

FIGURE 11.12 Incidence functions for supertramp species on islands with (**a**) low and (**b**) high species diversity (note the different *x*-axis scales). On low-diversity islands (**a**), incidence increases with species number and remains high. On islands on which diversity reaches much higher levels (**b**), incidence eventually declines. The elimination of supertramps from islands with many potential competitors is cited as evidence for the importance of competition in shaping community structure.

resources. D. R. Strong (1982) described such a situation for a group of hispine beetles that feed on the leaves of *Heliconia,* a perennial tropical herb. Up to 13 species of beetles coexist and can even be found on the same leaf. Only 1 of these 13 species is even weakly segregated from the others in its use of the resource—it feeds only on younger leaves. Strong hypothesized that in this case, coexistence of many microsympatric species is possible because the resource is not limiting.

In an important series of papers published in the late 1970s and early 1980s, Strong and D. S. Simberloff questioned the validity of many of the empirical studies cited as evidence for the role of competition in natural communities (Strong, Szyska, and Simberloff 1979; Simberloff 1983). Their criticisms were based on philosophical considerations. They argued that a scientist can never prove that a hypothesis is true but can only show that an alternative is false. Good science requires that we reject the null

hypothesis—that an observation is a result of chance alone—before we accept any alternative hypothesis. Strong and Simberloff asserted that too many studies sought evidence consistent with competition without first rejecting the null hypothesis that communities are structured by the chance occurrence of species.

How can we design studies of natural communities that test the null hypothesis? Let us examine an example from Strong and Simberloff's work that illustrates the kind of analysis that can be brought to bear on this problem. Strong, Szyska, and Simberloff (1979) examined the niche relationships of bird species on three sets of islands: the Tres Marías Islands

TABLE 11.2

Interspecific Bill Shape Ratios for Birds on the Tres Marías Islands, Mexico, Calculated from Adjacent Species in the Ranking from Largest to Smallest

Family	Bill Shape Ratio	
	Observed	Source Expected
Cathartidae	1.057	1.110
Accipitridae	1.708	1.306
	1.132	1.252
Falconidae	1.050	1.305
	1.568	1.314
	1.709	1.371
Scolopacidae	2.517	1.743
	1.399	2.378
Laridae	1.447	1.429
Columbidae	1.217	1.089
	1.071	1.068
	1.047	1.100
	1.433	1.297
Pstittacidae	2.409	1.907
Caprimulgidae	2.153	1.803
Alcedinidae	1.422	1.324
Tyrannidae	1.233	1.139
	1.113	1.160
	1.525	1.222
	1.272	1.202
	1.052	1.253
Mimidae	1.470	1.457
Turdidae	1.111	1.064
	1.069	1.161
	1.626	1.272
Vireonidae	1.112	1.203
Parulidae	1.061	1.107
	1.114	1.123
	1.043	1.122
Fringillidae	2.175	1.354

The uppermost ratio in each family is for the largest to the second-largest species, the second ratio is for the second-largest to the third-largest species, and so on.

Adapted from Strong, Szyska, and Simberloff 1979.

of Mexico, the Galápagos, and the Channel Islands off California (Table 11.2). Their quantitative measure of niches of coexisting species was the ratio of culmen length to bill depth, a measure of the shape of the bill. As we have seen, bill shape in birds is intimately related to the food resources utilized. The researchers asked the following question: Are the patterns of bill shapes among coexisting island species a random sample of the pool of species that might be found on the islands?

For each island they arranged the ratios of culmen length to bill depth for the coexisting species from highest to lowest. They then calculated the ratios of nearest neighbors in this sequence. This formed the "observed" portion of the comparison. To obtain the bill shape ratios expected under the null hypothesis, the researchers randomly selected from the mainland a sample of confamilial (same-family) species equal to the number of species on the island. This was based on the premise that the pool on the mainland was the ancestral group from which the current island community evolved. The researchers then calculated the ratios of bill shapes for nearest neighbors. They repeated the random draws and calculations of nearest neighbors 100 times and calculated the "mean expected ratio of nearest neighbors" for each island. The expected ratios from a random assemblage were compared to the actual ratios for species living on the islands. For birds in several families on the Tres Mariás Islands, the observed ratios were not significantly different from those of randomly assembled communities. Similar results were obtained for the Channel and Galápagos Islands. Thus Strong, Szyska, and Simberloff accepted the null hypothesis that the bird species inhabiting these islands were random samples of species from the mainland pool. There was no justification for invoking competition as the organizing force of the communities.

Another kind of analysis of the Galápagos data led to the same conclusion. In this analysis, for each Galápagos finch species they plotted the relative bill shapes for the populations on each island of

the archipelago as a function of culmen length to control for the differences in size among birds. In Figure 11.13, which in essence depicts the position of each species in "niche space," each letter represents the mean bill shape for a particular island; each species is enclosed by a polygon. On such a graph, the distances between the same letters represent the niche separations of species coexisting on a given island. For example, the distances between all letters "F" (dotted lines in Figure 11.13) represent the niche separation of all species on Santiago Island, and the pattern of separation of species represents the niche structure of the community. When the distances between species are small, niches are packed closely together; large distances suggest that only species with different niches can coexist.

Strong, Szyska, and Simberloff calculated the expected niche separation distances in much the same way as the ratios of bill shapes described previously. They generated expected distances for each island by randomly choosing the same number of species from the mainland pool and then compared these expected niche separation distances with those observed in

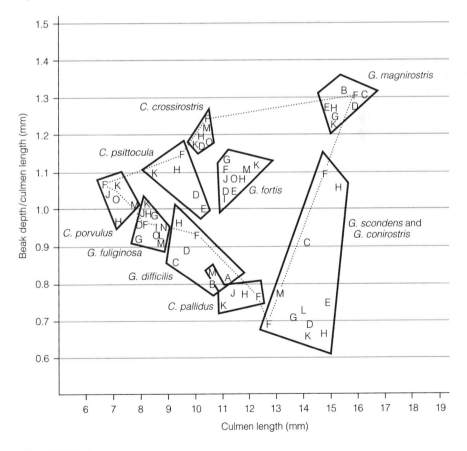

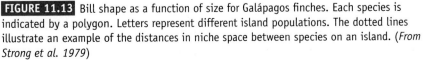
FIGURE 11.13 Bill shape as a function of size for Galápagos finches. Each species is indicated by a polygon. Letters represent different island populations. The dotted lines illustrate an example of the distances in niche space between species on an island. (*From Strong et al. 1979*)

actual communities. Again, the actual island communities could not be distinguished from the randomly constructed communities (Table 11.3). If competition were the explanation for these data, we would expect that the mean observed distances would be consistently greater than the mean expected distances. Given that the data in Table 11.3 contained approximately equal numbers of observed distances that were greater than and less than the expected distances, the conclusion was that competition alone does not explain the assemblage of these communities. The point is not that competition isn't occurring but, rather, that we should not infer its importance if we cannot first reject the null hypothesis.

Strong and Simberloff's position generated considerable controversy. R. K. Colwell and D. W. Winkler (1984) suggested that three important kinds of bias were possible in studies like theirs. The first is what they termed the **J. P. Morgan** effect, which arises when the pool group (that is, the mainland group) of species from which random guilds are generated is too large. Imagine as an extreme example that we include a flamingo in the pool with a group of warblers. This decreases the probability of finding patterns among the warblers because in any consideration of distribution in niche space, the flamingo is so different that the warblers will appear clumped (Figure 11.14). As a result, it will be difficult to detect competitive patterns among the warblers, even if they exist. Note in Figure 11.14 that the warbler niches are evenly spaced, as would be expected if competition limits their similarity, but that inclusion of the flamingo in the analysis causes the warblers to appear clumped, at least superficially; their interspecific niche differences are small relative to their difference from the flamingo.

The second bias in such studies is the **Icarus** effect. If species differ in dispersal ability, then the absence of one species from an island may have more to do with its inability to disperse long distances than with any competitive effects. Third is the **Narcissus** effect. If competition structured the mainland pool in the distant past, then the current island communities would appear to be random samples of the mainland pool, and we would conclude that the island communities are simply random assemblages. This conclusion would be incorrect, however, because competition really did structure these communities—it just happened a very long time ago. This effect is sometimes referred to as "the ghost of competition past."

Several attempts have been made to test the null hypothesis in such a way as to avoid these potential biases. For example, Bowers and Brown (1982) attempted to elucidate the role of competition by examining the patterns of coexistence of granivorous rodents in desert communities in the American Southwest. Their study was designed to investigate whether competition affects coexisting rodent assemblages. As noted earlier (see Figure 11.2), a number of

TABLE 11.3		
Observed and Expected Mean Niche Separation Distances for Finches on the Galápagos Islands		
	Mean Niche Separation Distance	
Island	Observed	Expected
Culpepper	3.71	4.20
Wenman	5.03	5.58
Tower	4.80	3.84
Pinta	3.29	3.15
Marchena	3.81	3.82
Santiago	3.13	3.19
Rabida	3.77	4.03
Santa Cruz	3.28	3.19
Pinzon	2.70	3.21
North Isabela	2.55	2.43
South Isabela	3.33	3.14
Santa Fe	5.60	5.79
San Cristobol	2.50	3.13
Española	6.60	5.72
Floreana	2.94	3.07

From Strong, Szyska, and Simberloff 1979.

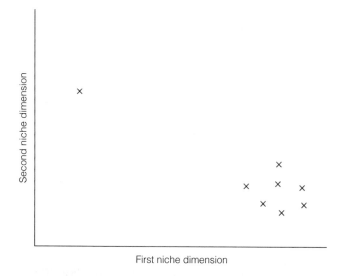

FIGURE 11.14 Hypothetical niche relationships of a group of warblers and a flamingo in two-dimensional niche space. The outlying point is the flamingo.

widely separated guilds of granivores are strikingly similar in the patterns of body size, even though the particular coexisting species differ. Bowers and Brown began with the hypothesis that competition structures these communities. If this hypothesis is true, then three predictions should hold:

1. Members of the same guild that coexist in the same local habitats should show nonrandom patterns of body size, because species of similar size should coexist less frequently than would be expected on the basis of chance alone.
2. Similar, nonrandom patterns of body size should be seen on a larger geographic scale. These patterns might be less strong than those observed locally because of the potential use of different habitats within this larger geographic range.
3. Inclusion of species from other guilds in the analysis should diminish or obliterate nonrandom patterns of occurrence, because species that are not competing should exhibit random patterns of body size.

Inherent in this design is the opportunity to accept the null hypothesis—that coexisting species are random assemblages of body sizes.

The researchers surveyed 95 sites for the presence or absence of granivorous rodents. The expected values of co-occurrence of species were based on the null hypothesis of random association of species. If species are associated at random, the frequency of co-occurrence should be the product of their independent frequencies of occurrence. (The product rule states that the probability of two independent events occurring together is the product of their independent probabilities.) If, for example, Merriam's kangaroo rat is found at 50 percent of the sites, and the spiny pocket mouse is found at 30 percent of the sites, then the null hypothesis predicts that the two species should coexist at $0.50 \times 0.30 = 0.15$ (15 percent) of the 95 sites. If they occur less (or more) frequently than this, we can conclude that something other than chance determines the co-occurrence of the two species.

Bowers and Brown compared the co-occurrences for pairs of species whose body mass ratios differed by either more or less than 1.5. They did this to take into account that species of similar body size are more likely to compete than those of very different sizes.

For each pair of species, the association was scored as either positive (+) or negative (−), depending on whether the frequency of co-occurrence across the 95 sites was greater or less than that predicted by chance.

We can evaluate the results of these analyses (Table 11.4) for each of the three predictions:

Prediction 1: For local assemblages, species with similar body sizes (ratios > 1.5) coexisted less frequently than expected by chance; that is, there were more negative associations than positive ones. For species with very different body sizes (ratios > 1.5), the co-occurrences were consistent with chance (equal numbers of positive and negative associations).

Prediction 2: On the geographic scale, the same general pattern was observed but less strongly; that is, species of similar size co-occurred more often in a large geographic area than they did in local habitats.

Prediction 3: When members of other guilds were included in the analysis, the occurrence of species was as predicted by chance: There were approximately equal numbers of positive and negative associations.

These results are consistent with the explanation that competition plays a role in structuring these communities. Species of similar size were less likely to occur together. Species of different guilds, among which competition would be unlikely, occurred together at random. The null hypothesis could be tested without the biases that Winkler and Colwell suggest are problematic on islands.

M. P. Moulton and J. L. Lockwood (1992) also attempted to surmount the problems that these biases would introduce. They reasoned that a study of introduced, closely related species would allow them to

TABLE 11.4

Patterns of Coexistence among Desert Rodents Categorized by Scale (Geographic versus Local), Body Mass Ratios, and Granivores versus All Guilds Combined.

		Granivores		All Guilds	
	Body Mass Ratio	Association		Body Mass Ratio	Association
		−	+		−	+
Geographic overlap	<1.5	44	13	<1.5	65	53
	>1.5	72	60	>1.5	162	176
	$\chi^2 = 8.60$	$P = .010$		$\chi^2 = 1.64$	$P = .231$
Local coexistence	<1.5	27	0	<1.5	93	15
	>1.5	65	28	<1.5	274	98
	$\chi^2 = 10.60$	$P = .008$		$\chi^2 = 5.55$	$P = 0.53$

Values are the number of associations that were more (+) or less (−) common than expected by chance.

From Bowers and Brown 1982.

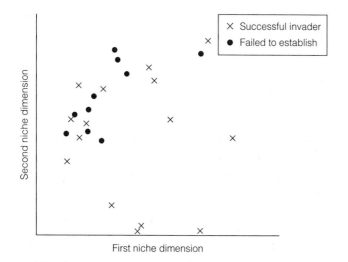

FIGURE 11.15 Two-dimensional niche spaces of finches introduced to the Hawaiian Islands. (*From Moulton and Lockwood 1992*)

avoid all three biases. In their study of finches introduced to the Hawaiian Islands, the J. P. Morgan effect does not arise because the guild includes finches only. Because these are introduced species, the Icarus effect is irrelevant. And "the ghost of competition past" does not haunt their study because the introduced species come from disparate places and thus have no evolutionary history together.

Of the total of 25 species of finches (in four families) introduced to the island of Oahu, 15 have established wild populations. For all 25 of these species, Moulton and Lockwood measured a series of 15 morphological characters that provided a measure of the niche of the species. They included such things as bill shape (related to diet) and wing morphology (related to flight ability and thus to microhabitat). A plot of each species in the two-dimensional morphological niche space is shown in Figure 11.15. Moulton and Lockwood compared the distributions of the niches of the 15 successful species and found that they were statistically "hyperdispersed" in niche space; that is, niches were widely separated. Note in Figure 11.15 that the niches of species that failed to establish wild populations are very similar to those of species that have established populations. Moulton and Lockwood reasoned that the failures were too similar to other species and were prevented by competitive exclusion from establishing populations.

Resolving the Role of Competition   Given these various studies, how do we resolve the issue of the role of competition in structuring communities? The very fact that we can demonstrate competition between two or more species makes it reasonable to

conclude that competition plays a role in the structure of communities. But Strong and Simberloff's criticisms are valid. We should no longer assume that just because we can demonstrate interspecific competition, it must be a major force in all communities. We should begin our empirical studies with a test of the null hypothesis. This is a sufficiently difficult task that we may not easily be able to do studies of the role of competition in just any community. As we saw, Moulton and Lockwood used the peculiar situation of species recently introduced to an island for their test.

It is also reasonable to suppose that competition will not be uniformly important across or within communities. In some communities, competition may be of paramount importance because of the importance of crucially limiting resources. In others it may play a less important role. In Chapter 2 we emphasized that evolution is an ongoing process and that different species are at different stages of the process. We should expect that in some communities, we will find guilds of species in which competition had its effects via competitive exclusion and character displacement long ago. Other communities may still be in the process of sorting out which species will be able to persist. Studies of extant communities will be confounded and confused by the fact that we expect them to differ.

J. S. Findley and H. Black (1983) proposed a model for the structure of communities that predicts some of these kinds of differences. Their model takes into account the importance of history in determining the kinds of niche relationships we see in communities. On the basis of their empirical studies of bat commu-

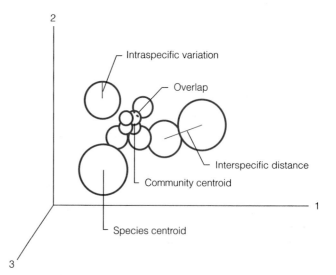

FIGURE 11.16 J. S. Findley and H. Black's model of a community in niche space. Each axis represents a different niche dimension. (*From Findley and Black 1983*)

nities in Africa, they proposed that long-standing communities will appear as shown in Figure 11.16. The older members of such a community may be tightly packed near the community centroid. Competitive exclusion has weeded out species that were too similar, and character displacement has shifted the niche centroids and niche breadths of the remaining species sufficiently far apart that coexistence is possible. New species can be added only on the periphery, where they overlap little with established species. These newer species are likely to have larger niches because they do not face much competition on the periphery of niche space. Although this model may not apply to all possible communities, it emphasizes the variation we expect to see among species within a community and across communities in different stages of the evolutionary ecological process.

Mathematical models of community assembly reveal other important factors that help shape the structure of a community. Some of these models suggest that sequential colonization and extinction gradually lead to a stable, persistent community structure (Drake 1991; Wilson 1992; Drake et al. 1993). Chance events, particularly in the early stages of community formation, lead to very different yet persistent structures. Recent modeling efforts by Lockwood et al. (1997) show that the number of species persisting in a community increases as the rate of invasion increases. One might thus think of invasion rate as a niche dimension that contributes to the structure of the community. Moreover, some species are shown in Lockwood's model to be able to invade only when others invade simultaneously.

Predation as a Structuring Force

Predation is also known to play an important role in organizing communities. Its contribution has been less controversial than that of competition, both because few ecologists claim that predation is as dominant an organizing force as competition and because rigorous experiments have demonstrated its importance. Interestingly, the mechanism by which predation is thought to structure communities includes a major role for competition.

R. T. Paine (1966) performed some of the original and classic work in this area. He studied the communities of invertebrates and algae in the rocky intertidal zone at Neah Bay, Washington (Figure 11.17). This

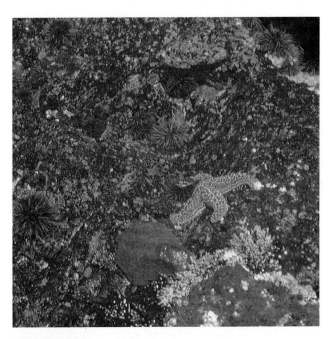

FIGURE 11.17 The rocky intertidal community, consisting of algae, a number of sessile and mobile invertebrates, and fish. Starfish, especially in the genus *Pisaster*, are keystone predators whose activity modifies the competitive interactions of many of the sessile invertebrates. (*Photo by David Krohne*)

system is composed of a large group of invertebrates and algae, with a single top carnivore, the starfish, *Pisaster ochraceous*. Paine was interested in the role the predator plays in organizing the community. He removed *Pisaster* from experimental plots and allowed them to remain on control plots. In plots from which starfish were removed, species diversity was significantly lower than in the controls; only 8 of 15 species remained in communities protected from starfish predation.

Paine believed that this change was mediated by competition. In the absence of starfish, barnacles in the genus *Balanus* increased on both sites. Later they were replaced by mussels (*Mytilus californicus*). Space became limiting on the substrate, and most species of algae were eliminated. Competitive exclusion by a few dominant species decreased community diversity.

This phenomenon has given rise to the **keystone predator** hypothesis. In this system, *Pisaster* is a keystone predator—the animal central to the organization of the community as it appears in undisturbed systems. The keystone predator's role is crucial in this system.

Other apparent examples of keystone predation have been described. J. H. Brown and E. J. Heske (1990) have described kangaroo rats (genus *Dipodomys*) as keystone species in a desert system. In their experiments, three species of kangaroo rats were consistently removed for a period of 12 years from a series of 0.25-hectare enclosed plots in the Chihuahuan Desert in southern Arizona. The effect of removing the kangaroo rats was apparent in significant changes in the vegetation. Inside the removal plots, the character of the vegetation changed from desert scrub to grassland (Figure 11.18). Annual and perennial grasses colonized the spaces between shrubs and increased threefold in the removal plots (Figure 11.19) compared to the controls. Brown and Heske believe that seed predation by these granivores and the soil disturbance associated with their foraging were responsible for the community changes observed.

In addition to its importance as an organizing process in communities, the keystone predator hypothesis has implications for the debate over the relationship between diversity and ecosystem stability (see Chapter 12). In that discussion, we consider the importance of the number of species to the normal functioning of an ecosystem. One concern about human impacts on biodiversity is that we may be destabilizing some systems by decreasing species diversity. The existence of keystone species indicates that we may not always be able to make predictions about stability and diversity without considering which species are being lost. Clearly, the elimination of keystone species like *Pisaster* or kangaroo rats may have greater consequences than the loss of some other species.

In our previous discussion of competition and community structure (Chapter 9), we reviewed an example of a situation in which competition determined

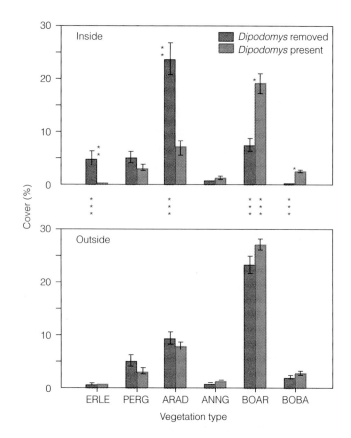

FIGURE 11.19 Effects of kangaroo rat (*Dipodomys*) removal from enclosures. The top figure represents data from inside the enclosures; the bottom figure represents data from transects immediately outside the plots. ERLE = *Eragrostis lehmanniana*; PERG = all other tall perennial grasses; ARAD = *Aristida adscencionis;* ANNG = other annual grasses; BOAR = the short annual grass *Bouteloua aristidoides* and BOBA = the short annual grass *B. barbata*. *, **, and *** indicate significant differences at the .05, .01, and .001 levels, respectively. (*From Brown and Heske 1990*)

FIGURE 11.18 Photograph across the fences surrounding plots from which kangaroo rats were removed for 13 years (left of fence). Note the increase in grass cover when the keystone rodents were absent. (*From Brown and Heske 1990. Photos by James H. Brown © 1989*)

the distribution of barnacles in the rocky intertidal zone. This classic work by J. H. Connell (1961) illustrated the role of community-level interactions in determining the patterns of species distribution.

Recently, D. J. Minchella and M. E. Scott (1991) reviewed an analogous study by G. G. E. Scudder in which parasitism (recall that we consider parasitism to be a form of predation) determines the distributions of two species of water boatmen in British Columbia. The situation is very similar to that of Connell's barnacles, in which both species could tolerate the lower intertidal zone but competition excluded one species from that region. In this case, both *Cenocorixa expleta* and *C. bifida* tolerate freshwater lakes, but only *C. expleta* can tolerate high salinity near coastal waters. Both species are potential hosts for a number of parasitic water mites, but because the mites cannot tolerate high salinity, *C. expleta* is relegated to waters with higher salt content, where it escapes parasitism. Minchella and Scott suggest that these kinds of community effects driven by parasitism may be extremely common in nature—we simply have not looked for them.

In addition to these kinds of effects on communities, predation can be responsible for complex webs of interactions that ultimately determine the nature of the community. T. J. Wooten (1992) described such a case in a marine intertidal community composed of a complex association of invertebrates and algae. Three species of limpets (*Lottia digitalis, L. pelta,* and *L. strigatella*) grow on barnacles or mussels. *L. digitalis* and *L. pelta* match the color of the species on which they grow: *L. digitalis* on the light-colored gooseneck barnacle (*Pollicipes polymerus*) and *L. pelta* on the darker California mussel (*Mytilus californianus*). Several species of algae are also members of this community. Predation on limpets is by birds—specifically, by oystercatchers and gulls.

Wooten excluded avian predation by enclosing the communities in cages. The responses of the enclosed limpets differed from those of the controls (Figure 11.20). *L. digitalis* increased significantly, *L. pelta* did not change (even though it is the most frequently consumed species), and *L. strigatella* decreased. Wooten suggested that bird predation competitively releases *L. strigatella* by decreasing the population sizes of one the other two limpets. Removal of the birds also caused a decrease in algal biomass and species composition (Figure 11.21). Wooten hypothesized that when birds were removed, the larger populations of *L. digitalis* decreased the algal populations. The barnacles increased as well, and this decreased the amount of space available to algae. As this example demonstrates, complex webs of predation and competition (Figure 11.22) determine the species diversity and the structure of communities.

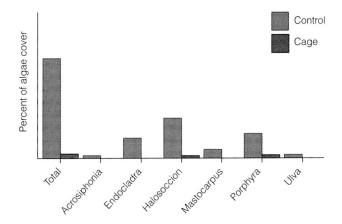

FIGURE 11.21 Percent cover of six fleshy algae in control plots with bird predation (hatched bars on left for each group) and caged plots without birds (open bars on right for each group). (*From Wooten 1992*)

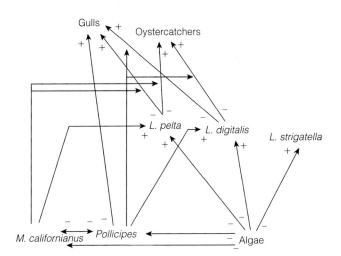

FIGURE 11.22 A diagram of the interactions affecting limpet abundance in the middle intertidal zone. The solid arrows indicate direct interactions among species (such as predation, competition, and habitat preference). The stippled arrows indicate indirect interactions that act by modifying a direct interaction between two other species. (*From Wooten 1992*)

Nonequilibrium Views of Community Structure

Thus far we have approached the problem of the structure and assembly of communities from an equilibrium point of view; that is, we have assumed that communities tend toward some equilibrium state and that community processes such as predation and competition determine the nature of the community that

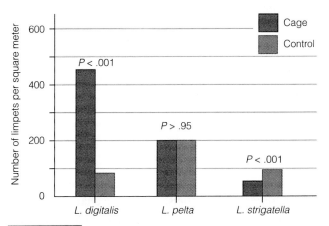

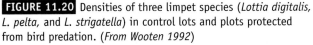

FIGURE 11.20 Densities of three limpet species (*Lottia digitalis, L. pelta,* and *L. strigatella*) in control lots and plots protected from bird predation. (*From Wooten 1992*)

emerges. In the past 20 years, however, ecologists have come to suspect that most of the time, most populations and communities are not at equilibrium (Egerton 1973).

A nonequilibrium concept of communities has profound effects on the way we think about niche relations and the assembly of communities (Sousa 1984). Combinations of species that would not be expected if the community had reached some equilibrium may be possible in a nonequilibrium situation. The most crucial aspect of the nonequilibrium view is that it suggests that we should observe temporary violations of some of the "rules" for communities. If change in the community is frequent and devastating, processes such as competition and predation will be less important as determinants of community structure. A number of nonequilibrium views of community organization are based on the idea that communities are infrequently at an equilibrium state. The first view we will examine involves disturbance.

The Role of Disturbance

A number of factors disturb communities in ways that affect the coexisting species. Abiotic factors such as fire, volcanic eruptions, floods, and storms have different impacts on the various species in a community and thus affect community organization.

J. A. Wiens and J. T. Rotenberry (1980) contributed important information to this view of communities. Using long-term studies of the bird community structures of shrub–steppe communities in the Great Basin, they concluded that at the largest spatial scale—the regional habitat—consistent community patterns of structure are recognizable. At the local level, however, there was so much variation among sites and among years within a site that Wiens and Rotenberry were unable to identify consistent community composition. They pointed out that it is at precisely this local level that competition-mediated structure should be most evident, as shown for mammal communities by Bowers and Brown (1982). Wiens and Rotenberry concluded that the severe and unpredictable climate characteristic of the Great Basin prevents the local species assemblages from achieving full ecological saturation and resource limitation. The result, in their view, is that the bird communities are decoupled from close biotic interactions.

J. H. Connell (1979) also emphasized the role of disturbance. He introduced the *intermediate disturbance hypothesis* to explain the number of species coexisting in a community. According to Connell, the frequency and intensity of disturbance determine the relative impor-

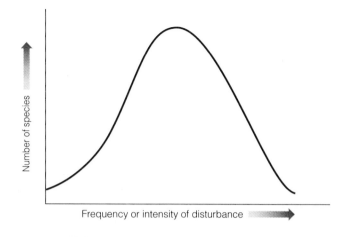

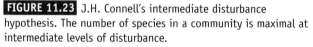

FIGURE 11.23 J.H. Connell's intermediate disturbance hypothesis. The number of species in a community is maximal at intermediate levels of disturbance.

tance of processes such as competition and predation in structuring communities. The general form of the hypothesis is shown in Figure 11.23. At high levels of disturbance, species diversity is low because few species can tolerate the repeated devastation. At low levels of disturbance, competition or predation—processes Connell termed "compensatory mechanisms"—reduces species diversity. Competitive exclusion or elimination of species by predation affects community structure because the absence of disturbance allows competition to become the predominant mechanism. Note that this hypothesis does not negate the operation of such equilibrium processes as predation and competition. Rather, it suggests that disturbance prevents them from organizing community structure in many communities. We shall return to this hypothesis in Chapter 12, where we will consider the evidence supporting it.

A related notion emphasizes the role of population instability. We saw in Chapter 8 that populations fluctuate and occasionally go extinct. By definition, this changes the structure of the community. R. M. May (1979) analyzed this problem theoretically and showed that for randomly generated communities, systems with high species diversity were relatively unstable and susceptible to the effects of disturbance.

Patch Dynamics Models

Another nonequilibrium approach is the development of patch dynamics models, which are characterized by emphasis on the following properties of communities:

1. *Spatial variation in the nature of the community.* Communities are "patchy" in distribution, and

these patches may reflect various abiotic or biotic conditions.

2. *Movement between patches via dispersal of individuals.* The dynamics of these movements depend on the scale of spatial variation and the vagility of organisms.

3. *Disturbance as an important feature of communities.* Physical factors such as storms, fire, and volcanic eruptions are superimposed on the biotic interactions among members of a community. The abiotic factors may prevent biotic interactions from reaching their predicted endpoints.

According to D. L. DeAngelis and J. C. Waterhouse (1987), who developed a model to synthesize these kinds of interactions, three mechanisms have traditionally been invoked to explain the persistence of communities on small spatial scales:

1. Functional relationships between species counteract biologically induced instability. If one species is absent, another may fulfill its role in the community.

2. Environmental disturbance decreases the effect of biotic instability. This process is similar to the effect of disturbance in the intermediate disturbance hypothesis in which abiotic disturbance prevents some species from dominating a community.

3. Species have compensatory mechanisms that operate at low population densities to prevent extinction. These may include rapid growth rates at low densities or related life history strategies.

DeAngelis and Waterhouse then propose that two additional factors are important:

4. The isolation of subpopulations determines the degree to which dispersal can lead to repopulation of extinct patches.

5. Spatial heterogeneity in physical and biological factors leads to movement of individuals to obtain important resources.

Their model is shown diagrammatically in Figure 11.24. Several sources of change destabilize the community, including abiotic disturbance, competitive exclusion, and stochastic demographic processes. Other factors, such as density-dependent reproductive

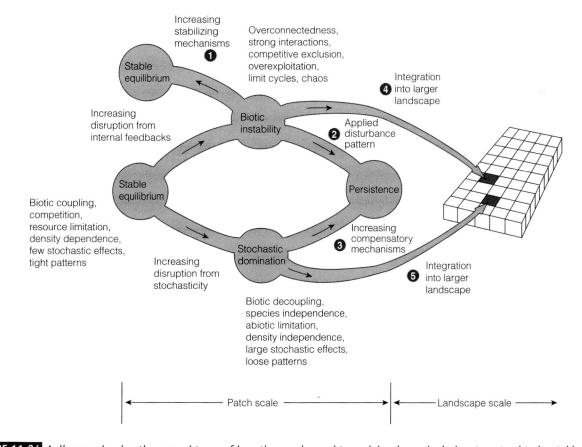

FIGURE 11.24 A diagram showing the general types of hypotheses advanced to explain why ecological systems tend to be stable despite the prevalence of biotic instabilities and environmental stochasticity. (*From DeAngelis and Waterhouse 1987*)

changes and dispersal to empty patches, lead to stability and equilibrium. The combined responses of individual species lead to community persistence. Note that this model assumes a metapopulation structure to populations (see Chapter 8) and, in essence, expands the concept to the community level.

R. T. Paine and S. A. Levin (1981) have documented the importance of patches of disturbance in structuring marine intertidal habitats. In the systems they studied on Tatoosh Island in Washington, the mussel (*Mytilus californianus*) competitively dominates the intertidal community. If left undisturbed long enough, mussels form a dense monoculture that eventually is disturbed by storm-generated waves. The extent of these disturbances ranges in size from the area occupied by a single mussel to an area as large as 38 square meters. Disturbances initiate the recruitment of other species, resulting in a temporal community pattern like that shown in Figure 11.25. In this scenario, the structure of the community over time is determined by the combined forces of competitive interactions and physical disturbance. Disturbance is sufficiently frequent that the community we observe at any moment in time is a mosaic of local sites, each with a slightly different disturbance and competitive history.

R. W. Graham and colleagues (1996) describe similar long-term patterns in North American mammal communities. This group of researchers analyzed fossil mammal faunas from 2945 sites in the United States. Their analysis shows that the geographic ranges of species shifted individually at different times and in different directions in response to the major climatic changes that occurred in the Pleistocene, and not as part of a shift in the entire community. Jablonski and Sepkoski (1996) report similar results for a wide variety of terrestrial and marine species in the Quaternary period.

The fossil record shows that most modern communities emerged very recently. These data support the Gleasonian view of communities as random, nonequilibrium assemblages of species.

The variety of studies described in this chapter suggests the complexity of the interactions that determine the structure of communities. This empirical work indicates that both predation and competition play a role in organizing communities. However, neither mechanism is sufficient to explain all the patterns we observe. The interaction of a rich, complex web of processes determines the nature and structure of biological communities.

Tallgrass Prairie, Bison, and Fire—A Synthesis

We have seen a number of mechanisms that organize community structure. These mechanisms employ deterministic, biotic factors such as competition and predation as well as stochastic influences such as disturbance in the assembly of communities. As in most ecological phenomena, the complete explanation must take into account a complex set of interacting systems. Let us illustrate the process with the community dynamics of tallgrass prairie.

The tallgrass prairie of North American occupied some 68 million hectares (see Chapter 16) before Europeans arrived. This ecosystem was composed of large, treeless tracts of grassland. The dominant plants (numerically and in biomass) were C_4 grasses. In addition to these grasses, a large number of C_3 forbs (nonwoody nongrasses) contributed to the species diversity of the system. Fire, started by lightning and consuming dead biomass, was an important abiotic factor in this system. Large portions of the original prairie burned at irregular intervals—a stochastic impact on the system. Bison (*Bos bison*), numbering between 30 and 60 million, were important grazers in this system (Figures 11.26 and 11.27).

A number of recent studies of tallgrass prairie community dynamics have shown that grazing (preda-

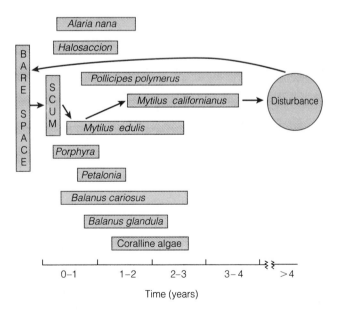

FIGURE 11.25 Generalized patterns of temporal persistence for typical and patch biota. The basic interactions are indicated by the arrows. All other species and interrelationships are of secondary importance. (*From Paine and Levin 1981*)

FIGURE 11.26 Bison are keystone grazers in tallgrass prairie. (*Photo by David Krohne*)

FIGURE 11.27 Fire is an important stochastic abiotic factor in tallgrass prairie. It interacts with grazing to determine the structure of the community. (*Photo by David Krohne*)

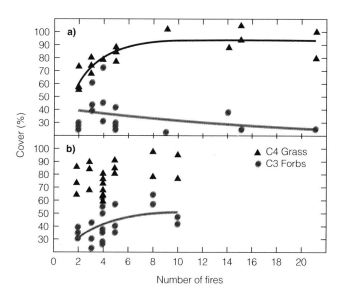

FIGURE 11.28 The effect of fire frequency on abundance measured as percent cover of C_4 and C_3 species in ungrazed (**a**) and grazed (**b**) watersheds. Number of fires refers to the number of fires occurring over a 20-year period. (*From Collins et al. 1998*)

tion), competition, and fire interact to determine the structure of this community (Knapp et al. 1999; Collins et al. 1998; Knapp et al. 1998) on the Konza Prairie in eastern Kansas. These researchers initiated a series of experiments on the effects of burning on species diversity. By burning a series of large watersheds at different intervals, they were able to analyze the role of this stochastic abiotic factor. In ungrazed plots, increasing fire frequency leads to increased cover of C_4 grasses and decreased cover of C_3 species (Figure 11.28). Frequent fire thus leads to lower numbers of species and a more homogeneous community (Figure 11.29). The mechanism for this effect is shown in Figure 11.30. Frequent fire favors the dominant C_4

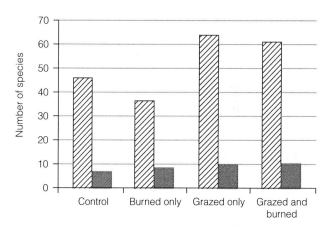

FIGURE 11.29 The number of species in watersheds with different annual treatments of grazing and burning. The solid portion of the bar represents C_4 grasses; the hatched portion represents C_3 species. (*Data from Collins et al. 1998*)

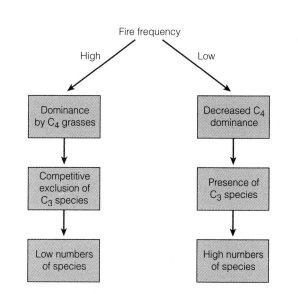

FIGURE 11.30 Mechanisms leading to high and low numbers of species under different fire frequencies in the absence of grazing.

grasses. These grasses competitively suppress the many C_3 forbs by shading and space preemption. When fire is infrequent, the competitive dominance of C_4 grasses decreases and the community contains more species. Thus, under certain fire regimes, competition determines the community structure.

The other important factor in this system is grazing by bison. Bison preferentially select burned areas because of their preference for C_4 grasses. Grazing reduces the dominance of these species (Figure 11.28) and allows the C_3 forbs to increase in numbers. The largest numbers of species were found in plots that were grazed or grazed and burned (Figure 11.29). Bison are considered to be keystone species in this system.

This example illustrates the web of interactions leading to the community structure in tallgrass prairie. Competition, predation (grazing), and stochastic abiotic effects at the patch and regional level—all influence the composition of the community. This system is turning out to be an important model for a number of community and ecosystem processes. We will return to it in our discussion of top-down and bottom-up control of biogeochemical cycles in Chapter 15.

SUMMARY

1. The nature of ecological communities was a focal point of debate through much of the twentieth century. The Clementsian view held that communities are highly organized, persistent entities that reach equilibrium. The Gleasonian view held that communities are random assemblages of species that result from the individual responses of species to the environment.

2. An important paradigm in community ecology has been that competition is the major force determining the structure of communities. According to this view, competition determines which species can coexist and the nature of the niche relationships among coexisting species. Competitive exclusion removes some species from the community, and character displacement modifies the position and shape of niches.

3. Empirical studies indicate that in communities with high species diversity, niche relations among coexisting species do indeed show the effects of competition. Diverse communities are characterized by decreased niche breadth, increased niche overlap, or increased range of resources used.

4. An important criticism of the competition model is that good science demands that before we accept the hypothesis that competition structures communities, we must first reject the null hypothesis that communities are random assemblages of species. Evidence suggests that some communities are random assemblages, but designing and testing a valid null hypothesis is difficult.

5. Predation also plays a role in structuring communities by altering the population densities of certain species. Predation is thought to operate via its effect on competitive relationships.

6. Nonequilibrium views of communities emphasize the importance of disturbance preventing the community from reaching equilibrium.

7. In patch dynamics models, communities represent nonequilibrium entities in which the heterogeneity of the environment, disturbance, and dispersal among patches are the primary structuring forces.

SELF-ASSESSMENT: CAN YOU ...?

1. Articulate the main differences between the Gleasonian and Clemenstian views of communities, and give an example of each.

2. Explain the differences between demonstrating the existence of competition between two species (as in Chapter 9) and demonstrating it among many species in a community.

3. Explain, with examples, the effects of adding species to a community on the niche relationships of species within that community.

4. Describe the nature of a null hypothesis (H_0) for community structure. What factors complicate empirical studies incorporating H_0?

5. Contrast the concept of a null hypothesis for community structure and a stochastic view of communities.

6. Relate patch dynamic models to metapopulation models (Chapter 5).

7. Discuss the relationship among predation, competition, and abiotic factors.

PROBLEMS AND STUDY QUESTIONS

1. Explain the concept of "the ghost of competition past."

2. The following data were recorded (by a graduate student) for the distribution of dung beetles on cow pies in pastures.

Species	Percentage of Sites Occupied
1	0.2
2	0.4
3	0.5
4	0.1
5	0.8
6	0.3
7	0.3

The following patterns of coexistence of species were recorded (this list is not complete).

Species	Percentage of Sites Co-occurring
1 and 2	0.08
2 and 3	0.25
3 and 4	0.05
6 and 7	0.09
5 and 7	0.30

Are these data more consistent with competition structuring the community or with a null model? Explain.

3. One criticism of the analysis of null hypotheses has been the possible effect of "the ghost of competition past." If this factor were operating, how would it affect the data in Problem 2? If you could go back to that previous time, what might you observe? Explain.

4. In Findley and Black's model of community structure, why do peripheral species have big niches whereas central species have small ones?

5. Explain the relationship among the Clementsian, Gleasonian, and patch dynamics models of community structure.

6. The Lotka–Volterra competition models introduced in Chapter 9 describe the conditions for stable species coexistence. How do these conditions apply to communities of many species? What do these models tell us about the relationships in high-diversity communities?

7. Design an experiment to test the hypothesis that disturbance is an important structuring force in a community.

8. Discuss the relationship between these two concepts: Communities are not at equilibrium, and evolution is an ongoing process.

9. For two communities of spiders that differ in species diversity, you find the arrangement of niches shown in the accompanying graph. How have the niches been modified in the high-diversity community? How can you test the idea that competition has caused these changes?

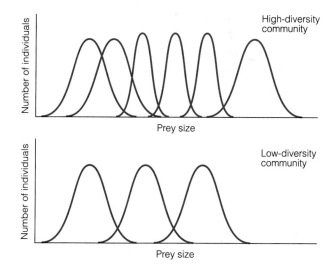

PROJECTS AND ADDITIONAL STUDY

1. Perform a literature search to identify the kinds (taxonomy, functional niches, types of predation) of predators that are keystone predators and the kinds of systems they dominate.

2. For a local community, perform an analysis like that of Bowers and Brown (1982) to test the null hypothesis that species co-occur at random. Measure the frequencies of occurrence of individual species, calculate the expected frequencies of co-occurrence, and compare the observed patterns to this expectation.

3. Use the following paper as the basis for a discussion of the role of keystone species in conservation biology: Power et al. 1996. Challenges in the quest for keystones. *BioScience* 46:609–620.

4. The paper by Knapp et al. (1999. *BioScience* 49:39–50.) sparked a letter by Givnish and Leach in a subsequent issue of *BioScience* and a response by Knapp et al. (*BioScience* 1999. Letters. 49(8):599–601. Read the original paper by Knapp et al., the letter by Givnish and Leach, and Knapp et al.'s response. Discuss and debate the issues raised in these publications.

Species Diversity

ommunities differ markedly in the number of species they contain. In fact, one of the most conspicuous ecological observations we can make is that the number of species varies widely from place to place and community to community. This simple observation has intriguing implications. It raises a number of fundamentally important research questions: Why do some communities have more species than others? Are there consistent geographic trends in the number of species? What are the ecological consequences of large or small numbers of species? How are species "packed" together in high-diversity communities? Interest in these and related questions has been so intense that much of the history of ecology over the past 30 years can be described as a search for explanations of patterns of species diversity. As we will see, some of the important issues have been resolved. Nevertheless, several major questions remain.

One of the reasons for the intense interest in species diversity is that its preservation is important to humans. We make extensive use of the diversity of life-forms on the planet for our own benefit. Obviously, domesticated animals and crops originated as wild members of the pool of species. In addition, we make extensive use of plant species for medicines. Plants are the source of fully 50 percent of the prescription drugs used in the United States, including such important drugs as quinine, digitalis, and morphine. More than 1000 antibiotics have been derived from wild species. We are continually searching for new species with medicinal value. Of course, some species are also of value as pesticides and herbicides. In addition to these direct practical uses, species diversity is, in effect, a barometer of how well we are treating the Earth. Anthropogenic extinctions of species are indicative of environmental degradation and thus measure our impact on the planet.

In Chapter 6 we explored the nature of intraspecific variation. We saw there (and in Chapter 2) that genetic variation is the raw material on which natural selection and the other forces of evolution act. The issue of species diversity is intimately tied to the issue of genetic diversity because the preservation of species also preserves the genetic variation they hold. Biotechnology and the revolution in molecular biology have underscored the potential uses of genetic variants for human purposes, but the concept was applied even before Mendel first articulated genetic principles. Long before the advent of molecular measures of variation, we understood the relationship between species and genetic variation. More than 60

species of potatoes have been described in nature, for example. Cultivated strains of potatoes are derived from only a handful of species and thus contain low levels of genetic variation. In 1846 and 1847, the potato blight, caused by the mold *Phytophthora infestans,* virtually destroyed the potato crop in Ireland, causing widespread starvation. Strains resistant to the blight were later developed by importing genetic diversity from other lines, some of which came from natural populations.

Our focus in this chapter is on the diversity of species in a community. As the potato example illustrates, however, this topic is intimately related to the more general subject of **biodiversity.** Even though this term may be assigned many different meanings by different ecologists, all are generally related to the concept of total biological diversity; thus the term encompasses species diversity, genetic variation and diversity, habitat diversity, and other patterns of ecological variation.

Patterns of Species Diversity

Measures of Species Diversity

If we are to search for patterns in the numbers of species in communities, we must first define the term **species diversity.** We have introduced the topic of species diversity as the *number* of species in a community. However, it turns out that assessing the number of species is only one of several ways of characterizing the species diversity of a community. As we will see, the measurement of species diversity can take on many dimensions.

The simplest measure of species diversity is the total number of species in a community, which is referred to as the **species richness.** In many cases it is an entirely appropriate and useful measure. Obtaining adequate measures of richness may be difficult, however.

One problem with the concept of species richness is that it incorporates no functional information. All we have done is tally the total number of species without regard to how individuals are distributed among species. Thus this simple number may not reflect important ecological information.

To illustrate, consider two communities, A and B, each with 100 individuals spread over 10 species as follows:

	Number of Individuals	
	Community A	Community B
Species 1	10	1
Species 2	10	1
Species 3	10	1
Species 4	10	1
Species 5	10	1
Species 6	10	1
Species 7	10	1
Species 8	10	1
Species 9	10	1
Species 10	10	91
	100	100

Even though both communities have the same number of species, they differ markedly in the distribution of individuals among the 10 species. Community A has a perfectly even distribution; Community B is heavily dominated by species 10, with minimal numbers of the other nine species. We might expect that there will be important functional differences between these two communities. Merely reporting the number of species in each community does not convey the possible functional differences.

Communities such as A have high evenness. We define **evenness** as the degree to which the number of individuals is evenly distributed among species in a community. Community A is much more heterogeneous (less homogeneous) than is Community B. We might expect that a heterogeneous community will be more functionally complex than a homogeneous one. Community B is so heavily dominated by a single species that ecologically it might function more as a single-species community does.

Clearly, then, ecologists need to take into account both the number of species and their heterogeneity. Accordingly, a number of measures of these properties have been derived.

Indices of Species Diversity One of the first measures to incorporate both species abundance and heterogeneity was Simpson's index (Simpson 1949). This index (D) is calculated using the equation

$$D = \frac{1}{\sum\limits_{i=1}^{n} p_i^2}$$

where p_i is the proportion of individuals in the ith species and n is the total number of species. D can vary from 1 to n. The values of this index for Communities A and B are 10.00 and 1.21, respectively, calculated as follows:

For Community A:

$$D = \frac{1}{\sum\limits_{i=1}^{n} p_i^2}$$

$$= \frac{1}{0.1}$$

$$= \frac{1}{\sum\limits_{1}^{10} (0.1)_1^2 + (0.1)_2^2 + \cdots + (0.1)_{10}^2}$$

$$= 10$$

For Community B:

$$D = \frac{1}{\sum\limits_{1}^{10} (0.1)_1^2 + (0.1)_2^2 + \cdots + (0.91)_{10}^2}$$

$$= \frac{1}{0.829}$$

$$= 1.21$$

Another widely used index was devised by C. E. Shannon and W. Weaver (1949) to measure message diversity in the field of information theory. R. H. MacArthur (1955) proposed its application to ecological diversity. The Shannon–Weaver index is based on the notion of "uncertainty." If we were to choose an individual at random from a very heterogeneous community such as Community A, we would be very *uncertain* which species we would get. The probability of choosing any one of the ten species is equal in Community A because each species in the community has an equal number of individuals. In Community B, however, we can be much more nearly certain which species we would choose in a random draw. Because 91 percent of all individuals belong to species 10, the probability that an individual from this species would be chosen is very high (91 percent). We use this notion of uncertainty as a measure of heterogeneity: Heterogeneous communities have low certainty; homogeneous communities have high certainty.

The Shannon–Weaver index quantitatively measures the degree of uncertainty and thus the heterogeneity (H). The index is calculated as follows:

$$H = -\sum\limits_{i=1}^{n} (p_i \ln p_i)$$

where p_i is the frequency of the ith species, n is the number of species, and ln is the natural log. This unitless index measures the degree of uncertainty, or heterogeneity, in a community: High values indicate that the community is very heterogeneous; low values indicate homogeneity. The values of H for Communities A and B are 2.30 and 0.50, respectively, calculated as follows:

For Community A:

$$H = -\sum_{i=1}^{n} (p_i \ln p_i)$$

$$H = -\sum_{1}^{10} [(0.1)(-2.3)] + [(0.1)(-2.3)]$$
$$+ \cdots + [(0.1)(-2.3)]$$

$$= -[(-0.23) + (-0.23) + \cdots + (-0.23)]$$

$$= 2.30$$

For Community B:

$$H = -\sum_{i-1}^{10} [(0.01)(-4.6)] + [(0.01)(-4.6)]$$
$$+ \cdots + [(0.91)(-0.009)]$$

$$= -[(-0.046) + (-0.046) + \cdots + (-0.085)]$$

$$= 0.50$$

The Relationship Between Abundance and Diversity

Thus far, we have assumed that we have complete information about both the number of species in a community and the distribution of individuals among those species. Of course, these data are generally estimates based on samples from a community rather than complete counts (censuses) of every individual. For some large, conspicuous taxa such as mammals and birds, complete counts may be possible and practical. For many invertebrate communities, however, samples are the only source of information. Consequently, any biases or inadequacies in sampling will affect the estimates.

In general, as the sample size increases, so does the number of species known to be in the community. Rare species will not be encountered until a large sample has been taken. An example of the effect of sample size on estimates of species richness is shown in Figure 12.1. It may be impractical to obtain complete samples from some environments. With the function shown in Figure 12.1, however, one can either make inferences about the total number of species present or compare species diversities in communities sampled equally.

In many thoroughly sampled communities, a graph of the number of species as a function of the number of individuals per species follows a log-normal curve (Figure 12.2) (Sugihara 1980). F. W. Preston (1948) was the first to describe this relationship. From a sam-

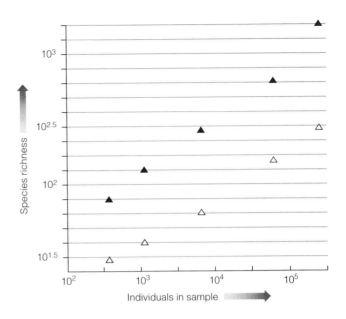

FIGURE 12.1 Species richness as a function of the number of individuals in samples from the deep ocean. Ocean symbols represent rare species; closed symbols represent all species. (*From Grassle 1991*)

ple of species in a community, each species is assigned to an abundance "class" on the basis of the log number of individuals in that species. Preston referred to these classes of species as "octaves"; each octave contains a group of species with similar abundances.

This curve is described by the equation

$$n_r = n_0 e^{-(1/2)(r/s)^2}$$

where n_0 is the modal number of species (the number of species in the most abundant class), s is the standard

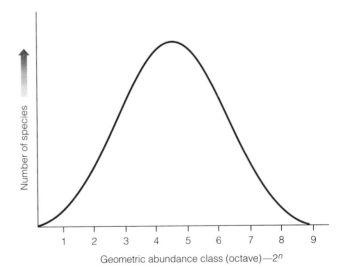

FIGURE 12.2 The log-normal species curve—the number of species as a function of the geometric abundance class. Each "octave" represents an abundance class (n) in which the number of individuals is 2^n.

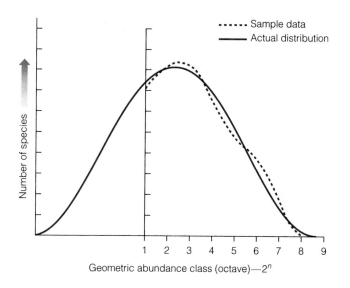

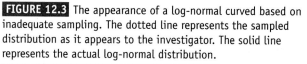

FIGURE 12.3 The appearance of a log-normal curved based on inadequate sampling. The dotted line represents the sampled distribution as it appears to the investigator. The solid line represents the actual log-normal distribution.

ber of species in the modal class, n_0, and the standard deviation, s, we can calculate the total number of species (N):

$$N = n_0\sqrt{2\pi s^2}$$

Preston's log-normal curve is based on the frequency distribution of abundance classes. Community structure can also be described with another method. Hanski's (1982) **core–satellite hypothesis** relates rarity and commonness of species to abundance. Hanski showed that local abundance and regional distribution are not independent. The distribution of rare and common species is bimodal—most species are either rare or common (Figure 12.4). Moreover, mean abundance and the number of sites occupied by a species are positively related (Figure 12.5). From these patterns, Hanski separates core and satellite species. The former are found in many sites at high abundance. In our grassland example, these are the dominant

deviation, and n_r is the number of species whose abundance is r octaves more or less than the mode.

In any sampling regime, the investigator cannot obtain samples of all individuals in all species. Incomplete sampling might generate a curve like that shown in Figure 12.3, which represents only a portion of the total log-normal curve for the species. In effect, the left side of the curve is hidden by the inadequate sampling because only the most abundant species have been recorded. As the sample size increases, more of the true curve is revealed.

A number of taxa, including those containing diatoms (Patrick 1968), soil arthropods (Hairston and Byers 1954), birds, and mammals (Preston 1962), fit the log-normal pattern. That so many taxa show this pattern is of interest for two reasons. First, it suggests that general principles may underlie the organization of communities. Second, it offers the practical benefit that we can estimate the total richness of a community from a sample. If we know both the num-

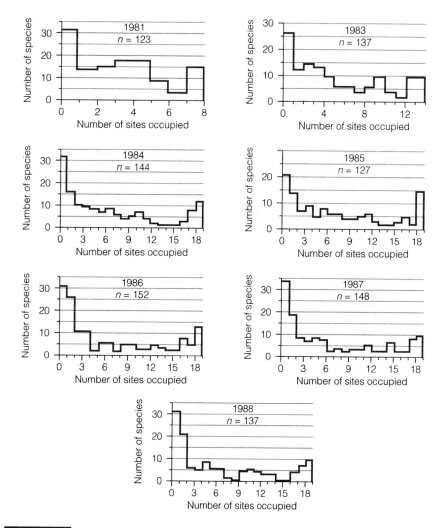

FIGURE 12.4 Bimodal distributions for the number of prairie plants occupying different numbers of sites at Konza Prairie, Kansas. (*From Collins and Glenn 1991*)

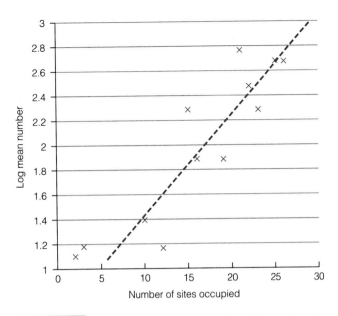

FIGURE 12.5 Number of species as a function of the number of sites occupied. Data are for prairie plants inhabiting a series of small, prairie remnants in Illinois and Indiana.

prairie grasses. Satellite species are sparse in any site and occur in few sites. In tallgrass prairie, the satellite species are the less abundant and rarer forbs (nonwoody, nongrass species). Aside from providing a framework against which to test community structure and the patterns of rarity, the core–satellite hypothesis asserts that stochastic factors are important in regional population dynamics. Indeed, it is these stochastic forces that lead to the bimodality of core and satellites (Hanski 1982; Collins and Glenn 1991).

Total Species Richness

Before we consider how species diversity varies among communities, we must consider a fundamental property of the ecology of the Earth: How many species exist on the planet? This is a question of great practical and theoretical importance, not just a matter of idle curiosity, for if we are to interpret the biological meaning of patterns of community diversity, we must be able to put them in the broad context of planetary diversity. In practical terms, we depend very heavily on thousands (if not millions) of species in the natural world for everything from food and medicine to the cycling of important nutrients. As species go extinct at ever-increasing rates, it is important to know how many species reside on the planet. Clearly, anthropogenic extinction is proceeding more rapidly than the formation of new species. Thus we need to know the magnitude of our impact on global diversity.

Interestingly, we do not know the number of species on the planet within even an order of magni-

tude! Estimates vary from several million to nearly 100 million (May 1988; Wilson 1992). Making and verifying such estimates is a daunting task. Obviously, we cannot actually count the number of species. We must rely on estimates that extrapolate from some sort of sampling scheme. Like any estimate, the results are sensitive to the assumptions we make. Let us consider some of the estimates currently being debated.

Most ecologists accepted a figure of 3–10 million species until some startling data emerged from the tropics in the 1980s (Erwin 1982, 1983, 1988). T. L. Erwin studied the insect species diversity of a relatively poorly known portion of the tropical rain forest: the canopy. Erwin sampled insects from the canopy by spraying it with insecticide and collecting the rain of insects that resulted. The diversity of insects, particularly beetles, was astounding. In a Peruvian rain forest, one sample contained 3099 individuals, and among these 3099 individuals were 1093 species! At another site in the Amazon Basin, Erwin found only 1 percent overlap of species among 1080 species collected from four different forest types. Two important results emerge from his data: (1) the species diversity in the tropical rain forest canopy is phenomenal, and (2) the distribution of insect species is extremely local. As a result, Erwin argued that the estimate of the total planetary diversity must be revised upward by an order of magnitude—to approximately 30 million.

Others place the total even higher. When N. E. Stork (1988) used techniques similar to Erwin's in Indonesia, he found as many as 42 million arthropods in a single hectare of rain forest. On the basis of these samples, Stork judges that there are between 10 and 80 million species on Earth. P. R. Ehrlich and E. O. Wilson (1991) suggest that the total may be as high as 100 million.

Clearly, each of these estimates is heavily dependent on crucial, untested assumptions. Most ecologists would probably agree that 3–5 million species is too low and that 100 million species is too high. More detailed information on the assumptions or altogether new methods of estimation are required to provide an estimate in which we can have confidence.

Even if it were possible to enumerate all the species on the planet or even in various selected habitats, the survey would take so long that many important habitats and species would be destroyed before it was complete. Thus we have to adopt different, sometimes indirect approaches to the problem. N. Myers et al. (2000) recognized that certain places contain disproportionately large numbers of species. The identification of such places, their inventory, and ultimately their preservation should be given high priority.

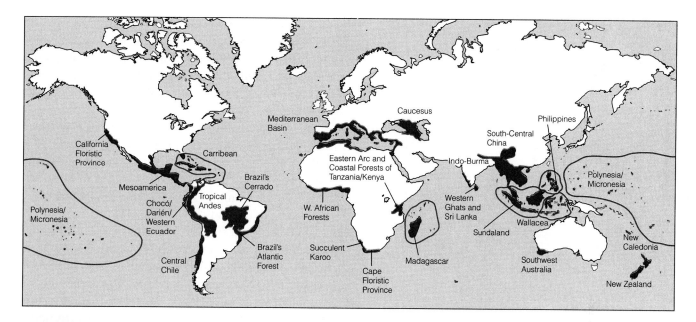

FIGURE 12.6 Twenty-five "hot spots" of species diversity—regions of exceptional concentration of endemic species and exceptional loss of habitat. The regions are important foci for conservation efforts. (*From Norman et al. 2000*).

Figure 12.6 shows the species diversity "hot spots" identified by Myers. Even though these regions make up only 0.5 percent of the Earth's land surface, they contain 20 percent of the world's species.

Hughes et al. (1997) recently attempted to make similar estimates for the number of *populations* on the planet, another related aspect of biodiversity. Using literature estimates of the area extent of populations and species's geographic ranges, these researchers estimate that there are approximately 220 populations per species—that is, between 1.1 and 6.6 billion populations, depending on the number of total species one assumes. As these unique populations go extinct, which Hughes et al. estimate is happening at a rate of 1800 per hour, an evolutionarily and ecologically important component of biodiversity is being lost.

Geographic Patterns of Diversity

We have alluded to the fact that the distribution of species is not random across the Earth. Here we will consider four species diversity patterns: diversity on islands, latitudinal trends, diversity in ecotones, and deep-ocean diversity.

Species Diversity on Islands One well-documented general pattern is that islands have fewer species than similar areas on the mainland. It is also well known that there is a relationship between species diversity and island area: Larger islands tend to have more species than smaller islands. In fact, this relationship is remarkably constant across island systems. The general equation describing the species/area relationship is $S = CA^z$, where S is the number of species, A is the island area, C is a constant, and z is the slope of the line. An example of this relationship is shown in Figure 12.7. The value of z ranges from 0.20 to 0.35 for many archipelagoes. This implies that there might be an underlying explanation for the pattern of island species richness. In addition to the size effect, there is a distance effect: Islands nearer to the mainland tend to have more species than more distant islands.

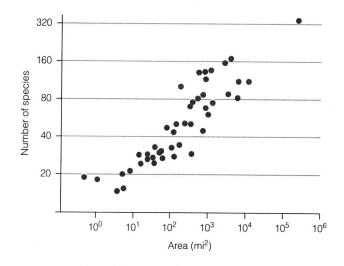

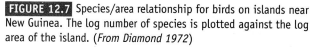

FIGURE 12.7 Species/area relationship for birds on islands near New Guinea. The log number of species is plotted against the log area of the island. (*From Diamond 1972*)

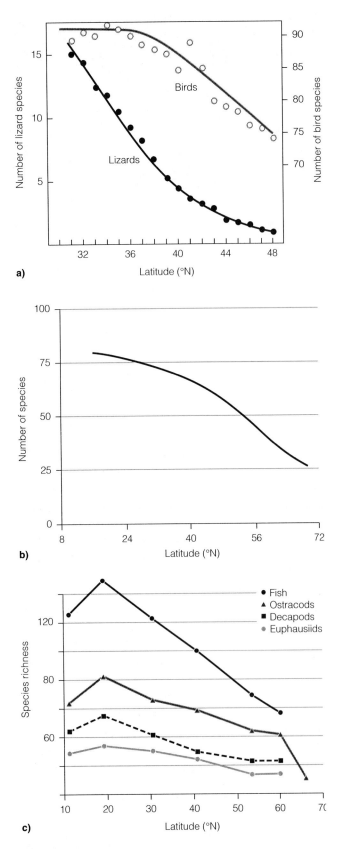

a)

b)

c)

FIGURE 12.8 Species richness as a function of latitude for (**a**) lizards and insectivorous birds in the United States, (**b**) quadrupedal mammals worldwide, and (**c**) pelagic organisms worldwide (*From Schall and Pianka 1978, Rosenzweig 1992, and Angel 1993*)

Latitudinal Trends in Diversity

We have already discussed one of the most intensely studied geographic trends in species diversity: the latitudinal trend. For more than a hundred years, researchers have been aware that the tropics are much richer in species than temperate or arctic regions.

For virtually every taxonomic group, species richness decreases with increasing latitude (Figures 12.8 and 12.9). For those of us with a "temperate zone" view of ecology, the number of species in tropical systems is overwhelming. We have already discussed the phenomenal diversity of insects in the rain forest canopy. The plant species diversity is no less stunning. In some tropical systems in the Far East, a single 1.0-hectare plot may contain more than 250 species of trees with individuals greater than 10 centimeters in diameter (Whitmore 1984). Plots like this may contain only a single individual of each species. On Barro Colorado, an island in the Panama Canal, there are more species of trees than in all of Canada.

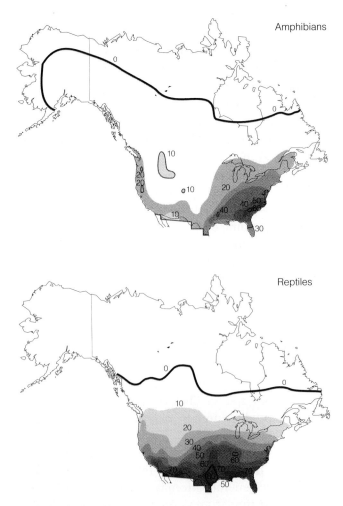

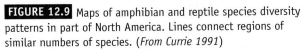

FIGURE 12.9 Maps of amphibian and reptile species diversity patterns in part of North America. Lines connect regions of similar numbers of species. (*From Currie 1991*)

There are exceptions to the general pattern of high diversity in the tropics. For example, portions of the Congo River basin are dominated by a single species of tree (Connell and Lowman 1989). On one 40-hectare plot on Trinidad, 84 percent of the trees were found to be a single species, *Mora excelsa* (Beard 1946). And in lowland Malaysia, many thousands of square kilometers are dominated by the single species *Dryobalanops aromatica* (Whitmore 1984). Any hypothesis that explains the general trend toward high species diversity in the tropics must also explain these exceptional situations.

Diversity in Ecotones Species diversity is also high in **ecotones,** the boundaries between two habitats. The defining feature of an ecotone, which can exist on nearly any spatial scale, is a sharp transition from one habitat type to another. For example, the boundary between a small woodlot and an adjacent meadow is an ecotone. If the transition is more gradual, it is appropriate to refer to it as an **ecocline** (van der Maarel 1990).

One of the most obvious features of ecotones is that they typically have very high species diversity relative to other kinds of habitats (Leopold 1933). As we have seen, species have specific habitat requirements. Thus one set of species is associated with forest habitats, and a different set is associated with grasslands. Ecotones typically contain more species than the sum of the two connecting habitats. In other words, the ecotone between a grassland and a forest supports both forest and grassland species, as well as an additional set of species. Generally, these additional species prefer the transition itself; they are not typically found in either forest or grassland. The effect is also seen in the pelagic ocean, where high species richness is associated with the boundaries between different types of oceanic water where different faunas intermix (Angel 1993).

Deep-Ocean Diversity The last pattern of diversity to consider is that found in deep-ocean habitats. For many years marine biologists believed that species diversity is low in deep-sea habitats. Recently, however, J. F. Grassle (1991) and Grassle and N. J. Maciolek (1992) have shown that the low species richness previously reported for these areas was an artifact of the difficulty of sampling the deep-sea environment. Quantitative sampling at depths greater than 1000 meters poses serious technical problems. New sampling equipment—in particular, dredges with doors that can be remotely controlled during retrieval—have allowed marine biologists to sample from great depths. Their studies have shown that the deep-ocean floor is

TABLE 12.1

Number of Species and Number of Families in Each Phylum from 233 0.09-Square-Meter Samples Taken Between 1500-Meter and 2500-Meter Depths on the Continental Slope Off New Jersey

Phylum	Number of Species	Number of Families
Cnidaria	19	10
Nemertea	22	1
Priapulida	2	1
Annelida	385	49
Echiurida	4	2
Sipuncula	15	3
Pogonophora	13	5
Mollusca	106	43
Arthropoda	185	40
Bryozoa	1	1
Brachiopoda	2	1
Echinodermata	39	13
Hemichordata	4	1
Chordata	1	1
Total	798	171

From Grassle 1991.

not the biological desert that we once thought. Grassle (1991), for example, sampled 233 sites between 1500 and 2500 meters below the surface off the coast of New Jersey. The samples revealed 798 species in 171 families of 14 phyla (Table 12.1).

Even more interesting was the finding that the proportion of rare species did not decline with increased sampling. As more individuals were captured, the number of species continued to increase. This means that the collections were *underrepresentations* of the total number of species. In other words, the new sampling techniques reveal high species diversity, but they have not yet adequately measured the extent of that diversity. The samples represent an example of the "hidden" left side of the log-normal curve in Figure 12.3. Grassle suggests that the actual species richness of the deep-sea environment may rival that of some tropical systems. In any case, we can no longer regard this portion of the biosphere as a species-poor desert. Certainly, we must not think that dumping human wastes of various kinds into the deep ocean will have no biological effects.

Determinants of Species Diversity

Now we consider the kinds of ecological processes and mechanisms that can account for these differences in species richness. Note that islands are characterized by lower species diversity, whereas species richness is unusually high in the other three systems we discussed—the tropics, ecotones, and the deep ocean. We will first consider the processes that lead to lower richness on islands. Then we will discuss mechanisms that can account for high diversity and consider how these mechanisms may apply to different kinds of high-diversity systems.

Mechanisms That Produce Low Diversity: Island Biogeography

Islands are typically depauperate in species richness relative to mainland areas of comparable size. Originally, this phenomenon was explained by a *nonequilibrium* theory of island biogeography. According to this notion, islands are depauperate because they have not had sufficient time since their formation to accumulate species via immigration. In other words, islands have not yet reached an equilibrium state. Given the reasonable assumption that species reach islands by dispersal from the mainland, it will take longer for islands to accumulate all the species present on the mainland. Imagine a newly formed volcanic island like those in the Hawaiian chain. Dispersal from the mainland will be unusual, and even after long periods of time we might expect fewer species to have arrived there.

The Equilibrium Theory of Island Biogeography

In 1963, R. H. MacArthur and E. O. Wilson published a new hypothesis to explain the patterns of species richness on islands. Described as an *equilibrium theory of island biogeography* (MacArthur and Wilson 1963, 1967), it proposed that the lower number of species on islands is not a result of insufficient time for species to disperse to the islands, but rather an equilibrium situation peculiar to all islands. The theory could be readily adapted to account for the size and distance effects just described.

MacArthur and Wilson's theory is based on the supposition that at a given time, the number of species present on an island is a result of two processes: colonization and extinction. On any island, new species colonize by immigrating from the nearest mainland. For animals that can fly, such as birds and insects, this process may be relatively easy. For others, it may be possible but improbable. Animals and plants with poor locomotory powers can travel great distances if carried by wind, by other organisms, or on rafts of vegetation that break away from the mainland. When the island of Krakatau exploded in 1883 and all life there was decimated, among the first organisms to

recolonize were spiders that reached the island by ballooning on silk threads. Obviously, the rate of colonization by these processes is generally much lower than it is for species with the power of flight.

As more and more species colonize an island, the rate of immigration by new species declines (Figure 12.10). This is due in large part to the fact that as diversity increases, more of the new immigrants are already represented on the island. When all the species from the mainland have colonized, the immigration rate must drop to zero; there are no species on the mainland that have not colonized.

MacArthur and Wilson also proposed that the extinction rate should change as a function of the number of species present on the island. As the island fills up with species, the extinction rate should increase (Figure 12.10) because there are more species to go extinct. Also, there is some evidence that competition for the limited resources of the island becomes more intense as species accumulate.

At some number of species, the immigration and extinction lines intersect (Figure 12.10). At that point, immigration exactly balances extinction, and the number of species on the island should stabilize. This equilibrium number of species is referred to as $\hat{S}$. Note how this theory differs from the nonequilibrium theory, which explains the depauperate nature of islands, a consequence of species' not having had enough time

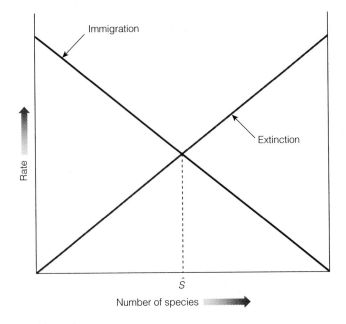

FIGURE 12.10 MacArthur and Wilson's equilibrium theory of island biogeography. The point where the lines for the immigration and extinction rates intersect represents the equilibrium number of species for the island. Note that the immigration and extinction functions are not necessarily straight lines.

to accumulate. MacArthur and Wilson's equilibrium theory suggests that there is something fundamentally different about islands—that no matter how long we let species accumulate, islands will never reach the same species richness as the mainland.

This equilibrium theory can also be used to explain the effects of island size and distance from other lands. Consider two islands of the same size but different distances from the mainland. Because extinction rates are dependent on the amount of resources available, they should be a function of island size, not distance. Thus the extinction rates should be the same on near and far islands (Figure 12.11a). The immigration rate, however, should be much lower on distant islands, because the probability of a colonist making it there is much lower. Accordingly, near and far islands should have different immigration rates. The colonization rate should always be less for the distant island (Figure 12.11a). The result will be a difference in the equilibrium numbers of species on the two islands. Note in Figure 12.11 that $\hat{S}_{near}$ is larger than $\hat{S}_{far}$.

A similar argument can be used to explain the effect of island size. Again consider two islands, this time a large island and a small island equidistant from the mainland. We might expect size to affect both the immigration rate and the extinction rate. A large island is a bigger target for potential colonists. Immigration events are random—an individual traveling over water by whatever means just happens to reach the island. If so, a large island will accumulate immigrants faster than a small one. This will affect the slope of the immigration rate (Figure 12.11b). We have said that extinction occurs because of competition for the limited resources on islands. A small island

will have fewer resources than a large one and consequently will have a higher extinction rate (Figure 12.11b). The equilibrium values for the large and small island will differ as shown in the figure; $\hat{S}_{sm}$ is smaller than $\hat{S}_{lge}$ in all cases.

Empirical Evidence for the Equilibrium Theory

The equilibrium and nonequilibrium theories make very different predictions about the nature of island species:

1. The equilibrium theory predicts that the number of species on an island will not change over time. Once the equilibrium between immigration and extinction has been reached, the value of $\hat{S}$ should be relatively constant. The nonequilibrium theory predicts that the value of $\hat{S}$ will increase as a function of time until it equals the number of species on the mainland.

2. The equilibrium theory predicts that although the value of $\hat{S}$ will be relatively constant over time, the actual species that make up $\hat{S}$ will change; that is, immigration and extinction are in a dynamic equilibrium. The nonequilibrium theory predicts that species will remain on the island; the increase in $\hat{S}$ over time depends on this.

A number of data sets support the equilibrium theory and discredit the nonequilibrium theory. In one classic study, J. L. Diamond (1969) studied bird species diversity on the Channel Islands off the coast of southern California. A. B. Howell had surveyed the islands in 1917 and recorded all species present on each

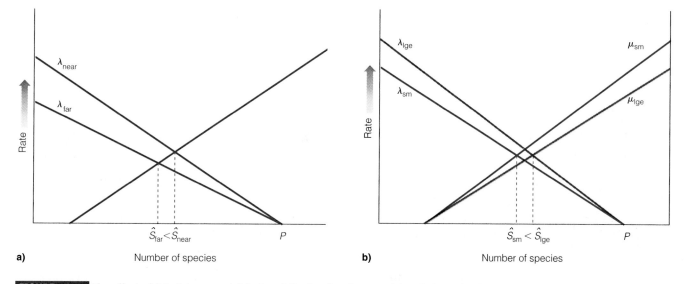

FIGURE 12.11 The effect of (**a**) distance and (**b**) size of the immigration rate (λ) and the extinction rate (μ) in MacArthur and Wilson's equilibrium theory.

island. Diamond resurveyed the islands in 1967 and compared the species diversities and compositions with Howell's data. As is clear from Table 12.2, species diversity did not increase appreciably between 1917 and 1967. Although islands gained between 1 and 9 species and lost between 3 and 12 species, the total number of species on each island was very similar to what it had been 50 years earlier. Moreover, the particular species that made up the total on each island had changed. Some species present in 1917 had gone extinct, and some new species had colonized. These data are consistent with both predictions of MacArthur and Wilson's equilibrium model.

Another confirmation of the theory was provided by D. S. Simberloff and E. O. Wilson (1969), who studied small mangrove islands in the Florida Keys. Mangrove trees produce seeds that drop into the water and float about until they reach shallow mud, where they germinate and grow into adult mangroves that grow vegetatively. The extensive root systems accumulate sediments, forming small islands that range in size from a few square meters up to several thousand square meters.

Simberloff and Wilson studied the biogeography of the arthropod fauna of these islands. First they surveyed a series of islands of different sizes and distances from shore; then they defaunated the islands by enclosing them in plastic and pumping in methyl bromide to kill all the arthropods. After allowing recolonization to occur, they resurveyed the islands. They found that species accumulated on the islands for a time and then reached an asymptote that was approximately equal to the original number of species on each island (Figure 12.12). However, *the actual species making up the total changed*. Again, both predictions of the equilibrium theory were confirmed. The researchers were also able to document effects of island size and distance on the composition of species.

Following the publication of MacArthur and Wilson's theory, a number of studies of island commu-

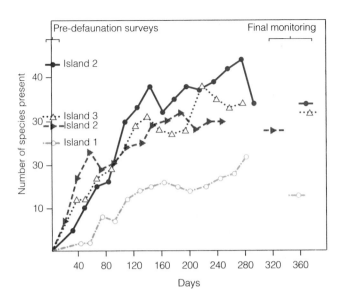

FIGURE 12.12 Pattern of recolonization of defaunated mangrove islands. In each curve the last two census points are not connected by a line to emphasize the greater period of elapsed time compared with times separating earlier censuses. (*From Simberloff and Wilson 1969*)

nities were undertaken. Researchers discovered that a number of factors associated with isolation and island size play a role. For example, Kohn and Walsh (1994) showed that the number of plant species on an archipelago of small islands near Shetland in northern Britain is affected by island size and isolation according to the stochastic effects leading to equilibrium as predicted by the MacArthur and Wilson theory. However, they also found indirect effects of size: Larger islands have greater habitat diversity, and this heterogeneity leads to more species.

Another factor is the nature of the island system—specifically, whether the islands are oceanic or land-bridge types. The former are islands that have arisen in the ocean de novo, often as a result of volcanic activity. The Galápagos and Hawaiian Islands are examples; there has never been a land connection between either archipelago and the mainland. Land-bridge islands are formed by a rise in sea level that isolates land formerly contiguous with the current mainland. The islands of the Aleutian chain in southwestern Alaska fall into this group.

Communities on these two kinds of islands have different histories. Organisms must colonize oceanic islands by long-distance dispersal events. These islands conform relatively well to the predictions of the equilibrium theory of island biogeography. On land-bridge islands, on the other hand, the full complement of mainland species becomes isolated when sea levels rise to create an island. Thus these islands begin with a

TABLE 12.2
Total Number of Bird Species on Selected Channel Islands

Island	1917	1967
Los Coronados	11	11
San Nicolas	11	11
San Clemente	28	24
Santa Catalina	30	34
Santa Barbara	10	6

Data from Diamond 1969.

higher number of species than we expect to see on an island. Because of their new insularity, the communities on these islands gradually shrink in species richness. The animals thus isolated are referred to as **relaxation faunas.**

Diamond (1972) provided data on the nature of relaxation bird faunas. For land-bridge islands near New Guinea, Diamond calculated the degree of relaxation from the current species diversity and the species present on the mainland. From information on the timing of sea level changes and the species diversity of islands of different size and distance, Diamond could calculate the time required for relaxation to occur.

The different histories of species diversity on land-bridge and oceanic islands are shown in Figure 12.13. The number of species on land-bridge islands gradually declines toward an equilibrium described by the balance between extinction and colonization; on oceanic islands, the species diversity builds over time until the MacArthur and Wilson equilibrium is attained. Note that the equilibrium number of species on oceanic islands may be slightly less than the number predicted solely on the basis of extinction and immigration curves. This is because some very remote islands will accumulate species at very low rates because the probability of long-distance immigration occurring is so low. Diamond (1972) provided data to substantiate this relationship (Figure 12.14) for birds on islands more than 300 miles from New Guinea. In this graph, the ratio of the actual number of species on the islands to the number expected on an island of equivalent area less than 300 miles from New Guinea is plotted as a function of distance. We see that the ratio of observed to expected by the MacArthur and Wilson model declines linearly with distance from New Guinea.

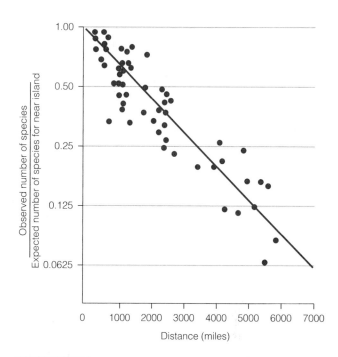

FIGURE 12.14 The ratio of observed number of species to the number expected from the MacArthur–Wilson equilibrium theory as a function of island distance for very distant islands. (*From Diamond 1972*)

Biogeography of Habitat Islands

The success of MacArthur and Wilson's model prompted ecologists to search for the operation of this mechanism in other island-like situations, especially habitat islands on the mainland. Many reasoned that some habitats are effectively islands because they are surrounded by inhospitable country that serves as a barrier to dispersal. In the western United States, for example, a geological province known as the Basin and Range lies between the Sierra Nevada to the west and the Rocky Mountains to the east. Extending between the two main ranges are many smaller ranges surrounded by

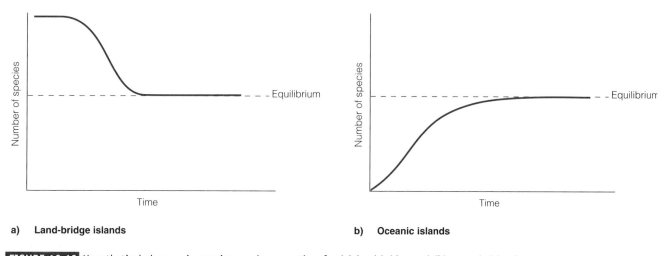

a)   **Land-bridge islands**

b)   **Oceanic islands**

FIGURE 12.13 Hypothetical changes in species number over time for (**a**) land-bridge and (**b**) oceanic islands.

Great Basin desert (Figure 12.15). The higher elevations of these ranges support coniferous forest and even alpine tundra habitat "islands" separated from large regions of similar habitat on the "mainlands" to the east and west by expanses of desert.

Ecologists have sought to explain the patterns of species diversity in a number of such island-like systems, including birds and mammals on mountaintops, specialist invertebrates in caves, fish in lakes, and even phytophagous insects on their host plants. In general, we can say that the strict application of the equilibrium theory to these habitat islands fails (Gilbert 1980).

In one investigation of island-like habitats, J. C. Hickman (1968) examined the species diversity of plants inhabiting the upper elevations of volcanic peaks in the Cascade Mountains of the Pacific Northwest. He found that whereas recently emerged volcanic cones supported 35 endemic species on average, older peaks supported an average of 130 species. Thus these habitat islands do not appear to behave according to the equilibrium theory of island biogeography. The number of species is not a function of immigration balanced by extinction, but rather of the amount of time during which species can accumulate.

Nevertheless, there are some island-like features of isolated mainland habitats to which portions of the equilibrium theory do apply. First, many of these habitat islands show species-area effects much like true islands, a fact recognized long ago (Grinnell and Swarth 1913). Examples of these effects are shown in Figure 12.16 for birds and mammals inhabiting the isolated forests of the Basin and Range province. Second, some habitat islands show isolation effects

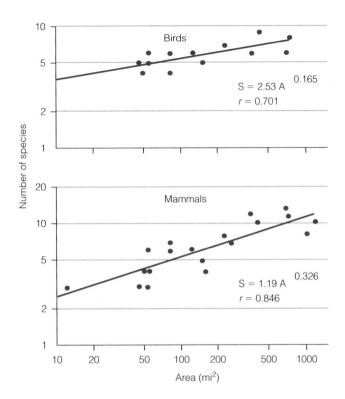

FIGURE 12.16 The number of species of resident boreal birds and boreal small mammals as a function of the size of isolated mountain ranges in the Basin and Range province. (*From Brown 1978*)

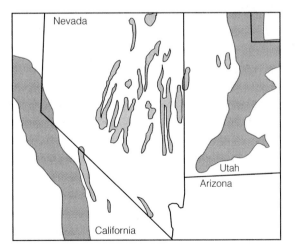

FIGURE 12.15 The Basin and Range province. The crosshatched regions are the Sierra Nevada to the west and the Rocky Mountains to the east. The shaded regions are mountain ranges with boreal habitats on their summits. (*From Brown 1971*)

(Figure 12.17). Third, many of the habitat island systems show patterns consistent with relaxation of species diversity. For example, M. V. Lomolino, J. H. Brown, and R. Davis (1989) present evidence that mammals on forest islands in the Basin and Range province are a relaxation fauna. Forest habitat was much more widespread prior to the Pleistocene. Afterward, a warming and drying climatic trend

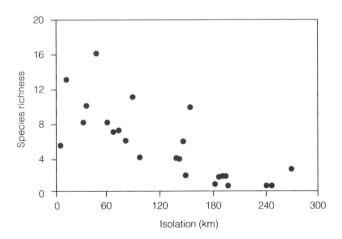

FIGURE 12.17 The effect of isolation (distance to nearest habitat island) on species richness for mammals on montane forest habitat islands. (*From Lomolino, Brown, and Davis 1989*)

caused these habitats to shrink to a few remnants on mountaintops. Colonization of such habitats is rare, particularly for mammals. Thus the low species diversity of mammals on forest islands relative to the Rockies or Sierra Nevada is not an equilibrium achieved by a balance of immigration and extinction, but rather occurs through a process of gradual extinction without colonization by new species.

Habitat islands do not conform as well to the equilibrium theory because two important effects confound the simple operation of MacArthur and Wilson's model. First, the relationship between species diversity and habitat diversity confounds the effect of area. It is not always clear whether an area effect is a direct result of the effect of area on extinction and immigration or is a consequence of the fact that large areas tend to have more kinds of habitats. This effect can be seen in true oceanic islands as well, but studies on habitat islands suggest it is especially confounding in mainland habitats. N. K. Johnson (1975) found that bird species diversity in Basin and Range mountaintop forests is more strongly correlated with the diversity of habitat types in the region than with area (Figure 12.18).

A second confounding factor is that the formation of a habitat island on the mainland creates what is known as an "edge effect." This is a change in the physical and biological parameters of a habitat at its boundaries relative to what is found in its center. For example, if a forest is clear-cut around a patch of remnant forest, the boundary of the forest will differ in important ways from the center. On a 100-hectare island of forest, the edge can differ from the interior by as much as 20 percent relative humidity and 4.5°C. (Lovejoy et al. 1986). This difference may have impor-

tant effects on the vegetation as well as on the suitability of the forest for many species of animals.

G. G. Whitney and J. R. Runkle (1981) have demonstrated the nature of the edge effect in woodlots in Ohio. They studied a series of woodlots with different logging histories and thus different successional status. After making full tallies of all tree species in the interior and edge of both old-growth and second-growth woodlots, they calculated the percent similarities of tree species for all possible comparisons of age and position. Position had important effects. The edges of old growth and second growth did not differ from each other, but the edges of old and second growth differed from the interiors of both old growth and second growth.

Conservation Applications of Island Theory

There is great interest among ecologists in the application of island theory to conservation problems. One of the most profound effects that humans have on the landscape is to fragment habitats into smaller and smaller "islands" (Figure 12.19). Fragmented habitats show relaxation effects like some islands. For example, A. R. Weisbrod (1976) studied the changes in

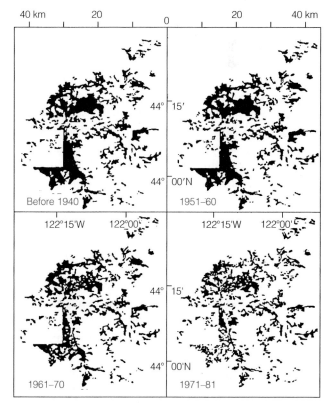

FIGURE 12.19 The distribution of old-growth forest in the Willamette National Forest over time. Tracts of old-growth forest have become increasingly fragmented by logging. (*From Harris 1984*)

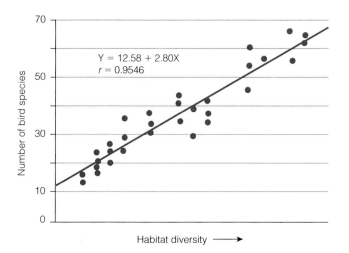

FIGURE 12.18 Bird species diversity as a function of habitat diversity in montane forest islands in the Great Basin. (*From Johnson 1975*)

mammal species in Mt. Rainier National Park over time. As logging has proceeded around the park, forest habitats have become increasingly island-like. Using historical records and modern surveys, Weisbrod showed that mammalian species diversity decreased by 26 percent between 1920 and 1976.

It would be useful to be able to apply the features of island biogeography theory to practical conservation problems. In particular, ecologists are interested in (1) using the theory to predict the effects of anthropogenic habitat fragmentation and (2) using our knowledge of species-area effects and relaxation faunas to design nature preserves that will maximize long-term species diversity (see the box on page 311). Because of the importance of habitat diversity and edge effects, however, we cannot simply transfer the MacArthur and Wilson equilibrium theory *in toto* to any habitat island on the mainland. Nevertheless, there are enough similarities—particularly the area effects—that some useful information on patterns of species diversity emerges from studying mainland habitat islands. In addition to the stochastic effects associated with reduced area, other biological factors associated with fragmentation may decrease species richness in fragmented habitat.

For example, Leach and Givnish (1996) examined the loss of species richness from small island-like patches of tallgrass prairie in Wisconsin. Prior to European settlement, Wisconsin had some 800,000 hectares of tallgrass prairie. Fragmentation of this habitat has been extreme: Less than 0.1% of this original grassland remains, mostly in small, isolated fragments. Leach and Givnish compared the current species composition of a series of 54 remnants with data from surveys done 32 to 52 years before. Up to 60 percent of the plant species have been lost from these remnants.

Mechanisms That Produce High Species Diversity

As we have seen, MacArthur and Wilson's equilibrium theory has been very useful in explaining the depauperate nature of island systems. Now we turn our attention to habitats that support large numbers of species.

As mentioned previously, there has been great interest in explaining the patterns of species abundance, particularly highly diverse systems such as the tropics. Ecologists would like to be able to explain how large numbers of species can be packed together in a community and also how highly diverse communities come to contain so many species. Why, for example, do tropical rain forests contain so many more species than temperate forests?

A large number of hypotheses have been suggested over the years to explain the existence of highly diverse systems. Even though many of these hypotheses were originally conceived as explanations of tropical diversity, some may apply to the other high-diversity communities we described: the deep sea and ecotones. As you consider all of these hypotheses, try to imagine experiments or observations that could result in rejection of each of them. Also ask yourself which of these hypotheses are logically exclusive. That is, which hypotheses, if deemed correct, logically preclude the operation of others? Which hypotheses could work together?

Evolutionary Time   According to the evolutionary time hypothesis, highly diverse communities have simply been in existence longer and thus have had more opportunities for species to evolve. Thus, because the temperate and arctic regions were disturbed relatively recently by extensive glaciations, whereas the tropics have remained climatically stable for many millions of years, there have been more opportunities for speciation and adaptive radiation in the tropics. As a result, they contain more species than higher latitudes.

Ecological Time   The ecological time hypothesis holds that high-diversity communities have been stable long enough for immigration to increase the species diversity relative to younger communities. Some ecologists believe that this hypothesis may be more applicable to small-scale disturbances, such as the effects of a windstorm, than to broad regional differences in diversity. In some tropical forests, the number of woody tree species increases with the amount of time since catastrophic disturbance.

Climatic Stability   According to the climatic stability hypothesis, climatic stability leads to increased diversity. When climate is highly stable, the rate of extinction from the random effects of weather will be lower. If two regions are similar except for the extinction rate, the one with the greater loss of species will have fewer species. In addition, a stable climate may allow species to specialize to a greater extent, providing resources for a larger number of species.

Another causal correlation related to climate is the relationship between species diversity and latitudinal range. The decrease in species richness with increasing latitude is paralleled by an increase in mean latitudinal range: As latitude increases, so does the range of

Habitat Fragmentation and Species Diversity

As we have seen, the species diversity of islands is best explained by the equilibrium theory of island biogeography: The number of species is lower on islands than on the mainland because of a dynamic equilibrium between colonization and extinction. Within certain limits, the concept is also applicable to habitat islands caused by fragmentation of large tracts of habitat.

The extent to which natural habitats are being fragmented is alarming. In the Pacific Northwest, clear-cutting of forests has resulted in a patchwork quilt of forest. The process is also occurring at an astounding rate in tropical forests, where approximately 4.5 million hectares are lost each year worldwide. Nearly 40 percent of the world's original tropical forests have been cleared.

Ecologists are concerned about this because the tropics contain so many of the world's species. The loss of species diversity occurs in several ways. First, an area effect is associated with fragmentation. A newly isolated forest fragment gradually decreases in species diversity (a relaxation fauna) until an equilibrium is reached between extinction and colonization. In addition to this statistical phenomenon, there are other interactive effects. For example, the creation of a forest fragment changes the physical conditions on the edge of the fragment. As the size of the fragments decreases, the proportional area of the edge increases. With a 10-hectare island, for example, if the edge effect extends 10 meters into the forest, edge habitat constitutes 14 percent of the total forest area. On the other hand, if the island is only 1 hectare, the 10-meter edge represents 36 percent of the total area.

In complex systems such as the tropics, the loss of species can be compounded by the web of interactions among species. For example, certain bird species specialize in foraging on insects and other small animals that are displaced in the wake of army ants as they move through a forest. Each colony of ants requires about 30 hectares to support it, so a 100-hectare forest fragment could support three colonies of ants. The ants do not forage continuously, however; they have a 35-day cycle of activity and are bivouacked and do not forage for part of that time. Because of these periods of inactivity, the small fragment of forest cannot support the birds that depend on the ants; in a large tract of continuous forest with more colonies of ants, the birds would be able to seek out a foraging colony. Thus, complex webs of interspecies dependence may be upset by habitat fragmentation.

We can estimate the rate of loss of species from the effects of fragmentation of tropical forest. Wilson (1992) calculates that for a value of $z = 0.30$ in the species-area curve (see the text), the current rate of forest fragmentation results in a loss of species at a rate of 0.54 percent per year. If we consider a conservative estimate in which $z = 0.15$ and there are 10 million species on the planet, species are being lost at a rate of 27,000 per year. From the fossil record we can calculate that this rate is *1000 to 10,000* times the "background," or natural, rate of extinction. New estimates refine these predictions (Primm and Raven 2000). The rate of loss depends on whether the "hotspots" in Figure 12.6 are preserved. If not, Pimm and Raven predict an increase in the extinction rate to increase to a value of 50,000 species *per million species in existence per decade* by 2060, followed by a decline in the extinction rate. Humans are generating a mass extinction event not unlike some of those detectable in the fossil record. We have seen that species diversity does play some role in the stability of ecosystems. We cannot at present predict the impact of this rate of loss of species diversity.

latitudes occupied by the species (Figure 12.20). This relationship is known as Rappaport's rule (Rappaport 1982).

G. C. Stevens (1989) suggested that Rappaport's rule might be explained as follows. Species at high latitudes face a greater range of climatic conditions; they are selected to be able to cope physiologically with a great range of abiotic conditions. One result is that high-latitude species have such great tolerance of conditions that their latitudinal range can be large relative to that of more specialized tropical species. The greater specialization of tropical species allows more species to "fit in" in the tropical environment. At high latitudes, a few generalists exclude other species and thus decrease the overall species richness.

Climatic Predictability   The climatic predictability hypothesis is based on the notion that species can specialize to exploit a particular set of environmental conditions. Accordingly, regions with predictable variation in climate may support more species because many of them are able to specialize on a particular daily or seasonal set of environmental conditions. It is important that the variation be predictable so that each species can count on the appearance of its favorable period.

On the surface, this theory appears to apply to tropical-temperate comparisons systems because we tend to think of the temperate zones as seasonal and of the tropics as highly stable. In actuality, however, many portions of the tropics are highly seasonal, particularly with respect to rainfall. The difference in moisture availability between the dry and rainy seasons in many tropical systems is conspicuous— and sometimes greater than the difference between summer and winter in temperate zones. This hypothesis requires additional testing and refinement of the variables. Scheiner and Rey-Benayas (1994) showed that climatic equability affects plant species richness only in temperate zones. Richness is greatest where temperatures are most variable and where precipitation is least variable.

Structural Heterogeneity   The strong correlation between the structural heterogeneity of a habitat and species diversity may suggest that in more structurally diverse systems, species can specialize to a greater extent on differences in microhabitat. As a result, more species can be packed into heterogeneous habitats. For example, considerable evidence suggests that species diversity among birds increases with the diversity of the foliage structure in the habitat (Figure 12.21). As habitat structure (as measured by variation in height of vegetation) becomes more complex, more species of birds are present. Even though this correlation is strong and can be demonstrated for a number of organisms, we must

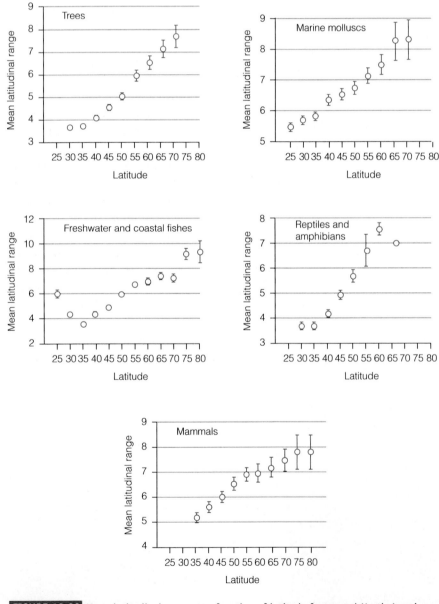

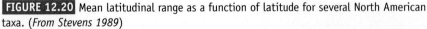

FIGURE 12.20 Mean latitudinal range as a function of latitude for several North American taxa. (*From Stevens 1989*)

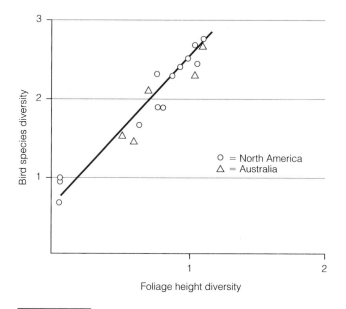

FIGURE 12.21 Index of bird species diversity as a function of foliage height diversity in North America and Australia. The regression line is based on the North America data. (*From Recher 1969*)

also note that this hypothesis simply pushes the question back a step. If a forest contains more species than a grassland, we must ask how the forest came to be present (and more structurally diverse).

Predation Certain predators have important effects on species diversity by reducing the population sizes of particular prey species. If one or a few prey species in the habitat are superior competitors and by virtue of their competitive ability cause the extinction of their competitors, a predator that reduces the density of these highly competitive species will have the effect of maintaining species diversity. The predator prevents the superior competitor species from dominating the community and causing the extinction of other species. This is the keystone predator hypothesis (Chapter 11).

Competition Increases in intensity of competition can increase species diversity. If competition in a community is very intense, there will be strong selection for species to specialize in order to avoid its effects. If many species are highly specialized, it may be possible for more species to coexist, and thus species diversity increases.

This raises the question of what causes intense competition. If populations are very large, near the carrying capacity, interspecific competition may be more acute. Thus it has been suggested that in tropical systems, many populations are near K and thus face intense competition. As we noted in Chapter 11, high-diversity communities often have highly specialized

species. The direction of causality is not clear, however. Are species specialized because they are in diverse communities, or are communities diverse because the species are specialized?

Primary Production A positive correlation between species diversity and primary productivity has led some ecologists to suggest that there might be a causal relationship between these two properties of communities. Two potential mechanisms may operate.

The first was suggested by Preston (1962), who argued that species with higher population densities have a lower risk of extinction; chance events are less likely to wipe out an abundant species. As primary productivity in a habitat increases, the population sizes of species throughout the food web also increase. Thus the least abundant species in a productive habitat may be more abundant than many species in less productive regions. The lower rate of extinction that results will lead to higher species diversity in the more productive community.

Another mechanism once again points to the importance of specialization. If primary productivity is high, food will be abundant all along the food chain. One consequence of abundant food is that organisms have more choice in terms of the food items taken and thus can specialize on particular food types. Again, specialization is thought to allow more species to coexist.

Although there are data to support the relationship between productivity and species richness (Scheiner and Rey-Banayas 1994), a number of authors have questioned the validity of this hypothesis. Rosenzweig (1992) has pointed out that the correlation between species diversity and primary productivity is not so clear as was once thought. It turns out that many graphs of species diversity versus productivity are bell-shaped (Figure 12.22), which calls for an explanation of why species diversity drops off at high productivity.

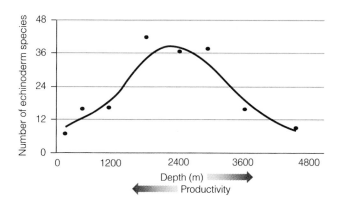

FIGURE 12.22 Species richness of echinoderms as a function of ocean productivity. (*From Rosenzweig 1992*)

In addition, some data suggest other potentially related correlations. In some cases, explanations of these correlations require logical arguments similar to those presented by Preston, but with a slightly different slant. For example, D. J. Currie (1991) showed that in terrestrial vertebrates, species richness is strongly correlated with potential evapotranspiration, or the amount of water that would evaporate from a saturated surface (Figure 12.23). In trees, it is correlated with actual evapotranspiration (Figure 12.24).

These correlations may indicate an effect of availability of energy on species richness similar to that suggested by the relationship between primary productivity and richness. If there is finite energy to apportion among species, the total amount of energy available potentially limits species richness. Potential evapotranspiration (PET) is dependent on the energy available to evaporate water (heat) and on the relative humidity; thus PET is a measure of crude ambient energy. As this quantity increases, more energy will be available to vertebrates. They will benefit from increased activity levels and foraging time in regions with high evapotranspiration. For homeotherms, energy use decreases with increased temperature (up to a point). Again, energy saving is the result. An analogous situation occurs in trees. Actual evapotranspiration (AET) is a measure of the amount of energy and water available for photosynthesis; hence it too is a measure of available energy. As AET increases, so does tree species richness.

The effects of energy limitation can be mediated through complex effects. Among plants, species diversity commonly decreases as nutrient availability increases (Pratt 1984). D. E. Goldberg and T. E. Miller (1990) hypothesize that this effect is mediated by primary productivity. As nutrients increase and are no

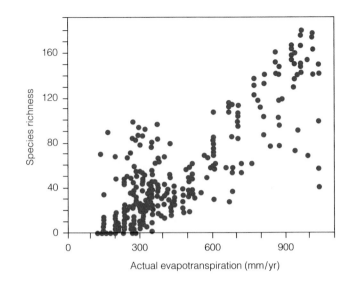

FIGURE 12.24 Species richness of trees as a function of actual evapotranspiration. (*From Currie 1991*)

longer limiting, plants grow vigorously and are more likely to compete for light. Because some species are competitively eliminated by this competition for light, energy limitation causes species diversity to decline.

Recent data suggest that production is predictive of species richness only where energy is limiting. Kerr and Packer (1997) showed that mammalian species richness is strongly correlated with potential evapotranspiration only in those North American regions that are energy-limited, specifically Alaska and much of Canada (Figure 12.25). In more southern regions where energy is abundantly available, the correlation

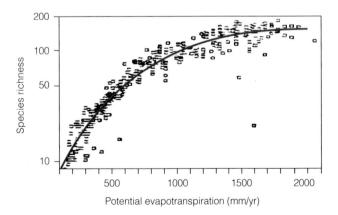

FIGURE 12.23 Species richness of terrestrial vertebrates as a function of potential evapotranspiration. Each point represents a different species. (*From Currie 1991*)

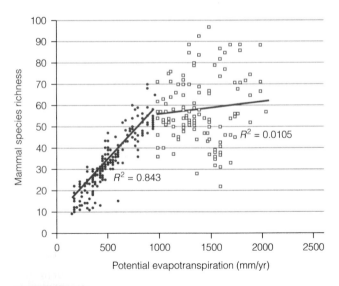

FIGURE 12.25 North American mammalian species as a function of potential evapotranspiration. Closed circles are for northern regions in Canada and Alaska; open circles are for southern portions of the continent where energy is not limiting. (*Kerr and Packer 1997*)

with PET is not significant. There, habitat heterogeneity is strongly associated with species richness.

Biotic Interactions

One of the most striking features of high-diversity tropical systems is the complexity of interactions among species. Each tropical species seemingly is involved in a complex web of parasitism, predation, mutualism, competition, and so on. Once diversity begins to increase, increased numbers of highly specialized niches may become available. Species richness increases still further as species evolve and specialize in an ever more complex system.

P. R. Ehrlich and P. H. Raven (1964) suggested a specific system in which this kind of specialization might have led to an increase in species richness. They presented a scenario in which plant evolution and insect species richness are intimately linked. Plants are known to protect themselves from insect herbivory by producing chemicals that deter insect feeding. Ehrlich and Raven hypothesize the following mechanism by which species richness expands:

1. A plant produces a novel compound that deters herbivory by insects.
2. This compound frees the plant from herbivory and thus increases its competitive advantage in the habitat.
3. The plant enters a new adaptive zone in which, free of competition, it evolves in novel directions.
4. An adaptive radiation of plants occurs in this new adaptive zone.
5. Insects evolve ways to circumvent the novel compound.
6. New species of insects evolve that specialize on these plants.
7. The insects enter a new adaptive zone and undergo adaptive radiation.

The result is an increased richness of both insects and plants. Ultimately, this radiation of new species is a result of the original interaction between a plant and its insect herbivore.

Givnish (1999) has proposed a different model that incorporates biotic interactions in the development of high species richness and illustrates the potential interaction of the large number of hypotheses advanced to explain tropical species richness. According to his hypothesis, the high rainfall and low relative seasonality in the tropics favor important plant enemies: insects and fungi. These in turn are responsible for high rates of density-dependent plant mortality and thus boost the potential for coexistence of greater numbers of tree species.

Intermediate Levels of Disturbance

J. H. Connell (1978) and Horn (1975) proposed that in some systems, species diversity is controlled by the frequency and intensity of disturbance (Figure 12.26). Connell proposed that some very diverse systems, such as certain tropical forests and coral reefs, owe their richness to disturbance. Few species can tolerate repeated disturbance, so frequent disturbance leads to low-diversity communities. In infrequently disturbed communities, one or a few species are competitively dominant and exclude other species, so species richness is low for these systems as well. At intermediate levels of disturbance, however, space is opened up by the disturbance and no species can competitively dominate. Consequently, species richness is maximal at these intermediate levels. The most diverse communities will be those with intermediate levels of disturbance.

W. P. Sousa (1979a, b) provided perhaps the best evidence of the operation of this model. In the rocky intertidal zone off the California coast, a succession of algae species grows on boulders. Waves and storms regularly disturb these boulders, creating new patches in which successional change occurs. The general pattern of succession is shown in Figure 12.27. The green alga, *Ulva*, is a pioneering species in this system; it colonizes boulders newly cleansed by disturbance. Next, boulders are colonized by several species of red algae. By the third year, a single species of red alga, *Gigartina caniculata*, dominates.

Sousa went on to examine the effects of the pattern of disturbance on the species diversity of algae. As you

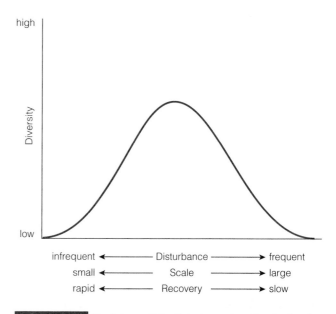

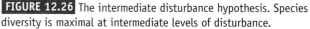

FIGURE 12.26 The intermediate disturbance hypothesis. Species diversity is maximal at intermediate levels of disturbance.

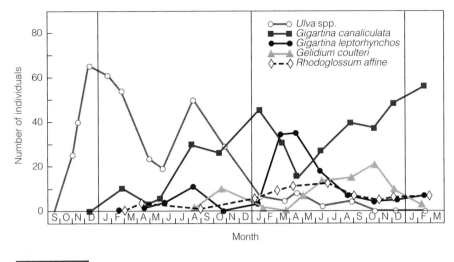

FIGURE 12.27 Colonization of boulders by algae over time in the intertidal zone of the California coast. (*From Sousa 1979a*)

might expect, large boulders are less likely to be rolled by waves and disturbed than are smaller ones. Sousa divided boulders into three classes on the basis of the force required to move them. The probability that the largest boulders would move was only 0.1 percent per month, whereas the probability of movement of the smallest boulders was 42 percent. The boulders in the three size classes thus represent low, intermediate, and high levels of disturbance. The species richness associated with each class over time is shown in Figure 12.28. Note that the intermediate disturbance class consistently had the highest species richness: Both the high- and the low-frequency disturbance classes had lower diversity.

The extent to which this theory accounts for tropical forest or coral reef species diversity is still debated. Certainly, some tropical forests are frequently dis-

turbed by hurricanes, and coral reefs are subject to the same effects of tropical storms. More experimental data will be required to assess the applicability of Connell's hypothesis to broad geographic differences in diversity.

Geographic Area The final hypothesis applies specifically to the latitudinal gradient on species diversity. Terborgh (1973) posited that tropical regions have more species simply because these regions contain more land area. The combined area between the Tropic of Capricorn and the Tropic of Cancer is approximately two times that of the other zones. Moreover, the tropics are contiguous, whereas the northern and southern temperate and polar zones are separated from each other. Thus there is no habitat barrier to the spread of species across the tropics.

According to Rosenzweig (1992), the large, contiguous area of the tropics has several consequences. First, the larger region allows greater population sizes and thus lower probabilities of species extinction. Second, the larger area provides more niches in which species can specialize. Third, the increased area provides a larger target for physical isolating barriers in the speciation process and thus higher levels of speciation. Together, these features result in tropical regions with higher speciation rates and lower extinction rates relative to the temperate and polar zones.

This hypothesis is intriguing in its simplicity. It depends on biological explanations for the differences, but underlying these factors is a simple fact of geography. It remains to be seen, however, whether this hypothesis can account for the huge diversity of the tropics, where there may be thousands of species per hectare. In fact, there are some ambiguities that make it difficult to interpret what at first glance seems to be a rather straightforward hypothesis. Implicit in Rosenzweig's assertion about the effect of area is that the crucial factor is the area of land masses *with relatively uniform temperature conditions*. In fact, the total land area is greater in the region from 23°27'N to 66°33'N, where it accounts for nearly 50 percent of the total global land mass, than it is in the tropics (Rohde 1997). The area of regions with small temperature fluctuations is greater in the tropics. Again, we must disentangle the multitude of biological, climatic, and geographic differences between the tropics and other regions if we are to understand the fantastic tropical species richness.

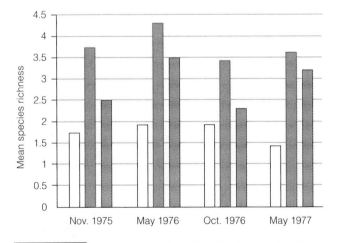

FIGURE 12.28 The average algal species richness associated with three different disturbance classes over time. Class 1 is the lowest frequency of disturbance. (*Data from Sousa 1979b*)

Consequences of Species Diversity

Thus far we have addressed the mechanisms that attempt to explain why some habitats have more species than others. Now we examine the properties that diversity per se confers on a system.

One early principle of community ecology held that diversity and stability are causally related. This principle, first espoused by C. Elton (1927) and further developed by MacArthur (1955), is based on the following reasoning:

1. Species diversity leads to complex ecological interactions; the more species that are present, the more levels of predation, competition, parasitism, and so on that can exist.
2. Complex systems are inherently more stable. The complex web of interactions lends stability because any one species in a complex network is less important than it would be in a simpler system. If we perturb the activity of a species in a complex system, the perturbation will have less effect than in a simple one. Functional redundancy is achieved via increased complexity.

This intuitive argument is attractive. In addition, it can be applied to some familiar ecological systems (Goodman 1975) including polar habitats, agricultural fields, and islands. The polar regions have low species diversity compared with the tropics; one characteristic of polar populations is that they are more likely to fluctuate widely than are tropical populations. It has also been suggested that agricultural systems exemplify this principle. Most agricultural fields are monocultures and are also inherently unstable. We must invest massive amounts of energy in the form of cultivation, pesticides, and herbicides to maintain them. According to this argument, these monocultures (low-diversity systems) are inherently unstable. Finally, islands are both depauperate in species richness and vulnerable to invasion.

This notion has also been applied to succession. Early in succession, species diversity is rather low, because only a few colonists can find the newly vacant habitat, and few of these colonists can survive there. This low-diversity community changes rapidly; by definition, it is not stable. As climax is approached, the community has much higher species diversity. We define the climax by its stability: It is a self-perpetuating state.

One of the first and most important challenges to the principle that diversity leads to stability was a theoretical paper by Robert May (1972). He was able to show that there is no mathematical reason why diversity in and of itself should *necessarily* lead to stability in ecological systems.

More recent models by McCann et al. (1998), incorporating more sophisticated nonlinear growth rates and foregoing the assumption of equilibrium, lead to a different conclusion. They show that complexity tends to stabilize systems by dampening population fluctuations and decreasing the rate of species loss.

Since MacArthur's original articulation of the concept that diversity and stability are related, additional formulations of the idea have appeared. MacArthur envisioned a direct, linear relationship between diversity and stability in ecosystem processes. Ehrlich and Ehrlich (1981) proposed the so-called **rivet hypothesis** in which species are analogous to the rivets holding an airplane together. If rivets are gradually lost, at first there will be little noticeable effect on the plane. However, at some point enough rivets have been lost that there is sudden and catastrophic failure of the structure. A related idea, known as the **redundancy hypothesis,** was proposed by Walker (1992). In this concept, some species can be lost with little consequence, because other species can expand their roles in the ecosystem to replace the lost functions. This hypothesis makes explicit the idea that particular species, not simply all species, are essential because of their ability to act as redundant components. The concept of the keystone species (Chapter 11) is related to this idea. Some species play a keystone role in which their effect is significant and greater than predicted from their abundance (Power et al. 1996). Finally, Lawton (1994) proposed the **idiosyncratic hypothesis** as a null model in which the relationship between diversity and ecosystem function is indeterminate.

A large number of empirical studies have attempted to test these hypotheses and the general notion that diversity, complexity, and stability are related. Table 12.3 summarizes a representative sample of such studies. Note that across a variety of habitat or ecosystem types, there is no consistent relationship between diversity and stability. How should we interpret these equivocal results? One thing should be clear. What appears to be a simple, intuitive argument about stability and diversity has inherent ambiguities that result from the fact that we are working at a high level of the ecological hierarchy across many spatial and temporal scales (Risser 1995). Consider the following questions: Is it reasonable to expect that all taxa should operate in a similar fashion? Does it matter whether we analyze stability at the ecosystem or population level? What do we really mean by *stability* and how can we measure it in nature? Let us look at empirical data relevant to each of these issues.

The taxonomic groups we include in the analysis may affect the results. Tilman and Downing (1994) and Tilman et al. (1996) showed that in tallgrass prairie, species richness plays a significant role in the stability of ecosystem function. Species richness was positively associated with drought tolerance, production, and nitrogen retention (Figure 12.29). In contrast, Finlay et al. (1997) showed that species richness and ecosystem function in microbes inhabiting a pond ecosystem operate in a fundamentally different way. They measured the patterns of carbon fixation and nutrient cycling in relation to microbial diversity. They found that most microbes are rare or cryptic until conditions change to favor their growth. When an ecosystem function changes to favor another species, one or more is generally available to take up the new role. For example, when conditions change from aerobic to anaerobic, a new group of anaerobic producers takes over. This is not the same as redundancy.

The level of the ecological hierarchy also leads to different results. As shown above, the stability of grassland ecosystem function is positively associated with species diversity. However, this was not true at the population level; that is, year-to-year variation in a species's abundance was not lower in more diverse plots. Apparently, the effect is mediated by competition: When abiotic factors depress competitively dominant species, their competitors increase in abundance. Ecosystem function and population processes respond differently to diversity and complexity.

Another source of ambiguity and conflicting results is the fact that the term *stability* is rather vague. Precisely what do we mean by this term? Certainly the concept was vague in the original formulation of the idea that diversity and stability are related. Some of the variation in the results presented in Table 12.3 is due to the fact that a number of different definitions of stability have been used: decreased fluctuation, recovery from disturbance, resistance to disturbance, and temporal persistence of the community or certain functions. We should not be surprised if the results of empirical studies vary with different measures of stability.

One of the first to incorporate an explicit, quantifiable definition of stability was VanVoris et al. (1980). They recognized that stability has two separate components: *resistance* and *resilience*. The first refers to the ability of a system to resist perturbation, whereas the

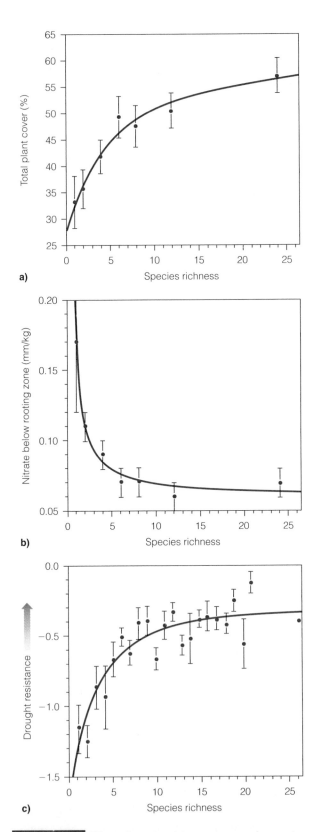

FIGURE 12.29 Effect of species richness on several ecosystem properties in tallgrass prairie. (**a**) Production (total cover) increases with richness. (**b**) Nitrate below the root zone indicates leeching of nitrogen to depths where it is unavailable to plants. Losses were less in diverse plots. (**c**) Drought resistance as a function of richness. (*From Tilman and Downing 1994 and Tilman et al. 1996*)

TABLE 12.3
Patterns of Stability and Diversity from Various Communities

Habitat	Disturbance	Stability Measure	Diversity–Stability Relationship
California annual grassland[1]	Annual variation	SC(NPP)$^{-1}$	0
Old field[2]	Annual variation	ΔNet P° Production	+
Serengeti[3] annual grassland	ΔPrecipitation	ΔSpecies composition	
Yellowstone[4] grasslands	Drought	ΔAbundance	+
British grassland[5]	Fertilization Mowing ΔPrecipitation	ΔComposition	−
Minnesota[6] grassland	Drought	ΔSpecies composition	+
Old field[7]	—	ΔDecomposition	−
Ponds[8]	—	Nutrient cycling Carbon fixation	−

[1,2]*McNaughton 1993*
[3]*McNaughton 1977*
[4]*Frank and McNaughton 1991*
[5]*Silvertown et al. 1994*
[6]*Tilman and Downing 1994; Tilman et al. 1996*
[7]*Wardke et al. 1997*
[8]*Finlay et al. 1997*

second alludes to its ability to return to normal following disturbance. They also suggested that functional complexity of the ecosystem is more important than simply the number of species.

VanVoris and co-workers (1980) designed an experiment to test the relationship between complexity and stability. On the floor of a deciduous forest, they set up a series of ten microcosms; each consisted of a sample of forest soil containing litter, soil invertebrates, and bacteria enclosed in PVC pipe. In these tiny ecosystems, they first measured the temporal pattern of carbon dioxide emissions as an index of complexity. (They reasoned that the timing and amounts of CO_2 emission are related to the complexity of ecological interactions and activities of the system's living components.) Then each microcosm was ranked on the basis of its complexity. Next, they measured the natural baseline rate of loss of calcium from each system as it functioned normally. Finally, each microcosm was perturbed by adding a sublethal dose of the heavy metal cadmium.

The pattern of calcium loss from each disturbed system was measured to quantify stability. Both the amount of calcium lost at the time of perturbation (resistance) and the time required for calcium levels to return to predisturbance levels (resilience) were measured (Figure 12.30). The area under the curve in each of these graphs represents stability as the

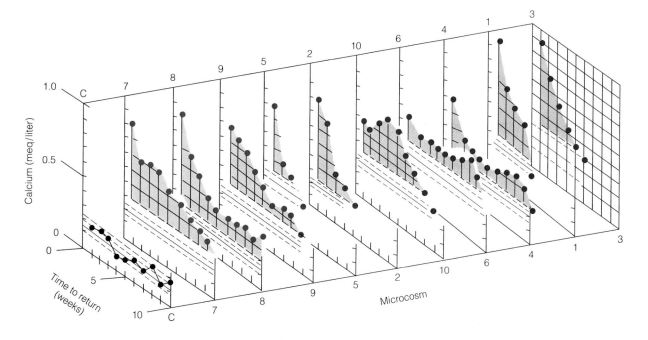

FIGURE 12.30 Patterns of microcosm stability for ten microcosms ranked by complexity. On one axis is time to return to preperturbation conditions; rapid return indicates high resilience. On the other axis is initial loss of calcium, a measure of resistance to perturbation; low loss indicates high resistance. The area under the resulting curves represents stability; smaller areas represent greater stability. Microcosm C was a control that was not perturbed. (*From VanVoris et al. 1980*)

combination of resistance and resilience. The micro-cosms were then ranked in order of stability. Clearly, there is a strong positive correlation between the complexity of the ecosystems and their stability (Figure 12.31).

D. A. Frank and S. J. McNaughton (1991) exam-ined the relationship between diversity and stability in a natural experiment created by the 1988 drought in Yellowstone Park. They too used a precise definition of stability. They measured the resistance (R_j) of the plant communities to the effects of drought (distur-bance) as:

$$R_j = 1 - \sum_{i=1}^{m} \frac{\Delta p_{ij}}{2}$$

where R_j is the resistance of community j, m is the number of species in the community, and Δp_{ij} is the change in relative abundance of the ith species in community j between 1988 and 1989. R_j is inversely related to the cumulative differences in species rela-tive abundances at a site between 1988 and 1989. In other words, if the abundances of species changed a great deal as a result of the 1988 drought, the resis-tance would be low. Frank and McNaughton then examined the relationship between the values of R_j and the Shannon–Weaver index of diversity for sev-eral communities. They found a strong positive corre-lation: Sites with high diversity as measured by the Shannon–Weaver index also had high resistance to perturbation caused by drought.

We should no longer hold as universal dogma the vague and simplistic statement

Diversity → complexity → stability

Sufficient counterexamples exist to reject this as a general and universal ecological "law." Nevertheless, there is strong evidence that for a number of precisely defined systems, there is an association among species

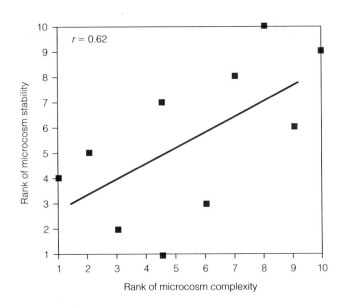

FIGURE 12. 31 Rank of microcosm stability as a function of rank complexity. Stability ranks were determined from the areas under the curves in Figure 12.30. Two values appear at complexity rank 4.5 because these microcosms had equal complexity values. (*Data from VanVoris et al. 1980*)

richness, complexity, and stability. We need an array of empirical studies at the ecosystem, population, and community levels, for a wide range of taxonomic groups, to assess the universality of the relationship. The clear connection between this topic and the importance of maintaining biodiversity argues for the immediate undertaking of such studies.

We have now considered the fundamental mecha-nisms that organize communities and account for the interactions among species. Our discussion has focused on mature communities, but along the way we have seen that disturbance plays an important role in community structure. In the next chapter we dis-cuss succession, the process of community develop-ment after disturbance.

SUMMARY

1. Species diversity, the number of species in a com-munity, is one component of biodiversity. There are several measures of species diversity that take into account the number of coexisting species and the het-erogeneity in species abundance.

2. Estimates of global species richness vary over more than an order of magnitude. We do not know how many species the Earth supports.

3. Islands are typically depauperate in species diver-sity. The MacArthur and Wilson equilibrium theory of island biogeography suggests that this pattern results because island biotas are in a dynamic equilibrium between size- and distance-dependent colonization and extinction rates.

4. The theory of island biogeography does not directly apply to habitat islands on the mainland. For many

island habitats, however, there is a species-area effect like that seen on oceanic islands. This has important consequences for conservation of species diversity in fragmented habitat.

5. A negative relationship exists between species diversity and latitude. A number of explanations of this relationship have been suggested, including the stability and predictability of the tropical climate, area effects, predation, competition, production, and time effects.

6. Ecotones, the boundary zones between two communities, typically contain more species than the two communities combined.

7. The deep ocean was once thought to contain few species, but recent research suggests that this habitat may be especially diverse.

8. There is a complex relationship between species diversity and ecological stability. There is evidence that in some systems, species diversity or functional diversity lends stability to certain aspects of the system.

SELF-ASSESSMENT: CAN YOU ...?

1. Explain the patterns of species distribution and abundance in communities.

2. Explain the hypotheses that (a) certain variables or conditions lead to lower species richness and (b) certain variables or conditions lead to higher species richness.

3. How does each of the theories in Question 2 explain high species diversity in the tropics? Which

are mutually exclusive? Which complement others? Which interact to account for increases in richness?

4. Explain why it has been so difficult to determine the relationship between diversity and stability.

5. Explain and discuss why we should preserve biodiversity.

6. What essential ecological features must we recognize if we are to maintain biodiversity?

PROBLEMS AND STUDY QUESTIONS

1. In a study of species diversity and productivity across a number of habitats, you find the relationship shown in the graph below. Do these data suggest that one or several factors determine species diversity? What could account for the fact that the curve is not a simple straight line?

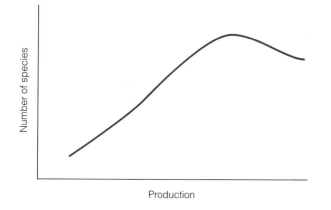

2. Compare the species diversities and heterogeneities of the following two communities:

	Community 1	Community 2
Species 1	12	23
Species 2	21	6
Species 3	18	45
Species 4	14	11
Species 5	19	36
Species 6	12	33
Species 7	17	8
Species 8	0	24
Species 9	0	6
Species 10	18	0

Which community is more diverse?

3. Which of the theories to explain high tropical species diversity are mutually exclusive?

4. Design an experiment to test one of the theories advanced to account for the latitudinal trend in species diversity.

5. Consider the following two graphs of colonization and extinction rates as functions of species diversity on an island. Which graph do you think represents real islands? Why?

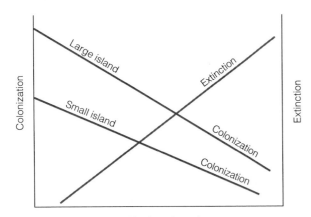

Number of species

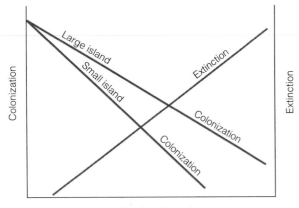

Number of species

6. Discuss the relationship between the area effect from island biogeography and the effect of sample size on estimates of species diversity.

7. Can you estimate the number of species in the community from the following data? Why or why not?

Area Sampled (m^2)	Number of Species
100	26
200	46
400	88
800	151
1600	280

8. Devise a hypothesis to explain the high diversity of species in ecotones. How could you test your hypothesis?

9. Does it matter whether we know how many species are on the Earth?

10. What are the components of ecological stability? How can we measure them?

11. Old-growth forest habitat in the Pacific Northwest is becoming fragmented by logging. A threatened species, the spotted owl, inhabits old-growth forests. Discuss the utility of the equilibrium theory of island biogeography in assessing the potential survival of the spotted owl. In what ways is the theory useful? What are its limitations?

PROJECTS AND ADDITIONAL STUDY

1. Perform a literature search on the relationship between diversity and stability for a particular taxonomic group or habitat. Analyze the differences or similarities in the results.

2. Using literature and Internet sites, determine whether some taxa (birds, for example) can be used as an index of species richness. That is, are there strong correlations in species among taxa that would allow us to use one group as in index in surveys of the patterns of species richness?

3. Choose a local habitat and taxonomic group. Design and carry out a sampling regime that allows you to test whether the community follows a log-normal distribution or supports the core–satellite hypothesis.

4. Read the following paper: Blumstein, D.T. and K.B. Armitage. 1997. *Amer. Nat.* 150:179–200. Analyze the use of the Shannon-Weaver diversity index in this system. How is this use related to, and how is it different from, its application to species diversity?

Succession

t 10:02 on the morning of August 27, 1883, a remarkable event took place. The Indonesian island of Krakatau exploded in the most violent volcanic eruption in modern history. The explosion, estimated to be equivalent to that of 100–150 tons of TNT, created a shock wave that circled the globe seven times. Eighteen cubic kilometers of rock were blasted into tiny pieces and ejected into the atmosphere, where they circled the planet and caused spectacular sunsets for years. In a moment, most of the island disappeared. All that remained was a small remnant island, Rakata (Figure 13.1), which was completely sterilized by the explosion.

Nine months later an expedition to the island discovered only a single living thing, a lone spider. Now, with the regular addition of species over the hundred

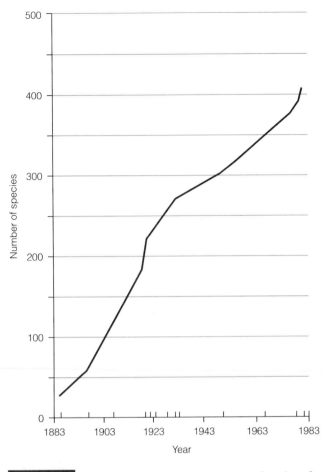

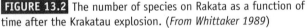

FIGURE 13.2 The number of species on Rakata as a function of time after the Krakatau explosion. (*From Whittaker 1989*)

years and more since the catastrophe (Figure 13.2), forest has returned.

The development of a community after a disturbance, the process that occurred following the explosion of Krakatau, is the subject of this chapter. We will explore several related questions: How does this process occur? What is the effect of the magnitude of the disturbance? Does the process have a definable endpoint? What mechanisms drive the process?

Succession Defined

Succession was originally defined by F. E. Clements as continuous, directional change in the species composition of a community by extinction and colonization, *leading to a single, ultimate community* (Clements 1916). Unlike Clements, who believed there was a single, repeatable successional pathway in each region of the country, modern ecologists no longer emphasize continuous, directional change to a single endpoint. Instead, we define **succession** as *sequential change in*

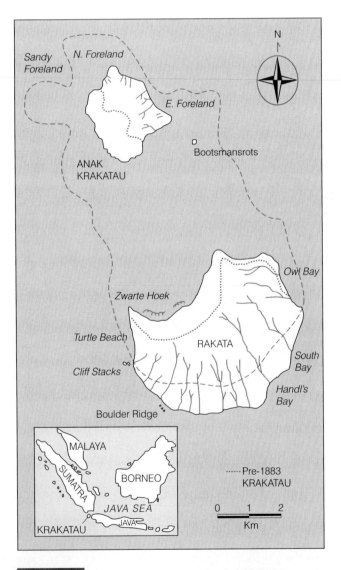

FIGURE 13.1 Old Krakatau (shown by the dashed line) was destroyed by a volcanic eruption in 1883, leaving only Rakata, a lifeless remnant, at the southern end. Anak Krakatau emerged from the sea as a volcanic cone in 1930. (*From Wilson 1992*)

the relative abundances of the dominant species in a community following a disturbance (Huston and Smith 1987). Because of variation in the biotic and abiotic factors associated with each case, the sequence may not be the same each time succession occurs. By *sequential,* we mean that as the community changes, once-dominant species will not predominate again unless there is another disturbance.

Encompassed by this concept are the processes of degradative succession and autotrophic succession. In **degradative succession,** organic material is degraded by detritivores and microorganisms. The organic material can be the body of a dead animal, a fecal deposit, a log, or any other bit of organic matter that has been deposited or that has died. In **autotrophic succession,** a new community develops in an empty habitat, usually one that has been emptied by some kind of disturbance. The succession can be a result of colonization by plants or, in some cases, by sessile animals.

There are two fundamental types of autotrophic succession. In **primary succession,** the community develops from an essentially abiotic situation following a cataclysmic disturbance. The disturbance may even remove or destroy the soil itself. Thus the process of primary succession may involve soil formation, potentially a very slow process. Examples of primary succession include the development of communities following glaciation, volcanic eruption, or receding waters that expose new substrate to colonization. In **secondary succession,** the community develops in a habitat that has been disturbed, but not so severely as to destroy all life. Examples include community development after a fire or hurricane. Because soil usually still remains, a community can develop in much less time than in primary succession. The later stages of primary succession and the stages of secondary succession, when they occur in the same region, are often very similar in terms of the appearance of the community and the nature of the mechanisms of change in community composition.

The entire sequence of stages in succession is referred to as a **sere;** any particular phase is a **seral stage.** The final stage of community change is called the **climax.** By definition, the climax persists and reproduces itself until the next disturbance moves succession back to an earlier seral stage. We will discuss the nature of the climax in greater detail later in this chapter. In some habitats, recurring disturbance continually returns succession to an earlier stage. As a result, this stage may appear to be self-perpetuating. An apparent climax that is in fact an earlier seral stage maintained by disturbance is called a **disclimax.**

The Role of Disturbance in Communities

We define **disturbance** as any relatively discrete event in time that disrupts ecosystem, community, or population structure and changes resource or substrate availability or the abiotic environment (Turner et al. 1997; White and Pickett 1985). Disturbances that initiate succession kill or remove organisms from the community. Consequently, there are changes in the biotic interactions of the system. In many cases, removal of species also has abiotic consequences. For example, in 1989 Hurricane Hugo removed leaves from the canopy of the tropical forest in Puerto Rico, causing a more than tenfold increase in light intensity on the forest floor (Walker, Volzow, and Akerman 1992). It is these transformations that lead to successional change.

A great variety of factors can disturb communities. Weather-related disturbances include thunderstorms, ice storms, tornadoes, and hurricanes. Volcanic eruptions are effective disturbances, whether they occur as catastrophic explosions, as on Krakatau and Mt. St. Helens, or as slowly emerging lava, as in Hawaii. Fire is a major disruptive force in many communities. Until the 1970s, the conventional wisdom among ecologists was that most ecological systems are in equilibrium; that is, they have reached some endpoint state that is persistent. Disturbance, though known to occur, was thought to represent an irregular and infrequent event. This view began to change in the 1970s (Egerton 1973), however, as ecologists gathered evidence of the regularity and frequency of disturbance. We now understand that disturbance is frequent and that, consequently, many communities are often in various stages of change following some sort of perturbation. Under such a nonequilibrium view, the process of succession takes on added importance.

Ecologists have collected ample evidence, from diverse communities, of the frequent occurrence of disturbance. In the taiga of interior Alaska, for example, fire is a recurrent disturbance. Prior to 1945 when aircraft contributed to fire suppression, 10 million hectares in that region burned each year, and no forest there was more than 200 years old. In the montane rain forests of Puerto Rico, landslides caused by the heavy rains and shallow soils account for considerable disturbance. It has been estimated that 1 percent of the forest on slopes slides each year. Thus any particular area can expect this form of disturbance at 100-year intervals. Satellite images depicting blowdowns of tropical rain forest by strong thunderstorm downdrafts indicate that 0.02 percent of the landscape

a)

b)

c)

FIGURE 13.3 (a) Extensive blast damage to the conifer forests surrounding Mt. St. Helens. (*Photo by John Lemker/Earth Scenes*) (b) Variation in fire intensity of the 1988 Yellowstone fire. In the dark center, hot fire destroyed all trees. The brown area surrounding the center represents hot fire but less severe damage. Other areas show a mix of living and dead trees in regions of cooler fire. Some stands were entirely unaffected by fire. (*Photo by David Krohne*) (c) Damage to the canopy of tropical forest on Puerto Rico by Hurricane Hugo. Note that many trees were denuded of leaves, whereas others, especially in the understory, were relatively unaffected. (*Photo by David Krohne*)

in the Amazon Basin is affected annually (Nelson, Kapos, and Adams 1994). Thus even environments that we think of as highly stable are subject to repeated and relatively frequent disturbance.

The effect of a disturbance depends on the nature of the disturbance and the type of habitat (Figure 13.3). Obviously, a cataclysmic explosion like the eruption of Mt. St. Helens has a greater effect than the fall of a few trees in a windstorm. We can characterize disturbances according to their severity, frequency, and spatial extent as depicted in Figure 13.4. Tree falls are quite frequent, but they are neither severe nor extensive in their effect. In contrast, hurricanes are relatively infrequent, but their effects can be severe and extensive. Even large, severe disturbances differ significantly in their effects. For example, the Yellowstone fires of 1988, Hurricane Hugo in Puerto Rico, and the explosion of Mt. St. Helens were all large and devastating disturbances, yet they differed in a number of important ways (Table 13.1). Note in Figure 13.3 the local variation in fire intensity. Turner et al. (1997) concluded that for large disturbances, severity affects the nature of succession more than the size of the disturbance.

The effect of disturbance factors varies widely across habitat types. Consider disturbance by wind. When Hurricane Andrew passed directly over the Everglades in 1993, the sawgrass community was virtually unaffected because of the plants' flexible stems. When Hurricane Hugo crossed the island of Puerto Rico, it damaged a large number of trees. Even so, the long-term effect was much less than in the pine forests

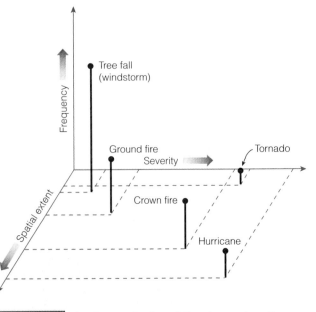

FIGURE 13.4 The characterization of disturbance along three dimensions: frequency, severity, and spatial extent.

TABLE 13.1
Characteristics of Three Types of Disturbance

Characteristic	Crown Fire (Yellowstone)	Hurricane (Hugo, Puerto Rico)	Volcano (Mt. St. Helens)
Area of impact	250,000 ha	All of Puerto Rico, 10^6 ha in South Carolina	700 km²
Resulting habitat heterogeneity	Coarse-grained mosaic	Fine-grained mosaic	Very coarse-grained at landscape level
Spatial predictability	Unpredictable because of topographic effects	Predictable	Predictable, based on blast direction
Return time of disturbance	100–500 years	60–200 years	100–1000 years
Differential direct mortality	No	Greater mortality of early successional species	Often no

of South Carolina. Seven weeks after the hurricane, 70 percent of the affected trees in Puerto Rico had produced new leaves (Walker, Volzow, and Akerman 1992). Similarly, four months after the 250-km/hr winds of Hurricane Joan passed over Nicaragua, more than 75 percent of the trees in primary forest had resprouted (Boucher 1990). In contrast, Hurricane Hugo caused extensive mortality in the pine forests of South Carolina. There the trees were snapped in two and could not recover. Position in the habitat also plays a role. Figure 13.5 shows the pattern of new

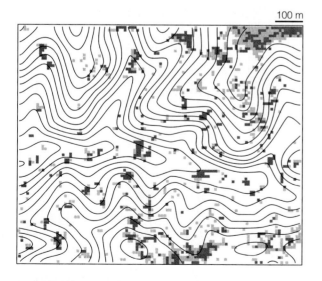

100 m

FIGURE 13.5 Map of gaps in a Japanese temperate deciduous forest. Light green squares represent gaps that existed during 1981. Dark green squares are gaps forming in the period 1981–1986. Note that the newly formed gaps tend to occur at the edges of the gaps existing in 1981. (*From Tanaka and Nakashizuka 1997*)

canopy gaps in temperate deciduous forest in Japan. Note that the majority of the newly created gaps occurred at the margins of existing gaps (Tanaka and Nakashizuka 1997).

The seral stages that follow disturbance depend on the type of habitat that was originally present. Each region and habitat type has a unique sequence of successional changes. We will not attempt to catalog all the patterns that have been described. Instead, we will discuss a few well-studied examples that can provide a basis for considering the mechanisms of successional change.

Patterns of Successional Change

This section presents two examples each of both primary and secondary succession. The understanding we gain here of the nature of the changes that take place in succession will prepare us to examine, in the next section, the mechanisms that drive the process. The differences between the patterns of change in different regions are also instructive.

Primary Succession

Dune Succession   One of the first systems of primary successional change to be described was the sequence that occurs on sand dunes. In fact, H. C. Cowles (1901) first described succession and understood it as a process of community development on a sere at the Indiana dunes on Lake Michigan. Cowles was able to infer that a sequence of changes had occurred on the dunes—a conclusion that other ecologists quickly appreciated and that led to descriptions of succession in other habitats. Cowles's insight was a major factor in the development of community ecology.

To understand the successional changes that occur at the Lake Michigan dunes, one must understand that the lake has not always had the same shoreline we see today. After the Pleistocene glaciers receded 12,000 years ago, a much larger lake, Lake Chicago, occupied a depression gouged by the glacier and filled with glacial meltwater. The shoreline of this lake extended much farther south than that of the present Lake Michigan. Unlike Lake Michigan, which drains to the north through the Straits of Mackinac and into Lake Huron, Lake Chicago drained to the south into the Illinois River and ultimately into the Mississippi. As water flowed out of the lake, its shoreline receded.

The river draining the lake cut down through the debris piled up by the glacier. At various times, this erosive water encountered layers of substrate that were harder and more resistant to erosion, and the lake level stabilized for a time, until the river cut into softer layers and again drained the lake more rapidly. Eventually, the lake dropped low enough so that it no longer drained to the south, and a northern outlet formed. At that time, the flow of water came to resemble what we see today.

Thus, on several occasions in the past, the lake level remained relatively constant for some time and then receded further, exposing new sand substrate in the process. One consequence of this geological history is that as we move inland from the present-day shoreline, we travel through progressively older habitats that at some time in the past were newly exposed shoreline. Cowles recognized that the dunes on the southern shore of Lake Michigan in effect represent a transect through time. The dunes near the shore represent areas undergoing the first stages of primary succession. Those just inland from these dunes have already passed through that first stage and have entered later stages of the process. As we continue inland, we reach older and older habitats. Of key importance is the fact that *all these habitats passed through similar earlier stages.* This was Cowles's great insight.

In this system there are five distinct zones of vegetation as one proceeds inland from the beach (Figure 13.6). Immediately adjacent to the shore is the foredune, a habitat battered by storms, waves, and wind. The substrate—sand newly washed up from the lake—contains little organic material. Any plants trying to grow within reach of the waves are regularly flooded and physically pounded, especially during winter storms. Just back from the breaking waves, the sand is loose and unstable. Plants that try to grow here face the problems of very high light intensities, sand with poor water-holding capacity, low levels of organic matter and nutrients, and a shifting substrate. Gradually, a few hardy species such as Marram grass (*Ammophila brevigulata*) colonize and spread by rhizomes, thus stabilizing the sand. If buried by sand, Marram grass can sprout a new stem near the surface (Figure 13.7).

Immediately inland from the foredune is the first stabilized dune. Here, Marram grass has stopped the sand from moving, and the root mass and dead plants have provided some organic matter in the initial stage of soil building. Other species can colonize, including little bluestem (*Andropogon scoparius*). Occasionally, a stabilized dune will be disturbed by an animal or humans. If the matrix holding the sand in place is

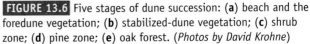

FIGURE 13.6 Five stages of dune succession: (**a**) beach and the foredune vegetation; (**b**) stabilized-dune vegetation; (**c**) shrub zone; (**d**) pine zone; (**e**) oak forest. (*Photos by David Krohne*)

FIGURE 13.7 Marram grass rhizomes can resprout if buried by sand. (*Photo by David Krohne*)

destroyed, then a **blowout**—an area of shifting sand devoid of vegetation—occurs, and the process of stabilization must begin again.

The next seral stage is the shrub stage, which includes species such as dune willow (*Salix syrticola*), sand cherry (*Prunus pumila*), and grape (*Vitus* spp.). These species require well-stabilized dunes and more soil than the earlier pioneers. They add more nutrients to the soil and further improve its ability to hold water. They also cast some shade, reducing the high light intensities characteristic of earlier stages.

A zone of pines is located inland from the shrub region. The main species inhabiting this area are white pine (*Pinus strobus*), jack pine (*Pinus banksiana*), and junipers (*Juniperus* spp.). These species are able to grow in the still rather harsh, xeric conditions. They require few nutrients, but more than the shrub stage they replace.

Finally, there is a region of deciduous trees dominated by black oak (*Quercus velutina*). Also present are northern red oak (*Quercus rubra*), various species of hickory (*Carya* spp.), and witch hazel (*Hamamelis virginiana*). This region looks much like the surrounding deciduous forest. Over the course of 12,000 years, a forest has developed on what was once the unstable sand of an ancient lake shore.

Succession After Recent Glaciation Another classic example of primary succession is the sequence at Glacier Bay in Alaska (Cooper 1939; Lawrence 1958; Bormann and Sidle 1990). Again, we need an understanding of the geology of the region to discuss this sere. Glacier Bay lies at latitude 58°N in southeastern Alaska in a region containing a number of active glacial systems. We tend to emphasize the major Pleistocene glaciations, such as the most recent Wisconsin Glacial Period that ended about 9000 years ago, but over the past million years, a series of lesser glacial advances and retreats have taken place, each with a period of about 10,000 years. There is evidence that nested within the most recent 10,000-year cycle was a series of minor advances and retreats; these "little ice ages" occurred at 1400–1300 B.C., 900–300 B.C., A.D. 400–750, and A.D. 1200–1300 (May 1979). Thus disturbance by glaciation and the subsequent periods of primary succession have been frequent and important influences on the northern landscape.

The glaciers of Glacier Bay are unusual in that they have receded very quickly over the past 200 years (Crocker and Major 1955). When Captain George Vancouver discovered the bay in 1794, he found a wall of glacial ice at Muir Inlet; in the 200 years since then, the glaciers have receded some 80 miles (Figure 13.8; 13.9). In the process they have exposed miles of

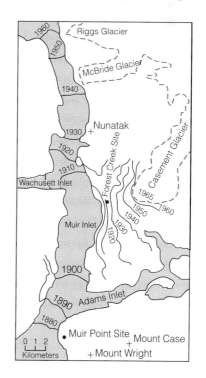

FIGURE 13.8 Migration of the ice front in Muir Inlet at the northeastern extremity of Glacier Bay, Alaska. Dates indicate the snout of the glacier from 1860 to 1960. (*From Bormann and Sidle 1990*)

FIGURE 13.9 Recession of the Muir Glacier from 1978 (**a**) to 1997 (**b**). Note how far the glacier front has receded in the 19 years between these photos. (*Photos by David Krohne*)

FIGURE 13.10 New granite exposed by a retreating glacier in Glacier Bay, Alaska. (*Photo by David Krohne*)

new substrate on which primary succession is taking place (Figure 13.10).

Much of the substrate in this region is granite. Thus, when a glacier recedes, the exposed substrate is hard and devoid of soil. This granite presents a different set of problems for pioneering plants than dunes present, with their unstable, shifting sand. At Glacier Bay, plants must colonize an area of smooth, hard rock with few crevices to hold seeds, water, or organic matter. Not surprisingly, the first colonists are usually lichens, which can cling to the rock surface. They provide some microtopographic relief so that dust and water can accumulate in small amounts. This material—particularly the water, which freezes (and expands) and thaws over the seasons—begins to break up the rock. Seeds of higher plants can cling to this poor "soil" and continue the process of succession. Jumpponen et al. (1999)

recently showed that early colonists of deglaciated sites are most successful if they colonize microsites that are concave, of coarse substrate, and near large rocks. Such sites not only are more likely to trap seeds, but they also protect seedlings from desiccation.

Among the first higher plants to colonize is *Dryas drummondi,* a small plant with reduced leaves and photosynthetic stems. This plant is particularly important because it is a nitrogen fixer, which provides much-needed nitrogen to the developing soil. Sitka alder (*Alnus sitchensis*) colonizes soon afterward and reaches its peak abundance after some 50 years (Figure 13.11). This species is also a nitrogen fixer and further fertilizes the soil. At about 60 years, Sitka willow (*Salix sitchensis*) and black cottonwood (*Populus trichocarpa*) invade. They increase the shade, and as they die, they add organic matter to the soil. At about 80 years, any alder that remains has been replaced by Sitka spruce (*Picea sitchensis*), which dominates until the site is 160 years old. At this time the final species, hemlock (*Tsuga heterophylla*), takes over and forms a dense climax forest.

Thus, as in the dunes, a transect from the mouth of Glacier Bay toward the active glaciers represents a transect through time. As one proceeds up the inlet, progressively younger habitats are encountered. Note that the pioneering species in this sere, like those in the dunes, must be hardy plants capable of withstanding desiccation, high light intensities, and low levels of soil nutrients and organic matter.

For some time we believed that Glacier Bay represented a sequence of communities of different ages that had passed through the same set of developmental changes, much like the sequence at the Lake Michigan dunes. Recently however, Fastie (1995)

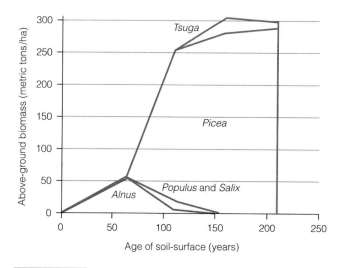

FIGURE 13.11 Changes in above-ground biomass over time for different plant species in succession in Glacier Bay, Alaska. (*From Bormann and Sidle 1990*)

showed that in fact, there are three different successional sequences represented in Glacier Bay. Fastie used tree-ring data to compare the vegetation histories and age structures at ten sites along a transect in Glacier Bay. He found that the oldest sites, all of which were deglaciated prior to 1840, differed from all the rest in that Sitka spruce and western hemlock established themselves there very early in succession. Nitrogen fixation by alder was a significant factor only in sites established after 1840. At the youngest sites (less than 90 years old), black cottonwood (*Populus trichocarpa*) dominated succession. The main factors leading to these differences were nearby seed sources. In the oldest sites, near the mouth of the bay, seeds from old-growth Sitka spruce and western hemlock forests were readily available, so these species colonized earlier there than at other sites. Fastie hypothesizes that in sites deeper in the bay where cottonwood dominated, nearby more xeric inland sites contain large source populations of cottonwood and their highly dispersive seeds. Thus Glacier Bay does not represent a single sequence of successional change. This study reminds us not to assume that, in every place where we can make a linear transect through progressively younger sites, all parts of that transect have experienced the same pattern of community development.

Secondary Succession

If the disturbance leaves some soil and living organisms, the development of a new community is called secondary succession. Often this process is not so lengthy as primary succession, and it may resemble the later stages of primary succession.

Old Field Succession One of the most thoroughly studied examples of secondary succession is the sequence called **old field succession,** a name that derives from the fact that it occurs on abandoned agricultural fields. Old field succession was one of the first examples of secondary succession described, and numerous studies have documented the patterns of change and, as we will see later, some of the important mechanisms of change.

Old field succession has been most thoroughly described in the midwestern and eastern United States, where agricultural land abounds. The exact species present in the various seres will, of course, vary with the local conditions and the nature of the climax deciduous forest common in the region. Nevertheless, the general sequence is quite consistent across the region, and we can describe it in general terms. An old field secondary successional sequence from Indiana is shown in Table 13.2.

One of the major differences between this type of succession and the examples of primary succession that we have discussed is that the physical conditions are much more benign early in secondary succession than early in primary succession. Although the habitat may have been depleted of nutrients by agriculture, there is soil present to provide some nutrients, water-holding capacity, and a reservoir of seeds. Consequently, secondary succession can often proceed far more rapidly than many primary successional sequences.

TABLE 13.2

Typical Time and Species Sequence in Midwestern Old Field Succession

Stage	Time Since Disturbance (yr)	Typical Species
Annuals	0–15	Foxtail (*Setaria* sp.) Ragweed (*Ambrosia* sp.) Queen Anne's lace (*Daucus carota*)
Perennials	16–25	Broomsedge (*Andropogon virginanus*) Aster (*Aster* sp.) Goldenrod (*Solidago* sp.)
Shrubs	26–50	Sumac (*Rhus* sp.) Multiflora rose (*Rosa multiflora*) Blackberry (*Rubus* sp.)
Softwood trees	51–100	Tulip poplar (*Liriodendron tulipifera*) Cottonwood (*Populus deltoides*) Red maple (*Acer rubrum*) Eastern red cedar (*Juniperus virginiana*)
Hardwood trees	1001	Sugar maple (*Acer saccharum*) American beech (*Fagus grandifolia*) Red oak (*Quercus rubra*)

Succession in Lodgepole Pine Forest   In recent years we have come to understand much more about the patterns of succession in coniferous forests. One impetus for this work was the series of major fires that burned in Yellowstone National Park in the summer of 1988. Approximately 33 percent of the park's 2.1 million acres experienced fire of some intensity during that summer. Because of Yellowstone's importance as a natural reserve, there is great interest in the nature of succession in this biome and at this particular locale. Consequently, a rather detailed description of the nature of succession has emerged (Romme and Despain 1989).

Much of Yellowstone National Park lies on a plateau of relatively uniform elevation (for a mountainous region, at least). Much of the region is an

FIGURE 13.12 The four stages of lodgepole pine succession in Yellowstone National Park. For an explanation of each stage, see the text. (*Photos by David Krohne*)

ancient volcanic caldera that sank after exploding. Although many life zones are represented in the park, much of the park lies at between 2500 and 3500 meters in elevation and consists of extensive lodgepole pine forests. By taking core samples from the trees, it is possible to obtain information on the age of a particular stand and (sometimes) something of its fire history. The latter is indicated by the presence of fire scars among the rings of the tree.

In extensive studies of this type across Yellowstone, W. H. Romme and D. G. Despain (1989) found that there is a pattern of recurrent small fires at 20- to 30-year intervals and that cataclysmic fires like those of 1988 are rare. But there is evidence that a fire on that scale occurred in 1690.

Romme and Despain describe a series of four successional stages after a fire in this region (Figure 13.12a). In Stage I (0–50 years), the stand is dominated by young lodgepole pines that germinated soon after the fire. Lodgepoles are well adapted for this system in that they produce two types of cones: those that open in the year they mature and serotinous cones. The latter are coated by a thin layer of wax that melts at 130°F, allowing the cone to open. Serotinous cones may remain closed on the tree for many years, until a fire causes them to open and shed their seeds. Thus the first stand of trees after a fire is often a monoculture of lodgepole pines. Growth rates are high because of the high light intensities. This first stage is relatively resistant to fire because there is little dead wood on the ground for fuel and the saplings are widely scattered.

Stage II (Figure 13.12b) is characterized by the maturation of these trees. A dense, closed canopy forms that excludes light, prevents the growth of seedlings of lodgepoles or other species, and thus maintains low species diversity. The shade kills the lower branches, and they are gradually pruned from the trees. This stage looks almost like a tree farm; it is a monoculture of same-aged lodgepoles with little or no understory. This stage is also relatively resistant to fire, again because there is little fuel on the ground (few trees have died and fallen). In addition, the lack of lower branches means that there is no conduit for flames up to the canopy. Consequently, it is difficult for a crown fire to begin. This stage lasts from about 50 to 150 years after fire.

As some of the trees in Stage II begin to die, Stage III begins. Trees mature and weaken, and many are killed by bark beetles. Now there are a few gaps in the canopy where lodgepoles have died and been windthrown. Light reaches the forest floor, and seedlings of

lodgepoles, as well as of Douglas fir (*Pseudotsuga menziessi*) and subalpine fir (*Abies lasiocarpa*), germinate and grow toward the canopy (Figure 13.12c). Fire is more likely because there is fuel on the forest floor (logs) and the young trees form a conduit for flames up to the canopy.

Stage IV begins after approximately 250 years. At this time lodgepoles are reaching their maximum life span, and death and windthrow are common. Douglas fir and subalpine fir are also maturing and dying. The resulting light gaps allow regeneration of spruce and fir. There is a great accumulation of litter and logs on the ground, and there are many standing dead trees (Figure 13.12d). This stage of succession is very vulnerable to fire, and cataclysmic fires can be propagated through such a stand.

Romme and Despain's studies at Yellowstone have shown that approximately 300 years ago, a major fire (or fires), probably much like those of 1988, burned across the Yellowstone landscape. By 1988, much of the Yellowstone ecosystem was in or near Stage IV— and thus was highly vulnerable to fire. It is likely that a cycle of fires like this has been operating in the region for some time (Figure 13.13; also see the box on page 334).

As in our previous examples, a number of successional pathways occur, depending on the local conditions, in this case especially the scale and severity of

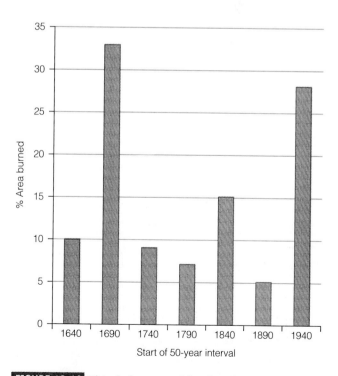

FIGURE 13.13 Historical pattern of fires in Yellowstone based on tree-ring data. (*Data from Romme and Despain 1989*)

The Yellowstone Fires and Fire Policy

For several weeks in the summer of 1988, television news showed us dramatic footage of the devastation a series of wildfires brought to Yellowstone National Park. Nearly a third of the park, some 700,000 acres, burned that summer. The public response was outrage: How could this happen to our most treasured national park? Politicians, including President Bush, demanded to know what policies and practices had led to this conflagration.

Understanding the principles of succession can help us put the origin and effects of those fires into perspective. As the text explains, there are four stages of lodgepole pine succession in the Yellowstone ecosystem. The probability of fire increases dramatically in the later stages. Data from fire scars in old trees indicate that extensive fires last occurred in Yellowstone around 1690. By 1988, much of the park was in the third or fourth stage of succession (Figure 13B.1). Thus, in the summer of 1988, Yellowstone was highly susceptible to fire.

Another factor that contributed to the fires was the unusual weather that summer. Although spring precipitation was normal, the summer was unusually hot and dry. The moisture content of dead wood on the forest floor had declined as low as 7 percent by late July. In midsummer a series of dry lightning storms passed over the park. On July 5, 1988, for example, over 2000 lightning strikes were recorded in Yellowstone, some of which started fires. Other fires started outside the park as a result of human activity. By late August the fires in the park could not be controlled. It was possible only to save buildings and people. The fires were not fully extinguished until the first snowfall in September.

The National Park Service came under heavy criticism for its "let-burn" policy in the park. This policy recognized fire as a natural and important factor in conifer forests. Under the policy, any fire caused by lightning was to be allowed to burn unless it threatened people or property. All fires caused by humans, and thus defined as unnatural, were to be put out immediately. In actuality, this policy had relatively little to do with the development of the fires that summer. Two of the major fires were started by humans, and attempts were made to extinguish them from the outset. Given the stage of succession in most of the park and the unusual climatic conditions, these efforts were futile. A major fire was nearly inevitable.

It is also unlikely that attempts at total fire suppression earlier in this century before the "let-burn" policy was adopted allowed fuel to build up to the point where a major conflagration resulted. Much of the Yellowstone backcountry is rather inaccessible, and efforts to suppress fires there were relatively unsuccessful until after aircraft were introduced following World War II. The 40-year period of fire suppression before 1988 was insufficient to allow large amounts of fuel to accumulate. Some interesting policy issues and questions were raised by the events of 1988 in Yellowstone. The fire policy in place in 1988 had been developed as a response to a growing appreciation of the natural role of fire in ecosystems. But one of the reasons why the park was set aside in the first place was that people appreciated the beauty of the Yellowstone landscape, and the miles of green forest accounted for much of this attraction. In essence, the public wanted to preserve a single moment in a long-term pattern of forest succession—Stages III and IV. But we know that the natural processes of succession and disturbance make such stages ephemeral; almost inevitably, those stages will pass. It is unlikely that any fire policy could have prevented the fires of 1988. Our park policies and the public that uses and supports the parks must recognize the difference between preserving a particular moment in the process of succession and preserving a system in which natural processes take place.

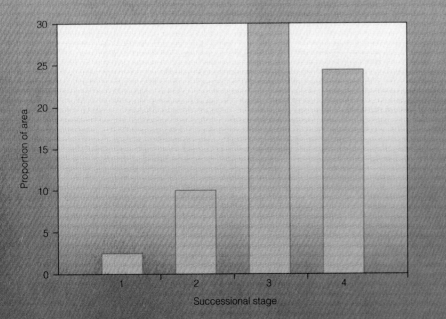

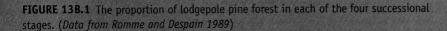

FIGURE 13B.1 The proportion of lodgepole pine forest in each of the four successional stages. (*Data from Romme and Despain 1989*)

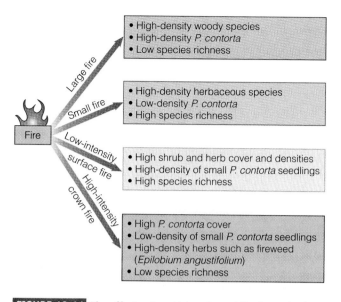

FIGURE 13.14 The effects of spatial scale and fire intensity in the Yellowstone fires of 1988.

the disturbance (Turner et al. 1997). Note in Figure 13.14 that extensive burned areas and high-intensity crown fires generally lead to high densities of lodgepole pine seedlings and low species richness. Smaller fires and cooler surface fires lead to lower numbers of lodgepoles and to extensive stands of many herbaceous species.

Variation in the Patterns of Succession

The examples just described have been presented as linear sequences of change in the vegetation over time; thus they closely fit our definition of succession. As we have learned more about succession, however, we have come to realize that simple, linear sequences like these are not complete descriptions of successional change. A number of important local factors influence the exact series of species replacements in a sere.

The sequence of habitat types that Cowles described as so clearly linear is in fact much more complex (Olson 1958). The exact set of species transitions depends on the initial conditions. Moisture levels as well as the aspect and degree of slope affect not only what the paths of change will be but also the final climax (Figure 13.15).

Another example of variation in the sequence of successional change is shown in Figure 13.16 for pathways of change in taiga in interior Alaska. Fire is the most common disturbance. Differences in slope have a major impact on the sequence of changes in succession. On north-facing slopes, where conditions are cooler and moister and where permafrost may remain, black spruce dominates immediately after

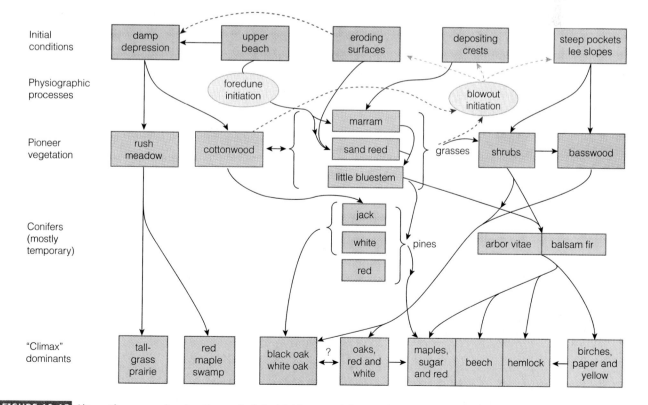

FIGURE 13.15 Alternative successional pathways in Lake Michigan sand dunes. The center of the diagram is a simplified outline of "normal" succession from pioneers through jack or white pine to black oak and white oak. (*From Olson 1958*)

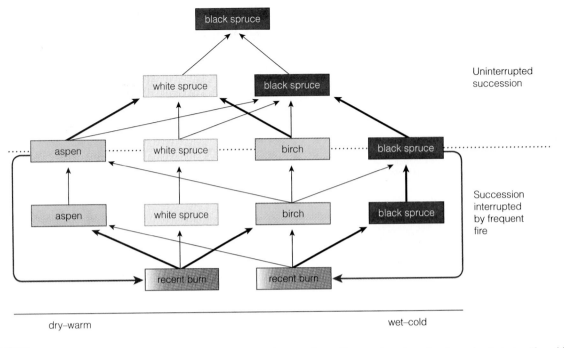

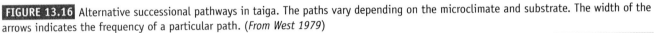

FIGURE 13.16 Alternative successional pathways in taiga. The paths vary depending on the microclimate and substrate. The width of the arrows indicates the frequency of a particular path. (*From West 1979*)

disturbance. On warmer, drier, south-facing slopes, succession passes through stages containing aspen, birch, and white spruce before reaching the black spruce climax.

These examples illustrate the kinds of variation that can occur in the sequence of seral stages. As we will see, the final climax is also potentially variable. Any descriptions of the mechanisms of successional change must be able to account for this important variation.

Mechanisms of Successional Change

Two things should be apparent from our examples of primary and secondary succession: (1) The kinds of species present (trees versus grass, softwoods versus hardwoods, and so on) change over the course of succession, and (2) the nature of the physical environment (development of soil, increasing shade, addition of nitrogen and organic matter to the soil, and so forth) also changes dramatically. Our explanation of successional change must account for these changes. Moreover, the mechanisms involved should answer an important question: How do the changes in the vegetation and the physical environment interact? We will explore several models of succession.

Connell–Slatyer Models

J. H. Connell and R. O. Slatyer (1977) organized the mechanisms of succession into three broad categories (Figure 13.17): facilitation, tolerance, and inhibition.

Facilitation   The first category is the **facilitation model.** In this model, the pioneering species modify the physical environment in such a way as to facilitate colonization by later succession species. From our consideration of primary succession, we can imagine the kinds of abiotic changes that would facilitate succession. As Marram grass stabilizes a sand dune, for example, a number of changes in the physical environment ameliorate the harsh conditions of the foredune. The rhizomes stabilize the sand; the grass casts some shade, which reduces the high light intensities, increases humidity near the ground, and modifies surface temperatures. As the Marram grass dies, its stems and roots add organic matter to the soil. As a result, the seeds of other grasses that colonize after the dune is stabilized find a more benign environment.

Lichter (1998) showed that a dune successional sequence on the eastern shore of Lake Michigan required approximately 300 years. In that sequence, a number of physical changes in the environment, caused by vegetation changes, greatly improved the abiotic conditions and permitted the development of a forest (Figure 13.18).

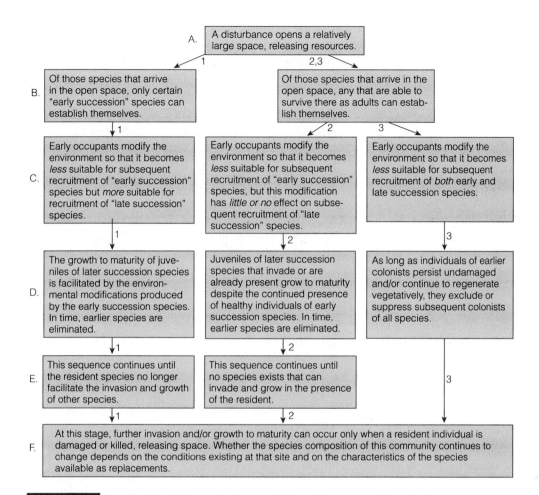

FIGURE 13.17 Three models of the mechanisms producing the sequence of species in succession: (1) facilitation, (2) tolerance, (3) inhibition. (*From Connell and Slatyer 1977*)

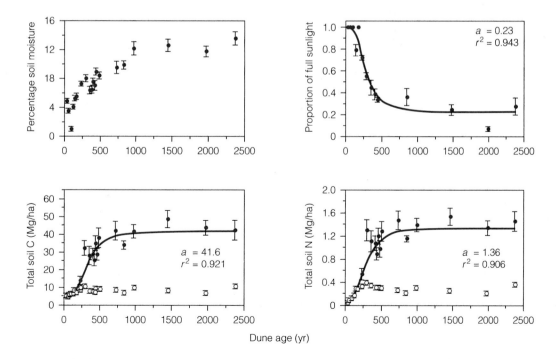

FIGURE 13.18 Changes in the abiotic environment associated with changes in vegetation during primary succession in Lake Michigan dune habitat. (*From Lichter 1998*)

In the Glacier Bay sere, lichens facilitate colonization by vascular plants by creating local microsites that hold organic matter and water and by physically cracking the granite surface. The nitrogen fixers, Sitka alder and *Dryas,* increase the soil nitrogen content, paving the way for the colonizations that occur in later stages (Chapin et al. 1994).

We might expect that facilitation processes are particularly important in primary succession, where pioneering species face an especially inhospitable abiotic environment. Nevertheless, there are examples of the role of facilitation in secondary succession as well. Experimental studies of old field succession by Li and Wilson (1998) showed that when the shrub *Symphoricarpos occidentalis* and white sprice (*Picea glauca*) were faced with competition from grasses, their growth and survival were both improved by facilitation effects of conspecifics. In semiarid parts of Spain, the legume *Retama sphaerocarpa* facilitates the growth of its conspecifics and of another species, *Marrubium vulgare.* Perhaps we will find that facilitation in secondary succession is more common in physically stressful environments like this one.

Tolerance   The second Connell–Slatyer model is the **tolerance model.** In this model, succession is also driven by the changes in the physical environment that each species causes. But in this case, one species makes it less likely that its offspring will be able to grow to maturity on the site, although other species are able to colonize and reproduce. According to this view, succession proceeds via the replacement of early succession species with others *more tolerant of* the physical conditions established by the early colonists. Perhaps the clearest example of this process is the set of replacements that occur late in old field succession. In parts of the Midwest, for example, sugar maple (*Acer saccharum*) and American beech (*Fagus grandifolia*), which are dominant in climax forest, are extremely shade-tolerant. Saplings of sugar maple can survive for 20 years in dense shade; beech saplings can survive for as long as 100 years under such conditions (Poulson and Platt 1989). The softwood stage (see Table 13.2) is replaced by beech and maple because the latter are more tolerant of the shade cast by the softwoods.

Tolerance may have more than one biological basis. In *active tolerance,* the presence of one species directly lowers the growth rate of others by reducing the availability of resources. In *passive tolerance,* the change is simply the result of differences in individual resource requirements of the species involved. In this type of tolerance, some species that differ in life history or other attributes are simply better able to tolerate the changing abiotic conditions.

P. V. McCormick and R. J. Stephenson (1991) showed that in an algal succession in a stream, both forms of tolerance operate. High levels of nitrogen in the water column result in a high probability of the early species attaching to the substrate and growing successfully. These algae become established first because their biology allows them to colonize rapidly (passive tolerance). The later species clearly decrease the reproductive rate of the early species via competition for resources; thus their effect is mediated via active tolerance.

There are numerous examples of species adapted in ways that allow them to compete for limiting resources, especially light, as succession proceeds. For example, F. A. Bazzaz (1979) has described the life cycle of an early successional species, common ragweed (*Ambrosia artemesifolia*). The life history of this species, which is common in old field succession soon after disturbance, is well adapted to this role (Figure 13.19). Common ragweed has evolved an innate seed dormancy that is broken when disturbance creates a high-light environment conducive to its growth. It can then germinate, grow, and reproduce very quickly. If

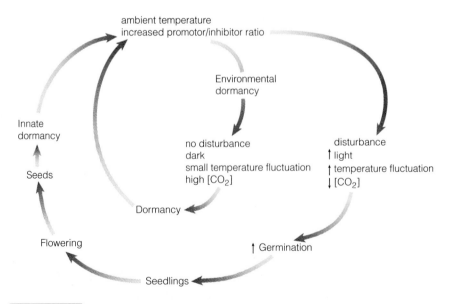

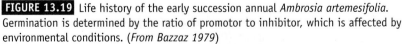

FIGURE 13.19 Life history of the early succession annual *Ambrosia artemesifolia.* Germination is determined by the ratio of promotor to inhibitor, which is affected by environmental conditions. (*From Bazzaz 1979*)

disturbance does not occur, individuals of this species can remain dormant in the soil for many years.

The architecture of the leaves and branches affects the physiology and photosynthesis of a tree, especially its energy balance and transpiration. A leaf does two important things: It photosynthesizes and it respires. The former produces energy for the tree; the latter uses some of that energy. If a leaf is to contribute to the growth of a tree, its rate of photosynthesis must exceed its rate of respiration. In shady environments, trees are selected for minimal leaf overlap and for perpendicular orientation of the leaves to the light source (Pearcy and Yang 1996). In high-light environments, leaf angle changes to distribute as much light as possible to all leaves of the tree. For example, in deciduous forest in Michigan, sugar maple (*Acer saccharum*) generally dominates American beech (*Fagus grandifolia*). Maple has strong apical dominance that enables it to grow rapidly in the vertical light gradient of a light gap. In contrast, beech has numerous side branches that support it better in dark understory, where it receives light from sun flecks filtering through the canopy or laterally from nearby light gaps (Poulson and Platt 1996).

Henry Horn (1971) provided a detailed example of the adaptations of later succession species. He was interested in the adaptive nature of the morphology of trees, especially the interaction of light intensity and the structure of trees. Horn noticed that the structure of early succession trees seemed different from that of later succession species. In particular, it seemed that early species had more layers of leaves than later species. This morphological difference could explain the replacement of early succession trees by those that appear later in the sere. Consider the two sapling trees in Figure 13.20. When the canopy is not closed and a great deal of light reaches these two trees, the early succession species (Figure 13.20a) will have a greater total photosynthetic area and a faster growth rate. When the canopy closes over, however, far less light reaches these two saplings, and a lower leaf on the early tree does not receive sufficient light to do enough photosynthesis to counterbalance its respiration. These lower branches and leaves thus become a detrimental net energy drain on the tree, and it dies. In contrast, the late-succession species has a single layer of leaves (Figure 13.20b). Although this tree absorbs less light than the tree in Figure 13.20a, there are no lower leaves acting as a drain on the tree; each leaf can "pay for itself" photosynthetically. Although this tree does not grow rapidly, it can survive under the low-light regime that exists later in succession, and it will replace the early species.

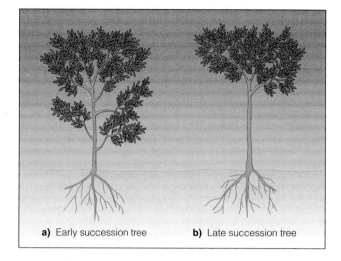

a) Early succession tree   b) Late succession tree

FIGURE 13.20 The diagrammatic structure of saplings of (**a**) early and (**b**) late succession trees. Early succession species have more layers of leaves than later species.

Horn attempted to quantify these structural differences among trees. This task was rather difficult because the differences among real trees are not so clear-cut as those depicted diagrammatically in Figure 13.20. Horn measured what he called the "effective number of layers of leaves" for several species of trees. He began by measuring the amount of light reaching the upper layer of leaves of a tree (point 1, Figure 13.21). Next he measured the amount of light passing through a *single* layer of leaves (point 2, Figure 13.21) and calculated the proportion of incident light that passes through a layer of leaves. He then measured

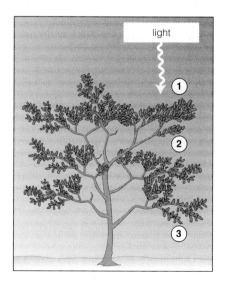

FIGURE 13.21 Design for the measurement of the number of effective layers of leaves. Light measurements at points 1 and 2 reveal the amount of light passing a single layer of leaves. The measurement indicates how much the ambient light is reduced by the number of leaf layers on the tree.

the amount of light reaching the base of the tree after passing through all the leaf layers of the tree (point 3, Figure 13.21). From these data he could calculate for each tree the number of effective layers of leaves, defined as the number of layers of single leaves that reduce the light by the amount it is reduced passing from the very top of the tree (incident sunlight) to the base of the tree. For example, suppose that his measurements showed that a single layer of leaves absorbs 10 percent of the incident sunlight (90 percent passes through) and that 30 percent of the incident sunlight reaches the base of the tree. We can calculate the number of effective layers of leaves as follows:

$$(0.90)^x = 0.30$$

In other words, how many layers of leaves, each of which passes 90 percent of the light, must we have to reduce the light to 30 percent of incident? Solving for x yields

$$\log (0.90)x = \log (0.30)$$
$$x = \log (0.30)/\log (0.90)$$
$$x = 11.4$$

This tree has 11.4 effective layers of leaves.

When Horn made measurements like these for a large number of early, mid-, and late succession trees, he found that the number of effective layers of leaves decreases from early to late succession trees (Table 13.3). Some of the most shade-tolerant species, such as sugar maple and American beech, have few effective layers of leaves.

It is important to note that this is a statistical rather than an absolute phenomenon. We do not observe tree morphologies as structurally simple as those in Figure 13.20. Moreover, we know that the shape of a tree is not entirely determined by a species's genetics. The light regime under which the tree grows also influences its shape and the number of effective layers; thus, trees grown in the open tend to have more lower branches than trees grown in dense shade. Finally, some species can modify their effective numbers of layers of leaves as the tree grows. In conifers, young trees growing in the open often have many effective layers of leaves. As the tree matures and grows taller, the lowest layers may be killed by shade.

From our knowledge of the different constraints imposed by the light regimes early and late in succession, we might predict that there would be differences in the photosynthetic rates of early and late succession species as well as in the responses of those rates to light levels. This is indeed what we find. Some photo-

TABLE 13.3

Effective Layers of Leaves for Selected Tree Species at Various Stages of Succession

Species	Number of Effective Layers
Early succession species:	
Gray birch (*Betula populifolia*)	4.3
Aspen (*Populus tremuloides*)	3.8
White pine (*Pinus strobus*)	3.8
Mid-succession species:	
White ash (*Fraxinus americana*)	2.7
Black gum (*Nyssa sylvatica*)	2.6
Red maple (*Acer rubrum*)	2.6
Late succession species:	
Sugar maple (*Acer saccharum*)	1.9
Hemlock (*Tsuga canadensis*)	1.6
American beech (*Fagus grandifolia*)	1.5

Data from Bazzaz 1979.

synthetic rates representative of early and late succession species are shown in Table 13.4. Note that the shade-tolerant late succession species have the lowest rates of photosynthesis. Summer annuals can clearly photosynthesize and grow far more rapidly. We would expect that in direct competition in strong sunlight, the early species would be able to outcompete the later ones.

Photosynthetic rates can be sensitive to light levels in ways that lead to successional change, however. The photosynthetic rates of early and late succession species change as a function of light intensity (Figure 13.22). At high light intensities, early species have an

TABLE 13.4

Photosynthetic Rates (mg CO_2/dm/hr) for Some Representative Early and Late Succession Species

Species	Rate
Early succession species:	
Common ragweed (*Ambrosia artemesifolia*)	35
Giant ragweed (*A. trifida*)	28
Foxtail (*Setaria faberii*)	38
Daisy fleabane (*Erigeron annuus*)	22
Aster (*Aster pilosus*)	20
Mid-succession species (softwood trees):	
Eastern red cedar (*Juniperus virginiana*)	10
Cottonwood (*Populus deltoides*)	26
Sassafras (*Sassafras albidum*)	11
Tulip poplar (*Liriodendron tulipifera*)	18
Late succession species (hardwood trees):	
Red oak (*Quercus rubra*)	7
American beech (*Fagus grandifolia*)	7
Sugar maple (*Acer saccharum*)	6
White oak (*Quercus alba*)	4

Data from Bazzaz 1979.

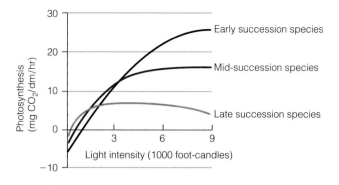

FIGURE 13.22 Photosynthetic rate as a function of light intensity for early, mid-, and late succession species. (*From Bazzaz 1979*)

advantage (as we saw in Table 13.4). In low light, as under a closed canopy later in succession, the situation changes. At light intensities of approximately 500–1000 foot-candles, so little light reaches leaves that photosynthesis does not compensate for the costs of respiration; the result is a net loss of carbon and energy (the "negative photosynthesis" in Figure 13.22). Note in this figure that the curve goes negative

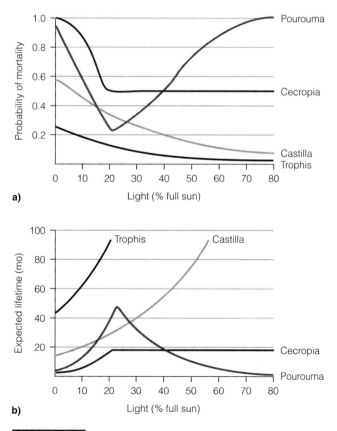

FIGURE 13.23 (a) Mortality rate for four species of tropical forest trees in Costa Rica as a function of light intensity. (b) Expected lifespan calculated from the mortality data suggests the pattern of successional change as light decreases during succession. (*From Kobe 1999*)

for late succession species at a lower light intensity than for early species; that is, late species have a positive energy budget at lower light intensities than early species do. As the canopy closes and light becomes limiting, the early succession species are at a greater disadvantage than the late species. These data demonstrate the physiological basis for the tolerance model of succession in which late succession species are more tolerant of the physical conditions late in succession.

Kobe (1999) demonstrated the ecological effects of such differences in the physiological response to the light regime. In experiments involving four related tree species inhabiting tropical forest in Costa Rica, Kobe showed that each species responded uniquely to the light environment, in effect partitioning the light resource niche (Figure 13.23a). The response of these species to differing light regimes after disturbance is clear. Kobe confirmed the ecological effect of their responses by modeling the expected lifetimes of the four species under different light conditions (Figure 13.23b.)

Inhibition   In the **inhibition model** of succession, the key element determining the outcome of succession is the nature of the initial colonization. Most species are considered capable of colonization, but by chance, there will be great variation in which species actually colonize. Those that do colonize attempt to exclude later species: They inhibit further colonization for the length of their life spans. This model precludes any precise, repeatable sequence of replacements; each site and each sere will be slightly different as a consequence of the vagaries of colonization, and each species that successfully colonizes will remain for its normal life span. Thus succession proceeds via the replacement of short-lived species with those with longer life spans.

Tilman's Resource Model

Another general mechanism of successional change was proposed by D. Tilman (1985). Tilman's model is based on the assumption that each species is a superior competitor for some particular limiting resource and that species composition should change whenever resources change. He hypothesizes that the abundances of two of the most important resources, light and soil nitrogen, are inversely related: Habitats with high light will tend to have low nitrogen, and vice versa. Succession is thus driven by the change in species composition dictated by the changes in resources over time.

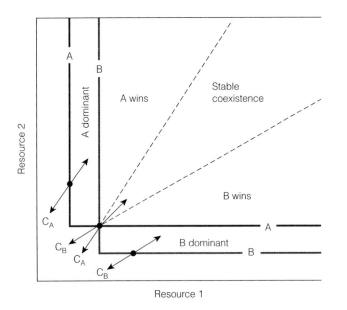

FIGURE 13.24 Tilman's model of succession. The solid line marked A is the isocline for one species; the solid line marked B is the isocline for the other. The vectors C_A and C_B represent the use of resources by A and B, respectively. The unmarked vectors represent the rate of production of resources. (*From Tilman 1985*)

In the graphical representation of this model shown in Figure 13.24, each axis represents a limiting resource. The two species, A and B, have different requirements for the two resources. Their population growth isoclines—the amounts of resources at which mortality equals reproduction—are shown by a solid line for each species. The right-angle bends in these isoclines indicate that each resource is limiting, and the population growth rate is determined by the resource in lower supply. Note that A has a lower requirement for resource 1 but a higher requirement for resource 2. The vectors labeled C_A and C_B represent the rate of use of the resources by the two species; the vectors without labels represent the rate of production of each resource by the habitat. By examining both the rate of resource production in each habitat and the rate of consumption by each species, we can project which species will dominate in any habitat. Note that in the figure there are regions where A wins, regions where coexistence is possible, and regions where B wins. As plant species modify the physical environment, A can be replaced by B, as shown in Figure 13.24.

Interactions Among the Succession Models

We should not consider the models of successional change to be mutually exclusive. Nothing inherent in any of them precludes the operation of the others simultaneously or in sequence. Indeed, the question of interest is how they interact. We next consider some examples of the interaction of mechanisms of succession.

W. F. Morris and D. M. Wood (1989) examined the roles of the facilitation and inhibition models in primary succession following the explosion of Mt. St. Helens in 1980. Specifically, they studied the effect of the nitrogen-fixing lupine (*Lupinus lepidus*) on two succeeding species, fireweed (*Epilobium angustifolium*) and the pearly everlasting (*Anaphalis margaritacea*). They found that both inhibition and facilitation occur, but at different life history stages of fireweed and everlasting. Morris and Wood prepared four plots: a control plot with no lupines, a plot with vigorous live lupines, a plot in which the lupines were killed and left as mulch, and a plot in which the lupines were killed and removed. Next, seedlings of fireweed and everlasting were transplanted into the four types of plots and observed over time. In the first year, the presence of live lupine plants in plots caused a decrease in the survivorship of both fireweed and everlasting seedlings (Figure 13.25), a finding consistent with the predictions of the inhibition model. In

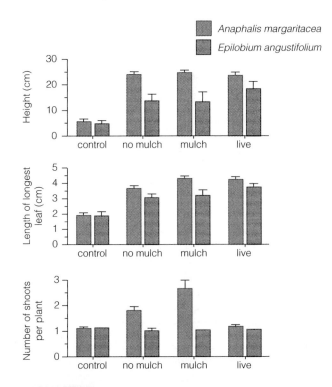

FIGURE 13.25 Growth of *A. margaritacea* and *E. angustifolium* in the second year of succession. No mulch = plots where lupines had been growing and were removed; mulch = plots where lupines had been growing and were then mulched into the soil; live = plots where lupines were growing with the other species. (*From Morris and Wood 1989*)

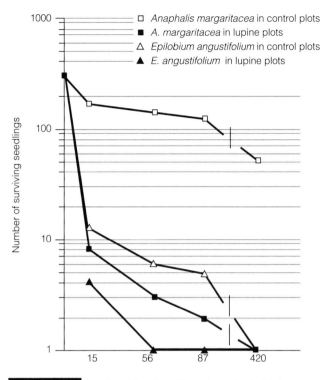

FIGURE 13.26 Survivorship of seedlings transplanted into a lupine patch and into a control plot in adjacent barren ground. (*From Morris and Wood 1989*)

the second year, however, in the plots in which lupines were growing (live) or had been growing (with or without mulch), both fireweed and everlasting grew more vigorously than the controls (Figure 13.26). Thus, in the second year there is a facilitation effect, probably associated with the addition of nitrogen to the soil by the lupine.

Among the first mechanistic studies of succession was an examination of the basis for successional change early in secondary succession in the Carolina Piedmont region (Keever 1950). In this habitat, the initial successional sequence in an old field sere is (1) horseweed, which dominates in the first year after disturbance; (2) aster, which dominates beginning in the second spring; and (3) broomsedge, which replaces asters in the third year. In a number of field and greenhouse experiments designed to elucidate the mechanisms responsible for these replacements, C. Keever found that fundamental aspects of the biology of each species account for the pattern. Horseweed, an annual forb, produces large numbers of seeds that have the ability to germinate in the year they are produced; in contrast, aster seeds must be cold-stratified in winter before they will germinate. Thus horseweed dominates in the first year. This species is self-allelopathic, however, and so is unable to replace itself. The aster seeds are capable of germinating in the second spring, so as the horseweed is declining, aster, a perennial,

comes to dominate the site. It can persist on the site, but it does not reproduce well in its own shade. Broomsedge, being more shade-tolerant, replaces the aster in the third year.

In this example, we see elements of both the tolerance and the inhibition models at work. The replacement of the aster by the more shade-tolerant broomsedge is an example of the former. The dominance of the site by horseweed and its replacement by a longer-lived species is consistent with elements of the inhibition model.

T. M. Farrell (1991) described a successional sequence in the rocky intertidal zone on the Oregon coast in which both facilitation and tolerance appear to operate. In the intertidal zone, disturbed sites are first colonized by the barnacle *Chthalamus dalli*, which is eventually replaced by another barnacle, *Balanus glandula*. In turn, three algal species, *Pelvetiopsis limitata*, *Fucus distichus*, and *Endocladia muricata*, follow the second barnacle species. *Balanus* colonizes sites containing *Chthalamus* because it can tolerate the presence of *Chthalamus*, although favorable weather does play a role in successful colonization by *Balanus*. *Balanus* clearly facilitated the colonization by the algae; when it was experimentally removed, colonization of all three algal species was prevented (Figure 13.27). Experiments in which epoxy-filled barnacle tests (shells) were applied to the substrate indicated that the facilitation effect results from alteration of the substrate, not from the activity of living barnacles.

The Role of Herbivores in Succession

Superimposed on the vegetational changes that occur as a result of competition and other interactions is the effect of consumers on the process of successional change. Given the complex interactions among herbivores and plants, we might expect that grazing or other consumer activities would affect the course of succession. Selective herbivory may favor one successional species over another, or consumers' activities may enhance the colonization abilities of certain species. The acorns of late succession oaks, for example, are so large and heavy they cannot be carried to new sites by wind or small animals; instead, they are carried there by squirrels.

Some of the best examples of the effects of grazers come from marine algal successional sequences. J. Lubchenko (1983) described a successional sequence in which herbivores play a key role. In the rocky intertidal zone in New England, disturbance of the substrate

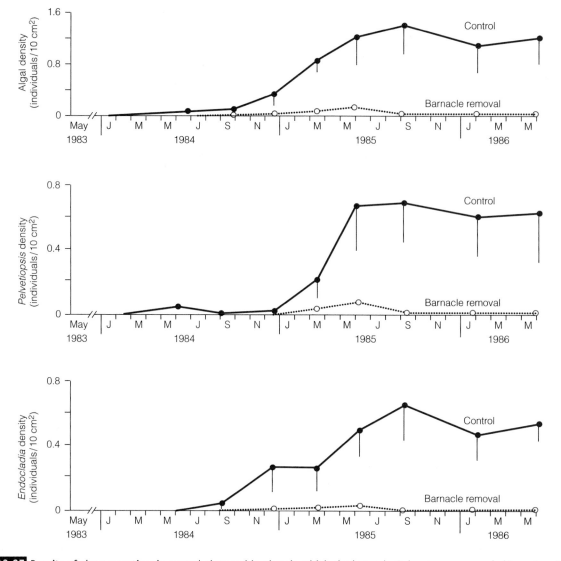

occurs from wave and storm action. Newly exposed sites are colonized first by early succession ephemeral algae such as *Ulva lactuea* L., *Enteromorpha* spp., and *Porphyra* spp. They are eventually replaced by later species, especially the brown alga, *Fucus vesiculosus.*

Lubchenko showed that gastropod herbivores, especially the periwinkle, *Littorina littorea*, play a crucial role in the succession. These snails preferentially graze on the early, ephemeral algae species. When the herbivores are excluded, the early succession species inhibit the colonization of *Fucus.* The snails also graze on small individuals of *Fucus*, but this alga can seek refuge in small crevices and grow large enough that the snails no longer graze on it. At that time the snails may actually enhance the growth of *Fucus* by grazing epiphytic algae from its blades.

These data suggest that local microtopography affects the nature and mechanisms of succession. W. P.

Sousa (1984) identified another example of complex herbivore effects on succession in the rocky intertidal zone. In Sousa's study sites in the intertidal on the California coast, extensive beds of the mussel *Mytilus californianus* are interrupted by patches of open space created by wave action. The open spaces are colonized by competitively inferior species of algae and sessile invertebrates. Eventually, these disturbed sites are covered by encroaching *Mytilus.*

Sousa experimentally created open patches of different sizes and found that grazing intensity by limpets was greater in small patches. This effect was manifest in a plot of the total cover of algal species in patches of different size, with and without limpet grazers (Figure 13.28). Small patches were dominated by competitively inferior but grazing-resistant species that would have been excluded from larger patches where limpet grazing was less intense.

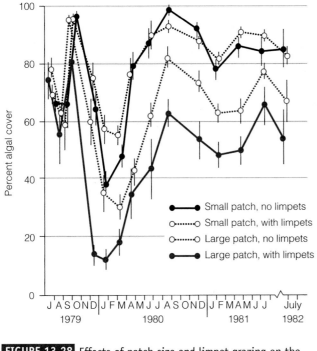

FIGURE 13.28 Effects of patch size and limpet grazing on the percent algal cover. *(From Sousa 1984a)*

Even in primary seres where abiotic factors and facilitation are predominant forces, herbivores modify the pattern of change. In Lake Michigan dune succession, a specialized herbivore, the flea beetle (*Altia subplicata*) attacks the sand willow (*Salix cordata*). When Bach (1994) excluded beetles from the willows, they grew approximately twice as fast. Survivorship in the exclosures was significantly higher as well. The indirect effects of herbivory spread beyond this one species: When beetles were excluded from the willows, other herbaceous species increased as well.

Probabilistic Models of Succession

Henry Horn also explored models of change for the system that are based on probabilities (Horn 1975). Rather than trying to discover morphological or physiological determinants of successional change, as previously discussed, Horn attempted to assess the degree to which succession could be described as a statistical process dependent on a series of transition probabilities from one state (which could be tree species composition or an individual tree) to another.

According to Horn (1975), "the most dramatic property of succession is its repeatable convergence on the same climax from any of many different starting points." He noted the resemblance of this property to statistical processes known as Markov chains. In Markov chains, the transitions from one "state" to another are stochastic processes that depend only on the current state, not on any previous state. An important characteristic of Markov chains is that they eventually settle into a pattern in which the states occur with frequencies that are independent of the initial states. Applied to succession, this property means that the climax state is characterized by a series of tree species frequency distributions that are independent of the initial distributions.

This approach to succession differs markedly from the models we have discussed thus far. It de-emphasizes specific biological determinants of the paths of succession and instead stresses the statistical nature of transitions from one stage of succession to another (or the probability distributions associated with individual tree replacements). The sum of these individual transitions describes successional change for the entire forest.

The models Horn developed are based on an analysis of tree-by-tree replacement in Institute Woods in Princeton, New Jersey. Horn began with the assumption that for any given canopy tree in a forest, the probability that a particular species will replace that tree is equal to the frequency of that species under the canopy. Consider the hypothetical situation in which a sugar maple in the canopy towers over four saplings growing underneath: two beeches, a sugar maple, and a black walnut. When the sugar maple dies, it will be replaced by one of these four saplings. Because 50 percent of them are beech, the probability of the sugar maple being replaced by beech is 50 percent. Similarly, the probabilities of replacement by sugar maple and by black walnut are both 25 percent. Horn extended this concept to the entire forest, so that if we want to know the probability that any given canopy species will be replaced by other species, we simply need to know the frequency distribution of sapling species under each canopy species for the entire forest.

The sapling distributions under various canopy tree species in Institute Woods are shown in Table 13.5. From these data, Horn calculated what the frequency distribution of species in the next generation would be. For example, if we wish to predict the number of big-tooth aspen in the next generation, we look at the distribution of big-tooth aspen saplings. Referring to Table 13.5, we see that there were 104 big-tooth aspens in the canopy. Of the sapling species underneath them, 3 percent were big-tooth aspen, 5 percent were gray birch, 9 percent were sassafras, and so on. According to our model, then, 3 percent of the 104 big-tooth aspen will be replaced by big-tooth aspen, 5 percent will be replaced by gray birch, and so on.

The same analysis applies to each of the canopy trees. The number of big-tooth aspen in the next generation will be determined by the frequency of big-tooth aspen saplings under the canopy trees. Note in the table that big-tooth aspen occur under only three canopy trees: big-tooth aspen, sassafras, and black gum. Thus, to predict the number of big-tooth aspen (BTA) in the next generation, we calculate as follows:

Number of BTA =
$$0.03(104) + 0.03(68) + 0.01(80) = 5.96$$

Similarly, the number of gray birch (GB) in the next generation is given by the equation

Number of GB =
$$0.05(104) + 0.01(68) + 0.01(80) = 6.68$$

Horn developed a computer model of succession based on these calculations. He predicted the frequency distribution of all species in the canopy in the next generation in the ways just shown. He then placed a similar sapling distribution under this new canopy and allowed those saplings to replace the canopy again. He allowed the process to continue, generation after generation, until the canopy species frequencies no longer changed. By definition, this represents the climax state.

To test the accuracy of his model, Horn compared the stable species distribution predicted by his model with a real climax forest that had remained undisturbed for 350 years. In Table 13.5 the hypothetical climax generated by Horn's computer model is compared with an actual climax forest. The two distributions are not statistically different. Thus Horn's simple model, in which the probability of replacement for a given species is based on the frequency distribution of saplings, accurately predicts the climax state.

Horn made an important and very interesting assumption that the probability of replacement of a canopy species is determined simply by the frequency distribution of sapling species. This implies that physiological and morphological differences among saplings are not important—each has an equal chance of reaching the canopy after the adult dies. Our previous discussions, including Horn's analysis of effective layers of leaves, would indicate that this assumption is not correct. The crucial feature of Horn's analysis, however, is that the process can be accurately modeled despite this assumption, which suggests that the deterministic features of the process may be less important than its probabilistic nature.

Of course, we also need to ask what deterministic factors govern the frequency distribution of sapling species. A number of deterministic factors, including morphology, physiology, and reproductive and dispersal characteristics, may determine the nature of that sapling distribution. Perhaps the process is essentially probabilistic once the sapling distribution has been determined. Horn's model is interesting because it gives us a clue about where to look for the important factors in the later stages of succession.

TABLE 13.5
The Species Distribution of Saplings in Institute Woods (Expressed as a Percentage of All Saplings) Under Various Canopy Species

Canopy Species	Number	Saplings (percentage of total)										
		BTA	GB	SF	BG	SG	WO	OK	HI	TU	RM	BE
Big-tooth aspen (BTA)	104	3	5	9	6	6	—	2	4	2	60	3
Gray birch (GB)	837	—	—	47	12	8	2	8	—	3	17	3
Sassafras (SF)	68	3	1	10	3	6	3	10	12	—	37	15
Black gum (BG)	80	1	1	3	20	9	1	7	6	10	25	17
Sweet gum (SG)	662	—	—	16	—	31	—	7	7	5	27	7
White oak (WO)	71	—	—	6	7	4	10	7	3	14	32	17
Red oak (OK)	266	—	—	2	11	7	6	8	8	8	33	17
Hickory (HI)	223	—	—	1	3	1	3	13	4	9	49	17
Tulip (TU)	81	—	—	2	4	4		11	7	9	29	34
Red maple (RM)	489	—	—	13	10	9	2	8	19	3	13	23
Beech (BE)	405	—	—	—	2	1	1	1	1	8	6	80
Predicted climax distribution		0	0	2	3	4	2	4	6	6	10	63
Observed climax distribution (350-year-old stand)		0	0	0	6	0	3	0	0	14	1	76

The last two rows provide a comparison of the climax state as predicted by Horn's model and in an actual old-growth forest.

After Horn 1975.

The Nature of the Climax

The Concept of Climax

We have defined the climax as the final, self-perpetuating stage in a successional sequence. In a general sense, this is a useful definition. For example, we can follow old field succession from annual weeds to a hardwood forest, which, if left undisturbed, will reproduce itself. None of the earlier stages of succession have this self-perpetuating property. As in many areas of ecology, however, broadly useful definitions are problematic when we look more closely and precisely at the situation.

The term *climax* was coined by F. E. Clements (1916, 1936). Clements chose the term because it shares a common root with the word *climate* and thus reflected his belief in the importance of the concept of the climatically determined biome. Clements believed that a biome is similar to an organic entity and that each major climatic region of a continent has its own climatically determined dominant vegetation type. It is this vegetation type that Clements (and many other plant ecologists) believed is the endpoint of succession. If an area in any part of the continent is left undisturbed long enough, its flora will eventually reach a single state characteristic of the biome. Clements's view of the climax is referred to as a **monoclimax.**

Nature does not appear to be that simple, however. For example, although the oldest dunes studied by Cowles did indeed support a hardwood forest, this forest is dominated by black oaks and hickories, not by the beeches and maples characteristic of nearby non-dune habitats. These dunes are several thousand years old, so it seems unlikely that they are still in the process of succeeding toward the "normal" beech–maple forest characteristic of the region.

In many old-growth forests in the deciduous forest biome of eastern North America, as well as in the conifer forests of the mountainous West, there are local assemblages of species that appear to be reproducing themselves. A. G. Tansley (1939) developed the concept of the **polyclimax** to account for this variation. According to this view, there is no such thing as a regional, climatically determined climax. Instead, there is a series of local climax states, each of which is determined by the local soil (edaphic) and microclimate conditions. The difference between this and the Clementsian view is that Clements expected succession eventually to take each edaphic climax to the single regional climax. A related version of the polyclimax theory is the **pattern climax** (Whittaker 1953), which extends the polyclimax theory to say that the edaphic climaxes may not be discrete entities but, rather, may grade into one another. The landscape is described as a mosaic of local, edaphic climaxes that merge gradually into one another.

These definitions are precise and quantifiable. In principle we should be able to distinguish among them empirically. Two related factors mitigate against this, however. First, we would need to study a system over an immense time scale. If the oak forests of the oldest dunes are in fact slowly modifying the soil so as to facilitate the eventual colonization of beech and maple, the relevant time scale is thousands of years. Second, we would need an area free of disturbance for an extremely long period of time. As we will see, this is an improbable occurrence. Even if we found such an area, the time scale is sufficiently long that other factors, including the possibility of climatic changes, would confound our conclusions.

As an example of the difficulty, consider the situation in many deciduous forests in the midwestern United States. In undisturbed forest, the current canopy is dominated by red oaks (*Quercus rubra*) and white oaks (*Q. alba*) that are 200–300 years old. The understory is dominated by sugar maple (*Acer saccharum*) that appear poised to replace the current canopy. Thus the oaks are not self-perpetuating. How should we interpret this situation in terms of the nature of climax? There are at least three plausible explanations:

1. The process of succession has not yet reached the monoclimax that Clements would suggest will eventually occur when the sugar maples and beech mature.
2. The oak forest did in fact represent a climatic climax in the past, but a climatic shift to a cooler, moister environment over the past 300 years has shifted the climax to beech–maple.
3. Beech and maple are in fact the climatic climax, but the current forest represents a fire disclimax. Prior to human settlement in the 1820s, fires were more common in these forests than they have been subsequently, and the oaks dominated the canopy because of their relative resistance to fire. Our recent fire suppression practices have selected for a beech–maple forest such as that manifest in the understory.

All three explanations are possible, but the time scale involved and the impossibility of replicating the natural experiment make hypothesis testing difficult.

Equilibrium and Climax

One of the recurring themes of this text is the question of the existence and nature of equilibrium in ecology. Succession is one of the most important focus areas for this question, because the definition of successional climax implies the existence of equilibrium. Does such an equilibrium state exist? If so, how common is it?

It is first important to understand that plant communities may *appear* to be at equilibrium when, in fact, long-term processes of change are in operation. In mature deciduous forest, which appears relatively stable, canopy gaps caused by tree mortality and windthrow cause local successional events. Tanaka and Nakashizuka (1997) used aerial photography to quantify the long-term patterns of gap formation. Figure 13.29 shows the rate of gap formation and closure over three 5-year time spans. Note that in two of the three, closure and formation are nearly equal, indicating an equilibrium number of gaps. In the third period (1981–1986), gap formation greatly exceeded closure. Thus the forest is in relative equilibrium over short time spans, but sporadic periods of canopy gap formation occur when a longer time scale is considered.

G. H. Aplet, R. D. Laven, and F. W. Smith (1988), in a study of Englemann spruce (*Picea engelmanii*)/subalpine fir (*Abies lasiocarpa*) forests in Colorado, demonstrated a similar long-term pattern of disturbance that at first glance appears to represent equilibrium. Extensive montane areas are covered with forests in which spruce and fir coexist, fir numerically dominating the stands. The traditional view was that such

forests represent equilibrium stands in which fir is numerically dominant.

Aplet, Laven, and Smith (1988) suggested that other dynamic processes requiring long periods of time could account for the appearance of these forests. They proposed three models to explain the commonly observed stand composition (Figure 13.30). All three models have a period (crosshatching) when, regardless of the model, the stands have a similar appearance and composition. The situation in Figure 13.30a represents a traditional equilibrium. The successional events leading to this state are unknown, but an equilibrium is reached in which spruce and fir coexist, with fir as the numerically dominant species. In Figure 13.30b, the situation is explained as an intermediate stage in the process of spruce exclusion as succession proceeds. After disturbance, both species can colonize. Fir maintains high rates of regeneration, whereas spruce, perhaps because of a factor such as shade intolerance, is excluded. In the third model, both species colonize after disturbance, but spruce is excluded when the canopy closes. As the stand ages, however, fir trees reach maximum age and begin to

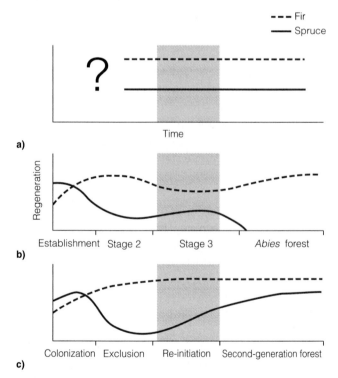

FIGURE 13.30 Three possible explanations for the development of spruce–fir forests in Colorado. The solid lines represent spruce; the dashed lines represent fir. The shaded areas show the situation in most forests today. (**a**) An equilibrium coexistence model. The relative numbers are unknown in the region marked with a question mark. (**b**) Nonequilibrium coexistence with competitive exclusion. (**c**) The model of stand development proposed by Aplet, Laven, and Smith (1988).

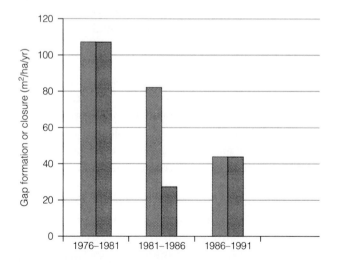

FIGURE 13.29 The rate of gap formation (solid bars) and gap closure (hatched bars) in a Japanese temperate deciduous forest for three 5-year periods. Note that in two of the periods, gaps are nearly at equilibrium but that in the period 1981–1986, many more gaps were formed than closed. (*Data from Tanaka and Nakashizuka 1997*)

die and fall, opening up light gaps. There is a second wave of spruce regeneration associated with these light gaps. Again, at intermediate stages, the stand resembles those produced by models a and b.

The key to Aplet's hypothetical models is that each makes a testable prediction about the composition of the stand during the process of succession:

Model a. We should observe constant ratios of spruce to fir over time—at least after succession has proceeded for a long time.

Model b. We should observe high rates of recruitment of both species immediately after disturbance. Subsequently, spruce regeneration should decline while fir regeneration holds relatively steady.

Model c. Both species should be recruited early in succession. Then spruce should decline while fir recruitment stays high. After several hundred years, spruce recruitment should increase again.

Aplet, Laven, and Smith surveyed five stands that ranged in age from 175 to 575 years since disturbance. In each stand, complete species tallies and increment cores were used to establish the numbers and age structures of each species. Inherent in the analysis is the assumption that the five sites experienced a common pattern of succession. The age class distributions for both species are presented in Table 13.6.

TABLE 13.6
Age Class Distributions for Spruce and Fir in Five Stands in Colorado

	Spruce					Fir				
Stand	1	2	3	4	5	1	2	3	4	5
Stand Age (yr)	175	275	375	575	575	175	275	375	575	575
Age Class					Stems in Age Class (no/ha)					
0–25		240	240	240	5040	1120	720	5790	4370	9600
26–50		480	240	240	4080	2440	5280	11820	7440	9360
51–75	20		490		2200	480	4560	1690	4100	4370
76–100	160				1720	1280	1680	720	2170	2040
101–125	340				770	2220	480	740	1000	680
126–150	1160	260			260	860	50	280	20	1370
151–175	280	10				140	520	360	490	370
176–200	20	30	10	10	20		390	320	540	430
201–225		160	10				630	780	630	340
226–250		610	10	10	10		190	130	130	170
251–275	20	370	30		10		190	100	340	80
276–300		20	50		10		10	70	80	40
301–325			120	30	10			100	120	30
326–350			140					50	40	10
351–375			160	30					70	10
376–400			20	30	10				60	
401–425				30	10					10
426–450				30	10				20	10
451–475				40	40				20	
476–500				60						
501–525				40	30					
526–550				20	50					
551–575				90	10					
576–600										
601–625					10					

From Aplet, Laven, and Smith 1988.

Clearly, the data do not support model a's prediction that both species should have constant recruitment rates over time. Compare the rate of regeneration of spruce in 575-year-old stands with that in 175-year-old stands. We thus reject the equilibrium model. Model b predicts that spruce recruitment should decline over time. Clearly, this is not the case, because the highest recruitment rates are in the oldest stands. Model c predicts an intermediate period when spruce recruitment is low, followed by a subsequent increase. This is supported by the data. Note that in 175-year-old stands there is no recruitment of spruce. This apparently occurred on all the sites; note the diagonal band of missing recruitment in the spruce data. Then, in the 575-year-old stands, spruce recruitment reaches very high levels. Fir recruitment is relatively constant over time. These data thus cause us to reject the first two models, including the equilibrium model. They support model c, in which the composition of the forest is dynamic over much of its successional history. Stand-devastating fire is sufficiently frequent (relative to the more than 500-year time scale of the process) that it is fair to say that these forests do not reach equilibrium.

Numerous other examples support the notion that succession rarely reaches long-term equilibrium. The original tallgrass prairies of the eastern Great Plains once extended into northeast Illinois and northwest Indiana, a region called the prairie peninsula (Transeau 1935). Here, at the ecotone with the deciduous forests, the eastern edge of the grassland pushed farther east into a region in which the climate can clearly support trees. It is likely that the prairie peninsula was a fire disclimax; the eastward extension of the prairie was maintained by recurring fires that did not have major river systems to impede their eastward progress. Efforts to maintain the few remnant prairies in this region or to restore the original prairie vegetation are heavily dependent on fire to prevent succession from proceeding to forest. Again, the natural appearance of the landscape was not a matter of an equilibrium vegetation type but was, rather, a product of recurrent disturbance.

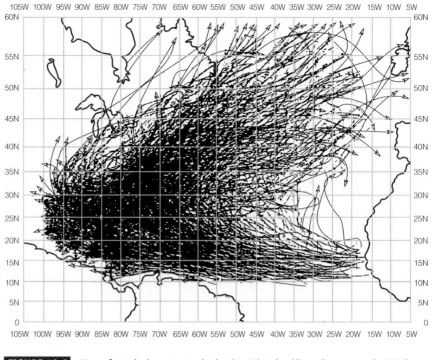

FIGURE 13.3 Map of tropical storm tracks in the Atlantic. (*From Duever et al. 1994*)

Mounting evidence suggests that even some of the systems we consider the most stable may rarely be at equilibrium. The tropical forests throughout the Caribbean region appear to be old and stable. They did not experience glaciation during the Pleistocene, as the northern forests did, and fire does not appear to be a major abiotic factor, as in many grasslands or northern forests. However, tropical storms and hurricanes have had frequent and major impacts on these forests. Figure 13.31 shows the tracks of major tropical storms in the Atlantic from 1886 to 1986. Note that few Caribbean forests (and not many temperate forests) have escaped hurricane damage. The average return time for a hurricane of any strength in the Caribbean is only 60 years; that is, any spot in the Caribbean will be hit by a hurricane approximately every 60 years. The average return time for a Category IV or V hurricane, the most severe storms, is only 100 years! Given the time scales we have been discussing for successional processes, it is clear that equilibrium is rare in these forests. Nearly all are probably at some preclimax successional stage as a result of disturbance by hurricanes.

These data suggesting that communities are nonequilibrium entities are profoundly important. As previously noted, in the 1970s many ecologists began to favor a nonequilibrium view that emphasizes the importance of successional patterns and processes because most communities will be subject to them

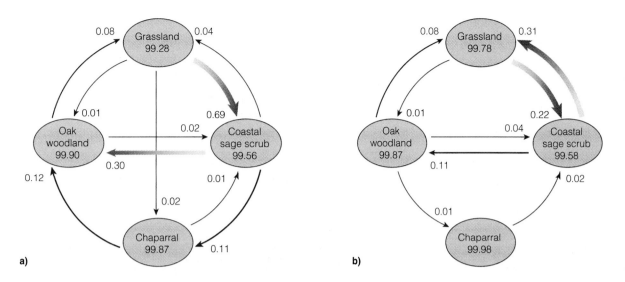

FIGURE 13.32 Annual transition rates among plant communities in a California vegetation mosaic: (**a**) unburned plots and (**b**) burned plots. The numbers in the ovals represent the probabilities that a given community will remain the same. The numbers beside the arrows estimate the probability that a community will change in the indicated direction. The thickness of the arrows is proportional to the probability of that change. (*Callaway and Davis 1993*)

much of the time. It also emphasizes the importance of the intermediate disturbance hypothesis (Connell 1978; see Chapter 12).

It is also important to recognize that a community may be undergoing succession at the same time that other biological interactions are affecting it. The result may be a complex sequence or pattern of community change. R. M. Callaway and F. W. Davis (1993) studied a mosaic of grassland, coastal sage scrub, chaparral, and oak woodland in California. This complex of communities is dynamic; there is no perfect correlation between physical factors and the vegetation types. Rather, the existence of a type in a particular spot is the result of succession after fire disturbance, plus competitive interactions among the species. The result is a cyclical pattern of change in community type (Figure 13.32), rather than the linear successional sequence previously described.

This chapter brings us to the end of our consideration of community ecology. We have discussed the important forces that structure mature communities, the patterns of species organization in communities, and the development of a community after disturbance. You should appreciate the dynamic nature of communities: They represent groups of species organized by long-term evolutionary forces such as character displacement and short-term effects such as competitive exclusion and disturbance. We turn our attention now to the highest level of ecological organization, the ecosystem.

SUMMARY

1. Succession is the process by which a community develops following disturbance. Primary succession follows a major disturbance that removes all living things from the region; secondary succession follows a less drastic disturbance that leaves soil and some living things.

2. At one time ecologists considered most communities to be stable, equilibrium entities in which disturbance is uncommon. Today, we recognize that most communities are disturbed frequently and are usually in various stages of succession. This makes succession a common and important ecological process.

3. Many factors, including fire, storms, glaciers, and volcanoes, disturb ecological communities. Disturbances can be classified according to their frequency, intensity, and spatial extent.

4. Primary succession begins with pioneering species that are able to colonize physically harsh environments. The harsh abiotic conditions may include high light intensity, high temperature, little or no organic matter, low moisture, and hard or unstable substrate.

5. Secondary succession typically proceeds faster than primary succession because soil and some organisms remain and the abiotic conditions are not so inhospitable.

6. Connell and Slatyer organized the mechanisms of successional change into three fundamental types: facilitation, tolerance, and inhibition. Many known sequences can be assigned to one of these models. However, interactions among the basic processes are probably common.

7. Grazing may play an important role in the development of a community by altering the competitive relationships among the species colonizing a disturbed site or changing the rate of colonization.

8. The frequency of disturbance in most communities suggests that the climax is an ephemeral stage. Even when communities mature, they are still dynamic and always changing.

SELF-ASSESSMENT: CAN YOU ...?

1. Explain the differences among various types of disturbance and their effect on the pattern of succession.

2. Describe and evaluate the evidence (from this and previous chapters) that suggests that communities are not at equilibrium.

3. Discuss the major differences between primary and secondary succession.

4. Describe the three major models of successional change described by Connell and Slatyer. Discuss the application of each to specific successional sequences.

5. Discuss the roles of other factors, such as nutrient availability, herbivores, and stochastic factors, in succession and in relation to the three mechanisms of Connell and Slatyer.

6. Explain the nature of the climax and the evolution of the concept.

PROBLEMS AND STUDY QUESTIONS

1. It is possible to consider succession from two standpoints:

> *Statistical/individual.* We can consider the process of change as a *community* process described by statistical changes in the composition of the community or as a matter of individual plant replacements over time.
>
> *Stochastic/deterministic.* We examined both stochastic and deterministic models for succession.
>
> Discuss the application of each of these views to succession.

2. Explain how effective layers of leaves differ between early and late succession species.

3. Is the climax state of secondary succession evidence to support the relationship between diversity and stability?

4. What fundamental assumption underlies Horn's stochastic model of succession? Explain.

5. Briefly present one example of the interaction of Connell and Slatyer's three models of succession. Indicate how and why each model is operating in that particular succession.

6. Why might we say that the Yellowstone fires in 1988 were "inevitable"?

7. Present an example showing how and why successional sequences in the same biome or general habitat type differ locally.

8. Consider the following data. The ages across the top are ages of stands since fire. The ages in the left column are the ages of the trees in the stands. The numbers represent the number of individuals of a given age in each stand.

Age of Individual Trees	Age of Stand Since Fire					
	Ponderosa Pine			Hemlock		
	100	200	500	100	200	500
0–25	15	23	0	475	616	456
25–50	9	46	5	310	523	415
50–100	1151	17	15	5	274	616
100–200	0	1063	8	9	516	589
200+	0	0	175	0	0	671

a. Is pine For hemlock an early succession species?

b. Is the 500-year-old stand at equilibrium?

c. Discuss and explain the successional events after a fire that are shown by these data.

d. What would you expect to see in a 600-year-old stand?

9. Using Tilman's resource model, explain how replacement of species occurs during succession.

10. Discuss the relationship between Horn's stochastic model of succession and the concept of the climax. Does Horn's model depend on a particular concept? Explain.

11. Discuss the concept of community stability and equilibrium in the context of your knowledge of succession, competition, and predation.

PROJECTS AND ADDITIONAL STUDY

1. Locate a series of local communities of different ages. Quantify the floristic differences among them and assemble a description of the successional sequence in that area. Address the assumptions and biases inherent in such a study. Or measure and compare a series of abiotic factors among such sites. You might include light intensity, soil moisture, soil nutrients, and so on.

2. Use the literature to assemble a description of the life history characteristics of species associated with different successional stages in a particular sere. Discuss the patterns that emerge.

3. Use historical data on hurricane tracks to determine the disturbance frequency for a particular set of sites in the Atlantic. Compare the patterns for different sites.

4. Gather size or age class data for a local forest. Develop a model (perhaps using a spreadsheet) of forest transitions for a local forest such as those used by Horn, assuming that the probability of replacement is proportional to the frequency of the species.

Ecosystems

Energy Flow and Trophic Structure

In this final section of the text, we turn our attention to the highest level of ecological organization, the ecosystem. At this level, the nature of the questions addressed by ecologists changes. The focus shifts to the movement of energy and materials within ecosystems.

We begin with a consideration of the energy that supports the living components of an ecosystem. It would be hard to overstate the importance of understanding the capture and transfer of energy. All life is energy-limited. Indeed, much of ecology follows directly from the need of every living thing to obtain energy. The size of a population is ultimately determined by energy limitations; a community develops as a result of change in the ability of plants to obtain solar energy; the reproductive strategy of an organism is constrained by energy; competition, a major organizing force in communities, is often a contest over energy. If we are to understand these topics, we must also understand the nature of energy transfer and the manner in which it becomes limiting.

In this chapter we will be concerned with the broad patterns of the movement of energy in ecosystems. Generally, we will group species into functional units such as herbivores, detritivores, and so on. As we summarize the general patterns of energy flow, remember that underlying them are species and individuals, each of which is adapting in the ways described in Chapters 2 and 3. These adaptations drive the more general ecosystem patterns we are trying to understand.

The eighteenth-century physiologist Antoine-Laurent Lavoisier was the first to study the role that energy plays in living organisms. He recognized that if organisms are to remain alive, they must obtain energy in some form and that most of this energy is gradually dissipated. It is this process of accumulating and dissipating energy, on an ecosystem scale, that is of interest to us in this chapter.

The Laws of Thermodynamics

The first two laws of thermodynamics are particularly relevant to this process. The first law states that

$$\Delta H = \Delta Q - \Delta W$$

where ΔH is the change in system enthalpy, the internal energy of a system; ΔQ is the net heat exchanged between the system and its surroundings; and ΔW is the net work exchanged with the environment. This equation describes the law of conservation of energy: The change in a system's energy is equal to the thermal energy added to the system minus the work done by the system. In other words, the first law states that *the total energy of an isolated system is constant, although within that system energy may change its form.*

The second law of thermodynamics is given by

$$\Delta H = \Delta G + T\Delta S$$

where ΔG is the net change in the Gibbs free energy of the system, T is temperature, and ΔS is the change in the entropy, or disorder, of the system. This equation describes the relationship between the forms of energy. When energy is spontaneously converted between forms, the conversion is always less than 100 percent efficient. Thus the second law states that *in any energy transfer, some energy is lost in the form of heat.* As we will see, this law imposes an important constraint on the organization of ecosystems.

Energy Capture in Ecological Systems

The energy to drive ecological systems ultimately comes from the sun. When life on Earth evolved some four billion years ago, the first organisms derived their energy from organic molecules in the "primordial soup." These organisms were **heterotrophs,** animals that derive energy either from organic compounds in the environment or from other organisms. As the supply of energy-containing molecules in the "soup" dwindled, the process of photosynthesis evolved, whereby organisms could capture solar energy directly. These photosynthesizers were **autotrophs,** organisms that derive energy from inorganic sources. Photosynthetic autotrophs, which can obtain energy from solar radiation and inorganic molecules, include higher plants, algae, and cyanobacteria. Chemoautotrophs derive energy from the energy stored in inorganic chemical bonds. Only certain bacteria are chemoautotrophic.

Primary Production

The fundamental processes of energy capture and use by organisms are photosynthesis and respiration.

Photosynthesis: $6CO_2 + 6H_2O \rightarrow C_6H_{12}O_6 + 6O_2$

Respiration: $C_6H_{12}O_6 + 6O_2 \rightarrow 6CO_2 + 6H_2O$

During photosynthesis, solar energy is captured and stored in the high-energy bonds of carbohydrates; oxygen is released in the process. The carbohydrate is used by the autotroph or is ingested by a heterotroph. The energy contained in the carbohydrate is released during respiration via glycolysis and the Krebs cycle; CO_2 and water are also released. In virtually all ecosystems, photosynthetic autotrophs provide energy for the rest of the system, so the ultimate source of energy for the system is the sun.

The rate of accumulation of energy in organic molecules by photosynthesis is known as **primary production.** This parameter is usually expressed in units of energy per unit area per unit time. For example, primary productivity might be expressed in terms of kcal per square meter per minute. Sometimes the energy unit is replaced with the quantity of organic matter in which the energy is stored. In this case we are using a measure of **biomass,** the total dry weight of organic matter. Biomass, which is correlated with energy content, is far easier to measure than energy content.

Biomass is measured in units of dry weight of organic matter per unit area (for example, kilograms per square meter). The fraction of the total biomass in living plant tissue is referred to as **standing crop biomass.** The nature of biomass changes constantly as plant material is produced and either dies or is consumed. In the prairies of North America, for example, many of the grasses are perennials. In the summer, the total standing crop increases as the plants grow. Approximately 20 percent of the standing crop is above ground in stems and leaves; 80 percent is below ground in root tissue. By fall, however, the amount and nature of the biomass have changed radically. Grazing has removed much of the plant material, and the remaining above-ground parts of the plant wither and die. At this point, the standing crop biomass is almost entirely in the roots.

The *total* rate of accumulation of energy is the **gross primary production.** It is a measure of the total rate of photosynthesis in the ecosystem. The **net primary production** is the gross primary production minus the energy used in respiration. Thus gross and net primary production and respiration are related as follows:

Net 1° production = gross 1° production − respiration

Net primary production is generally between 40 and 80 percent of gross primary production.

Some interesting ecosystems do not depend directly on photosynthetic organisms. The first was discovered deep beneath the sea in a region known as the Galápagos Rift, so called because it lies at the boundary between tectonic plates near the Galápagos archipelago in the Pacific. The Galápagos Rift system lies more than 2500 meters below the surface, far beyond the depth to which light penetrates. A feature of the rift system is a series of hydrothermal vents that release superheated water. Associated with these vents is a complex and diverse ecosystem composed of mussels, giant clams, and tube worms (Childress, Felbeck, and Somero 1981).

In most ecosystems a small but constant amount of the carbon fixed by photosynthesis is ^{13}C. In the rift systems, however, ^{13}C is relatively scarce, indicating that the carbon in the rift ecosystem is not photosynthetically derived. It turns out that the system is based entirely on chemoautotrophic bacteria. The vents release a constant stream of water containing high concentrations of hydrogen sulfide, which some bacteria can oxidize to derive energy. Most of the higher organisms at the vents are filter feeders that depend on these bacteria. The rift systems are not completely isolated from photosynthesis, however. The bacteria (and the other organisms) use dissolved oxygen that is photosynthetically derived in the upper layers of the ocean and carried to these depths by the circulation of the water.

Measuring Productivity   Productivity is a difficult quantity to measure in the field. To measure gross primary productivity in a plant community, we need to know the total rate of photosynthesis per unit area over the course of a year. Because it is difficult to measure photosynthesis directly in the field in terrestrial systems, we are often forced to use some indirect measure such as biomass. But this approach presents some obvious practical difficulties. It is hard to measure below-ground biomass without destroying the plant, but we cannot ignore that component because below-ground net production is estimated to range from 40 to 85 percent of total production (Fogel 1985). In addition, measuring new growth in a perennial presents problems. Imagine trying to quantify one year's addition of biomass in an oak tree. How do we measure the losses to respiration or to disease? Clearly, the measurement of gross production is not easy.

Although it is very difficult to measure gross production of a particular plant or group of plants in a community, remote sensing techniques have proved valuable in measuring the total rate of photosynthesis over a large area. Photosynthesizing plants characteristically reflect light in the visible (400–700 nanometers) and near-infrared (725–1100 nanometers) parts of the spectrum. Using satellites with high-resolution radiometers, we can measure the reflectance in each of these wavelengths. Plants' relative reflectances are highly correlated with primary production (Graetz, Fisher, and Wilson 1992).

Net primary production is somewhat easier to measure. We can harvest one year's growth and determine its biomass. This quantity is correlated with net production: The amount of new material left after respiration and herbivory have occurred represents the net production. Because assigning biomass accumulation to a particular year's growth can still be difficult, these measurements are simpler in certain kinds of communities such as annual grasslands. There the total biomass of an individual at the end of the growing season represents the net production of a single year.

In aquatic systems, the task is greatly simplified because we can easily use radioisotope markers to measure the rate of photosynthesis directly. The standard procedure is to use radioactively labeled carbon to measure the rate of carbon fixation by phytoplankton (the main photosynthetic organisms in aquatic systems). Because plants take up ^{12}C and the radioactive isotope ^{14}C at virtually the same rate, we can use the following relation to measure the total rate of carbon fixation by photosynthesis:

$$\frac{^{12}C \text{ assimilated}}{^{12}C \text{ available}} = \frac{^{14}C \text{ assimilated}}{^{14}C \text{ available}}$$

The quantity we are trying to measure is the amount of ^{12}C assimilated (the photosynthetic rate). We can readily measure the amount of ^{12}C available in the water. We then inoculate the water sample with a known quantity of ^{14}C-labeled bicarbonate (this is ^{14}C available). After a period of incubation, the amount of ^{14}C that has been assimilated is measured by placing the phytoplankton in a scintillation counter. We now know three of the four quantities in the equation, and it is a simple matter to solve for the unknown—the ^{12}C assimilated.

This method can be used in any system in which ^{14}C can be made available to the plants or phytoplankton. It is easiest in aquatic systems because carbon is taken up from the water, and it is relatively simple to give plants in a closed system (a transparent container) a known quantity of radioactive carbon. The incubation can be done in the field or in laboratory samples subjected to the same light regime as that in the field. Some attempts have been made to use a similar approach in terrestrial systems (Odum and Jordan 1970) by enclosing part of the community in transparent plastic and measuring changes in CO_2 concentration to monitor the uptake of carbon by the plants. This procedure presents considerable practical difficulties, however, including the task of ensuring that the measurement apparatus does not change important physical factors such as temperature and light intensity.

Patterns of Primary Production   Viewed from space, the Earth reflects a number of different shades of color (Figure 14.1). These shades in fact represent variation in primary productivity. At the poles, white

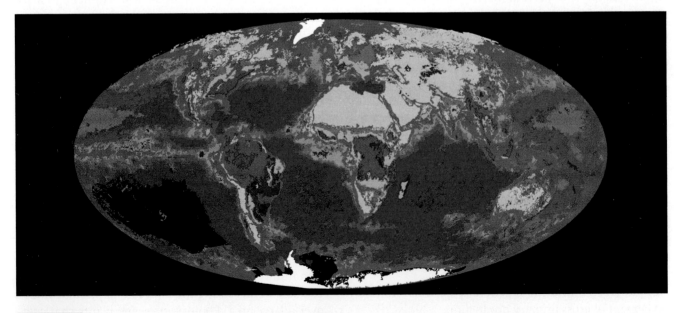

FIGURE 14.1 Composite satellite photograph of Earth primary production. In terrestrial systems, lighter areas indicate low production; intensity of green indicates level of primary production. In marine systems, phytoplankton, an indicator of production, ranges from lightest green (indicating highest level of production) to brighter green (indicating lowest). (*Gene Carl Feldman, Compton J. Tucker-NASA Goddard Space Flight Center*)

TABLE 14.1
Primary Productivity of Terrestrial and Marine Ecosystems

Ecosystem Type	Net Primary Productivity, per Unit Area (g m^{-2} or t km^{-2})				Biomass per Unit Area (kg m^{-2})		
	Area (10^6 km^2)	Normal Range	Mean	World Net Primary Production (10^{9}t)	Normal Range	Mean	World Biomass (10^{9}t)
Tropical rain forest	17.0	1000–3500	2200	37.4	6–80	45	765
Tropical seasonal forest	7.5	1000–2500	1600	12.0	6–60	35	260
Temperate evergreen forest	5.0	600–2500	1300	6.5	6–200	35	175
Temperate deciduous forest	7.0	600–2500	1200	8.4	6–60	30	210
Boreal forest	12.0	400–2000	800	9.6	6–40	20	240
Woodland and shrubland	8.5	250–1200	700	6.0	2–20	6	50
Savanna	15.0	200–2000	900	13.5	0.2–15	4	60
Temperate grassland	9.0	200–1500	600	5.4	0.2–5	1.6	14
Tundra and alpine	8.0	10–400	140	1.1	0.1–3	0.6	5
Desert and semidesert shrub	18.0	10–250	90	1.6	0.1–4	0.7	13
Extreme desert, rock, sand, and ice	24.0	0–10	3	0.07	0–0.2	0.02	0.5
Cultivated land	14.0	100–3500	650	9.1	0.4–12	1	14
Swamp and marsh	2.0	800–3500	2000	4.0	3–50	15	30
Lake and stream	2.0	100–1500	250	0.5	0–0.1	0.02	0.05
Total continential	149		773	115		12.3	1837
Open ocean	332.0	2–400	125	41.5	0–0.005	0.003	1.0
Upwelling zones	0.4	400–1000	500	0.2	0.005–0.1	0.02	0.008
Continental shelf	26.6	200–600	360	9.6	0.001–0.04	0.01	0.27
Algal beds and reefs	0.6	500–4000	2500	1.6	0.04–4	2	1.2
Estuaries	1.4	200–3500	1500	2.1	0.01–6	1	1.4
Total marine	361		152	55.0		0.01	3.9
Full total	510		333	170		3.6	1841

From Whittaker 1975.

snow and ice indicate regions of very low production. Through the temperate zones, and especially in the tropics, green indicates high productivity. There is relatively little biomass and production in the mid-latitude deserts. Clear water with low algal biomass reflects the bright green wavelengths. Thus the oceans and great lakes are, compared to other environments, unproductive. Even at this scale, the variation in production across the planet is clear.

The total primary production of the planet is approximately 170×10^9 metric tons of biomass per year (Whittaker 1975). About one-third of this is from marine systems, the rest from terrestrial ecosystems. Even though freshwater systems with emergent vegetation are among the most productive in the world, they contribute relatively little to total global production because they cover a relatively small percentage of the Earth's surface, compared to the productive parts of marine and terrestrial systems.

The broad patterns of primary production across the biomes of the world are shown in Figure 14.2 and Table 14.1. Note that for each ecosystem type, primary production varies more than an order of magnitude. Interestingly, the most productive terrestrial systems have about the same production as the most productive marine systems. Deserts are the least productive ecosystems, except for rock and ice. Even the open ocean exceeds desert production by almost 50 percent.

Note in Figures 14.1 and 14.2 that production tends to be higher near the equator than near the poles. However, primary production can vary tremendously at a given latitude. For example, following the 20° N and S latitude lines around the globe reveals that

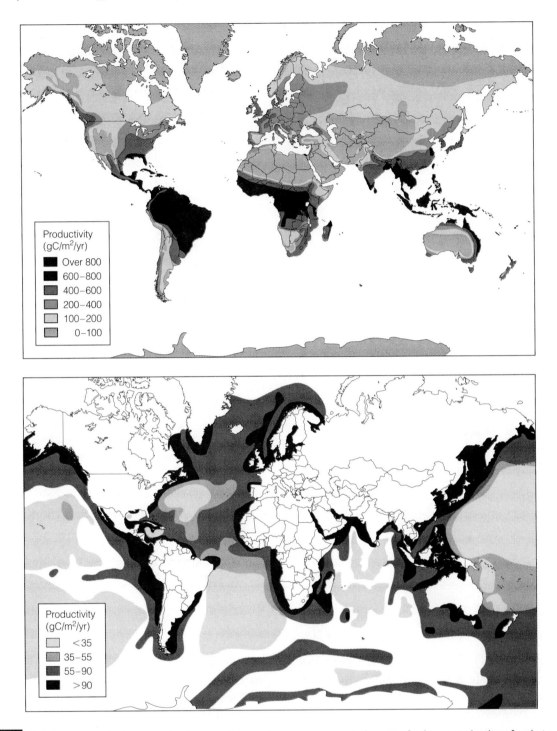

FIGURE 14.2 Global patterns of primary production. Shadings on the maps represent the rate of primary productions for that region.

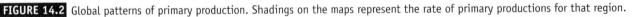

primary production varies as much on these latitudinal lines as it does between any two points on the globe.

Figure 14.3 shows the relationship between primary productivity and biomass. In general, biomes with high biomass also have high productivity. It is difficult to interpret this result. It may be that a large biomass is needed for production to occur at a high rate, or perhaps high production leads to high biomass.

If we calculate and plot productivity/biomass (P:B) ratios for various terrestrial and aquatic systems, some interesting patterns emerge (Figure 14.4). The ratios fall into three general groups. The first group, which has very low P:B ratios, corresponds to forest or shrubland habitats. The second group has slightly higher P:B ratios and is composed mostly of grassland-type habitats. The final group, which has very high P:B ratios, includes freshwater and marine habitats.

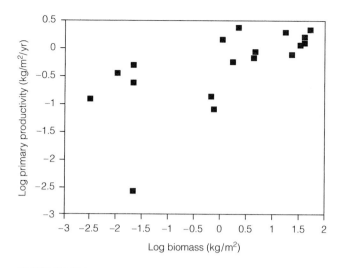

FIGURE 14.3 Log primary productivity as a function of log biomass for various biomes. (*Data from Whittaker 1975*)

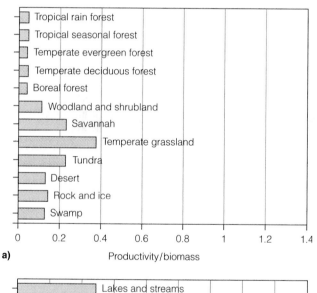

a)

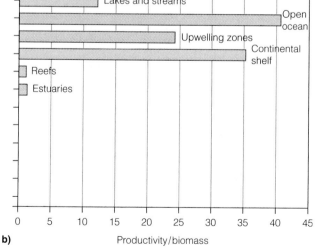

b)

FIGURE 14.4 The ratio of productivity to biomass for different vegetation regions and aquatic systems. (*Data from Whittaker 1975*)

These three groups correspond to three fundamental types of primary producers. In the forest and shrub systems, the trees and woody plants that do the bulk of the production maintain large amounts of nonphotosynthetic tissue in the form of trunks and massive root systems. Thus the amount of production, even in highly productive tropical forests, is low relative to the biomass. In the second group, a large fraction of the plant is photosynthetic—grasses have photosynthetic stems and no woody material. Finally, the aquatic systems have very high P:B ratios because their production is primarily due to single-celled phytoplankton. Note that among the aquatic systems, reefs and estuaries are exceptions: Their ratios are more like those of grasslands. The production in estuaries is more likely to be by multicellular plants. Reefs are exceptional because the primary producers are algae in symbiotic association with corals. Thus reefs contain more nonphotosynthetic tissue than other aquatic systems.

Factors That Determine Production

Figures 14.1 and 14.2 indicate the great spatial variation in productivity across the Earth. What factors lead to that variation? What factors stimulate or limit production? The latitudinal shifts in Figure 14.1 suggest that two factors that clearly vary with latitude—light intensity and temperature—are probably very important (Polis 1999). Nevertheless, the gradient is not perfectly correlated with latitude, and land and water differ. Thus other factors must also come into play.

We defined primary production as the *rate* of capture of solar energy by plants. Accordingly, the factors that limit the rate of photosynthesis limit production. In Chapter 3, we considered several of these factors in the context of abiotic limiting factors for plants. In describing the zone of tolerance for a plant, in effect we described the factors that determine photosynthetic rates. From our discussion in Chapter 3 we might surmise that the most important limiting factors for primary productivity are light, water, temperature, and nutrients.

How do we demonstrate which factors limit production? There are two fundamental approaches. First, we can examine the relationship between productivity and the natural variation in physical factors. A correlation between, say, primary production and average light intensity suggests that light is a limiting factor. These are correlational data, however, and as we know, correlation does not always imply cause and effect. Second, we might experimentally manipulate light intensity and measure its effect. If we shade a portion of a community and demonstrate decreased production, we can conclude that light is an important factor. Of course, experiments like this are difficult to perform on a large geographic scale. In the next paragraphs, we consider each of the putative limiting factors.

Light

If we compare a map showing the intensity of solar radiation across the globe (Figure 14.5) with the map of global primary production in Figure 14.2, we observe some strong similarities. This suggests not only that light is a limiting factor but also that it may interact with other parameters to determine local production.

The relationship between photosynthetic rate and light intensity (Figure 14.6) also suggests that light limits production. Note in Figure 14.6 that most of the C3 plants (exemplified by *Quercus rubra*) reach a point beyond which photosynthesis no longer increases. C4 plants (exemplified by *Chenopodium album*), however, do not reach an asymptote at higher light intensities, and this suggests that they are more likely to be light-limited.

As we saw in Chapter 13, light is an important force driving the pattern of plant community development or succession. In deciduous forests, saplings are suppressed in the shade of the canopy trees. If a canopy tree falls, saplings are "released" by the sudden influx of light and grow rapidly.

These data indicate that individual plants experience light limitation, but though the data are suggestive, they do not prove that ecosystem production is light-limited. There are, however, ecosystem-level

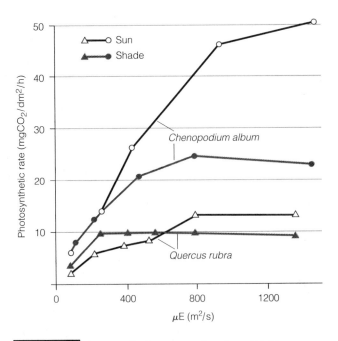

FIGURE 14.6 Photosynthetic rate as a function of light intensity (μE) in red oak (*Quercus rubra*), a C3 plant, and in pigweed (*Chenopodium albur*), a C4 plant. (*From Bazzaz 1979*)

data demonstrating that light limits primary productivity. In aquatic systems, for example, production drops to zero at depths where light intensity has significantly decreased. For much of the ocean, there is insufficient light for photosynthesis because water absorbs the key wavelengths of light so readily. In fact, more than 97.5 percent of the ocean lies below the photosynthetic zone (Polis 1999). Similarly, aquatic production varies over the course of the day according to the diurnal change in light intensity. Productivity is also lower on cloudy days than on bright, sunny days.

Water

There are ample data demonstrating that water limits primary productivity in some systems. As we saw in Chapter 3, water stress leads to a decrease in the photosynthetic rate because the stomata must be closed to conserve water. When they are closed, gas exchange is prevented and the plant becomes CO_2-limited, meaning that lack of CO_2 can limit photosynthesis even when light and other factors are abundant.

Ecosystems in extreme environments clearly demonstrate that water can limit production. For example, in many ecosystems

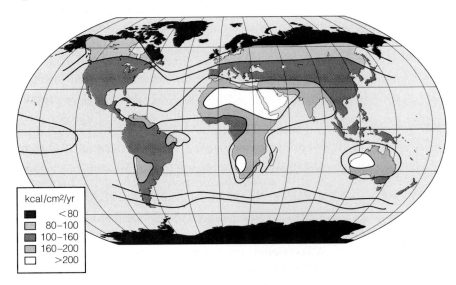

FIGURE 14.5 Annual average solar radiation (300–2200 nanometers) reaching the Earth's surface. (*From Geiger 1965*)

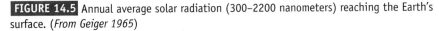

there is a strong correlation between primary production and annual precipitation. D. R. Cable (1975) showed that in desert perennial grasslands, the best predictors of net primary productivity are the precipitation that occurred the previous summer and the precipitation falling in the current season. These two factors explain between 64 and 91 percent of the annual variation in productivity. R. H. Whittaker and W. A. Niering (1975) demonstrated a correlation between precipitation and net productivity along a precipitation gradient representing habitats from desert to fir forest (Figure 14.7).

One of the tightest correlations between water and primary production is known from the McMurdo Dry Valleys in Antarctica (McKnight et al. 1999). This ecosystem is characterized by large expanses of barren ground, periglacial (freeze–thaw) processes, and large permanently ice-covered lakes (Figure 14.8). Streams in the dry valleys are ephemeral—they flow for only 4 to 10 weeks during the Austral summer when brief glacial thaw provides meltwater. The primary producers in this ecosystem are filamentous cyanobacteria and green algae. These algae form mats in the streambed that exist in a freeze-dried state for up to 11 months of the year. Even though summer temperatures are high enough to allow production, these algae do not function until they receive meltwater. Remarkably, they will begin photosynthesizing within minutes of contact with water from the first flows of the summer season.

As we might expect, the effect of water on productivity is greatest in systems where water is in particularly short supply. Indeed, W. Webb and colleagues

FIGURE 14.8 An antarctic dry valley: Taylor Glacier and the West Lobe of Lake Bonney, Taylor Valley, Antarctica. (*Photo by R. Price-OSF/Animals Animals*)

(1978) found that the above-ground net productivity of desert grassland and short-grass prairie is related linearly to water uptake, but the productivity of Arizona conifer forests is not.

Another indication that water is limiting is the correlation between production and **evapotranspiration,** the loss of water to the atmosphere from plants and soil. The **potential evapotranspiration (PET)** is the rate of loss judged possible on the basis of the system's abiotic factors, such as wind, temperature, and solar radiation. The **actual evapotranspiration (AET)** is the amount that is actually lost from plants and soil. The interpretation of evapotranspiration is more complex. AET includes a number of phenomena such as the amount of water available to evaporate from soil or transpire from the plants, the plants' efficiency of water use, and physical factors such as temperature and wind that increase the loss of water to the atmosphere.

For a number of ecosystems, there is a positive relationship between AET and productivity (Webb et al. 1978). Grasslands are an example of an ecosystem in which AET and productivity are correlated (Figure 14.9). As we might predict, these correlations are much stronger in biomes where precipitation is relatively low. In forest, where water is abundant, there is

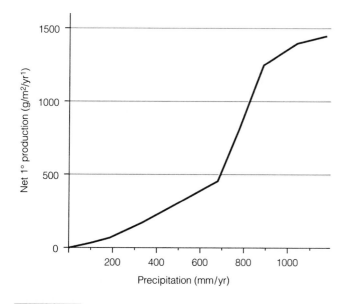

FIGURE 14.7 Change in net productivity along a precipitation gradient. (*From Whittaker and Niering 1975*)

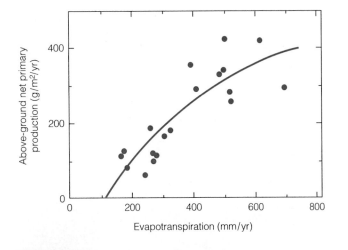

FIGURE 14.9 The rate of net primary production as a function of actual evapotranspiration measured in several grassland sites in the United States. (*From Webb et al. 1978*)

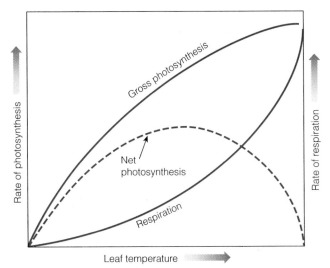

FIGURE 14.10 Photosynthetic rate and rate of respiration as functions of leaf temperature. The difference between the two curves is the net photosynthesis. (*After Fitter and Hay 1987*)

no correlation between AET and primary production (Webb et al. 1978).

Temperature   As we saw in Chapter 3, plants have a zone of tolerance for temperature, so we would expect high and low temperatures to decrease production. Within the zone of tolerance, we expect gross production to increase with temperature as the result of a Q_{10} effect. Respiration should also show an increase with temperature. Thus we expect a slightly different effect of temperature on *net* and *gross* primary production. Within the range of tolerable temperatures, gross productivity will increase with temperature. But because of higher respiration losses, net photosynthesis will peak at a lower temperature than gross production (Figure 14.10).

The effects of temperature on primary productivity are difficult to disentangle from the effects of water. Deserts and grasslands are characterized by *both* high temperature and low precipitation. The correlation between production and AET is indicative of this interaction between water and temperature effects.

Nutrients   The final major factor in plant productivity that we will consider is nutrients. Any of the macro- or micronutrients shown in Table 3.3 can limit plant growth and thus production. Nutrients can originate within the system (**autochthonous nutrients**) or from outside the system (**allochthonous nutrients**). Allochthonous inputs may come from great distances. For example, much of the phosphorus responsible for high biomass in the Amazon rain forest comes via dust from Africa (Swap et al. 1992).

The best data available on the role of nutrients in limiting production come from aquatic systems.

Productivity varies among lakes and is clearly related to their nutrient contents. Oligotrophic lakes, in which productivity is low, tend to be young and located in regions where runoff deposits few nutrients into them. Such lakes can be converted into eutrophic lakes, which are characterized by high productivity and by the natural or anthropogenic input of nutrients (such as the runoff of fertilizer from agricultural land). This process, called **eutrophication,** provides clear evidence that nutrients can limit productivity.

In the open ocean the relationship between nitrogen and productivity is so consistent that it can be used to estimate new primary production. S. Sathyendranath and colleagues (1991) define a variable *f*, which is the ratio of new to total production (P_n/P_t). The total production can be estimated indirectly because of a series of correlations: (1) The temperature of the open ocean can be measured from its color; (2) temperature is strongly correlated with nitrogen concentration; and (3) the value of *f* is a function of the nitrogen concentration. Thus the new primary production (P_n) can be estimated as $f \times P_t$.

There is also experimental evidence for the limiting role of nutrients in primary production. C. R. Goldman (1960) experimentally studied the role of nutrients as limiting factors in lakes in southwestern Alaska. Preliminary studies had indicated considerable variation in the concentrations of nutrients important to the growth of phytoplankton. At several times during the growing season, Goldman suspended in the lake bottles containing lake water, to which a specific amount of one of the putative limiting nutrients was added. Productivity (photosynthesis) was measured in terms of radioactive ^{14}C uptake. As Figure 14.11 shows, nitrogen, magnesium, and phosphorus were all limiting: Adding any of these nutrients increased production relative to controls.

Nutrient limitation of aquatic production has also been demonstrated on a larger scale. In a classic experiment, D. W. Schindler (1974) partitioned a lake in northwestern Ontario. One portion received phosphorus, nitrogen, and carbon supplements; the other received only nitrogen and carbon. Within two months, the portion given supplemental phosphorus was covered with a spectacular algal bloom. Such experiments are more difficult in the open ocean, with its huge expanses of unenclosed ecosystem and its currents and tides to move materials great distances. However, some recent experiments have shown that enrichment experiments can be conducted even in this more difficult environment. Coale et al. (1996) seeded 72 km^2 with iron in the form of acidic iron sulfate. The input of iron triggered a massive phytoplankton bloom, which certainly suggests

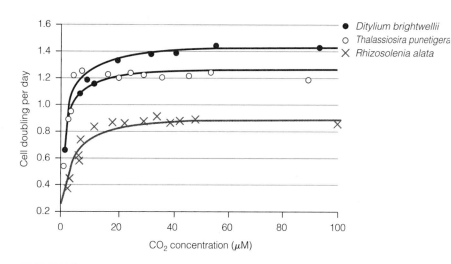

FIGURE 14.12 Productivity (rate of cell doubling) as a function of dissolved CO_2 concentration in three species of diatoms. (*From Riebesell, Wold-Gladrow, and Smetacek 1993*)

that iron is a significant limiting factor in the open ocean.

Sometimes nutrients that are seemingly in ample supply are limiting. For example, it is generally believed that dissolved inorganic carbon, or dissolved CO_2, is not limiting in marine systems because its concentration in the ocean is two to three orders of magnitude greater than that of any other plant macronutrient. However, recent work by V. Riebesell, D. A. Wold-Gladrow, and V. Smetacek (1993) with diatoms, whose main source of carbon is dissolved CO_2, indicates that under experimental conditions in which light and nutrients were not limiting, productivity was affected by the concentration of CO_2 (Figure 14.12).

Clearly, then, light, water, temperature, and nutrients are major factors that limit primary production in most systems. However, because any factor that limits plant growth affects production, we should not be surprised to find a number of new factors emerging as limiting factors. For example, viral pathogens of producers such as diatoms and cyanobacteria have recently been implicated as limiting factors in marine production. In the open ocean, each milliliter of seawater contains 10^6–10^9 virus particles, some of which are known pathogens of phytoplankton. Addition of virus-sized particles from seawater to laboratory cultures reduced production by as much as 78 percent (Suttle, Chan, and Cottrell 1990).

Interactions Among Factors Although we have discussed light, water, temperature, and nutrients separately, these factors surely interact in complex ways to determine local primary production. Simple

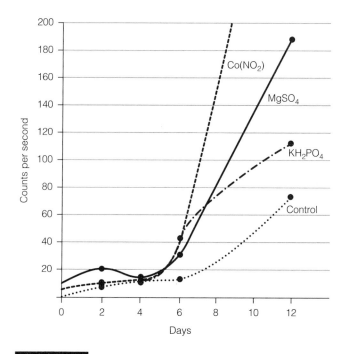

FIGURE 14.11 Photosynthetic rates (as reflected in the uptake of radioactive carbon) in samples of lake water to which nitrogen, magnesium, or phosphorus was added. (*From Goldman 1960*)

experiments that test the role of a single factor such as iron in the open ocean as discussed above are valuable because they identify key factors. But we should expect that second-order effects resulting from interactions among factors should be significant as well—and more difficult to ascertain.

One might expect that factors become limiting only when all other factors are in sufficient supply—in other words, the factor in shortest supply limits production. In aquatic systems, light and nutrients can both be limiting. Sterner et al. (1997) examined the relationship between these two factors in a series of lakes in Ontario. The ratio of light:phosphorus was strongly correlated with the C:P ratio of the seston (suspended particulate matter consisting of living algae, bacteria, protozoa, and abiotic material) in the water column. The authors suggest that the balance between light and nutrients controls the nutrient use efficiency. When light is high relative to nutrient availability, the producers are carbon-rich and phosophorus-poor. On the other hand, when light is reduced (as at greater depth), the producers are phosphorus-rich. Recently, a mode of analysis known as **meta-analysis** has been developed to examine such interactive effects (Adams et al. 1997). This procedure quantifies the effects of a number of factors examined in a series of experimental studies. Downing et al. (1999) undertook such a meta-analysis of marine nutrient experiments to determine the relative roles of various nutrients in limiting productivity. They calculated a measure, Δr, that estimates the change in growth rate of phytoplankton caused by specific nutrients. Figure 14.13 shows the result of their meta-analysis for nitrogen, phosphorus, and iron in several marine habitats. This kind of analysis quantifies the

relative limitation effects in a way that allows comparison across nutrients and habitats.

Abiotic factors other than those that contribute directly to photosynthesis can control interacting webs of light, water, temperature, and nutrients. In tallgrass prairie, production is determined in part by the regional and local climatic patterns. At a landscape level, the relatively low precipitation of the Great Plains leads to grass as the dominant form of vegetation, which significantly affects the rate of production (see Table 14.1 and Figure 14.1). Locally, annual net primary production (ANPP) is strongly correlated with precipitation. However, abiotic factors such as fire that don't directly affect photosynthesis can control interacting webs of light, water, temperature, and nutrients in ways that control ANPP. A series of studies (Knapp et al. 1998) at Konza Long-term Ecological Research site in Kansas has elucidated the role of fire in annual net primary productivity (ANPP). The dominant grass, big bluestem (*Andropogon gerardii*) responds physiologically to fire in ways that increase its production (Figure 14.14). Fire removes the dead biomass (detritus) at the soil surface. This increases light intensity for young shoots near the soil surface. Because light can penetrate to the soil surface, the soil warms more rapidly in the spring, and shoot growth is faster. Nitrogen uptake also seems to be more rapid after a burn. The only negative effect of fire occurs in a drought year when removal of the detritus exacerbates the drought conditions. Under dry conditions, the dead biomass near the soil surface holds more moisture and decreases surface temperature. Thus the availability of several important limiting factors is determined in large measure by the frequency of fire (Figure 14.15).

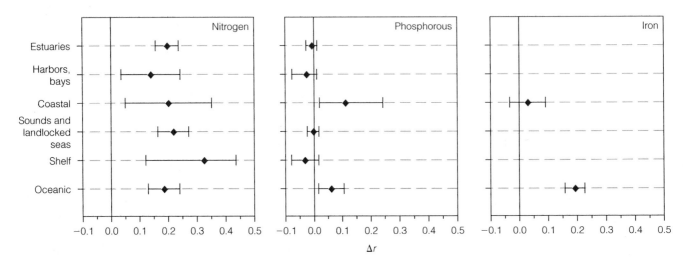

FIGURE 14.13 The variation in nitrogen, phosphorus, and iron limitation in various marine habitats. Δr is an estimate of the change in growth rate of phytoplankton caused by specific nutrients. Error bars represent 95% confidence intervals. (*From Downing et al. 1999*)

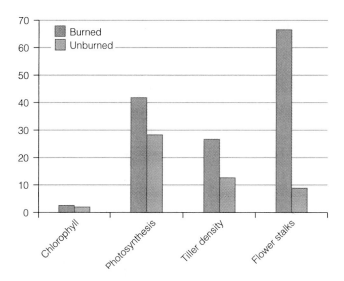

FIGURE 14.14 Effects of fire on several aspects of the dominant grass, big bluestem (*Adropogon gerardii*). Units are as follows: chlorophyll (mg/g); net photosynthesis (μmole/m^2/sec); tiller density (number/m$^2 \times 0.1$); and flower stalks (number/m^2). (*Data from Knapp et al. 1998*)

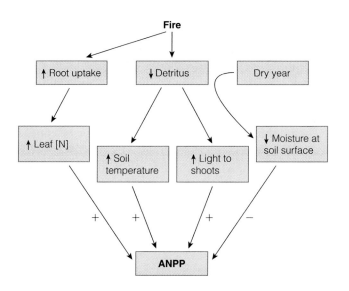

FIGURE 14.15 The indirect effect of fire on annual net primary productivity (ANPP) in tallgrass prairie. Fire affects nitrogen uptake and the amount of dead biomass (detritus), which in turn affect ANPP positively (+) or negatively (−).

Trophic Structure

We now turn our attention to the movement of energy through ecosystems. The first and second laws of thermodynamics constrain the nature of energy flow. The first law says that only the form of energy can be changed, not the amount. Thus solar energy (radiation) is captured via the photosynthetic process and stored as chemical energy in the bonds of organic molecules such as glucose and starch. The chemical energy is utilized by the plant (or by an herbivore) via its respiratory machinery to do work—to transport ions, to move, to synthesize tissue from organic and inorganic molecules, and so on. The second law states that none of these energy transfers is 100 percent efficient; some energy is lost as heat at each transfer.

These two laws lead to the fundamental generalization we can make about energy transfer in ecosystems: Energy flow is linear. Materials can be cycled and reused by organisms (see Chapter 15), but energy flows unidirectionally through ecosystems. It enters the system in the form of solar radiation, is captured by plants, and then is transferred to one or more heterotrophs. Along the way energy is gradually lost as heat, so no recycling of energy is possible. This principle organizes the structure of the biotic components of ecosystems.

Food Chains

Let us consider the path that energy takes through an ecosystem. The solar radiation that strikes the plants in a community has several possible fates (Figure 14.16). Some sunlight is immediately reflected. The percentage reflected varies greatly with the nature of the community. In forests, for example, 10–12 percent is reflected from the leaves, whereas about half that amount reflects off herbaceous vegetation (Larcher 1980). Some sunlight is absorbed by the leaves to drive photosynthesis. The chloroplasts absorb 60–80 percent of the light that reaches them (Spanner 1963);

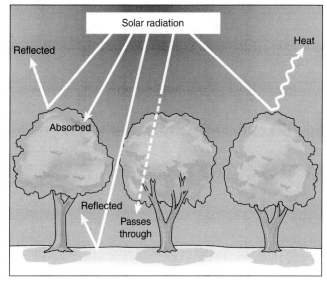

FIGURE 14.16 A diagrammatic representation of the fate of solar radiation in a forest.

the rest is reflected, passes through the leaf, or is converted to heat. The remaining light strikes the ground or other objects in the environment.

The portion of the energy that is captured by the plant is converted into chemical energy. This energy is available to the plant for its life functions and is also available to heterotrophs that consume the plant. These heterotrophs can in turn be consumed by others (carnivores), and so on. Eventually, the energy reaches a carnivore that is not preyed upon by any other animals. This animal is the **top carnivore.**

The pathway of energy transfer just described is a **food chain** (Figure 14.17). On the first level of the food chain are the primary producers, usually plants. The **primary consumers** feed directly on the plants. They in turn are fed upon by the **secondary consumers,** which are eaten by the tertiary consumers, and so on to the top carnivore. **Secondary production** is the term applied to the rate of accumulation of biomass by heterotrophs. Each of the steps in a food chain is called a **trophic** (feeding) **level.**

In a simple food chain like that depicted in Figure 14.17, the notion of a trophic level is straightforward. In practice, however, the concept is not so clear. There are several difficulties (Pimm 1988). The first problem is omnivory (Polis and Strong, 1996). Raccoons (*Procyon lotor*), for example, consume both plant and animal material. To which trophic level, primary or secondary consumer, do we assign them? What about carnivorous plants? Some species of carnivorous plants have lower photosynthetic rates associated with the higher nitrogen concentrations derived from their prey (Méndez and Karlsson 1999). Others use the carbon gained from the prey to supplement photosynthesis (Adamed 1997). Second, a number of species, such as some fish, feed at different trophic levels at different times in their lives. Third, how do we describe parasites? Do they constitute an additional trophic level? Finally, it is difficult to determine precisely where a food chain ends. Because matter cycles through an ecosystem, the energy it contains may pass through an organism more than once (without being used or lost as heat). How do we count these additional transfers?

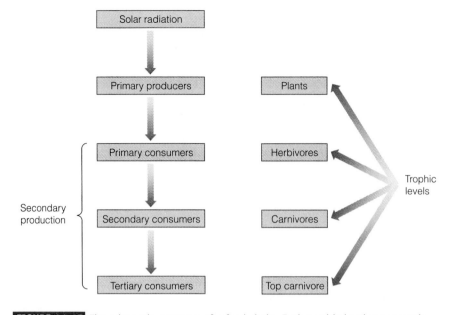

FIGURE 14.17 The schematic structure of a food chain. Each trophic level may contain many species.

Food Webs

In the simple food chain in Figure 14.17, we have not identified the species of plants and animals in each trophic level. Of course, many species may be feeding at any trophic level, and there may be multiple pathways of energy transfer: Food chains interact to form food webs (Figure 14.18). Indeed, the pathways of energy transfer may become exceedingly complex. When we depict this complexity by indicating all the species that participate in the energy transfer at each trophic level, we have described a **food web.** Figure 14.19 conveys the complexity of the interactions in these webs; even a seemingly simple community like that found in a pitcher plant exhibits complex sets of trophic relationships. In such a complex web it can be difficult even to determine what the various species feed on. Recently, chemical clues have been used to identify the paths of energy transfer when direct observation is difficult or impossible. Vander Zanden and Rasmussen (1999) have shown that the stable isotopes of carbon and nitrogen in consumers are similar to those of their food supply. Thus these isotopes can be used as indicators of trophic position and structure in complex food webs.

It is useful to have a measure of the complexity of a web because if we can quantify the complexity of interactions in a food web, we can make comparisons among systems, design experiments to test hypotheses about differences, and measure the effects of perturbations, whether anthropogenic or natural. We measure complexity with a parameter called **connectance:**

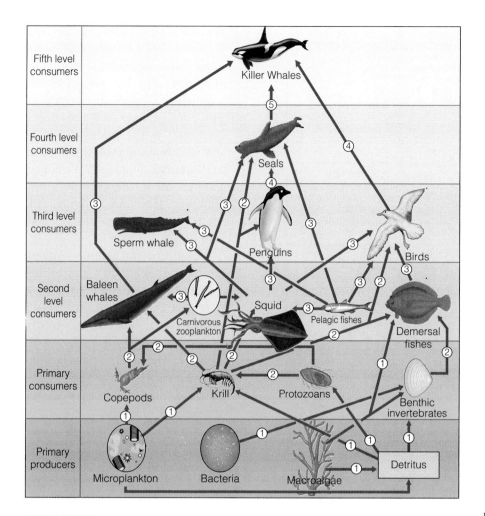

FIGURE 14.18 A simplified Antarctic food web. Note that instead of a simple, linear sequence of trophic levels, there are complex webs of interaction, including species that feed at more than one trophic level. (*From Segar 1998*)

seem stunningly intricate at first glance, but compartmentalization reduces its actual complexity. There is evidence that microhabitat variation increases the likelihood of compartmentalization (Pimm 1982) because species' preferences for a specific microhabitat isolate them from one another. The degree to which compartmentalization like this is common in nature is not well understood.

In describing food webs, we must delineate some boundary to the system of interest. This is simply a practical consideration: Additional knowledge of trophic interactions leads to bigger and more complex webs, until ultimately the entire planet is united in a single web. Even when the boundaries are clearly defined, the complexity can be stunning. When G. A. Polis (1991) described the intricate food webs of the desert sand community in the Coachella Valley of California, he found that the interacting biota included 174 species of plants, 138 species of invertebrates, more than 55 arachnids, and more than

Connectance = actual number of interspecific interactions ÷ potential number of interspecific interactions

In a web with *n* species, the potential number of interactions is given by

$$\frac{n(n-1)}{2}$$

We then count up the number of interactions that actually take place between species in the web (the number of connecting lines) and compare this with the maximum possible.

Finally, we can ask whether the food web is compartmentalized. Groups of trophic interactions may be organized into compartments, or subsets of the total species composition of the web, which may act more or less independently of one another. A web may

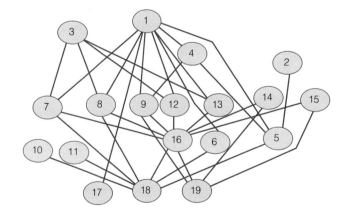

FIGURE 14.19 A food web of the insects inhabiting a pitcher plant. Each number represents a different insect species, except 16, 17, 18, and 19, which represent bacteria and protozoa, live insects, recently drowned insects, and older organic debris, respectively. Each line represents a trophic linkage; predators are higher in the figure than their prey. (*From Pimm, Lawton, and Cohen 1991*)

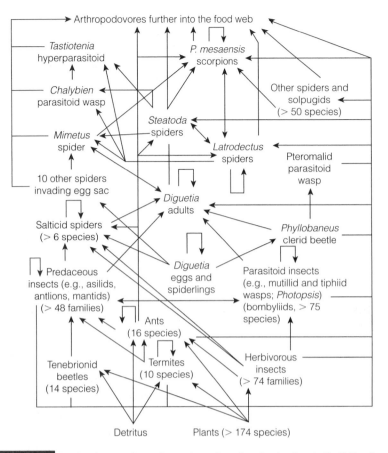

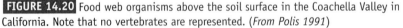

FIGURE 14.20 Food web organisms above the soil surface in the Coachella Valley in California. Note that no vertebrates are represented. (*From Polis 1991*)

Often a succession of decomposing organisms feed sequentially on the dead bodies of plants and animals (Figure 14.21). This energy is also eventually dissipated. Decomposers play a vital role in every ecosystem by preventing the gradual accumulation of dead plant and animal biomass. Moreover, they release nutrients to the environment that can re-enter the biotic component of the ecosystem.

Recall our discussion of diversity and stability in Chapter 12. Some of the early work in this area was based on the diversity and complexity of food webs. Robert May's pioneering theoretical work (May 1971; 1973) demonstrated that complex food webs are inherently *less* stable. Recent work on the stability of food webs (McCann et al. 1998) demonstrates that weak trophic interactions play an important role in the stability of food webs. Weak interactions are defined as those in which one species ingests another with low probability. The effect of such links is to dampen the oscillations in population density, thus decreasing the probability of stochastic extinction and stabilizing the system. Not surprisingly, empirical studies of food webs show that many are characterized by numerous weak interactions.

2000 species of microorganisms. Some of the trophic interactions occurring at just the soil surface of this community are depicted in Figure 14.20.

In all food webs, one energy pathway is worthy of special note. In each trophic level, some individuals die without being consumed by the organisms in the next trophic level. Moreover, eventually the top carnivores die. The energy in the top carnivores and the unconsumed portion of other trophic levels is utilized by a complex class of organisms called **decomposers**.

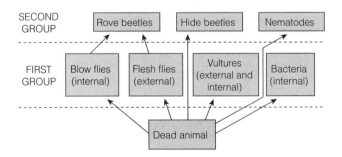

FIGURE 14.21 A decomposer food web based on a dead vertebrate animal. Decomposition is based on the activities of two sequential groups of organisms. Some in the second group feed on animals in the first group.

Efficiency of Energy Transfer

We can empirically measure the proportion of energy passed to the next trophic level in each energy transfer. These proportions are referred to as **Lindeman efficiencies.** In many systems, Lindeman efficiencies are on the order of 10 percent, but sometimes they run as high as 30 percent.

What determines the nature of the energy loss between trophic levels? We can break down the potential losses of energy into three fundamental components. Consider the energy transfer between a primary and a secondary consumer—say, a grasshopper eaten by a bird (Figure 14.22). Of the total amount of energy in grasshoppers, some will immediately be lost to the next trophic level because some grasshoppers will not be ingested by birds. This rate of transfer

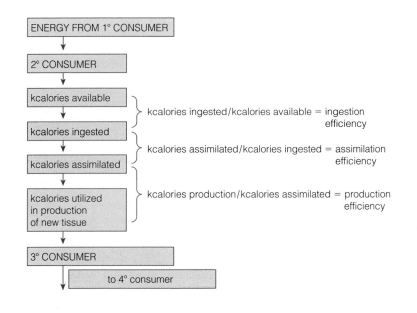

FIGURE 14.22 A diagrammatic representation of the efficiencies that determine the loss of energy in a transfer between trophic levels.

is called the **ingestion efficiency.** Of the energy ingested by the birds, some will not be digested and absorbed; this is the **assimilation efficiency.** Finally, of the energy assimilated by the bird, some will not end up as new bird biomass in the form of new tissue or reproduction. This rate of transfer is called the **production efficiency.**

Now let us consider in detail the biological bases of each of these efficiencies.

Ingestion Efficiency

The ingestion efficiency is the proportion of available energy eaten by a heterotroph. For an herbivore consuming plants, more biomass may be available than can be eaten. In the prairies of North America, grazers consumed vast quantities of plant biomass, but they never came close to consuming all the available forage. Other factors prevented the herbivore populations from reaching high enough densities across the entire continent to consume all the available plant material. In addition, plant adaptations that deter grazing, such as toxins, spines, and thick cuticles, ensured that the ingestion efficiency of herbivores remained low.

For a carnivore and its prey, the situation is similar. The predator must locate its prey, capture it, and then consume it. The predator is involved with its prey in a coevolutionary race (Chapters 2 and 10). Cryptic coloration or other adaptations prevent the detection of prey, and some prey are formidable enemies, difficult to capture and subdue if detected. Some parts of the prey may be impossible for the predator to ingest. All of the coevolutionary aspects of predator/prey adaptations considered in Chapter 10, including optimal for-

aging, affect ingestion efficiency. Other factors limit carnivore densities and thus further lower ingestion efficiency.

Assimilation Efficiency

The assimilation efficiency is determined by the nature of the material that is ingested. Once the food item has been ingested, some fraction is digested and absorbed, whereas the remainder is passed through the gut and eliminated. There is great variation in the availability of energy from different food items.

The consumption of plant tissue generally leads to assimilation efficiencies in the range of 30 percent. In many plant tissues, the energy is "locked up" by cellulose or lignin. Animals that digest plants tend to have very long guts so the plant material is processed slowly. Many, such as the ruminants, have symbiotic bacteria that can digest cellulose and lignin, releasing the energy of the plant cells for absorption by the animal.

Carnivores generally have higher assimilation efficiencies than herbivores. The exact value depends on the nature of the prey item. Prey with tough exoskeletons are difficult to digest; because much of the chitinous exoskeleton of insects passes through the gut of insectivorous vertebrates that consume them, this energy is unavailable to insectivores or higher trophic levels. Soft-bodied prey items or those in which the predator can ingest simply the muscles, organs, and gut result in much higher assimilation efficiencies.

We sometimes refer to the availability of energy in food items as **food quality.** This term refers in a general way to the relative ease of assimilation, or how readily the barriers to digestion that a prey item or plant contains can be overcome. We can rank potential food items on the basis of digestibility (quality in this sense). Carrion ranks high because bacteria and other decomposers have begun the process of breaking down the barriers to assimilating the energy content of the dead animal. Items such as fleshy fruits and soft-bodied animals are higher in quality than a seed within a hard shell or an insect protected by an exoskeleton. Leafy plant material with much cellulose is of very low quality but is still higher in quality than stems or woody material.

Of course, many adaptations by the heterotroph improve its assimilation efficiency. Certainly, there would be strong selection in favor of adaptations that enhance an organism's ability to derive energy from the food items it captures. Dentition for mastication, digestive enzymes, symbiotic relationships with other organisms, and the length and shape of the gut all represent potential features of heterotrophs that can be modified by selection. These features are much easier for selection to change than the fundamental nature of the food material. It is far easier for a grazer to develop a longer gut than to undergo the adaptations that becoming a carnivore would entail.

Production Efficiency The production efficiency is the proportion of the assimilated energy that is actually converted into new biomass by the herbivore or carnivore. Values of production efficiency range over nearly two orders of magnitude (Table 14.2). The nature of an animal's life history determines its production efficiency.

For example, one of the most important variables is whether the animal is poikilothermic or homeothermic. The latter expend considerable energy maintaining a high body temperature (Chapter 3), so this energy is unavailable for producing new tissue. Production efficiency is also affected by body size and reproductive rate. Small animals lose heat more rapidly than large ones, which decreases their produc-

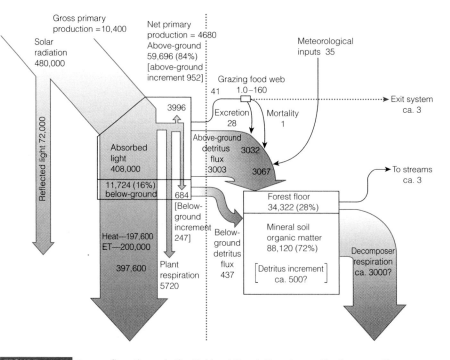

FIGURE 14.23 Energy flow through the Hubbard Brook Forest over the four-month summer period. Pool sizes are in kcalories per square meter; energy flows are in kcalories per square meter over the four-month period. (*From Gosz et al. 1978*)

tion efficiency. Animals with high reproductive rates convert a high proportion of assimilated energy into new tissue in the form of offspring.

Figure 14.23 depicts a general model of the nature of energy loss in the transfer between two trophic levels in a food chain in a forest system. It indicates the nature and magnitude of the energy losses associated with ingestion, assimilation, and production. The amount of energy in any component of the system is indicated, as are the rates of transfer between components. The higher trophic levels are amalgamated into a single block. Note that a huge amount of energy exists in the detritus component. It is also significant that only minute amounts of energy enter higher trophic levels.

Eltonian Pyramids

One way to depict the trophic structure of communities graphically is shown in Figure 14.24. The pyramids in this figure are called **Eltonian pyramids** after C. Elton (1927), who first represented trophic structure in this way. In these diagrams all the species feeding at a particular trophic level are lumped together.

Eltonian pyramids can represent the numbers, biomass, or energy content of each trophic level. Figure 14.24 shows two types of hypothetical pyramids for a deciduous forest and a grassland. The pyramids at the

TABLE 14.2	
Production Efficiencies (kcalories Used for Production/kcalories Assimilated) for Different Groups of Animals	
Taxon	Production Efficiency (%)
Insectivores	0.86
Birds	1.29
Small mammals	1.51
Other mammals	3.14
Fish	9.77
Social insects	9.77
Noninsect invertebrates	25.0
Nonsocial insects	40.7

From Humphreys 1979.

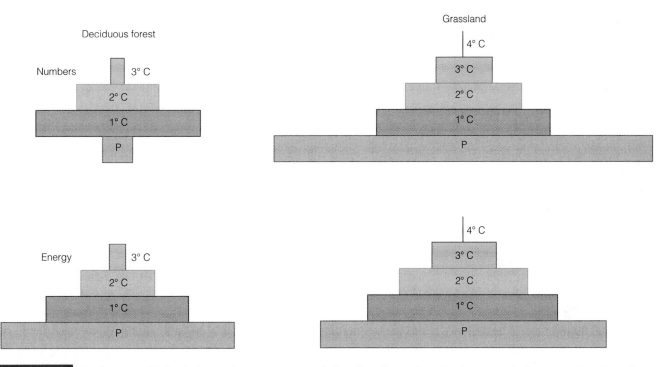

FIGURE 14.24 Eltonian pyramids for deciduous forest and grassland. Note that the producer level can contain fewer organisms than the consumer level.

top are based on numbers of organisms. Note that in the forest, a relatively few primary producers (mostly trees) support larger numbers of insect 1° consumers. In the grassland, the number of primary producers (grasses) is huge relative to the number of consumers. The forms of pyramids of numbers and biomass vary greatly among and within habitat types. In energy pyramids, the energy content must be greatest in the producer level, as demanded by the laws of thermodynamics. Energy pyramids can never be inverted.

The structure of energy pyramids has important practical consequences when toxins are introduced into the ecosystem (see the box on page 376).

The Length of Food Chains

The number of levels in the pyramids in Figure 14.24 indicates the lengths of the food chains in those communities. What are the general patterns of food chain length? Can we discern the factors that govern it?

The most straightforward measure of a food chain is simply the number of trophic levels. Because food chains are connected in complex webs, however, the number of trophic levels does not measure the length of the various pathways by which energy flows. One statistical method of measuring the pathways of energy flow is illustrated in Figure 14.25. It enables us to determine the number of trophic steps between each top carnivore and the primary producers. For

example, $3° C_2$ is connected to P_3 and P_4 by four different pathways. For three of the four food chains, there are four trophic levels; for the fourth, there are three. We describe food chain length by reporting the mode or the mean number of trophic linkages between the producer and the top carnivore. In this

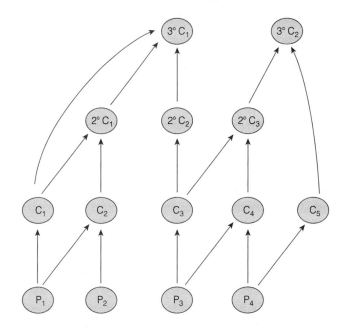

FIGURE 14.25 A diagram of food web interactions for calculating the length of the food chain. Subscripts refer to different species at a given trophic level.

Bioaccumulation

The energy relationships depicted in Figure 14.24 have important consequences. The fact that each trophic level must gather energy widely from lower trophic levels results in the phenomenon of **bioaccumulation.** Toxic pollutants that degrade slowly, or not at all, can accumulate throughout the food chain even though organisms lower in the food chain may not be affected.

Two groups of organisms are particularly vulnerable. Those high in the food chain, particularly the top carnivores, are likely to accumulate significant concentrations of toxins. In aquatic systems, piscivorous birds such as loons, eagles, pelicans, and cormorants are especially likely to accumulate toxins concentrated in fish. The other vulnerable group includes the filter feeders—organisms that glean their nutrition by processing large amounts of water. Consequently, some economically important shellfish can contain toxic levels of some elements and compounds.

An example of bioaccumulation is shown in Table 14B.1. Note that as one moves from the sediments to the vegetation and finally up the food chain to birds and mammals, the concentration increases by several orders of magnitude.

One of the most carefully documented examples of this phenomenon is known from Clear Lake in California (Southwick 1976). Even though the insecticide DDT was applied to the lake to control gnats at a low rate (0.05 ppm), the pesticide accumulated in insect-eating fish to levels of 40–2,500 ppm. Western grebes, which feed on these fish, suffered severe mortality rates. Analysis of their tissues indicated DDD concentrations of 1600 ppm; the concentration of toxin had increased to 32,000 times the application rate.

The *bioconcentration factor* (BCF) is an important parameter that measures the degree to which pollutants in the water are concentrated along the food chain. This factor enables us to estimate the concen-

tration of elements in fish tissue given the concentration in water:

$$\text{Concentration in fish} = \text{concentration in water} \times \text{BCF}$$

The BCF is simply a function of the kind of trophic relationships and energy constraints shown in Figure 14.24. The persistence and stability of the pollutant also play an important role. Toxic elements such as lead and zinc cannot degrade, and thus their toxicity persists. Some compounds such as DDT degrade very slowly and also persist as they pass through the food chain. The life span of the concentrating organisms also plays a role. Obviously, the longer the organism lives, the more toxin that can accumulate within it. Consequently, there is a broad range of BCFs for different elements. For example, cadmium and chromium have BCFs greater than 300,000, whereas the value for molybdenum is only 90 (Goudie 1993).

TABLE 14B.1

Mean Methyl Mercury Concentrations in Organisms in a Salt Marsh

Organism	Parts per Million
Sediments	<0.001
Spartina	<0.001–0.002
Echinoderms	0.01
Annelids	0.13
Bivalves	0.15–0.26
Gastropods	0.25
Crustaceans	0.28
Fish muscle	1.04
Fish liver	1.57
Mammal muscle	2.2
Bird muscle	3.0
Mammal liver	4.3
Bird liver	8.2

Data from Gardner et al. 1978.

TABLE 14.3

Distribution of Food Chain Lengths from Cohen, Briand, and Newman (1986) and Pimm (1982)

	Trophic Level						
	1	2	3	4	5	6	Source
Frequency	2	45	37	22	5	3	(1)
	0	25	44	25	7	2	(2)
	0	2	23	24	5	2	(3)

Those from Cohen et al. (1) are the single most common length over all the top predators in each web. There was no attempt to exclude data where detritus or plants were omitted from the published web or where the top predator was unlikely to be the end of the food chain. In the studies compiled by Pimm (2), each top predator was counted separately, and webs were excluded if detritus or plants were not at the base of each chain. Webs for which small invertebrates were not the end of the food chain (and thus webs that were less likely to be incomplete) are shown in the third line of the table (3).

From Pimm 1988.

example, the modal value is 4; the mean number of trophic levels is 3.75.

The Range of Food Chain Lengths

Data on the frequency distribution of the number of trophic levels, compiled using three different criteria and gleaned from a large number of published reports, are presented in Table 14.3. Note the general similarity among the three frequency distributions despite the variation in the data sets from which they were derived. The interesting feature of these frequency distributions is that there is a rather narrow range of food chain lengths across a tremendous array of community types—aquatic, terrestrial, detritus-based, and so on. No community has more than six trophic levels, and most have three or four. These data suggest that there may be some very general mechanism that determines food chain length.

We can make a few modest generalizations about the patterns of food chain length. It appears that communities in patchy habitats tend to have shorter food chains (Schoener 1989). This may be because patches are more likely to be missing one of the top predators. It also seems that two-dimensional habitats such as grasslands and intertidal zones have shorter food chains (Cohen, Briand, and Newman 1990).

Hypotheses to Explain Food Chain Length

From our knowledge of the mechanics of food webs, we can devise some hypotheses to identify the factors that determine the length of a food chain. We know that energy is gradually dissipated as it passes down a food chain. Possibly, the food chain ends (and thus the length is determined) when there is insufficient energy to support another trophic level. This suggests two basic hypotheses to explain the variation in food chain length:

1. Primary production may determine the length of food chains. Given that energy is lost all along the pathway, it may be that the amount of energy entering the food chain determines its ultimate length. This hypothesis predicts that food chains should be longer in more productive habitats.
2. The efficiencies of energy transfer within a food chain may vary such that some chains "run out" of energy sooner than others. This could occur in a number of ways. Any of the three efficiencies—ingestion, assimilation, or production—could be responsible for the energy limitation. For example, in forests, most energy in the plants is not consumed (low ingestion efficiency); instead, it enters the detritus chain. Thus we might predict that forest food chains will be shorter than others. The amount of unused energy that passes through the digestive tracts of grazers varies considerably. In some systems, production efficiency is low. If a food chain contains a number of homeotherms that dissipate large amounts of energy maintaining constant body temperatures, less energy will be available to maintain higher trophic levels, and we might then predict that the length of a food chain will be negatively correlated with the number of homeotherms it contains.

There is little evidence to support either of these hypotheses (Pimm 1988). Consider the production hypothesis first. The variation in primary productivity among aquatic ecosystems is shown in Table 14.4. Despite the broad range of energy input, there is little variation in food chain length. For example, even though Lake Myvatn has very low productivity, it supports four trophic levels (Larsson et al. 1979). Some tropical lakes have 60 times Lake Myvatn's primary productivity but no more trophic levels.

Grasslands also have a wide range of productivities and thus offer us an opportunity to test the production hypothesis. Production differs by a factor of 20 between temperate and tropical systems without differences in the number of trophic levels (Pimm 1982). Again, primary production does not correlate with food chain length.

R. L. Kitching and S. L. Pimm (1986) tested the production hypothesis experimentally in food webs found in water-filled tree holes in subtropical forest in Australia. These food chains are detritus-based; leaf

TABLE 14.4
Studies to Illustrate the Range of Primary Productivity in Aquatic Ecosystems

Productivity	Kind of System	Source
Marine		
mg C/m^3/day		
0.1–100	Various marine planktonic	Koblenz-Mishke et al. (1970)
200	Coastal upwellings	Barber and Smith (1981)
1000	Coastal upwelling	Barber and Chavez (1986)
mg C/m^2/yr	(i.e., productivity integrated over water column and over year)	
50	Open oceans	Ryther (1969)
82	Sargasso Sea	Platt and Harrison (1985)
50	Sargasso Sea	Jenkins and Goldman (1985)
690–720	NW Africa, Peru Coastal upwellings	Barber and Smith (1981)
100	Coastal zones	Ryther (1969)
200–300	Three shelf-sea ecosystems from Bering Sea, Oregon, and New York	Walsh (1981)
>1000	Coastal zone, Peru	Walsh (1981
150	Upwellings	Ryther (1969)
Freshwater Lakes		
mg C/m^{-2}/yr	(i.e., productivity integrated over water column and over year)	
5–6400	Various lakes	Likens (1975)
9	Lake Myvatn, Iceland	Johanssen (1979)
22	Ovre Heimdalsvatn, Norway	Larsson et al. (1978)
430	Lake Suwa, Japan	Mori and Yamamoto (1975)
450	Lake George, Uganda	Burgis et al. (1973)

Estimates based on daily rates

From Pimm 1988.

litter is the primary food source for a group of detri-vores, which in turn are fed upon by mites, a chirono-mid (a midge), and tadpoles of the frog *Lechriodus fletcheri*. When Kitching and Pimm provided artificial tree holes with amounts of energy inputs equal to, twice as great as, and half as great as those of natural holes, they found that the density of the tadpoles decreased with increased energy input. Holes with abundant energy input did not have longer food chains.

The second hypothesis, which essentially states that food chain length is associated with variation in sec-ondary production, is also not supported by the evi-dence. There is tremendous variation in secondary production that is not associated with differences in trophic structure. In fish populations, for example, secondary production ranges from 22 to 2034 kcalo-ries per square meter per year in herbivores and from 0.3 to 450 kcalories per square meter per year in car-nivores (Odum 1971). Thus there is as much variation at trophic levels two and three as there is in primary production, but food chain length is not associated with the differences.

If neither energy input (primary production) nor the efficiency of energy transfer (secondary produc-tion) accounts for the variation in the length of food chains, what does? One result from Kitching and Pimm's (1986) experiments is suggestive. They found that in the successional development of their artidicial tree holes, the predatory chironomid did not colonize until rather late. The length of the food chain was shorter in "disturbed" holes. Pimm (1988) suggests that food chains that are disturbed frequently tend to be shorter. Empirical data on the nature of disturbance and recovery in natural systems, as well as experi-mental disturbance of food chains, will be needed to test this hypothesis. The narrow range of food chain lengths across such a wide array of community types remains one of the most fascinating puzzles of ecosys-tem ecology.

Top-Down versus Bottom-Up Control

In Chapter 10, we discussed the control of populations by top-down or bottom-up mechanisms. That is, in top-down control, a population is regulated by preda-tion—in trophic level terms, from *above*. In bottom-up control, the population is regulated by its food resource—from *below*. Ecologists debate whether pri-marily top-down or bottom-up processes control food chains and webs (Polis and Winemiller, 1996; Moran and Hurd, 1998; Scheu and Schaefer, 1998). The classic form of the bottom-up hypothesis states that predators and plants are resource-limited: plants by nutrients, water, and light; predators by the number of available prey. The bottom-up hypothesis is based on the laws of thermodynamics that constrain the energy availability as one moves up the food chain. This hypothesis predicts that the attenuation of energy at each trophic level ultimately limits the biomass or numbers at the next level up.

Also associated with the top-down/bottom-up controversy is the concept of a **trophic cascade.** Trophic cascades are indirect effects of one trophic level on either lower levels (top-down cascade) or higher level (bottom-up cascade). Dyer and Letourneau (1999) provide an example of trophic cascades in a tropical system. They experimentally enhanced light and nutrient conditions for the primary producers. This led to increased plant biomass, a result that suggested bottom-up control at this level. However, there was no evidence that these effects cascaded upward to the herbivores or predators. When these investigators examined top-down effects, they found that ants have indirect (top-down cascade) effects on the herbivores because they decreased herbivory on the plants. Figure 14.26 summarizes these interactions.

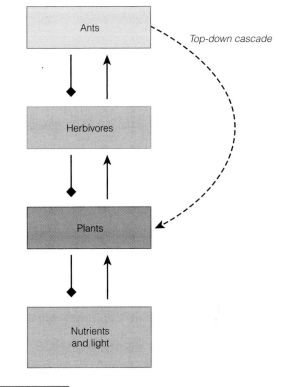

FIGURE 14.26 Top-down and bottom-up forces in a simple, tropical food chain. Arrows indicate a positive effect on the next level; diamonds indicate a negative impact. Ants exert a positive impact on plants via an indirect effect—a top-down cascade—by decreasing the density of herbivores.

SUMMARY

1. Energy can be neither created nor destroyed, and no energy transfer is 100 percent efficient. These two laws of thermodynamics constrain the operation of natural ecosystems.

2. Energy to drive ecosystems is derived from the sun. A few deep-sea systems are exceptions; there, chemoautotrophic bacteria convert chemical energy into forms useful for living organisms.

3. Communities differ over more than an order of magnitude in primary production.

4. Ecosystems fall into three fundamental groups on the basis of their productivity biomass ratios. The nature of the producers—trees, herbaceous plants, or algae—determines this ratio.

5. Primary production is most often limited by light, water, temperature, or nutrients. These four factors thus constitute important abiotic limiting factors, as discussed in Chapter 3. Their limiting role constitutes an important selective force on producers.

6. Energy flow is linear. Because of the laws of thermodynamics, energy gradually dissipates as it moves through the ecosystem.

7. The pathway of energy flow is called the food chain. Food chains are connected into food webs as multiple species occupy the same trophic level. Some species feed on more than one trophic level.

8. The proportion of energy in one trophic level that is available to the next is determined by three fundamental efficiencies of transfer: ingestion efficiency, assimilation efficiency, and production efficiency. Complex aspects of the biology of the species and its food determine the values of these efficiencies.

9. Two logical hypotheses to explain the variation in food chain length—the role of increased production and the role of increased transfer efficiency—have little experimental support. Some recent evidence suggests that disturbance may explain some of the variation.

SELF-ASSESSMENT: CAN YOU ...?

1. Explain how the laws of physics (thermodynamics) are manifest in the biology of food chains.

2. Explain the biological basis of the different shades shown on the Earth in Figure 14.1.

3. Relate the efficiencies below to the related concepts.

   ingestion efficiency → predation
   optimal foraging theory
   predator and prey adaptations

   production efficiency → life history strategies
   temperature balance and homeostasis

4. Explain, using appropriate examples, why it is difficult to identify the nature of a food web and the links inherent in it.

5. Explain why there is so little variation in food chain length.

6. Relate top-down and bottom-up control of food chains to the mechanisms of population regulation.

PROBLEMS AND STUDY QUESTIONS

1. Explain the importance of the first two laws of thermodynamics to the functioning of ecosystems. How do they constrain the organisms that constitute an ecosystem? How do they determine the nature of an ecosystem?

2. Why is it easier to measure primary production in aquatic systems? What technical problems are involved?

3. If C4 photosynthesis is advantageous under hot, dry conditions, why do you suppose more plants have not adopted the C4 pathway? Wouldn't it always be advantageous to be able to continue production during the occasional drought?

4. How do the patterns of energy flow influence the nature of predation? Consider the various types of predators described in Chapter 10. How are their adaptations for predation affected by the nature of energy flow?

5. Describe the relationship between ingestion efficiency and optimal foraging.

6. Does your understanding of energy flow influence your position on the "top-down/bottom-up" controversy described in Chapter 10?

7. What crucial factors from ecosystem energetics and production are important aspects of agricultural systems as ecosystems?

8. Discuss the relationship between species diversity as defined in Chapter 12 and the functional diversity of an ecosystem food web. Are the two concepts synonymous?

9. Would you expect the Eltonian pyramid of an early succession community to differ from that of a late succession community? Why or why not?

10. Design laboratory or field experiments to test the following hypotheses:
    a. Increased production leads to increased food chain length.
    b. Increased species diversity leads to higher primary production.
    c. Increased species diversity leads to lower variation in primary production among years.

PROJECTS AND ADDITIONAL STUDY

1. For a local ecosystem, attempt to identify the major components of the food chain and their relationships.

2. Devise an aquarium or laboratory culture experiment using algae and protozoa to test the hypothesis that the length of the food chain is dependent on primary production. Develop a specific protocol to test the predictions of this hypothesis

3. Survey the literature for examples of top-down and bottom-up control of food chains. Attempt to identify the features of the ecosystem that are associated with these two kinds of control. Is it possible to draw any generalizations about the relative importance of top-down and bottom-up control or about the contexts in which they occur?

CHAPTER 15

Biogeochemical and Nutrient Cycles

We turn our attention now to the other fundamental process in ecosystems: the movement of materials. Because most such movements are at least potentially cyclic, we refer to them as **biogeochemical cycles** or **nutrient** (or **element**) cycles, depending on their geographic scale. The movement of nutrients on a global scale is a biogeochemical cycle. The time scales on which biogeochemical cycles operate tend to be long—in some cases they require geological time. The movement of materials in a local ecosystem is a nutrient cycle. In this case, the time scale approximates the life span of the longer-lived organisms in the ecosystem.

Biogeochemical cycles are ecosystem-level processes with important effects on individuals, populations, and communities. In Chapter 14 we noted the ways in which nitrogen and phosphorus limit primary production. Nutrients often represent a critical limiting resource (Chapter 3) for individual plants. Similarly, animals seek out specific nutrients that are in short supply; for example, ungulates often seek nutrients available in mineral licks. We also saw in Chapter 3 that nutrients are thought to play a role in the population biology of lemmings in the Arctic.

The movement of materials through the ecosystem is an area of basic ecology that is closely related to environmental science (see the box on page 390 later in the chapter). One of the most conspicuous—and perhaps consequential—effects of human activity is our influence on the major patterns of the movement of materials. There are two major reasons for this. First, the movement of materials is an area of ecology in which interactions occur on a global scale. Thus, for example, the amount of carbon in the Antarctic Ocean is related to the amount of carbon in the atmosphere over the Northern Hemisphere. Second, our industrial and agricultural activities are so extensive and so intensive that their ecological effects are global in scale. Indeed, the emissions of certain substances by industrial processes are as large as or larger than the emissions produced by natural processes. If the natural processes of the ecosystem cannot accommodate these massive inputs, there will be ecological consequences.

The movement of materials through the ecosystem is driven by the movement of energy. In the previous chapter, in which we described the patterns of energy transfer, we saw that the food chain represents a linear sequence of energy transfers from primary producers to the top carnivore. The essence of this pattern is *consumption:* Energy transfer is driven by the consumption of carbon by primary producers and the consumption of other organisms by heterotrophs. In the process,

materials such as carbon, nitrogen, phosphorus, oxygen, and water are transferred along with the energy.

In many cases, nutrient cycles depend on the activity of decomposers to release materials back into the ecosystem, where they can be used again. In Chapter 13, we alluded to **degradative succession,** the sequential replacement of decomposer species. These sequences are important for the cycling of nutrients because different organisms in the succession have the ability to decompose different kinds of material.

One of the best-studied examples of degradative succession occurs on pine (*Pinus sylvestris*) needles (Kendrick and Burgess 1962). The primary agents of decomposition are fungi of various species that colonize the needles in a predictable sequence (Figure 15.1). The needles remain in the leaf litter layer for as long as seven years, but ultimately the chemical products of decomposition leach into the underlying soil and rejoin the various nutrient cycles.

There is a fundamental difference between the patterns of energy flow and those of material flow in ecosystems. As we saw in Chapter 14, *energy flow is linear.* Energy cannot be recycled: At the end of the food chain, insufficient energy remains to support another trophic level. Ecosystems require the constant input of solar energy to the Earth. In contrast, *the movement of materials is at least potentially cyclic;* that is, elements and compounds can move back and forth from the biotic to the abiotic components of the ecosystem essentially forever. Elements may form new compounds but they are neither created *de novo* nor destroyed. For example, the carbon that is fixed by photosynthesis as car-

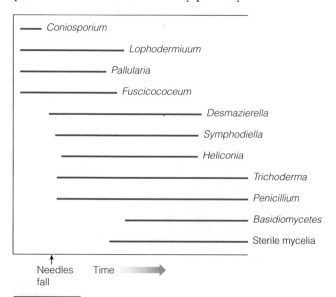

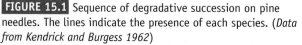

FIGURE 15.1 Sequence of degradative succession on pine needles. The lines indicate the presence of each species. (*Data from Kendrick and Burgess 1962*)

bohydrate passes through the food chain in several forms and ultimately is released back to the atmosphere as CO_2 by the process of respiration. It can then be taken up again by a plant.

It is important to realize that the movement of materials is *potentially but not necessarily cyclic.* As we will see, on a global scale, there is a general cyclicity of many important elements. But we will also see that under certain circumstances, elements may remain for long periods of time in one component of the ecosystem, where they are essentially unavailable to other parts of the system.

Ecosystem ecologists are concerned with several elements of material flow. Their research focuses on questions like the following:

1. What are the patterns of movement of materials through the ecosystem? This is first a qualitative question: What are the reservoirs of the material, and what are the pathways and mechanisms of transfer? Once those parameters have been identified, we wish to know what quantities of materials are present in various components of the system and how rapidly they are changing.
2. What is the nutrient budget for the system? This question extends the quantitative part of the previous question to include an assessment of the net gains and losses of materials from the system. Here we are concerned with the extent to which the system is in equilibrium.
3. How is the flow of material controlled? Are certain species crucial to the movement of materials? Do certain physical processes regulate the cycle? What are the crucial transfer points?

As we address these questions, a model of the movement of materials can be helpful. Figure 15.2 shows a highly simplified *compartment model* of material flow in an ecosystem. The first step in developing such a model is the qualitative description of where the materials are found in the system. The reservoirs for materials are called **pools** or **sinks,** depending on whether they are sources of material for other parts of the system (pools) or reservoirs with a net input (sinks). The atmosphere, primary producers, and the soil are possible examples of pools and sinks. Next, we need to be able to assign quantitative values to the size of the pools and the rates of transfer between pools. The rates of flow are called **fluxes.** A complete ecosystem model for a cycle consists of a series of pools or sinks with known quantities of the material and a series of equations that describe the fluxes. The latter are essentially differential equations that describe the

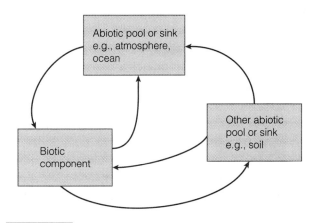

FIGURE 15.2 A simplified compartment model for biogeochemical cycles.

amount of material moved over time. An example of a nutrient cycle for a tallgrass prairie ecosystem, showing data on the content of pools and sinks and the magnitude of fluxes, is shown in Figure 15.3.

Models like this provide important insights into the workings of the system. If, for example, the flow from one reservoir to another is exceedingly slow, we have an important clue that this might be the rate-limiting, or controlling, step in the cycle. In addition, we can begin to ask mechanistic questions of the system. What if we increase one of the flow rates? What if we artificially add more material to a certain pool? These kinds of models are particularly useful for predicting anthropogenic effects on an ecosystem.

Obviously, a number of factors constrain the utility and accuracy of such models. We must know, for

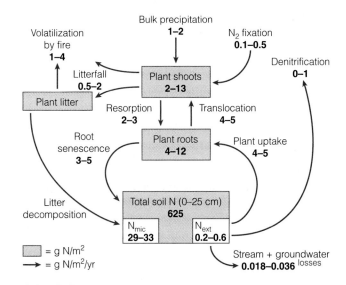

FIGURE 15.3 A quantitative compartment model of the nitrogen cycle in tallgrass prairie, showing pools (boxes) and fluxes (arrows). N_{mic} = microbial biomass N; N_{ext} = NH_4^+ and NO_3^-. (*From Blair et al. 1998*)

example, all the important sinks and fluxes. Identifying and quantifying these can be a daunting task because the number of potential interactions is so great—and not all are intuitively expected. For example, recent evidence documents the *interspecific* transfer of nutrients in tallgrass prairies. Fischer-Walter et al. (1996) have shown that phosphorus is transferred among neighboring species over considerable distances in prairie. A role for mycorrhizal interconnection is suspected in such transfers. We know that plants compete for nutrients, so we might not be surprised to discover that interspecific transfer must be incorporated in our model.

In addition, when we construct a model like that in Figure 15.2, we implicitly assume a set of boundaries to the ecosystem. The model depicted in that figure could apply to a single square meter of forest floor, to the entire North American taiga, or to the entire planet. The choice of an appropriate scale is crucial in ecosystem studies. If the scale is too small, important processes may occur outside the system (and then be represented in the model as input and output), and the model will produce little insight into the important mechanisms of nutrient cycling. If the scale is too large, it may be difficult (if not impossible) to obtain the detailed quantitative information necessary to describe the pools and fluxes in a meaningful way.

The choice of scale, then, depends on a number of factors. Foremost is the nature of the question being asked. If we are concerned with, say, the effect of increasing inputs of atmospheric CO_2 from the burning of fossil fuels, our scale is necessarily global, and we must be prepared to deal with the uncertainties associated with a scale of this magnitude. If, on the other hand, we are concerned with the role of fire in recycling nutrients in grasslands, then our scale can be much smaller—perhaps even a relatively small portion of a much larger grassland.

In this chapter we will first consider four global biogeochemical cycles: the carbon, nitrogen, phosphorus,

and sulfur cycles. We will then consider how local nutrient cycles operate and, in particular, how they differ from global cycles. Finally, we will discuss the nature of nutrient budgets and the factors involved in controlling local cycles.

Biogeochemical Cycles

In this section we consider examples of cycles of elements on a planetary scale. Such cycles exist for virtually every element or compound important to living organisms. There are two fundamental types of cycles. Cycles involving gaseous materials such as oxygen, nitrogen, and carbon dioxide are **volatile cycles.** Nonvolatile materials such as phosphorus are transferred through the ecosystem in **sedimentary cycles.** We will consider the cycles of four elements crucial to life: the carbon cycle, the nitrogen cycle, the phosphorus cycle, and the sulfur cycle. Each cycle differs from the others in important ways. Taken together, they illustrate the essential features of global cycles. Their differences and similarities are summarized in Table 15.1.

The Carbon Cycle

Life on Earth is based on carbon. Indeed, the special branch of chemistry called *organic chemistry* involves the study of carbon compounds.

Carbon molecules form the structural basis for living things; carbohydrates, lipids, nucleic acids, and proteins all contain carbon atoms. The fundamental energy currency of living organisms is based on molecules that contain carbon:

$$6CO_2 + 6H_2O \underset{\text{respiration}}{\overset{\text{photosynthesis}}{\longleftrightarrow}} C_6H_{12}O_6 + 6O_2$$

TABLE 15.1
Some Major Features of Four Important Biogeochemical Cycles

Element	Cycle Type	Reservoirs	Special Features
Carbon	Volatile	Large amounts in atmosphere and ocean	Intimately related to energy processes through photosynthesis and respiration; no specialized organisms required for chemical conversion.
Nitrogen	Volatile [?]	Large amounts in atmosphere	Complex chemistry involved; specialized organisms required for several chemical conversions.
Phosphorus	Sedimentary	Large amounts in phosphate rocks	Availability depends on pH of soil or water; can be lost in deep sediments, so cycling times are extremely long.
Sulfur	Volatile and sedimentary	Large amounts in atmosphere and sediments	Specialized bacteria required for conversion of some forms of sulfur.

Photosynthesis is the process whereby solar energy is stored in the form of carbohydrates, and respiration is the process whereby that energy is released for use by the organisms. These fundamental processes of energy capture and use drive the movement of carbon in the carbon cycle.

Over geological time, these two processes have not been in exact balance. In the ocean especially, net photosynthesis has exceeded respiration, leading to the accumulation of oxygen and depletion of carbon in the atmosphere and to the accumulation of carbon in marine sediments (Falkowski 1997). A number of authors have suggested that the coevolutionary race between plants and their grazers initiated major climatic changes over the past several hundred million years (Zimmer 1993). There have been important shifts in plant and animal evolution and hence climatic change. The balance between herbivores and plants is important because it determines in large part the carbon dioxide content of the atmosphere: Plants use CO_2 and animals release it. The incorporation of CO_2 from the atmosphere during photosynthesis has two effects. First, it decreases the CO_2 content of the atmosphere and thus decreases the greenhouse warming of the atmosphere. Second, the activity of plants increases weathering of rock to produce soil. When atmospheric CO_2 taken up by plants is respired by the roots, the CO_2 reacts with water in the soil to form carbonic acid (H_2CO_3), a weak acid that dissolves bicarbonate from rocks, contributing to the weathering process. If the coevolutionary race between plants and their herbivores is not always in balance, the CO_2 content of the atmosphere has varied considerably—with significant effects on climate and weathering.

Approximately 450 million years ago plants invaded the land; within 50 million years, vascular plants that could colonize drier terrestrial habitats had evolved. In the absence of terrestrial animals, these plants escaped much grazing pressure, and thus the balance of the carbon cycle shifted toward stored carbon and away from atmospheric CO_2. Another result was a 30-million-year period of climatic cooling as a result of the diminished greenhouse effect.

When the first terrestrial animals appeared 360 million years ago, most were carnivores, but soon herbivores evolved to exploit the terrestrial plants. Their adaptations included long guts with symbiotic bacteria to facilitate digestion of plant material, as well as changes in dentition to facilitate grinding coarse plant material. The communities that evolved over the next 100 million years came to have trophic relationships similar to those that modern communities exhibit. As animals consumed more of the abundant plant resources, the CO_2 balance shifted back toward the atmosphere, and a warming trend ensued.

When angiosperms appeared 100 million years ago, their more rapid reproductive rate shifted the carbon balance back to the plants, and mean temperatures declined again. The demise of the dinosaurs (many of which were herbivores) about 65 million years ago opened up opportunities for mammals to enter the herbivorous niche. When this adaptive shift occurred, CO_2 was again shunted back to the atmosphere by mammalian respiration, and temperatures increased once again.

The evidence for this hypothesis is correlational; many other explanations might account for the pattern of climatic change and the CO_2 content of the atmosphere. Clearly, though, this hypothesis requires additional testing. Regardless of the outcome of such studies, its proposal is an indication of how important ecologists deem the relationship between herbivores and the plant producers.

The basic pattern of carbon flow is illustrated in Figure 15.4. Plants move carbon from abiotic pools (atmosphere and water) into the biotic component of the ecosystem. There carbon constitutes structural elements and energy sources for organisms all the way up the food chain. Respiration releases CO_2 back to the abiotic environment. Some carbon is stored for varying lengths of time in biomass, but eventually this too is respired by decomposers. During geological periods in which photosynthesis exceeds respiration, carbon accumulates in the form of fossil fuels.

The CO_2 concentration in the atmosphere is approximately 0.03 percent. Although this number seems small, it translates into about 730 million metric tons (730 gigatons). Consequently, carbon is abundantly

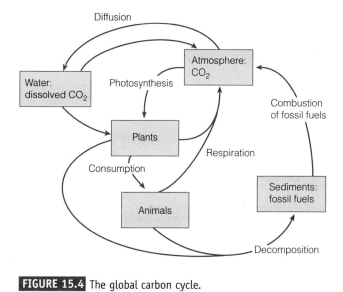

FIGURE 15.4 The global carbon cycle.

available to terrestrial plants. As plants and consumers die, their carbon becomes part of the reservoir of dead organic matter. As this material decomposes, carbon is released to the atmosphere or held in organic molecules in the soil. Measuring the size of the biomass and soil reservoirs of carbon is difficult, so there is considerable uncertainty in the values. Other studies put the quantity at 1200–1600 gigatons (Post et al. 1990).

Aquatic systems, particularly the ocean, have a much higher carbon content. At least 30,000 gigatons of carbon are held in the ocean. This is also a difficult quantity to measure accurately, and estimates are likely to go higher as more refined techniques become available.

The distribution of carbon in the ocean is not uniform. Exchange of carbon with the atmosphere occurs at the surface. Carbon enters the ocean when atmospheric CO_2 diffuses into the water at the surface. The CO_2 then reacts with the water to form a weak acid, carbonic acid, which dissociates:

$$CO_2 + H_2O \rightleftharpoons H_2CO_3 \rightleftharpoons H^+ + HCO_3^- \rightleftharpoons 2H^+ + CO_3^{-2}$$

The form that the carbon takes in the water is determined by the pH, which determines the degree to which carbonic acid dissociates (Figure 15.5). At low pH, the reactions above are driven to the left; at high pH, they are driven to the right.

This chemistry is important because, as we saw in Chapter 14, even though carbon is found at relatively high concentrations in aquatic systems, it occasionally limits primary production. Free CO_2 is used by virtually all algae and higher plants. Some plants can use carbon from bicarbonate (HCO_3^-) when CO_2 is limiting. Apparently, plants cannot use the carbonate ion (CO_3^{-2}) as a source of carbon for photosynthesis (Raven 1970).

The rate at which CO_2 dissolves in the water is a function of surface area, wind action, and the partial pressures of CO_2 (pCO_2) in the air and water. As CO_2 dissolves, the partial pressure of CO_2 in the surface water comes to equal that of the atmosphere. Photosynthesis is confined to this upper water layer by the availability of light. As carbon is taken up by phytoplankton, the surface pCO_2 decreases, and more carbon can dissolve at the water–air interface. Thus photosynthesis in effect pumps carbon into the ocean. Because dead plankton and fecal material sink and carry the carbon to deeper layers, the carbon concentration may be considerably higher at greater depths. Carbon is further distributed by ocean currents, convection, and other physical processes. Large pools of carbon also exist in carbonate rocks. New carbonate rock that forms in the ocean may be a sink for atmospheric carbon.

In the centuries since the Industrial Revolution, the flux of carbon from storage in fossil fuels back to the atmosphere has increased markedly. This process is of concern because CO_2 is one of the most important greenhouse gases. The ability of the carbon cycle to accommodate such changes is not yet clear.

The Nitrogen Cycle

Like carbon, nitrogen is a fundamental component of living things. It is an integral element of the structure of amino acids (and thus proteins), and nitrogenous bases form the backbone of nucleic acids. We saw in Chapter 14 that nitrogen often limits primary productivity, particularly in aquatic systems. As is the case with carbon, human activities are altering the nature of the reservoirs of nitrogen on the planet, with some potentially significant consequences.

The chemistry of the nitrogen cycle is complex. Nitrogen in the ecosystem exists in a number of chemical forms, each with a different chemistry (Table 15.2). One of the most important features of the nitrogen cycle is that specialized bacteria are needed to carry out many of the interconversions of nitrogen that occur during its movement through the ecosystem. This contrasts with the carbon cycle, in which fundamental metabolic pathways that are universal to all life-forms accomplish the conversion of the various forms of carbon. The crucial sets of reactions of the nitrogen cycle did not all evolve at the same time (Falkowski 1997). Nitrogen fixation occurred in the ancient, reducing atmosphere. The sequences of genes coding for nitrogenase, the enzyme responsible for

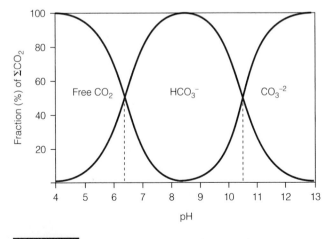

FIGURE 15.5 Dissociation curve for carbonic acid, H_2CO_3, as a function of pH.

TABLE 15.2
The Important Chemical Reactions in the Nitrogen Cycle

Process	Organisms
Denitrification	
$C_6H_{12}O_6 + 24KNO_3 \rightarrow 6CO_2 + 3H_2O + 6KOH + 3N_2O$	*Pesudomonas denitrificans*
$5C_6H_{12}O_6 + 24KNO_3 \rightarrow 30CO_2 + 18H_2O + 24KOH + 12N_2$	*Pseudomonas denitrificans*
Ammonification	
$C_2H_5NO_2 + 1\frac{1}{2}\,O_2 \rightarrow 2CO_2 + H_2O + NH_3$	Many bacteria, plants, and animals
Nitrification	
$NH_3 + 1\frac{1}{2}\,O_2 \rightarrow HNO_2 + H_2O$	*Nitrosomonas* bacteria
$KNO_2 + 1\frac{1}{2}\,O_2 \rightarrow KNO_3$	
Nitrogen fixation	
$N_2 + 3H_2 \rightarrow 2NH_3$	Cyanobacteria, azotobacteria

nitrogen fixation, are highly conserved across the bacteria and cyanobacteria, which suggests a single, ancient origin of this enzyme. Ammonification and nitrification probably arose next. Both require molecular oxygen, which would not have been present until evolution had produced an oxidizing atmosphere. Denitrification probably arose last, and seems to have evolved independently more than once (Zumpft 1992).

The basic elements of the nitrogen cycle are shown in Figure 15.6. A major pool of nitrogen exists in the atmosphere, which is approximately 78 percent nitrogen. Primary producers in terrestrial and aquatic systems acquire nitrogen from the atmosphere via associated nitrogen fixers or dissolved nitrogen in the soil; animals obtain nitrogen via the consumption of producers. Nitrogen is released from dead plants, feces, and dead animals in forms that can be taken up again by plants. Thus a major portion of the nitrogen cycle flows from the living component of the ecosystem to the soil and back. Nitrogen also enters aquatic systems from the soil via runoff.

Nitrogen is found in the atmospheric pool in the form of molecular nitrogen, N_2. Although cycling of nitrogen occurs once it enters the biotic component of the ecosystem, all nitrogen ultimately came from this atmospheric pool. Most organisms cannot use molecular nitrogen—it must be converted into usable forms by other

organisms. The crucial reaction is nitrogen fixation, the conversion of molecular nitrogen (N_2) into hydroxylamine (NH_2OH). Then many species can convert this compound into nitrite and nitrate, which enter a number of biochemical pathways.

The fixation of atmospheric nitrogen is carried out by cyanobacteria and by some bacteria. Often nitrogen-fixing microorganisms are symbiotically associated with higher plants, especially legumes, where they are located in nodules on the roots. Other important plant species with symbiotic nitrogen fixers include alder (*Alnus*) and some members of the buckthorn family in the genus *Ceanothus*. These plants are important colonizers of nutrient-poor soils (see Chapter 13).

The crucial enzyme in nitrogen fixation is nitrogenase, which accomplishes the initial reduction of N_2. This is an energetically expensive reaction that must occur anaerobically. In cyanobacteria, nitrogenase is sequestered in the heterocysts, structures that apparently serve to protect the enzyme and the reaction from oxygen. Ammonia, the end product of nitrogen fixation (see Table 15.2), is rather toxic and generally is converted to nitrate (see below), a form of nitrogen that is readily used by plants.

Obviously, the rate of nitrogen fixation is crucial to ecosystem production. The high production in coral reefs, which are among the most productive ecosystems on Earth, is made possible by rates of nitrogen fixation that exceed those of any other marine ecosystem (Capone 1983). The reasons for such high rates of fixation are not clear. However, grazing by fish may be important, because it may reduce competition for

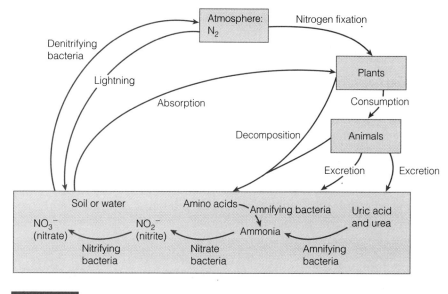

FIGURE 15.6 The global nitrogen cycle.

space between nitrogen-fixing cyanobacteria and other algae.

Ammonia from nitrogen fixation or from the decomposition of amino acids by detritivores is oxidized to form nitrite (NO_2^-) and nitrate (NO_3^-) in the process called **nitrification** (see Table 15.2). The bacteria *Nitrosomonas* and *Nitrococcus* nitrify ammonia to nitrite; *Nitrobacter* and *Nitrococcus* nitrify nitrite to nitrate. Nitrate may be then released into the soil, from which it can be taken up by plants and recycled; it may also enter aquatic systems (via runoff from the land), where it can be taken up by phytoplankton. Nitrate and nitrite that are not utilized in this way may be reduced by denitrifying bacteria to form molecular nitrogen that is released into the atmosphere.

Human activities have had an impact on the global nitrogen cycle as well. The addition of massive amounts of nitrogenous fertilizers to agricultural systems has produced an excess of nitrogen in surface waters. The resulting algal blooms and eutrophication of lakes are a significant problem in much of the world. In addition, the combustion of oil and gasoline releases nitrogen dioxide (NO_2) into the atmosphere, where it can form nitric acid, which significantly lowers the pH of precipitation.

The Phosphorus Cycle

Phosphorus, like nitrogen and carbon, is essential for all organisms. An important element of nucleic acids, it is fundamental to the structure of cell membranes and also plays a major role in the structure and function of ATP, the basic energy currency of life. We noted in Chapter 14 that phosphorus is often a limiting element in primary productivity. As we will see, this is because certain processes associated with the phosphorus cycle sometimes render the element unavailable.

The essential pools and fluxes of phosphorus are shown in Figure 15.7. The phosphorus cycle is considerably simpler than the nitrogen cycle. There is no atmospheric pool; the phosphorus cycle is a sedimentary cycle in which phosphate rock constitutes a major reservoir.

Phosphorus enters the cycle via the weathering and erosion of phosphate deposits. Dissolved phosphorus enters plants and then animals and decomposers, from which it returns to the soil. Eventually, runoff following precipitation carries the phosphorus to the ocean, where it again can cycle between plants and animals. Zooplankton, which graze on phytoplankton, play an important role in this cycling because they excrete excess phosphorus into the water, where it

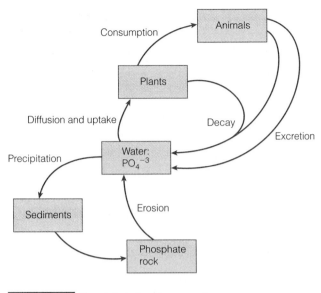

FIGURE 15.7 The global phosphorus cycle.

can be taken up again by primary producers. This cycling occurs in the relatively warm surface layers. As organisms in that layer die and sink to the bottom, they carry phosphorus with them. Some are returned to the surface layers by upwelling, whereas some become buried in sediments.

Once particulate phosphorus is deposited in sediments, it is relatively unavailable to living organisms. Eventually, the sediments are buried deeply, and the phosphorus they contain may not enter the biotic component of the ecosystem for tens or hundreds of millions of years—not until the sediments are uplifted by geological process and the phosphate rock weathers. As a result, the complete phosphorus cycle is exceedingly slow. If the element remains buried in sediments for long periods of time, the movement of phosphorus becomes essentially acyclic.

The useful form of phosphorus, the phosphate ion (PO_4^{-3}), can be readily absorbed from either the soil or water and used by plants. However, phosphate also forms a number of insoluble compounds with aluminum, iron, and calcium. The availability of PO_4^{-3} is determined by the pH and the concentrations of other ions present. The highest levels of dissolved $H_2PO_4^-$ occur at intermediate pH. Under alkaline or acid conditions, relatively insoluble complexes form in water or in soil. Thus one of the concerns about acid precipitation and the acidification of aquatic systems is their effect on phosphorus availability.

The Sulfur Cycle

Sulfur is a component of two amino acids important to plants and animals, cysteine and methionine. In addition, it forms reduced compounds such as hydrogen sulfide (H_2S) that some chemoautotrophs can utilize as an energy source.

Some of the ecologically important forms of sulfur are gases and others are solids; thus the sulfur cycle contains both volatile and sedimentary elements. Like nitrogen, sulfur exists in several chemical forms that differ in their oxidation states. And, as in the nitrogen cycle, some of the interconversions of the forms of sulfur require specialized bacteria.

The sulfate ion (SO_4^{-2}), the most oxidized form, is the form that is taken up by plants. Thus, this ion lies at the center of the cycle (Figure 15.8), with several pathways leading to it. Hydrogen sulfide (H_2S) enters the atmosphere from volcanic eruptions and biological processes. There it is oxidized to sulfur dioxide (SO_2), which moves by rain into the soil and water, where it is oxidized to SO_4^{-2} and taken up by plants.

Sulfur is then passed along the food chain and is eventually released by detritivores as sulfate ion, which can then be taken up by plants again. Alternatively, it can sink to the bottom of aquatic systems. Under the anaerobic conditions common in mud and sediments, the bacteria *Desulfovibrio* and *Desulfomonas* respire using the oxygen of sulfate. In the process they reduce SO_4^{-2} to H_2S and sulfide (S^{-2}). S^{-2} is then used as a reducing agent by photoautotrophic bacteria, which use sulfur instead of oxygen as an electron donor. (In photosynthesis, O_2 is used as an electron donor in the light reactions.) The green and purple photoautotrophic bacteria use H_2S as an oxygen receptor in their photosynthetic machinery. The purple bacteria produce SO_4^{-2}, which can be assimilated by higher plants.

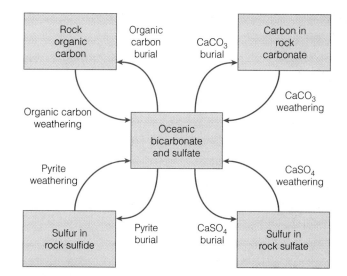

FIGURE 15.9 The interaction of the carbon and sulfur cycles with respect to oxidation state. Long-term shifts from the oxidized to the reduced state in each element require a complementary shift in the other element if atmospheric oxygen is to remain relatively constant. (*From Bernes and Petsch 1998*)

If iron is present in the sediments under anaerobic conditions, the elemental sulfur precipitates in the form of ferrous sulfide (FeS_2), a very insoluble compound that may remain in sediments for long periods of time. If this material, which is often found in association with coal deposits, is suddenly exposed to the air, the sulfur is oxidized by the bacterium *Thiobacillus* to form SO_3^{-2} and SO_4^{-2}. These ions combine with water to form sulfuric acid (H_2SO_4), a common pollutant emanating from mines.

Anthropogenic addition of SO_2 to the atmosphere is an important source of pollution. Coal and other fossil fuels with a high sulfur content release SO_2 when burned. In the upper atmosphere, a series of reactions take place that generate sulfuric acid (see the box on page 390).

We have considered four examples of biogeochemical cycles. Keep in mind that although we described them as separate entities, they are in fact interrelated in complex ways. Figure 15.9 represents an example of this kind of interaction among organic and inorganic forms of carbon and sulfur. Sulfur and carbon both exist in oxidized forms ($CaSO_4$ and $CaCO_3$) as well as reduced forms (FeS_2 and organic carbon).

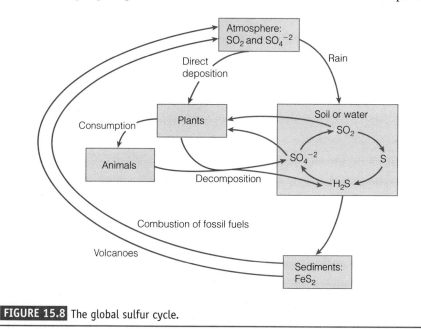

FIGURE 15.8 The global sulfur cycle.

Acid Precipitation

Our descriptions of biogeochemical cycles in this chapter have focused primarily on the natural patterns of movement of materials. Human activity can also have profound effects on the nature of some of these cycles. Excess input of materials, alteration of the flow rates between compartments, and extraction of material from an ecosystem can influence the cycle. If the anthropogenic effects are relatively small, the system can compensate rather easily. However, some industrial impacts are large enough to have a global effect. Acid precipitation is an example. The origins of acid precipitation are found in disruption of natural biogeochemical cycles. In addition, the input of large amounts of acid disrupts other biogeochemical cycles, with significant biological effects.

We begin with some rain chemistry. Rain has a slight natural acidity. In the atmosphere, CO_2 diffuses into water droplets and comes to equilibrium with gas in the atmosphere. The dissolved CO_2 reacts with water to form carbonic acid, a weak acid:

$$CO_2 + H_2O \rightarrow H_2CO_3$$

If carbonic acid is the only acid present in raindrops, the pH of the water is approximately 5.4, slightly acidic. Human industrial activity significantly lowers the pH of precipitation through anthropogenic disruptions of the sulfur and nitrogen cycles.

Let us consider first the addition of acid via the sulfur cycle (Figure 15.8). When fossil fuels, especially coal, are burned, large amounts of sulfur dioxide (SO_2) are emitted. This gas reaches the upper atmosphere, where it undergoes a series of chemical reactions. In the presence of light, SO_2 reacts with the oxygen in the atmosphere:

$$2SO_2 + O_2 \rightarrow 2SO_3$$

SO_3 is a gas that diffuses into water droplets where it reacts with water:

$$SO_3 + H_2O \rightarrow H_2SO_4$$

H_2SO_4 (sulfuric acid) is a strong acid that dissociates readily to produce very acidic precipitation.

The second major pathway leading to acid rain is an alteration of the nitrogen cycle. The combustion of gasoline in automobiles releases nitrogen dioxide (NO_2) into the atmosphere. At the high temperatures of an internal combustion engine, atmospheric nitrogen (N_2) is oxidized to form nitric oxide (NO), which is quickly converted to NO_2. In the upper atmosphere, a series of reactions takes place involving naturally occurring ozone (O_3) and NO_2:

$$O_3 \rightarrow O^- + O_2$$
$$O^- + H_2 \rightarrow 2HO^{-2}$$
$$2HO^- + 2NO_2 \rightarrow 2HNO_3$$

HNO_3 is nitric acid, another strong acid, which dissolves in water droplets and contributes to the acidity of precipitation.

Even though NO_2 and SO_2 are also released from natural sources such as volcanoes, they have become contributors to pollution because human industrial activity generates them in amounts far beyond normal levels. In the United States alone, approximately 21 million metric tons of SO_2 and 20 million metric tons of NO_2 are produced each year. Attempts to decrease local pollution by increasing the height of smokestacks has resulted in the injection of these compounds into the general atmospheric circulation. Thus midwestern industrial emissions are carried to the northeastern United States, where acid precipitation is a serious problem. Similarly, pollution from countries to the south contributes to acid precipitation in Scandinavia. In parts of the northeastern United States, the average pH of precipitation is approximately 4.1. At several sites in the United States, precipitation has been

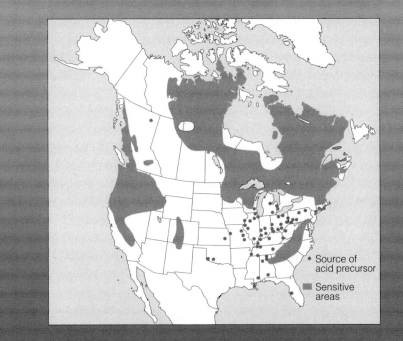

FIGURE 15B.1 Acid-sensitive areas in the United States and Canada (shaded areas) and major sources of acid precursors (dots). (*From Chiras 1994*)

Acid Precipitation, continued

recorded with a pH below 3.0, nearly 1000 times more acidic than natural rainfall.

The effects of acid precipitation are manifold and are heavily dependent on the local biogeochemistry. Soil is rather well buffered so that acid precipitation has a relatively small effect on it. Direct deposition of acid precipitation on plants can damage plant tissues, however.

In aquatic systems, the effects of acid precipitation depend heavily on the buffering capacity of the water. Lakes are buffered primarily by bicarbonate (HCO_3^-), an ion formed from the carbonic acid that occurs naturally in water:

$$H_2CO_3 \longleftrightarrow H^+ + HCO_3^-$$

As hydrogen ions are added via acid precipitation, they are removed as the reaction moves to the left. No pH change occurs until all the carbonate ion is removed, but then the pH drops quickly upon the further addition of acid. The lakes of Scandinavia, the northeastern United States, and much of eastern Canada lack natural buffering capacity (Figure 15B.1) and thus are at risk of acidification.

As the pH of a lake drops, aquatic organisms face increasing problems. Many species of fish succumb at pH values less than 5.0. In addition to direct toxicity, acid increases the solubility of certain toxic elements, especially aluminum and some heavy metals. It also decreases the availability of necessary nutrients such as phosphorus. At low pH, phosphate ion (PO_4^{-3}), the form generally assimilated by plants, forms insoluble compounds with ferric ion (Fe^{-3}) and aluminum. The problem is particularly acute in the spring, when many animals are reproducing. At this time of year, a sudden pulse of acid enters the water as snow melts, decreasing the survival of various life history stages of a wide variety of animals. The effect can occur throughout the food chain.

Moreover, they are both involved in redox reactions. As a result, there is a balance via reciprocal redox conversion. Not surprisingly, this also affects and is affected by atmospheric oxygen—if atmospheric oxygen is to remain relatively constant, oxidation of one element must be balanced by the opposite shift in the other (Berner and Petsch 1998).

Nutrient Cycles

In the previous section, we described the global patterns of nutrient cycling for four important elements. At this most inclusive level of ecosystem organization, nutrient budgets must be balanced: There is no gain or loss of these elements from the planet. But as we noted, there are sinks in which the nutrients are relatively unavailable.

We turn our attention now to studies of ecosystems conducted on a smaller geographic scale. The same fundamental biological processes—consumption, defecation, decomposition, and so forth—operate on this smaller scale. Indeed, the broad outlines of each of the cycles described for the planet can often be observed on a smaller geographic scale. However, one difficulty in the study of nutrient cycles is that our effects on global biogeochemical cycles is so pervasive, particu-

larly in the Northern Hemisphere, that analyses of local cycles is confounded by inputs of pollutants that affect the movement of materials (Hedin et al. 1995). Also, at the smaller scale, there is the possibility that nutrients leave or enter the system from elsewhere (Figure 15.10); as a result, nutrients may simply flow through the system rather than cycle.

Nutrient cycles vary enormously because of the idiosyncrasies of local ecosystems. For example, lakes in the McMurdo region of Antarctica have unique patterns of carbon flux that are strongly affected by the fact that these lakes are permanently ice-covered (Priscu et al. 1999). Unlike all other lakes on Earth, these systems contain no metazoan plankton or fish. Still, they maintain a complex carbon cycle in which bacteria and protozoa play a key role in cycling carbon from phytoplankton. In addition, viruses play a significant role as carbon sinks in this system. Moreover, the carbon cycle depends on contemporaneously produced organic carbon as well as "legacy carbon" derived from past biogeochemical events.

Indeed, an extreme example of an ecosystem with no nutrient cycling has been described by J. R. Ehrlinger and colleagues (1992). In some regions of the Atacama Desert of Chile, no precipitation has been recorded in historical time (see Chapter 16). Two introduced species of mesquite, *Prosopis tamarugo* and *P. alba,* inhabit these desert basins, where they are able

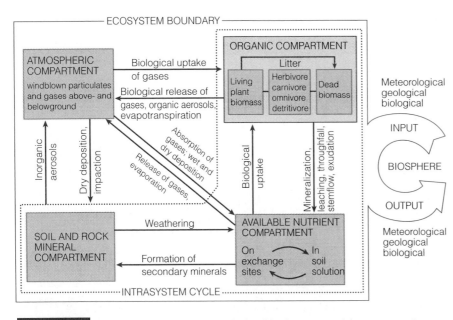

FIGURE 15.10 A model depicting nutrient relationships in a terrestrial ecosystem. Inputs and outputs to the ecosystem are moved by meteorological, geological, and biological vectors. (*From Likens et al. 1977*)

Northern Hardwood Forests

One of the most intensively studied ecosystems is the hardwood forest in the Hubbard Brook Experimental Forest (HBEF) in the White Mountains of New Hampshire. This site, which has been the subject of long-term studies of ecosystem function (Likens et al. 1977; Bormann and Likens 1979), is typical of northern hardwood forests. It is dominated by species such as American beech (*Fagus grandifolia*), sugar maple (*Acer saccharum*), yellow birch (*Betula alleghaniensis*), white ash (*Fraxinus americana*), and basswood (*Tilia americana*). Portions of the forest were logged between 1910 and 1920, but it has not been disturbed since then except for experimental manipulations.

Two important aspects of the topography of the HBEF are crucial. First, the forest is located in a mountainous region dissected by a number of valleys that serve to define small watersheds, each of which can be considered a separate ecosystem. Second, the forest is underlain by granitic bedrock that is impervious to water. This means that water does not percolate through the rock lying beneath the soil. As a result, nutrients cannot escape in this way, which makes it far easier to measure the nutrient flow in and out of the system.

to survive because their taproots penetrate up to 12 meters into the substrate to pull water to the surface. Ehrlinger and colleagues found that these trees are underlain by a layer of dead leaves as much as 45 centimeters thick. In the hot, dry air, these leaves do not decompose. In fact, the ^{14}C content of the leaves found at a depth of 20 centimeters in the litter indicates that they were deposited before 1950—that is, before atmospheric atomic testing. In the intervening 40 years, they have not decomposed.

Thus, in this system, nitrogen is not recycled through decomposition of plants; it simply accumulates in the detritus layer. There is virtually no nitrogen in the soil. The mesquite trees also produce a mat of subsurface roots that contain nodules in which nitrogen fixation occurs. Thus the nitrogen to support these plants is derived entirely from the atmosphere.

We will consider examples of three well-studied ecosystems: northern temperate forests, tropical forests, and coral reefs. We will see that the local pattern of nutrient cycling is specific to each of these systems. Then we will see how representative these examples are of local ecosystems in general. Finally, we will examine the kinds of processes that can control local nutrient cycles.

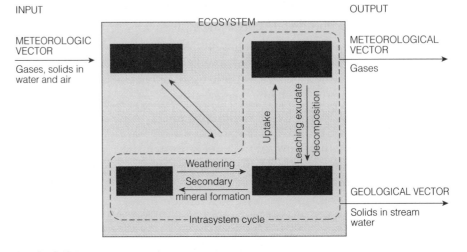

FIGURE 15.11 A generalized local nutrient cycle in the Hubbard Brook Experimental Forest. (*From Bormann and Likens 1979*)

The basic structure and fluxes of small ecosystems such as these are depicted in Figure 15.11. Note that material can enter the ecosystem in water from the watershed or atmosphere (precipitation) or in solid or gaseous form from the atmosphere. Exports are in the form of gases, dissolved material in stream flow, or solids suspended in the stream. New materials can become available from within the system by the weathering of rock.

Like most New England forests, the HBEF is a second-growth forest that is redeveloping after logging. Experimental studies on the effects of ecosystem disturbance were carried out by clear-cutting some sections to compare ecosystem processes with those in uncut control plots. As the forest develops after disturbance (see Chapter 13), the ecosystem proceeds through several phases (Figure 15.12).

Because of the relatively recent and extensive logging, the **aggradation** period is particularly important. Aggradation is characterized by increased production—in this case, as the forest regrows—and the accumulation of biomass and nutrients (Figure 15.13).

Let us consider the nitrogen cycle in the HBEF in some detail. During the first part of the aggradation phase, nitrogen accumulates rapidly (Figure 15.14), although eventually the rate of accumulation slows. This pattern occurs because early in aggradation, nitrogen concentrations in both the forest floor and live biomass are increasing. Later, the accumulation in live biomass levels off, reducing the overall rate of increase.

The essential features of the local nitrogen cycle in an aggrading forest are shown in Figure 15.15.

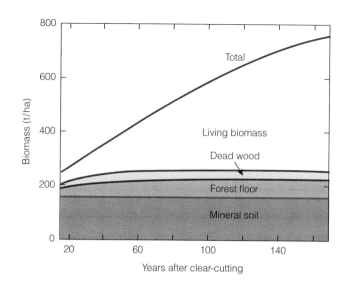

FIGURE 15.13 The accumulation of biomass during the aggradation phase of ecosystem development in the HBEF. (*From Bormann and Likens 1979*)

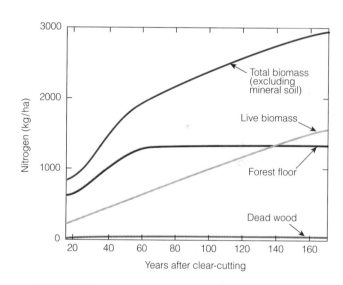

FIGURE 15.14 Estimated accumulation of nitrogen during the aggradation phase in the HBEF, based on projections of living and dead biomass multiplied by appropriate nitrogen concentrations. The estimate of total nitrogen does not include nitrogen n the mineral soil. (*From Bormann and Likens 1979*)

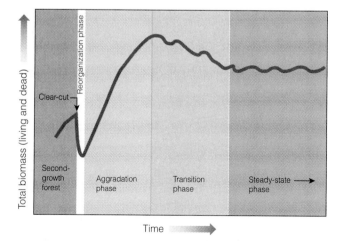

FIGURE 15.12 Phases of forest ecosystem development after clean-cutting of second-growth northern hardwood forest. Phases are delineated by changes in total biomass accumulation. (*From Bormann and Likens 1979*)

Approximately 70 percent of the nitrogen added to the system comes from nitrogen fixation. Much of this nitrogen input remains within the system; nitrogen cycling is very tight. Only about 5 percent of mineralized nitrogen is lost as geological output.

This kind of detailed quantitative budget allows us to draw some conclusions about the potential effects of human activity. Note in Figure 15.15 that a total of 119.4 kilograms of nitrogen per hectare per year (79.6 + 39.8) is used by plant growth. Fully 33 percent of this comes from resorption from storage in plant

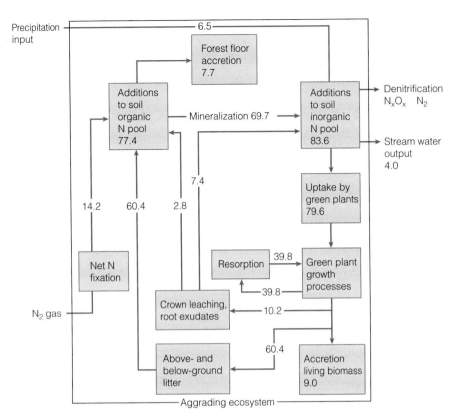

Precipitation input

FIGURE 15.15 Major nitrogen transfers into, out of, and within a 55-year-old aggrading northern hardwood ecosystem. Numbers are kilograms of nitrogen per hectare per year. (*From Bormann and Likens 1979*)

tissues. This is significant because it tells us that clear-cutting destroys a large portion of the pool of nitrogen used in plant growth.

Removal of the trees significantly increased stream flow. The uptake and transpiration of water by plants were reduced, and rainwater was no longer intercepted by plants. Coincident with the excess stream flow was a major increase in the concentration of dissolved nutrients in the water. Losses of nutrients were approximately eight times those of intact forest.

Bormann and Likens propose that after a long period of regeneration, the forest enters a steady-state phase (see Figures 15.12 and 15.14). At this time, no net accumulation of biomass occurs. Primary production balances respiration, and the rate of accumulation of nutrients slows greatly. The existence of a true steady state is a matter of debate among ecosystem ecologists. It may be that no system truly achieves such a condition. As detailed in Chapter 13, disturbance can take many forms. For forests like the HBEF, factors such as fire and storms are important disturbances. Because these factors operate on many spatial and temporal scales, it is perhaps more likely that disturbance is so frequent that a true steady state does not develop.

It is important to keep in mind that although we have discussed nutrient cycles as separate entities, the connection of nutrient cycles to production and consumption leads to the interactive effects. For example, in terrestrial ecosystems like these forests, rapid plant growth, associated with high efficiency of nitrogen use, leads to litter with a high ratio of carbon to nitrogen and subsequently a low mineralization rate of nitrogen as plants and animals decompose (Tateno and Chapin 1997). Consequently, there is a negative feedback between plant growth and production mediated by nutrients and the pools of many nutrients of the system.

Some scientists have hoped that as increasing global CO_2 concentrations lead to increased temperatures, those higher temperatures will concomitantly lead to increased rates of nitrogen mineralization. The latter could increase plant production and pull CO_2 from the atmosphere, slowing the rate of global warming. However, Anser et al. (1997) argue that this effect will be temporary, because the carbon and nitrogen cycles are becoming decoupled as a result of massive changes in land use and cover. As more and more forest is converted to agriculture, the nitrogen and carbon cycles become less tightly coupled. Agricultural lands typically grow crops with high turnover rates and a low C:N ratio. Consequently, agricultural land use does not stimulate the storage of excess carbon in biomass that occurs in natural vegetation.

Tropical Forests

The tropical forests of the world represent some of the most productive habitats on the planet (see Table 14.1). Thus, it is of interest to consider the patterns of nutrient flow in such systems and the role that nutrient cycling plays in the productivity of such forests.

For some time, the tropics, with their lush vegetation and high production, were thought to represent potentially important agricultural areas for humans. Indeed, H. D. Thurston (1969) stated, "The most realistic short-range method of solving the world's food crisis is to increase food production through conventional agriculture in the tropics."

The promise of high agricultural production in the tropics has not been realized, however. Many factors have contributed to this failure, including the prevalence of crop pests and the lack of fossil fuels to support advanced agricultural practices. Furthermore, there are nutrient limitations.

Many tropical soils are nutrient-deficient. This is particularly true of extremely old soils known as **laterite.** Because the tropics have not been glaciated for many millions of years, the soil is highly weathered, and many of the nutrients have been lost. Laterites are composed of extensive sheets of crust made up of oxides of aluminum and iron; the iron gives the soil its characteristic reddish color. Most of the silica has been removed by leaching, as has most of the sodium, potassium, calcium, and magnesium. The result is a very nutrient-poor soil. More than 50 percent of all tropical soils have these characteristics (Sanchez 1976). In addition, the high rainfall that is characteristic of the tropics results in high rates of leaching.

There is intense selection on tropical plants to trap the nutrients that are in scarce supply (Jordan 1985). Indeed, there is ample evidence that tropical plants are nutrient traps. Tropical plants tend to have higher root-to-shoot ratios, and their roots also tend to be concentrated near the soil surface so that they can appropriate nutrients before other plants absorb them. Mats of epiphytic bryophytes, lichens, club mosses, ferns, and other plants occur on tropical forest trees. Adventitious tree roots penetrate these mats to secure nutrients there that may never enter the soil. In addition, most tropical plants have mycorrhizal associations (Chapter 3) to increase their rates of nutrient uptake.

These characteristics of tropical forest nutrient dynamics help explain the failure of traditional slash-and-burn agriculture in these systems. Removal of the natural nutrient traps results in rapid leaching of the remaining nutri-

ents (Figure 15.16). In addition, lateritic soil causes other problems once the vegetation has been removed. When exposed directly to the heavy tropical rains, the layers of aluminum and iron oxides form a cement-like crust that greatly increases runoff.

This pattern of rapid nutrient loss predominates in much of the lowland tropics. Nutrient cycling varies among tropical forests, however (Whitmore 1990; Harcombe 1977a, 1977b); not all forests have nutrient-poor soils. C. F. Jordan and R. Herrera (1981) propose that there are two fundamental types of tropical forests. *Oligotrophic forests* have the characteristics just described. They are extensive in lowland regions, particularly the Amazon Basin. *Eutrophic tropical forests* are found in mountainous regions where bedrock has been exposed by geological uplifting. The weathering of this rock provides the system with a source of new nutrients.

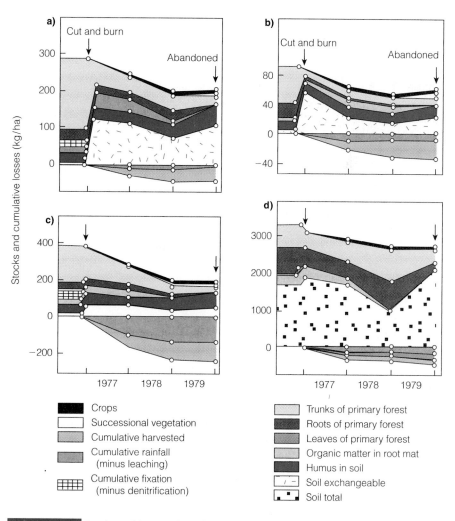

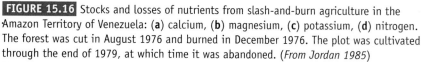

FIGURE 15.16 Stocks and losses of nutrients from slash-and-burn agriculture in the Amazon Territory of Venezuela: (**a**) calcium, (**b**) magnesium, (**c**) potassium, (**d**) nitrogen. The forest was cut in August 1976 and burned in December 1976. The plot was cultivated through the end of 1979, at which time it was abandoned. (*From Jordan 1985*)

Coral Reefs

We have already seen (Chapter 14) that coral reefs have high primary production. However, net community production is low (D'Elia 1988). Ecologists generally infer from this that the nutrients that support primary production move rapidly. Indeed, compared to many other shallow-water marine ecosystems, nutrient flux is rapid.

Many coral reefs, particularly atolls in the Pacific, are rather isolated from large landmasses. Thus they represent islands of high productivity surrounded by vast expanses of relatively nutrient-poor waters. Therefore, there must be some mechanism of accu-

mulating sufficient nutrients to support production. *Advection,* large-scale horizontal movements of ocean water, carries large volumes of water to the reef. Even though this water has a relatively low nutrient concentration, the volume may be sufficient to carry a large absolute amount of nutrient to the reef. It has also been suggested (D'Elia 1988) that "geothermal endo-upwelling" brings a variety of nutrients to the reef. In this process, geothermally heated deep-ocean water, rich in nutrients, is carried to the surface and through the reef structure, where the nutrients are quickly absorbed.

Two features distinguish the nitrogen cycle on coral reefs. First, reefs are notable for their high rates of

TABLE 15.3

Input and Output of Calcium and Phosphorus in Various Forest Ecosystems

		Calcium (kg/ha/yr)			Phosphate Phosphorus (kg/ha/yr)		
		Precipitation Runoff			Precipitation Runoff		
Author	Forest Type	Input (P)	(R)	P − R	Input (P)	(R)	P − R
Shugart et al. 1976	Mixed mesophytic, eastern Tennessee	10.51	27.40	−16.89	***	***	***
Jordan et al. 1972, Jordan 1969	Montane tropical rain forest, Puerto Rico	21.80	43.10	−21.30	***	***	***
Woodwell and Whittaker 1967; Woodwell et al, 1975	Oak–pine, Long Island	3.30	9.70	−6.40	***	***	***
Likens et al. 1977	Northern hardwoods, New Hampshire	2.20	13.90	−11.70	.036	.019	+.017
Cole et al. 1967	Douglas fir, Washington	2.80	4.50	−1.70	Trace	.020	−.020
Turvey 1974	Tropical rain forest, New Guinea	0	24.75	−24.75	***	***	***
Heinrichs and Mayer 1977	Spruce on sandstone, Germany	12.80	13.50	−0.70	.80	.020	+0.78
Heinrichs and Mayer 1977	Beech on sandstone, Germany	12.80	12.70	+0.10	.80	0.10	+0.70
Wright 1976	Coniferous, north Minnesota	3.10	4.50	−1.40	.18	.03	+.15
Bernhard-Reversat 1975	Evergreen forest, Ivory Coast	1.90	3.80	−1.90	0.15	.03	+0.12
Kenworthy 1970	Tropical rain forest, Malaysia	14.00	2.10	+11.90	***	***	***
Golley et al. 1975	Tropical, moist forest, Panama	29.30	163.20	−133.90	1.0	.70	+0.30
Swank and Douglass 1975; Johnson and Swank 1973	Hardwoods, North Carolina	6.16	6.92	−0.76	0.11	0.02	+0.09
Swank and Douglass 1975; Johnson and Swank 1973	Pine, North Carolina	6.51	4.10	+2.42	0.12	0.02	+0.10
Richardson and Lund 1975	Aspen, Michigan	8.30	19.4 to 38.8	−11.1 to 30.5	***	***	***
Gosz 1975	Mixed conifer, New Mexico	7.61	4.88	+2.74	***	***	***
Herrera 1979	Rain forest, Amazon Basin, Spodosol	16.00	2.80	+13.20	16.7	16.00	+0.70
Jordan, 1982	Rain forest, Amazon Basin, Oxisol	11.61	3.90	+7.71	26.91	30.00	−3.09

From Jordan 1982.

nitrogen fixation; they import large amounts of nitrogen from the atmosphere. Second, they are also characterized by high rates of nitrogen export to the open ocean, mostly in the form of nitrate.

The most important feature of the phosphorus cycle on coral reefs is the dynamics of phosphorus within the corals. Many corals contain endosymbiotic zooxanthellae—photosynthetic dinoflagellates that account for much of the primary production of reefs. L. R. Pomeroy and E. J. Kuenzler (1969) have shown that corals with such symbionts are very effective at assimilating phosphorus from the water. In fact, these symbiotic associations can remove dissolved inorganic phosphate from waters containing extremely low phosphorus concentrations. Corals with endosymbionts have very long phosphorus retention times. Internal cycling of phosphorus supports the high productivity of the reefs.

As in the case of nitrogen, there is general throughput of phosphorus in reefs. Although it may be cycled internally in corals for a time, eventually phosphorus is exported. Thus, on reefs, true nutrient cycling does not occur. The conventional view has been that cycling within the biotic component accounts for much of the flux:

$$\text{autotrophs} \xleftarrow{\text{internal cycling}} \text{heterotrophs}$$

According to C. F. D'Elia (1988), a more realistic model includes a higher degree of throughput of nutrients:

$$\begin{array}{c} \text{internal cycling} \\ \text{external} \rightarrow \text{autotrophs} \leftrightarrow \text{heterotrophs} \rightarrow \text{external} \\ \text{inputs} \qquad\qquad\qquad\qquad\qquad\qquad\qquad \text{outputs} \end{array}$$

In other words, nutrients do not cycle as tightly on reefs as they do in forest systems, especially tropical forest (D'Elia 1988); that is, reefs function more as open systems (Webb et al. 1975).

Local Nutrient Budgets

The three ecosystems we have considered—hardwood forests, tropical forests, and coral reefs—do not appear to be in a steady state with respect to nutrient cycles. Even though some degree of internal cycling occurs in each of these systems, each is also characterized by net gain or loss of nutrients. These systems exemplify one of the fundamental distinctions between the global biogeochemical cycles and local nutrient cycles: Local cycles are generally not in a steady-state condition. Global cycles must be balanced, but as soon as we

define an ecosystem boundary, we observe some net gain or loss of nutrients. Inputs from the atmosphere, from stream flow, or from the weathering of parent materials add nutrients; outputs in the form of leaching and runoff remove nutrients.

The degree to which nutrient budgets are not in a steady state varies with the ecosystem type and the particular nutrient. The net input and output of calcium and phosphorus for a number of forest ecosystems are shown in Table 15.3. For calcium, the net balance is negative across a wide variety of systems. The implication is that calcium must be replaced in these systems by weathering of parent materials. In contrast, phosphorus appears to be in approximate balance in many forest systems, because the ratio of input to output is close to 1.0.

Control of Nutrient Cycles

We have seen that there are complex patterns of nutrient movement and flow in ecosystems. But what sorts of mechanisms and factors control the pattern of nutrient flux? Are nutrient cycles analogous to biochemical cycles such as the Krebs cycle and photosynthesis, with crucial points of control analogous to rate-limiting enzymes? Are there feedback mechanisms that control the rate of nutrient flux?

There is ample evidence that certain steps are crucial to the operation of nutrient cycles. We have already examined a few of them. The high rate of nitrogen fixation in coral reefs is obviously an important part of the pattern of nitrogen flux there. The rapid leaching of nutrients in tropical forests constrains the components of nutrient cycles there.

In some cases, the activity of a particular trophic level exerts important control. For example, N. T. Hobbs and colleagues (1991) have shown that primary consumers (grazers) play a significant role in the nitrogen budgets of tallgrass prairie ecosystems. In the prairie nitrogen cycle, fire releases large quantities of nitrogen stored in plant biomass, much of it into the atmosphere. Experimental plots from which cattle were excluded lost twice as much nitrogen to combustion as control plots on which cattle were permitted to graze (Figure 15.17). The amount of nitrogen lost was proportional to the amount of standing crop biomass before fire.

It is even possible that a single species plays a critical role in nutrient cycling. One classic example comes from salt marsh ecosystems in Georgia (Kuenzler 1961), in which a flush of growth of the major grass,

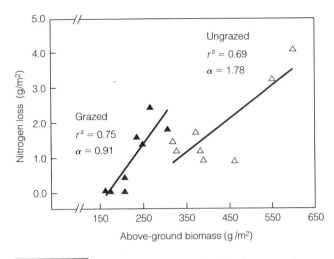

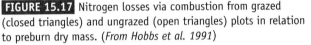

FIGURE 15.17 Nitrogen losses via combustion from grazed (closed triangles) and ungrazed (open triangles) plots in relation to preburn dry mass. (*From Hobbs et al. 1991*)

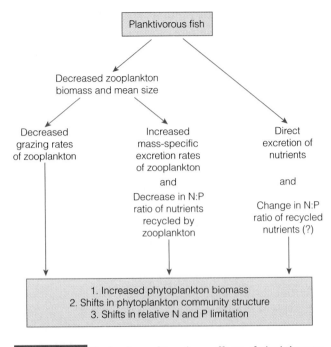

FIGURE 15.19 Mechanisms of top-down effects of planktivorous fish in a lake ecosystem. (*From Vanni and Layne 1997*)

Spartina, occurs in the summer. To support this growth, the plants' roots penetrate deep into the sediments to pump phosphorus to the surface of the marsh. Later, when the grass dies and decomposes, the phosphorus it contains is released into the water in the form of small particles.

E. J. Kuenzler showed that one species of mussel, *Modiolus demissus,* plays a key role in pulling this phosphorus from the water for recycling into the biotic component of the ecosystem. This filter feeder extracts large quantities of phosphorus from the water. In fact, every two and a half days, the mussels cycle as much phosphorus as is contained in all the water of the

marsh! Clearly, this one species is a major control point in salt marsh phosphorus cycling; any change in the population size or activity of *Modiolus* has a major impact on the ecosystem.

H. Jenny (1980) developed a useful paradigm for the control of ecosystem nutrient cycling. In this paradigm, five external conditions, called state factors, regulate all ecosystem functions including the flow of

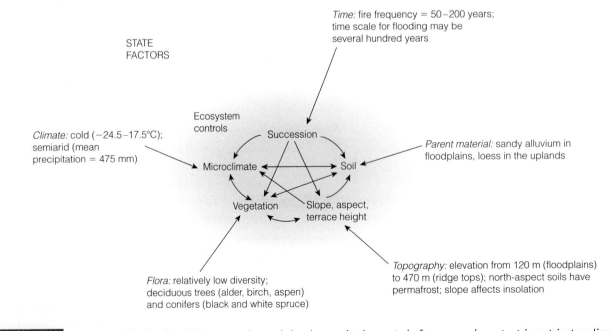

FIGURE 15.18 The pattern of taiga floodplain succession and the changes in the control of processes important in nutrient cycling in taiga. (*From Van Cleve et al. 1991*)

nutrients. The five state factors are the regional climate, flora, topography, parent material of the soil, and time (development of the community after disturbance). Some of these state factors, such as topography and parent material, are static; that is, they do not change within a region. Others, such as the climate and flora, are active.

K. Van Cleve and colleagues (1991) have used this approach to understand the control of nutrient cycling in taiga floodplain forests in Alaska. The state factors for this system and their interactions are depicted in Figure 15.18. From this set of state variables, one can model the flow of nutrients in this system.

We can also consider regulation of nutrient cycles from the perspective of the top-down and bottom-up patterns of control. For example, plants and their associated symbionts are responsible for the fixation of atmospheric nitrogen and hence exert bottom-up control over the movement of nitrogen into many ecosystems, particularly those that do not have closed or balanced nitrogen budgets.

Recent experiments by Vanni and his colleagues (Vanni and Layne 1997; Vanni et al. 1997) demonstrate top-down control of nutrient dynamics in a lake ecosystem. By enclosing portions of a lake ecosystem and selectively allowing input or output from various components (such as sediments and atmosphere), Vanni and his co-workers were able to elucidate the mechanisms of top-down control of nutrient dynamics (Figure 15.19). When planktivorous fish were present, phytoplankton biomass increased, and phosphorus and nitrogen concentrations increased in the water column (Figure 15.20). These results were due to three mechanisms of predator effects: reduction in herbivory rates, changes in the rates at which nutrients were excreted from herbivores, and nutrient excretion by the predators themselves (Figure 15.20).

Our consideration of nutrient cycles provides convincing evidence of the importance of having complete ecosystems for normal function. One of the most important principles to emerge from ecosystem studies of nutrient cycles is the complexity of interactions that occur within and between cycles. Only for a few ecosystems do we have enough information to identify all the crucial steps, such as the importance of *Modiolus* in a salt marsh, the role of alders in taiga floodplains, and the coral/zooxanthellae interaction in coral reefs. We will not be able to predict accurately the effects of human activities on ecosystems until we have more knowledge of the key points of nutrient cycles.

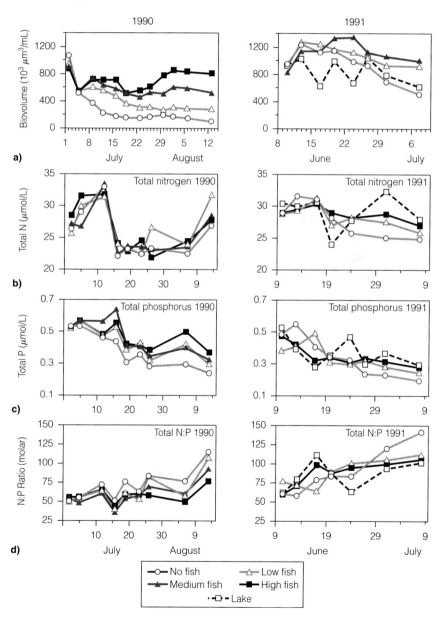

FIGURE 15.20 Total phytoplankton volume (**a**), total nitrogen concentration (**b**), total phosphorus concentration (**c**), and total N:P (**d**) in systems from which fish were excluded, or were low, medium, or high in density. (*From Vanni et al. 1997*)

SUMMARY

1. Unlike energy flow, the movement of materials in ecosystems is at least potentially cyclic. Elements can move in cyclic fashion among biotic components of the ecosystem or between abiotic and biotic components.

2. Biogeochemical cycles can be represented with compartment models. These models are based on the amount of material in various biotic and abiotic reservoirs and on the pathways and rates of transfer among pools.

3. Biogeochemical cycles are driven by the fundamental energy capture processes of ecosystems—that is, by the activity of primary producers and heterotrophic consumption.

4. The carbon cycle is intimately related to the basic energy capture and use processes. Carbon dioxide is removed from a large atmospheric pool by photosynthesis. It is returned to this pool by respiration.

5. The nitrogen cycle is characterized by cycling among a number of chemical forms of nitrogen. Many of the reactions, such as nitrogen fixation, require specialized organisms.

6. The phosphorus cycle is a sedimentary cycle. Because phosphorus can be lost as a result of changes in solubility, the potential exists for loss of phosphorus from the rapidly cycling portions of the system.

7. The sulfur cycle is characterized by a large atmospheric pool, transfer into the biotic component of the system by rainfall and plant uptake, and the cycling of various redox states of sulfur.

8. Disturbance affects biogeochemical cycles. Loss of nutrients may occur after a disturbance, followed by aggradation and (eventually) a steady-state concentration.

9. Tropical forests are characterized by adaptations to trap nutrients before they are leached from the system. Oligotrophic forests are nutrient-poor because of low rates of nutrient input; eutrophic systems are nutrient-rich.

10. Coral reefs are known for their high gross primary production, which apparently depends on high rates of nitrogen fixation. The phosphorus cycle on reefs depends on the rapid cycling of phosphorus made possible by mutualistic zooxanthellae.

11. Most local nutrient budgets are not in steady-state balance. Most are aggrading or degrading.

12. Nutrient cycles are controlled by rate-limiting steps such as nitrogen fixation or by the activities of certain species such as *Modiolus* in salt marsh phosphorus cycles. State factors such as climate, topography, time since disturbance, flora, and parent material of soil interact to control the movement of nutrients in ecosystems.

SELF-ASSESSMENT: CAN YOU ...?

1. Distinguish between global biogeochemical cycles and local nutrient cycles. Discuss the important differences and their significance.

2. Outline the essential pools, sinks, and fluxes of the carbon, nitrogen, sulfur, and phosphorus cycles. (Try to do this not just from memory but, rather, by reasoning the logical pathways and forms of these elements from your general knowledge of physiology and ecology. Some aspects have to be memorized, but many can be logically worked out).

3. Outline the fundamental differences and similarities between biogeochemcial cycles and the movement of energy in ecosystems.

4. Explain the role of disturbance in local nutrient cycles.

5. Explain the role of successional processes (Chapter 13) in local nutrient cycles.

6. Make an annotated list of the effects of top-down and bottom-up control in as many ecological processes as possible (from this and any previous chapters).

PROBLEMS AND STUDY QUESTIONS

1. Why are regional nutrient budgets generally not balanced?

2. What is the general pattern of change in the nutrient budget when a forest is disturbed?

3. Explain the relationship between biogeochemical cycles and trophic structure.

4. Review and outline the role of nitrogen, phosphorus, and sulfur in important biomolecules.

5. Discuss the following statement: "Nutrient cycles are not controlled or regulated. The patterns of flow are simply the result of recovery from the most recent disturbance."

6. You hypothesize that *Modiolus* regulates the phosphorus cycle in a salt marsh ecosystem. How would you test that hypothesis?

7. What are the important differences between biogeochemical cycles and local nutrient cycles?

8. Relate your understanding of biogeochemical cycles to the importance of conserving biodiversity.

9. In what ways are coral reef ecosystems unusual in terms of production and nutrient cycling? Generate a list of hypotheses to explain these unusual features. How could you test these hypotheses?

10. We have stated that understanding the nature of interactions among nutrient cycles is an important challenge for ecologists. Speculate on the nature of these interactions and their consequences.

PROJECTS AND ADDITIONAL STUDY

1. Measure the pH of precipitation in your area. Keep track of changes associated with the direction of the storm track, amount of rainfall, temperature, season, nearby pollution sources, and other variables that might affect precipitation pH.

2. Using values from a survey of the literature, generate a graphical summary of the nitrogen and phosphorus content of key ecosystem pools for temperate deciduous forest, temperate grassland, desert, tundra, and tropical rain forest. How do these systems differ? What biological factors are correlated with these differences?

3. Design (and carry out if possible) a microcosm experiment to test the hypothesis that carbon or nitrogen cycling is controlled by top-down forces. Begin by making predictions about the relative sizes of the pools under this hypothesis. Use a simple ecosystem containing just three trophic levels.

Patterns of
Terrestrial Vegetation

O ne of the simplest and most fundamental eco- logical observations that one can make is that there is geographic variation in the nature of the vege- tation. Both Darwin and Wallace were intrigued by the variations they observed during their nineteenth- century travels. This observation, almost trivial in its simplicity, has profound implications for much of eco- logical science. A series of fundamental questions fol- lows from this simple observation. For example:

1. What are the root causes of these changes in vege- tation?
2. How do animals respond to the changes in the veg- etation?
3. Are the same groups of plants concentrated in sim- ilar climatic regions?
4. Are the vegetation units we observe actual biologi- cal entities or simply useful constructs that ecolo- gists can use to classify natural systems?

Our purpose in this and the next chapter is to out- line the basic nature of terrestrial and aquatic habitats. In Chapter 3 we examined the kinds of adaptations that allow some organisms to inhabit environments with various abiotic challenges. In these two chapters, our emphasis will be on the relationship between biotic communities, especially the vegetation, and the physical environment. Our approach will be largely descriptive in this chapter (and in the next on aquatic systems).

Climate

The climate is one of the most important factors influ- encing the distribution of plants. Temperature and precipitation are two climatic factors of paramount importance to plants (Chapter 3). Plants are fre- quently limited by available water. Recall that the process by which plants obtain CO_2 for photosynthesis (opening the stomata) is also a major avenue for the loss of water by the plant. Each species has water-tol- erance limits and preferences for how much water is available (see Chapter 3). Plants also differ in their tol- erance of temperature extremes. The effects of tem- perature interact with moisture stress. At high temperature, water loss is likely to be higher as well. At temperatures below freezing, the plant faces two difficulties. One is the direct effect of the formation of ice crystals in the tissue. The other is that frozen water is physiologically unavailable to the plant, so moisture stress becomes a problem.

Other climatic factors may occasionally be impor- tant as well, although these often operate through temperature and moisture effects. For example, wind may be an important physical factor. High winds may have direct physical effects, such as the damage they can cause to woody plants. Often, however, the effect is indirect. Strong winds contribute significantly to desiccation.

In this section, we will consider the major determi- nants of climate, including altitude and latitude, the proximity of large bodies of water, and the presence of mountain ranges.

Altitude and Latitude

One of the major determinants of temperature is lati- tude. As Figure 16.1 shows, the nature of solar input changes from the equator toward the poles: Solar radiation strikes the Earth nearly perpendicularly at the equator but strikes at a smaller angle at the poles. Consequently, solar radiation heats the atmosphere and soil less at the poles than in the tropics.

The variation in the orientation of the Earth to the sun at the poles relative to the equator also affects the seasonality in temperatures (Figure 16.2). At the poles, the difference between summer and winter tempera- ture is far greater than it is closer to the equator.

Latitude and altitude have similar effects on the vegetation. This is due in part to the fact that an increase in either latitude or altitude results in a

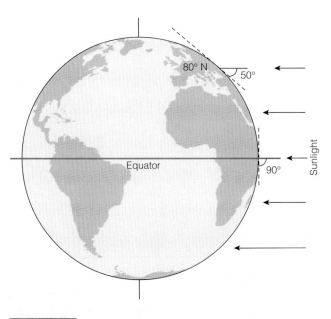

FIGURE 16.1 The angle of incidence of solar radiation at different latitudes. The axis of the Earth is tilted relative to the sun. In this figure, however, the tilt has been omitted to emphasize the effect of latitude.

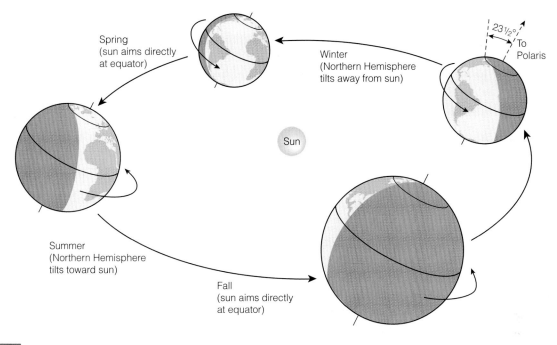

FIGURE 16.2 Changes in the orientation of the Earth relative to the sun at different seasons in the Northern Hemisphere.

decrease in temperature. However, the reasons for these changes are different. Altitudinal temperature changes are a result of the decreasing air pressure at higher altitude. As the atmospheric gases expand, the temperature of the gases drops, an effect known as **adiabatic cooling.** We observe a similar phenomenon when we let air out of a tire. The emerging stream of air is cool because of the release of energy associated with the expanding gases. The latitudinal change, by contrast, results from the lower angle of incidence of solar radiation.

Obviously, altitude and latitude interact as determinants of a region's vegetation. At high latitude, a slight increase in altitude can have a profound effect on the mean temperature. Near sea level at the equator, we might expect to find plants adapted to warm climates, but the vegetation on mountain peaks in that region is often more similar to that found closer to the poles than to that found at lower altitudes near the equator.

Maritime and Continental Climates

Temperatures are also affected by the presence of large bodies of water. Water, which has a very high heat capacity, acts as a heat reservoir. Compared to the land, water tends to change temperature slowly: It warms slowly and, once warm, loses that heat slowly.

The regions near an ocean are said to have a **maritime climate;** inland regions have a **continental climate.** The proximity of a large body of water tends to have two effects. First, temperatures in maritime climates are milder than one might expect on the basis of latitude or altitude alone. For example, Juneau, Alaska, is located at latitude 58° N, where one might expect very cold winter temperatures like those in other northern cities such as Edmonton, Alberta. Juneau is on the shore of the Pacific Ocean, however, and the presence of the ocean ameliorates the temperature. Mean winter temperature in Juneau is −6°C, a value similar to that of Chicago, much farther south. Second, the range of fluctuations is smaller in maritime climates. Continental climates are characterized by much wider seasonal changes in temperature. Arcata, California, on the Pacific Ocean, has a mean summer temperature of 11.5°C and a mean winter temperature of 10.1°C. At the same latitude but farther inland, Denver, Colorado, has mean summer and winter temperatures of 23°C and −1.2°C, respectively.

Rain Shadow

An important determinant of precipitation is the presence and arrangement of mountain ranges. A mountain range running perpendicular to the prevailing wind creates regions of high precipitation on the windward slope and regions of significantly less moisture on the leeward side. This mechanism is a result of the fact that warm air can hold more moisture than cool air. As moisture-laden winds strike the mountain

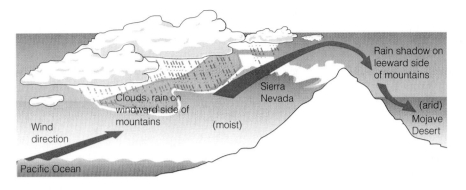

FIGURE 16.3 The rain shadow effect. The prevailing winds carry moisture-laden air upslope, where it cools and precipitates. An area with much drier conditions—the rain shadow—occurs on the leeward side of the mountain.

range, the air rises and cools adiabatically (Figure 16.3). As a result, the air can hold less moisture than before, and water precipitates and falls to the Earth. The air that reaches the leeward side contains much less moisture, so precipitation there is greatly decreased. In addition, as the air moves downslope and warms, it is able to hold more moisture, thus precluding precipitation. The net result is that the windward sides of mountains tend to be wetter; the lee sides tend to be dry. A **rain shadow** occurs on the lee side. The gradient in precipitation across western North America is largely due to this phenomenon

(Figure 16.4). The Sierra Nevada, Cascades, and Rocky Mountains all pull moisture from the air on their western slopes, creating rain shadows to their east. As the air moves east, it picks up moisture again from the Great Lakes and the Gulf of Mexico, and precipitation totals increase.

Slope Effects

Mountains have a third important influence on climate (in addition to adiabatic and rain shadow effects). The direction a particular slope faces—its aspect—influences the temperature regime and, indirectly, the amount of precipitation the slope receives. In the Northern Hemisphere, north- and east-facing slopes tend to be cooler and wetter than south- and west-facing slopes. North-facing slopes receive less total sunlight, whereas east-facing slopes receive less intense sunlight. The latter phenomenon occurs because as temperatures rise over the course of the day, direct sunlight has a more important effect. Thus east-facing slopes that receive direct sunlight only in the morning tend to have cooler mean temperatures than their west-facing counterparts. Evapotranspiration is reduced, and humidity and soil moisture are greater. South-facing slopes are generally hotter and drier than north-facing ones. Of course, this phenomenon is reversed in the Southern Hemisphere, where the north- and west-facing slopes are warmer and drier.

At first glance, this slope effect may seem to contradict the rain shadow effect, which creates wetter regions on the west sides of mountain ranges. The two phenomena differ in scale, however. The rain shadow effect is a broad, regional pattern: the west slopes of the Sierra Nevada are wetter than the east side. In contrast, the slope effects occur locally. For any single mountain, on either side of a range, west- and south-facing slopes will be hotter and drier than the east- and north-facing slopes on the same mountain.

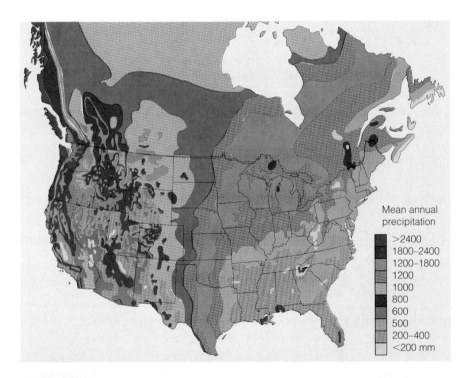

Mean annual precipitation

	>2400
	1800–2400
	1200–1800
	1200
	1000
	800
	600
	500
	200–400
	<200 mm

FIGURE 16.4 The change in precipitation from west to east in the United States is due in large part to the rain shadow effects of the Sierra Nevada, Cascade, and Rocky Mountains. (*Data from Chapman and Sherman 1975*)

FIGURE 16.5 An east-facing slope and a west-facing slope in Wyoming. The east-facing slope maintains montane conifer forest, whereas the west-facing slope is covered with sagebrush, a much more xeric-adapted community. (*Photo by David Krohne*)

The impact on the vegetation can be profound (Figure 16.5). As we will see, the vegetation on north-facing slopes may be more like that on flat ground several hundred miles north, whereas plants sensitive to frost damage may extend to higher altitudes on south-facing slopes.

Midlatitude Desertification

Many of the world's deserts are located at the midlatitudes, at approximately 30° north or south of the equator (Figure 16.6). This phenomenon is a consequence of the interaction between direct solar radiation at the equator and adiabatic cooling. At the equator, solar radiation strikes the Earth at close to a 90° angle, warming the Earth and the air near the surface. As this warm air rises, it cools adiabatically. Like the air that cools as it passes over a mountain range, this air can hold less water, and precipitation results. This is one of the reasons why the tropics have such high levels of precipitation.

The air rising over the tropics reaches a layer of air in the upper atmosphere that prevents it from rising further. It then spreads away from the equator toward the poles. As it does so, the cooler air sinks back to the Earth's surface at about latitudes 30° north and south. Much of its moisture, however, has already dropped out over the equatorial regions. Thus the surface air contains little moisture, and precipitation is greatly reduced at those latitudes. Consequently, many of the world's deserts are concentrated in this latitudinal band.

Classification of Communities

One of the tasks that early ecologists undertook was the categorization of ecological communities. Their goal was to organize the obvious spatial variation in plant and animal life so that ecological hypotheses might emerge from the patterns.

One of the first scientists to attempt such a classification was Alfred Russel Wallace, who organized the biota of the world into six realms (Figure 16.7). Major physiographic features such as oceans and mountain ranges provide obvious boundaries between the biota of most regions, but the boundary between the Oriental and Australian regions is less obvious. Based on zoogeographic affinities, Wallace placed the boundary between the Molucca Islands extending southeast of Borneo, splitting the islands of Bali and Lombok. This has come to be known as Wallace's line (see Figure 16.7).

One of the classifications that emerged early in the twentieth century and is still in use today is the concept of the biome. In this chapter we will use this scheme to organize the vegetation of the world into a set of "natural" units. We define a **biome** as the major terrestrial unit of vegetation. It is characterized by a specific form of vegetation—shrub, coniferous forest, grassland, and so on. The fundamental characteristic is the

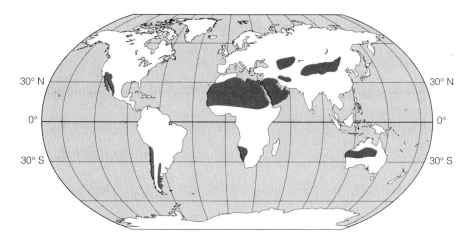

FIGURE 16.6 Many of the world's deserts (shaded area) are located at approximately 30° north or south of the equator.

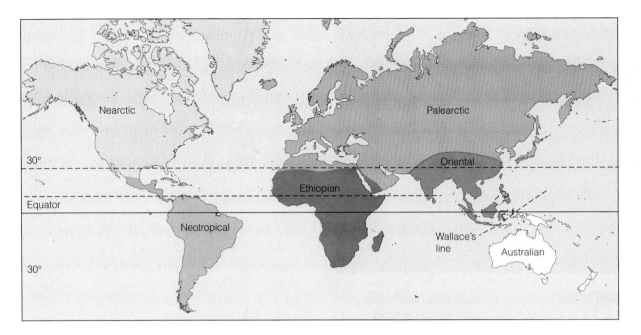

FIGURE 16.7 Wallace's six biotic realms. Wallace's line divides the Australian Biotic Realm from the Oriental Biotic Realm.

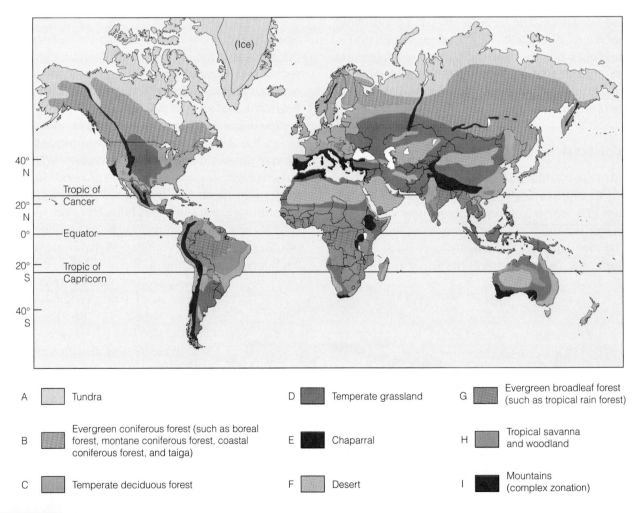

A Tundra

B Evergreen coniferous forest (such as boreal forest, montane coniferous forest, coastal coniferous forest, and taiga)

C Temperate deciduous forest

D Temperate grassland

E Chaparral

F Desert

G Evergreen broadleaf forest (such as tropical rain forest)

H Tropical savanna and woodland

I Mountains (complex zonation)

FIGURE 16.8 The location of the world's biomes. Note that in this figure, taiga has been combined with coniferous forest because the boundary between these vegetation types is indistinct. The region designated as mountains contains a complex pattern of vegetation types arranged according to altitude.

growth form of the plants. The **flora** of a region is the set of plant species present. Thus the same biome (vegetation type) in different regions can have the same or different flora species present. Each biome is associated with a particular climatic regime and general soil type. As you might expect, certain animals are associated with each biome.

The locations of the world's biomes are shown in Figure 16.8. Note that we have identified a total of eight biomes. One of the difficulties with the concept of the biome is that the vegetation in one region often grades into another over a climatic cline. Thus the boundaries are sometimes indistinct, and broad transitional regions occur that some ecologists prefer to classify as distinct biomes. For example, in the southwestern United States, grassland merges into true desert, but this occurs over such a broad region that some ecologists classify this transition zone as a distinct biome. Another such region characterizes the transition from tundra to boreal forest in the northern latitudes of Europe and North America. This region, called the taiga, contains many boreal forest species that gradually diminish in stature or number as one travels north. At or near the northern treeline, boreal forest species are replaced by those of the tundra. Similar transitional regions occur between virtually all biomes. We must make a somewhat arbitrary decision about whether or not to call such regions separate biomes.

This text uses a classification scheme based on the presence of some recognizable growth form of the vegetation associated with a particular climatic region. The student should be aware, however, that any such scheme is in a sense doomed to fail. The ecological world is too complex to categorize so simply. And we should keep in mind that our goal is really to develop a language of description that will facilitate further discussion of the ecological processes in different regions.

Tundra

The **tundra** biome is distributed across the northern fringes of the Old and New Worlds (Figure 16.9). Note that in the Southern Hemisphere, the landmasses of both Africa and South America taper off before true tundra can occur at high latitudes.

The term is derived from the Finnish word *tunturi*, which means "treeless plain." This is an apt description of the tundra, for it is a land beyond the northern treeline where the growth form of the plants is low to the ground. The flora of the tundra is quite varied across its vast extent. Many species contribute to the vegetation, but all are characterized by a low growth form. Some are species that are also small in stature

FIGURE 16.9 The tundra in northern Alaska. (*Photo by David Krohne*)

elsewhere, such as some of the grasses and sedges; others are dwarf forms of species that grow as trees in more southern latitudes. One example is shown in Figure 16.10. The low, matted vegetation in this photograph is actually a birch tree that grows prostrate on the ground in the severe northern climate.

As in virtually all the biomes we will discuss, the nature of the vegetation varies across the tundra. We can recognize regions dominated by grasses, others

FIGURE 16.10 This mat of vegetation is actually a single, prostrate birch tree whose growth habit results from harsh conditions in the tundra. The blue baseball cap is shown for scale. (*Photo by David Krohne*)

with a shrub—steppe vegetation, and still others with extremely sparse vegetation and extensive areas of rocks and boulders called fell fields.

The climate of the tundra is indeed severe. Winter temperatures may drop below −30°C during the six months of darkness. Of course, in summer the sun never sets, and temperatures may exceed 25°C. Many arctic regions receive less than 25 centimeters of precipitation per year. This has an important influence on the plants because, as we have seen, one of the crucial effects of low temperature is that water becomes unavailable, or at least limiting. If precipitation is low as well, the plants must be well adapted to xeric (dry) conditions.

The Pleistocene glaciations had a profound effect. The most recent glaciation event, the Wisconsin, receded less than 10,000 years ago. Consequently, the ground a few centimeters or meters below the surface is still frozen solid. The first few centimeters of soil may thaw each summer, but below that is a zone of **permafrost,** or permanently frozen ground. This too limits plant growth, because roots cannot penetrate easily and the groundwater there is frozen and unavailable. The permafrost also prevents drainage of surface water. Although the tundra receives relatively low levels of precipitation, water cannot easily percolate into the ground. For this reason, and because glaciation scoured out many depressions, the tundra is dotted with ponds and lakes of various sizes.

In more southerly regions, tundra vegetation is found at high altitudes (Figure 16.11). We must distinguish between arctic tundra and **alpine tundra.** Although they have similar growth forms of plants and share many species, there are three important physical differences between them: (1) Alpine tundra does not have permafrost. (2) Alpine tundra does not have the long periods of winter darkness found in the Arctic. (3) Alpine tundra often has much higher precipitation than arctic tundra characteristically experiences. Usually, a large portion of this precipitation comes in the form of winter snows. These great accumulations of snow melt gradually over the summer, and even if summer precipitation is infrequent, there may be a constant input of water from snowmelt.

Taiga

Immediately south of the arctic tundra lies a band of forest known as the **taiga.** The term is derived from a Russian word meaning "land of little sticks." This is an appropriate description, because in many places the taiga resembles a stunted, sickly forest (Figure 16.12). Coniferous trees with an open canopy are the dominant form of vegetation. Although the predominant species are black spruce (*Picea mariana*) and white spruce (*Picea glauca*), a few deciduous trees such as paper birch (*Betula papyrifera*) and quaking aspen (*Populus tremuloides*) are found mixed with the conifers and in disturbed sites. Many of these species are also found in the coniferous forests farther south, which leads some ecologists to regard taiga as simply a transition between the tundra and the boreal coniferous forest to the south. However, the consistent physical structure of the vegetation (sparse and stunted

FIGURE 16.11 Alpine tundra in northern Montana. (*Photo by David Krohne*)

FIGURE 16.12 Taiga vegetation in interior Alaska. (*Photo by David Krohne*)

trees with an open canopy even in old, mature forests) leads other ecologists to consider taiga a separate biome.

The taiga has a circumpolar distribution: It stretches across Alaska, northern Canada, northern Scandinavia, and Russia. Many of the species found in taiga, including the dominant black and white spruce, are also distributed in circumpolar fashion; if a person familiar with the flora of interior Alaska were plunked down in the taiga of Siberia, he or she would recognize many of the species.

The climate of the taiga is one of the most severe on the planet. Unlike many parts of the tundra, which have maritime influences, much of the taiga has a continental climate. Thus it experiences a tremendous annual range of temperatures. Summer temperatures may exceed 30°C, whereas taiga winters are characterized by some of the lowest recorded temperatures outside Antarctica. For example, Fort Yukon in interior Alaska has recorded temperatures of −50°C. The cold can be so intense that spruce trees sometimes shatter as water inside them freezes and expands.

The substrate in the taiga is characterized by two important features: a young, recently glaciated soil and discontinuous permafrost. The Wisconsin glaciations left this region recently, so the soil tends to be thin and young. Portions of the substrate a few centimeters below the surface are still frozen (permafrost). Unlike the tundra, however, the permafrost of the taiga is discontinuous, being found primarily on north- and east-facing slopes. On the warmer west- and south-facing slopes, the high summer temperatures have melted the substrate. These patterns of permafrost have profound effects on the vegetation. Areas with permafrost tend to be dominated by black spruce, whereas those without permafrost are more likely to support deciduous trees such as aspen or birch.

Fire constitutes a major abiotic factor in taiga forests. High summer temperatures and the flammability of spruces lead to a high incidence of natural fires in taiga. It has been estimated that as many as 10 million hectares of Alaskan taiga burned annually before World War II, when serious attempts at fire suppression began in Alaska. Studies of the fire patterns in Alaska have shown that few forests in the interior are older than 200 years, indicating a high frequency of fire disturbance. A few islands in the Yukon River that are protected from fire support trees as old as 350 years, which indicates that spruces can indeed live longer than the ages they are observed to reach throughout most of Alaska.

Coniferous Forests

Broad bands of coniferous forests extend through much of the Northern Hemisphere (see Figure 16.8). Forests across this huge range are similar in general appearance: dense, sturdy conifers with relatively little understory and a closed canopy (Figure 16.13). Unlike the taiga vegetation, however, the actual species that make up this vegetation vary geographically. In North America we can divide the coniferous forests into three broad groups: the boreal forests, montane conifer forests, and coastal conifer forests.

Boreal forests are characterized by a lusher, more vigorous growth of many of the same species that constitute the taiga to the north. As previously mentioned, the gradation from boreal forest to taiga is subtle and accounts for the reluctance of some ecologists to distinguish between them. Further south, black and white spruce develop forests with a closed canopy. Other conifer species are also common in these forests. Eastern white pine (*Pinus strobus*), an important timber tree, reaches great stature and relatively high densities in these forests.

FIGURE 16.13 Coniferous forest in Ontario, Canada. (*Photo by David Krohne*)

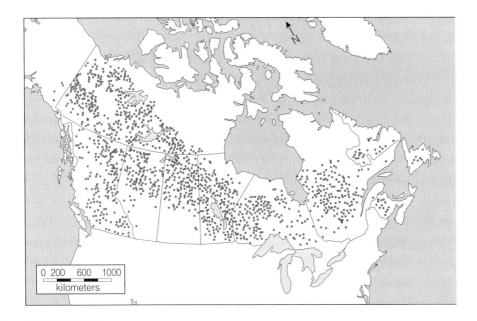

FIGURE 16.14 The distribution of forest fires burning more than 200 hectares in Canada from 1980 to 1989. (*From Johnson 1992*)

As in the taiga, fire is an important component of the boreal forest landscape. Lightning-caused fires are frequent in this region (Figure 16.14). Jack pine (*Pinus banksiana*) occupies a particular niche in boreal forests because it is well adapted to fire: Its cones open when heated by a forest fire. Consequently, it is one of the first to seed a recently burned area, and extensive stands of jack pine delimit the boundaries of old fires in the north.

The montane conifer forests are found at higher elevations in North America, Europe, and Asia (Figure 16.15). Though structurally similar to the boreal forests, they are dominated by different species (Table 16.1). Among the montane forests of North America, some species are widespread, whereas others are endemic to certain mountain chains. In all the montane forests, the trees become progressively more stunted and widely scattered (as in taiga) as elevation increases, until the altitudinal treeline is reached and alpine tundra begins.

The third zone of North American conifer forest is the coastal zone. This band of coniferous forest extends up the Pacific coast of the continent to southeastern Alaska (see Figure 16.16). These forests, which are also characterized by closed canopies and lush, vigorous growth, are among the most magnificent on the continent, particularly the old-growth stands that have not yet been logged.

Climatically, this region differs from the other two in that it has a strong maritime influence. Thus the coastal forests are characterized by a much milder temperature regime and significantly greater rainfall than the montane or boreal forests. Indeed, temperate rain forest, in which annual precipitation may exceed 380 centimeters (150 inches), occurs in this region. Snowfall may occur, but even in southeastern Alaska it does not persist on the ground for long periods. In northern California, Oregon, and Washington, signifi-

FIGURE 16.15 Montane conifer forest in Wyoming. (*Photo by David Krohne*)

TABLE 16.1

Important Species in Three Types of Coniferous Forests in North America

Forest	Important Conifers
Northern boreal	Eastern white pine (Pinus strobus) Black spruce (Picea mariana) White spruce (Picea glauca) Jack pine (Pinus banksiana)*
Sierra Nevada (montane)	Ponderosa pine (Pinus ponderosa) Douglas fir (Pseudotsuga menziesii) White fir (Abies concolor) Sequoia (Sequoiadendron giganteum)* Lodgepole pine (Pinus contorta) Jeffrey pine (Pinus jeffreyi)
Rocky Mountains (montane)	Ponderosa pine (Pinus ponderosa) Lodgepole pine (Pinus contorta) Blue spruce (Picea pangens)* Engelmann spruce (Picea engelmanni) Limber pine (Pinus flexilis)

*Endemic to the region.

FIGURE 16.16 Temperate rain forest in southeast Alaska. (*Photo by David Krohne*)

cant amounts of precipitation are added to the system by fog drip: Low clouds and fog moving past the needles of conifers condense to water that drips to the soil, where it is absorbed. Fog drip may provide an additional 30 centimeters of precipitation per year.

Coastal coniferous forests are dominated by such species as Douglas fir (*Pseudotsuga menziesii*) and western hemlock (*Tsuga heterophylla*). In addition, a number of **endemics** (species found in no other place) occur in response to local climatic and substrate conditions. Among the most well known of these are the redwoods (*Sequoia sempervirens*) of California and Oregon (Figure 16.17). These trees, known for their great height and age, are restricted to a small band

near the coast in northwestern California and southwestern Oregon, where summer fog adds significantly to the available moisture.

Desert

We have seen that the deserts of the world (see Figure 16.8) generally result from the effects of rain shadows or the general phenomenon of midlatitude desertification. This biome is somewhat different from the others we have discussed in that it is perhaps better characterized by climate than by vegetation. Even though the vegetation of most deserts consists of small, widely scattered shrubby plants, the nature and amount of vegetation are sufficiently variable that no single structural aspect of deserts serves to identify them in the way that tundra or coniferous forests can be characterized. Some deserts are completely barren of vegetation (Figure 16.18), whereas others show remarkable densities of plants and can even support

FIGURE 16.17 Redwood coastal conifer forest in northern California. (*Photo by David Krohne*)

FIGURE 16.18 The range of variation in desert vegetation: (**a**) salt flats in Utah; (**b**) Sonoran Desert near Tucson, Arizona. (*Photos by David Krohne*)

large trees (such as Joshua trees) or cacti (such as saguaro).

We generally define deserts as those regions in which evaporation exceeds precipitation. Beyond the general statement that deserts are typically hot and dry, there is considerable variation in temperature and precipitation. As we have noted, parts of the Atacama Desert in Chile have never experienced rain. On the other hand, some parts of the Kalahari Desert in Africa benefit from moisture derived from fog. One of the outstanding features of desert climate is the unpredictability of rainfall. Often precipitation comes in the form of local thunderstorms. Rainfall may be intense at one site in the Sonoran Desert of Arizona, while just a few kilometers away, no rain reaches the Earth.

The substrate of deserts is highly variable. Some soils are young, composed primarily of small stones of various sizes that have been weathered very little. Others are older and more fully developed. Substrates range from pure white, as in the gypsum sands of White Sands in New Mexico, to black, lava-derived soils in volcanic regions like the Ubehebe Crater of Death Valley. Another important feature of the substrate is the concentration of salt or alkali. Many valley floors in the American Southwest were filled with water during the Pleistocene. As temperatures increased and precipitation declined, these lakes evaporated, leaving behind high concentrations of salt. The salt flats to the west of Great Salt Lake in Utah are the best-developed example of this phenomenon. Because salt makes it difficult for plants to absorb water, the presence of high salt concentrations exacerbates the problems of desiccation due to heat and low rainfall.

Large areas of desert in many parts of the world are covered with a layer of stones referred to as *desert pavement.* Such substrates are found on low-angle slopes and in valley bottoms. Stones may be so dense on the surface that few plants can grow through them. Another substrate phenomenon of interest is "desert varnish." In many regions with a rocky substrate, the surface of the stones is covered with a dark, shiny veneer that contains clay and oxides of iron and manganese. In sandy regions the surface may be covered by a thin crust formed from chemical interactions and

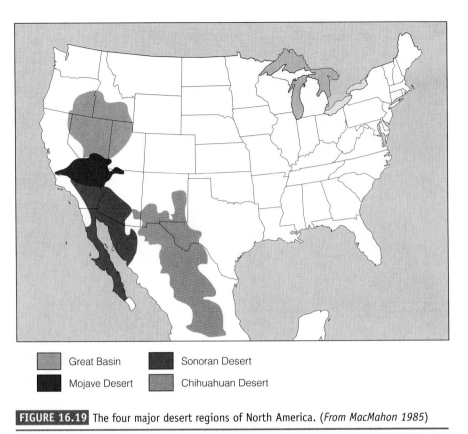

| Great Basin | Sonoran Desert |
| Mojave Desert | Chihuahuan Desert |

FIGURE 16.19 The four major desert regions of North America. (*From MacMahon 1985*)

a network of algae and lichen. It serves to prevent the finer materials in the substrate from being carried away by wind. Some algae live just beneath the soil surface and subsist on light that penetrates the upper few millimeters of soil. This crust is extremely fragile and can be broken down merely by walking on it. It is one of the reasons why deserts are often more fragile and vulnerable to degradation than they appear at first glance.

The deserts of North America are divided into four distinct regions (Figure 16.19): the Mojave, the Sonoran, the Chihuahuan, and the Great Basin Deserts. The first three are low-elevation, hot deserts distinguished by differences in the dominant species that inhabit them. The Great Basin Desert is found at elevations exceeding 1300 meters and is characterized by cooler temperatures and some persistent winter snow. Sagebrush is a dominant plant in this desert.

Chaparral

Chaparral is a shrub-dominated community (Figure 16.20) generally associated with a Mediterranean climate, which is characterized by winter rainfall and summer drought. The most extensive representation of this climatic type and of chaparral is, as you might expect, found in southern Europe and northern Africa, along the coast of the Mediterranean Sea.

FIGURE 16.20 Chaparral, a shrub-dominated community, is found in regions with a Mediterranean climate. (*Photo by Nigel J. H. Smith/Animals Animals*)

Other examples are found in California, southern Australia, and Chile. Most chaparral systems are found in areas with insufficient precipitation to support trees and a substrate that is too rocky to support extensive stands of grass.

Chaparral is typically a dense community in which the dominant plant growth form is the shrub. In fact, the word chaparral is from the Spanish word for a dense, shrub thicket. The shrubs have extremely dense foliage, the "canopy" is closed, and most species grow less than 5 meters high. In North American chaparral, species such as madrone (*Arbutus menziesii*), manzanita (*Arctostaphylos patula*), and chamise (*Adenostema fasciculatum*) are characteristic.

As we might expect, chaparral species are adapted to the long period of summer drought. Among the adaptations that have evolved in this context is sclerophyllous foliage—leaves that are small and leathery to avoid water loss. The shrubs become semidormant during the summer dry season. Many typically have a deep taproot to bring up water from lower soil strata during the summer, but they also have extensive lateral roots near the surface so that water can be absorbed quickly when rain does come. Most species are also evergreen; thus their leaves are present for photosynthesis during the wet winter months.

Fire is one of the most crucial abiotic factors in chaparral. The long, hot, dry summers of these regions leave the vegetation vulnerable to fire, and most chaparral species have become highly adapted to recurrent fire. Indeed, many chaparral species need fire either to reproduce or to maintain vigor. Chaparral is a closed vegetation community; that is, as plants mature, little or no open space is available for seedlings or vegetative growth. Because the plants release allelopathic chemicals that inhibit the growth of other plants, fire plays an important role; it opens space for reproduction and growth and volatilizes the toxic compounds that inhibit growth.

Many species in the chaparral contain volatile chemicals that, not unlike gasoline, are highly flammable. When chaparral ignites, the combustion of these chemicals can create temperatures exceeding 500°C. Only those species most highly adapted to such fires survive. After fire, they find an open community in which growth and reproduction are now possible. Sprouting of mature plants is common, and many seeds germinate quickly in soil that contains few noxious allelochemicals and has recently received a massive input of nutrients formerly tied up in dead plant material. In the first few years after a fire, extensive growth of herbaceous plants ensues from germination. After about five years, these plants become secondary to the shrubs that were not killed but have resprouted. These shrubs have the competitive advantage of a large and intact root system to obtain water and nutrients.

Grassland

In its pristine state, nearly half of the terrestrial land area of the planet was grassland. Before agriculture, this vegetation type covered large areas of the central regions of North America, the steppes of Asia, the pampas region of South America, and the African veldt. Europe is the only temperate continent without extensive development of this vegetation type. Because the fertile soils of the grasslands are so important economically, most grasslands have been significantly altered, perhaps more so than any other biome. Certainly, in North America, grassland has been all but replaced by row crops or pastures with species very different from those found there when Europeans first arrived.

Most of the world's grasslands are found in arid or semiarid regions with annual precipitation in the range of 25 to 75 centimeters. Above this amount, trees can flourish; below it, desert vegetation is likely to predominate. Many grasslands are also characterized by long periods of drought on both seasonal and multiyear scales. Most precipitation comes in the summer months in the form of thunderstorms.

As we might expect from the huge geographic coverage, the world's grasslands are diverse in species composition and function. The grasslands of Africa are characterized by annual grasses supported by seasonal tropical monsoons. The North American grasslands are dominated by perennial grasses. The steppes of

Asia are characterized by very arid grasslands that gradually merge into shrub steppes similar to North American high desert.

The North American grasslands were called prairies (from the French word for "meadow") by the Europeans who first saw them (Figure 16.21). The perennial grasses that dominated these grasslands exist in two growth forms, depending on the species: sod formers like big bluestem (*Andropogon gerardii*) and bunch grasses like northern dropseed (*Sporobolis heterolepis*). Sod formers produce extensive lateral shoots (stolons) from which new individuals can grow, creating a dense matrix of plant material within 1 meter of the soil surface. Bunch grasses grow in small clumps that gradually become larger as the plant ages and continues to grow. The other component of the prairie is the **forbs**—nonwoody, nongrass species. Among the plants in this group are the hundreds of species of wildflowers that competed with the grasses in prairies.

The North American prairies have soils whose fertility is exceeded by only a few other systems in the world. A unique set of factors and processes combined to produce these soils. First, the Pleistocene glaciations extended southward throughout the Rocky Mountains, grinding the mineral-rich Rockies into a fine flour-like material called **loess.** As the glaciers retreated, this loess was carried eastward by the prevailing winds and deposited on the Great Plains in depths as great as 3 meters. Next, the climate became warmer and drier with long, sunny summers, resulting in a highly productive grassland vegetation. Each year the above-ground parts of the perennial grasses died and contributed their organic matter to the mineral-rich loess. Because the climate was relatively arid, few compounds were leached away by rainfall, and

deep root systems kept minerals near the surface. Finally, a number of the forbs in this system were legumes that fixed nitrogen and contributed this crucial nutrient to the soil when the foliage died each year. The result was deep, fine, loamy soil that is very high in essential plant nutrients.

Like the chaparral, the prairies were shaped by fire. Because each year's above-ground production dies and remains on the surface, considerable fuel accumulates. The hot, dry climate with frequent thundershowers and their accompanying lightning is also conducive to fire. Most prairie plants are highly adapted to fire, because most are perennials with meristems, or growing points, that are underground, where fire cannot kill them. Fire increases light penetration to the ground in this closed community and thereby enhances production. It also releases nutrients that have been locked up in dead plant tissue, thus stimulating the growth of many plants.

A special variant on grassland is the **savanna,** a grassland on which woody plants are evenly and widely distributed. One of the most extensive such regions is found in Africa. Obviously, the density of woody plants and trees increases with precipitation. Therefore, we observe a gradation from pure grassland through savanna to forest across a precipitation gradient. In Africa, savanna is so extensive and ecologically important that it is shown separately in Figure 16.8.

In the Great Plains of North America, savannas were locally important. Often this habitat represented a dynamic balance between the forests of the east or in river bottoms and the prairies of the west and upland sites. Fire played a significant role in the ecocline from forest to prairie, with savanna habitats representing the shifting boundaries.

FIGURE 16.21 Variation in grassland: (**a**) tallgrass prairie in Iowa, (**b**) shortgrass prairie in eastern Colorado. (*Photos by David Krohne*)

FIGURE 16.22 Deciduous forest in Indiana. (*Photo by David Krohne*)

Deciduous Forests

Deciduous forests (Figure 16.22), composed of trees that drop their leaves seasonally, develop where seasonal dry periods occur. Both temperate and tropical deciduous forests occur (they are combined in Figure 16.8). In the temperate zone, deciduous forests develop where winter temperatures are relatively mild and annual precipitation exceeds 80 centimeters. The ability to drop the leaves in the fall protects the broad-leaved trees characteristic of deciduous forests from winter desiccation. In most tropical deciduous forests, temperatures are mild all year, but precipitation is strongly seasonal, selecting for deciduous trees that can avoid water loss during the dry season.

In North America there are broad regions of transition from deciduous forest to the neighboring biomes. In the northern parts of the United States and southern Canada is a belt of transitional forest in which conifers and deciduous species intermingle. To the west, the boundary between the prairie and forest is indistinct. In fact, this boundary was probably dynamic during pre-settlement

times. Because deciduous trees are certainly capable of growing throughout much of the Great Plains where prairie dominated, it is likely that the forests were relegated to the east by the fires that must have been more frequent in the arid lands to the west of the deciduous forest boundary. Deciduous forest trees are relatively intolerant of fire compared to many of the boreal species—and certainly compared to the prairie grasses.

Deciduous forest in North America is characterized by a relatively high diversity of tree species, as well as by regional variation in the dominant species according to local climate and substrate. Most of the major subregions of deciduous forest (Figure 16.23) are characterized by associations of two or more species that have similar moisture and substrate requirements. Thus the western reaches of deciduous forest are dominated by oaks (*Quercus*) and hickories (*Carya*), species that prefer a drier microclimate. Immediately to the east, where precipitation is higher, an association of American beech (*Fagus grandifolia*) and sugar maple (*Acer saccharum*) replaces oak–hickory forest.

The deciduous nature of these forests plays a significant role in the distribution and abundance of other,

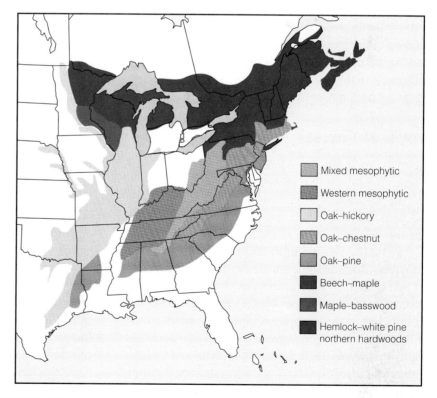

- Mixed mesophytic
- Western mesophytic
- Oak–hickory
- Oak–chestnut
- Oak–pine
- Beech–maple
- Maple–basswood
- Hemlock–white pine northern hardwoods

FIGURE 16.23 Some of the major tree associations within the deciduous forest of eastern North America. The mesophytic deciduous forests are relatively moist forest communities. One of these associations, the hemlock–white pine northern hardwoods, occurs in the ecotone between deciduous forest and boreal coniferous forest and thus contains both deciduous trees and conifers.

nonwoody species in the system. Only limited sunlight reaches the forest floor during the summer months, when the trees bear leaves. Thus many of the forest-floor herbs flower in the early spring, when sufficient sunlight can reach them to support reproduction. These forests are noted for their luxurious displays of spring-blooming herbs.

The forests of this biome are characterized by a physical structure unlike that seen in conifer forests. Temperate deciduous forests are composed of essentially four discrete layers. Uppermost is the canopy, consisting of the interlocking branches of the mature trees on the site. Often only a few widely scattered trees are tall enough to constitute this layer. Beneath the canopy is a second layer, the understory, consisting of saplings of potential canopy species and a few species that never reach canopy height. Among the latter group are species such as dogwood (*Cornus canadensis*) and American hornbeam (*Carpinus caroliniana*). The saplings of canopy species are often suppressed by the shade cast by the canopy. They must await the death of a mature tree before they can grow into the canopy. The other understory trees, however, must be able to photosynthesize at low light levels if they are to carry out their life histories. The third layer consists of a shrub stratum composed of woody shrubs such as leatherwood (*Dirca palustris*) and pawpaw (*Asimina triloba*), as well as young trees. Finally, there is an herbaceous layer of species that can complete their reproduction during the spring, before tree leaves appear, or can photosynthesize at low light levels.

Tropical Forests

One of the most extensive forest systems on the planet includes the forests that grow near the equator. When we think of tropical forest, we often think of *lowland rain forest* (Figure 16.24), which is found in a zone near the equator where rainfall is high. Areas in this zone generally receive more than 10 centimeters of rain per month and sometimes more than 50 centimeters in a month. Temperatures are high and relatively constant over the course of the year. In Java, for example, the mean monthly temperature is 24.3°C in February and 25.3°C in October.

Lowland tropical rain forest is characterized by luxuriant plant growth. One of the most striking features of the canopy in tropical forest is the astounding diversity of tree species (Chapter 12). In some tropical systems, more than 250 species of canopy trees can be identified in a single 1-hectare plot! The diversity of shrubs and understory trees is similarly high. Moreover, tropical rain forest is characterized by more

FIGURE 16.24 Lowland tropical rain forest. (*Photo by David Krohne*)

extensive growth of epiphytic plants than is found in any other biome. A huge number of species grow directly on other plants (Figure 16.25). Some are parasitic; others simply use the host tree as support in their attempt to compete for sunlight and nutrients.

Like deciduous forests, rain forests are organized into discrete vegetation strata. The uppermost layer consists of a few giant trees, sometimes more than 60 meters tall, that protrude above the general canopy. The middle layer is composed of the lush growth of many species whose upper branches intersect to form a dense canopy. A lower layer of smaller trees may or may not be present, depending on how much light penetrates the canopy.

The soils of tropical forests are very old, because this part of the planet did not experience the Pleistocene glaciations that generated new soils in more northerly and more southerly regions. Tropical soils are typically highly weathered, so much of the nutrient content of bedrock has long since been released. In addition, the high levels of precipitation have leached out many nutrients and carried them from the system. Thus, even though tropical forests support diverse and luxurious vegetation, the soil is not particularly fertile (nevertheless, the forests are being cleared for agriculture at an alarming rate, as the box on page 420 discusses). Tropical plants are adept at gathering nutrients before they are leached

FIGURE 16.25 Epiphytic species in tropical rain forest in Puerto Rico. (*Photo by David Krohne*)

away or captured by another plant. The many epiphytes are examples of plants adapted for this characteristic of the forest.

Although lowland rain forest is quite extensive in Amazonia, Africa, and Australia, a number of other forest types are also present in equatorial regions. The trade winds and mountain slopes interact to produce a great variety of climatic regimes, particularly with respect to rainfall. Each supports a particular type of vegetation. Two forest types—cloud forest and seasonal deciduous forest—are common enough in the tropics to warrant further description.

Cloud forests are found between 1000 and 2500 meters in elevation. As the name implies, these forests are restricted to a band of near-permanent clouds that envelop tropical mountains. Clouds and the associated forests develop where wet air rises and cools. The altitude at which permanent clouds develop depends on the humidity of the air at the base of the mountain: The higher the humidity, the lower the elevation at which the clouds begin. The vegetation in cloud forests is less luxurious than in lowland forests. Wind, cooler temperatures, and well-drained soil combine to support gnarled, stunted trees. One of the most conspicuous features of cloud forests is the lush growth of epiphytes. Although the diversity of epiphytic species declines with elevation, their abundance increases markedly.

Tropical deciduous forests occur where seasonality in precipitation is more conspicuous. The gradation from evergreen rain forest to deciduous forest may be subtle. In some parts of the world, including Africa, deciduous forests are uncommon. Where they do occur, the degree to which leaf loss is synchronous varies with the length of the dry season. Where the dry period is prolonged, fire is an important selective force.

The Biome Concept and the Nature of Communities

The classification of vegetation types and the description of biomes were based on early efforts to catalog the major plant communities of North America. At one level, this process is simply a means of ensuring that we have a common set of terms to describe habitat types. However, it led to one of the first controversies in twentieth-century ecology. F. E. Clements (1916) described the plant communities associated with different biomes as highly organized, integrated entities. H. A. Gleason (1926), who was the major spokesperson for the alternative view, proposed that communities are not "natural entities" but, rather, random assemblages of species. Each member of a community is there as a result of its own evolutionary, biogeographic, and ecological history.

Few ecologists accept the extreme Clementsian view that the biomes represent integrated entities with superorganismal properties. Nevertheless, the debate continues in a slightly different form. A fundamental aspect of this debate is the relative importance of biotic interactions. If communities are aggregations of species independently responding to the abiotic environment according to species-specific tolerance limits, then biological interactions among species are secondary effects. If, on the other hand, biotic interactions, both positive and negative, play a role in species' presence and absence, then community composition becomes vastly more complicated. This issue underlays much of the discussion of community structure (Chapter 11) and succession (Chapter 13). In both chapters, we discussed the role of stochastic and deterministic effects.

The fact that the same biome in different parts of the world may or may not share the same actual species of plants or animals raises some interesting ecological and evolutionary questions. For example, the tundra and taiga biomes that stretch across the northern parts of North America, Europe, and Asia share a number of plant and animal species. In

Habitat Destruction

The biomes described in this chapter represent the major habitat types for the world's plants and animals. The alteration and destruction of these habitats have a profound effect on our global ecology. The rate of loss of some habitats is frightening; other habitats are on the verge of elimination.

Considerable attention is paid by the media and the public to the loss of tropical rain forest. Indeed, this habitat is being destroyed at a startling rate. Even though tropical forests originally covered approximately 10 percent of the Earth's surface, they contained up to 50 percent of all the Earth's species (see Chapter 12). Currently, deforestation is reducing the planet's tropical forests at a rate of 17 million hectares per year. At this rate, these forests will disappear in less than 50 years.

A large proportion of the loss of tropical forest occurs when rain forest is cleared and burned to produce agricultural land, particularly grazing lands for cattle. The vast majority of rain forest occurs in less developed countries, where poverty and the need for foreign investment drive the development of these lands.

Once rain forest has been cut, it may be difficult or impossible, for several reasons, to return the land to its original forested state. Most of the nutrients in tropical forest are stored in the living biomass (see Chapter 15); thus, when a forest is cleared and burned, much of its nutrient content is lost. The extent of these forests and their tremendous biomass mean that they play a significant role in the Earth's biogeochemical cycles. As much as 13 percent of the annual input of atmospheric CO_2 may be a result of the elimination of tropical forest. Lateritic soils exposed to the heavy rainfall of the tropics are quickly washed away, silting streams and rivers and leaving a concrete-like surface that can support few plants.

As important as this loss of habitat is, we should not limit our concern to these dis-

tant lands. In the United States, we also face a number of challenges to our habitat diversity. Our grasslands have been severely altered and reduced in area by agriculture. The tallgrass prairies of the Midwest have essentially disappeared—less than 1 percent of the original prairie remains. Most of it is now relegated to tiny remnants in old pioneer cemeteries and along railroad rights-of-way that managed to escape the plow.

In addition, we are rapidly losing our own temperate rain forests and old-growth conifer forests in the Pacific Northwest. Virtually no old-growth forest remains on private land. The remaining patches on publicly owned land, such as national forests and national parks, represent less than 10 percent of the original extent of these forests.

Old-growth forests are tremendously valuable both economically and ecologically. They are being cut for their wood, which is prized for lumber. Because trees in old-growth forests have grown slowly and attained great size, they produce a great

deal of lumber of the highest quality. (The slow growth rate results in lumber with extremely fine grain and high tensile strength.) In addition to their value for lumber, old-growth forests are great reservoirs of species diversity. The spotted owl, a threatened species, breeds only in large tracts of old-growth forest. But this bird is only an indicator. Hundreds of species depend on the unusual structure and physical factors associated with old-growth forests: huge, tall trees casting deep shade, many decomposing logs that support complex communities of small animals and plants, and snags that provide habitat and nesting sites for many species.

In addition to these examples of threatened biomes, many specialized habitats within some of the major biomes are threatened as well. For example, one variant of the chaparral region of California, coastal sage scrub, is disappearing rapidly. More than 70 percent of California coastal scrub is gone because the slopes of the Coast Range in southern California where this habitat is found are prime real estate. Urban sprawl and housing developments threaten this habitat and the species it supports.

The destruction of habitats has two significant ecological consequences. First, our understanding of the operation of ecological systems depends on the existence of functioning natural systems. It is from these habitats that we elucidate the principles that govern ecological processes. The information we glean from them provides the foundation for understanding the environment and our effects on it. Second, the destruction of habitat is the single most important factor leading to species extinction. In the case of rain forest, habitat loss is causing the extinction of as many as 17,000 species each year (see Chapter 12). In other systems, such as the ancient forests of the Pacific Northwest, individual species of concern, such as the spotted owl, are at risk of extinction.

contrast, the rain forests of Australia are very different from those of South America or Africa. Many Australian species are endemic. Often, similar *but unrelated* species occur in the rain forests on each continent. These represent examples of convergence (Chapter 2).

Spatial Patterns of Terrestrial Communities: Landscape Ecology

The vegetation types we have just described were identified and classified early in this century (Clements 1936). The study of patterns of habitat and vegetation continues in a more sophisticated form as **landscape ecology**—the study of spatial patterns in ecological phenomena and processes. Landscape ecology developed in parallel with mathematical techniques to study spatial patterns and the availability of remote sensing data and satellite imagery (and sufficient computer power to analyze these large digital data sets). As a result, landscape ecologists can address a number of important questions regarding spatial and temporal habitat patterns. Landscape ecology is a young and developing branch of ecology. Because so much of ecology is based in some way on spatial variation, much of what we have discussed in this text could in fact be considered landscape ecology. Any patterns or processes with important spatial variation—metapopulation dynamics, shifts in nutrient dynamics, mosaics of climax and successional processes, and so forth—are landscape issues. Moreover, as anthropogenic changes in the landscape become more widespread and pronounced, landscape ecology is playing an increasing role in conservation biology.

Spatial Habitat Patterns   Landscape ecology, particularly when it relies on high altitude or satellite imagery and remote sensing data, allows us to quantify patterns of spatial variation in habitat. In addition, it allows us to track temporal change in the environment.

One fundamental aspect of landscape structure is the size and scale of habitat patches. For example, we saw in Chapter 13 and in several of the biomes discussed in this chapter that fire is a recurring influence on the vegetation. In addition, the temporal pattern of fire varies enormously across habitat types. Thus we expect that a landscape analysis of vegetation will also require a spatial analysis of fire frequency and pattern. Brown et al. (1999) examined the patterns of fire in Ponderosa pine–Douglas fir forests in central Colorado. Trees that are exposed to fire but are not killed record the event as fire scars that are visible in certain growth rings. By cross-dating trees from a number of areas, these researchers could identify the timing of fires and their spatial extent. For example, compare the spatial extent of fires in 1841 and 1851 (Figure 16.26). Fires occurred in both years but were much more widespread in 1851. The pattern of fire frequency since

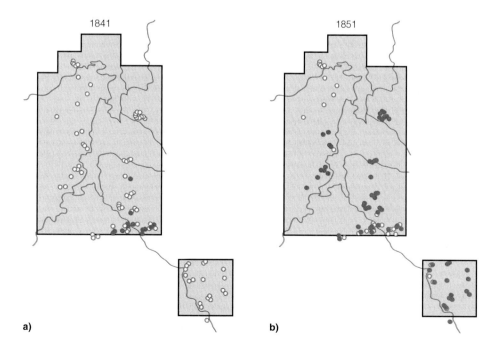

a)   b)

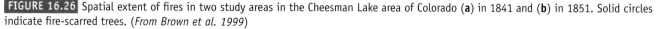

FIGURE 16.26   Spatial extent of fires in two study areas in the Cheesman Lake area of Colorado (**a**) in 1841 and (**b**) in 1851. Solid circles indicate fire-scarred trees. (*From Brown et al. 1999*)

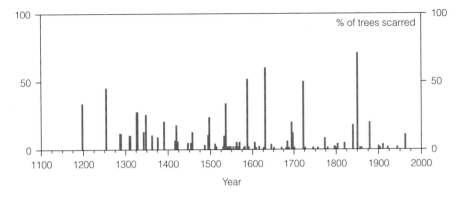

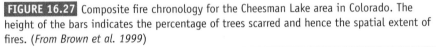

Composite fire chronology for the Cheesman Lake area in Colorado. The height of the bars indicates the percentage of trees scarred and hence the spatial extent of fires. (*From Brown et al. 1999*)

1100 A.D. is shown in Figure 16.27. Note that the interval between fires varies considerably. More important, the relative size of the fires, as indicated by the number of scarred trees, also varies. Thus this landscape represents a mosaic of habitats that differ in age since disturbance.

In addition to the importance of habitat size and spatial distribution, habitat shape has significant ecological consequences. The western trillium (*Trillium ovatum*) is a common herb of western conifer forests. In the Siskiyou Mountains of Oregon, clear-cutting has reduced much of the original old-growth forest to a series of patches surrounded by early successional clearings. Jules (1998) studied the effect of forest fragmentation on this species. He showed that recruitment of young individuals is greatly reduced in "edge" habitat, where exposure to increased light and wind and other abiotic changes occur. Although the reasons for the species's decline at the edge of fragments are complex and not fully understood, it is clear that fragmentation leads to significant changes in habitat shape that, in turn, affect the demography of this species. As large forests are broken into smaller ones, the relative amount of edge habitat increases (Figure 16.28). This aspect of shape is often quantified with the edge/area ratio or the perimeter/area ratio.

A related aspect of habitat shape is the complexity of the edge. We can measure this property of the habitat by using fractal geometry (Mandelbrot 1983). Consider the curve in Figure 16.29. If we want to measure the length of this curve, we can apply a ruler of a certain length as shown in the figure. Note, however, that the measurement we get will depend on the length of the ruler we use. As shown in the figure, a large ruler will result in a smaller measurement than a small ruler. The latter will measure more of the little nooks and crannies, and our end result will be different.

We can describe this relationship mathematically according to a simple power law. The length (*L*) of any object depends on the measurement scale (δ):

$$L(\delta) = K(\delta)^{1-D}$$

where *K* is a constant, and δ is the measurement scale (the length of the "ruler"). The *D* in the exponent is the **fractal dimension**, which varies between 1 and 2 (Sugihara and May 1990).

What is the significance of this to ecology? First, organisms that perceive the habitat with a different "ruler" will be affected differently by the habitat. If the curve in Figure 16.29 is the boundary of a coastline, large organisms will "measure it," or perceive it, differently than small ones. For example, for a coastline with *D* = 1.5, a ten-fold reduction in the measurement scale will result in an increase in the perceived length by a factor of about 3. Large organisms, with greater vagility and senses that perceive greater distances, will "see" this coastline as *smaller* than a small organism with more limited measuring perception. For the smaller organism, a greater length of coastline is available. This is more than the trivial

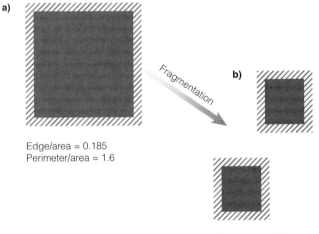

The effect of habitat fragmentation on size relationships. (**a**) In the large block of habitat, the effect of abiotic factors associated with the "edge habitat" extend in as far as shown by the cross-hatching. (**b**) When this large block of habitat is broken down into a series of smaller habitat fragments, the importance of that edge habitat increases. Note the increase in the edge/area and perimeter/area ratios. Any ecological factors associated with edge or perimeter are more pronounced in smaller blocks of habitat.

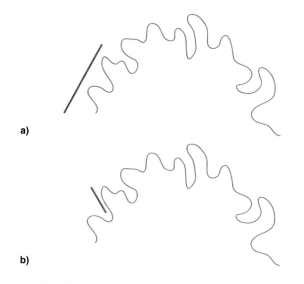

a)

b)

FIGURE 16.29 The measured length of a complex curve depends on the scale ("ruler length") of the measurement. Measuring this complex shape with a long "ruler" (**a**) will result in a shorter measurement than if we use a short "ruler" that measures more of the complexity of the shape (**b**).

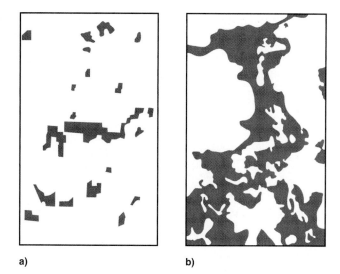

a)   **b)**

FIGURE 16.30 Patterns of deciduous forest in Mississippi that illustrate the spatial patterns in agricultural regions (**a**) and in relatively undisturbed forest (**b**). (*From Krummel et al. 1987*)

matter of the smaller organism's size relative to the perimeter of the coast—it is a matter of a fundamental difference in the habitat pattern.

Second, the fractal dimension reveals something about the complexity of a habitat's shape. Higher values of *D* signify more complex shapes. In addition, for larger values of *D*, the apparent length changes more rapidly with the scale of the measurement (the "ruler length"). It should be intuitively obvious that for more complex and convoluted shapes, the effect of the ruler length will be more pronounced. Krummel et al. (1987) used this property to analyze the spatial patterns of deciduous forest in regions of Mississippi that were experiencing rapid conversion from forest to agricultural use (Figure 16.30). They measured the perimeter of forest patches from aerial-map information and calculated fractal dimensions for scales of different sizes. Note that there is a rapid shift in *D* at a patch size of approximately 70 ha (Figure 16.31). Smaller forests tend to be relatively smooth (*D* = 1.20) compared to larger patches (*D* = 1.4–1.5). Larger forests are relatively undisturbed, and natural properties such as soil and slope determine their smoother boundaries. Smaller patches are those that are delineated by human activity; their boundaries tend to be simpler, as revealed by the smaller fractal dimension.

Satellite imagery can be used to monitor habitat changes as well. Satellites can record the intensity of reflected light in discrete segments of the electromagnetic spectrum. Different land cover types and vegeta-

tion reflect these spectra differently, giving habitat and vegetation types a unique pattern. One can even identify differences in production from spectral reflectance because actively growing plants reflect different patterns of light than dormant or nongrowing plants.

This technique provides landscape ecologists with a valuable tool to measure spatial habitat variation. The use of remote sensing data makes it possible to quantify the changes across a much larger landscape than one could document using field methods on the ground. For example, over the last 30 years, the mid-continent population of snow geese (*Chen hyperborea*) increased from approximately 700,000 in 1969 to nearly 3 million in 1999. This species congregates in

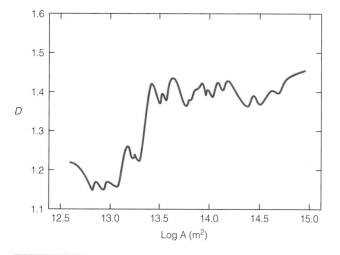

FIGURE 16.31 Changes in fractal dimension (*D*) as a function of spatial scale (log area). Note the sharp increase in *D* at approximately 13.5. (*From Krummel et al. 1987*)

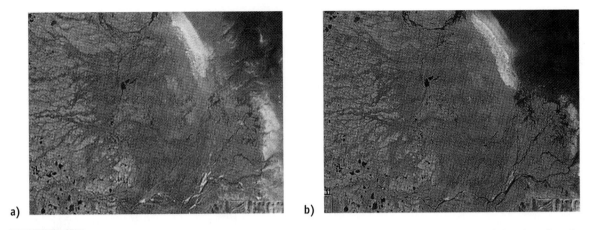

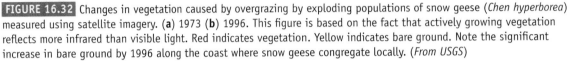

a)   b)

FIGURE 16.32 Changes in vegetation caused by overgrazing by exploding populations of snow geese (*Chen hyperborea*) measured using satellite imagery. (**a**) 1973 (**b**) 1996. This figure is based on the fact that actively growing vegetation reflects more infrared than visible light. Red indicates vegetation. Yellow indicates bare ground. Note the significant increase in bare ground by 1996 along the coast where snow geese congregate locally. (*From USGS*)

relatively small breeding areas in Canadian tundra along Hudson Bay (Alisauska, 1992). There the increased numbers have had a significant local impact on the vegetation (Figure 16.32).

It is clear that the snow geese are responsible for the habitat change in Figure 16.32. In other cases, the root causes must be determined from other kinds of landscape data. For example, Hoch (2000) examined invasion of grassland by eastern red cedar (*Juniperus virginiana*) in the Flint Hills region of Kansas. To assess woody invasion, Hoch used satellite measurements of the patterns of reflected light in seven portions of the electromagnetic spectrum (blue, green, red, near infrared, thermal infrared, and two mid-infrared bands) to identify red cedar forest. By identifying which reflectance patterns represent closed-canopy red cedar forest and verifying on the ground that the pattern indeed correctly identifies red cedar, Hoch was able to measure the spatial extent of these invading forests, as well as the proportion of closed-canopy areas. The results are illustrated in Figure 16.33. Note the significant increase in the proportion of closed-canopy forest in the interval 1983–1997. It would be virtually impossible to obtain data like these on a landscape scale without remote-sensing information.

What causes underlie the invasion of red cedar? The land cover shift could be a result of local land management practices, such as changes in grazing intensity or fire frequency. Or it may be that large-scale changes such as global climate change or even increases in CO_2 centrations are responsible. One can support or reject these hypotheses by examining correlates of red cedar invasion. Hoch found that red cedar invasion was not random with respect to soil

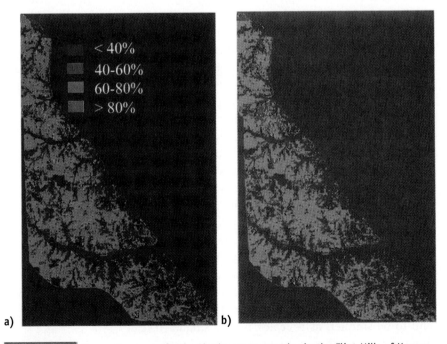

a)   b)

FIGURE 16.33 Change in cover of red cedar in seven counties in the Flint Hills of Kansas, measured using remote sensing techniques from 1983 (**a**) to 1997 (**b**). Reflectance patterns were used to measure the spatial extent of red cedar forest with different amounts of canopy closure. Note the significant increase in the proportion of closed canopy forests. (*From Hoch 2000*)

type. It occurred more frequently on thin, poor soils classified as unsuitable for trees—a pattern consistent with the influence of local land management. The fact that red cedar density is highly correlated with local human population growth also supports the hypothesis of local change in land management. As people move into grassland habitats, they typically reduce the frequency and spatial extent of fires, thus favoring the growth of red cedar. Moreover, the shapes of the new forests are telling. They tend to be rectangular with square corners and a low edge area ratio—a pattern consistent with abrupt changes in land management practices at property boundaries. If large-scale climatic or atmospheric change were important, forest boundaries would be correlated more strongly with soil type, slope aspect, or other topographic features, rather than with land ownership boundaries.

The terrestrial biomes are, of course, only a portion of the habitat types in the world. The next chapter reviews the important habitat types and basic ecology of aquatic (marine and freshwater) systems.

SUMMARY

1. Abiotic factors play a crucial role in the distribution of plant species. Water and temperature are among the most important regional, effects on vegetation. In many regions such as grassland, chaparral, and taiga, fire is an important abiotic factor.

2. The regional climate determines the nature of the precipitation and temperature regime.

3. North American climates and vegetation types result from a number of important factors: altitude, latitude, the proximity of large bodies of water, rain shadows, slope effects, and midlatitude desertification.

4. The plant life in a region is described as the flora, the species of plants present, and the vegetation, the structure of the plant life (trees, grassland, etc.).

5. Ecologists have attempted to organize the major terrestrial units of vegetation into a series of categories sometimes called biomes. These are based on the structural form of the vegetation.

6. The boundaries between vegetation types are sometimes indistinct. For example, in the eastern regions of the tallgrass prairie, grassland and forest overlapped in a mosaic vegetation.

7. The classification of vegetation types is somewhat arbitrary. For example, some ecologists do not consider taiga to be separate from the boreal coniferous forests to the south.

8. Landscape ecology focuses on spatial patterns of ecological processes. Any phenomenon or process with spatial variation falls under the rubric of landscape ecology. In this chapter we focused on patterns of spatial habitat and variation in vegetation.

SELF-ASSESSMENT: CAN YOU . . . ?

1. Outline the relationships among climate, geography, and vegetation for each of the major biomes.

2. Explain the physical factors that limit, act as selective agents on, and determine each specific vegetation type.

3. Discuss the concept of the biome in relation to the concept of the community.

4. Explain the variation in vegetation *within* biomes.

5. Explain the importance of habitat shape in landscape ecology.

PROBLEMS AND STUDY QUESTIONS

1. Discuss the nature of the climax and the concept of biomes, or major terrestrial vegetation units.

2. Prepare a table outlining the major climatic and vegetation differences in North America.

3. Choose a field site you are familiar with. Discuss the variation in vegetation and flora within that region. Do your own observations lead you to think of plant communities as highly integrated units or random assemblages of plants?

4. Relate the patterns of primary production discussed in Chapter 14 to the global patterns of vegetation.

5. Temperature and moisture are two critical abiotic variables that affect the nature of the vegetation and flora. Design a study (perhaps choose a specific region with which you are familiar) to elucidate the relative importance of temperature and moisture. Would your study be experimental or observational?

6. Originally, tallgrass prairie occurred in parts of Indiana and Illinois, yet trees grow well in this region.

Why was this area prairie before the European settlement?

7. Habitat destruction is strongly linked to species extinction. Drawing on the material in Chapter 12, discuss the consequences of habitat loss and loss of biodiversity.

8. In Chapter 3 we explored the role of abiotic factors in determining where plants can live. If we hypothesize that each species has tolerance limits that define its geographic range, we might conclude that plant communities are simply made up of the species that, by chance, have overlapping ranges. What elements of communities from Chapters 9–13 suggest that this explanation is too simple?

9. How could you test the ideas you propose in Problem 8?

10. Discuss the role of the Pleistocene glaciations in shaping the plant communities of North America.

PROJECTS AND ADDITIONAL STUDY

1. Use topographic maps (showing water and forest) or high-altitude photographs to measure landscape characteristics of a particular region, including (a) percent of total area in various habitat types and (b) perimeter/area ratios for important habitat types.

2. Calculate the fractal dimension (D) from the data in Project 1, using the methods outlined in Sugihara and May, 1990. *Trends in Ecology and Evolution* 5:79.

3. The biome concept was developed in North America and applied first to the vegetation of this continent. Use primary and secondary literature to compare the nature of the "biomes" on another continent to those of North America. Do the biomes of North America translate directly to other vegetation systems? What is the significance of this?

Freshwater and Marine Systems

In this chapter we continue the description of the basic habitat types that we began in Chapter 16. We turn our attention now from the land to freshwater and marine environments. As before, our purpose is to develop a working knowledge of the nature of these communities, especially the variation within habitats. We should appreciate that not all watery habitats are the same—there is tremendous variation in the physical factors and hence in the numbers and kinds of organisms in these habitats.

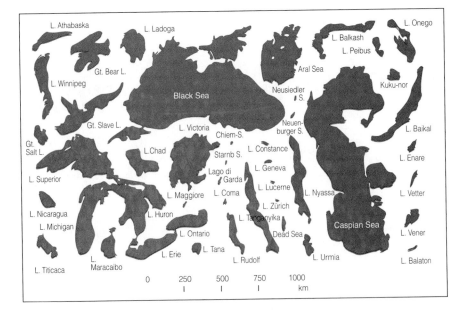

FIGURE 17.1 An approximate comparison of the surface areas of many of the world's largest inland bodies of water. (*From Ruttner 1963*)

Freshwater Systems

We make a fundamental distinction between saline waters, such as are found in the oceans, and freshwater, which has a very low salt content. As we will see, however, the two basic types of aquatic systems merge at or near the coastline, and the boundaries are not always distinct. There are ecologically important regions in which the salt content gradually increases as one moves closer to the ocean.

We divide freshwater systems into two fundamental types: lentic systems and lotic systems. **Lentic systems** contain water that is not flowing, as in lakes or ponds. **Lotic systems** are flowing streams and rivers. **Wetlands** do not fit neatly into either type but represent a composite habitat—the region of transition from aquatic to terrestrial.

Lentic Systems

Inland lakes cover approximately 2.25 million square kilometers around the world. As large as this area is, the volume of water in such lakes constitutes only a tiny fraction of the water on Earth, more than 97 percent of which is contained in the oceans. A few large lakes contain huge amounts of freshwater (Figure 17.1), although most lakes are much smaller. The largest purely inland body of water is the Caspian Sea (436,000 square kilometers). The deepest lake on the planet, Lake Baikal in Russia (maximum depth 1620 meters), contains fully 20 percent of the world's freshwater.

Lakes are formed by a number of processes. One of the most important is glacial activity. An advancing glacier gouges depressions in the Earth's surface, and when it retreats, its meltwater may fill the depressions. Many extant lakes and ponds at high latitude have a glacial origin because the last glacial event, the Wisconsin, occurred only 12,000 years ago. Perhaps in other periods of the Earth's geological history, other processes were more important.

Another important lake-creating process is geological activity associated with major movements in the Earth's crust. Earthquakes occurring along or near fault lines can cause significant amounts of land to drop to lower levels. Tectonic activity has formed some of the deepest lakes on the planet, particularly in the Rift Valley of East Africa. There a series of large lakes formed in a region of intense seismic activity near the boundaries between two diverging tectonic plates (Figure 17.2).

Lakes can also form if the course of moving water is changed in some way. Floods may shift the current into new channels, leaving lakes in the old river basin. In slow-moving waters in relatively flat depressions, silt accumulating in bends can cut off a section of the river to form an oxbow lake. High flow rates are likely to break through any dams that form or to prevent material from settling out of the current.

Lakes, both large and small, are characterized by great heterogeneity in physical factors. Abiotic factors vary greatly among locations and depths within a lake. This variation has profound effects on the plants and animals in the system and plays a key role in their distribution and abundance in the lake.

As in any ecological system, the patterns of light intensity play a crucial role by virtue of their influence

FIGURE 17.2 The locations of African rift lakes along a seismically active line in East Africa.

on the distribution of plants and algae. Water absorbs sunlight, so the amount of solar input available to plants decreases with depth (Figure 17.3). Any suspended material in the water may also absorb or scatter light and thus decrease the amount available at greater depths.

The upper layer of a lake, the layer with enough light for photosynthesis, is called the **trophogenic zone.** Below this region is the **tropholytic zone,** where light intensities are insufficient to support plants. At a certain depth, called the **compensation point,** the rate of photosynthesis by plants exactly balances the rate of respiration by plants and animals. Photosynthesis balances respiration when the absorption of sunlight has reduced the light available to approximately 1 percent of the sunlight reaching the surface of the lake. The exact depth of the compensation point varies with the amount of incident sunlight, the turbidity of the water, and the rate of respiration.

Temperature also changes with depth in lentic systems because of the relationship between density and temperature. Like most substances, water becomes denser as it cools. However, it reaches its maximum density at 4°C. Below this temperature, its density *decreases* again. This is why ice cubes float in a drink and why lakes freeze from the surface down. Waters of different temperature, and thus different density, do not mix readily. As a result, the water in lakes becomes stratified by temperature. In deep lakes, the degree of stratification and mixing undergoes seasonal changes as a result of the pattern of temperature changes in the air.

In the spring, the surface water warms, and temperature stratification begins. As the season progresses, this gradient becomes more pronounced. A layer of water at 4°C lies near the bottom, and the water gets progressively warmer above it. A sharp temperature and density gradient, called the **metalimnion,** develops (Figure 17.4). Within the metalimnion the plane of maximum temperature change is called the thermocline. The upper layers of the lake still mix as a result of wind action and the absence of large density differences there. Because of its higher density, the 4°C water at the bottom does not mix

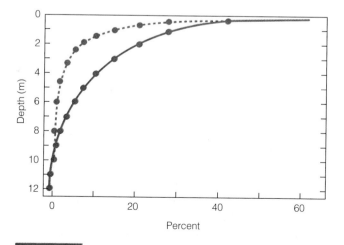

FIGURE 17.3 The solid line represents water with relatively low sediment content; the dashed line represents sediment-laden water.

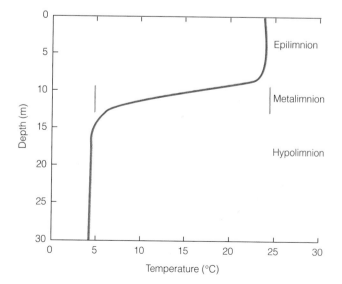

FIGURE 17.4 Summer temperature stratification in a lake. (*From Wetzel 1983*)

with the water above it. The lake has become **thermally stratified.** The layer above the thermocline is called the **epilimnion;** the layer below is called the **hypolimnion.**

The thermal stratification of lakes affects other physical factors as well. There may be a strong oxygen concentration gradient in the lake associated with the summer temperature stratification. Three processes determine the oxygen content of water in a lake: (1) Oxygen enters the lake at the surface via exchange with the air; (2) the activity of photosynthetic algae and other plants add oxygen to the water; and (3) respiration by animals, plants, and microorganisms removes oxygen from the water.

In the summer, oxygen concentrations are high in the epilimnion as a consequence of exchange and photosynthesis. At the same time, however, dead and decaying material sinks into the hypolimnion, where it is decomposed by bacteria. Here, oxygen may be greatly depleted, especially because this lower layer does not mix with the oxygen-rich epilimnion. Therefore, the hypolimnion typically has low oxygen concentrations during the summer.

Nutrients may also be stratified as a result of these processes. As algae photosynthesize, they use the nutrients in the epilimnion. When the algae die and sink into the hypolimnion, those nutrients are removed from circulation.

In the fall, as the surface of the lake cools, the temperature becomes more uniform, and mixing occurs. This is called the **fall turnover.** During the winter, the temperature of the lake is uniform, and a layer of ice may form on the surface. Because of the ice, winds do not circulate the water. When the ice melts in the spring, the surface waters warm, and spring winds mix the water again.

In addition to the zonation in a lake's physical parameters, there is zonation in terms of the organisms that inhabit the lake (Figure 17.5). Near the shore, light can penetrate to the bottom, and rooted plants can survive. This region is referred to as the **littoral zone.**

Water too deep for rooted plants is inhabited by **plankton,** tiny free-floating or swimming organisms in the **pelagic** region. The waters of a lake beneath the littoral zone and above the compensation point make up the **limnetic zone.** In many lakes, the density of phytoplankton is stratified by depth by virtue of both physical processes (buoyancy characteristics) and active movement to regions of optimal growth conditions.

Below the compensation point, photosynthesis is greatly diminished. The organisms that live here are generally not green plants but animals or microorganisms that depend primarily on the rain of organic matter (dead organisms) from above as an energy source. This region is the **profundal** zone. Finally, the bottom layer of a lake in both the littoral and the profundal zones is the **benthic zone.**

Lake systems can be broadly classified as oligotrophic or eutrophic on the basis of their nutrient contents and rates of photosynthesis. **Oligotrophic lakes** are characterized by low nutrient content, especially of phosphorus. They have small populations of photosynthetic algae and hence low rates of photosynthesis. The low densities of algae result in less material for decomposition in the hypolimnion and hence higher oxygen concentrations there. Oligotrophic lakes are often young and have not yet accumulated sufficient nutrients from the land to support high levels of photosynthesis. They also frequently have low surface-to-volume ratios; that is, they tend to be deep lakes with small surface areas. This results in relatively low solar input to support photosynthesis.

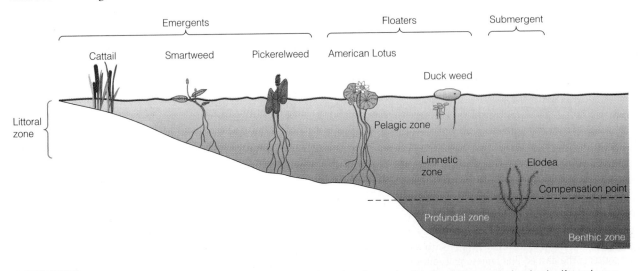

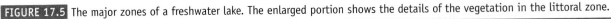

FIGURE 17.5 The major zones of a freshwater lake. The enlarged portion shows the details of the vegetation in the littoral zone.

Eutrophic lakes are characterized by high levels of nutrients and high rates of photosynthesis. Large populations of algae develop and provide a steady supply of organic material for decomposing in the profundal zone. Thus the hypolimnion may become oxygen-depleted in such lakes. Indeed, during the summer months, oxygen depletion may be so severe that it causes the death of bottom-dwelling fish. Although eutrophication of a lake can occur through natural processes as a lake ages and accumulates nutrients, the process is accelerated when humans greatly increase the nutrient content. This often happens as a result of fertilizer runoff from agricultural land.

Lotic Systems

Physical factors play a crucial role in the ecology of flowing water. Among the most important of these is current. The rate of flow is determined by the stage of the streambed, the volume of water in the stream, and the nature of the resistance downstream. The volume of water in a stream or river can change dramatically over a very short period of time. Storms or rapid snowmelt can greatly increase the rate of flow. If a stream encounters resistance, as it does when it enters a lake, flow decreases near the point of resistance. The density and diversity of life-forms in the water and on the bottom are determined largely by flow rate.

Fast-flowing streams and rivers have rocky bottoms because smaller particles of dirt and silt are carried away. Of course, within a stream the current speed changes as the water enters riffles and pools, and the nature of the bottom substrate changes accordingly. Slow-moving streams are more likely to have muddy bottoms.

Most rivers and streams are too shallow to have temperature stratification, but variation can occur along the length of a lotic system. Narrow, deep portions of streams are cooler than broad, shallow portions. The amount of shading from streamside vegetation also affects water temperature. In addition, moving water is less likely to freeze during winter.

The oxygen content of lotic systems is generally higher than in lentic systems. A greater surface-to-volume ratio in streams and rivers than in lakes, as well as aeration from churning, accounts for this.

Streams differ fundamentally from lentic systems in the nature of their energy source. Most of the energy for stream ecosystems comes from outside sources (**allochthonous sources**). Streams receive large amounts of litter and detritus from the stream basin, and for most streams this source provides a far larger proportion of the energy available to the system than

that derived *in situ* by photosynthesis (**autochthonous sources**) (Moss 1980).

In lotic systems, most of the photosynthesis is done by plants attached to the substrate. These plants, known as **periphyton,** reach their highest densities in riffles where the water is moving rapidly over rocks. They are susceptible to scouring, however, if the current increases rapidly, such as after a storm. Stream periphyton generally have very high reproductive rates and can recolonize and grow quickly after scouring.

The bulk of the primary consumers in a stream are detritivores that obtain energy from the allochthonous inputs. Typically, a degradative successional sequence, similar to that described for pine needles in Chapter 15, breaks down this organic material.

Stream ecosystem ecology changes markedly over the course of its length, a concept known as the **river continuum.** Near the source of the river, autotrophic production is limited (often by shade from terrestrial vegetation) such that respiration exceeds production. Much of the organic input to the stream is from allochthonous sources. Consequently, most stream heterotrophs are adapted to use such materials—many are detritivores. Downstream, the relative importance of terrestrial vegetation decreases in the wider stream. Accordingly, temperature increases. Higher levels of *in situ* primary production may exceed respiration. The wider stream, with open swifter water as well as near-shore eddies and pools, has a greater variety of microhabitats and thus a greater total species diversity. As the size of the system continues to increase, another ecological shift occurs. The current slows and sediments accumulate, and production declines again.

Wetlands

At the interface between freshwater and terrestrial systems are a number of semiaquatic habitats collectively termed **wetlands.** Wetlands, habitats that are perpetually or periodically flooded, are of crucial importance to a number of vertebrate and invertebrate organisms as breeding grounds, wintering areas, and feeding sites. Humans often covet these lands for various kinds of development, and thus only a tiny fraction of the original wetlands of the United States remains (see the box on page 432).

In wetlands, which are transitional habitats between terrestrial and lentic or lotic systems, the water table is near the surface, and shallow water (less than 2 meters) of variable depth periodically inundates the area. We distinguish a number of different types of wetlands on the basis of the nature of the drainage and other physical characteristics, especially

Wetlands

Ecologists have begun to appreciate the tremendous value of wetland systems. First, wetlands are crucial habitat for a number of organisms. For example, many species of waterfowl nest in the upper Great Plains in potholes, small ponds formed by the Pleistocene glaciers that hold water during the early summer. Along the southern coasts of North America, mangrove swamps provide crucial habitat for the reproduction of many species of marine fish and invertebrates. The trees' dense, interconnecting roots trap nutrients and slow the flow of water, thus creating ideal habitat for the development of eggs and larvae.

In addition to their importance as habitat, wetlands often play a crucial hydrological role. In freshwater systems they absorb excess precipitation and thus prevent flooding downstream. Much of the disastrous flooding along the Mississippi River in the summer of 1993 resulted from the replacement of natural wetlands along the river by levees that contain the river in its channel. In a year of unusually high precipitation, there were few natural wetlands to hold and gradually release the excess water.

Wetlands also serve as important filtering systems. The vegetation of wetlands slows water flow, allowing sediment to settle out and thereby reducing the siltation of streams and rivers. In addition, the plants absorb excess nutrients such as phosphorus and nitrogen, which slows the eutrophication of lakes and rivers. Plants even accumulate some toxins and heavy metals.

The environmental issues surrounding the Everglades in southern Florida are representative of those associated with many wetland systems in North America. The Everglades, a huge wetlands complex that comprises much of the southern third of Florida, has been referred to as a "river of grass" because water flows through miles of sawgrass (actually a

sedge) from the southern shore of Lake Okeechobee across a wide swath of the southern tip of the state to reach the ocean at Florida Bay (Figure 17B.1).

Beginning in the late nineteenth century, this natural water flow was altered by dams and canals. In 1948 Congress approved a complex system of water diversion and management that included building some 1400 miles of canals. The Kissimmee River, which flows into Lake Okeechobee, was changed from a meandering 100-mile-long river to a 48-mile-long canal. The results of these projects were devastating. The complex wetlands were habitat to thousands of water birds, such as egrets, herons, ibis, and wood storks, whose nesting success depends on suitable water levels. These species have declined by as much as 90 percent because of the diversion of water

from the Everglades. As a result of pollution input and the altered water flow, mercury contaminates virtually all the fish in the Everglades, and phosphorus levels in the water are extremely high. The altered hydrology and pollution have allowed the invasion of a number of noxious non-native species, such as water hyacinth and Brazilian pepper, that are crowding out native species.

There is great interest in improving the ecological balance in this wetlands system. The U.S. Army Corps of Engineers has begun a project to re-create the meandering flow of the Kissimmee River, and the Florida legislature has enacted laws requiring increased flow of water through the glades. These actions will not suffice to restore this system to normal function, but they are steps in the right direction.

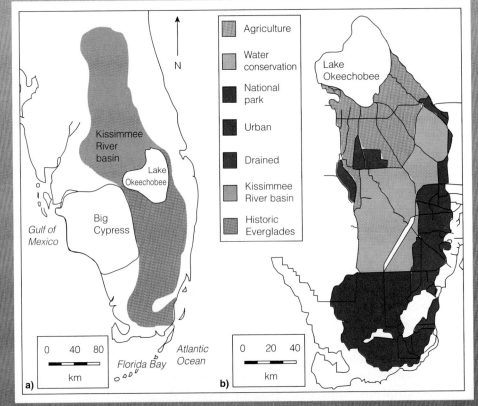

FIGURE 17B.1 (a) The historic Everglades ecosystem in the late 1800s. (b) Land uses of the region as of 1990. (*From Gunderson, Light, and Holing 1995*)

pH. These wetland types sometimes grade into one another, making precise definitions difficult.

Marshes contain soft-stemmed herbaceous plants (Figure 17.6). The most conspicuous marsh plants are **emergents,** plants that are rooted in the substrate but protrude above the shallow water. Cattails (*Typha* spp.) are a prominent example. Other members of the community, such as pond lilies, float on the surface. Marshes originate in a number of ways. They may be formed when lakes or ponds fill in with vegetation. The edges of lakes or rivers where water flow is reduced may also develop marsh vegetation.

Marshes may represent spatially extensive, continuous systems or small, discrete wetlands. For example, one of the most extensive marshes in the world is the Everglades of southern Florida (Figure 17.7). This marsh, which is dominated by a sedge called sawgrass (*Cladium jamaicense*), was formed as the flat southern portion of Florida was exposed by dropping sea levels in the last million years. Water flowing out of Lake Okeechobee in central Florida slowly drains over the width of the entire state to the sea, nourishing the sawgrass marsh with nutrients and freshwater. In contrast, prairie of the Great Plains is dotted with numerous **prairie potholes,** small marshes formed from depressions gouged by the advance of the Wisconsin glaciation (Figure 17.8). These potholes are critical habitat for waterfowl production. Their suitability as breeding sites is a result of their food-rich waters and the protection they provide from predators.

Swamps also occur in flat basins through which water flows slowly, saturating the soil (Figure 17.9). They differ from marshes in that they are dominated by woody plants. The Okefenokee Swamp in southern Georgia is typical. This swamp occupies a basin formed by the remnant of a marine sand bench to the east and

FIGURE 17.7 The Everglades, a large freshwater marsh in southern Florida, is dominated by sawgrass. (*Photo by David Krohne*)

FIGURE 17.8 A prairie pothole in North Dakota surrounded by agriculture fields of sunflowers. (*Photo by David Krohne*)

FIGURE 17.6 A freshwater marsh in northern Wisconsin. (*Photo by David Krohne*)

FIGURE 17.9 A bald cypress swamp in southern Illinois. (*Photo by David Krohne*)

higher land to the west. As the sea level dropped, the ocean receded eastward, exposing the sand bench. Freshwater accumulated in the basin.

The species that dominate swamps differ regionally. In the northeastern United States, red maple (*Acer rubrum*) dominates. Farther south, bald cypress (*Taxodium distichum*) is a common dominant. The vegetation in a swamp can vary markedly with the depth of standing water. In the Okefenokee, for example, bald cypress occupies the wettest areas, and pines are found in the driest.

The last major type of wetland is the **bog,** which is typically found in glaciated regions of the northern United States and Canada. The archetypal bog does not have an outlet for water. Often bogs are formed in small depressions (kettleholes) scoured out and exposed by a receding glacier.

After a kettlehole bog is formed and filled with water, the bog gradually begins to fill in with sediments. Such a young system is, of course, highly oligotrophic. Sphagnum moss (*Sphagnum* spp.) is an early colonist of such sites in northern regions, where it can tolerate the low nutrient levels. Sphagnum has the ability to pull nutrients from the water, even if they are at low concentrations, via a cation exchange system that dumps hydrogen ions into the water, gradually lowering its pH.

The sphagnum grows from the edges of the bog toward the center, gradually covering the surface (Figure 17.10). It thus forms a floating mat that rains organic matter down to the cold, acid waters below, where peat gradually accumulates. The outer edges of the bog are the oldest sections of this successful sequence. There a layer of sphagnum and sediment

FIGURE 17.10 A bog in northern Alaska. Note the circular nature of the bog, the concentric rings of vegetation, and the mat of vegetation in the center. (*Photo by David Krohne*)

extends from the surface to the bottom of the bog. Often the sphagnum closer to the center of the bog lies directly over water and forms a springy mat across the surface.

As the bog gradually fills in, other plants such as cottongrass (*Eriophorum polystachion*) colonize. The nutrient content of the water and developing soil is still low. Consequently, carnivorous plants such as sundews (*Drosera* spp.) and pitcher plants (*Serracenia* spp.) that obtain nutrients from insects begin to colonize the margins of what was once open water. Eventually, woody plants, including poison oak (*Rhus toxicodendron*) and tamarack (*Larix laricina*), move in. The result is a wetland system composed of concentric rings of different vegetation types surrounding a small center of open water.

Marine Systems

The world's oceans cover nearly three-fourths of the surface of the planet, and so they constitute one of the largest potential habitats for plants and animals. Of course, life arose and diversified in the ocean, and virtually every major taxonomic group of organisms has representatives adapted for life in salt water. Given the ocean's tremendous size, we might expect great variation among marine habitats, and indeed this is what we find. In this section we describe the basic nature of oceans and the fundamental kinds of habitats associated with them.

The Physical Nature of Oceans

The principal chemical component of seawater is salt in the form of sodium chloride. In addition to these two elements, a number of other ions, including magnesium, sulfur, calcium, and potassium, are important constituents of seawater. The number of cations in seawater exceeds the number of anions, and this leads to two other important properties: Seawater is slightly alkaline (pH 8.2), and the excess of cations buffers the pH of seawater.

There is a huge range of temperature across the world's oceans. Polar waters are often below freezing (and may remain liquid because of the salt content), whereas some equatorial seas approach 30°C. As in freshwater, temperature decreases with depth because the main factor that increases temperature is solar input. The presence of salt in ocean water changes the relationship between density and temperature, however. Maximum density is not achieved at 4°C as in freshwater. The deepest water in oceans may be as cold as 2°C to 3°C.

As in freshwater, light intensity decreases with depth. In the open ocean, this property determines the maximum depth at which photosynthesis can occur. In addition to changes in the quantity of light available, important changes in the quality of light occur with depth. Blue light (short wavelengths) penetrates deeper than red light. Different groups of marine algae utilize different photosynthetic pigments, each of which absorbs a particular wavelength of light.

The oceans reach such great depths that pressure becomes a significant factor for organisms living on or near the bottom. At the bottom of the deep oceans, the pressure may be a thousand times that at the surface. Bottom-dwelling fish must maintain incredible internal pressure to counteract the external forces. In fact, many of these animals explode when brought to the surface. Of great interest to physiological ecologists are animals that range from the surface to the bottom and thus are adapted to a tremendous range of pressure.

Ocean currents are generated by a variety of factors. Changes in temperature and salinity lead to density changes, resulting in the vertical movement of water. These movements lead to horizontal flow as water moves to replace the water that has been displaced up or down. In addition, the rotation of the Earth generates the Coriolis forces that cause east–west movements of oceans in both hemispheres. Some of the major ocean currents are shown in Figure 17.11.

Near coastlines, these currents tend to move water away from the shore. As this water moves, it is replaced by water from below through upwelling. This process has crucial effects on the productivity of coastal environments, because upwelling brings important nutrients to the surface, where they support the growth of phytoplankton. The Pacific Ocean off the coast of Peru is the best-known site of this phenomenon; upwelling of nutrients there leads to high production and to one of the richest fisheries in the world.

A number of other patterns of water movement affect the flow of materials in the oceans. **Langmuir cells** are patterns of circulation caused by winds of

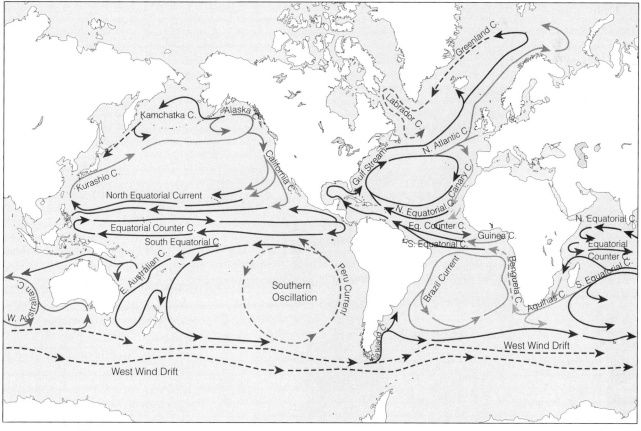

— Warm water current   ----- Cold water current

FIGURE 17.11 The general pattern of the flow of ocean currents, showing the Southern Oscillation, which reverses during El Niño events.

speed greater than 3 meters per second. The wind generates circular rotations of the water just below the surface, as shown in Figure 17.12. Adjacent cells flow in opposite directions, establishing areas of convergence and divergence. Materials accumulate at convergences but are absent from divergences. The **Coriolis Forces** is the deflection of water (or other objects) by the rotation of the Earth. The movement is to the right in the Northern Hemisphere and to the left in the Southern Hemisphere. Near the equator, the effect is minimal. The patterns of ocean currents are determined at least partly by this effect. The **Ekman spiral** is the product of wind in combination with the Coriolis forces. As wind sets the surface water in motion, the Coriolis forces deflect the angle of the surface current (to the right in the Northern Hemisphere, to the left in the Southern Hemisphere). Surface water also transfers its energy by friction to successive layers of water at greater and greater depths. However, the energy dissipates with depth, so the deflection changes from the surface to greater depths (Figure 17.13) where the current flow is opposite the flow on the surface.

Tides are important physical factors at the shore. The tides are caused by the gravitational attraction of the moon (and, to a lesser extent, the sun) on the

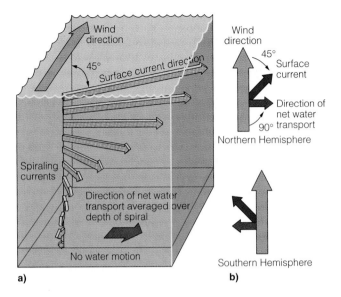

FIGURE 17.13 The operation of the Ekman spiral. The Coriolis forces deflect the path of movement away from that of the wind direction to varying degrees, depending on the depth. The result is a spiraling current. (*From Segar 1998*)

oceans. The nature of the shore determines the tidal range (the difference in height between high tide and low tide) that it experiences; narrow channels, in which the advancing tide is compressed, experience great tidal ranges. The most spectacular example of this is the Bay of Fundy in Nova Scotia, where the tidal range is more than 16 meters. Where the shore is wide and flat, the tidal range will be much less.

Marine Environments

We define a series of regions within marine systems analogous to those in freshwater (Figure 17.14). Near the shore is the **littoral zone.** As the bottom drops away, we enter the **pelagic zone.** The open ocean waters in this zone are divided into two main parts, the neritic and the oceanic. The **neritic zone** is the portion of the pelagic that lies above the continental shelf; the **oceanic zone** lies over the deep ocean. The **benthic region** is the bottom from the lower intertidal out to the greatest depths of the ocean.

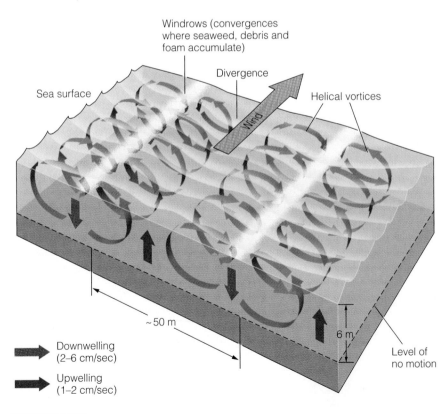

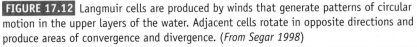

FIGURE 17.12 Langmuir cells are produced by winds that generate patterns of circular motion in the upper layers of the water. Adjacent cells rotate in opposite directions and produce areas of convergence and divergence. (*From Segar 1998*)

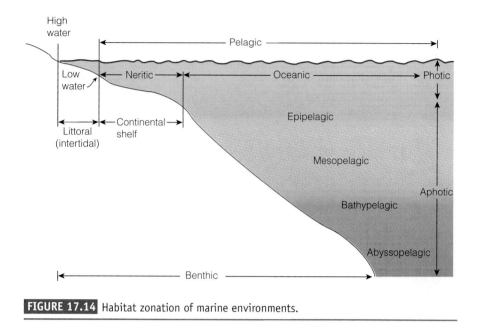

FIGURE 17.14 Habitat zonation of marine environments.

suggests that this serves to make these predators targets for other predatory species.

Because of its great depth, the bottom of the deep oceans (the **bathypelagic**) is cold, dark, and experiences tremendous pressures. As forbidding as this region seems to us, it receives a constant rain of nutrients and debris from above, and a complex web of life exists there. Some of the most bizarre fish on the planet come from this region (Figure 17.15).

Hydrothermal Vents Marine biologists once believed that the abyssal oceans were relative biological deserts, dependent on the rain of material from above. In 1977, however, a complex community of life was discovered on the Galápagos Ridge in the deep Pacific Ocean. Subsequently, many other such systems were found. These communities have grown up around hydrothermal vents located at midocean ridges (Figure 17.16). There are two types of vents in these systems. Some discharge warm water (2°–23°C.) containing hydrogen sulfide, oxygen, and nitrate. Others, known as **black smokers**, release much hotter water (270°–380°C.). Although these vents do not discharge oxygen or nitrate, they do release hydrogen sulfide and many dissolved metals. The "black smoke" is

Open Ocean Biologists have long thought that the most biologically active and interesting regions of the oceans were the shallow-water and photic regions. There was a tendency to regard the large proportion of the ocean's volume found at middle depths far from shore as a biological desert. Recent studies of the midocean depths, however, have revealed that they exhibit great diversity and interesting ecology (Robison 1995).

Relatively little light penetrates into this aphotic zone, but studies in submersibles have shown that there is great biodiversity in this region as well. In the dark, bioluminescence is ubiquitous among species from plankton to vertebrates. It is used for a variety of novel purposes. Some species bioluminesce faintly on their undersides so that they are countershaded when viewed from below against the lighter upper waters (Robison 1995). Among the most intriguing uses of bioluminescence is thwarting predators. Some invertebrates shoot bioluminescent spheres onto their potential predators, lighting them up. B. H. Robison

FIGURE 17.15 One of the many bizarre bathypelagic fish.

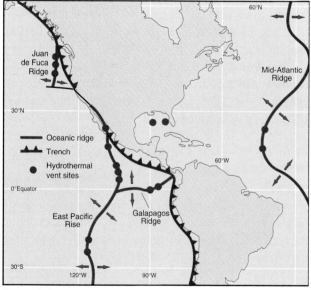

FIGURE 17.16 The locations of hydrothermal vents in the world's oceans. (*From Segar 1998*)

FIGURE 17.17 Photograph of the giant tube worm (*Riftia pachyptila*), an important component of a hydrothermal vent ecosystem. (*From Segar 1998*)

FIGURE 17.18 Low tide in Southeastern Alaska. The organisms like the popcorn kelp in this photograph must be able to tolerate relatively long periods of complete exposure to the air. (*Photo by David Krohne*)

derived from hydrated iron and manganese oxides that precipitate under the high pressure as clouds of small particles.

Diverse and complex biological communities surround these vents. Among the most striking members of this community are the giant tube worms (*Riftia pachyptila*; Figure 17.17) and several species of clams that live close to the vent. Located at varying distances from the vent are many other invertebrate groups, including shrimp, plumeworms, crabs, snails, and limpets. The vent is clearly the center of this ecosystem; the biomass and diversity of organisms decline rapidly with distance from the hole.

The primary producers in this ecosystem are chemosynthetic bacteria, some of which live symbiotically in *Riftia*. The bacteria oxidize the sulfide and use this energy to produce organic matter:

$$6H_2O + 6CO_2 + 6H_2S + 6O_2 \rightarrow C_6H_{12}O_6 + 6H_2HSO_4$$

It is not entirely clear what the tube worms receive from this relationship, but perhaps they obtain waste products or the bacteria themselves. Other free-living bacteria are grazed on by many of the other associated invertebrates.

Rocky Intertidal Zone   This habitat, as its name implies, occurs on rock shores that experience a range of tidal fluctuation. The zone is delimited by the mean high and low tidelines. Marine organisms that grow in

the intertidal zone must be able to deal with periodic exposure to the air and the associated effects of heat and desiccation. The shape of the shoreline determines the pattern of exposure and coverage during the tidal cycle. A relatively flat shore (Figure 17.18) has a lower tidal range and greater variation in the amount of coverage by water. On a steep shore, the tides rise higher and fall lower. Also associated with tides are major changes in the intensity of wave action that are due, in part, to water depth. Waves have an important scouring effect on the substrate. Finally, as tides rise, organisms experience greater salinity than they do at low tide, when salt water does not reach as far shoreward. Because these factors vary, there is sharp zonation of plants and animals in the intertidal zone (Figure 17.19).

Tide pools are important features of rocky shores. These depressions in the substrate trap seawater when the tide retreats, thus providing a refuge for organisms that cannot tolerate exposure to the air and allowing them to persist higher in the intertidal zone than they otherwise might. Salinity and temperature in the pools may fluctuate drastically relative to more open water. As the pool dries up, its salinity increases; an input of freshwater, from rain perhaps, lowers it. Similar effects increase or decrease the temperature as well. Species that inhabit tide pools must have evolved physiological mechanisms to cope with these extremes (see Chapter 3).

Salt Marshes   Salt marshes are important soft-bottom systems along many coasts. They tend to form in regions where the topography meets an exacting set of criteria: There must be an offshore barrier island to moderate the waves that otherwise would

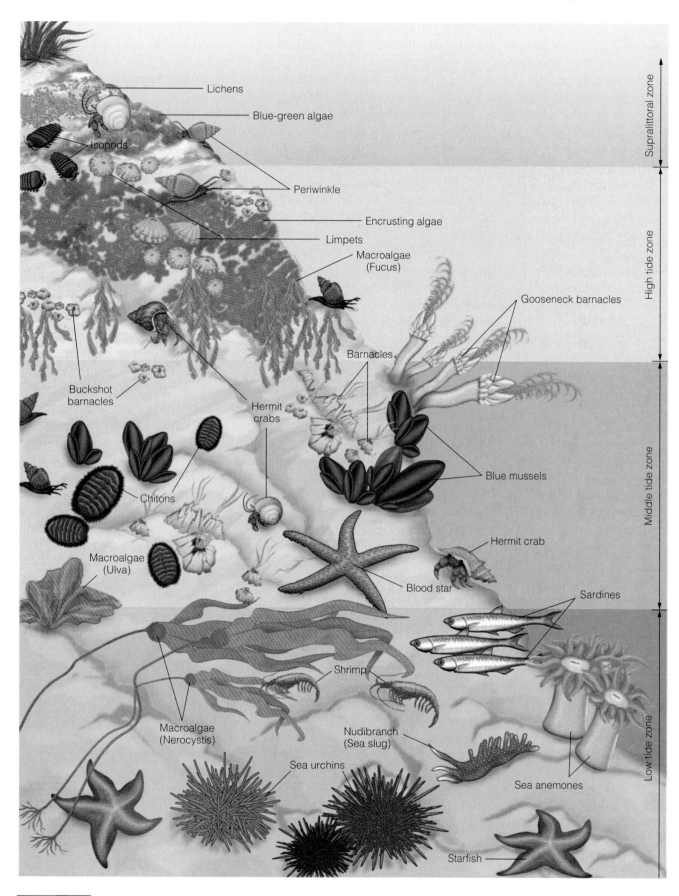

FIGURE 17.19 The four zones of a typical rocky intertidal, defined by the amount of exposure to air. Typical organisms are shown for each zone. (*From Segar 1998*)

physically destroy the community, and the bottom must slope gradually upward so that tidal waters can spread and deposit the silt and mud they import into the system.

The physical conditions in a salt marsh can be harsh for plants. Species must be able to tolerate water-logged soil, inundation with salt water, and wave damage, which can be severe during storms.

Salt marshes are at least superficially depauperate in species; that is, the diversity of emergent plant life there is not conspicuous. The flora of many salt marshes consists of members of only a few genera, including *Salicornia, Spartina,* and *Juncus.*

The vegetation of a salt marsh shows conspicuous zonation according to topography and distance from the sea (desiccation and salinity). Most salt marshes have three zones. The first, or lowest, zone is known as the "pioneer zone" because new colonists must continually replace species removed by storms and tides. The middle zone is often the most diverse. The upper zone contains a mix of species with different salinity tolerances and requirements, grading, of course, to less salt-tolerant species farther from the shore.

Pools of water somewhat analogous to the tide pools of rocky shores form in salt marshes. These depressions, which are called **pannes,** arise from variation in the substrate topography or differences in the colonization patterns of the vegetation as the marsh develops. Like tide pools, pannes exhibit great variation in temperature and salinity, depending on their depth and on whether channels connect them to the sea. Deep pannes retain salt water as the tide ebbs; shallow ones may partially evaporate at low tide, causing salinity to increase. Pannes often support distinctive vegetation that is visible as pockets dominated by a particular species.

Estuaries

Estuaries are ecologically important nearshore habitats located where rivers enter the ocean. They are defined as partially enclosed bodies of water that are directly connected with the ocean, but where seawater is diluted by freshwater entering from a river system.

The key physical factor in an estuary is the degree of salinity. Water entering from the river (0 ppt NaCl) tends to float on top of the denser saline waters from the ocean (37 ppt). A number of factors determine the estuarine salinity gradient. Evaporation of freshwater from the surface increases salinity. In tropical systems, evaporation rate may be so great that highly saline water extends farther into the river mouth. Tidal or storm-driven influx of salt water also increases salin-

ity. River flooding flushes salt out of the system. Because of the rotation of the Earth, seawater entering an estuary on an east coast in the Northern Hemisphere is deflected clockwise, causing one side of the bay to be saltier than the other.

The salinity gradient has several functional consequences. Of course, plants and animals must be able to cope with the salinity regime. Most estuarine animals are marine organisms that are tolerant of salinity. However, many forms are dwarfed as a result of the lower-than-optimal salinity they encounter. There is also zonation according to their tolerance of the freshwater inputs.

For some species, such as the oyster, the salinity regime provides a refuge from predation. The eastern oyster (*Crassostrea virginica*) tolerates wide fluctuations in salinity, but two of its main predators, Forbes's common sea star (*Asterias forbesi*) and the Atlantic oyster drill (*Urosalpinx cinerea*), are less tolerant of freshwater. Consequently, the oyster can escape predation in the estuarine environment.

In addition to these direct salinity effects, the merging of salt water and freshwater has indirect effects. Because the freshwater floats, it flows seaward on the surface. Below the outflow of freshwater, an influx of saline water occurs. This countercurrent forms a circulation nutrient trap that prevents nutrients and particulate matter from being washed out to sea.

The land form of most estuaries is dynamic. If stream flows are slow, material imported by the river is deposited to form a delta at its mouth. Opposing this deposition is the action of ocean currents, wind, and waves. The substrate of estuaries is thus in a constant state of flux as these forces interact.

Many estuarine systems are associated with extensive salt marshes and include *Spartina* grass, pickleweed (*Salicornia*), eelgrass (*Zostera marina*), and turtle grass (*Thalassia testudinum*). A significant fraction of the production is from phytoplankton, especially diatoms. Oysters are central components of the animal life of estuaries. The gradual accumulation of oysters, both living and dead, forms reefs that serve as substrate for a variety of other marine invertebrates.

Mangroves

Another important nearshore community associated with soft bottoms is the **mangrove swamp** (Figure 17.20), a habitat distributed across much of the tropics and subtropics. We can consider the mangrove swamp the tropical analog of the temperate salt marsh system. Although they do not constitute a large fraction of all shore communities, mangrove swamps, like salt marshes, are extremely important ecologically. They serve as nurseries for

FIGURE 17.20 A red mangrove at the edge of a mangrove swamp in southern Florida. (*Photo by David Krohne*)

FIGURE 17.21 The seed of the red mangrove germinates while still on the tree. (*Photo by David Krohne*)

many marine species, vertebrate and invertebrate alike. A large number of species of commercial importance, such as shrimp and several species of snappers, develop in these protected and productive habitats before entering the open ocean as adults. Moreover, like salt marshes, mangrove swamps filter materials, including pollutants, from water that enters the ocean.

Mangrove swamps are characterized by bands of mangrove species, each associated with a different position relative to the ocean based on its tolerance of salt and the water saturation of the substrate. In much of the New World, the species occur in the following sequence from the shore inland: red mangrove (*Rhizophora mangle*), black mangrove (*Acicennia gerinans*), white mangrove (*Laguncularia racemosa*), and finally buttonwood (*Conocarpus erectus*). The red mangrove is the most tolerant of salt and saturated soil; buttonwood is least tolerant of these conditions.

The red mangrove pioneers the formation of a mangrove swamp. This species, which is highly adapted for life in warm, shallow, saline waters, produces salt-tolerant seeds that have the unusual property of germinating while still attached to the tree (Figure 17.21). A small root forms before the seed drops from the parent. When released, the seedling floats on the currents, often for great distances and long periods of time. When the seedling reaches shallow water or runs aground in mud, it begins rapid growth. The plant produces many arching prop roots that vegetatively propagate the individual to form a mass of interconnecting roots and above-water growth.

Small islands composed almost entirely of mangroves are formed by this process. One of the most important features of red mangrove stands is the extensive prop roots that serve to slow the flow of water around the stands. This causes silt and other particles to drop to the bottom and gradually build the substrate. A stable mud flat develops that is rich in nutrients and well protected from wave and wind action. The other species of mangroves may then be able to colonize. Most of these islands begin as nearly pure stands of mangroves, but as they age, some other species colonize as well.

Coral Reefs   **Coral reefs** are important nearshore habitats that are confined to shallow waters in tropical and subtropical latitudes where water temperature does not drop below 18°C (Figure 17.22). They are composed of a skeleton of colonial corals, both living and dead. Coral reefs are divided into three fundamental types. **Fringing reefs** develop on rock substrate near islands and continents. This is the main type of reef in the Florida Keys. **Barrier reefs** are found along the shore of continents; an example is Australia's Great Barrier Reef. Finally, **atolls** are island reefs that develop as corals and sediments accumulate and approach the ocean's surface. Changes in sea level may bring the reef closer to the surface. Atolls usually surround a lagoon of shallow water. Coral reefs are among the most productive habitats in the ocean; they average more than two orders magnitude higher production than the surrounding open ocean. What accounts for this production? Several factors probably play a role. The reef-building corals have associated with them zoox-

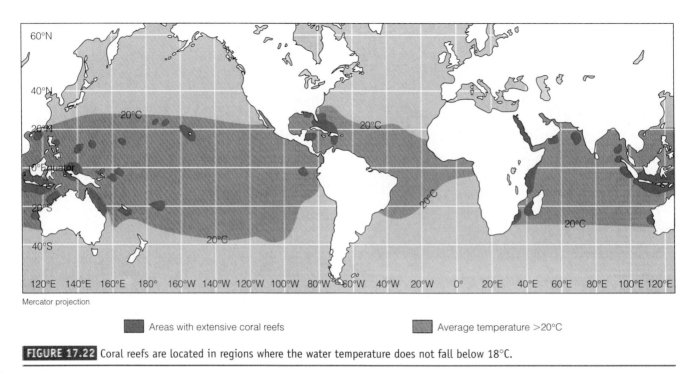

Mercator projection

■ Areas with extensive coral reefs   ■ Average temperature >20°C

FIGURE 17.22 Coral reefs are located in regions where the water temperature does not fall below 18°C.

anthellae, a photosynthetic dinoflagelate. Other producing algae live on or near the surface of the reef-building corals. The symbiotic association between the zooxanthellae and the reef-building corals plays a key role in production. The association is highly efficient. The zooxanthellae use solar radiation passing through the transparent coral tissue. In addition, they efficiently use the coral's waste CO_2 and nutrients. A high percentage of the photosynthate produced by the zooxanthellae is absorbed directly by the coral.

The location of reefs relative to land masses also plays a role. All forms of reefs—fringing reefs, barrier reefs, and atolls—are located near large land masses. Ocean currents are deflected by these masses, causing upwelling of nutrients. Finally, the organisms inhabiting reefs, from the grazers on the algae to the predatory fish, are restricted to the immediate reef vicinity. Nutrients are not lost from the system; rather, they are quickly recycled in a manner reminiscent of tropical forests.

FIGURE 17.23 Coral reefs maintain very high local species diversity. (*Photo by Joyce and Frank Burek/Animals Animals*)

Coral reefs are also unique in the extreme species diversity they support (Figure 17.23). Reefs contain a bewildering array of niches for both plants and animals. In addition to the symbiotic dinoflagellates, many species of algae incrust the surfaces of the corals or live in the interstices of the coral and the limestone it builds. A wide range of invertebrate and vertebrate herbivores graze the algae. Many species of filter feeders live on suspended organic matter or other organisms. Deposit feeders, including several mollusks and worms, sift through the sediment on or around the reef. Hundreds of species of predators live higher in the food chain. Perhaps most remarkable is the scale over which this ecosystem functions. A single square meter of coral reef may support hundreds of species of organisms.

This completes our survey of the variety of habitat types. Freshwater and marine systems constitute an important set of habitats, in part because they are so extensive, but also because they are connected to the ecosystem processes of terrestrial systems. They have also been important to the science of ecology because certain kinds of experimental studies are more easily accomplished in aquatic systems.

SUMMARY

1. Aquatic systems are divided into two fundamental groups: freshwater and marine systems.

2. Freshwater systems consist of lentic systems (nonflowing), lotic systems (flowing), and wetlands (the interface between terrestrial and aquatic systems).

3. Lentic systems are characterized by zonation according to light penetration and the density of water. Deep lakes become thermally stratified, a phenomenon that affects the distribution of nutrients in the lake.

4. The ecology of lotic systems is strongly affected by flow rate. Most of the production in lotic systems is a result of periphyton.

5. Important abiotic factors in marine systems are salinity, currents, tides, and depth.

6. In marine systems as in lakes, the shallow water near shore and deep offshore areas differ markedly in ecology.

7. A number of ecologically important habitats occur at the boundary between marine and freshwater or terrestrial habitats: tide pools, salt marshes, estuaries, mangrove swamps, and reefs.

8. Wetlands and other aquatic systems play a key role in processes important to humans. For example, marshes may slow the rate of flooding, absorb pollutants, and provide breeding habitat for animals of importance to humans. Salt marshes and mangrove swamps are vital breeding grounds for many commercially important species.

SELF-ASSESSMENT: CAN YOU . . . ?

1. Outline the important physical features of freshwater and salt water habitats, and explain the fundamental differences between them.

2. Explain what environmental benefits aquatic habitats provide that are important to humans.

3. Explain the role of outside energy sources or forces in aquatic habitats (such as gravity in a stream).

4. Compare the ecology of aquatic systems at *boundaries*—that is, near shore or at a habitat discontinuity—to the ecology of large, relatively homogeneous systems.

5. Compare the nature of primary productivity and its limiting factors in the various aquatic habitats.

PROBLEMS AND STUDY QUESTIONS

1. Is there an aquatic system or entity analogous to the biome in terrestrial systems?

2. Using the information on biogeochemical cycles in Chapter 15, discuss the interface of aquatic and terrestrial ecosystems with respect to the movement of materials. Which cycles involve both terrestrial and aquatic systems?

3. Discuss the development of a bog in terms of succession: the generation of a terrestrial plant community after a disturbance (glaciation).

4. Which physical and chemical properties of water are crucial to the ecology of aquatic systems?

5. Compare the abiotic limiting factors in terrestrial systems with those in freshwater and marine systems.

6. How could you test the hypothesis that algae are distributed in aquatic systems according to the nature of their photosynthetic pigments?

7. Compare the tolerance limits of species inhabiting tide pools, the open ocean, and a freshwater lake.

8. What patterns of genetic variation would you expect to see in red mangroves? (See Chapter 6.)

9. Summarize the major types of wetland habitats. Of what importance are these regions to humans?

10. Identify and classify the major factors that can act as disturbance aquatic systems. Are there examples of successional sequences in aquatic systems like those in terrestrial systems?

PROJECTS AND ADDITIONAL STUDY

1. Research local laws and zoning regarding wetland habitats. Determine how well protected such habitats are in your area.

2. Perform a literature search to identify the important abiotic factors in a local aquatic habitat. Measure the spatial and temporal patterns of change in these factors.

3. Choose a taxonomic group or ecological guild that inhabits a nearby aquatic system. Make collections (the techniques will vary with the systems) to map the patterns of distribution and abundance in the system. For example, what is the pattern of algal species distribution as a function of depth in a lake? Or what is the distribution of emergent aquatic plants in a marsh?

Epilogue

Natural History and Ecological Theory
Quality of Information

The Boundaries of Ecological Science
Ecology and Our Environment

In the first chapter of this text, a description of a prairie fire illustrated the concept of interaction as the most fundamental characteristic of ecological systems. The purpose of that scenario was simply to provide a natural history description of a web of interactions. At that point we had no ecological theory in which to place those interactions.

By this point you should have developed a rather sophisticated appreciation of the nature of interactions in ecological systems. The discussion of ecological interaction with which we shall conclude is possible now because of the theoretical and factual background you have developed. It illustrates the kinds of analysis in which you should now be able to engage.

Natural History and Ecological Theory

In the first chapter, we discussed the differences between natural history and ecology, noting that natural historical observations underlie much of modern ecology. We saw that the science of ecology places the descriptive observations of natural history in an analytic and theoretical context.

We turn now to another very different natural history. It too illustrates the richly complex web of interactions in ecological systems, but it also shows that the interactions can be taken as examples of some fundamental theoretical principles that we have studied in the science of ecology.

In the tropical forests of Costa Rica, a wonderful web of interactions surrounds a brilliant red flower, *Hamelia*; its pollinator, the magnificent hummingbird *Eugenes fulgens*; and an array of tiny pollen-eating mites in the genus *Proctolaelaps* (Colwell 1995). *Hamelia* is a typical hummingbird-pollinated flower, with a long, red corolla (Figure 18.1) in which nectar is produced as a reward for the hummingbirds that visit and pollinate the species. The mites travel from flower to flower in the bills of the hummingbirds. In the few seconds during which the bird visits a flower, the mites charge down the bill at a relative rate equivalent to the speed of a cheetah and leap onto the new flower, where they consume nectar and pollen.

The complexity of the system is deeper still. The hummingbirds visit several species of red flowers, each of which is utilized by a particular species of mite. The mites must disembark on the correct species or face two problems: They may be killed by other species of mites, or they may be stranded without hope of finding a mate and breeding. In the few sec-

FIGURE 18.1 The hummingbird-pollinated tropical flower *Hamelia*. (*Photo by Michael Fogden/Animals Animals*)

onds of the hummingbird's visits, the mite ascertains whether the flower is the correct species by the fragrance it emits, which is carried to the mite via the breathing of the hummingbird.

Robert K. Colwell of the University of Connecticut has spent the last 20 years working out the details of this remarkable relationship among flowers, hummingbirds, and mites. No doubt the story is not yet complete. What Colwell has discovered thus far, however, in this one tiny portion of the enormous biological diversity of the rain forest, illustrates some of the most fundamental ecological and evolutionary principles we have studied:

1. *Coevolution.* The hummingbird pollinators and their red-flowered partners offer a classic example of coevolution. The hummingbird forages for nectar produced as a reward by the flowers. In the process, it pollinates plants that may be separated by great distances.
2. *Interspecific competition.* Colwell has shown that the mites, in fact, are significant competitors with the hummingbirds for the nectar and pollen resources of the host plants. The mites consume as much as 50 percent of the nectar and 33 percent of the pollen produced by the plant.
3. *Mating systems.* The mites have a polygynous mating system in which the males defend harems of females.
4. *Dispersal.* For *Hamelia*, the hummingbirds are a dispersal vector as they transfer pollen and thus affect gene flow. For the mites, a complex set of dispersal issues also pertains. The flowers of *Hamelia* are ephemeral, sometimes lasting only a day. If the

mites do not leave before the flower wilts and falls from the plant, they will be stranded. If a female mite disperses to another flower via a hummingbird, she may have the opportunity to lay her eggs in a virgin flower where her offspring will face no competition from other mites. She might, however, land on an already inhabited flower. Clearly, she must make an "evolutionary decision" about the most adaptive dispersal strategy. Similarly, dispersal can carry a male mite to a flower with new virgin females, in which case his fitness will be high. On the other hand, he might land on a flower with females that have already been inseminated, in which case his fitness will be lower than if he had not dispersed.

5. *Niche specialization.* As many as five species of mites may be transported by the same individual hummingbird. The mites are, in fact, herbivores of a kind; each species has specialized on a particular species of flower, thereby decreasing competition. In general, herbivores' plant specificity is a response to plant defenses; in this case, the specificity and niche specialization are also related to the mites' need to find flowers that harbor potential mates.

The natural history of this system is fascinating in and of itself, and we need not know any sophisticated ecological theory to find it interesting. However, you should now also be able to place each of these interactions in the larger context of the applicable ecological principles, such as competition theory, the competitive exclusion principle, dispersal strategies, and the theory of sexual selection. And you should be able to compare this system to others you have studied.

Quality of Information

Data, analysis, and scientific interpretations are accumulating at an increasing rate in all fields of science. Consequently, a science textbook must offer more than simply a catalog of information and interpretations, a great many of which will be outdated in a short time. For example, the paradigm that competition structures *all* ecological communities was held as dogma 25 years ago. Now we understand that that interpretation was not entirely correct. We also once assumed that most populations are at equilibrium most of the time. Thus we could search for those mechanisms of population regulation that determine the equilibrium population density. Now we under-

stand that the equilibrium view of populations is insufficient to explain the dynamics of all natural populations.

Given the changes that will inevitably occur in the science of ecology, what can we single out as enduring? We can depend on the temporal stability of our *methods*—the means by which we analyze and understand ecological systems. The scientific method, with all its varied applications, is fundamentally the same today as it has been for a century or more, and we can count on its continued utility long after some of the interpretations explained in this text have been discarded.

That is why this text has emphasized the data on which our current understanding of ecology is based. It is my hope that as you have surveyed the actual data and their analysis on which modern ecology rests, you have developed an appreciation for the methodology of science as applied to ecology. I hope that you have questioned the validity of some of the studies discussed in this text. Through the discussion of the examples cited, you should have come to appreciate the range of certainty in the studies on which the science is based. If you have found yourself wondering about the scatter of points through which a regression line is drawn, you have begun the process. If you have asked about the assumptions underlying a study, you're on the right track. If you have considered alternative interpretations of the data presented, you have gone farther still. A healthy skepticism should be your goal. In the words of Whitehead (1920), "Seek simplicity and distrust it." These are skills that will be of permanent value, even as our current understanding of ecology evolves.

The Boundaries of Ecological Science

There is another level at which we can now discuss interaction. In the first chapter, we placed the science of ecology at the most inclusive end of the biological hierarchy. That hierarchy is a useful construct with which to discuss the relationships among the biological sciences. If anything characterizes science at the beginning of the new century, however, it is the blurring of traditional boundaries between disciplines. If ecology is in fact a science of interaction, it is most assuredly a science that is connected in important ways to all the rest of biology. Indeed, Grant (2000) has reiterated that the American Society of Naturalists, a prominent ecological society, is "devoted to the conceptual unification of the biological sciences." The modern ecologist draws on all areas of the biological hierarchy—from the molecular level to the ecosystem. If we are to understand the ecological interactions between the organism and its physical environment, the elements of animal and plant physiology naturally come into play. If we are interested in the effects of hot, dry conditions on primary production, we must delve into the intricacies of the C4 biochemical pathway. If we are interested in the adaptive morphology of early and late succession trees, we must understand the fundamentals of plant morphology and development. And nowhere has the interaction between ecology and other aspects of biology changed more profoundly than in the application of molecular techniques to ecological problems. The revolution in molecular biology has transformed the way we approach all biology, including ecology.

Ecology and Our Environment

In the first chapter I drew a distinction between environmental science, the applied science, and ecology, the basic science. This distinction, though useful in a "taxonomy" of related sciences, is becoming less pronounced all the time. Ecologists have an obligation to apply their basic knowledge to environmental problems. Anthropogenic environmental problems are complex, and we will need sophisticated ecological insights to address them. Many of the simpler problems have been solved—or at least we understand the scientific components of the solution. For example, we now have a good understanding of the ways in which DDT becomes concentrated in ecological systems, its physiological effects, and how top carnivores are particularly affected. However, many of the most pressing current environmental problems are very complex. Consider the problem of maintaining the planet's biodiversity. Conservation biologists devising strategies for species preservation must include at least the following in their analyses:

Island biogeography theory and habitat fragmentation

Metapopulation dynamics

The details of species-specific population regulation

Patterns of genetic variation and gene flow

Dispersal strategies and ability

Behavioral responses to ecological stress

This is a far more complex ecological problem than that posed by DDT.

There is another reason why the distinction between basic and applied ecology has become blurred. Anthropogenic effects are now measurable in virtually all areas of the planet, even the most pristine habitats in the polar regions. Indeed, the ecology of our planet is now dominated by humans (Vitousek et al. 1997). As a result, the very nature of scientific inquiry in ecology is forever altered. Every year it is more difficult to devise ecological experiments that do not include, as a potential confounding variable, the effects of human activity and disturbance on the system. Dayton (1998), in a discussion of the effects of the fishing industry on benthic marine systems, summarized the issue:

> One inescapable consequence of this widespread damage is the loss of the opportunity to study and understand intact communities. In most cases there are no descriptions of the pristine habitats. The damage is so pervasive that it may be impossible to ever know or reconstruct the ecosystem. In fact, each succeeding generation of biologists has markedly different explanations of what is natural because they study increasingly altered systems that bear less and less resemblance to the former, pre-exploitation versions. This loss of perspective is accompanied by fewer direct human experiences (or even memories) of once undisturbed systems.

This situation poses two challenges for the ecologists of the new century. First, we must continue to fine-tune our methods of analysis to account for the confounding effects of our activities and the loss of pristine systems. We must devise ways to isolate from the effects of humans, at least for purposes of analysis, the fundamental ecological processes we hope to investigate. Second, ecologists must contribute to the maintenance and restoration of the environmental health of the planet. We have an obligation to our fellow citizens, to the animals and plants whose habitats we share, and to the ecosystems that sustain us. Our well-being and our science depend on it.

Glossary

Abiotic extrinsic school A school of population regulation that emphasizes the role of abiotic factors, such as weather, as important limiting or regulating factors for populations.

Absolute fitness The expected number of offspring produced by a particular genotype. This measure incorporates both the survival ability and the fecundity of the genotype.

Acclimatization The process of physiological adaptation by an individual to the environment it experiences.

Actual evapotranspiration (AET) The amount of water that is actually lost from plants and soil.

Adaptation (1) An evolutionary solution to an ecological problem. (2) Features of the organism that increase its fitness.

Adiabatic cooling A decrease in air temperature at higher elevations as a result of decreasing air pressure.

Aggradation A phase of ecosystem development after disturbance, in which nutrients or biomass is accumulating.

Aggregation Groups of conspecifics that include more than one family unit (parents and offspring) localized in a particular area. There is no internal organization or cooperative behavior.

Aggression An act by one individual that achieves dominance over another by physical violence or the threat of it.

Aggressive mimicry A mimicry system in which a noxious or dangerous species resembles a benign one.

Allee effect A population phenomenon in which populations have lower or negative growth rates at very low densities.

Allele frequency The frequency, measured in proportion to the total number of alleles at a locus, of a particular allele.

Allochthonous sources Sources of energy outside an aquatic system.

Allometry A phenomenon in which the relative sizes of various morphological traits of an organism change at different rates.

Allopatry The occurrence of organisms in different places or localities.

Altruistic behavior Behavior that enhances the fitness of other individuals in the population at the expense of the individual performing it.

Aposematic coloration Bright or distinctive coloration that indicates that the species is dangerous or noxious.

Assimilation efficiency The fraction of ingested energy that is assimilated by an organism.

Atolls Island reefs that develop as corals and sediments accumulate and approach the ocean's surface.

Autecology A branch of ecology that focuses on the relationship and interactions between the individual and its environment.

Autochthonous sources *In situ* sources of energy in an aquatic system.

Autotrophic succession The development of a new community in an empty habitat, usually one that has been made so by some kind of disturbance.

Autotrophs Organisms that derive energy from inorganic sources.

Barrier reef A type of coral reef found along the shores of continents.

Batesian mimic A benign species that resembles a noxious or dangerous one.

Bathypelagic zone In oceans, the bottom regions of the deep ocean, characterized by great depth, high pressure, low temperature, and darkness.

Benthic zone The bottom layer of a lake or ocean in both the littoral and profundal zones.

Big-bang reproduction See **Semelparity.**

Bioaccumulation The accumulation of toxins or other materials to high concentrations at the higher levels of a food chain.

Biodiversity Biological diversity, including species diversity, genetic variation and diversity, and the diversity of ecological interactions.

Biogeochemical cycle The global pattern of movement of materials (nutrients).

Biomass The total dry weight of organic matter.

Biome A major unit of terrestrial vegetation, characterized by the structure of the plant community.

Black smoker A type of hydrothermal vent characterized by extremely hot water and black "smoke" consisting of hydrated iron and manganese oxides.

Bog An aquatic system characterized by the presence of sphagnum moss, by high acidity, and often by having no water outlet.

Bottleneck A pattern of population change in which the population is reduced to small numbers and then expands again.

Browser An herbivore that consumes woody material and bark.

Carnivory The predatory interaction in which an organism captures, kills, and consumes an animal.

Carrying capacity (*K*) The maximum number of individuals that the resources in a particular environment can support.

Chaos A pattern of population fluctuation that appears to be random but is in fact underlain by deterministic factors. The pattern of fluctuation is particularly sensitive to the initial conditions.

Character displacement A long-term, evolutionary response to competition in which one species's niche shifts to avoid competition with other species.

Classical conditioning A process of behavioral learning in which certain environmental stimuli come to be associated with particular responses.

Climax The final, self-perpetuating stage of community change or succession.

Coarse-grained species A species inhabiting a landscape in which the habitat patches are large relative to the vagility of the species. An individual is likely to experience only one habitat patch in its travels.

Coefficient of relationship (*r*) The average number of genes shared by two individuals.

Coevolution An intimate and interactive evolutionary relationship between two or more species in which direct genetic change in one species is attributable to genetic change in the other(s).

Coevolutionary arms race A series of escalating adaptations to counteradaptations in another species, often in predator–prey relations.

Cohort life table A life table derived by following a single cohort from birth until the death of the last individual.

Colony A conspecific group composed of a society of highly integrated individuals. There may be division of labor and/or the organization of individuals into specialized castes.

Commensalism An interaction in which one species benefits but the other is neither helped nor harmed.

Communication Any action on the part of one organism that alters the probability of a behavior in another individual.

Community A group of species occurring in a defined region.

Compensation point The depth in a lake at which the rate of photosynthesis by plants exactly balances the rate of respiration by plants and animals.

Competition The interaction between two species over a limiting resource that negatively affects one or both of their population growth rates.

Competitive exclusion The local extinction of a species as a result of interspecific competition.

Competitive exclusion principle (Gausse's Law) The principle that no two species can long coexist on the same limiting resource; that is, they cannot occupy the same niche.

Connectance A measure of the interaction among trophic levels or the complexity of a food chain.

Conservation biology The science of the maintenance and management of biodiversity in natural systems affected by human activity.

Continental climate A climate of large land masses, as distant from large bodies of water, characterized by large annual temperature fluctuations.

Convergent evolution Similar adaptations in organisms with different phylogenetic histories.

Coral reef A shallow-water tropical marine ecosystem characterized by complex communities associated with the growth of corals.

Core-satellite hypothesis A pattern of community structure in which mean abundance and the number of sites occupied by a species are positively related and the distribution of rare and common species is bimodal (that is, most species are *either* rare *or* common).

Coriolis force A force resulting from the rotation of the earth that causes objects to veer to the right in the Northern Hemisphere and to the left in the Southern Hemisphere.

Crude density The population density measured across a defined area without regard to microhabitat suitability.

Crypticity An adaptation through which prey avoid detection by appearing to their predators as a random sample of the environmental background.

Current reproductive value The relative genetic contribution of an individual from birth to its current age.

Decomposers Heterotrophs that obtain energy from dead organic matter.

Deme A local, genetically defined population characterized by random mating within the group.

Demography The quantitative description of a population, including parameters such as density, age structure, and sex ratio.

Diapause A genetically determined resting stage characterized by cessation of development and protein synthesis and by suppression of the metabolic rate.

Diffuse competition The combined competitive effects of several species.

Directional selection A form of natural selection in which individuals in one tail of the phenotypic distribution are favored and increase in frequency, shifting the mean phenotype in one direction.

Disclimax A seeming climax that is in fact an earlier seral stage maintained by disturbance.

Dispersal The movement of individuals from the natal area or home area.

Dispersion The pattern of spatial distribution of individuals in the environment.

Disruptive selection A form of natural selection in which individuals in both tails of the phenotypic distribution are favored and increase in frequency, leading to a bimodal phenotypic distribution.

Disturbance any relatively discrete event in time that disrupts ecosystem, community, or population structure and changes resource or substrate availability or the abiotic environment.

Ekman spiral A pattern of flow in the ocean caused by wind in combination with the Coriolis force. Water is deflected to different angles at different depths.

Ecocline A gradual transition from one habitat type to another.

Ecological density The population density, measured in appropriate microhabitats for the species.

Ecology The study of the interactions between the organism and the physical and biological components of its environment.

Ecosystem The most inclusive ecological level. The sum of the abiotic and biotic components of a defined system or region.

Ecotone The boundary between two habitats.

Ecotypes Genetically distinct geographic variants that result from adaptation to local conditions.

Ectomicorrhizal associations Micorrhizal associations in which the fungal hyphae do not penetrate the root of the plant.

Effective population size (N_e) The number of individuals whose allele frequencies behave as expected of a randomly mating group.

Eltonian pyramid Diagrammatic representation of the energy relationships among trophic levels in an ecosystem, in which the numbers, biomass, or energy content of each trophic level is represented as a portion of a pyramid.

Emergent properties Features of a system not directly predictable from the properties of the component parts.

Emergent vegetation Plants that extend above the surface of the water in aquatic systems.

Endemic species Species found in a particular region and nowhere else.

Endomicorrhizal associations Micorrhizae whose hyphae enter the root of the plant.

Energy maximizer In optimal foraging theory, a species for which obtaining the maximum amount of energy is crucial.

Environmental science (1) The study of the ecological effects of human activities on the environment. (2) The application of basic ecological principles to human environmental concerns.

Environmental toxicology The study of the effects of pollutants on individuals and populations.

Epilimnion The layer of water in a lake above the thermocline.

Eusocial A complex and highly evolved social system in which extensive cooperation, communication, and division of labor are present.

Eutrophic Systems characterized by high nutrient content and production.

Evapotranspiration The loss of water to the atmosphere from plants and soil.

Evolution Genetic change in a population over time. The process of long-term development and change in living organisms as a result of processes such as natural selection, genetic drift, and speciation.

Evolutionarily stable strategy (ESS) A genotype (and phenotype) that cannot be replaced by competition with another invading genotype.

Exploitation competition Competition in which the actions of one species (or individual, in the case of intraspecific competition) significantly reduce the amount or availability of the resource for another.

Facilitation A model of successional change in which pioneering species modify the environment in a way that enhances the colonization of later species.

Fall turnover The autumnal mixing of hypolimnion and epilimnion in a lake.

Female defense polygyny A polygynous system that occurs because males defend groups of females.

Fine-grained species A species that inhabits a landscape in which the habitat patches are small relative to the vagility of the species. Such a species is likely to experience more than one patch type in its life or in its travels.

Fitness The relative genetic contribution of an individual's descendants to future generations. Fitness is enhanced by the individual's ability to survive and reproduce.

Fixed action pattern (FAP) A behavior that is highly stereotyped. It appears in the same form each time it is performed.

Folivore See **Grazer.**

Food chain The pathway of energy transfer from producers to consumers to the top carnivore.

Food quality The availability of energy from food for assimilation.

Food web The interaction of a number of food chains in an ecosystem.

Forb A nonwoody, nongrass species of plant.

Fractal dimension A property of a geometric figure that indicates the relationship between the perimeter and the measurement scale; an index of the perimeter complexity of an area.

Frequency-dependent selection A form of natural selection in which the fitness of a particular genotype varies with its frequency in the population.

Fringing reef A type of coral reef that develops on a rock substrate near islands and continents.

Frugivore An herbivore that consumes fruit.

Fugitive species A species that inhabits a transient environment.

Functional niche An aspect of a species's niche that emphasizes the ecological role of the species in the community.

Functional response A predator response to increased prey density in which the rate of consumption increases.

Fundamental niche The Hutchinsonian niche for a species in the absence of any potential competitors.

Gausse's Law. See **Competitive exclusion principle.**

Gene flow The net movement of alleles into or out of a population.

Gene pool The sum total of all the alleles in a population.

Genet In plants, the developmental product of a single zygote; a clone.

Genetic distance A measure of the genetic differences between two individuals based on the number of alleles they share.

Genetic drift Stochastic (random) shifts in allele frequencies.

Genetic neighborhood The local region in which mating occurs at random.

Genetic polymorphism hypothesis A hypothesis of population regulation in which two interacting populations (such as a parasite and a host) constitute selective forces on one another. As a result of adaptations in each species, the fluctuations in population density gradually decline.

Genotype The genetic constitution of an individual that, along with the environment, determines the phenotype.

Granivore An herbivore that consumes seeds.

Grazers Herbivores that consume leafy material; also called **folivores.**

Gross primary production The total rate of accumulation of energy by photosynthesis.

Group A band, assemblage, or party of individuals. Sometimes the term is used to refer to an assemblage that is intermediate between an aggregation and a colony.

Habitat (1) The physical location of an organism in the environment. (2) The type of environment—vegetation, climate, and so forth—inhabited by an organism.

Haplodiploid A sex-determining system in which males are haploid (derived from unfertilized eggs) and females are diploid.

Hard selection A form of natural selection in which the fitness of an individual is unrelated to the other genotypes, usually as a result of important mortality effects due to the abiotic environment.

Hardy–Weinberg equilibrium Genetic equilibrium where the genotype frequencies in a population are described by the equation $p^2 + 2pq + q^2 = 1$. A population in Hardy–Weinberg equilibrium is not evolving.

Herbivory The consumption of plant tissue by an animal.

Heterosis See **Heterozygote advantage.**

Heterotrophs Animals that derive energy either from organic compounds in the environment or from other organisms.

Heterozygote advantage A phenomenon in which the heterozygote has higher fitness than either homozygote.

Holism A scientific approach that emphasizes the totality of the interactions, and the processes that depend on the sum of the interactions.

Home range The area an individual normally traverses in its daily activities.

Hutchinsonian niche A concept of a species's niche that is based on the species's use and tolerance of a series of *n* resources and abiotic factors.

Hydrothermal vent A phenomenon on the deep ocean floor in which superheated water is released from vents. A rich and unique biological community surrounds these vents.

Hypertonic A condition in which an organism or other entity has a higher solute concentration than that on the other side of a semipermeable membrane.

Hypolimnion The layer of water in a lake below the thermocline.

Hypotonic A condition in which an organism or other entity has a lower solute concentration than that on the other side of a semipermeable membrane.

Icarus effect A difficulty in generating a null community when species differ in dispersal ability; the absence of one species from an island may have more to do with its inability to disperse long distances than with any competitive effects.

Identical by descent The property of two alleles that are derived from the same ancestral copy of the gene.

Idiosyncratic hypothesis A hypothesis regarding the effect of species loss on ecosystem function. This hypothesis is a null model in which the relationship between diversity and ecosystem function is indeterminate and ecosystem-specific.

Inclusive fitness The relative ability of an organism to get its genes *or copies of them* into the next generation.

Index of relative abundance An indirect measure of relative population size, using evidence of an animal's presence—calls, feces, tracks, or other indicators.

Ingestion efficiency The fraction of the energy available to a trophic level that is actually ingested.

Inhibition model A model of succession in which the pioneering species inhibit colonization by other species. They exclude other species for their life span.

Innate behavior A behavior that is genetically determined or programmed.

Instinctual behavior See **Innate behavior.**

Integrated pest management (IPM) A program of carefully selected control techniques tailored to address each particular insect pest problem while minimizing the application of chemicals.

Interference competition Competition in which one species actively interferes with the ability of one or more others to use a resource.

Interspecific competition Competition between two or more different species.

Intertidal zone The region between mean high tide and mean low tide.

Intraspecific competition Competition between members of the same species.

Irruption A pattern of population fluctuation in which long periods of low density are interspersed with irregular outbreaks of high density.

Iteroparity More than one reproductive effort in an individual's life span.

J. P. Morgan effect A difficulty in generating a null community that arises when the pool group (that is, the mainland group) of species from which random guilds are generated is too large.

Karyotype The number and morphology of the chromosomes of a species.

Key factor analysis A means of analyzing populations that is based on using the information in a life table to identify the life cycle stages most vulnerable to mortality and thus crucial to achieving high density in an invading species.

Keystone predator A predator that, because of its predation, is central to the organization of the community, often through its effects on competition in the community.

Landscape ecology A branch of ecology that focuses on spatial ecological patterns such as the spatial distribution of populations or communities over the landscape.

Langmuir cells Patterns of circulation caused by winds of speed greater than 3 meters per second. The wind generates circular rotations of the water just below the surface.

Laterite A type of soil common in tropical ecosystems, characterized by advanced weathering, low nutrient content, and high aluminum and iron content.

Lek A communal male display on a small plot of ground to which females are attracted for mating.

Lentic systems Aquatic systems in which the water is not flowing, such as lakes or ponds.

Leslie matrix A method of modeling population growth and change that is based on the transition from one age class to the next, as predicted by age-specific birth and death rates; also called a **transition matrix.**

Life history The pattern of a species's development, growth, life span, and reproduction.

Life table A table of vital statistics for a population, including age-specific mortality and birth rates and life expectancy.

Life zone A plant community type characteristic of a certain elevation in montane regions.

Limiting factor Abiotic parameters, such as salinity, pH, and temperature, whose physical or chemical effects delimit a zone in which an organism can survive.

Limnetic zone The waters of a lake or ocean beneath the littoral zone and above the compensation point.

Lindeman efficiency The percentage of energy present in one trophic level that is transferred to the next level in a food chain.

Littoral zone The region near shore where light can penetrate to the bottom and rooted plants can survive.

Loess Fine-grained mineral particles scoured from rock by a glacier and accumulated as a massive soil deposit elsewhere.

Lotic systems Aquatic systems in which the water flows, such as streams and rivers.

Macronutrients Nutrients required by plants or animals in relatively large amounts.

Mangrove swamp A wetland found on the edge of tropical oceans, characterized by zones of several species of mangroves.

Maritime climate Climate of a region near a large body of water, characterized by small annual temperature fluctuations.

Marshes Aquatic systems whose plant life consists of soft-stemmed herbaceous species.

Mating system The mating interactions of a species, consisting of the number of mates an individual takes, whether the male and female form a pair bond, and the duration of the pair bond.

Matriline A female, her female offspring, and other female relatives (sisters, her mother, and so on).

Mediterranean climate A climate characterized by winter rainfall and summer drought.

Metalimnion A region in a lake where the temperature drops rapidly.

Metamorphosis A developmental pattern in which the organism undergoes a complete change of body form.

Meta-analysis An analytical technique developed to examine interactive effects by quantifying the effects of a number of factors examined in a series of experimental studies.

Metapopulation A "population of populations." A group of discrete populations connected by potential and actual routes of dispersal.

Micorrhizae Symbiotic fungi associated with plant roots that increase the absorption of important nutrients.

Micronutrients Nutrients required by plants or animals in relatively small amounts.

Mimic A species that resembles another as part of a mimicry system.

Mimicry The physical resemblance of two or more species, resulting from inherent advantages of having similar appearance.

Model (in mimicry) The species that a mimicking species resembles.

Monoclimax A concept of the climax in which succession proceeds to a single state characteristic of the biome.

Monogamy A mating system in which a male mates with a single female.

Mullerian mimicry The physical resemblance of various members of a group of noxious or dangerous species. Each species gains advantage from the fact that potential predators avoid their collective appearance.

Mutualism An interaction between two or more species in which all species benefit.

Narcissus effect A difficulty in generating a null community in situations where competition structured the mainland pool in the distant past.

Natural history The descriptive study of the habits, behavior, and interactions of species in their environment.

Natural selection An evolutionary process by which allele frequencies change from one generation to the next to reflect the differential success of different genotypes in surviving and reproducing.

Neoteny A life cycle in which the larvae of some populations or races become sexually mature and no longer metamorphose into adults.

Neritic zone In marine systems, the portion of the pelagic region that lies above the continental shelf.

Net primary production The amount of the gross primary production that remains after all the losses to herbivores, parasites, and respiration.

Net reproductive rate (R_o) The average number of individuals produced by a female over her lifetime.

Niche The ecological role of the species in the community, including the interactions in which it participates.

Nitrification The oxidation of ammonia to form nitrite.

Null hypothesis The hypothesis that an observation is a result of chance.

Numerical response A predator response to increased prey density in which the number of predators increases locally.

Nutrient (element) cycling The movement of materials (nutrients) in a local ecosystem.

Nutrients Compounds or elements required by an organism for its growth, reproduction, and survival.

Oceanic zone The portion of a marine system that lies over the deep ocean, beyond the continental shelf.

Oligotrophic systems Systems in which productivity and nutrient content are low.

Optimal foraging theory A body of ecological theory that relates the actual behavior of a predator as it searches for and consumes its prey to some "optimal" foraging pattern predicted by a model.

Osmosis The movement of water across a semipermeable membrane, from regions of low solute concentration to regions of higher solute concentration.

Panne A small depression in a salt marsh, analogous to a tide pool on a rocky shore.

Parasite An organism that forms an association with one or more individuals of another species (called *hosts*) in order to obtain nutrients or energy from them.

Parasitoid A parasite that specifically limits its parasitic attack to the larval stages of the host.

Pattern climax A concept of the climax in which the multiple climax states are not discrete entities but grade into one another.

Pelagic region The region of a lake or ocean too deep for rooted plants.

Periphyton Aquatic plants attached to the substrate.

Permafrost Permanently frozen soil.

Phenotype The manifestation of a trait in an individual.

Physical resources Abiotic factors that an organism must assimilate if it is to survive.

Plankton Small free-floating or swimming organisms in a lake or ocean.

Polyandry A mating system in which one female mates with more than one male.

Polyclimax A concept of the climax in which the endpoint of succession is a series of local climax states, each of which is determined by local soil (edaphic) and microclimate conditions.

Polygyny A mating system in which one male mates with more than one female.

Population A group of individuals of one species in a specified area.

Population control The ecological mechanisms that limit population density.

Population density The number of individuals per unit area.

Population regulation The processes by which population density is governed by factors internal or external to the population; bounded fluctuations in abundance.

Potential evapotranspiration (PET) The rate of water loss based on the system's abiotic factors such as wind, temperature, and solar radiation.

Prairie potholes Small wetlands and marshes formed from depressions gouged by glaciers in the northern Great Plains.

Predation The capture, killing, and consumption of one animal by another.

Preemptive competition The competition among individual plants for space.

Presaturation dispersal Dispersal that occurs prior to a population's attainment of carrying capacity.

Primary consumer A heterotroph that feeds directly on plants (primary producers).

Primary production The rate of accumulation of energy in organic molecules by photosynthesis.

Primary succession The development of a community from an essentially abiotic setting following a cataclysmic disturbance.

Production efficiency The fraction of assimilated energy that is converted into new tissue (somatic or reproductive).

Profundal zone The region of a lake or ocean below the compensation point.

Proximate explanation The immediate cause of a phenomenon.

Race See **Subspecies.**

Rain shadow Lower precipitation on the lee side of a major mountain range. This happens because, as moist air passes over the mountains, most of the precipitation occurs at high elevations on the windward side.

Ramet The above-ground "individuals" of a plant; the physiological plant unit.

Realized niche The subset of the fundamental niche used when competitors are present.

Reciprocal replacement A phenomenon in succession in which one individual of a particular species is replaced in the climax by another, and vice versa.

Reductionism A scientific approach in which each part of a system is described, studied, and understood in terms of its component parts.

Redundancy hypothesis A hypothesis regarding the effect of the loss of species on ecosystem stability. Some species can be lost with little consequence because other species can expand their roles in the ecosystem to replace the lost functions.

Relative fitness The ability of a genotype to obtain representation in the next generation relative to other genotypes.

Relaxation fauna (or flora) The community that results from fragmentation of habitat and the consequent loss of species diversity.

Reproductive value (V_x) The relative expected genetic contribution of an individual of age x to the next generation.

Residual reproductive value The relative expected genetic contribution of an individual of age x over the remainder of its life.

Resource defense polygyny A polygynous system in which some environmental resource is clumped and males defend the resource. Females are attracted to the territory for its resources, and males gain access to multiple females in this way.

River continuum The concept that stream ecosystem ecology changes markedly over the course of a river's length.

Rivet hypothesis A hypothesis regarding the effect of the loss of species on ecosystem stability. The loss of species is analogous to the loss of rivets from an airplane. There is little effect until many rivets (species) are lost. Then, at some threshold level of loss, there are major, catastrophic effects.

Saturation dispersal Dispersal that occurs when the population is at or near carrying capacity.

Savanna A grassland on which woody plants are evenly and widely distributed.

Search image A pattern of sensory input from the environment that is recognized by a predator as a prey item.

Secondary compounds Chemical compounds (produced by plants) that deter herbivores.

Secondary consumers Heterotrophs that feed on herbivores.

Secondary production The rate of accumulation of biomass by heterotrophs.

Secondary succession The development of a community in a habitat that has been disturbed, but not so severely as to destroy all life.

Self-thinning A phenomenon in plants in which a population gradually decreases in density, but the size of the remaining individuals increases.

Semelparity A single reproductive event in an individual's life span. Also termed *big-bang reproduction.*

Senescence The degenerative changes that result in an increase in expected mortality with age. Eventually the probability of survival reaches zero.

Seral stage A particular stage in succession.

Sere The entire sequence of stages in succession.

Serpentine soil A peculiar soil type characterized by low concentrations of Ca, N, and P; high concentrations of Mg, Ni, Al, and Fe; and low water-holding capacity.

Seston Suspended particulate matter in an aquatic system consisting of living algae, bacteria, protozoa, and abiotic material.

Sexual selection A form of natural selection in which the fitness of one sex (usually males) depends on their attractiveness to the other sex.

Sign stimulus The stimulus that elicits a fixed action pattern.

Soft selection A form of natural selection in which certain genotypes have an advantage because they are able to obtain resources in competition with other members of the population.

Speciation The evolutionary process of species formation.

Species A group of actually or potentially interbreeding individuals, reproductively isolated from other such groups.

Species diversity The number of species and the heterogeneity of their abundances in a community.

Species heterogeneity The degree to which the distribution of the number of individuals among species in a community is uneven.

Species richness The total number of species in a community.

Stabilizing selection A form of natural selection in which intermediate phenotypes have the highest fitness and lead to a narrowing of the phenotypic distribution.

Stable age distribution An unchanging frequency distribution of ages in a population, which results from constant birth and death rates in the population.

Standing crop biomass The fraction of the total biomass that contains living plant tissue.

Static life table A life table derived from a sample of a population taken at one moment in time. The age-specific mortality rate is inferred from the age distribution.

Subspecies A morphologically or genetically distinct group within a species, occupying a distinct geographic range.

Superstimulus An exaggerated sign stimulus that elicits an exaggerated behavioral response.

Survivorship curve A plot of the log number (of a cohort) surviving, as a function of age.

Sympatric In the same place or locality.

Systems ecology A branch of ecology dealing with the mathematical description of interactions among component parts of an ecological system.

Territory An area or region to which an individual has exclusive access as a result of aggression or advertisement.

Tertiary consumers Heterotrophs that feed on secondary consumers in a food chain.

Thermal stratification The phenomenon in a lake in which warm surface water is separated from—and does not mix with—lower layers of cold water.

Thermocline The plane of maximum temperature change within the metalimnion.

Tide pools Depressions in the substrate that trap seawater when the tide retreats.

Time maximizers In optimal foraging theory, a species for which foraging efficiently (so as to maximize the time available for other activities) is crucial.

Tolerance model A model of succession in which species modify the environment in ways that make it difficult for the offspring to replace the adults. Replacement of these species occurs by other species more tolerant of the new conditions.

Top carnivore The highest trophic level in a food chain.

Transition matrix See **Leslie matrix.**

Trophic cascade An indirect effect of one trophic level on either lower levels (top-down cascade) or higher levels (bottom-up cascade).

Trophic level A link in the food chain.

Trophogenic zone The upper layer of a lake, where enough light is present to permit photosynthesis.

Tropholytic zone The region in a lake where light intensity is insufficient to support plants.

Ultimate explanation A general, overarching reason for a phenomenon, often expressed in terms of evolutionary principles.

Upwelling The upward movement of water and nutrients in oceans, often near coasts.

Vagility The relative mobility of a species, determined by its mode of locomotion.

Wetlands Habitats that are perpetually or periodically flooded.

Wildlife management The science of the control and manipulation of wildlife populations to achieve for human objectives.

Literature Cited

Adams, D., J. Gurevitch, and M. S. Rosenberg. 1997. Resampling tests for meta-analysis of ecological data. *Ecology* 78:1277–1283.

Alcock, J. 1989. *Animal behavior.* Sunderland, Mass.: Sinauer.

Alford, R. A., and R. N. Harris. 1988. Effects of larval growth history on anuran metamorphosis. *American Naturalist* 131:91–106.

Alisauska, R. T. 1992. Spring habitat use and diets of midcontinent adult lesser snow geese. *Journal of Wildlife Management* 56:43.

Allee, W. C., A. E. Emerson, O. Park, T. Park, and K. P. Schmidt. 1949. *Principles of animal ecology.* Philadelphia: W. B. Saunders.

Allison, V., D. W. Dunham, and H. H. Harvey. 1992. Low pH alters response to food in the crayfish, *Cambarus bartoni. Canadian Journal of Zoology* 70:2416–2420.

Anderson, D. J. 1982. The home range: A new non-parametric estimation technique. *Ecology* 63:103–112.

Anderson, T. R. 1977. Reproductive responses of sparrows to a superabundant food supply. *Condor* 79:205–208.

Andren, H. 1994. Effects of habitat fragmentation on birds and mammals in landscapes with different proportions of suitable habitat: A review. *Oikos* 71:355–366.

Andrewartha, H. G., and L. C. Birch. 1954. *The distribution and abundance of animals.* Chicago: University of Chicago Press.

Angel, M. V. 1993. Biodiversity of the pelagic ocean. *Conservation Biology* 7:760–772.

Antonovics, J., and A. D. Bradshaw. 1970. Evolution in closely adjacent plant populations, VIII. Clinal patterns at a mine boundary. *Heredity* 25: 349–362.

Aplet, G. H., R. D. Laven, and F. W. Smith. 1988. Patterns of community dynamics in Colorado Engelmann spruce–subalpine fir forests. *Ecology* 69:312–319.

Archambault, D. J., and K. Winterhalder. 1994. Metal tolerance in *Agrostis scabra* from the Sudbury, Ontario area. *Canadian Journal of Botany* 73:766–775.

Armitage, K. B. 1962. Social behavior of a colony of the yellow-bellied marmot (*Marmota flaviventris*). *Animal Behavior* 10:319–331.

Asner, G. P., T. R. Seastedt, and A. R. Townsend. 1997. The decoupling of terrestrial carbon and nitrogen cycles. *BioScience* 47: 226–234.

Avise, J. C. 1993. The evolutionary biology of aging, sexual reproduction and DNA repair. *Evolution* 47:1293–1301.

Ayala, F. J. 1969. Experimental invalidation of the principle of competitive exclusion. *Nature* 224:1076.

Ayala, F. J. 1972. Competition between species. *American Scientist* 60:348–357.

Bach, C. E. 1994. Effects of a specialist herbivore (*Altica subplicata*) on *Salix cordata* and sand dune succession. *Ecological Monographs* 64:423–445.

Baker, R. J., L. W. Robbins, F. B. Stangl, and E. C. Birney. 1983. Chromosomal evidence for a major subdivision in *Peromyscus leucopus. Journal of Mammalogy* 64:356–359.

Balda, R. O. 1980. Recovery of cached seeds by a captive *Nucifraga caryocatactes. Zeitschrift fir Tierpsychologie* 52:331–346.

Baptista, L. F., and L. Petrinovich. 1984. Song development in the white-crowned sparrows: Social factors and sex differences. *Animal Behavior* 32:172–181.

Barkalow, F. S., Jr. 1962. Latitude related to reproduction in the cottontail rabbit. *Journal of Wildlife Management* 26:32–37.

Barnes, B. M. 1989. Freeze avoidance in a mammal: Body temperature below 0°C in an arctic hibernator. *Science* 244:1593–1595.

Bay, J. C. 1894. On compass plants and the twisting of leaves. *Botanical Gazette* 19:251–252.

Bazzaz, F. A. 1979. The physiological ecology of plant succession. *Annual Review of Ecology and Systematics* 10:351–371.

Beard, J. S. 1946. The mora forests of Trinidad, British West Indies. *Journal of Ecology* 33:173–192.

Beaulieu, J., G. Gauthier, and L. Rochefort. 1996. The growth response of graminoid plants to goose grazing in a high arctic environment. *Journal of Ecology* 84:905–914.

Bell, G. 1984. Evolutionary and non-evolutionary theories of senescence. *American Naturalist* 124:600–603.

Belsky, A. J. 1986. Does herbivory benefit plants? *American Naturalist* 127:870–888.

Benzie, J. A. H., and J. A. Stoddart. 1992. Genetic structure of outbreaking and non-outbreaking crown-of-thorns starfish (*Acanthaster planci*) populations on the Great Barrier Reef. *Journal of Marine Biology* 112:119–130.

Berg, E. E., and J. L. Hamrick. 1994. Spatial and genetic structure of two sandhills oaks: *Quercus laevis* and *Quercus magaretta* (Fagacae). *American Journal of Botany* 81:7–14.

Bergerud, A. T. 1967. The distribution and abundance of arctic hares in Newfoundland. *Canadian Field Naturalist* 81:242–248.

Berner, R. A., and S. T. Petsch. 1998. The sulfur cycle and atmospheric oxygen. *Science* 282:1426–1427.

Bertness, M.D., L. Gought, and S. W. Shumway. 1992. Salt tolerances and the distribution of fugitive salt marsh plants. *Ecology* 73:1842–1851.

Billick, I., and T. J. Case. 1994. Higher order interactions in ecological communities: What are they and how can they be detected? *Ecology* 75:1529–1543.

Bjorkman, O., and J. Berry. 1973. High efficiency photosynthesis. *Scientific American* 229:80–93.

Blumstein, D. T., and K. B. Armitage. 1997. Does sociality drive the evolution of communicative complexity? A comparative test with ground-dwelling sciurid alarm calls. *American Naturalist* 150:179–200.

Bohn, H. B., B. McNeal, and G. O'Connor, 1985. *Soil chemistry.* New York: John Wiley & Sons.

Bonnel, M. L., and R. K. Selander. 1974. Elephant seals: Genetic variation and near extinction. *Science* 184:908–909.

Bonner, J. T. 1965. *Size and cycle.* Princeton, N.J.: Princeton University Press.

Bormann, B. T., and R. C. Sidle. 1990. Changes in productivity and distribution of nutrients in a chronosequence at Glacier Bay National Park, Alaska. *Journal of Ecology* 78:561–578.

Bormann, F. H., and G. E. Likens. 1979. *Pattern and process in a forested ecosystem.* New York: Springer-Verlag.

Botkin, D. B. 1990. *Discordant harmonies: A new ecology for the twenty-first century.* New York: Oxford University Press.

Boucher, D. H. 1990. Growing back after hurricanes: Catastrophe may be critical for rain forest dynamics. *BioScience* 40:163–166.

Boutin, S. 1992. Predation and moose population dynamics: A critique. *Journal of Wildlife Management* 56:116–127.

Bowen, B. S. 1982. Temporal dynamics of microgeographic structure of genetic variation in *Microtus californicus. Journal of Mammalogy* 63:625–638.

Bowers, M. A., and J. H. Brown. 1982. Body size and coexistence in desert rodents: Chance or community structure? *Ecology* 63:391–400.

Boyce, M. S., and C. M. Perrins. 1987. Optimizing great tit clutch size in a fluctuating environment. *Ecology* 68:142–157.

Brimley, C. S. 1923. Breeding dates of small mammals at Raleigh, North Carolina. *Journal of Mammalogy* 4:253–264.

Brodie, E. D., III, and E. D. Brodie, Jr. 1990. Tetrodotoxin resistance in garter snakes: An evolutionary response of predators to dangerous prey. *Evolution* 44:651–660.

Brodie, E. D., III, and E. D. Brodie, Jr. 1999. Predator–prey arms races. *BioScience* 49:557–568.

Brower, L. P. 1969. Ecological chemistry. *Scientific American* 220:22–29.

Brown, C. H., and P. M. Waser. 1984. Hearing and communication in blue monkeys (*Cercopithecus mitis*). *Animal Behavior* 32:66–75.

Brown, J. H. 1971. Mammals on mountaintops: Nonequilibrium insular biogeography. *American Naturalist* 105:467–478.

Brown, J. H. 1978. The theory of insular biogeography and the distribution of boreal birds and mammals. *Great Basin Naturalist Memoirs* 2:209–227.

Brown, J. H. 1995. *Macroecology.* Chicago: University of Chicago Press.

Brown, J. H., and D. W. Davidson. 1977. Competition between seed-eating rodents and ants in desert ecosystems. *Science* 196:880–882.

Brown, J. H., and E. J. Heske. 1990. Control of a desert-grassland by a keystone rodent guild. *Science* 250:1705–1707.

Brown, J. H., P. A. Marquet, and M. L. Taper. 1993. Evolution of body size—consequences of an energetic definition of fitness. *American Naturalist* 142:573–584.

Brown, J. H., and B. A. Maurer. 1989. Macroecology: The division of food and space among species on continents. *Science* 243:1145–1150.

Brown, J. H., D. W. Mehlman, and G. C. Stevens. 1995. Spatial variation in abundance. *Ecology* 76:2028–2043.

Brown, J. L. 1964. The evolution of diversity in avian territorial systems. *Wilson Bulletin* 76:160–169.

Brown, L. D. 1964. *Deciduous forests of Eastern North America.* New York: Hafner Publishing Co.

Brown, P. M., M. R. Kaufmann, and W. D. Shepperd. 1999. Long-term landscape patterns of past fire events in a montane ponderosa pine forest of central Colorado. *Landscape Ecology* 14:513–532.

Buchner, P. 1965. *Endosymbiosis of animals with plant microorganisms.* New York: John Wiley & Sons.

Buckley, D. S., T. L. Sharik, and J. G. Isebrands. 1998. Regeneration of northern red oak: Positive and negative effects of competitor removal. *Ecology* 79:65–78.

Cable, D. R. 1975. Influence of precipitation on perennial grass production in the semi-desert Southwest. *Ecology* 56:981–986.

Cain, A. J., and P. M. Sheppard. 1952. The effects of natural selection on body colour in the land snail, *Cepaea nemoralis. Heredity* 6:217.

Cain, A. J., and P. M. Sheppard. 1954. Natural selection in *Cepaea. Genetics* 39:89–116.

Callaway, R. M., and F. W. Davis. 1993. Vegetation dynamics, fire, and the physical environment in coastal central California. *Ecology* 74:1567–1578.

Cameron, G. N. 1973. Effect of litter size on postnatal growth and survival in the desert woodrat. *Journal of Mammalogy* 54:489–493.

Canfield, R. H. 1957. Reproduction and life span of some perennial grasses of southern Arizona. *Journal of Range Management* 10:199–203.

Capone, D. G. 1983. Benthic nitrogen fixation. In *Nitrogen in the marine environment,* ed. E. J. Carpenter and D. G. Capone, pp. 105–137. New York: Academic Press.

Carbyn, L. N. 1987. Gray wolf and red wolf. In *Wild furbearer management and conservation in North America,* ed. M. Novak, J. A. Baker, M. E. Obard, and B. Malloch, pp. 358–377. Toronto, Ontario: Ministry of Natural Resources.

Carpenter, R. C. 1986. Partitioning herbivory and its effects on coral reef algal communities. *Ecological Monographs* 56:345–363.

Carpenter, S. R., and J. F. Kitchell. 1988. The temporal scale of variance in lake productivity. *American Naturalist* 129:417–433.

Casper, B. B., and R. B. Jackson. 1997. Plant competition underground. *Annual Review of Ecological Systems* 28:545–570.

Caughley, G. 1966. Mortality patterns in mammals. *Ecology* 47:906–918.

Chapin, F. S., L. R. Walker, C. L. Fastie, and L. C. Sharman. 1994. Mechanisms of primary succession following deglaciation at Glacier Bay, Alaska. *Ecological Monographs* 64:149–175.

Chapman, J. D., and J. C. Sherman. 1975. *Oxford regional economic atlas: The United States and Canada.* Oxford: Oxford University Press.

Chapman, R. N. 1931. *Animal ecology.* New York: McGraw-Hill.

Chappell, M. A., and G. W. Bartholomew. 1971. Activity and thermoregulation of the antelope ground squirrel, *Ammospermophilus leucurus,* in winter and summer. *Physiological Zoology* 54:215–223.

Charnesky, E. A. 1989. Repeated reversals during spatial competition between corals. *Ecology* 70(4):843–855.

Charnov, E. L. 1976. Optimal foraging: The marginal value theorem. *Theoretical Population Biology* 9:129–136.

Childress, J. J., H. Felbeck, and G. N. Somero. 1981. Symbiosis in the deep sea. *Scientific American* 256:114–120.

Chiras, D. D. 1994. *Environmental science.* Redwood City, Calif.: Benjamin Cummings.

Chitty, D. 1967. The natural selection of self-regulatory behavior in animal populations. *Proceedings of the Ecological Society of Australia* 2:51–78.

Christian, J. J., and D. E. Davis. 1971. Endocrines, behavior and populations. In *Natural regulation of animal populations,* ed. I. A. McLaren, pp. 69–98. New York: Atherton Press.

Clausen, J., D. D. Keck, and W. M. Hiesey. 1948. Experimental studies on the nature of species. III. Environmental responses of climatic races of *Achillea.* Carnegie Institute Publication no. 581. Washington, D.C.

Clegg, M. T., and R. W. Allard. 1972. Patterns of genetic differentiation in the slender wild oat species, *Avena barbata. Proceedings of the National Academy of Science* 69:1820–1825.

Clements, F. E. 1916. Plant succession: Analysis of the development of vegetation. *Carnegie Institute of Washington Publications* 242:1–512.

Clements, F. E. 1936. Nature and structure of the climax. *Journal of Ecology* 24:252–284.

Clutton-Brock, T. H., S. D. Albon, and F. E. Guiness. 1986. Great expectations: Dominance, breeding success and offspring sex ratios in red deer. *Animal Behavior* 34:460–471.

Clutton-Brock, T. H., F. E. Guiness, and S. D. Albon. 1982. *Red deer: Behavior and ecology of two sexes.* Edinburgh: Edinburgh University Press.

Coale, K. H., et al. 1996. A massive phytoplankton bloom induced by an ecosystem-scale iron fertilization experiment in the equatorial Pacific Ocean. *Nature* 383:495–501.

Cohen, J. E., F. Briand, and C. M. Newman. 1986. A stochastic theory of community food webs. II. Predicted and observed lengths of food chains. *Proceedings of the Royal Society of London, Series B* 228:317–328.

Cohen, J. E., F. Briand, and C. M. Newman. 1990. *Community food webs: Data and theory.* New York: Springer-Verlag.

Cole, L. C. 1954. The population consequences of life history phenomena. *Quarterly Review of Biology* 29:103–137.

Coley, P. D., J. P. Bryant, and F. S. Chapin. 1985. Resource availability and plant antiherbivore defense. *Science* 230:895–899.

Collins, S. L., and S. M. Glenn. 1991. Importance of spatial and temporal dynamics in species regional abundance and distribution. *Ecology* 72:654–664.

Collins, S. L., A. K. Knapp, J. M. Briggs, J. M. Blair, and E. M. Steinauer. 1998. Modulation of diversity by grazing and mowing in native tallgrass prairie. *Science* 280:745–747.

Colwell, R. K., and D. W. Winkler. 1984. A null model for null models in biogeography. In *Ecological communities: Conceptual issues and the evidence,* ed. D. R. Strong and D. S. Simberloff, pp. 344–359. Princeton, N.J.: Princeton University Press.

Connell, J. H. 1961. The influence of intraspecific competition and other factors on the distribution of the barnacle *Chthalamus stellatus. Ecology* 42:710–723.

Connell, J. H. 1978. Diversity in tropical rain forests and coral reefs. *Science* 199:1302–1310.

Connell, J. H. 1979. Tropical rain forests and coral reefs as open non-equilibrium systems. In *Population dynamics,* ed. R. M. Anderson, B. D. Turner, and L. R. Taylor, pp. 141–163. Oxford: Blackwell.

Connell, J. H., and M. D. Lowman. 1989. Low-diversity tropical rain forests: Some possible mechanisms for their existence. *American Naturalist* 133:240–257.

Connell, J. H., and R. O. Slatyer. 1977. Mechanisms of succession in natural communities and their role in community stability and organization. *American Naturalist* 111:1119–1144.

Cooke, M. T. 1928. The spread of the European starling in North America (to 1928). Circular. U.S. Department of Agriculture 40:1–9.

Coomes, D. A., and P. J. Grubb. 1998. Responses of juvenile trees to above- and belowground competition in nutrient-starved Amazonian rain forest. *Ecology* 79:768–782.

Cooper, W. S. 1939. A fourth expedition to Glacier Bay, Alaska. *Ecology* 20:130–155.

Cope, E. D. 1896. *Primary factors of organic evolution.* Chicago: Open Court.

Cowles, H. C. 1901. The physiographic ecology of Chicago and vicinity, a study of the origin, development, and classification of plant societies. *Botany Gazette* 31:170–177.

Craig, C. L., and G. D. Barnard. 1990. Insect attraction to ultraviolet-reflecting spider webs and web decorations. *Ecology* 71:616–624.

Craig, C. L., R. S. Weber, and G.D. Bernard. 1996. Evolution of predator–prey systems: Spider foraging plasticity in response to the visual ecology of prey. *American Naturalist* 147:205–229.

Crocker, R. L., and J. Major. 1955. Soil development in relation to vegetation and surface age at Glacier Bay, Alaska. *Journal of Ecology* 43:427–448.

Cronin, E. W., Jr., and P. W. Sherman. 1977. A resource-based mating system: The orange-rumped honey guide. *Living Bird* 15:5–32.

Cronin, G., and M. E. Hay. 1996. Induction of seaweed chemical defenses by amphipod grazing. *Ecology* 77:2287–2301.

Currie, D. J. 1991. Energy and large-scale patterns of animal- and plant-species richness. *American Naturalist* 137:27–50.

Cyr, H., and M. L. Pace. 1993. Magnitude and patterns of herbivory in aquatic and terrestrial ecosystems. *Nature* 361:148–150.

D'Elia, C. F. 1988. The cycling of essential elements in coral reefs. In *Concepts of ecosystem ecology,* ed. L. R. Pomeroy and J. J. Alberts, pp. 195–230. New York: Springer-Verlag.

Dalke, P. D. 1942. The cottontail rabbits in Connecticut. *Connecticut State Geological and Natural History Survey Bulletin* 65:1–97.

Dane, B., and W. G. vander Kloot. 1964. An analysis of the display of the goldeneye duck (*Bucephala clangula* L.). *Behaviour* 22:283–328.

Darling, F. F. 1938. *Bird flocks and the breeding cycle: A contribution to the study of avian sociality.* Cambridge: Cambridge University Press.

Daubenmire, R. F. 1974. *Plants and the environment: A textbook of autecology.* New York: John Wiley & Sons.

Davidson, D. W. 1978. Size variability in the worker caste of a social insect (*Veromesser pergandei* Mayr) as a function of the competitive environment. *American Naturalist* 112:523–532.

Davies, N. B., and M. Brooke. 1991. Coevolution of the cuckoo and its hosts. *Scientific American* (January):92–98.

Dawkins, R., and J. R. Krebs. 1979. Arms races between and within species. *Proceedings of the Royal Society of London, Series B* 205:489–501.

Dayton, P. K. 1998 Reversal of the burden of proof in fisheries management. *Science* 279:821–822.

DeAngelis, D. L., and J. C. Waterhouse. 1987. Equilibrium and nonequilibrium concepts in ecological models. *Ecological Monographs* 57:1–21.

Deevey, E. S., Jr. 1947. Life tables for natural populations of animals. *Quarterly Review of Biology* 22:283–314.

Diamond, J. L. 1969. Comparison of faunal equilibrium turnover rates on the Channel Islands of California. *Proceedings of the National Academy of Science* 67:1715–1721.

Diamond, J. L. 1972. Biogeographic kinetics: Estimation of relaxation times for avifaunas of Southwest Pacific Islands. *Proceedings of the National Academy of Science* 69:3199–3203.

Diamond, J. M. 1975. Assembly of species communities. In *Ecology and evolution of communities,* ed. M. L. Cody and J. M. Diamond, pp. 342–444. Cambridge, Mass.: Belknap Press.

Dixon, A. F. G. 1971. The role of aphids in wood formation. I. The effect of the sycamore aphid *Drepanosiphum plantaoides* (Schr.) (Aphididae) on the growth of sycamore, *Acer pseudoplantanus* (L.). *Journal of Applied Ecology* 8:165–179.

Doak, D. F. 1992. Lifetime impacts of herbivory for a perennial plant. *Ecology* 73:2086–2099.

Dobson, A., and M. Meagher. 1996. The population dynamics of brucellosis in the Yellowstone National Park. *Ecology* 77(4):1026–1036.

Dobzhansky, T. 1970. *Genetics of the evolutionary process.* New York: Columbia University Press.

Dodge, K. L., and P. W. Price. 1991. Eruptive versus noneruptive species—a comparative study of host plant use by a sawfly, *Euura exiguae* and a leaf beetle, *Disonycha pluriligata. Environmental Entomology* 20:1129–1133.

Doherty, P., and T. Fowler. 1994. An empirical test of recruitment limitation in a coral reef fish. *Science* 263:935–939.

Downhower, J. F., and K. B. Armitage. 1971. The yellow-bellied marmot and the evolution of polygyny. *American Naturalist* 105:355–370.

Drake, J. A. 1991. Community assembly mechanics and the structure of an experimental species ensemble. *American Naturalist* 137: 1–26.

Drake, J. A., T. E. Flum, G. J. Witteman, T. Voskul, A. M. Hoylman, C. Creson, D. A. Kenny, G. R. Huxel, C. S. LaRue, and J. R. Duncan. 1993. The construction and assembly of an ecological landscape. *Journal of Animal Ecology* 63:117–130.

Duever, M. J., J. F. Meeder, L. C. Meeder, and J. M. McCollum. 1994. The climate of South Florida and its role in shaping the Everglades ecosystem. In *The Everglades: The ecosystem and its restoration,* ed. S. H. Davis and J. C. Ogden, pp. 225–248. Delray Beach, Fl.: St. Lucie Press.

Duke, N. C., J. A. H. Benzie, J. A. Goodall, and E. R. Ballment. 1998. Genetic structure and evolution of species in the mangrove genus *Avicennia* (Avicenniaceae) in the indo-west Pacific. *Evol.* 5:1612–1626.

Dumbacher, J. P., B. M. Beehler, T. F. Spande, H. M. Garraffo, and J. W. Daly. 1992. Homobatrachotoxin in the genus Pitohiu: Chemical defense in birds? *Science* 258:799–801.

Dyer, L. A., and D. K. Letourneau. 1999. Relative strengths of top-down and bottom-up forces in a tropical forest community. *Oecologia* 119:265–274.

Eanes, W. F., and R. K. Koehn. 1978. An analysis of genetic structure in the monarch butterfly, *Danaus plexippus* L. *Evolution* 32:784–797.

Edwards, P. J. 1989. Insect herbivory and plant defense theory. In *Toward a more exact ecology,* ed. P. J. Grubb and J. B. Whittaker, pp. 275–297. Oxford: Blackwell.

Egerton, F. H. 1973. Changing concepts of the balance of nature. *Quarterly Review of Biology* 48:322–350.

Ehrlich, P. R., and A. H. Ehrlich. 1981. *Extinction: The causes and consequences of the disappearance of species.* New York: Random House.

Ehrlich, P. R., and P. H. Raven. 1964. Butterflies and plants: A study in coevolution. *Evolution* 18:586–608.

Ehrlich, P. R., and E. O. Wilson. 1991. Biodiversity studies: Science and policy. *Science* 253:758–762.

Ehrlinger, J. R., H. A. Mooney, P. W. Rundel, R. D. Evans, B. Palma, and J. Delatorre. 1992. Lack of nitrogen cycling in the Atacama Desert. *Nature* 359:316–319.

Eldridge, N. 1974. Character displacement in evolutionary time. *American Zoologist* 14:1083–1097.

Elton, C. 1927. *Animal ecology.* New York: Macmillan.

Emlen, S. T. 1994. Benefits, constraints and the evolution of the family. *Trends in Ecology and Evolution* 9:282–285.

Emlen, S. T., and L. W. Oring. 1977. Ecology, sexual selection, and the evolution of mating systems. *Science* 197:215–223.

Endler, J. A. 1993. The color of light in forests and its implications. Ecological Monographs 63:1–27.

Ennos, A. R. 1997. Wind as an ecological factor. *Trends in Ecology and Evolution.* 12:108–111.

Enquist, B. J., J. H. Brown, and G. B. West. 1998. Allometric scaling of plant energetics and population density. *Nature* 395:163–165.

Eriksson, O., and B. Bremer. 1993. Genet dynamics of the clonal plant *Rubus saxitilis. Journal of Ecology* 81:533–542.

Erwin, T. L. 1982. Tropical forests: Their richness in Coleoptera and other arthropod species. *Coleoptera Bulletin* 36:74–75.

Erwin, T. L. 1983. Beetles and other arthropods of the tropical forest canopies at Manaus, Brazil, sampled with insecticidal fogging techniques. In *Tropical rainforests: Ecology and management,* ed. S. L. Sutton, T. C. Whitmore, and A. C. Chadwick, pp. 59–75. Oxford: Blackwell.

Erwin, T. L. 1988. The tropical forest canopy: The heart of biotic diversity. In *Biodiversity,* ed. E. O. Wilson, pp. 123–126. Washington, D.C.: National Academy Press.

Etter, R. J. 1988. Asymmetrical developmental plasticity in an intertidal snail. *Evolution* 42:322–334.

Falconer, D. S. 1981. *Introduction to quantitative genetics.* London: Longman.

Falkowski, P. G. 1997. Evolution of the nitrogen cycle and its influence on the biological sequestration of CO_2 in the ocean. *Nature* 387:272–275.

Farrell, T. M. 1991. Models and mechanisms of succession: An example from a rocky intertidal community. *Ecological Monographs* 61:95–113.

Fastie, C. L. 1995. Causes and ecosystem consequences of multiple pathways of primary succession at Glacier Bay, Alaska. *Ecology* 76:1899–1916.

Faulkes, C. G., D. H. Abbott, and A. L. Mellor. 1990. Investigation of genetic diversity in wild colonies of naked mole-rats by DNA fingerprinting. *Journal of Zoology London* 221:87–89.

Fautin, R. W. 1941. Development of nestling yellow-headed blackbirds. *Auk* 58:215–232.

Feeny, P. 1976. Plant apparency and chemical defense. *Recent Advances in Phytochemistry* 10:1–40.

Fenner, F., and F. N. Ratcliffe. 1965. *Myxomatosis.* Cambridge, England: Cambridge University Press.

Finch, C. E. 1990. *Longevity, senescence and the genome.* Chicago: University of Chicago Press.

Findlay, C. S., and F. Cooke. 1982. Synchrony in the lesser snow goose (*Anser caerulescens*). II. The adaptive value of reproductive synchrony. *Evolution* 36:786–799.

Findley, J. S., and H. Black. 1983. Morphological and dietary structuring of a Zambian insectivorous bat community. *Ecology* 64:625–630.

Finlay, F. J., S. C. Maberly, J. I. Cooper. 1997. Microbial diversity and ecosystem function. *Oikos* 80:209–213.

Fisher, R. A. 1930. *The genetical theory of natural selection.* Oxford: Oxford University Press.

Fischer Walter, L. E., D. C. Hartnett, B. A. D. Hetrick, and A. P. Schwab. 1996. Interspecific nutrient transfer in a tallgrass prairie plant community. *American Journal of Botany.* 83:180–184.

Fitter, A. H., and R. K. M. Hay. 1987. *Environmental physiology of plants.* London: Academic Press.

Fogel, R. 1985. Roots as primary producers in below ground ecosystems. In *Ecological interactions in soil,* ed. A. H. Fitter, D. Atkinson, D. Read, and M. B. Usher, pp. 23–35. British Ecological Society Special Publication no. 4. Oxford: Blackwell.

Ford, R. G., and F. A. Pitelka. 1984. Resource limitation in the California vole. *Ecology* 65:122–136.

Foster, S. A. 1999. The geography of behaviour: An evolutionary perspective. *Trends in Ecology and Evolution* 14:190–195.

Foufopoulos, J. and A. R. Ives. 1999. Reptile extinctions on land-bridge islands: Life-history attributes and vulnerability to extinction. *American Naturalist.* 153:1–25.

Fowler, A. C., R. L. Knight, T. L. George, and L. C. McEwen. 1991. Effects of avian predation on grasshopper populations in North Dakota grasslands. *Ecology* 72:1775–1781.

Fox, B. 1981. Niche parameters and species richness. *Ecology* 62:1415–1425.

Francis, W. J. 1970. The influence of weather on population fluctuations in California quail. *Journal of Wildlife Management* 34:249–266.

Frank, D. A., and S. J. McNaughton. 1991. Stability increases with diversity in plant communities—empirical evidence from the 1988 Yellowstone drought. *Oikos* 62:360–362.

Frank, D. A. and S. J. McNaughton. 1992. The ecology of plants, large mammalian herbivores, and drought in Yellowstone National Park. *Ecology* 73:2043–2058.

Frank, D. A., S. J. McNaughton, and B. F. Tracy. 1998. The ecology of the earth's grazing ecosystems. *BioScience* 48:513–521.

Frank, L. G. 1986. Social organization of the spotted hyaena *Crocuta crocuta.* II. Dominance and reproduction. *Animal Behavior* 34:1510–1527.

Frank, L. G. 1991. Fatal sibling aggression, precocial development, and androgens in neonatal spotted hyenas. *Science* 252:704.

Franklin, I. R. 1980. Evolutionary change in small populations. In *Conservation biology: An evolutionary-ecological perspective,* ed. M. E. Soule and B. A. Wilcox, pp. 135–150. Sunderland, Mass.: Sinauer.

Fredga, K. 1988. Aberrant chromosome sex-determining mechanisms in mammals, with special reference to species with XY females. *Proceedings of the Royal Society of London, Series B* 322:83–95.

Fredga, K., A. Gropp, H. Winking, and F. Frank. 1976. Fertile XX- and XY-type females in the wood lemming *Myopus schisticolor. Nature* 261:225–237.

Frisch, O. von. 1967. *The dance language and orientation of bees.* Trans. by L. E. Chadwick. Cambridge, Mass.: Belknap Press.

Futuyma, D. J. 1986. *Evolutionary biology.* Sunderland, Mass.: Sinauer.

Futuyma, D. J., and M. Slatkin (Eds.). 1983. *Coevolution.* Sunderland, Mass.: Sinauer.

Gadgil, M., and O. T. Solbrig. 1972. The concept of "r" and "K" selection: Evidence from wildflowers and some theoretical considerations. *American Naturalist* 106:14–31.

Gaines, M. S., and C. J. Krebs. 1971. Genetic changes in fluctuating vole populations. *Evolution* 24:702–723.

Gardezi, T., and J. Dasilva. 1999. Diversity in relation to body size in mammals: A comparative study. American Naturalist 153:110–123

Gasaway, W. C., R. D. Boertje, D. C. Grangaard, D. G. Kelleyhouse, R. O. Stephenson, and D. G. Larsen. 1992. The role of predation in limiting moose at low densities in Alaska and Yukon and implications for conservation. *Wildlife Monographs* 120:1–59.

Gasaway, W. C., R. O. Stephenson, J. L. Davis, P. E. K. Shepherd, and O. E. Burre. 1983. Interrelationships of wolves, prey and man in interior Alaska. *Wildlife Monographs* 84:1–50.

Gause, G. F. 1934. *The struggle for existence.* New York: Williams & Wilkins.

Geiger, R. 1965. *Die Atmosphare der Erde.* Darmstadt, Germany: Perthes.

Gensler, H. L., and H. Bernstein. 1981. DNA damage as the primary cause of ageing. *Quarterly Review of Biology* 56:279–303.

Gibbs, H. L., and P. R. Grant. 1987. Oscillating selection on Darwin's finches. *Nature* 327:511–513.

Gibson, R. M., and J. W. Bradbury. 1986. Male and female mating strategies on sage grouse leks. In *Ecological aspects of social evolution: Birds and mammals,* ed. D. I. Rubenstein and R. W. Wrangham. Princeton, N.J.: Princeton University Press.

Gilbert, D. A., N. Lehman, S. J. O'Brien, and R. K. Wayne. 1990. Genetic fingerprinting reflects population differentiation in the California Channel Island fox. *Nature* 344:764–766.

Gilbert, F. S. 1980. The equilibrium theory of island biogeography: Fact or fiction? *Journal of Biogeography* 7:209–235.

Giles, R. H., Jr. 1971. *Wildlife management techniques.* Washington, D.C.: The Wildlife Society.

Gill, F. B., and L. L. Wolf. 1975a. Foraging strategies and energetics of East African sunbirds at mistletoe flowers. *American Naturalist* 109:491–510.

Gill, F. B., and L. L. Wolf. 1975b. Economics of feeding territoriality in the golden-winged sunbird. *Ecology* 56:333–345.

Givnish, T. J. 1999. On the causes of gradients in tropical tree diversity. *Journal of Ecology* 87:193–210.

Gleason, H. A. 1926. The individualistic concept of the plant association. *Torrey Botanical Club Bulletin* 53:7–26.

Gleason, H. A. 1939. The individualistic concept of the plant association. *American Midland Naturalist* 21:92–110.

Glynn, P. W., and W. H. de Weerdt. 1991. Elimination of two reef-building hydrocorals following the 1982–83 El Niño warming event. *Science* 253:69–71.

Goldberg, D. E., and T. E. Miller. 1990. Effects of different resource additions on species diversity in an annual plant community. *Ecology* 71:213–226.

Goldman, C. R. 1960. Primary productivity and limiting factors in three lakes of the Alaska Peninsula. *Ecological Monographs* 30:207–230.

Goodman, D. S. 1975. The theory of diversity–stability relationships in ecology. *Quarterly Review of Biology* 50:237–266.

Gordon, M. S., K. Schmidt-Nielsen, and H. M. Kelley. 1961. Osmotic regulation in the crab-eating frog (*Rana cancrivora*). *Journal of Experimental Biology* 42:437–445.

Gosz, J. R., R. T. Holmes, G. E. Likens, and F. H. Bormann. 1978. The flow of energy in a forest ecosystem. *Scientific American* 238:92–103.

Gotelli, N. J., and M. Pyron. 1991. Life history variation in North American freshwater minnows—effects of latitude and phylogeny. *Oikos* 62:30–40.

Goudie, A. 1993. *The human impact on the environment.* Oxford: Blackwell.

Gould, S. J. 1973. Positive allometry of antlers in the "Irish elk," *Megaloceros giganteus. Nature* 244:375–376.

Gould, S. J. 1980. *The panda's thumb.* London: Penguin.

Gould, S. J., and R. C. Lewontin. 1979. The spandrels of San Marco and the Panglossian paradigm: A critique of the adaptationist programme. *Proceedings of the Royal Society of London, Series B* 205:581–598.

Graetz, R. D., R. Fisher, and M. Wilson. 1992. *Looking back: The changing face of the Australian continent, 1972–1992.* Canberra, Australia: CSIRO.

Graham, R. W., E. L. Lundelius, M. A. Graham, E. K. Schroeder, R. S. Toomey III, E. Anderson, A. D. Barnosky, J. A. Burns, C. S. Churcher, D. K. Grayson, R. D. Guthrie, L. R. Harington, G. T. Jefferson, L. D. Martin, H. G. McDonald, R. E. Morlan, H. A. Semken Jr., S. D. Webb, L. Werdelin, and M. C. Wilson. 1996. Spatial response of mammals to late Quaternary environmental fluctuations. *Science* 2722:1601–1606.

Grant, P. R. 1986. *Ecology and evolution of Darwin's finches.* Princeton, N.J.: Pinceton University Press.

Grant, P. R. 2000. What does it mean to be a naturalist at the end of the twentieth century? *American Naturalist* 155:1–12.

Grant, P. R., and B. R. Grant. 1992. Demography and the genetically effective sizes of two populations of Darwin's finches. *Ecology* 73:766–784.

Grassle, J. F. 1991. Deep-sea biodiversity: The ocean bottom supports communities that may be as diverse as those of any habitat on Earth. *BioScience* 41:464–470.

Grassle, J. F., and N. J. Maciolek. 1992. Deep-sea species richness: Regional and local diversity esti-mates from quantitative bottom samples. *American Naturalist* 139:313–341.

Greene, H. W., and R. W. McDiarmid. 1981. Coral snake mimicry: Does it occur? *Science* 213:1207–1211.

Greenwood, P. J. 1983. Mating systems and the evolu-tionary consequences of dispersal. In *The ecology of animal movement,* ed. I. R. Swingland and P. J. Greenwood, pp. 100–131. Oxford: Clarendon Press.

Grinnell, J. 1917. The niche relationships of the California thrasher. *The Auk* 21:364–382.

Grinnell, J., and H. S. Swarth. 1913. An account of the birds and mammals of the San Jacinto area of southern California, with remarks upon the behavior of geographic races on the margins of their habitats. *University of California Publications in Zoology* 10:197–406.

Gross, A. O. 1928. The heath hen. *Memoirs of the Boston Society of Natural History* 6:491–588.

Gross, M. R., R. M. Coleman, and R. M. McDowell. 1988. Aquatic productivity and the evolution of diadromous fish migration. *Science* 239:1291–1293.

Grubb, P. J. 1992. A positive distrust to simplicity— lessons from plant defenses and from competition among plants and among animals. Presidential address to the British Ecological Society, University of Manchester. *Journal of Ecology* 80:585–610.

Gubernick, D. J., and J. R. Alberts. 1987. The biparental care system of the California mouse, *Peromyscus californicus. Journal of Comparative Psychology* 101:169–177.

Gunderson, L. H., S. S. Light, and C. S. Holling. 1995. Lessons from the Everglades. *BioScience Supplement* 66–73.

Hainsworth, F. R. 1995. Optimal body temperatures with shuttling: Desert antelope ground squirrels. *Animal Behavior* 49:107–116.

Hairston, N. G., J. D. Allan, R. K. Colwell, D. J. Futuyma, J. Howell, M. D. Lubin, J. Mathias, and J. H. Vandermeer. 1968. The relationship between species diversity and stability: An experimental approach with protozoa and bacteria. *Ecology* 49:1091–1111.

Hairston, N. G., and G. W. Byers. 1954. The soil arthropods of a field in southern Michigan: A study in community ecology. *Contributions of the Laboratory of Vertebrate Biology, University of Michigan* 64:1–37.

Hairston, N. G., F. E. Smith, and L. B. Slobodkin. 1960. Community structure, population control, and competition. *American Naturalist* 44:421–425.

Hall, E. R., and K. R. Kelson. 1959. *The mammals of North America.* New York: Ronald Press.

Hamilton, W. D. 1964. The genetical evolution of social behavior. *Journal of Theoretical Biology* 7:1–52.

Hamilton, W. D. 1971. Geometry for the selfish herd. *Journal of Theoretical Biology* 31:295–311.

Hamrick, J. L., and R. W. Allard. 1972. Microgeographical variation in allozyme frequen-cies in *Avena barbata. Proceedings of the National Academy of Science* 69:2100–2104.

Hansen, L. P., and G. O. Batzli. 1979. Influence of sup-plemental food on local populations of *Peromyscus leucopus. Journal of Mammalogy* 60:331–342.

Hanski, I. 1982. Dynamics of regional distribution: Core and satellite species hypothesis. *Oikos* 38:210–221.

Hanski, I., T. Pakkala, M. Kuussaari, and G. Lei. 1995. Metapopulation persistence of an endangered butterfly in a fragmented landscape. *Oikos* 72:21–28.

Harcombe, P. A. 1977a. The influence of fertilization on some aspects of succession in a humid tropical forest. *Ecology* 58:1375–1383.

Harcombe, P. A. 1977b. Nutrient accumulation by vegetation during the first year of recovery of a tropical forest ecosystem. In *Recovery and restoration of damaged ecosystems,* ed. J. Cairns, K. Dickson, and E. Herricks, pp. 347–378. Charlottesville: University Press of Virginia.

Harcourt, A. H., P. H. Harvey, S. G. Larson, and R. V. Short. 1981. Testis weight, body weight and breeding system in primates. *Nature* 293:55–57.

Hardy, G. H. 1908. Mendelian proportions in a mixed population. *Science* 28:49–50.

Hardy, S. B. 1979. Infanticide among mammals: A review, classification, and examination of the implications for the reproductive strategies of females. *Ethology and Sociobiology* 1:13–40.

Harris, L. D. 1984. *The fragmented forest.* Chicago: University of Chicago Press.

Harvell, C. D. 1986. The ecology and evolution of inducible defenses in a marine bryozoan: Cues, costs and consequences. *American Naturalist* 128:810–823.

Hasler, J. F., and E. M. Banks. 1975. Reproductive performance and growth in captive collared lemmings (*Dicrostonyx groenlandicus*). *Canadian Journal of Zoology* 53:777–787.

Hassell, M. P., J. H. Lawton, and R. M. May. 1976. Patterns of dynamical behavior in single-species populations. *Journal of Animal Ecology* 45:471–486.

Hastings, J. R., and R. M. Turner. 1965. *The changing mile.* Tucson: University of Arizona Press.

Hatcher, B. G., and A. W. D. Larkin. 1983. An experimental analysis of factors controlling the standing crop of epilithic algal community on a coral reef. *Journal of Experimental Marine Biology and Ecology* 69:61–84.

Hatchwell, B. J. 1999. Investment strategies of breeders in avian cooperative breeding systems. *American Naturalist* 154:205–219.

Hatchwell, B. J. and A. F. Russell. 1996. Provisioning rules in cooperatively breeding long-tailed tits *Aegithalos caudatus:* An experimental study. *Procceedings of the Royal Society of London B. Biological Sciences* 263:83–88.

Haukioja, E., and S. Neuvonen. 1985. Induced long-term resistance of birch foliage against defoliators: Defensive or incidental? *Ecology* 66:1303–1308.

Hauser, M. D. 1996. The evolution of communication. Cambridge, Mass.: MIT Press.

Hay, M. E., W. Fenical, and K. Gustafson. 1987. Chemical defense against diverse coral reef herbivores. *Ecology* 68:1581–1591.

Heard, D. C. 1992. The effect of wolf predation and snow cover on musk-ox group size. *American Naturalist* 139:190–205.

Heard, D. C., and J. P. Ouellet. 1994. Dynamics of an introduced caribou population. *Arctic* 47: 88–95.

Hedin, L. O., J. J. Armesto, and A. H. Johnson. 1995. Patterns of nutrient loss from unpolluted, old-growth temperate forests: Evaluation of biogeochemical theory. *Ecology* 76:493–509.

Hendrix, S. D. 1979. Compensatory reproduction in a biennial herb following insect defloration. *Oecologia* 42:107–118.

Herbold, B., and P. B. Moyle, 1986. Introduced species and vacant niches. *American Naturalist* 128:751–760.

Heschel, M. S., and K. N. Paige. 1994. Inbreeding depression, environmental stress, and population size variation in scarlet gilia (*Ipomopsis aggregata*). *Conservation Biology* 9:126–133.

Heschel, M. S., and K. N. Paige. 1995. Inbreeding depression, environmental stress, and population size in scarlet gilia (*Ipomopsis aggregata*). *Conservation Biology* 9:126–133.

Hickman, J. C. 1968. Disjunction and endemism in the flora of the central western Cascades of Oregon: A historical and ecological approach to plant distributions. Ph.D. dissertation. University of Oregon, Eugene.

Hobbs, N. T., D. S. Schimel, C. E. Owensby, and D. S. Ojima. 1991. Fire and grazing in the tallgrass prairie: Contingent effects on nitrogen budgets. *Ecology* 72:1372–1382.

Hodges, L. 1973. *Environmental pollution.* New York: Holt.

Holling, C. S. 1959. Some characteristics of simple types of predation and parasitism. *Canadian Entomologist* 91:385–398.

Holmes, W. G., and P. W. Sherman. 1982. The ontogeny of kin recognition in two species of ground squirrels. *American Zoologist* 22:491–517.

Honeycutt, R. L. 1992. Naked mole-rats. *American Scientist* 80:43–53.

Horn, H. S. 1971. *The adaptive geometry of trees.* Princeton, N.J.: Princeton University Press.

Horn, H. S. 1975. Markovian properties of forest succession. In *Ecology and evolution of communities,* ed. M. L. Cody and J. M. Diamond. Cambridge, Mass.: Harvard University Press.

Horvitz, C. C., and D. W. Schemske. 1995. Spatio-temporal variation in demographic transitions of a tropical understory herb: Projection matrix analysis. Ecology Monographs 65:155–192.

Howe, H. F. 1977. Sex-ratio adjustment in the common grackle. *Science* 198:744–745.

Huffaker, C. B. 1958. Experimental studies on predation: Dispersion factor and predator–prey oscillations. *Hilgardia* 27:343–383.

Hughes, J. B., G. C. Daily, and P. R. Ehrlich. 1997. Population diversity: Its extent and extinction. *Science* 278:689–692.

Hughes, J. W. 1992. Effect of removal of co-occurring species on distribution and abundance of *Erythronium americanum* (Liliaceae), a spring ephemeral. *American Journal of Botany* 79:1329–1336.

Huisman, J., R. R. Jonker, C. Zonneveld, and F. J. Weissing. 1999. Competition for light between phytoplankton species: Experimental tests of mechanistic theory. *Ecology* 80: 211–222.

Humphrey, S. R. 1985. How species become vulnerable to extinction, and how we can meet the crises. In *Animal extinctions,* ed. R. Hoage. Washington, D.C.: Smithsonian Institution Press.

Humphreys, W. F. 1979. Production and respiration in animal populations. *Journal of Animal Ecology* 48:427–454.

Hunter, M. D. and J. N. McNeil. 1997. Host-plant quality influences diapause and voltinism in a polyphagous insect herbivore. *Ecology* 78: 4:977–986.

Huston, M., and T. Smith. 1987. Plan succession: Life history and competition. *American Naturalist* 130:168–198.

Hutchinson, G. E. 1951. Copepodology for the ornithologist. *Ecology* 32:570–577.

Hutchinson, G. E. 1957. Concluding remarks. *Cold Spring Harbor Symposium on Quantitative Biology* 22:415–427.

Hutchinson, G. E. 1965. *The ecological theater and the evolutionary play.* New Haven, Conn.: Yale University Press.

Hutchinson, G. E. 1978. *An introduction to population ecology.* New Haven, Conn.: Yale University Press.

Ishikawa, M., and L. V. Gusta. 1996. Freezing and heat tolerance of *Opuntia* cacti native to the Canadian prairie provinces. *Canadian Journal of Botany* 74:1890–1895.

Jablonski, D., and J. J. Sepkoski, Jr. 1996. Paleobiology, community ecology, and scales of ecological pattern. *Ecology* 77: 1367–1378.

Jacobsen, K., and K.E. Erikstad. 1995. An experimental study of the costs of reproduction in the kittiwake *Rissa triactyla. Ecology* 76:1636–1642.

Janzen, D. H. 1976. Why bamboos wait so long to flower. *Annual Review of Ecology and Systematics* 7:347–391.

Janzen, D. H. 1979. New horizons in the biology of plant defenses. In *Herbivores: Their interaction with secondary plant metabolites,* ed. G. A. Rosenthal and D. H. Janzen, pp. 331–350. New York: Academic Press.

Janzen, D. H. 1980. When is it coevolution? *Evolution* 34:611–612.

Janzen, F. J. 1994. Vegetational cover predicts the sex ratio of hatchling turtles in natural nests. *Ecology* 75: 1593–1599.

Jarvis, J. U. M., and N. C. Bennett. 1993. Eusociality has evolved independently in two genera of bathyergid mole-rats but occurs in no other subterranean mammal. *Behavioral Ecology and Sociobiology* 33:253–260.

Jeffreys, A. J. 1987. Highly variable minisatellites and DNA fingerprints. *Biochemistry Society Transactions* 15:309–317.

Jeffreys, A. J., V. Wilson, and S. L. Thein. 1985. Hypervariable "minisatellite" regions in human DNA. *Nature* 314:67–73.

Jenkins, D. A., A. Watson, and G. R. Miller. 1963. Population studies on red grouse, *Lagopus lagopus sciticus* (Lath.), in northeast Scotland. *Journal of Animal Ecology* 32:317–376.

Jenkins, D. A., A. Watson, and G. R. Miller. 1964. Current research on red grouse in Scotland. *Scottish Birds* 3:3–13.

Jenny, H. 1980. *The soil resource: Origin and behavior.* New York: Springer-Verlag.

Jerison, H. J. 1955. Brain to body ratios and the evolution of intelligence. *Science* 121: 447–449.

Johnson, E. A. 1992. *Fire and vegetation dynamics: Studies from the North American boreal forest.* Cambridge: Cambridge University Press.

Johnson, N. K. 1975. Controls of number of bird species on montane islands in the Great Basin. *Evolution* 29:545–574.

Jones, J. S., J. A. Coyne, and L. Partridge. 1987. Estimation of the thermal niche of *Drosophila melanogaster* using a temperature-sensitive mutation. *American Naturalist* 130:83–90.

Jones, W. T., P. M. Waser, L. F. Elliott, and N. E. Link. 1988. Philopatry, dispersal, and habitat saturation in the banner-tailed kangaroo rat, *Dipodomys spectabilis. Ecology* 69:1466–1473.

Jordan, C. F. 1985. *Nutrient cycling in tropical forest ecosystems.* New York: John Wiley & Sons.

Jordan, C. F., and R. Herrera. 1981. Tropical rain forests: Are nutrients really critical? *American Naturalist* 117:167–180.

Jules, E. S. 1998. Habitat fragmentation and demographic changes for a common plant: Trillium in old-growth forest. *Ecology* 79:1645–1656.

Jumpponen, A., H. Vare, K. G. Mattson, R. Ohtonen, and J. M. Trappe. 1999. Characterization of "safe sites" for pioneers in primary succession on recently deglaciated terrain. *Journal of Ecology* 87:98–105.

Kalela, O., and T. Oksala. 1966. Sex ratio in the wood lemming, *Myopus schisticolor* (Lilljeb.), in nature and in captivity. *Annales Universitatis Turkuensis Series A, II, Biologica–Geographica–Geologica* 37:5–24.

Kaplan, R. H., and S. N. Salthe. 1979. The allometry of reproduction: An empirical view in salamanders. *American Naturalist* 113:671–689.

Kareiva, P. 1994. Higher order interactions as a foil to reductionist ecology. *Ecology* 75:1527–1528.

Kaufman, D. G., and C. M. Franz. 1993. *Biosphere 2000.* New York: HarperCollins.

Keeler, K. H. 1981. A model of selection for facultative nonsymbiotic mutualism. *American Naturalist* 118:488–498.

Keever, C. 1950. Causes of succession on old fields of the Piedmont, North Carolina. *Ecological Monographs* 20:231–250.

Keith, L. B. 1963. *Wildlife's ten-year cycle.* Madison: University of Wisconsin Press.

Keith, L. B. 1974. Some features of population dynamics in mammals. *Proceedings of the International Congress of Game Biologists* 11:17–58.

Keith, L. B. 1983. Role of food in hare population cycles. *Oikos* 40:385–395.

Keith, L. B. 1987. Dynamics of snowshoe hare populations. *Current Mammalogy* (Plenum, New York) 2:119–195.

Kendrick, W. B., and A. Burgess. 1962. Biological aspects of the decay of *Pinus sylvestris* leaf litter. *Nova Hedwiga* 4:313–342.

Kenyon, C., J. Chang, E. Gensch, A. Rudner, and R. Tabtiang. 1993. A *C. elegans* mutant that lives twice as long as the wild type. *Nature* 366:461–464.

Kerr, J. T., and L. Packer. 1997. Habitat heterogeneity as a determinant of mammal species richness in high-energy regions. *Nature* 385: 252–254.

Kirk, K. L., and J. J. Gilbert. 1992. Variation on herbivore response to chemical defenses: Zooplankton foraging on toxic cyanobacteria. *Ecology* 73:2208–2217.

Kitching, R. L., and S. L. Pimm. 1986. The length of food chains: Phytotelmata in Australia and elsewhere. *Proceedings of the Ecological Society of Australia* 14:123–139.

Kleiman, D. G. 1980. The sociobiology of captive propagation. In *Conservation biology,* ed. M. E. Soule and B. A. Wilcox, pp. 243–262. Sunderland, Mass.: Sinauer.

Kobe, R. K. 1999. Light gradient partitioning among tropical tree species through differential seedling mortality and growth. *Ecology* 80:187–201.

Koenig, W. D., and R. L. Mumme. 1987. *Population ecology of the cooperatively breeding acorn woodpecker.* Princeton, N.J.: Princeton University Press.

Komdeur, J. 1992. Importance of habitat saturation and territory quality for evolution of cooperative breeding in the Seychelles warbler. *Nature* 358:493–495.

Knapp, A. K., J. M. Briggs, D. C. Harnett, and S. L. Collins. 1998. *Grassland dynamics: Long-term ecological research in tallgrass prairie.* LTER Network Series. New York: Oxford University Press.

Knapp, A. K., J. M. Blair, J. M. Briggs, S. L. Collins, D. C. Harnett, L. C. Johnson, and E. G. Towne. 1999. The keystone role of bison in North American tallgrass prairie. *BioScience* 49: 39–50.

Kohn, D. D., and D. M. Walsh. 1994. Plant species richness—the effect of island size and habitat diversity. *Journal of Ecology* 82:367–377.

Konishi, M. 1965. The role of auditory feed-back in the control of vocalization in the white-crowned sparrow. *Zeitschrift fir Tierpsychologie* 22:770–783.

Kraak, W. K., G. L. Rinkel, and J. Hoogenheide. 1940. Oecologishce bewerking van de Europese ringgegevens van der Kievit (*Vanellus vanellus* L.). *Ardea* 29:151–157.

Krebs, C. J. 1970. Genetic and behavioral studies on fluctuating vole populations. *Proceedings of the Advanced Institute on the Dynamics of Numbers in Populations,* September 7–18 (Oosterbeek, Netherlands), 243–256.

Krebs, C. J. 1979. A review of the Chitty hypothesis of population regulation. *Canadian Journal of Zoology* 56:2463–2480.

Krebs, C. J. 1992. Population regulation revisited. *Ecology* 73:714–715.

Krebs, C. J., and K. T. DeLong. 1965. A *Microtus* population with supplemental food. *Journal of Mammalogy* 46:566–573.

Krebs, C. J., M. S. Gaines, B. L. Keller, J. H. Myers, and R. H. Tamarin. 1973. Population cycles in small rodents. *Science* 179:35–41.

Krebs, J. R., R. Ashcroft, and M. Webber. 1978. Song repertoires and territory defense in the great tit. *Nature* 271:539–542.

Krohne, D. T., and R. Baccus. 1985. Genetic and ecological structure of a population of *Peromyscus leucopus. Journal of Mammalogy* 66:529–537.

Krohne, D. T., J. T. Couillard, and J. R. Riddle. 1991. Population responses of *Peromyscus leucopus* and *Blarina brevicauda* to emergence of periodic cicadas. *American Midland Naturalist* 126:317–321.

Krohne, D. T., and G. A. Hoch. 1999. Demography of populations of *Peromyscus leucopus* on habitat patches: The role of dispersal. *Canadian Journal of Zoology* 77:1247–1253.

Kruk, H. 1972. *The spotted hyena: A study of predation and social behavior.* Chicago: University of Chicago Press.

Krummel, J. R., R. H. Gardner, G. Sugihara, R. V. O'Neill, and P. R. Coleman. 1987. Landscape patterns in a disturbed environment. *Oikos* 48:321–324.

Kuenzler, E. J. 1961. Phosphorus budget of a mussel population. *Limnology and Oceanography* 6:400–415.

Kukal, O., M. P. Ayres, and J. M. Scriber. 1991. Cold tolerance of the pupae in relation to the distribution of swallowtail butterflies. *Canadian Journal of Zoology* 69:3028–3037.

Kurten, B. 1953. On the variation and population dynamics of fossil and recent mammal populations. *Acta Zoologica Fennica.* 76:1–122.

Lack, D. 1943a. *The life of the robin.* London: H. F. & G. Witherby.

Lack, D. 1943b. The age of the blackbird. *British Birds* 36:166–175.

Lack, D. 1943c. The age of some more British birds. *British Birds* 36:193–221.

Lack, D. 1947. The significance of clutch size. *Ibis* 89:302–352.

Lack, D. 1948a. Notes on the ecology of the robin. *Ibis* 90:252–279.

Lack, D. 1948b. The significance of litter size. *Journal of Animal Ecology* 17:45–50.

Lack, D. 1954. *The natural regulation of animal numbers.* Oxford: Clarendon Press.

Lack, D. 1956. Variations in the reproductive rate of birds. *Proceedings of the Royal Society Series B, Biological Sciences* 145:329–333.

Lack, D. 1966. *Population studies of birds.* Oxford: Clarendon Press.

Lack, D. 1968. *Ecological adaptations for breeding in birds.* London: Methuen.

Lahaye, W. S., R. J. Guitierrez, and H. R. Akcakaya. 1994. Spotted owl metapopulation dynamics in Southern California. *Journal of Animal Ecology* 63:775–785.

Lande, R. 1988. Genetics and demography in biological conservation. *Science* 241:1455–1460.

Lande, R. 1993. Risks of population extinction from demographic and environmental stochasticity and random catastrophes. *American Naturalist* 142:911–927.

Lande, R. 1994. Risk of population extinction from fixation of new deleterious mutations. *Evolution* 48:1460–1469.

Lande, R., and G. F. Barrowclough. 1987. Effective population size, genetic variation, and their use in population management. In *Viable populations for conservation,* ed. M. Soule, pp. 87–123. New York: Cambridge University Press.

Larcher, W. 1980. *Physiological plant ecology.* Berlin: Springer-Verlag.

Larsson, P., J. E. Brittain, L. Lein, A. Lillehammer, and K. Tangen. 1979. The lake ecosystem of Ovre Heimdalsvatn. *Holarctic Ecology* 1:304–320.

Law, R. 1975. Colonisation and the evolution of life histories in *Poa annua.* Unpublished Ph.D. thesis. University of Liverpool.

Lawrence, D. B. 1958. Glaciers and vegetation in southeastern Alaska. *American Scientist* 46:89–122.

Lawton, J. H. 1992. There are not 10 million kinds of population dynamics. *Oikos* 63:337–338.

Lawton, J. H. 1994. What do species do in ecosystems? *Oikos* 71:367–374.

Lawton, J. H. 1999. Are there general laws in ecology? *Oikos* 84:177–192.

Leach, M. K., and T. J. Givnish. 1996. Ecological determinants of species loss in remnant prairies. *Science* 273:1555–1558.

Leberg, P. L. 1991. Influence of fragmentation and bottlenecks on genetic divergence of wild turkey populations. *Conservation Biology* 5:522–530.

LeBoeuf, B. J. 1972. Sexual behavior in the northern elephant seal *Mirounga angustirostris. Behavior* 41:1–26.

LeBoeuf, B. J. 1974. Male–male competition and reproductive success in elephant seals. *American Zoologist* 14:163–176.

Ledig, F. T., M. T. Conkle, B. Bermejovelazquez, T. Eguiluzpiedra, P. D. Hodgskiss, D. R. Johnson, and W. S. Dvorak. 1999. Evidence for an extreme bottleneck in a rare Mexican pinyon: Genetic diversity, disequilibrium, and the mating system in *Pinus maximartinezii. Evolution* 53: 91–99.

Lee, A., and R. Martin. 1990. Life in the slow lane: For Australia's koalas surviving on eucalyptus leaves means taking things easy. *Natural History* August, pp. 34–43.

Lehman, N., and R. K. Wayne. 1991. Analysis of coyote mitochondrial DNA genotype frequencies: Estimation of the effective number of alleles. *Genetics* 128:405–416.

Leopold, A. 1933. *Game management.* New York: Scribner.

Lescop-Sinclair, K., and S. Payette. 1995. Recent advances of the arctic treelid along the eastern coast of Hudson Bay. *Journal of Ecology* 83:929–936.

Leslie, P. H. 1945. On the use of matrices in certain population mathematics. *Biometrika* 33:183–212.

Leverich, W. J., and D. A. Levin. 1979. Age-specific survivorship and reproduction in *Phlox drummondi. American Naturalist* 113:881–893.

Levin, D. A., and H. W. Kerster. 1969. Density-dependent gene dispersal in *Liatris. American Naturalist* 103:61–74.

Levins, R. 1968. *Evolution in changing environments: Some theoretical explorations.* Princeton, N.J.: Princeton University Press.

Levins, R. 1971. *Evolution in changing environments.* Princeton, N.J.: Princeton University Press.

Levins, S. A. (ed.) 1975. *Ecosystem analyses and prediction.* Society for Industrial Applications of Applied Mathematics. New York.

Lewontin, R. C. 1965. Selection for colonizing ability. In *The genetics of colonizing species,* ed. H. G. Baker and G. L. Stebbins. New York: Academic Press.

Li, X.D., S. D. Wilson. 1998. Facilitation among woody plants establishing in an old field. *Ecology* 79:2694–2705.

Licht, L. E. 1992. The effect of food level on growth rate and frequency of metamorphosis and paedomorphosis in *Ambystoma gracile. Canadian Journal of Zoology* 70:87–93.

Lichter, J. 1998. Primary succession and forest development on coastal Lake Michigan sand dunes. *Ecological Monographs* 68:487–510.

Lidicker, W. Z., Jr. 1973. Regulation of numbers in an island population of the California vole, a problem in community dynamics. *Ecological Monographs* 43:271–302.

Lidicker, W. Z., Jr. 1975. The role of dispersal in the demography of small mammals. In *Small mammals: Their productivity and population dynamics,* ed. F. B. Golley, K. Petrusewicz, and L. Ryskowski, pp. 101–128. Cambridge: Cambridge University Press.

Lidicker, W. Z., Jr. 1985. Dispersal. In *The biology of New World Microtus,* ed. R. H. Tamarin, pp. 420–454. American Society of Mammalogists Special Publication no. 8.

Lidicker, W. Z., Jr. 1986. An overview of dispersal in non-volant small mammals. In *Migration mechanisms and adaptive significance,* ed. M. A. Rankin, pp. 359–375. Contributions in Marine Science (University of Texas) 27.

Lidicker, W. Z., Jr., and P. K. Anderson. 1962. Colonization of an island by Microtus californicus, analyzed on the basis of runway transects. *Journal of Animal Ecology* 31:503–517.

Likens, G. E., F. H. Bormann, R. S. Pierce, J. S. Eaton, and N. M. Johnson. 1977. *Biogeochemistry of a forested ecosystem.* New York: Springer-Verlag.

Lincoln, R. J., G. A. Boxshall, and P. F. Clark. 1982. *A dictionary of ecology, evolution and systematics.* Cambridge: Cambridge University Press.

Linström, J., E. Ranta, and H. Lindén. 1996. Large-scale synchrony in the dynamics of capercaillie, black grouse and hazel grouse populations in Finland. *Oikos* 76:221–227.

Livdahl, T. P., and M. S. Willey. 1991. Prospects for an invasion: Competition between *Aedes albopictus* and *A. triseriatus. Science* 253:189–191.

Lloyd, A. H., and L. J. Graumlich. 1997. Holocene dynamics of treeline forests in the Sierra Nevada. *Ecology* 78:1199–1210.

Lloyd, J. E. 1966. Studies on the flash communication system in *Photinus* fireflies. *Miscellaneous Publications of the Museum of Zoology, University of Michigan.* 130.

Lockwood, J. L., R. D. Powell, M. P. Nott and S. L. Pimm. 1997. Assembling ecological communities in time and space. *Oikos* 80:549–553.

Lomolino, M. V., J. H. Brown, and R. Davis. 1989. Island biogeography of montane forest mammals in the American Southwest. *Ecology* 70(1):180–194.

Lonsdale, W. M., and A. R. Watkinson. 1983. Light and self-thinning. *New Phytologist* 90:399–418.

Lord, R. D., Jr. 1960. Litter size and latitude in North American mammals. *American Midland Naturalist* 64:488–499.

Lotka, A. J. 1922. The stability of the normal age distribution. *Proceedings of the National Academy of Science* 8:339–345.

Lotka, A. J. 1925. *The elements of physical biology.* Baltimore: Williams & Wilkins.

Lovatt, L. 1911. Moor management. In *The grouse in health and in disease,* ed. L. Lovatt, pp. 372–391. London.

Lovegrove, B. G. 1991. The evolution of eusociality in mole-rats (*Bathyergidae*): A question of numbers and cost. *Behavioral Ecology and Sociobiology* 28:37–46.

Lovejoy, T. E., R. O. Bierregaard Jr., A. B. Rylands, J. R. Malcolm, C. E. Quintela, L. H. Harper, K. S. Brown Jr., A. H. Powell, G. V. N. Powell, H. O. R. Shubart, and M. B. Hays. 1986. Edge and other effects of isolation on Amazon forest fragments. In *Conservation biology: The science of scarcity and diversity,* ed. M. E. Soule, pp. 257–285. Sunderland, Mass.: Sinauer.

Lowe, C. E. 1958. Ecology of the swamp rabbit in Georgia. *Journal of Mammalogy* 39:116–127.

Lowe-McConnell, R. H. 1987. *Ecological studies in tropical fish communities.* Cambridge: Cambridge University Press.

Lubchenko, J. 1983. *Littorina* and *Fucus:* Effects of herbivores, substratum heterogeneity and plant escapes during succession. *Ecology* 64:1116–1123.

Lynch, M., and R. Lande. 1993. Evolution and extinction in response to environmental change. In *Biotic interactions and global change,* ed. P. M. Kareiva, J. G. Kingsolver, and R. B. Huey, pp. 234–250. Sunderland, Mass.: Sinauer.

MacArthur, R. H. 1955. Fluctuations of animal populations, and a measure of community stability. *Ecology* 36:533–536.

MacArthur, R. H. 1958. Population ecology of some warblers of northeastern coniferous forests. *Ecology* 39:599–619.

MacArthur, R. H. 1968. The theory of the niche. In *Population biology and evolution,* ed. R. C. Lewontin. Syracuse, N.Y.: Syracuse University Press.

MacArthur, R. H., and E. R. Pianka. 1966. On optimal use of a patchy environment. *American Naturalist* 100:603–609.

MacArthur, R. H., and E. O. Wilson. 1963. An equilibrium theory of insular zoogeography. *Evolution* 17:373–387.

MacArthur, R. H., and E. O. Wilson. 1967. *The theory of island biogeography.* Monographs in Population Biology No. 1. Princeton, N.J.: Princeton University Press.

MacDonald, G. M., T. W. D. Edwards, K. A. Moser, and R. Pienitz. 1993. Rapid response of treeline vegetation and lakes to past climate warming. *Nature* 361:243–246.

MacMahon, J. A. 1985. Deserts. New York: Knopf.

Magnus, D. B. E. 1958. Experimental analysis of some "overoptimal" sign stimuli in the mating behavior of the fritillary butterfly, *Argynnis paphia* L. (*Lepidoptera nymphalidae*). *Proceedings of the 10th International Congress of Entomology, Montreal* 2:405–418.

Majors, N. 1955. Population and life history of the cottontail rabbit in Lee and Tallapoosa Counties, Alabama. Unpublished master's thesis, Alabama Polytechnic Institute, Auburn, Alabama.

Malthus, T. R. 1826. *An essay on the principles of population,* 6th ed. London.

Mandlebrot, B. B. 1983. The Fractal Geometry of Nature. W. H. Freeman. San Francisco.

Marler, P. 1970. Bird song and speech development: Could there be parallels? *American Scientist* 58:669–673.

Marler, P., and M. Tamura. 1964. Culturally transmitted patterns of vocal behavior in sparrows. *Science* 146:1483–1486.

Marshall, D. R., and S. K. Jain. 1969. Interferences in pure and mixed populations of *Avena fatua* and *A. barbata. Journal of Ecology* 57:251–270.

Martin, A., and C. Simon. 1990. Differing levels of among-population divergence in the mtDNA of periodical cicadas related to historical biogeography. *Evolution* 44:1066–1080.

Martin, M. M., and J. Harding. 1981. Evidence for the evolution of competition between two species of annual plants. *Evolution* 35(5):975–987.

Martin, T. E. 1996. Fitness costs of resource overlap among coexisting bird species. *Nature* 380:338–340.

Maschinski, J., and T. G. Whitham. 1989. The continuum of plant responses to herbivory: The influence of plant association, nutrient availability and timing. *American Naturalist* 134:1–19.

Massey, A. B. 1925. Antagonism of the walnuts (*Juglans nigra* L. and *J. cinerea* L.) in certain plant associations. *Phytopathology* 15:773–784.

May, R. M. 1971. Stability in multi-species community models. *Mathematical Bioscience* 12:59–79.

May, R. M. 1972. Will a large complex system be stable? *Nature* 238:413–414.

May, R. M. 1973. *Stability and complexity in model ecosystems.* Princeton U. Press, Princeton, NJ.

May, R. M. 1974. Biological populations with nonoverlapping generations: Stable points, stable cycles and chaos. *Science* 186:645–647.

May, R. M. 1979a. Arctic animals and climatic changes. *Nature* 282:177–178.

May, R. M. 1979b. The structure and dynamics of ecological communities. In *Population dynamics,* ed. R. M. Anderson, B. D. Turner, and L. R. Taylor, pp. 385–407. Oxford: Blackwell.

May, R. M. 1988. How many species are there on Earth? *Science* 24:1441–1449.

Maynard-Smith, J. 1982. *Evolution and the theory of games.* Cambridge: Cambridge University Press.

Maynard-Smith, J. 1989. *Evolutionary genetics.* Oxford: Oxford University Press.

Maynard-Smith, J., and G. R. Price. 1973. The logic of animal conflict. *Nature* 246:15–18.

McCabe, T. T., and B. D. Blanchard. 1950. *Three species of Peromyscus.* Santa Barbara, Calif.: Rood.

McCann, K., A. Hastings and G. R. Huxel. 1998. Weak trophic interactions and the balance of nature. *Nature* 395:794–795.

McCormick, P. V., and R. J. Stephenson. 1991. Mechanisms of benthic algal succession in lotic environments. *Ecology* 72:1835–1848.

McKnight, D. M., D. K. Niyogi, A. S. Alger, A. Bomblies, P. A. Conovitz, and C. M. Tate. 1999. Dry valley streams in Antarctica: Ecosystems waiting for water. *BioScience* 49:995.

McNab, B. K. 1963. Bioenergetics and the determination of home range size. *American Naturalist* 97:133–140.

McNab, B. K. 1971. On the ecological significance of Bergmann's rule. *Ecology* 52(5):845–854.

McNaughton, S. J. 1976. Serengeti migratory wildebeest: Facilitation of energy flow by grazing. *Science* 191:92–94.

McNaughton, S. J. 1977. Diversity and stability of ecological communities: A comment on the role of empiricism in ecology. *American Naturalist* 111:515–525.

McNaughton, S. J. 1993. Biodiversity and function of grazing ecosystems. In *Biodiversity and ecosystem function,* ed. E. D. Schulze and H. A. Mooney, pp. 361–383. New York: Springer-Verlag.

McNaughton, S. J., R. W. Ruess, and S. W. Seagle. 1988. Large mammals and process dynamics in African ecosystems. *BioScience* 38:794–800.

McNaughton, S. J., J. L. Torrents, M. M. McNaughton, and R. H. Davis. 1985. Silica as a defense against herbivory and a growth promoter in African grasses. *Ecology* 66:528–535.

McQueen, D. J., J. R. Post, and E. L. Mills. 1986. Trophic relationships in freshwater pelagic ecosystems. *Canadian Journal of Fish and Aquatic Science* 43:1571–1581.

Mech, L. D. 1966. The wolves of Isle Royale. Fauna of the national parks of the United States. Fauna Series 7.

Mech, L. D. 1970. *The wolf: The ecology and behavior of an endangered species.* Minneapolis, Minn.: University of Minnesota Press.

Mehrhoff, L. A., and R. Turkington. 1990. Microevolution and site-specific outcomes of competition among pasture plants. *Journal of Ecology* 78:745–756.

Méndez, M., P. S. Karlsson. 1999. Costs and benefits of carnivory in plants: Insights from the photosynthetic performance of four carnivorous plants in a subarctic environment. *Oikos* 86:105–112.

Mennema, J. A., J. Quene-Bogerenbrood, and C. L. Plate. 1985. *Atlas van de Nederlands flora, deel 2. Bohn.* Utrecht, The Netherlands: Schelgema en Holkema.

Merriam, C. H. 1898. *Life zones and crop zones.* U.S. Department of Agriculture, Division of Biology, Survey Bulletin no. 10.

Messier, F., and M. Crete. 1984. Body condition and population regulation by food resources in moose. *Oecologia* 65:44–50.

Miller, R. S. 1964a. Ecology and distribution of pocket gophers (Geomyidae) in Colorado. *Ecology* 45:256–272.

Miller, R. S. 1964b. Larval competition in *Drosophila melanogaster* and *D. simulans. Ecology* 45:132–148.

Minchella, D. J., and M. E. Scott. 1991. Parasitism: A cryptic determinant of animal community structure. *Trends in Ecology and Evolution* 6:250–254.

Monson, G. 1943. Food habits of the banner-tailed kangaroo rat in Arizona. *Journal of Wildlife Management* 7:98–102.

Moran, M. D. and L. E. Hurd. 1998. A trophic cascade in a diverse arthropod community caused by a generalist arthropod predator. *Oecologia* 113:126–132

Moran, N. A., and T. G. Whitham. 1990. Interspecific competition between root-feeding aphids mediated by host-plant resistance. *Ecology* 71(3):1050–1058.

Morris, R. F. 1957. The interpretation of mortality data in studies on population dynamics. *Canadian Entomology* 89:49–69.

Morris, W. F., and D. M. Wood. 1989. Ecological constraints to seedling establishment on the Pumice Plains, Mount St. Helens, Washington. *Ecology* 70:697–704.

Mosby, H. S. 1949. The present status and future outlook of the eastern and Florida wild turkey. *Transactions of the North American Wildlife Conference* 12:346–358.

Moss, B. 1980. *Ecology of freshwaters.* New York: John Wiley & Sons.

Moulton, M. P., and J. L. Lockwood. 1992. Morphological dispersion of introduced Hawaiian finches—evidence for competition and a narcissus effect. *Journal of Evolutionary Ecology* 6:45–55.

Muller, C. H. 1970. Phytotoxins as plant habitat variables. *Recent Advances in Phytochemistry* 3:105–121.

Mumme, R. L., W. D. Koenig, and F. A. Pitelka. 1988. Costs and benefits of joint nesting in the acorn woodpecker. *American Naturalist* 131:654–677.

Mumme, R. L., W. D. Koenig, R. M. Zink, and J. A. Marten. 1985. Genetic variation and parentage in populations of acorn woodpeckers. *The Auk* 102:305–312.

Murdoch, R. 1994. Population regulation in theory and practice. *Ecology* 75:271–287.

Murdoch, W. W., J. Chesson, and P. L. Chesson. 1985. Biological control in theory and practice. *American Naturalist* 129:263–282.

Murie, A. 1944. The wolves of Mt. McKinley. Fauna of the National Parks of the U.S., Fauna Series No. 5. U.S. Department of Interior, National Park Service.

Murphy, G. I. 1968. Patterns in life history and the environment. *American Naturalist* 102:390–404.

Murray, B. G. 1999. Can the population regulation controversy be buried and forgotten? *Oikos* 84:148–152

Myers, J. H. 1993. Population outbreaks in forest lepidoptera. *American Scientist* 81:241–251.

Myers, N., R. A. Mittermeir, C. G. Mittermeir, G. A. B. da Fonseca, and J. Kent. 2000. Biodiversity hotspots for conservation priorities. *Nature* 403:853–858.

Nei, M. 1972. Genetic distance between populations. *American Naturalist* 106:283–292.

Nei, M., and K. Roychoudhury. 1982. Genetic relationship and evolution of human races. *Evolutionary Biology* 14:1–59.

Nelson, B. W., V. Kapos, and J. B. Adams. 1994. Forest disturbance by large blowdowns in the Brazilian Amazon. *Ecology* 75:853–858.

Nevo, E. 1978. Genetic variation in natural populations: Patterns and theory. *Theoretical Population Biology* 13:121–177.

Newman, D., and K. Pilson. 1997. Increased probability of extinction due to decreased genetic effective population size: Experimental populations of *Clarkia pulchella.* *Evolution* 51:354–362.

Newton, I., and P. Rothery. 1997. Senescence and reproductive value in sparrowhawks. *Ecology* 78:1000–1008.

Nicholson, A. J. 1933. The balance of animal populations. *Journal of Animal Ecology* 2:131–178.

Nicholson, A. J. 1954. Compensatory reactions of populations to stress and their evolutionary significance. *Australian Journal of Zoology* 2:1–8.

Nicholson, A. J. 1957. The self-adjustment of populations to change. *Cold Spring Harbor Symposium on Quantitative Biology* 22:153–172.

Nicholson, A. J. 1958. Dynamics of insect populations. *Annual Review of Entomology* 3:107–136.

O'Brien, S. J., and E. Mayr. 1991. Bureaucratic mischief: Recognizing endangered species and subspecies. *Science* 251:1187–1188.

O'Brien, S. J., M. E. Roelke, L. Marker, A. Newman, C. A. Winkler, D. Meltzer, L. Colly, J. F. Evermann, M. Bush, and D. E. Wildt. 1985. Genetic basis for species vulnerability in the cheetah. *Science* 227:1428–1434.

O'Brien, S. J., D. E. Wildt, and M. Bush. 1986. The cheetah in genetic peril. *Scientific American* 254:84.

O'Brien, S. J., D. E. Wildt, D. Goldman, C. R. Merril, and M. Bush. 1983. The cheetah is depauperate in genetic variation. *Science* 221:459–462.

Odum, E. P. 1971. *Ecology.* New York: Holt.

Odum, H. T., and C. F. Jordan. 1970. Metabolism and evapotranspiration of the lower forest in a giant cylinder. In *A tropical rainforest: A study of irradiation and ecology at El Verde, Puerto Rico,* ed. H. T. Odum and R. F. Pigeon. Washington, D.C.: U.S. Atomic Energy Commission.

Olson, J. S. 1958. Rates of succession and soil changes on southern Lake Michigan sand dunes. *Botany Gazette* 119:125–170.

Olsson, M., and R. Shine. 1997. The limits to reproductive output: Offspring size versus number in the sand lizard (*Lacerta agilis*). *American Naturalist* 149:179–188.

Orians, G. 1974. Tropical population ecology. In *Fragile Ecosystems,* ed. E. G. Farnsworth and F. B. Golley, pp. 5–65. New York: Springer-Verlag.

Orians, G. L. 1969. On the evolution of mating systems in birds and mammals. *American Naturalist* 103:589–604.

Oring, L. W. 1985. Avian polyandry. *Current Ornithology* 3:309–351.

Oring, L. W., and M. L. Knudson. 1973. Monogamy and polyandry in the spotted sandpiper. *The Living Bird* 11:59–73.

Orive, M. E. 1995. Senescence in organisms with clonal reproduction and complex life histories. *American Naturalist* 145:90–108.

Orr, W. C., and R. S. Sohal. 1994. Extension of life-span by overexpression of superoxide dismutase and catalase in *Drosophila melanogaster. Science* 263:1128–1130.

Owen, D. F., and R. G. Wiegert. 1976. Do consumers maximize plant fitness? *Oikos* 35:230–235.

Packer, C. 1986. The ecology of sociality in felids. In *Ecological aspects of social evolution,* ed. D. I. Rubenstein and R. W. Wrangham. Princeton, N.J.: Princeton University Press.

Paige, K. N. 1999. Regrowth following ungulate herbivory in *Ipomopsis aggregata:* Geographic evidence for overcompensation. *Oecologia* 118:316.

Paige, K. N., and T. G. Whitham. 1987. Overcompensation in response to mammalian herbivory: The advantage of being eaten. American Naturalist 129:407–416.

Paine, R. T. 1966. Food web complexity and species diversity. *American Naturalist* 100:65–75.

Paine, R. T., and S. A. Levin. 1981. Intertidal landscapes: Disturbance and the dynamics of pattern. *Ecological Monographs* 51:145–178.

Parish, J. K. 1989. Re-examining the selfish herd: Are central fish safer? *Animal Behavior* 38: 1048–1054.

Park, T. 1954. Experimental studies of interspecific competition. II. Temperature, humidity and competition in two species of *Tribolium. Physiological Zoology* 27:177–238.

Park, T. 1962. Beetles, competition and populations. *Science* 138:1369–1375.

Partridge, L., and K. Fowler. 1992. Direct and correlated responses to selection on age at reproduction in *Drosophila melanogaster. Evolution* 46:76–91.

Patrick, R. 1968. The structure of diatom communities in similar ecological conditions. *American Naturalist* 102:173–183.

Patton, J. L., and P. V. Brylski. 1987. Pocket gophers in alfalfa fields: Causes and consequences of habitat-related body size variation. *American Naturalist* 130:493–506.

Patton, J. L., and J. H. Feder. 1981. Microspatial genetic heterogeneity in pocket gophers: Non-random breeding and drift. *Evolution* 35:912–920.

Pearcy, R.W., and W. M. Yang. 1996. A three-dimensional crown architecture model for assessment of light capture and carbon gain by understory plants. *Oecologia* 108:1–12.

Pearl, R. 1940. *Introduction to medical biometry and statistics.* Philadelphia: W. B. Saunders.

Pearson, O. P. 1960. Habits of *Microtus californicus* revealed by automatic photographic records. *Ecological Monographs* 30:231–249.

Pearson, O. P. 1963. History of two local outbreaks of feral house mice. *Ecology* 44:540–549.

Peet, R. K., and N. L. Christensen. 1987. Competition and tree death; most trees die young in the struggle for the forest's scarce resources. *BioScience* 37:586–595.

Pennings, S. C., and R. M. Callaway. 1992. Salt marsh plant zonation—the relative importance of competition and physical factors. *Ecology* 73:681–690.

Perrins, C. M. 1964. Survival of young swifts in relation to brood-size. *Nature* 201:1147–1149.

Peterson, J. J., H. C. Chapman, and D. B. Woodward. 1968. The bionomics of a mermithid nematode of larval mosquitoes in southwestern Louisiana. *Mosquito News* 28:346–352.

Peterson, R. O. 1999a. Ecological studies of wolves on Isle Royale. Annual Report Michigan Tech. University.

Peterson, R. O. 1999b. Wolf–moose interaction on Isle Royale: The end of natural regulation? *Ecological Applications* 9:10–16.

Peterson, R. O., and R. E. Page. 1983. Wolf–moose fluctuations at Isle Royale National Park, Michigan, USA. *Acta Zooligi Fennici* 174:251–253.

Phillips, M., and S. N. Austad. 1990. Animal communication and social evolution. In *Interpretation and explanation in the study of animal behavior,* I. *Interpretation, intentionality, and communication,* ed. M. Bekoff and D. Jamieson, pp. 254–268. Boulder, Colo.:Westview.

Pianka, E. R. 1973. The structure of lizard communities. *Annual Review of Ecology and Systematics* 4:54–74.

Pianka, E. R. 1975. Niche relations of desert lizards. In *Ecology and evolution of communities,* ed. M. L. Cody and J. M. Diamond, pp. 221–314. Cambridge, Mass.: Harvard University Press.

Pianka, E. R. 1988. *Evolutionary ecology.* New York: Harper & Row.

Picker, M. D., B. Leon, and J. G. H. Londt. 1991. The hypertrophied hindwings of *Palmipena aeloeptera* Picker. *Animal Behavior* 42:821–826.

Pielou, E. C. 1969. *An introduction to mathematical ecology.* New York: John Wiley & Sons.

Pielou, E. C. 1974. *Population and community ecology.* New York: Gordon & Breach.

Pimentel, D. 1968. Population regulation and genetic feedback. *Science* 159:1432–1437.

Pimm, S. L. 1982. *Food webs.* London: Chapman & Hall.

Pimm, S. L. 1988. Energy flow and trophic structure. In *Concepts of ecosystem ecology,* ed. L. R. Pomeroy and J. J. Albers, pp. 263–278. Berlin: Springer-Verlag.

Pimm, S. L., J. H. Lawton, and J. E. Cohen. 1991. Food web patterns and their consequences. *Nature* 350:669–674.

Pimm, S. L. and P. R. Raven. 2000. Extinction by numbers. *Nature* 403:843–845.

Pitelka, F. A. 1964. The nutrient-recovery hypothesis for arctic microtine cycles. I. Introduction. In *Grazing in terrestrial and marine environments,* ed. D. J. Crisp, pp. 55–56. Oxford: Blackwell.

Pitkow, R. B. 1960. Cold death in the guppy. *Biology Bulletin* 119:231–245.

Platt, W. J. 1976. The natural history of a fugitive prairie plant (*Mirabilis hirsuta* [Pursh]). *Oecologia* 22:399–409.

Pleasants, J. M. 1989. Optimal foraging by nectarivores: A test of the marginal-value theorem. *American Naturalist* 134:51–71.

Pockman, W. T. and J. S. Sperry. 1997. Freezing-induced xylem cavitation and the northern limit of *Larrea tridentata. Oecologia* 109:19–27.

Polis, G. A. 1991. Complex trophic interactions in deserts: An empirical critique of food-web theory. *American Naturalist* 138:123–155.

Polis, G. A. 1999. Why are parts of the world green? Multiple factors control productivity and the distribution of biomass. *Oikos* 86:3–15.

Polis, G. A., and D. R. Strong. 1996. Food web complexity and community dynamics. *American Naturalist* 147:813–846.

Polis, G. A., and K. O. Winemiller. 1996. Food webs: integration of patterns and dynamics. New York: Chapman & Hall.

Pomeroy, L. R., and E. J. Kuenzler. 1969. Phosphorus turnover by coral reef animals. *Proceedings of the Second National Symposium on Radioecology no. 2,* ed. D. J. Nelson and F. S. Evans, pp. 474–482.

Poole, R. W. 1974. *An introduction to quantitative ecology.* New York: McGraw-Hill.

Porsild, A. E., C. R. Harrington, and G. A. Mollisu. 1967. *Lupinus arcticus* Wats grown from seeds of Pleistocene age. *Science* 158:113–114.

Porter, J. W., and N. M. Targett. 1988. Allelochemical interactions between sponges and corals. *Biological Bulletin* 175:230–239.

Post, W. M., T-H. Peng, W. R. Emanuel, A. W. King, V. H. Dale, and D. L. DeAngelis. 1990. The global carbon cycle. *American Scientist* 78:310–326.

Poulson, T. L. and W. J. Platt. 1996. Replacement patterns of beech and sugar maple in Warren Woods, Michigan. *Ecology* 77:1234–1253.

Poulson, T. L., and J. Platt. 1989. Gap light regimes influence canopy tree diversity. *Ecology* 70:553–555.

Power, M. E., D. Tilman, J. A. Estes, B. A. Menge, W. J. Bond, L. S. Mills, G. Daily, J. C. Castilla, and J. Lubchenco, 1996. Challenges in the quest for keystones. *BioScience* 46:609–623.

Pratt, C. R. 1984. Response of *Solidago graminifolia* and *Solidago juncea* to nitrogen fertilization applications: Changes in biomass allocation and implications for community structure. *Bulletin of Torrey Botanical Club* 111:469–478.

Prescott, L. M., J. Harley, and D. H. Klein. 1990. *Microbiology.* Dubuque, Iowa: W. C. Brown.

Preston, F. W. 1948. The commonness, and rarity, of species. *Ecology* 29:254–283.

Preston, F. W. 1962. The canonical distribution of commonness and rarity: Part I. *Ecology* 43:185–215.

Priscu, J. C., C. F Wolf, C. D. Takacs, C. Fritsen, J. Laybourn-Parry, E. C. Roberts, B. Sattler, and W. B. Lyons. Carbon transformations in a perennially ice-covered Antarctic lake. *BioScience* 49:997–1008.

Pugnaire, F. I., P. Haase, and J. Puigdefabregas. 1996. Facilitation between higher plant species in a semiarid environment. *Ecology* 77:1420–1426.

Raijman, L. E., N. C. van Leeuwen, R. Kersten, H. C. M. Den Nijs, and S. B. Menken. 1994. Genetic variation and outcrossing rate in relation to population size in *Gentiana pneumonanthe* L. *Conservation Biology* 8:1014–1026.

Rappaport, E. H. 1982. *Areography: Geographical strategies of spacing.* 1st English ed., trans. B. Dravsal, *Publ. Fundacion Bariloche,* vol. 1. New York: Pergamon.

Raskin, I., and H. Kende. 1985. Mechanism of aeration in rice. Science 205:327–329.

Raven, J. A. 1970. Exogenous inorganic carbon sources in plant photosynthesis. *Biological Review* 45:167–221.

Recher, H. F. 1969. Bird species diversity and habitat diversity in Australia and North America. *American Naturalist* 103:75–80.

Reeve, H. K. 1992. Queen activation of lazy workers in colonies of the eusocial naked mole-rat. *Nature* 358:147–149.

Reeve, H. K., D. F. Westneat, W. A. Noon, P. W. Sherman, and C. F. Aquadro. 1990. DNA "fingerprinting" reveals high levels of inbreeding in colonies of the eusocial naked mole-rat. *Proceedings of the National Academy of Science* 87:2496–2500.

Rhoades, D. F. 1985. Offensive-defensive interactions between herbivores and plants: Their relevance in herbivore population dynamics and ecological theory. *American Naturalist* 125:205–238.

Ribble, D. O. 1991. The monogamous mating system of *Peromyscus californicus* as revealed by DNA fingerprinting. *Behavioral Ecology and Sociobiology* 29:161–166.

Ricklefs, R. E. 1968. On the limitation of brood size in passerine birds by the ability of adults to nourish their young. *Proceedings of the National Academy of Science* 61:847–851.

Ricklefs, R. E. 1998. Evolutionary theories of aging: Confirmation of a fundamental prediction, with implications for the genetic basis and evolution of life span *American Naturalist* 152:24–44.

Riebesell, V., D. A. Wold-Gladrow, and V. Smetacek. 1993. Carbon dioxide limitation of marine phytoplankton growth rates. *Nature* 361:249–251.

Risser, P. G. 1995. Biodiversity and ecosystem function. *Conservation Biology* 9:742–746.

Robison, B. H. 1995. Light in the ocean's midwaters. *Scientific American* 273:60–64.

Rohde, K. 1997. The larger area of the tropics does not explain latitudinal gradients in species diversity. *Oikos* 79:169–172.

Roldan, E. R. S., and M. Gomendio. 1999. The Y chromosome as a battle ground for sexual selection. *Trends in Ecology and Evolution* 14:58–62.

Romme, W. H., and D. G. Despain. 1989. Historical perspectives on the Yellowstone fires of 1988. *Bioscience* 39:695–699.

Root, R. B. 1996. Herbivore pressure on goldenrods (*Solidago altissima*): Its variation and cumulative effects. *Ecology* 77:1074–1087.

Rose, R. K. 1979. Levels of wounding in the meadow vole, *Microtus pennsylvanicus. Journal of Mammalogy* 60:37–45.

Rose, R. K., and M. S. Gaines. 1976. Levels of aggression in fluctuating populations of the prairie vole, *Microtus ochrogaster,* in eastern Kansas. *Journal of Mammalogy* 57:43–57.

Rosenthal, J. P., and P. M. Kotanen. 1994. Terrestrial plant tolerance to herbivory. *Trends in Ecology and Evolution* 9:145–148.

Rosenzweig, M. L. 1969. Why the prey curve has a hump. *American Naturalist* 103:81–87.

Rosenzweig, M. L. 1973. Evolution of the predator isocline. *Ecology* 27:84–94.

Rosenzweig, M. L. 1992. Species diversity gradients: We know more and less than we thought. *Journal of Mammalogy* 73:715–730.

Rummel, J. D., and J. Roughgarden. 1985. A theory of faunal buildup for competition communities. *Evolution* 39:1009–1033.

Rusterholz, K. A. 1981. Competition and the structure of an avian foraging guild. *American Naturalist* 118:173–190.

Ruttner, F. 1963. *Fundamentals of limnology.* Toronto: Toronto University Press.

Saether, B. E. 1997. Environmental stochasticity and population dynamics of large herbivores: A search for mechanisms. *Trends in Ecology and Evolution* 12:143–149.

Saitoh, T., O. N. Bjornstad, and N. C. Stenseth. 1999. Density dependence in voles and mice: A comparative study. *Ecology* 80:638–650.

Sanchez, P. A. 1976. *Properties and management of soils in the Tropics.* New York: John Wiley & Sons.

Sanders, T. B., J. L. Hamrick, and L. R. Holden. 1979. Allozyme variation in *Elymus canadensis* from the tallgrass prairie region: Geographic variation. *American Midland Naturalist* 101:1–12.

Sathyendranath, S., T. Platt, E. P. W. Horne, W. G. Harrison, O. Ulloa, R. Outerbridge, and N. Heopffner. 1991. Estimation of new production in the ocean by compound remote sensing. *Nature* 353:129–133.

Schaeffer, W. M. 1974. Selection for optimal life histories: The effects of age structure. *Ecology* 55:291–303.

Schall, B. A. 1974. Isolation by distance in *Liatris cylindracea. Nature* 252:703.

Schall, B. A. 1975. Population structure and local differentiation in *Liatris cylindracea. American Naturalist* 109:511–528.

Schall, J. J., and E. R. Pianka. 1978. Geographical trends in numbers of species. *Science* 201:679–686.

Schaller, G. W. 1972. *The Serengeti lion.* Chicago: University of Chicago Press.

Scheiner, S. M., and J. M. Reybenayas. 1994. Global patterns of plant diversity. *Evolutionary Ecology* 8:331–347

Scheu, S., and M. Shaffer. 1998. Bottom-up control of the soil macrofaunal community in a beechwood on limestone: Manipulation of food resources. *Ecology* 79:1573–1585.

Schindler, D. W. 1974. Eutrophication and recovery in experimental lakes: Implications for lake management. *Science* 184:897–899.

Schluter, D., and P. R. Grant. 1984. Determinants of morphological patterns in communities of Darwin's finches. *American Naturalist* 123:175–196.

Schmidt-Nielsen, K. 1964. *Desert animals: Physiological problems of heat and water.* Oxford: Oxford University Press.

Schmidt-Nielsen, K. 1979. *Animal physiology: Adaptation and environment.* Cambridge: Cambridge University Press.

Schoener, T. W. 1968. Sizes of feeding territories among birds. *Ecology* 49:123–141.

Schoener, T. W. 1989. Food webs from the small to the large. *Ecology* 70:1559–1589.

Schultz, A. M. 1964. The nutrient-recovery hypothesis for arctic microtine cycles. II. Ecosystem variables in relation to arctic microtine cycles. In *Grazing in terrestrial and marine environments,* ed. D. J. Crisp, pp. 57–68. Oxford: Blackwell.

Schultz, A. M. 1969. A study of an ecosystem: The arctic tundra. In *The ecosystem concept in natural resource management,* ed. G. VanDyne. New York: Academic Press.

Selander, R. K. 1965. On mating systems and sexual dimorphism. *American Naturalist* 105:400–437.

Shaffer, W. M., and M. Kot. 1986a. Chaos in ecological systems: The coals that Newcastle forgot. *Trends in Ecology and Evolution* 1:58–63.

Shaffer, W. M., and M. Kot. 1986b. Differential systems in ecology and epidemiology. In *Chaos,* ed. A. V. Holden, pp. 158–178. Princeton, N.J.: Princeton University Press.

Shannon, C. E., and W. Weaver. 1949. *The mathematical theory of communication.* Urbana: University of Illinois Press.

Shapiro, J., V. Lamarra, and M. Lynch. 1975. Biomanipulation: An ecosystem approach to lake restoration. In *Water quality management through biological control,* ed. P. L. Brazonik and J. L. Fox, pp. 85–96. Gainesville: University of Florida Press.

Sharitz, R. R., and J. F. McCormick. 1973. Population dynamics of two competing annual plant species. *Ecology* 54:723–739.

Shelford, V. E. 1945. The abundance of the collared lemming (*Dicrostonyx groenlandicus* (Tr.) var *richarsoni Mer*) in the Churchill area, 1929 to 1940. *Ecology* 24:472–484.

Sherman, P. W. 1977. Nepotism and the evolution of alarm calls. *Science* 197:1246–1253.

Sherman, P. W., J. U. M. Jarvis, and R. D. Alexander. 1991. *The biology of the naked mole-rat.* Princeton, N.J.: Princeton University Press.

Sherman, P. W., J. U. M. Jarvis, and S. H. Braude. 1992. Naked mole-rats. *Scientific American* 267:72–79.

Silvertown, J., et al. 1994. Rainfall, biomass variation and community composition in the Park Grass Experiment. Ecology 75:2430–2437.

Simberloff, D. S. 1983. Competition theory, hypothesis testing and other community ecological buzzwords. *American Naturalist* 122:626–635.

Simberloff, D. S., and E. O. Wilson. 1969. Experimental zoogeography of islands: The colonization of empty islands. *Ecology* 50:278–296.

Simpson, E. H. 1949. Measurement of diversity. *Nature* 163:688.

Simpson, G. G. 1965. *The geography of evolution.* Philadelphia: Chilton Books.

Sinclair, A. R. E., and J. M. Gosline. 1997. Solar activity and mammal cycles in the northern hemisphere. *American Naturalist* 149:776–784.

Slobodchikoff, C. N., C. Fisher, and J. Shapiro. 1986. Predator-specific alarm calls of prairie dogs. *American Zoologist* 26:557.

Slobodchikoff, C. N., J. Kiriazis, C. Fisher, and E. Creef. 1991. Semantic information distinguishing individual predators in the alarm calls of Gunnison's prairie dogs. *Animal Behavior* 42:713–720.

Smallwood, P. D., and J. A. Smallwood. 1998. Seasonal shifts in sex ratios of fledgling American kestrels (*Falco sparverius* paulus): The Early Bird Hypothesis. *Evolutionary Ecology* 12:839–853.

Smedley, S. R., and T. Eisner. 1995. Sodium uptake by puddling in a moth. *Science* 270:1816–1818.

Smith, C. C. 1968. The adaptive nature of social organization in the genus of tree squirrels *Tamiasciurus. Ecological Monographs* 38:31–63.

Smith, M. H., and J. T. McGinnis. 1968. Relationship of latitude, altitude, and body size to litter size and mean annual production of offspring in *Peromyscus. Researches in Population Ecology* 10:115–126.

Smith, M. L., J. N. Bruhn, and J. B. Anderson. 1992. The fungus *Armillaria bulbosa* is among the largest and oldest living organisms. *Nature* 356:428–431.

Smith, N. G. 1968. The advantage of being parasitized. *Nature* 219:690–694.

Smith, T. B. 1991. A double-billed dilemma. *Natural History* (January):14–19.

Smith-Gill, S. J., and D. E. Gill. 1978. Curvilinearity in the competition equations: An experiment with ranid tadpoles. *American Naturalist* 110:849–860.

Somero, G. N., and A. L. DeVries. 1967. Temperature tolerance of some antarctic fishes. *Science* 156:257–258.

Sousa, W. P. 1979a. Experimental investigation of disturbance and ecological succession in a rocky intertidal algal community. *Ecological Monographs* 49:227–254.

Sousa, W. P. 1979b. Disturbance in marine intertidal boulder fields: The non-equilibrium maintenance of species diversity. *Ecology* 60:1225–1239.

Sousa, W. P. 1984a. Intertidal mosaics: Patch size, propagule availability and spatially variable patterns of succession. *Ecology* 65:1918–1935.

Sousa, W. P. 1984b. The role of disturbance in natural communities. *Annual Review of Ecology and Systematics* 15:353–392.

Southwick, C. H. 1976. *Ecology and the quality of our environment.* New York: Van Nostrand.

Southwood, T. R. E. 1966. *Ecological methods.* London: Chapman & Hall.

Spanner, D. C. 1963. The green leaf as a heat engine. *Nature* 198:934–937.

Spencer, A. W., and H. W. Steinhoff. 1968. An explanation of geographic variation in litter size. *Journal of Mammalogy* 49:281–286.

Sprules, W. G. 1974. The adaptive significance of peadogenesis in North American species of *Ambystoma* (Amphibia:Caudata). *Canadian Journal of Zoology* 52:393–400.

Stacey, P. B., and W. D. Koenig. 1984. Cooperative breeding in the acorn woodpecker. *Scientific American* 251:114–121.

Stamp, N. E. 1992. Theory of plant–insect herbivore interactions on the inevitable brink of re-synthesis. *Bulletin of the Ecological Society of America* 73:28–34.

Stapp, P., P. J. Pekins, and W. W. Mautz. 1991. Winter energy expenditure and the distribution of southern flying squirrels. *Canadian Journal of Zoology* 69:2548–2555.

Stearns, S. C. 1976. Life history tactics: A review of the ideas. *Quarterly Review of Biology* 51:3–47.

Stearns, S. C. 1977. The evolution of life history traits: A critique of the theory and a review of the data. *Annual Review of Ecology and Systematics* 8:145–171.

Stephens, D. W., and J. R. Krebs. 1986. *Foraging theory.* Princeton, N.J.: Princeton University Press.

Sterner, R. W., J. J. Elser, E. J. Fee, S. J. Guildford, and T. H. Chrzanowski. 1997. The light:nutrient ratio in lakes: The balance of energy and materials affects ecosystem structure and process. *American Naturalist* 150:663–684.

Stevens, G. C. 1989. The latitudinal gradient in geographical range: How so many species coexist in the tropics. *American Naturalist* 133:240–257.

Storey, K. B. 1990. Life in a frozen state: Adaptive strategies for natural freezing tolerance in amphibians and reptiles. *American Journal of Physiology* 258:R559–R568.

Stork, N. E. 1988. Insect diversity: Facts, fiction and speculation. *Biological Journal of the Linnean Society* 35:321–337.

Strickberger, M. W. 1990. *Evolution.* Boston: Jones & Bartlett.

Strong, D. R. 1982. Harmonious coexistence of hispine beetles on *Heliconia* in experimental and natural communities. *Ecology* 63:1039–1049.

Strong, D. R., L. A. Szyska, and D. S. Simberloff. 1979. Tests of community-wide character displacement against null hypotheses. *Evolution* 33:897–913.

Sugihara, G. 1980. Minimal community structure: An explanation of species abundance patterns. *American Naturalist* 116:770–787.

Sugihara, G., and R. M. May. 1990. Applications of fractals in ecology. *Trends in Ecology and Evolution* 5:79–86.

Suominen, O., K. Danell, and R. Bergström. 1999. Moose, trees, and ground-living invertebrates: Indirect interactions in Swedish pine forests. *Oikos* 84:215–226.

Suttle, C. A., A. M. Chan, and M. T. Cottrell. 1990. Infection of phytoplankton by viruses and reduction of primary productivity. *Nature* 347:467–470.

Swap, R., M. Garstang, S. Greco, R. Talbot, and P. Kallberg. 1992. Saharan dust in the Amazon Basin. Tellus. Series B. *Chemical and Physical Meteorology.* 44B:133–149.

Taber, R. D., and R. F. Dasmann. 1958. The dynamics of three natural populations of the deer *Odocoileus hemionus columbianus. Ecology* 38:233–246.

Tadros, T. M. 1957. Evidence of the presence of an edapho-biotic factor in the problem of serpentine tolerance. *Ecology* 38:14–23.

Taitt, M. J. 1981. The effect of extra food on small rodent populations. I. Deermice (*Peromyscus leucopus*). *Journal of Animal Ecology* 50:111–124.

Takens, F. 1981. Strange attractors in turbulence. In *Dynamical systems and turbulence,* ed. D. A. Rand and L. S. Young, pp. 366–381. New York: Springer-Verlag.

Tamarin, R. H. 1978. Dispersal, population regulation and *K*-selection in field mice. *American Naturalist* 112:545–555.

Tamarin, R. H. 1985. *The biology of New World Microtus.* American Society of Mammalogists Special Publication no. 8.

Tamarin, R. H., and C. J. Krebs. 1969. Microtus population biology II. Genetic changes at the transferrin locus in fluctuating populations of two vole species. *Evolution* 23:183–211.

Tanaka, H., and T. Nakashizuka. 1997. Fifteen years of canopy dynamics analyzed by aerial photographs in a temperate deciduous forest. Japan. *Ecology* 78:612–620.

Tansley, A. G. 1939. *The British islands and their vegetation.* Cambridge: Cambridge University Press.

Tatar, M., D. W. Gray, and J. R. Carey. 1997. Altitudinal variation for senescence in *Melanoplus* grasshoppers. *Oecologia* 111: 3:357–364.

Tateno, M., and F. S. Chapin. 1997. The logic of carbon and nitrogen interactions in terrestrial ecosystems. *American Naturalist* 149:723–744.

Terborgh, J. 1973. On the notion of favorability in plant ecology. *American Naturalist* 107: 481–501.

Thompson, J. N. 1999. The raw material for ecoevolution. *Oikos* 84:5–16.

Thomson, J. D., and B. A. Thomson. 1989. Dispersal of *Erythronium grandiflorum* pollen by bumblebees: Implications for gene flow and reproductive success. *Evolution* 43:657–661.

Thornhill, R. 1981. *Panorpa (Mecoptera:Panorpidae)* scorpionflies: Systems for understanding resource-defense polygyny and alternative male reproductive tactics. *Annual Review of Ecology Systematics* 12:335–386.

Thurston, H. D. 1969. Tropical agriculture: A key to the world food crises. *BioScience* 19:29–34.

Tilman, D. 1985. The resource-ratio hypothesis of plant succession. *American Naturalist* 125:827–852.

Tilman, D., and J. A. Downing. 1994. Biodiversity and stability in grasslands. *Nature* 367:363–365.

Tilman, D., D. Wedin, and J. Knops. 1996. Productivity and sustainability influenced by biodiversity in grassland ecosystems. *Nature* 379: 718–720.

Tinbergen, N. 1951. *The study of instinct.* New York: Oxford University Press.

Transeau, E. N. 1935. The prairie peninsula. *Ecology* 16:423–437.

Trivers, R. L. 1974. Parent–offspring conflict. *American Zoologist* 14:249–264.

Trivers, R. L., and D. E. Willard. 1973. Natural selection of parental ability to vary the sex ratio of offspring. *Science* 191:249–263.

Trostel, K., A. R. E. Sinclair, C. J. Walters, and C. J. Krebs. 1987. Can predation cause the 10-year hare cycle? *Oecologia* 74:185–192.

Trussell, G. C. 1997. Phenotypic plasticity in the foot size of an intertidal snail. *Ecology* 78:1033–1048.

Turchin, P. 1999. Population regulation: A synthetic view. *Oikos* 84: 153–159.

Turner, M. G., V. H. Dale, E. H. Everham. 1997. Fires, hurricanes, and volcanoes: Comparing large disturbances. *BioScience* 47:758–768.

Turner, M. G., W. H. Romme, R. H. Gardner, and W. W. Hargrove. 1997. Effects of fire size and pattern on early succession in Yellowstone National Park. *Ecological Monographs* 67:411–433.

Tuttle, M. D. 1982. The amazing frog-eating bat. *National Geographic* 161:78–91.

U.S. Department of Agriculture. 1965. Sylvics of forest trees of the United States. Agriculture Handbook no. 271.

Urquhart, F. A. 1987. *The monarch butterfly: International traveler.* Chicago: Nelson-Hall.

Valverde, T., and J. Silvertown. 1997. A metapopulation model for *Primula vulgaris,* a temperate forest understorey herb. *Journal of Ecology* 85: 193–210.

Valone, T. J., J. H. Brown, and E. J. Heske. 1994. Interactions between rodents and ants in the Chihuahuan Desert—an update. *Ecology* 75: 252–255.

Van Ballenberghe, V. 1987. Effects of predation on moose numbers: a review of recent North American studies. *Swedish Wildlife Research Supplement* 1:431–460.

Van Cleve, K., F. S. Chapin III, C. T. Dyrness, and L. A. Viereck. 1991. Element cycling in taiga forests: State-factor control. *BioScience* 41:78–88.

Van Cleve, K., and L. A. Viereck. 1972. Forest succession in relation to nutrient cycling in the boreal forest of Alaska. In *Forest succession: Concepts and application,* ed. D. West, H. Shugart, and D. Botkin, pp. 185–211. New York: Springer-Verlag.

vandenBosch, R., P. S. Messenger, and A. P. Gutierrez. 1982. *An introduction to biological control.* New York: Plenum Press.

van der Maarel, E. 1990. Ecotones and ecoclines are different. *Journal of Vegetative Science* 1: 135–138.

Vander Zanden, M. J., and J. B. Rassmussen. 1999. Primary consumer δ^{13}C and δ^{15}N and the trophic position of aquatic consumers. *Ecology* 80:1395–1404.

Vanni, M. J., and C. D. Layne. 1997. Nutrient recycling and herbivory as mechanisms in the "top-down" effect of fish on algae in lakes. *Ecology* 78:21–40.

Vanni, M. J., C. D. Layne, and S. E. Arnott. 1997. "Top-down" trophic interactions in lakes: Effects of fish on nutrient dynamics. *Ecology* 78:1–20.

VanValen, L. 1973. A new evolutionary law. *Evolutionary Theory* 1:1–30.

Vanvoorhies, W. A. 1992. Production of sperm reduces nematode lifespan. *Nature* 360:456–458.

VanVoris, P., R. V. O'Neil, W. R. Emanuel, and H. H. Shugart Jr. 1980. Functional complexity and ecosystem stability. *Ecology* 61:1352–1360.

Varley, G. C. 1949. Population changes in German forest pests. *Journal of Animal Ecology* 18:117–122.

Varley, G. C., and G. R. Gradwell. 1960. Key factors in population studies. *Journal of Animal Ecology* 26:251–261.

Vaughan, T. A. 1972. *Mammalogy.* Philadelphia: Saunders.

Verboom, J., A. Schotman, P. Opdam, and A. J. Metz. 1991. European nuthatch metapopulations in a fragmented agricultural landscape. *Oikos* 61:149–161.

Veit, R. R., and M. A. Lewis. 1996. Dispersal, population growth, and the Allee effect: Dynamics of the house finch invasion of eastern North America. *American Naturalist* 148:255–274.

Vermeij, G. J. 1983. Intimate associations and coevolution in the sea. In *Coevolution,* ed. D. J. Futuyma and M. Slatkin, pp. 311–327. Sunderland, Mass.: Sinauer.

Vinton, M. A., and D. C. Hartnett. 1992. Effects of bison grazing on *Andropogon gerardii* and *Panicum virgatum* in burned and unburned tallgrass prairie. *Oecologia* 90:374–382.

Vitala, J., E. Korplmäkl, P. Palokangas, and M. Kolvula. *Nature* 373:425–427.

Vitousek, P. M., et al. 1997. Human domination of the earth's ecosystems. *Science* 277:494–499.

Voight, D. R., and W. E. Berg. 1987. Coyote. In *Wild furbearer management and conservation in North America,* ed. M. Novak, J. A. Baker, M. E. Obard, and B. Malloch, pp. 345–356. Toronto, Ontario: Ministry of Natural Resources.

Volterra, V. 1926. Variations and fluctuations of the numbers of individuals in animal species living together. Reprinted in 1931 in R. N. Chapman, *Animal ecology.* New York: McGraw-Hill.

Vorhies, C. T., and W. P. Taylor. 1922. Life history of the kangaroo rat, *Dipodomys spectabilis spectabilis* (Merriam). U.S. Department of Agriculture Bulletin no. 1091.

Wagner, F. H., and L. C. Stoddart. 1972. Influence of coyote predation on black-tailed jackrabbit populations in Utah. *Journal of Wildlife Management* 36:329–342.

Walker, L. R., J. Volzow, and J. D. Akerman. 1992. Immediate impact of Hurricane Hugo on a Puerto Rican rain forest. *Ecology* 73:691–694.

Wallace, B. 1981. *Basic population genetics.* New York: Columbia University Press.

Walls, S. C. 1990. Interference competition in post-metamorphic salamanders: Interspecific differences in aggression by coexisting species. *Ecology* 71(1):307–314.

Walker, B. 1992. Biodiversity and ecological redundancy. *Conservation Biology* 6:18–23.

Walter, H. 1973. *Vegetation of the earth.* London: English Universities Press.

Wardle, D. A., K. I. Bonner and K. S. Nicholson. 1997. Biodiversity and plant litter: Experimental evidence which does not support the view that enhanced species richness improves ecosystem function. *Oikos* 79: 247–258.

Waser, P. M., and W. T. Jones. 1983. Natal philopatry among solitary mammals. *Quarterly Review of Biology* 58:355–390.

Watson, A., and D. Jenkins. 1968. Experiments on population control by territorial behavior in red grouse. *Journal of Animal Ecology* 37:595–614.

Watt, K. E. F. 1964. Comments on fluctuations of animal populations and measures of community stability. *Canadian Entomologist* 96:1434–1442.

Wayne, R. K., N. Lehman, M. C. Allard, and R. L. Honeycutt. 1992. Mitochondrial DNA variability of the gray wolf: Genetic consequences of population decline and habitat fragmentation. *Conservation Biology* 6:559–569.

Webb, K. L., W. D. DuPaul, W. J. Wiebe, W. Sottile, and R. E. Johannes. 1975. Enewetak (Eniwetok) Atoll: Aspects of the nitrogen cycle on a coral reef. *Limnology and Oceanography* 20:198–210.

Webb, W., S. Szarek, W. Lavenroth, R. Kenerson, and M. Smith. 1978. Primary production and water use in native forest, grassland and desert ecosystems. *Ecology* 59:1239–1247.

Weber, E., and B. Schmid. 1993. Das neophytenproblem. *Dissertationes Botanicae* 196:209–227.

Weinberg, W. 1908. Fiber den Nachweis der Vererburg Geim Menschen. *Jaresh. Verein. f. vater. Naturk. Wurtten* 64:368–380.

Weisbrod, A. R. 1976. Insularity of mammal species numbers in two national parks. In *Proceedings of 1st Conference on Scientific Research in National Parks,* ed. R. M. Linn, pp. 83–87. U.S. National Park Service Transactions and Proceedings, Series 5.

Wellington, W. G. 1960. Qualitative changes in natural populations during changes in abundance. *Canadian Journal of Zoology* 38:290–314.

Werner, E. J., and D. J. Hall. 1974. Optimal foraging and the size selection of prey by the bluegill sunfish *Lepomis macrochirus. Ecology* 55:1042–1052.

West, G. B., J. H. Brown, and J. H. Enquist. 1997. A general model for the origin of allometric scaling laws in biology. *Science* 276:122–123.

West, S. D. 1979. Habitat responses of microtine rodents to central Alaskan forest succession. Ph.D. dissertation. University of California, Berkeley.

Wetzel, R. G. 1983. *Limnology.* New York: Saunders.

White, J. 1980. Demographic factors in populations of plants. In *Demography and evolution in plant populations,* ed. O. T. Solbrig, pp. 21–48. Oxford: Blackwell.

Whitehead, A. N. 1920. The concept of nature. Tarner Lectures delivered in Trinity College, November 1919. Cambridge University Press, London.

Whitmore, T. C. 1984. *Tropical rain forests of the Far East.* 2d ed. Oxford: Clarendon Press.

Whitmore, T. C. 1990. *Introduction to tropical forests.* Oxford: Oxford University Press.

Whitney, G. G., and J. R. Runkle. 1981. Edge versus age effects in the development of a beech–maple forest. *Oikos* 37:377–381.

Whittaker, R. H. 1953. A consideration of climax theory: The climax as a population and pattern. *Ecological Monographs* 23:41–78.

Whittaker, R. H. 1975. *Communities and ecosystems.* London: Macmillan.

Whittaker, R. H. 1989. Plant recolonization and vegetation succession on the Krakatau Islands. *Ecological Monographs* 59:59–123.

Whittaker, R. H., and W. A. Niering. 1975. Vegetation of the Santa Catalina Mountains, Arizona. V. Biomass, production and diversity along the elevation gradient. *Ecology* 56:771–790.

Wiens, J. A., and J. T. Rotenberry. 1980. Bird community structure in cold shrub deserts: Competition or chaos? In *Acta XVII Congressus Internationalis Ornithologica,* pp. 1063–1070. Berlin, Germany.

Wilbur, H. M. 1980. Complex life cycles. *Annual Review of Ecology and Systematics* 11:67–93.

Wiley, R. H. 1973. Territoriality and non-random mating in sage grouse (*Centrocercus urophasianus*). *Animal Behavior Monographs* 6:85–169.

Williams, C. F., and R. P. Guries. 1994. Genetic consequences of seed dispersal in three sympatric forest herbs. I. Hierarchical population genetic structure. *Evolution* 48:791–805.

Williams, G. C. 1957. Pleiotropy, natural selection and the evolution of senescence. *Evolution* 11:398–411.

Willis, E. O. 1963. Is the zone-tailed hawk a mimic of the turkey vulture? *Condor* 65:313–317.

Wilson, D. S. 1992. Complex interactions in meta-communities, with implications for biodiversity and higher levels of selection. *Ecology* 73:1984–2000.

Wilson, E. O. 1971. *The insect societies.* Cambridge, Mass.: Harvard University Press.

Wilson, E. O. 1975. *Sociobiology.* Cambridge, Mass.: Belknap Press.

Wilson, E. O. 1992. *The diversity of life.* New York: W. W. Norton.

Witte, F. 1984. Ecological differentiation in Lake Victoria haplochromines: Composition of cichlid species flocks in African lakes. In *Evolution of fish species flocks,* ed. A. A. Echell and I. Kornfield, pp. 155–167. Orono, Maine: University of Maine Press.

Woddell, S. R. J., H. A. Mooney, and A. J. Hill. 1969. The behavior of *Larrea divaricata* (creosote bush) in response to rainfall in California. *Journal of Ecology* 57:37–44.

Woiwood, I. P., and I. Hanski. 1992. Patterns of density dependence in moths and aphids. *Journal of Animal Ecology* 61:619–629.

Wolda, S. J. 1992. Trends in abundance of tropical forest insects. *Oecologia* 89:47–52.

Wolfe, L. M. 1995. The genetics and ecology of seed size variation in a biennial plant, *Hydrophyllum appendiculatum* (Hydrophyllaceae). *Oecologia* 101:343–352.

Wolff, J. O. 1986. The effects of food on midsummer demography of white-footed mice, *Peromyscus leucopus. Canadian Journal of Zoology* 64:855–858.

Wolff, J. O., and D. M. Cicerello. 1989. Field evidence for sexual selection and resource competition infanticide in white-footed mice. *Animal Behavior* 38:637–642.

Wooten, T. J. 1992. Indirect effects, prey susceptibility and habitat selection: Impact of birds on limpets and algae. *Ecology* 73:981–992.

Wooten, J. T. 1994. Putting the pieces together: Testing the independence of interactions among organisms. *Ecology* 75:1544–1551.

Wooten, J. T., and D. A. Bell. 1992. A metapopulation model of the peregrine falcon in California: Viability and management strategies. *Ecology Applications* 2:307–321.

Wright, S. 1931. Evolution in Mendelian populations. *Genetics* 16:97–159.

Wright, S. 1965. The interpretation of population structure by *F*-statistics with special regard to systems of mating. *Evolution* 9:395–420.

Yosef, R., and D. W. Whitman. 1992. Predator exaptations and defensive adaptations in evolutionary balance—no defence is perfect. *Evolutionary Ecology* 6:527–536.

Young, A. G., and H. G. Merriam. 1994. Effects of forest fragmentation on the spatial genetic structure of *Acer saccharum* Marsh (sugar maple) populations. *Heredity* 72:201–208.

Young, Z. 1981. *The life of vertebrates.* New York: Oxford University Press.

Zahavi, A. 1975. Mate selection—a selection for a handicap. *Journal of Theoretical Biology* 67:603–605.

Zimmer, C. 1993. The war between plants and animals. *Discover* 14 (July):16–17.

Zimmerman, J. K. 1991. Ecological correlates of labile sex expression in the orchid *Catasetum viridiflavum. Ecology* 72:597–608.

Zumpft, W. G. In *The prokaryotes: A handbook on the biology of bacteria: Ecophysiology, Isolation, Applications.* ed. A Ballow et al., pp. 554–582. New York: Springer-Verlag.

Zwölfer, J. 1976. The golden rod problem: Possibilities for a biological weed control project in Europe. EPPO Publication Series B no. 81. European and Mediterranean Plant Protection Organization. Paris.

Credits

CHAPTER 2. 19: Fig. 2.1 from *Introduction to Quantitative Genetics*, by D. S. Falconer. Copyright © D. S. Falconer 1975, 1989. Reprinted by permission of Pearson Education Limited. **23**: Fig. 2.4 from "Dispersal of *Erythronium grandiflorum* Pollen by Bumblebees: Implications for Gene Flow and Reproductive Success," by J. D. Thomson and B. A. Thomson, *Evolution*, 1989, 43:657–661. Reprinted by permission. **24**: Fig. 2.5 reprinted with permission from "Oscillating Selection on Darwin's Finches," by H. L. Gibbs and P. R. Grant, *Nature*, 1987, 327:511–513. Copyright © 1987 Macmillan Magazines Limited. **25**: Fig. 2.6 from *Evolution*, by M. W. Strickberger. Copyright © 1990 Jones & Bartlett. Reprinted by permission. **28**: Fig. 2.9 from © J. Z. Young 1981. Redrawn from *Life of Vertebrates* by J. Z. Young (3rd ed., 1981) by permission of Oxford University Press. **30**: Fig. 2.12 from *The Panda's Peculiar Thumb*, by S. J. Gould. With permission from *Natural History* (November 1978), copyright 1978, the American Museum of Natural History. **32**: Fig. 2.14 a & c reprinted with permission from "Positive Allometry of Antlers in the Irish Elk, *Megaloceros giganteus*," by S. J. Gould, *Nature*, 1973, 244:375–376. Copyright ©1973 Macmillan Magazines Limited. **34**: Fig. 2.15 reprinted with permission from "Coral Snake Mimicry: Does it Occur?" by H. W. Greene and R. W. McDiarmid, *Science*, 1981, 213:1207–1211, Figure 1. Copyright ©1981 American Association for the Advancement of Science.

CHAPTER 3. 48: Fig. 3.3 from "Recent Advances of the Arctic Treelid Along the Eastern Coast of Hudson Bay," by K. Lescop-Sinclair and S. Payette, *Journal of Ecology*, 1995, 83:929–936. Copyright © 1995 Blackwell Science Ltd. Reprinted by permission. **49**: Fig. 3.5 from "Winter Energy Expenditure and the Distribution of Southern Flying Squirrels," by P. Stapp, P. J. Pekins, and W. W. Mautz, 1991, *Canadian Journal of Zoology*, 69: 2548–2555. Reprinted by permission. **51**: Fig. 3.11 data from "Salt Marsh Plant Zonation—the Relative Importance of Competition and Physical Factors," by S. C. Pennings and R. M. Callaway, *Ecology*, 1992, 73:681–690. Copyright © 1992 The Ecological Society. Reprinted by permission. **57**: Fig. 3.19 from "Activity and Thermoregulation of the Antelope Ground Squirrel, *Ammospermophilus leucurus*, in Winter and Summer," by M. A. Chappell, and G. W. Bartholomew, *Physiological Zoology*, 1981, 54:215–223. Copyright © 1981 University of Chicago Press. Reprinted by permission. **60**: Fig. 3.21 from "Freezing-induced Xylem Cavitation and the Northern Limit of *Larrea tridentata*," by W.T. Pockman and J. S. Sperry, *Oecologia*, 1997, 109:19–27. Reprinted by permission. **67**: Fig. 3.29, 3.30 from "The Color of Light in Forests and its Implications," by J. A. Endler, *Ecological Monographs*, 1993, 63:1–27. Copyright © 1993 The Ecological Society of America. Reprinted by permission. **71**: Fig. 3.32 from *Microbiology*, by L.M. Prescott, J. Harley, and D. H. Klein. Copyright © 1990 W. C. Brown.

CHAPTER 4. 81: Fig. 4.1 from "Spatial Variation in Abundance," by J. H. Brown, D. W. Mehlman, & G. C. Stevens, *Ecology*, 1995, 76:2028–2043. Copyright © 1995 The Ecological Society of America. Reprinted by permission. **83**: Fig. 4.4 from "Genetic Dynamics of the Clonal Plant *Rubus saxitilis*," by O. Eriksson and B. Bremer, *Journal of Ecology*, 1993, 1:533–542. Copyright © 1993 Blackwell Science Ltd. Reprinted by permission. **83**: Fig. 4.5 adapted from "Differing Levels of Among-population Divergence in the mtDNA of Periodical Cicadas Related to Historical Biogeography," by A. Martin and C. Simon, *Evolution*, 1990, 44:

1066–1080. Reprinted by permission. **87**: Fig. 4.8 from *An Introduction to Quantitative Ecology*, by R. W. Poole. Copyright © 1974 McGraw-Hill Book Companies. Reprinted by permission. **89**: Fig. 4.11 from *Wildlife Management Techniques*, by R. H. Giles. Copyright © 1971 The Wildlife Society. Reprinted by permission. **90**: Fig. 4.12 from "An 85-year Study of Saguaro (*Carnegiea gigantea*) Demography," by E. A. Pierson and R. M. Turner, *Ecology*, 1998, 79:2676–2693. Copyright © 1998 The Ecological Society of America. Reprinted by permission. **95**: Fig. 4.13 from (a) data from Deevey, 1947; (b) data from Caughey, 1966; (c) data from Allee, et al., 1949. Reprinted by permission of The Ecological Society. **95, 96, 97**: Figs. 4.14–4.16, 4.18 from *An Introduction to Population Ecology*, by G. E. Hutchinson. Copyright © 1978 Yale University Press. Reprinted by permission. **99**: Fig. 4.20 from "Vegetational Cover Predicts the Sex Ratio of Hatchling Turtles in Natural Nests," by F. J. Janzen, *Ecology*, 1994, 75:1593–1599. Copyright © 1994 The Ecological Society of America. Reprinted by permission. **100**: Fig. 4.22 from "Great Expectations: Dominance, Breeding Success and Offspring Sex Ratios in Red Deer, by T. H. Clutton-Brock, et al., *Animal Behavior*, 1986, 34:460–471. Reprinted by permission of Academic Press Ltd. **102**: Fig. 4.24 from "Dynamics of an Introduced Caribou Population," by D. C. Heard, J. P. Ouellet, *Arctic*, 1994, 47:88–95. Copyright © 1994 Arctic Institute of North America. Reprinted by permission. **103**: T. 4.9 from "Spatiotemporal Variation in Demographic Transitions of a Tropical Understory Herb: Projection Matrix Analysis," by C. C. Horvitz and D. W. Schemske, *Ecological Monographs*, 1995, 155–192. Copyright © 1995 The Ecological Society of America. Reprinted by permission.

CHAPTER 5. 111: Fig. 5.3 adapted from "Population Changes in German Forest Pests," by G. C. Varley, *Journal of Animal Ecology*, 1949, 18:117–122. Copyright © 1949 Blackwell Science Ltd. Reprinted by permission. **112**: Fig. 5.4 from "The Abundance of the Collared Lemming (*Dicrostonyx groenlandicus* (Tr.) *var richarsoni* Mer) in the Churchill Area, 1929 to 1940," by V. E. Shelford, *Ecology*, 1945, 24: 472–484. Copyright © 1945 The Ecological Society. Reprinted by permission. **112**: Fig. 5.5 from *Wildlife's Ten-Year Cycle*, by L. B. Keith. Copyright © 1963 The University of Wisconsin Press. Reprinted by permission. **113**: Fig. 5.6 from "History of Two Local Outbreaks of Feral House Mice," by O. P. Pearson, *Ecology*, 1963, 44:540–549. Copyright © 1963 The Ecological Society. Reprinted by permission. **115**: Fig. 5.8 adapted from "Resource Limitation in the California Vole," by R. G. Ford and F. A. Pitelka, *Ecology*, 1984, 65:122–136. Copyright © 1984 The Ecological Society. Reprinted by permission. **116**: Fig. 5.9 data from "Competition and Tree Death; most Trees Die Young in the Struggle for the Forest's Scarce Resources," by R. K. Peet and N. L. Christensen, *Bioscience*, 1987, 37:586–595. Copyright © 1987 American Institute of Biological Sciences. Reprinted by permission. **116**: Fig. 5.10 from "Influence of Coyote Predation on Black-tailed Jackrabbit Populations in Utah," by F. H. Wagner and L. C. Stoddart, *Journal of Wildlife Management*, 1972, 36:329–342. Copyright © 1972 Wildlife Society. Reprinted by permission. **118**: Fig. 5.12 from "The Population Dynamics of Brucellosis in the Yellowstone National Park," by A. Dobson and M. Meagher, *Ecology*, 1996, 77(4):1026–1036. Copyright © 1996 The Ecological Society of America. Reprinted by permission. **118, 119**: Figs. 5.14–5.16 adapted from "Population Outbreaks in Forest *Lepidoptera*," by J. H. Myers, *American Scientist*, 1993, 81:241–251. Copyright © 1993 American Scientist. Reprinted by permission. **120**: Fig. 5.17 reprinted with permission from "Population Regulation and Genetic Feedback," by D. Pimentel, *Science*, 1968, 159:1432–1437. Copyright © 1968 American Association for the Advancement of Science. **121**: Fig. 5.18 from "Genetic Structure of Outbreaking and Non-outbreaking Crown-of-Thorns Starfish (*Acanthaster planci*) Populations on the Great Barrier Reef," by J. A. H. Benzie and J. A. Stoddart, *Journal of Marine*

1972, 106:14–31. Copyright © 1972 The University of Chicago Press. Reprinted by permission. **186**: Fig. 7.18 reproduced with permission from "The Evolution of Life History Traits: A Critique of the Theory and a Review of the Data," by S. C. Stearns, *Annual Review of Ecology and Systematics*, 1977, 8:145–171. Copyright © 1977 Annual Reviews, Inc. **187**: Fig. 7.19 from "The Allometry of Reproduction: An Empirical View in Salamanders," by R. H. Kaplan and S. N. Salthe, *American Naturalist*, 1979, 113:671–689. Copyright © 1979 The University of Chicago Press. Reprinted by permission.

CHAPTER 8. 194: Fig. 8.1 from *Animal Behavior*, by J. Alcock. Copyright © 1989 Sinauer. Reprinted by permission. **195, 203**: Figs. 8.2, 8.10, 8.11 reprinted by permission of the publishers from *Sociobiology: The New Synthesis*, by E. O. Wilson, Cambridge, Mass.: The Belknap Press of Harvard University Press. Copyright © 1975 by the President and Fellows of Harvard College. **197**: Fig. 8.4 from "Studies on the Flash Communication System in Photinus Fireflies," by J. E. Lloyd, *Miscellaneous Publications of the Museum of Zoology*, University of Michigan, 1966, vol. 130. Reprinted by permission. **197**: Fig. 8.5 from *The Dance Language and Orientation of Bees*, by O. von Frisch, trans. by L. E. Chadwick. Copyright © 1967 Belknap Press. Reprinted by permission. **198**: Fig. 8.6 from "Does Sociality Drive the Evolution of Communicative Complexity? A Comparative Test with Ground-dwelling Sciurid Alarm Calls," by D. T. Blumstein and K. B. Armitage, *American Naturalist*, 1997, 150:179–200. Copyright © 1997 University of Chicago Press. Reprinted by permission. **200**: Fig. 8.8 adapted from "The Home Range: A New Nonparametric Estimation Technique," by D. J. Anderson, *Ecology*, 1982, 63:103–112. Copyright © 1982 The Ecological Society. Reprinted by permission. **201**: Fig. 8.9 from "Sizes of Feeding Territories Among Birds," by T. W. Schoener, *Ecology*, 1968, 49:123–141. Copyright © 1968 The Ecological Society. Reprinted by permission. **205**: Fig. 8.12 Reprinted with permission from "Ecology, Sexual Selection, and the Evolution of Mating Systems," by S. T. Emlen and L. W. Oring, *Science*, 1977, 197:215–223. Copyright © 1977 American Association for the Advancement of Science. **210**: Fig. 8.14 from "Cooperative Breeding in the Acorn Woodpecker," by P. B. Stacey and W. D. Koenig, *Scientific American*, 1984, 251:114–121. Copyright © 1984 Scientific American, Inc. All rights reserved. **211**: Fig. 8.16 from "Naked Mole-rats," by R. L. Honeycutt, *American Scientist*, 1992, 80:43–53. Copyright © 1992 American Scientist. Reprinted by permission. **212, 213**: Figs. 8.17 & 8.18 from "Investment Strategies of Breeders in Avian Cooperative Breeding Systems," by B. J. Hatchwell, *American Naturalist*, 1999, 154:205–219. Copyright © 1999 The University of Chicago Press. Reprinted by permission.

CHAPTER 9. 224: Fig. 9.6 from *Introduction to Ocean Sciences*, by D. A. Segar, 1998. Belmont, Calif.: Wadsworth. All rights reserved. **224**: Fig. 9.7 from "The Influence of Intraspecific Competition and other Factors on the Distribution of the Barnacle *Chthalamus stellatus*," by J. H. Connell, *Ecology*, 1961, 42:710–723. Copyright © 1961 The Ecological Society. Reprinted by permission. **225**: Fig. 9.8 reprinted by permission of the publishers from *Ecology and Evolution of Communities* edited by Martin L. Cody and Jared M. Diamond, Cambridge, Mass.: The Belknap Press of Harvard University Press. Copyright © 1975 by The President and Fellows of Harvard College. **227**: Fig. 9.10 from "Light and Self-thinning," by W. M. Lonsdale and A. R. Watkinson, *New Phytologist*, 1983, 90:399–418. Reprinted by permission. **227**: Fig. 9.11 from "Demographic Factors in Population of Plants." In *Demography and Evolution in Plant Populations*, by O. T. Solbrig (ed.), 1980, pp. 21–48. Copyright © 1980 Blackwell Science. Reprinted by permission. **228**: Fig. 9.12 from "Interferences in Pure and Mixed Populations of *Avena fatua* and *A. barbata*," by D. R. Marshall and S. K. Jain, *Journal of Ecology*, 1969, 57:251–270. Copyright © 1969 Blackwell Science, Ltd. Reprinted by permission. **229**: Fig. 9.14 from "On the Ecological Significance of Bergmann's Rule," by B. K. McNab, *Ecology*, 1971, 52(5):845–854. Copyright © 1971 The Ecological Society. Reprinted by permission. **230**: Fig. 9.15 from "Size Variability in the Worker Caste of a Social Insect (*Veromesser pergandei Mayr*) as A Function of the Competitive Environment," by D. W. Davidson, *American*

Naturalist, 1978, 112:523–532. Copyright © 1978 The University of Chicago Press. Reprinted by permission. **230**: Fig. 9.16 from "Size Variability in the Worker Caste of a Social Insect (*Veromesser pergandei Mayr*) as a Function of the Competitive Environment," by D. W. Davidson, *American Naturalist*, 1978, 112:523–532. Copyright © 1978 The University of Chicago Press. Reprinted by permission. **231**: Fig. 9.18 from "Evidence for the Evolution of Competition Between Two Species of Annual Plants," by M. M. Martin and J. Harding, *Evolution*, 1981, 35(5):975–987. Reprinted by permission. **232**: Fig. 9.19 from *The Struggle for Existence*, by G. F. Gause. Copyright © 1934 Dover Publications, Inc. Reprinted by permission. **232**: Fig. 9.20 from *Mammalogy*, by T. A. Vaughan. Copyright © 1972 Saunders. Reprinted by permission. **240**: Fig. 9.28 reprinted with permission from "Prospects for An Invasion: Competition between *Aedes albopictus* and *A. triseriatus*," by T. P. Livdahl and M. S. Willey, *Science*, 1991, 253:189–191. Copyright © 1991 American Association for the Advancement of Science.

CHAPTER 10. 248: Fig. 10.4 reprinted with permission from "Flock-feeding on Fish Schools Increases Individual Success in Gulls," by F. Gotmark, D. W. Winkler and M. Anderson, *Nature*, 1986, 319:589–591. Copyright © 1986 Macmillan Magazines Limited. **250**: Fig. 10.6 from "The Effect of Wolf Predation and Snow Cover on Musk-ox Group Size," by D. C. Heard, *American Naturalist*, 1992, 139:190–205. Copyright © 1992 The University of Chicago Press. Reprinted by permission. **251**: Fig. 10.8 from "Synchrony in the Lesser Snow Goose (*Anser caerulescens*). II. The Adaptive Value of Reproductive Synchrony," by C. S. Findlay and F. Cooke, *Evolution*, 1982, 36:786–799. Reprinted by permission. **255**: Fig. 10.10 from "Chemical Defense Against Diverse Coral Reef Herbivores," by M. E. Hay, W. Fenical, and K. Gustafson, *Ecology*, 1987, 68:1581–1591. Copyright © 1987 The Ecological Society. Reprinted by permission. **261**: Figs. 10.14–10.16 from "Some Characteristics of Simple Types of Predation and Parasitism," by C. S. Holling, *Canadian Entomologist*, 1959, 91:385–398. Copyright © 1959 Canadian Entomology. Reprinted by permission.

CHAPTER 11. 272: Fig. 11.1 from *Ecological Studies in Tropical Fish Communities*, by R. H. Lowe-McConnell. Copyright © 1987 Cambridge University Press. Reprinted with the permission of Cambridge University Press. **273**: Fig. 11.2 from "Body Size and Coexistence in Desert Rodents: Chance or Community Structure?" by M. A. Bowers and J. H. Brown, *Ecology*, 1982, 63:391–400. Copyright © 1982 the Ecological Society. Reprinted by permission. **275**: Fig. 11.4 adapted from "Population Ecology of Some Warblers of Northeastern Coniferous Forests," by R. H. MacArthur, *Ecology*, 1958, 39:599–619. Copyright © 1958 The Ecological Society. Reprinted by permission. **276**: Figs. 11.6, 11.7 from "Niche Parameters and Species Richness," by B. Fox, *Ecology*, 1981, 62:1415–1425. Copyright © 1981 The Ecological Society. Reprinted by permission. **277**: Fig. 11.10 from *Evolutionary Ecology*, by E. R. Pianka. Copyright © 1988 HarperCollins Publishers. Reprinted by permission. **278**: Fig. 11B.1 from "Conservation of Hawaii's Vanishing Avifauna," by J. M. Scott, et al., *Biological Sciences*, 1988, 38:238–257. Copyright © 1988 American Institute of Biological Sciences. Reprinted by permission. **281**: Fig. 11.13 from "Test of Community-wide Character Displacement Against Null Hypotheses," by D. R. Strong, L. A. Szyska, and D. S. Simberloff, *Evolution*, 1979, 33: 897–913. Reprinted by permission. **284**: Fig. 11.16 from "Morphological and Dietary Structuring of a Zambian Insectivorous Bat Community," by J. S. Findley and H. Black, *Ecology*, 1983, 64:625–630. Copyright © 1983 The Ecological Society. Reprinted by permission. **285**: Fig. 11.17 from *Introduction to Ocean Sciences*, by D. A. Segar, 1998. Belmont, Calif.: Wadsworth Publishing Company. All rights reserved. **286**: Fig. 11.19 reprinted with permission from "Control of a Desert-grassland Guild by A Keystone Rodent Guild," by J. H. Brown and E. J. Heske, *Science*, 1990, 250:1705–1707. Copyright © 1990 American Association for the Advancement of Science. **287**: Figs. 11.20–11.22 from "Indirect Effects, Prey Susceptibility and Habitat Selection: Impact of Birds on Limpets and Algae," by T. J. Wooten, *Ecology*, 1992, 73:981–992. Copyright © 1992 The Ecological Society. Reprinted by permission.

289: Fig. 11.24 from "Equilibrium and Nonequilibrium Concepts in Ecological Models," D. L. DeAngelis and J. C. Waterhouse, *Ecological Monographs*, 1987, 57:1–21. Copyright © 1987 Ecological Monographs. Reprinted by permission. **290**: Fig. 11.25 from "Intertidal Landscapes: Disturbance and the Dynamics of Pattern," by R. T. Paine and S. A. Levin, *Ecological Monographs*, 1981, 51:145–178. Copyright © 1981 Ecological Monographs. Reprinted by permission. **291**: Fig. 11.28 reprinted with permission from "Modulation of Diversity by Grazing and Mowing in Native Tallgrass Prairie," by S. L. Collins, et al., *Science*, 1998, 280:745–747. Copyright © 1998 American Association for the Advancement of Science.

CHAPTER 12. 298: Fig. 12. 1 from "Deep-sea Biodiversity: The Ocean Bottom Supports Communities That May Be As Diverse as Those of Any Habitat on Earth," by J. F. Grassle, *BioScience*, 1991, 41: 464–470. Copyright © 1991 American Institute of Biological Sciences. Reprinted by permission of the publisher and author. **299**: Fig. 12.4 from "Importance of Spatial and Temporal Dynamics in Species Regional Abundance and Distribution," by S. L. Collins and S. M. Glenn, *Ecology*, 1991, 72:654–664. Copyright © 1991 The Ecological Society of America. Reprinted by permission. **301**: Fig. 12.6 reprinted with permission from "Biodiversity Hotspots for Conservation Priorities," by N. Myers, et al., *Nature*, 2000, 403:853–858. Copyright © 2000 Macmillan Magazines Limited. **301, 307**: Figs. 12.7, 12.14 from "Biogeographic Kinetics: Estimation of Relaxation Times for Avifaunas of Southwest Pacific Islands," by J. L. Diamond, *Proceedings of the National Academy of Science*, 1972, 69:3199–3203. Reprinted by permission of the author. **302**: Fig. 12.8a reprinted with permission from "Geographical Trends in Numbers of Species," by J. J. Schall and E. R. Pianka, *Science*, 1978, 201:679–686. Copyright © 1978 American Association for the Advancement of Science. **302, 313**: Figs. 12.8b, 12.22 from "Species Diversity Gradients: We Know More and Less Than We Thought," by M. L. Rosenzweig, *Journal of Mammalogy*, 1992, 73:715–730. Copyright © 1992 Journal of Mammalogy. Reprinted by permission. **302**: Fig. 12.8c from "Biodiversity of the Pelagic Ocean," by M. V. Angel, *Conservation Biology*, 1993, 7:760–772. Copyright © 1993 Blackwell Science Ltd. Reprinted by permission. **302, 314**: Figs. 12.9, 12.23, 12.24 from "Energy and Large-scale Patterns of Animal and Plant-species Richness," by D. J. Currie, *American Naturalist*, 1991, 137:27–50. Copyright © 1991 The University of Chicago Press. Reprinted by permission. **303**: T. 12.1 from "Deep-sea Biodiversity: The Ocean Bottom Supports Communities That May Be As Diverse as Those of Any Habitat on Earth," by J. F. Grassle, *BioScience*, 1991, 41:464–470. Copyright © 1991 American Institute of Biological Sciences. Reprinted by permission of the publisher and author. **306**: Fig. 12.12 from "Experimental Zoogeography of Islands: The Colonization of Empty Islands," by D. S. Simberloff and E. O. Wilson, *Ecology*, 1969, 50:278–296. Copyright © 1969 The Ecological Society. Reprinted by permission. **308**: Fig. 12.15 from "Mammals on Mountaintops: Nonequilibrium Insular Biogeography," by J. H. Brown, *American Naturalist*, 1971, 105:467–478. Copyright © 1971 The University of Chicago Press. Reprinted by permission. **308**: Fig. 12.16 from "The Theory of Insular Biogeography and the Distribution of Boreal Birds and Mammals," by J. H. Brown, *Great Basin Naturalist Memoirs*, 1978, 2:209–227. Copyright © 1978 Great Basin Naturalist, Brigham Young University. Reprinted by permission. **308**: Fig. 12.17 from "Island Biogeography of Montane Forest Mammals in The American Southwest," by M. V. Lomolino, J. H. Brown, and R. Davis, *Ecology*, 1989, 70(1):180–194. Copyright © 1989 The Ecological Society. Reprinted by permission. **309**: Fig. 12.18 from "Controls of Number of Bird Species on Montane Islands in the Great Basin," by N. K. Johnson, *Evolution*, 1975, 29:545–574. Reprinted by permission. **309**: Fig. 12.19 from *The Fragmented Forest*, by L. D. Harris. Copyright © 1984 The University of Chicago Press. Reprinted by permission. **312**: Fig. 12.10 from "The Latitudinal Gradient in Geographical Range: How so Many Species Coexist in the Tropics," by G. C. Stevens, *American Naturalist*, 1989, 133:240–257. Copyright © 1989 The University of Chicago Press. Reprinted by permission. **313**: Fig. 12.21 from "Bird Species Diversity and Habitat Diversity in Australia and North America," by H. F. Recher, *American Naturalist*, 1969, 103:75–80. Copyright © 1969 The University of Chicago Press. Reprinted by permission. **314**: Fig. 12.25 reprinted with permission from "Habitat Heterogeneity as a Determinant of Mammal Species Richness in High-energy Regions," by J. T. Kerr and L. Packer, *Nature*, 1997, 385:252–254. Copyright © 1997 Macmillan Magazines Limited. **316**: Fig. 12.27 from "Experimental Investigation of Disturbance and Ecological Succession in a Rocky Intertidal Algal Community," by W. P. Sousa, *Ecological Monographs*, 1979a, 49:227–254. Copyright © 1979 Ecological Monographs. Reprinted by permission. **316**: Fig. 12.28 from "Disturbance in Marine Intertidal Boulder Fields: The Non-equilibrium Maintenance of Species Diversity," by W. P. Sousa, *Ecology*, 1979b, 60:1225–1239. Copyright © 1979 The Ecological Society. Reprinted by permission. **318**: Figs. 12.29a, b, reprinted with permission from "Productivity and Sustainability Influenced by Biodiversity in Grassland Ecosystems," by D. Tilman, D. Wedin, and J. Knops, *Nature*, 1996, 379:718–720. Copyright © 1996 Macmillan Magazines Limited. **318**: Fig. 12. 29c reprinted with permission from "Biodiversity and Stability in Grasslands," by D. Tilman and J. A. Downing, *Nature*, 1994, 367:363–365. Copyright © 1994 Macmillan Magazines Limited. **319**: Fig. 12.30 from "Functional Complexity and Ecosystem Stability," by P. VanVoris, et al., *Ecology*, 1980, 61:1352–1360. Copyright © 1980 The Ecological Society. Reprinted by permission.

CHAPTER 13. 324: Fig. 13.1 reprinted with permission from *The Diversity of Life*, by E. O. Wilson. Copyright © 1992 by Edward O. Wilson. **324**: Fig. 13.2 from "Plant Recolonization and Vegetation Succession on the Krakatau Islands," by R. H. Whittaker, *Ecological Monographs*, 1989, 59:59–123. Copyright © 1989 The Ecological Society. Reprinted by permission. **327**: Fig. 13.5 from "Fifteen Years of Canopy Dynamics Analyzed by Aerial Photographs in a Temperate Deciduous Forest, Japan." by H. Tanaka and T. Nakashizuka, *Ecology*, 1997, 78:612–620. Copyright © 1997 The Ecological Society. Reprinted by permission. **329, 331**: Figs. 13.8, 13.11 from "Changes in Productivity and Distribution of Nutrients in a Chronosequence at Glacier Bay National Park, Alaska," by B. T. Bormann and R. C. Sidle, *Journal of Ecology*, 1990, 78:561–578. Copyright © 1990 Blackwell Science Ltd. Reprinted by permission. **335**: Fig. 13.15 from "Rates of Succession and Soil Changes on Southern Lake Michigan Sand Dunes," by J. S. Olson, *Botany Gazette*, 1958, 119:125–170. Copyright © 1958 The University of Chicago Press. Reprinted by permission. **336**: Fig. 13.16 modified from "Wildfire in the Taiga of Alaska," by L. A. Viereck, *Journal of Quaternary Research*, 1973, 3(3):465–495. In S. West, *Habitat Responses of Microtine Rodents to Central Alaskan Forest Succession* (Berkeley, CA: University of California, PhD. Thesis, 1979). **337**: Fig. 13.17 from "Mechanisms of Succession in Natural Communities and Their Role in Community Stability and Organization," by J. H. Connell and R. O. Slatyer, *American Naturalist*, 1977, 111:1119–1144. Copyright © 1977 The University of Chicago Press. Reprinted by permission. **337**: Fig. 13.18 from "Primary Succession and Forest Development on Coastal Lake Michigan Sand Dunes," by J. Lichter, *Ecological Monographs*, 1998, 68:487–510. Copyright © 1998 The Ecological Society. Reprinted by permission. **338, 341**: Figs. 13.19, 13.22 from "The Physiological Ecology of Plant Succession," by F. A. Bazzaz, *Annual Review of Ecology and Systematics*, 1979, 10:351–371. Copyright © 1979 Annual Reviews, Inc. Reprinted by permission. **340**: T. 13.3, 13.4 from "The Physiological Ecology of Plant Succession," by F. A. Bazzaz, *Annual Review of Ecology and Systematics*, 1979, 10:351–371. Copyright © 1979 Annual Reviews, Inc. Reprinted by permission. **341**: Fig. 13.23 from "Light Gradient Partitioning Among Tropical Tree Species Through Differential Seedling Mortality and Growth," by R. K. Kobe, *Ecology*, 1999, 80:187–201. Copyright © 1999 The Ecological Society. Reprinted by permission. **342**: Fig. 13.24 from "The Resource-ratio Hypothesis of Plant Succession," by D. Tilman, *American Naturalist*, 1985, 125:827–852. Copyright © 1985 The University of Chicago Press. Reprinted by permission. **342, 343**: Figs. 13.25, 13.26 from "Ecological Constraints to Seedling Establishment on the Pumice Plains, Mount St. Helens,

Washington," by W. F. Morris and D. M. Wood, *Ecology*, 1989, 70:697–704. Copyright © 1989 The Ecological Society. Reprinted by permission. **344**: Fig. 13.27 from "Models and Mechanisms of Succession: An Example from a Rocky Intertidal Community," by T. M. Farrell, *Ecological Monographs*, 1991, 61:95–113. Copyright © 1991 The Ecological Society. Reprinted by permission. **345**: Fig. 13.28 from "Intertidal Mosaics: Patch Size, Propagule Availability and Spatially Variable Patterns of Succession," by W. P. Sousa, *Ecology*, 1984a, 65:1918–1935. Copyright © 1984 The Ecological Society. Reprinted by permission. **348**: Fig. 13.30 from "Patterns of Community Dynamics in Colorado Engelmann Spruce-subalpine Fir Forests," by G. H. Aplet, R. D. Laven, and F. W. Smith, *Ecology*, 1988, 69:312–319. Copyright © 1988 The Ecological Society. Reprinted by permission. **349**: T. 13.6 from "Patterns of Community Dynamics in Colorado Engelmann Spruce-subalpine Fir Forests," by G. H. Aplet, R. D. Laven, and F. W. Smith, *Ecology*, 1988, 69:312–319. Copyright © 1988 The Ecological Society. Reprinted by permission. **351**: Fig. 13.32 from "Vegetation Dynamics, Fire, and the Physical Environment in Coastal Central California," by R. M. Callaway and F. W. Davis, *Ecology*, 1993, 74:1567–1576.

CHAPTER 14. 364: Fig. 14.6 from "The Physiological Ecology of Plant Succession," by F. A. Bazzaz, *Annual Review of Ecology and Systematics*, 1979, 10:351–371. Copyright © 1979 Annual Reviews, Inc. Reprinted by permission. **365**: Fig. 14.7 from "Vegetation of the Santa Catalina Mountains, Arizona. V. Biomass, Production and Diversity Along the Elevation Gradient," by R. H. Whittaker and W. A. Niering, *Ecology*, 1975, 56:771–790. Copyright © 1975 The Ecological Society. Reprinted by permission. **366**: Fig. 14.9 from "Primary Production and Water Use in Native Forest, Grassland and Desert Ecosystems," by W. Webb, et al., *Ecology*, 1978, 59:1239–1247. Copyright © 1978 The Ecological Society. Reprinted by permission. **367**: Fig. 14.11 from "Primary Productivity and Limiting Factors in Three Lakes of the Alaska Peninsula," by C. R. Goldman, *Ecological Monographs*, 1960, 30:207–230. Copyright © 1960 The Ecological Society. Reprinted by permission. **367**: Fig. 14.12 reprinted with permission from "Carbon Dioxide Limitation of Marine Phytoplankton Growth Rates," by V. Riebesell, D. A. Wold-Gladrow and V. Smetacek, *Nature*, 1993, 361:249–251. Copyright © 1993 Macmillan Magazines Limited. **371**: Fig. 14.19 reprinted with permission from "Food Web Patterns and Their Consequences," by S. L. Pimm, J. H. Lawton and J. E. Cohen, *Nature*, 1991, 350:669–674. Copyright © 1991 Macmillan Magazines Limited. **371**: Fig. 14.18 from *The Living Resources of the Southern Ocean*, by I. Everson, 1977. U.N. Development Program, Rome. **372**: Fig. 14.10 from "Complex Trophic Interactions in Deserts: An Empirical Critique of Food-web Theory," by G. A. Polis, *American Naturalist*, 1991, 138:123–155. Copyright © 1991 The University of Chicago Press. Reprinted by permission.

CHAPTER 15. 390: Fig. 15B.1 from *Environmental Science*, by D. D. Chiras. Copyright © 1994 Benjamin/Cummings Publishing Company. Reprinted by permission. **392**: Fig. 15.10 from *Biogeochemistry of a Forested Ecosystem*, by G. E. Likens, et al. Copyright © 1977 Springer-Verlag. Reprinted by permission. **395**: Fig. 15.16 from *Nutrient Cycling in Tropical Forest Ecosystems*, by C. F. Jordan. Copyright © 1985 John Wiley & Sons Ltd. Reprinted by

permission. **398**: Fig. 15.17 from "Fire and Grazing in the Tallgrass Prairie: Contingent Effects on Nitrogen Budgets," by N. T. Hobbs, et al., *Ecology*, 1991, 72:1372–1382. Copyright © 1991 The Ecological Society. Reprinted by permission. **398**: Fig. 15.18 from "Element Cycling in Taiga Forests: State-factor Control," by K. Van Cleve, et al., *BioScience*, 1991, 41:78-88. Copyright © 1991 American Institute of Biological Sciences. Reprinted by permission. **398**: Fig. 15.19 from "Top-down Trophic Interactions in Lakes: Effects of Fish on Nutrient Dynamics," by M. J. Vanni, et al., *Ecology*, 1997, 78:1–20. Copyright © 1997 The Ecological Society. Reprinted by permission. **399**: Fig. 15.20 from "Nutrient Recycling and Herbivory as Mechanisms in the "Top-down" Effect of Fish on Algae in Lakes," by M. J. Vanni and C. D. Layne, *Ecology*, 1997, 78:21–40. Copyright © 1997 The Ecological Society. Reprinted by permission. **392, 393, 394**: Figs .15.11–15.15 from *Pattern and Process in a Forested Ecosystem*, by F. H. Bormann and G. E. Likens. Copyright © 1979 Springer-Verlag. Reprinted by permission.

CHAPTER 16. 405: Fig. 16.2 adapted from *Oceanography*, 2nd Ed., by T. Garrison, 1996. Belmont, Calif.: Wadsworth. **408**: Fig. 16. 8 adapted from *Biology: The Unity and Diversity of Life*, 7th ed., by C. Starr and R. Taggart, 1995. Belmont, Calif.: Wadsworth. **412**: Fig. 16.14 from *Fire and Vegetation Dynamics: Studies from the North American Boreal Forest*, by E. A. Johnson. Copyright © 1992 Cambridge University Press. Reprinted by permission. **423**: Figs. 16.30, 16.31 from "Landscape Patterns in a Disturbed Environment," by J. R. Krummel, et al., *Oikos*, 1987, 48:321–324. Reprinted by permission of the publisher. **406, 408**: Figs. 16.3, 16.7 adapted from *Biology: Concepts and Applications*, 3rd Ed., by C. Starr, 1997. Belmont, Calif.: Wadsworth. **421, 422**: Figs. 16.26, 16.27 from "Long-term Landscape Patterns of Past Fire Events in a Montane Ponderosa Pine Forest of Central Colorado," by P. M. Brown, et al., *Landscape Ecology*, 1999, 14:513–532. Copyright © 1999 Kluwer Academic Publishers. Reprinted by permission.

CHAPTER 17. 428: Fig. 17.1 from *Fundamentals of Limnology*, by F. Ruttner. Copyright © 1963 University of Toronto Press, Inc. Reprinted by permission of the publisher. **429**: Fig. 17.4 from *Limnology*, by R. G. Wetzel. Copyright © 1983 W. B. Saunders. Reprinted by permission of the publisher. **432**: Fig. 17B.1 from "Lessons from the Everglades," by L. H. Gunderson, S. S. Light and C. S. Holling, *BioScience Supplement*, 1995, 66–73. Copyright © 1995 American Institute of Biological Sciences. Reprinted by permission. **435**: Fig. 17.11 adapted from *Oceanography*, 2nd Ed., by T. Garrison, 1996. Belmont, Calif.: Wadsworth. **437, 436, 439**: Figs. 17.12, 17.13, 17.19 from *Introduction to Ocean Sciences*, by D. A. Segar, 1998. Belmont, Calif.: Wadsworth.

ADDITIONAL PHOTO CREDITS. 2: David Krohne; **17**: Merlin D. Tuttle/Bat Conservation International; **45, 75**: David Krohne; **79**: Art Wolfe; **109, 143**: David Krohne; **163**: John Pontier/Animals Animals; **167**: David Krohne; **191**: John Gerlach/Visuals Unlimited; **207**: David Krohne; **219**: Dave B. Fleetham/Visuals Unlimited; **233**: John Lemker/Animals Animals; **243**: Darren Bennett/Animals Animals; **271**: David Hall/Photo Researchers; **295, 311, 323, 354, 357, 376, 381, 403, 420, 427**: David Krohne; **445**: NASA.

Subject Index

Species Index